Physical Chemistry

for Engineering and Applied Sciences

Physical Chemistry
for Engineering and
Applied Sciences

Frank R. Foulkes

CRC Press
Taylor & Francis Group
Boca Raton   London   New York

CRC Press is an imprint of the
Taylor & Francis Group, an **informa** business

Library of Congress Cataloging-in-Publication Data

Foulkes, F. R.
  Physical chemistry for engineering and applied sciences / Frank R. Foulkes.
  pages cm
  Includes bibliographical references and index.
  ISBN 978-1-4665-1846-9 (hardback)
  1. Chemistry, Physical and theoretical--Textbooks.. I. Title.

QD453.3.F68 2013
541--dc23   2012024640

DEDICATION & APPRECIATION

I dedicate this book to Irene, Bryan, and Michael,
without whom I am nothing.

and

I wish gratefully to acknowledge Lance Wobus, my editor at
Taylor and Francis. Lance has been very supportive; it has
been a pleasure working with him on this project.

CONTENTS

PREFACE

Welcome to Physical Chemistry for Engineering and Applied Sciences!

This course has been running for many years (I took it myself as a first-year engineering student in 1961, and, in spite of the fact that I wasn't the sharpest tool in the shed, I seem to have passed it, so I guess it can't be all that tough).

Most first year university physical chemistry textbooks have been designed more for students in chemistry than for students in engineering and applied sciences. These books tend to be more theoretical than what the rest of us require. Frankly, freshman students in engineering and the applied sciences don't need to know too much at this stage about quantum mechanics, atomic structure, and molecular spectroscopy. But they do need to know about melting points, how to balance a chemical reaction, and how to calculate the voltage of a car battery.

For years my colleagues had been saying that we really ought to write our own textbook. So..... here it is! It may not be perfect,[1] but at least it doesn't cost $200, *and* it doesn't contain a lot of stuff that's not relevant to what you need to know, *and* it's small enough that it can almost be carried around and read on the subway.

When I was an undergraduate student there was one thing that especially bugged me about almost all the assigned textbooks for our various courses: I could almost never follow the derivations of the equations! The authors of these books would write down some equation, and then, skipping about 20 steps, say something like: "It is readily shown that, after simplification, equation [1] reduces to equation [2]." Huh? I remember wasting whole days on the weekends trying to figure out how we get to equation [2] from equation [1]. I vowed that if I ever wrote a textbook, I would *not* leave out all the intermediate steps. I have tried to live up to this promise in these course notes. If you are willing to take the time to read them carefully, I believe that it is possible for you (or for any average first year student[2]) to follow all the derivations, and therefore, to really understand— should you wish—where the equations come from.

This does not, unfortunately, mean that this course is particularly easy. If you want to pass it, you will have to be prepared to do a significant amount of work. Specifically, you are expected to (here

---

[1] Yet.
[2] And, of course, *you* probably are above average!

it comes) read the appropriate chapters and attempt the exercises *before* coming to the lectures.[3] In class, your professor probably will discuss and try to clarify points of confusion—there will be many—and possibly will take up the exercise problems (the ones that *you* already have tried to do before coming to class). Don't be surprised if your professor doesn't want to spend the lecture time rehashing the material given in the text.

It is really important that you start working right from the beginning. If you don't understand lecture n, then you will not be able to understand lecture $n + 1$. Please don't make the common undergraduate mistake of falling behind in your work. To encourage you to keep up, you probably will be given regular quizzes based on the example problems, exercise problems, and the extra problems found at the end of the chapters. And remember, *attend the lectures*! Many instructors give problems in class that are not in the text; and, of course, sometimes they give hints regarding the exams.

This book includes:

- Two extensive chapters (Appendices 8 and 9) in which I explain integral calculus for the mathematically impaired.
- Extensive tables of Standard Electrode Potentials (Appendix 10).
- Extensive chapter reviews at the end of each chapter.
- Problems and Exercises graded by difficulty as follows:

  ☺ = Easy to do; more or less plugging numbers into formulas.

  ☺ = Of average difficulty; the kind of problem you might find on the final examination.

  ☹ = Difficult; either in concepts or in the amount of work required.

  ☠ = These are really tough; not to be attempted by the weak of heart.

  An asterisk (*) indicates that some (usually pretty easy) integral calculus is required to solve this problem.

- A very detailed index at the end of the book!

Good luck, and hang in there! You can do it. Remember: "What one fool can do, so can another."

Frank Foulkes (a.k.a.,"Uncle Frank")

Department of Chemical Engineering and Applied Chemistry,
University of Toronto.

February, 2012

---

[3] At the very least, try to read the Key Points at the end of each chapter. These reviews are much more detailed than those found in most other textbooks, making them useful for studying for examinations.

ABOUT THE AUTHOR

Frank R. Foulkes is a Professor Emeritus in the Department of Chemical Engineering and Applied Chemistry at the University of Toronto, where he taught both undergraduate and graduate courses in physical chemistry, thermodynamics, and electrochemistry for almost forty years. His consulting activities and research publications are primarily in the areas of electrochemistry and corrosion. In 1969 he was the first Canadian to publish research in the area of solid polymer electrolyte fuel cells and was the leading author of the well-known *Fuel Cell Handbook* (1989). Born in Toronto, Professor Foulkes earned his B.A.Sc., M.A.Sc., and Ph.D. degrees at the University of Toronto. He is a registered professional engineer in the province of Ontario. In addition to his academic and scientific interests, he is a karate instructor, holding a seventh degree black belt in Shotokan karate. He lives in Toronto with his wife Irene, two grown sons, Bryan and Michael, and his motorcycle and Morgan sports car.

NOMENCLATURE

LIST OF SYMBOLS

Symbol	Quantity	Units
a	acceleration	$m\ s^{-2}$
a	constant in the van der Waals equation	$m^6\ Pa\ mol^{-2}$
a_i	relative activity of species i; $a_i = a_i'/(a_i')^o$	(unitless)
a_i'	absolute activity of species i	(concn units)
$(a_i')^o$	absolute activity of species i in its reference state	(concn units)
$a_i^{(m)}$	relative activity of species i on the molal scale	(unitless)
a_W	relative activity of water	(unitless)
A	pre-exponential factor in expression for rate constant k	(same as k)
A	area	m^2
atm	standard atmospheric pressure (= 1.01325 bar)	atm
b	constant of proportionality between $\bar{e}_{transl}$ and T	$J\ K^{-1}$
b	constant in the van der Waals equation	$m^3\ mol^{-1}$
B(T)	second virial coefficient	$m^3\ mol^{-1}$
$(B_m)_i$	partial molar value of property B: $(B_m)_i \equiv \left(\partial B/\partial n_i\right)_{T,P,n_j}$	
c	speed of light in vacuum	$m\ s^{-1}$
c^2	mean square speed of gas molecules	$m^2\ s^{-2}$
c	molar heat capacity	$J\ mol^{-1}\ K^{-1}$
$\bar{c}$	specific heat capacity	$J\ kg^{-1}\ K^{-1}$
c_P	molar heat capacity at constant pressure	$J\ mol^{-1}\ K^{-1}$
$\bar{c}_P$	specific heat capacity at constant pressure	$J\ kg^{-1}\ K^{-1}$
c_V	molar heat capacity at constant volume	$J\ mol^{-1}\ K^{-1}$
$\bar{c}_V$	specific heat capacity at constant volume	$J\ kg^{-1}\ K^{-1}$
c_i	concentration of species i	$mol\ L^{-1}$

c_o, c_i^o	initial concentration of species i at time $= 0$	mol L^{-1}
c^o	standard concentration ($= 1$ mol L^{-1})	mol L^{-1}
c_{RMS}	root mean square speed of a gas molecule	m s^{-1}
C	Celsius	°C
C	total heat capacity	J K^{-1}
C	number of components in a system	(unitless)
C(T)	third virial coefficient	m^6 mol^{-2}
C_i	concentration of species i (SI units)	mol m^{-3}
C_P	total heat capacity at constant pressure	J K^{-1}
C_V	total heat capacity at constant volume	J K^{-1}
ΔC_P	defined as $\sum C_P(products) - \sum C_P(reactants)$	J K^{-1}
∂	partial differential operator	
e	energy	J
$\bar{e}_{transl}$	average translational kinetic energy of a particle	J
E	total energy of a system	J
E	electric field strength	V m^{-1}
E_{cell}	cell voltage	V
E°	standard reduction potential (*vs.* standard hydrogen electrode)	V
E_{an}^o	standard reduction potential for anode reaction	V
E_{cat}^o	standard reduction potential for cathode reaction	V
E_{cell}^o	standard cell potential (cell potential under standard conditions)	V
f	fraction of gas molecules having speed between s and $(s + \Delta s)$	(unitless)
f_W	rational (mole fraction scale) activity coefficient	(unitless)
F	force	N = kg m s^{-2}
F	Faraday's constant	C mol^{-1}
F	Fahrenheit	°F
F	number of degrees of freedom in a system in phase equilibrium	$F = C - P + 2$
g	acceleration due to gravity	m s^{-2}
g	molar Gibbs free energy, defined as $g = h - Ts$	J mol^{-1}
$\bar{g}$	specific Gibbs free energy, defined as $\bar{g} = \bar{h} - T\bar{s}$	J kg^{-1}
g°	standard state (1 bar pressure) molar Gibbs free energy	J mol^{-1}
Δg_f^o	standard molar Gibbs free energy of formation	J mol^{-1}
G	Gibbs free energy, defined as $G = H - TS$	J
$(G_m)_i$	partial molar Gibbs free energy of species i $(= \mu_i)$	J mol^{-1}
$\Delta G°$	standard Gibbs free energy change	J

ΔG_R°	standard Gibbs free energy change of reaction	J
ΔG_a^{*}	activation energy for a chemical reaction	J mol^{-1}
$\Delta G_{T,P}$	ΔG for a process at constant temperature and constant pressure	J
$\Delta G_{298.15}^{\circ}$	standard Gibbs free energy change at 1 bar and 25°C	J
h	molar enthalpy, defined as $h = u + Pv$	J mol^{-1}
$\bar{h}$	specific enthalpy, defined as $\bar{h} = \bar{u} + P\bar{v}$	J kg^{-1}
Δh_f°	molar enthalpy (heat) of formation	J mol^{-1}
H	enthalpy, defined as $H = U + PV$	J
ΔH_P	change in enthalpy for a constant pressure process	J
ΔH_V	change in enthalpy for a constant volume process	J
ΔH_R°	standard enthalpy change of reaction	J
$\Delta H_{T_2}^{\circ}$	standard enthalpy change of reaction at temperature T_2	J mol^{-1}
$\Delta H_{298.15}^{\circ}$	standard enthalpy change at 298.15 K (25°C)	J mol^{-1}
$[H^+]_W$	hydrogen ion concentration from self-dissociation of water	mol L^{-1}
HA	generic term for any weak acid	
HAc	abbreviation for acetic acid, CH_3COOH	
[i]	molar concentration of species i	mol L^{-1}
I	molarity-based ionic strength of solution: $I = \frac{1}{2}\Sigma c_i z_i^2$	mol L^{-1}
$I^{(m)}$	molality-based ionic strength of solution: $I^{(m)} = \frac{1}{2}\Sigma m_i z_i^2$	mol kg^{-1}
k	Boltzmann's constant	J K^{-1}
k	chemical reaction rate constant	(units vary)
k_{obs}	experimentally observed apparent rate constant of ionic reaction	(units vary)
k°	rate constant of ionic reaction at zero ionic strength	(units vary)
K	thermodynamic equilibrium constant	(unitless)
K_a	acid dissociation constant	(unitless)
K_b	base dissociation constant	(unitless)
K_h	hydrolysis constant	(unitless)
K_{SP}	solubility product (solubility equilibrium constant)	(unitless)
K_W	self-dissociation constant for water: $K_W = a_{H^+}^{(m)} \cdot a_{OH^-}^{(m)}$	(unitless)
K_i	Henry's law constant for volatile solute i	(pressure units)
K_1	first dissociation constant for an acid or a base	(unitless)
K_2	second dissociation constant for an acid or a base	(unitless)
K_3	third dissociation constant for an acid or a base	(unitless)
KE	kinetic energy	J
L	length (also a unit, the *liter*)	m

$ln\ x$	natural logarithm of x, defined as $ln\ x = \int \frac{1}{x}dx$	(unitless)
$log\ x$	logarithm of x to the *base-10*: $log\ x = ln\ x/ln\ 10$	(unitless)
m	mass	kg
m_i	molal concentration of species i: mol solute/kg solvent	mol kg^{-1}
M_i	molar mass of species i	kg mol^{-1}
n	number of moles of electrons transferred per mole of reaction	(unitless)
n_i	number of moles of species i	mol
N	total number of particles in a system	(unitless)
N_A	Avogadro's number	mol^{-1}
N_i	number of molecules or atoms of species i	(unitless)
Ox	denotes oxidized form of a chemical species	
$[OH^-]_W$	hydroxyl ion concentration from self-dissociation of water	mol L^{-1}
P	total pressure	(pressure units)
P	number of phases in a system	(unitless)
P_c	critical pressure	(pressure units)
P_i	partial pressure of component i or species i	(pressure units)
P_W	partial pressure of water	(pressure units)
P^o	standard pressure (= 1 bar = 100 000 Pa)	(pressure units)
$P_i^\bullet$	equilibrium vapor pressure of pure component i	(pressure units)
PE	potential energy	J
pH	$pH \equiv -log_{10}\ a_{H^+}$	(unitless)
pOH	$pOH \equiv -log_{10}\ a_{OH^-}$	(unitless)
pX	$pX \equiv -log_{10}\ X$	(unitless)
Q	heat	J
Q_{rev}, Q^{rev}	reversible heat	J
Q	electric charge	C (coulombs)
Q	reaction quotient	(unitless)
r	rate of chemical reaction	mol L^{-1} s^{-1}
R	universal gas constant	J mol^{-1} K^{-1}
Red	denotes reduced form of a chemical species	
s	speed of an individual gas molecule	m s^{-1}
s	solubility of species i in aqueous solution	(concn units)
s	molar entropy	J K^{-1} mol^{-1}
$\bar{s}$	specific entropy	J K^{-1} kg^{-1}
$s^o_{298.15}$	standard (absolute) entropy at 25°C	J K^{-1} mol^{-1}
S	entropy	J K^{-1}

$\Delta S°$	standard entropy change	J K^{-1}
ΔS_R^o	standard reaction entropy change	J K^{-1}
STP	standard temperature and pressure (= 0°C, 1 atm)	
t	time	(units of time)
$t_{1/2}$	half-life of a chemical or nuclear reaction	(units of time)
T	temperature	K
T_c	critical temperature	K
u	molar internal energy	J mol^{-1}
$\bar{u}$	specific internal energy	J kg^{-1}
U	internal energy	J
$(U_m)_i$	partial molar internal energy of species i	J mol^{-1}
ΔU_P	change in internal energy for a constant pressure process	J
ΔU_V	change in internal energy for a constant volume process	J
v	molar volume	m^3 mol^{-1}
v_c	molar volume at the critical point	m^3 mol^{-1}
V	volume	m^3
$(V_m)_i$	partial molar volume of species i	m^3 mol^{-1}
W	total mass of gas particles in a system	kg
W	work	J
W_{rev}, W^{rev}	reversible work	J
w_{elec}	electrical work	J = V C
W_{PV}	PV-work	J
$W°$	"other" work" (work *other* than PV-work)	J
$W_{rev}°$	reversible (maximum or minimum) "other" work	J
x_i	mole fraction of species i or component i	(unitless)
y_i	mole fraction of species i in the gas or vapor phase	(unitless)
z	height or elevation	m
z_i	valence of species i	(unitless)
z_+	valence of cation (positive number)	(unitless)
z_-	valence of anion (negative number)	(unitless)
Z	compressibility factor ($Z = Pv/RT$)	(unitless)
$\rightleftharpoons$	denotes equilibrium condition for a process or reaction	
↑	increases	
↑	evolved gas	
↓	decreases; precipitate, deposit	

GREEK LETTERS

α	fractional degree of dissociation or of decomposition	(unitless)
δ	infinitesimal incremental amount of a path function	
Δ	change in a property (= *final value – initial value*)	
$^1\Delta^2B$	$\equiv B_2 - B_1$	
Δz	elevation	m
ΔT_b	boiling point elevation (increase of boiling point above that of pure solvent)	K
$|\Delta T_f|$	freezing point depression (lowering of freezing point below that of pure solvent)	K
ΔON	change in oxidation number or oxidation state	(unitless)
ϕ	voltage	V
$\Delta \phi$	potential difference, single electrode potential difference	V
γ_i	single ion molal ionic activity coefficient	(unitless)
$\gamma_\pm$	mean molal stoichiometric ionic activity coefficient	(unitless)
γ_+	molal ionic activity coefficient for a cation	(unitless)
γ_-	molal ionic activity coefficient for an anion	(unitless)
γ^*	molal ionic activity coefficient of activated complex	(unitless)
$\gamma_\pm$	mean molal ionic activity coefficient for an electrolyte	(unitless)
μ_i	chemical potential of species i ; $\mu_i \equiv \left(\partial G / \partial n_i \right)_{T,P,n_j}$	J mol^{-1}
μ_w^L	chemical potential of water in the liquid phase	J mol^{-1}
μ_w^V	chemical potential of water in the vapor phase	J mol^{-1}
μ_i^o	chemical potential of species i in the standard state	J mol^{-1}
μ_L^o	chemical potential of liquid solvent in standard state (pure liq.)	J mol^{-1}
μ_w^*	chemical potential of pure liquid solvent	J mol^{-1}
ν	$\nu_+ + \nu_-$	(unitless)
ν_+	stoichiometric coefficient for cation	(unitless)
ν_-	stoichiometric coefficient for anion	(unitless)
π	osmotic pressure	(pressure units)
θ	temperature	°C
ρ	density	kg m^{-3}
υ	velocity	m s^{-1}

SUBSCRIPTS

AB	denotes a change in some property in going from state A to state B; e.g., W_{AB} is the work done in passing from state A to state B
an	refers to anode
atm	of the atmosphere
bp	boiling point
c	critical state
cat	refers to cathode
comb	combustion
eqm	denotes equilibrium value
ext	denotes "external" value of a property; i.e., the value in the surroundings, external to the system
f	denotes final state
f	denotes the process of formation; e.g., Δg_f^o
fp	freezing point
fus	fusion (melting)
g	gas or vapor
gas	denotes a property of a gas; e.g., V_{gas} is the volume of the gas or of the gas phase
i	denotes initial state
irrev	for an irreversible process
L	liquid
liq	liquid
max	denotes maximum value
min	denotes minimum value
n	at constant number of moles
nbp	normal boiling point
nfp	normal freezing point
n_j	denotes that the moles of all other species except species i are held constant
P	at constant pressure
rev	for a reversible process
s	solid
soln	refers to solution phase
SP	solubility product
STP	standard temperature and pressure ($= 0°C$, 1 atm pressure)
sub	sublimation
surr	denotes the surroundings

syst	denotes the system
T	at constant temperature
t	refers to value at *time* $= t$
univ	denotes the universe (*system + surroundings*)
V	at constant volume
v	vapor
vap	vaporization
W	water, solvent
x	in the x-direction
y	in the y-direction
z	in the z-direction
12	denotes a change in some property in going from state *1* to state *2*; e.g., Q_{12} is the heat transferred in the transformation from state *1* to state *2*
o	denotes "stoichiometric concentration"; e.g., c_o, $[NaCl]_o$
o	refers to initial value; e.g., r_o = initial rate of chemical reaction
+	denotes a cationic species
–	denotes an anionic species

SUPERSCRIPTS

L	liquid phase
V	vapor phase
S	solid phase
soln	solution phase
m	order of a chemical reaction
n	order of a chemical reaction
(m)	refers to molal concentration scale
α	refers to phase α
β	refers to phase β
dot ($\cdot$)	denotes a *rate* (e.g., $\dot{W}$ is the rate of doing work, in J s^{-1})
macron	denotes the specific value of a property; i.e., the value per unit mass (e.g., $\bar{v}$ is the specific volume in m^3 kg^{-1})
o	degrees Celsius
o	denotes standard conditions
o	"other"; e.g., W^o denotes work "other" than PV-work

PREFIXES

Submultiple	Prefix	Symbol		Multiple	Prefix	Symbol
10^{-1}	deci-	d		10	deka-	da
10^{-2}	centi-	c		10^2	hecto-	h
10^{-3}	milli-	m		10^3	kilo-	k
10^{-6}	micro-	μ		10^6	mega-	M
10^{-9}	nano-	n		10^9	giga-	G
10^{-12}	pico-	p		10^{12}	tera-	T
10^{-15}	femto-	f		10^{15}	peta-	P
10^{-18}	atto-	a		10^{18}	exa-	E
10^{-21}	zeyto-	z		10^{21}	zetta-	Z
10^{-24}	yocto-	y		10^{24}	yotta-	Y

PHYSICAL CONSTANTS

		Currently accepted accurate value	Value to be used when solving problems in this text
e_o	elementary charge	$1.60217733(49) \times 10^{-19}$ C	1.602×10^{-19} C
F	Faraday's constant	$96{,}485.309(29)$ C mol^{-1}	$96{,}487$ C mol^{-1}
g	acceleration due to gravity	9.80665 m s^{-2} (exact)	9.806 m s^{-2}
k	Boltzmann's constant	$1.380658(12) \times 10^{-23}$ J K^{-1}	1.381×10^{-23} J K^{-1}
m_p	rest mass of the proton	$1.6726231(10) \times 10^{-27}$ kg	1.673×10^{-27} kg
m_n	rest mass of the neutron	$1.6749286(10) \times 10^{-27}$ kg	1.675×10^{-27} kg
m_e	rest mass of the electron	$9.1093897(54) \times 10^{-31}$ kg	9.109×10^{-31} kg
N_A	Avogadro's number	$6.0221367(36) \times 10^{23}$ mol^{-1}	6.022×10^{23} mol^{-1}
R	Universal gas constant	$8.314510(70)$ J mol^{-1} K^{-1}	$8.314(5)$ J mol^{-1} K^{-1}

CONVERSION FACTORS

1 cal (calorie) = 4.184 J

1 lb (pound) = 453.59 g

1 ft (foot) = 12 in (inches)

1 in = 2.54 cm

1 L $= 10^{-3}$ m^3

1 cm$^3 = 1$ mL $= 10^{-6}$ m^3

1 US gal = 3.785 L

1 Cdn gal = 4.546 L

$\ln x = \ln 10 \cdot \log x$

1 atm = 101 325 Pa

1 atm = 1.01325 bar

1 atm = 760 Torr

THINGS YOU SHOULD KNOW
BUT PROBABLY FORGOT

1.1 BASIC DEFINITIONS

Force—is an external agency capable of altering the state of rest or of motion of a body; defined by Newton's Law: $F = ma$, where F is the force in newtons (N), m is the mass in kilograms (kg), and a is the acceleration in m s^{-2}. Note the equivalency of units: N = kg m s^{-2}.

Mass—from observation, the acceleration induced in a body is directly proportional to the force applied to the body: $a \propto F$; i.e., $a = kF$ where k is the constant of proportionality. We call the ratio of the applied force to the induced acceleration (i.e., k^{-1}) the *mass m* of the body; thus, $m \equiv F/a$, where m is the mass in kg.

Electron—an elementary particle of rest mass $9.1093897 \times 10^{-31}$ kg (1/1836.152701 times the rest mass of a proton), having a negative charge of 1.602177×10^{-19} coulombs, and a classical radius of 2.81794092 fm (fm = femtometer = 10^{-15} m). This is very small! The size of an electron is to the size of a person what the latter is to the size of the Milky Way galaxy. Of all the subatomic particles that have mass, the electron is the smallest in size and the lightest in mass. Discovered in 1897 by Joseph John Thomson, the electron has never yielded any indication of having any further internal structure; it seems truly to be an elementary dot.

Electricity—a general term for all phenomena caused by electric charge.

Electric charge—Although we can describe, measure, and predict with considerable accuracy the properties and behavior of charged matter, we still don't know what "charge" is! The basic unit of negative charge is the charge of the electron, while the basic unit of positive charge is the charge of the proton, the latter having the same magnitude but opposite sign to that of the former. It is observed that like charges repel each other, whereas negative charges attract. These positive and negative conventions are purely arbitrary, even though much of science is based on them. In 1752 Benjamin Franklin called the charge—i.e., the static electricity—on a rubbed glass rod *positive* and that on a rubbed amber rod *negative*. Based on this convention, the charge of an electron turns out to be negative.

Electric field—the region in which these electric forces act.

Ion—an electrically charged atom or group of atoms.

Atom—the smallest portion of an element that can take part in a *chemical* reaction.

Element—a substance consisting entirely of atoms each having the same number of protons in the nucleus (i.e., the same *atomic number*). There are slightly more than 100 naturally occurring or artificially produced elements.

Isotope—Isotopes are forms of an element with very nearly the same chemical properties but different numbers of neutrons in their nuclei and hence different atomic masses. For example, in the case of carbon (atomic number 6), $^6C^{12}$ represents a carbon atom with 6 protons and 6 neutrons in its nucleus; this atom is a C-12 isotope. On the other hand, $^6C^{13}$ represents a carbon-13 isotope, having 6 protons and 7 neutrons in its nucleus.

Molecule—the smallest portion of a substance capable of existing independently while retaining the properties of the original substance.

Elementary particles—include electrons, protons, neutrons, photons, positrons, antiprotons, antineutrons, neutrinos, quarks, pions, mesons, gluons, gravitons, etc. In chemistry we deal primarily with protons, neutrons, electrons, and photons.

1.2 SI UNITS

The International System (*Système International*, SI) of units has been adopted for use internationally because of the following advantages: First, it is both metric and decimal. Second, fractions have been eliminated, multiples and submultiples being indicated by a system of standard prefixes, greatly simplifying calculations. SI provides a direct relationship between mechanical, electrical, nuclear, chemical, thermodynamic, and optical units, thereby forming a coherent system. There is no duplication of units for the same physical quantity, and all derived units are obtained by direct one-to-one relationships between base or other derived units. The seven base units are defined below in Table 1:

Table 1. *Base SI Units and Their Definitions*

Physical Quantity	Unit	Symbol	Definition
mass	kilogram	*kg*	The *kilogram* is equal to the mass of the international prototype of the kilogram, which is a cylinder of iridium-platinum alloy, 39 mm in diameter and 39 mm high, kept at Sèvres, France. [1901][1]
length	meter	*m*	The *meter* is the length of the path travelled by light in vacuum during a time interval of 1/299,792,458 s. [1983]
time	second	*s*	The *second* is the duration of 9,192,631,770 periods of the radiation corresponding to the transition between the two hyperfine levels ($F = 4$, $m_F = 0$ to $F = 3$, $m_F = 0$) of the ground state of the cesium-133 atom. [1967]
temperature	kelvin	*K*	The *kelvin* is the unit of thermodynamic temperature, and is the fraction 1/273.16 of the thermodynamic temperature of the triple point of water. [1967]

[1] Date of definition.

amount of substance	mole	*mol*	The *mole* is the amount of substance of a system that contains as many elementary entities as there are atoms in exactly 0.012 kg of carbon-12. When the mole is used, the elementary entities must be specified and may be atoms, molecules, ions, electrons, other particles, or specified groups of such particles. In this definition, it is understood that the carbon atoms are unbound, at rest, and in their ground state. [1971]
electric current intensity	ampere	*A*	The *ampere* is that constant current which, if maintained in two straight parallel conductors of infinite length, of negligible circular cross-section, and placed one meter apart in vacuum, would produce between these conductors a force equal to 2×10^{-7} newton per meter of length. [1948]
luminous intensity	candela	*cd*	The *candela* is the luminous intensity, in a given direction, of a source that emits monochromatic radiation of frequency 540×10^{12} Hz and that has a radiant intensity in that direction of 1/683 watts per steradian. [1979]

1.3 DALTON'S ATOMIC THEORY (1830)

John Dalton's Atomic Theory forms the basis of chemistry. The three postulates of the theory are:

(1) There exist indivisible atoms.
(2) Each atom of a given element has the same atomic mass.
(3) Atoms combine in a variety of simple, whole number ratios to form compounds.

Although generally valid, there are some notable exceptions to these postulates:

Non-stoichiometric compounds—Some substances, such as titanium oxide, actually have a *range* of compositions. In the case of titanium oxide, the compositions range from $Ti_{0.69}O$ to $Ti_{0.75}O$. This compound does not appear to obey the third postulate. In fact, titanium oxide consists of a homogeneous mixture of TiO and Ti_2O_3 in varying proportions. Another non-stoichiometric compound is zinc oxide (*ZnO*), whose actual composition may correspond to something such as $Zn_{1.02}O$. In this case, there are some extra neutral Zn atoms wedged into the crystal lattice of the material.

Isotopes—have natural variations in the number of neutrons in the nucleus, and therefore show that the same element is capable of having different atomic masses. This is in disagreement with the second postulate.

Atom splitting—it is possible to shatter atoms by bombarding their nuclei with high energy particles such as protons that have been accelerated to velocities approaching the speed of light. This violates the first postulate; however, the energies of particles involved in *chemical* reactions are not high enough to produce such phenomena.

1.4 STOICHIOMETRY

Stoichiometry is the branch of chemistry that deals with the combining proportions of elements or compounds involved in chemical reactions, and the methods of calculating them. The word comes from the Greek words *stoicheion*, meaning "element" plus *-metry*, "to measure." Stoichiometry makes use of three combining laws:

(1) The Law of Definite Proportions—This law deals with the *mass composition* of compounds, and states:

> *"The mass percent of the elements in a given compound is the same, regardless of how the compound is made."*

Thus, water is always 11.1% by mass hydrogen and 88.9% by mass oxygen, regardless of the method of preparation. (Strictly speaking, there may be small errors arising from non-stoichiometric compounds or isotopic variations; but such errors usually are too small to be significant.)

(2) The Law of Multiple Proportions—this law deals with the *formation of compounds* from atoms, and states:

> *"If two elements form more than one compound, then the different masses of one element that combine with the same mass of the other are in the ratio of small, whole numbers."*

Example 1-1

Consider FeO and Fe_2O_3: In FeO, 55.8 g of iron combines with 16.0 g of oxygen. In Fe_2O_3, 111.7 g of iron combines with 48.0 g of oxygen. Thus, in the case of FeO, the amount of iron combining with *one* gram of oxygen is 55.8/16.0 = 3.5; while in the case of Fe_2O_3 it is 111.7/48.0 = 2.3. According to the *Law of Multiple Proportions*, the numbers 3.5 and 2.3 should be in the ratio of simple, whole numbers. Is this true? Yes! *3.5/2.3* is the same as *3/2*, which is a ratio of simple, whole numbers.

(3) The Law of Equivalent Proportions—this law deals with elements and compounds *reacting* with each other, and states:

> *"The masses of various elements or compounds that combine with some fixed mass of some other element or compound taken arbitrarily as a standard, also react with one another in the same ratio or some simple multiple or submultiple thereof."*

This sounds more complicated than it really is. Thus, if *x* grams of *A* reacts with *y* grams of *B* to form *C*, and *x* grams of *A* also reacts with *z* grams of *D* to form *E*, then, when *B* reacts with *D*, it should react in a multiple (or submultiple) of *y:z*.

Example 1-2

Consider the following two reactions:

$$3H_2 + N_2 \rightarrow 2NH_3 \qquad \text{and} \qquad H_2 + \tfrac{1}{2}O_2 \rightarrow H_2O$$

$$\begin{array}{ccc} 1\text{ g} & 4.66\text{ g} & 5.67\text{ g} \end{array} \qquad\qquad \begin{array}{ccc} 1\text{ g} & 8\text{ g} & 9\text{ g} \end{array}$$

One gram of hydrogen reacts with 4.66 g of *nitrogen* to produce ammonia; one gram of hydrogen also reacts with 8 g of *oxygen* to produce water; thus, according to the *Law of Equivalent Propor-*

tions, when *nitrogen* reacts with *oxygen*, we expect these substances to combine in multiples or submultiples of N/O = 4.66/8 = 0.583. Consider the mass ratios of the reactants in some of the possible reactions between N_2 and O_2:

$$2N_2 + 3O_2 \rightarrow 2N_2O_3 \qquad \frac{N}{O} = \frac{2(28)}{3(32)} = 0.583 \qquad \left. \frac{0.583}{0.583} = \frac{1}{1} \right. \quad \text{(same)}$$

$$2N_2 + O_2 \rightarrow 2N_2O \qquad \frac{N}{O} = \frac{2(28)}{32} = 1.75 \qquad \left. \frac{1.75}{0.583} = \frac{3}{1} \right. \quad \text{(simple multiple)}$$

$$N_2 + 2O_2 \rightarrow 2NO_2 \qquad \frac{N}{O} = \frac{28}{2(32)} = 0.438 \qquad \left. \frac{0.438}{0.583} = \frac{3}{4} \right. \quad \text{(simple submultiple)}$$

1.5 EQUIVALENT WEIGHT (*EW*)

The *equivalent weight* of a substance is defined as *that mass of the substance that reacts with one mole of electrons*. The equivalent weight also is referred to as the *combining weight*. Although not an SI quantity—SI deals only with *moles* of substance—the *equivalent weight* is a useful concept still widely used in analytical chemistry, because one equivalent of one substance always combines with exactly one equivalent of another substance; the same cannot be said for moles. For example, one *equivalent* of H_2SO_4 (= 49 g) combines with one *equivalent* of NaOH (= 40 g); but 0.5 *moles* of H_2SO_4 combines with 1.0 *mole* of NaOH. Note that one mole of electrons is just *Avogadro's number* (N_A) of electrons: 6.022×10^{23} electrons).

Example 1-3

24.305 g of Mg reacts with 16.00 g of O_2 to form MgO according to $Mg + \frac{1}{2}O_2 \rightarrow MgO$.

The magnesium is oxidized from an oxidation state of *0* to an oxidation state of *+2*; i.e., $Mg \rightarrow Mg^{2+} + 2\ e^-$. One mole of Mg (= 24.305 g) releases 2 moles of electrons; therefore, the weight of Mg that releases one mole of electrons is 24.305/2 = 12.15 g. In this reaction the equivalent weight of Mg is 12.15 g. Also, in the same reaction, 16.00 g of O_2 reacts with 2 moles of electrons: $\frac{1}{2}O_2 + 2e^- \rightarrow O^{2-}$, therefore the equivalent weight of oxygen in this reaction is 16.00/2 = 8.00 g.

$$\boxed{\ Equivalent\ Weight\ = \ \frac{molar\ mass}{z}\ }$$

where z is a whole number.

Since oxides are so abundant in nature, sometimes we use oxygen as a secondary definition of equivalent weight: *The equivalent weight of an element is that weight which combines with or displaces 8.000 g of oxygen.*

Example 1-4

11.19 g of hydrogen reacts with 88.81 g of oxygen to form 100.00 g of water according to the reaction $H_2 + \frac{1}{2}O_2 \rightarrow H_2O$. Therefore, $(8.00/88.81) \times (11.19) = 1.008$ g of hydrogen combines with 8.00 g of oxygen; therefore, the equivalent weight of hydrogen is 1.008 g.

Note that the equivalent weight of a substance can *vary*, depending on the reaction. Thus, for $Fe^{2+} + 2e^- \rightarrow Fe$, the equivalent weight of iron is (molar mass)/2 = 58.85/2 = 29.43 g; but for $Fe^{2+} \rightarrow Fe^{3+} + e^-$, the equivalent weight of iron is (molar mass)/1 = 58.85/1 = 58.85 g.

1.6 AMOUNT OF SUBSTANCE: THE MOLE

The number of particles in one mole of substance is given by **Avogadro's number** N_A, which is defined as the number of carbon-12 atoms in exactly 0.012 kg (i.e., 12.000... g) of carbon-12. The currently accepted value of Avogadro's number is $N_A = 6.0221367 \times 10^{23}$ mol^{-1}. The **molar mass**, M (formerly called the *molecular weight*), of a substance is *the mass of one mole of the substance*. In SI units the molar mass is given in kg mol^{-1}; thus, water has a molar mass of 0.018015 kg mol^{-1}. For convenience, however, we still often refer to the "molecular weight" of water as being 18.015 (the units being understood to be g mol^{-1}). To avoid calculational errors, care must be taken when using SI units to remember to express the molar mass of a substance in kg mol^{-1}, and not in g mol^{-1}.

The number of moles of a substance often is represented by the symbol n:

$$No.\ moles = n = \frac{mass\ of\ sample}{molar\ mass\ of\ sample} = \frac{m}{M}$$

Molecular Formulas

The *molecular formula* of a substance gives the actual composition of one molecule of the substance: e.g., P_4S_{10}, C_6H_6, H_2.

Empirical Formulas

In the *empirical formula* of a substance, the subscripts are the smallest whole numbers that describe the ratios of the atoms in the substance: e.g., P_2S_5, CH, H. Empirical formulas are less informative than the molecular formulas.

1.7 AVOGADRO'S HYPOTHESIS (1811)

Avogadro's hypothesis originally stated: *"I gas a pari condizioni di pressione e di temperatura in eguali volumi contengono un egual numero di molecole."*

A more useful statement is: 22.41410×10^{-3} m^3 (= 22.41410 L) of an ideal gas at *Standard Temperature and Pressure (STP)* contains exactly one mole of gas (i.e., 6.0221367×10^{23} particles of gas). STP is defined as 0°C (273.15 K) and one **standard atmosphere** pressure (101 325 Pa). If one **bar** (= 10^5 Pa) is taken as the standard pressure, the volume is 22.71108×10^{-3} m^3.

Example 1-5

Silicon combines with fluorine to form two different gaseous fluorides. The silicon content of fluoride A is 26.9% by weight, and that of fluoride B is 32.9% by weight.

(a) Determine the empirical formula for each compound.
(b) Show that the *Law of Multiple Proportions* is obeyed.
(c) If 22.4 L at STP of compound A as a vapor weighs 104 g and the same volume at STP of compound B weighs 170 g, determine the molecular formula for each compound.

Solution

(a) Molar masses (g mol^{-1}): Si = 28.09; F = 19.00.   Basis: 100 g of each compound.
In 100 g of compound A there are 26.9 g of Si and 100 – 26.9 = 73.1 g of F.

$$\text{No. mol of Si} = \frac{26.9 \text{ g}}{28.09 \text{ g mol}^{-1}} = 0.958 \text{ mol}$$

$$\text{No. mol of F} = \frac{73.1 \text{ g}}{19.00 \text{ g mol}^{-1}} = 3.85 \text{ mol}$$

$$\frac{\text{mol F}}{\text{mol Si}} = \frac{3.85}{0.958} = 4.02 \approx \frac{4}{1}$$

Therefore, in compound A there are 4 moles of fluorine for each mole of silicon, and the simplest (empirical) formula is SiF_4.

Similarly, for 100 g of compound B there are 32.9 g of Si and 100 – 32.9 = 67.1 g of F.

$$\text{No. mol of Si} = \frac{32.9 \text{ g}}{28.09 \text{ g mol}^{-1}} = 1.17 \text{ mol}$$

$$\text{No. mol of F} = \frac{67.1 \text{ g}}{19.00 \text{ g mol}^{-1}} = 3.53 \text{ mol}$$

$$\frac{\text{mol F}}{\text{mol Si}} = \frac{3.53}{1.17} = 3.02 \approx \frac{3}{1}$$

Therefore, in compound B there are 3 moles of fluorine for each mole of silicon, and the simplest (empirical) formula is SiF_3.

Ans: $\boxed{SiF_4 \text{ and } SiF_3}$

(b) In compound A, 73.1 g of fluorine combines with 26.9 g of silicon. Therefore, the mass of fluorine that combines with *one* g of silicon is 73.1/26.9 = 2.72 g of fluorine. Similarly, in compound B, 67.1 g of fluorine combines with 32.9 g of silicon; therefore the mass of fluorine that combines with 1.00 g of silicon is 67.1/32.9 = 2.04 g. According to the *Law of Multiple Proportions* these two masses—2.72 and 2.04—should be in the ratio of simple whole numbers.

Thus, $\dfrac{2.72}{2.04} = 1.333 = \dfrac{4}{3}$, which is the ratio of two simple whole numbers. The Law is obeyed!

(c) Since one mole of any ideal gas or vapor occupies 22.4 L at STP, and 22.4 L of compound A at STP weighs 104 g, it follows that 104 g of compound A is one mole of compound A. Therefore the molar mass of compound A is 104 g. The formula weight of SiF_4—the empirical formula for the compound—is $28.09 + 4(19.00) = 104.09$ g, which is the same as the actual molar mass; therefore the molecular formula is the same as the empirical formula, namely, SiF_4. Similarly, one mole of compound B weighs 170 g. The formula weight of compound B is $28.09 + 3(19.00) = 85.09$ g, which is *one-half* the actual molar mass; therefore, the molecular formula for compound B must be *twice* the empirical formula, i.e., Si_2F_6.

Ans: $\boxed{SiF_4 \text{ and } Si_2F_6}$

Example 1-6

The molecular formula for vitamin B_1 is $C_{12}H_{18}N_4OCl_2S$ (molar mass = 337.27).
(a) How many moles of nitrogen atoms are there in 55.0 g of vitamin B_1?
(b) What is the mass of elemental chlorine in 75.0 g of vitamin B_1?
(c) How many moles of vitamin B_1 contain 25.0 g of carbon?
(d) What mass of vitamin B_1 contains 2.7×10^{23} nitrogen atoms?

Solution

(a) 337.27 g (= 1 mole) of vitamin B_1 has 4 moles of N atoms;

therefore 55 g of vitamin B_1 has $\left(\dfrac{55}{337.24}\right)(4) = 0.652$ moles of N atoms.

Ans: $\boxed{0.652 \text{ mol}}$

(b) 337.27 g of vitamin B_1 contains $2(35.453) = 70.906$ g of Cl;

therefore 75.0 g of vitamin B_1 contains $\left(\dfrac{75.0}{337.24}\right)(70.906) = 15.769$ g of Cl.

Ans: $\boxed{15.8 \text{ g}}$

(c) $12(12.01) = 144.12$ g of C is contained in 1.000 mole of vitamin B_1;

therefore 25 g of C is contained in $\left(\dfrac{25.0}{144.12}\right)(1.000) = 0.1735$ moles of vitamin B_1.

Ans: $\boxed{0.174 \text{ mol}}$

(d) 2.7×10^{23} nitrogen atoms is $\left(\dfrac{2.7 \times 10^{23}}{6.022 \times 10^{23} \text{ mol}^{-1}}\right) = 0.4484$ moles of N atoms.

4 moles of N atoms is contained in 337.27 g of vitamin B_1;

therefore 0.4484 moles of N atoms is contained in $\left(\dfrac{0.4484}{4}\right)(337.27) = 37.80$ g of vitamin B_1.

Ans: $\boxed{38 \text{ g}}$

1.8 CONSERVATION OF MASS

The total mass on the left-hand side (*LHS*) of a chemical equation is equal to the total mass on the right-hand side (*RHS*). This "law" of conservation of mass neglects minor gains or losses of mass resulting from Einstein's mass-energy relationship ($e = mc^2$). Because of the relatively low energies involved in the making and breaking of bonds in chemical reactions (in comparison with the energies involved in *nuclear* reactions), such minor violations of the *Law of Conservation of Mass* are almost always insignificant in chemical reactions. The law of conservation of mass as used in chemistry also ignores the mass of *electrons*; e.g., we assume that one mole of calcium *ion* (Ca^{2+}) has the same mass as one mole of calcium *atom* (Ca), namely, 40.08 g. In fact, because it has lost two electrons, the mass of one mole of calcium ions is less than that of one mole of calcium atoms by the mass of two moles of electrons, which is equal to about 0.0011 g, representing an error of only about 0.03%, which usually can be ignored.

A *balanced* chemical equation must have the same *number* of *atoms* of each element on each side of the equation:

Balanced: $3C_2H_6O + 2Na_2Cr_2O_7 + 8H_2SO_4 \rightarrow 3C_2H_4O_2 + 2Cr_2(SO_4)_3 + 2Na_2SO_4 + 11H_2O$

Unbalanced: $C_2H_6O + Na_2Cr_2O_7 + H_2SO_4 \rightarrow C_2H_4O_2 + Cr_2(SO_4)_3 + Na_2SO_4 + H_2O$

Example 1-7

Hydrogen sulfide gas (the gas that smells like rotten eggs) is a common contaminant of natural gas. After H_2S is separated from natural gas it can be converted to sulfur by the following two-step process:

$H_2S + O_2 \rightarrow H_2O + SO_2$ followed by $H_2S + SO_2 \rightarrow S\downarrow + H_2O$ (Both equations unbalanced!)

If air is 20.0 mol % oxygen, what is the minimum number of cubic meters of air at STP required to produce one tonne of sulfur by this process? Air may be assumed to behave as an ideal gas.

Solution

First, the two steps of the reaction are balanced, then combined to give one overall reaction:

$H_2S + \frac{3}{2}O_2 \rightarrow H_2O + SO_2$ From the overall reaction it can be seen that $\frac{3}{2}$ moles of O_2 is

$2H_2S + SO_2 \rightarrow 3S\downarrow + 2H_2O$ required to produce *3* moles of S; that is, $\frac{1}{2}$ mole of O_2 is re-

$\overline{3H_2S + \frac{3}{2}O_2 \rightarrow 3S\downarrow + 3H_2O}$ quired for each *one* mole of S produced.

Therefore, mol of O_2 required $= \left(\dfrac{0.5 \text{ mol } O_2}{\text{mol S}}\right)(1 \text{ tonne S})\left(\dfrac{1000 \text{ kg S}}{\text{tonne S}}\right)\left(\dfrac{1 \text{ mol S}}{0.03206 \text{ kg S}}\right)$

$= 15\ 596 \text{ mol } O_2$

At STP one mole of air occupies 22.414 L = 0.022414 m^3.

But one mole of air contains 0.20 moles of O_2; therefore,

No. m^3 of air required $= \dfrac{15{,}596 \ \text{mol} \, O_2}{\left(\dfrac{0.20 \ \text{mol} \, O_2}{0.022414 \ \text{m}^3 \, \text{air}}\right)} = 1748 \ \text{m}^3$ air.

Ans: $\boxed{1.75 \times 10^3 \ \text{m}^3}$

Exercise 1–1

Ammonia (NH_3) is made by the Haber process: $\quad N_2 + 3H_2 \xrightarrow[\substack{high \ pressure}]{catalyst} 2NH_3$

The nitrogen required for this process is obtained from air by using the air to burn natural gas (mostly CH_4), thereby removing the oxygen from the air, leaving behind only nitrogen. The hydrogen required is obtained by steam reforming natural gas to carbon monoxide and hydrogen, followed by a shift conversion reaction to increase the yield of hydrogen:

$$CH_4 + H_2O_{(g)} \xrightarrow{\ catalyst\ } CO + 3H_2 \qquad \text{(Steam Reforming)}$$

$$CO + H_2O_{(g)} \xrightarrow{\ catalyst\ } CO_2 + H_2 \qquad \text{(Shift Conversion)}$$

How many cubic meters of natural gas at STP are required to produce one tonne of ammonia? Air may be assumed to consist of 80.0 mol % nitrogen and 20.0 mol % oxygen. For the purpose of calculation, you may assume the complete stoichiometric combustion of methane in air to produce carbon dioxide and water, and that no nitrogen oxides are formed. ☺

1.9 CONSERVATION OF CHARGE

It is one of the fundamental principles of physics that electric charge can be neither created nor destroyed. For chemistry this means that the total charge on the left-hand side of a chemical reaction must be the same as the total charge on the right-hand side, as illustrated in the following examples:

Reaction	Charge on LHS	Charge on RHS
$H_2 \rightarrow 2H^+ + 2e^-$	0	$+2 + 2(-1) = 0$
$Zn + Cu^{2+} \rightarrow Zn^{2+} + Cu$	$0 + 2 = +2$	$+2 + 0 = +2$
$H^+ + Cl^- \rightarrow HCl$	$+1 + (-1) = 0$	0
$H_2O \rightarrow H^+ + OH^-$	0	$+1 + (-1) = 0$
$2Cl^- \rightarrow Cl_2 + 2e^-$	$2(-1) = -2$	$0 + 2(-1) = -2$

1.10 ATOMIC MASS SCALES

Atomic mass scales are scales of the relative masses of the various elements. Historically, since hydrogen was the lightest element, it was arbitrarily chosen as the standard, and assigned a relative mass of unity: thus, H ≡ 1, or H_2 ≡ 2. From the chemical analysis of water, it is found that oxygen (O) is 16 times heavier than hydrogen (H); therefore, on this scale of relative weights it follows that O = 16. H and O both became standards; but, gradually oxygen became the more prevalent standard since more elements combine with oxygen than with hydrogen. Therefore, up until 1961, by *definition*, chemists assigned naturally occurring oxygen a value of exactly 16 for its molar mass: i.e., O $\equiv$ 16.0000 and $O_2 \equiv$ 32.0000. However, naturally occurring oxygen also contains 0.04% $^8O^{17}$ and 0.20% $^8O^{18}$ isotopes in addition to 99.78% $^8O^{16}$. Physicists, on the other hand, preferred to deal with single atoms, rather than mixtures of isotopes; therefore they based *their* atomic mass scale on arbitrarily assigning $^8O^{16} \equiv 16.0000$. Therefore, for the physicists, naturally occurring oxygen—because it also contains some $^8O^{17}$ and $^8O^{18}$—had a molar mass that was *greater* than 16.0000! This situation

Up until 1961	Chemists	Physicists
Natural O	16.0000	16.0044
Isotope O^{16}	15.9956	16.0000

caused a lot of confusion, so in 1961 *both* scales were abandoned and a uniform scale was adopted by both the chemists and the physicists; this new scale is based on the most abundant isotope of carbon, carbon-12. In the new (1961) atomic mass scale, carbon-12 is assigned a value of *exactly* 12.0000: $^6C^{12} \equiv 12.0000$. Since naturally occurring carbon contains 98.892% $^6C^{12}$ (molar mass 12.0000) and 1.108% $^6C^{13}$ (molar mass 13.00335), its molar mass is therefore

$$(0.98892)(12.0000) + (0.01108)(13.00335) = 12.01115.$$

This is the value that is listed in tables of atomic masses. Note that atomic masses are only *relative* masses, not the *actual* masses of atoms. The *actual* mass of a single carbon atom is about 1.994×10^{-26} kg.

Example 1-8

In tables of molar masses, the molar mass of chlorine is listed as Cl = 35.453 g mol^{-1}. Naturally occurring chlorine consists of a mixture of two isotopes, Cl^{35} and Cl^{37}, having molar masses of 34.969 and 36.966 g mol^{-1}, respectively. What is the mole percent of these two isotopes in naturally occurring chlorine?

Solution

Consider 100 moles of chlorine:

Let the mole percent of Cl^{35} be x %; therefore, the mole percent Cl^{37} will be $(100 - x)$ %.

$\{$Mass of 100 mol of Cl$\} = \{$mass of x mol $Cl^{35}\} + \{$mass of $(100 - x)$ mol $Cl^{37}\}$

$$(100 \text{ mol Cl})\left(\frac{35.453 \text{ g Cl}}{1 \text{ mol Cl}}\right) = (x \text{ mol } Cl^{35})\left(\frac{34.969 \text{ g } Cl^{35}}{1 \text{ mol } Cl^{35}}\right) + \left\{(100 - x) \text{ mol } Cl^{37}\right\}\left(\frac{36.966 \text{ g } Cl^{37}}{1 \text{ mol } Cl^{37}}\right)$$

$$3545.3 = 34.969x + 3696.6 - 36.966x$$

$$1.997x = 151.3$$

$$x = \frac{151.3}{1.997} = 75.76.$$

Therefore, natural chlorine consists of 75.76 mol % Cl^{35} and $(100 - 75.76) = 24.24$ mol % Cl^{37}.

Ans: $\boxed{75.76 \text{ mol } \% \ Cl^{35}, 24.24 \text{ mol } \% \ Cl^{37}}$

Exercise 1–2

What is the *mass* percent of the two isotopes in naturally-occurring chlorine? ☺

Limits to Reliability of Atomic Masses

Accurate atomic masses are determined from measurements of gas densities, accurate combining weights, mass spectroscopy, mass spectrometry, etc. Although refinements are continually being made in these analytical techniques, because of small variations in the isotopic contents of naturally occurring substances, for general purpose calculations, 5 or 6 significant figures are probably the best accuracy that can be expected for values of atomic masses. For example, the atomic mass of boron (B) found in the USA is 10.826, while the boron found in Asia Minor and in Italy has an atomic mass of 10.821. Similar isotopic variations are found for Pb, H, C, S, Ge, and O, as well as for some other elements.

Example 1-9

Exactly one kilogram of carefully purified calcium carbonate was heated strongly at 900°C until there was no further weight loss. The weight of the residue was 560.3 g. Given that the molar masses of carbon and oxygen are 12.011 g mol^{-1} and 15.9994 g mol^{-1}, respectively, calculate the molar mass of calcium.

$$CaCO_{3(s)} \rightarrow CaO_{(s)} + CO_{2(g)}\uparrow$$

Solution

$CO_2 = 12.011 + 2(15.9994) = 44.0098$ g mol^{-1}

No. grams of CO_2 formed $= 1000 - 560.3 = 439.7$ g.

No. mol CO_2 released $= \dfrac{439.7 \text{ g}}{44.0098 \text{ g}/\text{mol}} = 9.991$ mol

From the stoichiometry of the decomposition reaction it can be seen that the number of moles of CO_2 formed is equal to the number of moles of $CaCO_3$ that decomposes.

Thus,   1000 g $CaCO_3 = 9.991$ mol $CaCO_3$

And, therefore: molar mass of $CaCO_3 = \dfrac{1000 \text{ g } CaCO_3}{9.991 \text{ mol } CaCO_3} = 100.09$ g mol^{-1}

(mass of 1 mol $CaCO_3$) = (mass of 1 mol Ca) + (mass of 1 mol C) + 3(mass of 1 mol O)

$$100.09 \text{ g} = x + 12.011 + 3(15.9994)$$
$$x = 100.09 - 12.011 - 47.9982 = 40.08 \text{ g mol}^{-1}$$

KEY POINTS FOR CHAPTER ONE

1. **Newton's Law**: $F = ma$, where the force F is in newtons (N), the mass m is in kilograms (kg), and the acceleration a is in m s^{-2}. Note that the newton also can be expressed as kg m s^{-2}. That is, 50 N is exactly the same as 50 kg m s^{-2}.

2. **Ion**: An electrically charged atom or group of atoms.

    Atom: The smallest portion of an element that can take part in a *chemical* reaction.

    Element: A substance consisting of atoms, each of which has the same number of protons in the nucleus (i.e., the same *atomic number*).

    Isotope: Forms of an element with almost the same chemical properties but different numbers of *neutrons* in their nuclei and hence different atomic masses.

    $^6C^{13}$ = 6 protons and 7 neutrons (C-13 isotope).

    Molecule: The smallest portion of a substance capable of existing independently while retaining the properties of the original substance.

3. **Basic SI (*Système International*) units**: **mass** = kilogram (kg); **length** = meter (m); **time** = second (s); **temperature** = kelvin (K); **amount of substance** = mole (mol); **electric current** = ampere (A = C s^{-1}, where the coulomb C is the basic unit of charge).

4. **1 mole** = the amount of substance of a system that contains as many elementary entities as there are atoms in exactly 0.012 kg of carbon-12.

    Avogadro's Number N_A: The number of carbon atoms in exactly 0.012 kg of carbon-12 (= 6.0221367×10^{23} mol^{-1}).

    Molar Mass M: the mass of one mole of the substance (kg mol^{-1} in SI units).

    Number of Moles n = mass of sample/molar mass = m/M.

5. **Dalton's Atomic Theory**: (1) There exist indivisible atoms. (2) Each atom of a given element has the same atomic mass. (3) Atoms combine in a variety of simple, whole number ratios to form compounds. Exceptions: (i) Non-stoichiometric compounds contradict postulate 3. (ii) Isotopes contradict postulate 2. (iii) Atoms can be split into constituent particles with bombardment by high energy particles; this contradicts postulate 1.

6. **Stoichiometry**: Deals with the combining proportions of elements or compounds involved in chemical reactions, and the methods of calculating them ("chemical arithmetic").

7. **The Law of Definite Proportions**: The mass percent of the elements in a given compound is the same, regardless of how the compound is made.

    The Law of Multiple Proportions: If two elements form more than one compound, then the different masses of one element that combine with the same mass of the other are in the ratio of small, whole numbers.

    The Law of Equivalent Proportions: The masses of various elements or compounds that combine with some fixed mass of some other element or compound taken arbitrarily as a standard, also react with one another in the same ratio or some simple multiple or submultiple thereof.

8. **Equivalent Weight**: The mass of a substance that reacts with one mole of electrons (= *molar mass*/z, where z is a whole number).

9. **Molecular Formula**: Shows the actual composition of one molecule of the substance.
    Empirical Formula: Shows the smallest whole numbers that describe the ratios of the atoms in the substance.

10. **Avogadro's Hypothesis**: 22.41410×10^{-3} m^3 (= 22.41410 L) of an ideal gas at *Standard Temperature and Pressure (STP)* contains exactly one mole of gas.
    STP = 0°C (273.15 K) and one **standard atmosphere** pressure (101 325 Pa).
    1 bar = 10^5 Pa = 0.9869 atm.

11. **Conservation of Mass**: The total mass on the left-hand side (*LHS*) of a balanced chemical equation is equal to the total mass on the right-hand side (*RHS*). Neglects minor gains or losses resulting from Einstein's mass-energy relationship, which usually are insignificant for the energy changes involved in chemical reactions.

12. **Conservation of Charge**: Electric charge can be neither created nor destroyed. For stoichiometry this means that the total charge on the left-hand side of a chemical reaction must be the same as the total charge on the right hand side.

13. **Atomic Mass Scales**: Scales of the *relative* masses of the various elements, not the *actual* masses of the atoms. Based on $^6C^{12}$ = 12.0000... (exactly).

PROBLEMS

1. How many tonnes of limestone ($CaCO_3$) are needed to prepare 5.00 tonnes of "dry ice" (solid CO_2), assuming that 30% of the CO_2 produced is wasted in converting it to solid, and that limestone is only 95% by weight $CaCO_3$? ☹

2. Chlorine can be produced by the reaction:
    $$2KMnO_4 + 16HCl \rightarrow 2KCl + 2MnCl_2 + 5Cl_2 + 8H_2O$$

    (a) What mass of $KMnO_4$ is required to produce 2.5 L of Cl_2 gas measured at STP?
    (b) Calculate the volume of commercial hydrochloric acid required. (Commercial grade hydrochloric acid is 36.0% HCl by weight, and has a density of 1.18 g mL^{-1}.) ☹

3. A gasoline mixture containing 60.0% by weight octane (C_8H_{18}) and 40.0% by weight hexane (C_6H_{14}) has a density of 0.660 g mL^{-1}. Calculate the number of cubic meters of air at STP drawn through the carburetor when one imperial gallon of this gasoline is combusted in an automobile engine, assuming complete combustion of the fuel to carbon dioxide and water

vapor. Air contains 21.0% by volume of oxygen. One pound (*lb*) = 453.59 g; one imperial gallon (*IG*) of water weighs 10.0 lb; liquid water has a density of 1.00 g mL^{-1}. ☹

4. Nicotine, which has a molar mass of 162.23 g mol^{-1}, is 74.03% by weight carbon, 8.70% by weight hydrogen, and 17.27% by weight nitrogen. What is the molecular formula of nicotine? ☺

5. Phosphorus trichloride (*PCl$_3$*) reacts with water to form phosphorous acid (*H$_3$PO$_3$*) and hydrochloric acid:

$$PCl_{3(liq)} + 3H_2O_{(liq)} \rightarrow H_3PO_{3(aq)} + 3HCl_{(aq)}$$

   (a) Which is the limiting reactant when 12.4 g of phosphorus trichloride is mixed with 10.0 g of water?

   (b) What masses of phosphorous acid and hydrochloric acid are formed? ☺

6. Solution *A* contains a mixture of iron (II) and iron (III) sulfates. When 25.0 mL of this solution is acidified and titrated with a potassium permanganate solution containing 3.12 g of KMnO$_4$ per liter, 27.6 mL of the permanganate solution is required. Next, a second sample of Solution *A* is taken, and all the iron (III) in it is reduced to iron (II); the resulting solution is labelled Solution *B*. A 25.0 mL sample of Solution *B* required 42.4 mL of the permanganate solution. The reaction involved is

$$2KMnO_4 + 10FeSO_4 + 8H_2SO_4 \rightarrow K_2SO_4 + 2MnSO_4 + 8H_2O + 5Fe_2(SO_4)_3$$

   Calculate the concentrations—expressed in grams of salt per liter of solution—of the iron (II) and iron (III) sulfates in Solution *A*. ☹

7. A 0.596 g sample of a gaseous compound containing only boron and hydrogen occupied 484 cm^3 at STP. When ignited in excess oxygen all the hydrogen was recovered as 1.17 g of water, and all the boron was present as B$_2$O$_3$. What is the empirical formula? The molecular formula? The molar mass? What mass of B$_2$O$_3$ was produced? ☺

8. A sample of pure lead weighing 2.07 g was dissolved in nitric acid to give a solution of lead nitrate. This solution was treated with hydrochloric acid, chlorine gas, and ammonium chloride, to yield a precipitate of ammonium hexachloroplumbate, (NH$_4$)$_2$PbCl$_6$. What is the maximum amount of this product that could be made from the lead sample? ☺

9. Equal masses of zinc metal and iodine were mixed together and the iodine was reacted completely to ZnI$_2$. What fraction by weight of the original zinc remained unreacted? ☺

10. When an alloy of aluminum and copper was treated with aqueous hydrochloric acid the following reaction occurred: $Al + 3H^+ \rightarrow Al^{3+} + \frac{3}{2}H_2$. The aluminum dissolved but the copper remained behind as the pure metal. A 0.350 g sample of the alloy gave 415 cm^3 of hydrogen gas at STP. What was the weight percent aluminum in the alloy? ☺

11. A 4.22 g sample of a mixture of calcium chloride (*CaCl$_2$*) and sodium chloride (*NaCl*) was treated to precipitate all the calcium as calcium carbonate (*CaCO$_3$*), which then was heated

and converted to calcium oxide (CaO). The final weight of the calcium oxide was 0.959 g. What was the percent by weight of calcium chloride in the original sample? ☺

12. When 0.210 g of a gaseous compound containing only hydrogen and carbon was burned, 0.66 g of carbon dioxide was obtained. What is the empirical formula of the compound? The density of the compound at STP was 1.87 g L^{-1}; what is the molecular formula of the compound? ☺

13. A beaker contained a solution consisting of 30.0 g of HCl dissolved in 100 g of water. When a 10.0 g piece of pure metallic aluminum (Al) was dropped into this solution, a chemical reaction took place in which aluminum trichloride ($AlCl_3$) and hydrogen gas (H_2) were produced. The gas was collected at a pressure of 99.0 kPa and a temperature of 24.0°C.

   (a) Write the balanced chemical reaction for the process.

   (b) What mass of hydrogen gas was produced? ☺

14. (a) Calculate the molar mass of dry air, which may be assumed to consist of 20.0% by volume oxygen gas and 80.0% by volume nitrogen gas.

   (b) A piece of pure tin (Sn) weighing 7.490 g was placed into a rigid vessel initially containing air. The air in the vessel was displaced by pure oxygen at the same temperature and pressure as that of the surroundings, namely 16.85°C and 101.0 kPa. The volume of the oxygen gas in the vessel was 2.895 L. After accurately weighing the vessel and its contents, it was heated until all the tin had irreversibly oxidized to a white, stable powder consisting of SnO_2. Next, the vessel and its contents were allowed once more to cool down to the temperature of the room, after which a valve in the vessel was opened slightly, and air—of the same composition as in part (a)—from the surroundings was permitted to enter the vessel until pressure equilibrium was reached between the surroundings and the contents of the vessel. At this point the valve again was shut and the vessel and its contents were re-weighed. It was found that the vessel and its contents had increased in weight by 1.818 grams. Neglecting any small differences in the densities of Sn and SnO_2, and assuming that the ideal gas law ($PV = nRT$) holds, use the above data to determine the molar mass of Sn. ☹

15. A 73.3 cm^3 sample of a slurry of impure ZnS ore of average density 1.365 g cm^{-3} and containing 50.5% by weight of impure ore was analyzed as follows: First the sample was digested with concentrated nitric acid until all the sulfur was converted to sulfuric acid:

$$ZnS + 10HNO_3 \rightarrow Zn(NO_3)_2 + H_2SO_4 + 8NO_2 + 4H_2O$$

Then the sulfate was completely precipitated as insoluble $BaSO_4$.

$$H_2SO_4 + BaCl_2 \rightarrow BaSO_4\downarrow + 2HCl$$

The barium sulfate was removed, washed, dried and weighed, yielding 58.4 g of $BaSO_4$. Calculate the percentage by weight of ZnS in the crude ore. ☺

16. Naturally occurring zinc contains five isotopes, each present at the following mole percentages:

$$^0Zn^{64} = 48.89\% \quad ^{30}Zn^{66} = 27.81\% \quad ^{30}Zn^{67} = 4.11\%$$
$$^{30}Zn^{68} = 18.57\% \quad ^{30}Zn^{70} = 0.62\%$$

The relative molar masses of each isotope are as follows:

$$^{30}Zn^{64} = 63.9291 \quad ^{30}Zn^{66} = 65.9260 \quad ^{30}Zn^{67} = 66.9271$$
$$^{30}Zn^{68} = 67.9249 \quad ^{30}Zn^{70} = 69.9253$$

Calculate the relative molar mass of naturally occurring Zn. ☺

17. What is the empirical formula of a compound that is, by mass, 52.1% carbon, 13.2% hydrogen, and 34.7% oxygen? ☺

18. For the reaction $\quad Ca(OH)_{2(s)} + 2HCl_{(aq)} \rightarrow CaCl_{2(aq)} + 2H_2O_{(liq)}$

what mass of $Ca(OH)_2$ in kg is required to neutralize (react completely with) 464.0 L of an aqueous HCl solution that is 30.12% HCl by mass and has a density of 1.15 g mL^{-1}? ☺

19. What volume of commercial hydrochloric acid solution and what mass of limestone are required to produce 2.00 kg of carbon dioxide if commercial hydrochloric acid solution contains 35% by mass HCl and has a density of 1.18 g mL^{-1}, and the limestone available consists of 90% by mass calcium carbonate plus 10% inert material?

$$CaCO_3 + 2HCl \rightarrow CaCl_2 + CO_2 + H_2O$$

20. A 5.00 g sample known to contain a mixture of potassium chloride (*KCl*) and lithium chloride (*LiCl*) was dissolved in water and all the chloride ion was precipitated as 15.8 g of silver chloride using an excess of silver nitrate (*AgNO$_3$*) solution. Calculate the weight fraction of lithium chloride in the unknown sample. ☺

21. Using the data provided and assuming complete combustion, estimate the mass of CO_2 and of SO_2 produced per day by a 1000 MW electric power plant burning bituminous coal from West Virginia. Mass composition of coal: 78% C, 6% H, 7% O, 1% N, 5% S, 3% inorganics. Heating value of coal = 29 MJ kg^{-1}. Overall efficiency of plant = 35%. ☹

22. A gaseous mixture entering a steady state continuous flow reactor at 300°F and 740 mm Hg has the following composition by volume: N$_2$ = 50%, SO$_2$ = 25%, O$_2$ = 25%. In the reactor, some of the SO$_2$ is oxidized to SO$_3$. If the gaseous mixture leaving the reactor is 55 mole % N$_2$ and there is no reaction of the N$_2$, calculate: (a) the percent yield of SO$_3$, and (b) the mass, in kg, of SO$_3$ produced per kg of entering mixture. ☺

23. A compound is known to consist of only carbon, hydrogen, and possibly oxygen. A 0.9467 g sample of this material was burned completely with excess air to form carbon dioxide and water vapor. The combustion products were collected by passage through a U-tube containing magnesium perchlorate, which absorbs water vapor, and a second U-tube containing calcium oxide, which absorbs carbon dioxide. If the weight of the magnesium perchlorate tube increased by 0.6812 g and that of the calcium oxide tube by 2.0807 g, what is the empirical formula of the compound? ☺

24. A sample containing NaCl, Na$_2$SO$_4$, and NaNO$_3$ gives the following elemental analysis in percent by mass: Na = 32.08%, O = 36.01%, Cl = 19.51%. Calculate the percent by mass of each compound in the sample. ☺

STATES OF MATTER AND THE PROPERTIES OF GASES

2.1 THE THREE STATES OF MATTER

Physical Chemistry explains the structure of matter and the changes in structure in terms of fundamental concepts such as atoms, electrons, and energy; accordingly, physical chemistry provides the basic framework for all other branches of chemistry (inorganic chemistry, organic chemistry, biochemistry, geochemistry, chemical engineering, etc.).

The Three States of Matter

Gas—a fluid state, highly compressible; a gas expands to occupy the total container. In a gas, the particles are in continuous, rapid, chaotic motion, usually travelling many atomic diameters before colliding with other gas particles or with the walls of the container. The interactions between gas particles are weak.

Liquid—a fluid state, with a well-defined surface; when in a gravitational field a liquid fills the lower part of its container; a liquid is almost incompressible. The particles of a liquid are in contact with each other and the motion about each other is restricted. Individual liquid particles travel only a fraction of a molecular diameter before colliding with other particles.

Solid—retains its shape (independent of the shape of the container); essentially incompressible. The particles in a solid are in contact, and are unable to move past one another; instead, they oscillate about a fixed position in the crystal lattice.

Gases

Gases are important because:

• Many elements and compounds exist as gases under ordinary conditions of temperature and pressure (e.g., O_2, H_2, N_2, Cl_2, CO_2, CO, SO_2, CH_4) and many others are easily volatilized.

• The concepts of temperature and pressure are defined in terms of gases.

• The planet (including us) is surrounded by a blanket of gas (the atmosphere); processes in this atmosphere determine the weather.

• We breathe gas to stay alive.

• Many industrial and natural processes involve gases.

• The interior behavior of the stars is described by the gas laws.

• If you don't understand gases you won't be able to pass this course.

The State of a System

In addition to the "states" of matter, we also talk about the "state" of a system. A *system* is usually some sample of matter, and its *state* refers to its condition as described by its pressure (P), volume (V), temperature (T), and amount of substance (n) or composition (x). Thus, for the kinds of systems that we will be dealing with, the states are functions of their P, V, T, and n (or x). For example, one mole of H_2 can have different P, V, and T values. Any given set of P, V, and T completely defines its "state." Two samples of the same substance are in the same "state" if each has the same P, V, T, and *mass*.

Volume (V)—a measure of the *space* occupied by the system. Units: mL, cm^3, dm^3, L, m^3, etc.

2.2 PRESSURE

Static pressure is just force divided by the area over which the force acts: $P = F/A$. In SI units, the pressure is in pascals (Pa), the force is in newtons (N) and the area is in square meters (m^2). Force is defined by Newton's law: $F = ma$; thus, the units of pressure are

$$P = N/m^2 = Pa = (kg)(m\ s^{-2})/m^2 = kg\ m^{-1}\ s^{-2}.$$

Pressure of a Gas—Pressure (P) is a force acting on an area (i.e., acting on a *surface*). For a *gas*, the only area is *the wall of the container*; therefore the pressure of a gas is a force acting on the wall of the container. This force results from the collision of gas molecules with the wall of the container. It turns out that the pressure of a gas is *a measure of the average (macroscopic) rate of change of momentum of the molecules or atoms per unit area of wall* (Fig. 1). If the change in momentum is *fast*, then the pressure is *greater* (more frequent bombardment of the wall).

change in momentum

Fig. 1

Confirm the units: $P \equiv \dfrac{d(mv)}{dt} \Big/ A = \dfrac{(kg)(m/s)}{(s)} \Big/ m^2 = kg\ m^{-1}\ s^{-2} = Pa.$

Note that the units are the same as those for static pressure. Table 1 lists some pressures:

Table 1   *Some Representative Pressures, in Pa*

Intergalactic space	2×10^{-16}	Standard atmosphere	101 325
Best laboratory vacuum	10^{-12}	At bottom of deepest ocean	1.1×10^8
Vapor pressure of Hg at 25°C	0.25	Center of the Earth	4×10^{11}
Vapor pressure of H_2O at 25°C	3167	Center of the Sun	2×10^{16}

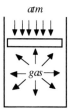

Consider a gas inside the piston/cylinder arrangement shown in Fig. 2: If the piston is weightless and frictionless, it can be seen that if $P_{gas} = P_{atm}$, then the piston doesn't move and we have a *mechanical equilibrium* (pressure equilibrium) between the internal pressure exerted by the gas and the outside pressure exerted by the atmosphere.

Pressure Units: 1 standard atmosphere $\equiv$ 101 325 Pa = 1.01325 bar = 101.325 kPa
= 14.6959 psi (pounds per square inch) = 760 Torr (1 torr = 1 mm Hg).

Fig. 2   It is observed experimentally that at any given point in a fluid (i.e., in a liquid or in a gas) at rest, a pressure sensor gives the same reading regardless of how it is oriented. Thus, pressure is a scalar quantity having no directional properties. That is, at any given point in the fluid, the pressure is the same in all directions. It also is observed that a *change* ΔP in pressure applied to an enclosed incompressible fluid is transmitted undiminished to every portion of the fluid and to the walls of the containing vessel. This is known as **Pascal's Principle**. We make use of this principle when we brush our teeth, since, when one end of a tube of toothpaste is squeezed, the change in pressure is transmitted to the other end.

Example 2-1

A room at 25°C and one atm pressure is 4.0 m wide, 5.0 m long, and 3.0 m high.

(a) What force is exerted on the floor by the air contained in the room?

(b) What is the total force exerted on the floor by the atmosphere?

The density of air at 25°C and one atm pressure is 1.18 kg m^{-3}, and the acceleration due to gravity is $g = 9.80$ m s^{-2}.

Solution

(a) The force exerted by the air contained in the room is obtained from Newton's law $F = ma$, where m is the mass of the air, and a is g, the acceleration due to gravity.

$$F = ma = mg = \left\{ \left(1.18 \; \frac{\text{kg}}{\text{m}^3} \right) \left(4.0 \times 5.0 \times 3.0 \; \text{m}^3 \right) \right\} \left\{ 9.80 \; \frac{\text{m}}{\text{s}^2} \right\} = 693.8 \; \text{N}$$

Ans: | 694 N |

(b) The total pressure in the room is one atm, i.e., 101 325 Pa (= N m^{-2});

therefore, $F_{total} = PA = \left\{ \left(101\ 325 \; \frac{\text{N}}{\text{m}^2} \right) \left(4.0 \times 5.0 \; \text{m}^2 \right) \right\} = 2.03 \times 10^6 \; \text{N} = 2.03 \; \text{MN}$

Ans: | 2.0 MN |

This force corresponds to a mass of $\quad m = \dfrac{F}{g} = \dfrac{2.03 \times 10^6}{9.8} = 207 \times 10^3$ kg = 207 tonne, and is equal to the force exerted by a column of air covering the floor and extending all the way to the top of the atmosphere.

A **barometer** (Fig. 3) measures the pressure of the atmosphere.

Fig. 3(a) on the next page shows Fig. 3 in more detail.

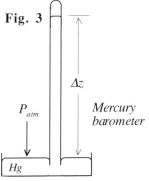

Fig. 3

P_{atm}

Hg

Mercury barometer

Δz

For mechanical equilibrium at point A, the pressure P_{atm} of the atmosphere must be balanced by the pressure P_{Hg} generated by the weight of the column of mercury *plus* the pressure $P_{Hg}^{\bullet}$ exerted by the vapor pressure of the mercury vapor in the space above the surface of the mercury column:

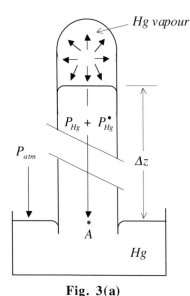

Fig. 3(a)

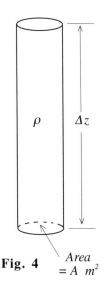

Fig. 4

$$P_{atm} = P_{Hg} + P_{Hg}^{\bullet} \qquad \ldots [1]$$

At ordinary temperatures the vapor pressure of mercury is negligible and can be ignored: for example, at 25°C it is only 0.25 Pa. Therefore the condition for mechanical equilibrium becomes

$$P_{atm} = P_{Hg} \qquad \ldots [2]$$

What is the pressure exerted by the column of liquid mercury?

Consider a column of mercury of cross-sectional area A m^2, height Δz meters, and density ρ kg m^{-3} (Fig. 4). The mass of the column is

$$m = \left(A \text{ m}^2 \right)\left(\Delta z \text{ m} \right)\left(\rho \, \tfrac{\text{kg}}{\text{m}^2} \right)$$

$$= A \rho \, \Delta z \quad \text{kg} \qquad \ldots [3]$$

The force of gravity acting on this mass is

$$F = ma$$

$$= mg$$

$$= A \rho \, \Delta z \, g \quad \text{newtons} \qquad \ldots [4]$$

The pressure exerted by the column of mercury at its base is

$$P_{Hg} = \frac{force}{area} = \frac{F}{A}$$

$$= \frac{A \rho \, \Delta z \, g}{A}$$

$$= \rho g \, \Delta z \quad \text{pascals} \qquad \ldots [5]$$

Substituting Eqn [5] into Eqn [2] gives

$$\boxed{P_{atm} = \rho g \Delta z} \qquad \ldots [6]$$

where P_{atm} is the pressure of the atmosphere, ρ is the density of the liquid in the barometer, Δz is the height of the column of liquid in the barometer tube, and g is the acceleration due to gravity. Mercury (Hg) is commonly used as the liquid. At 0°C and 25°C its density is 13 595.5 kg m^{-3} and 13 534.0 kg m^{-3}, respectively. The standard value of g is 9.80665 m s^{-2}.

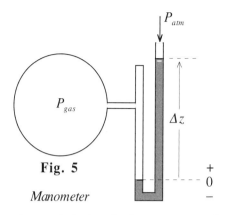

Fig. 5

Manometer

A **manometer** measures the pressure of a gas inside a container. The most common type of manometer is the U-tube manometer, shown in Fig. 5. At pressure equilibrium, the manometer formula for the pressure of the contained gas is

$$P_{gas} = P_{atm} + \rho g \Delta z \qquad \ldots [7]$$

Gauge Pressure—many pressure gauges measure the pressure in *excess* of that of the atmosphere, often expressed the units of psi ("pounds per square inch"). For this kind of device

$$psia = psig + P_{atm} \qquad \ldots [8]$$

where *psia* is the *absolute* pressure and *psig* is the **gauge** pressure.

How big is 1 atm pressure? $F = ma = mg$. If $F = 1$ newton, then $m = F/g = 1/9.81$ $= 0.102$ kg $= 102$ g. Thus, one newton is the force of gravity acting on a mass of 102 g. Since one atm is equal to 101 325 N/m², this corresponds to a mass of $(101\ 325)(0.102) = 10\ 332$ kg acting on an area of one square meter. This is approximately the weight of 10 automobiles per square meter! One atm is a pretty *big* pressure!

Example 2-2

At 25°C a certain pressure P can support a column of pure water 80 cm high. The same pressure will support a column of ethanol 102 cm high. What is the density of the alcohol? The density of water at 25°C is 0.997 g cm⁻³.

Solution

Let the density of the ethanol be ρ_E. Referring to the diagram at the right, a force balance at $z = 0$ inside the liquid gives

$$P = P_{atm} + \rho g \Delta z$$

that is,

$$P - P_{atm} = \rho g \Delta z$$

We are told that $\Delta z = 80$ cm when $\rho = 0.997$, and that $\Delta z = 102$ cm when $\rho = \rho_E$. In each case, $P - P_{atm}$ is the same.

Thus,

$$P - P_{atm} = \rho_E g \Delta z$$

$$= \rho_E g (102)$$

$$= (0.997) g (80)$$

and

$$\rho_E = \frac{(0.997) g (80)}{g (102)}$$

$$= \frac{(80)}{(102)}(0.997)$$

$$= 0.782 \text{ g cm}^{-3}$$

$$= 782 \text{ kg m}^{-3}$$

Ans: $\boxed{782 \text{ kg m}^{-3}}$

Example 2-3

Consider the mechanical equilibrium in the open U-tube shown at the right, which contains water and an unknown oil that is immiscible with water. If a = 100 cm and b = 125 cm, what is the density of the oil? The density of the water is 997 kg m^{-3}.

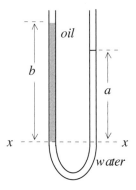

Solution

The pressure in each arm of the tube must be the same at the level x–x, otherwise movement would occur. The pressure in each arm is given by

$$P = P_{atm} + \rho g \Delta z$$

Equating the two pressures gives

$$P_{atm} + \rho_{oil} g \Delta z_{oil} = P_{atm} + \rho_{water} g \Delta z_{water}$$

$$\rho_{oil} g \Delta z_{oil} = \rho_{water} g \Delta z_{water}$$

$$\rho_{oil} = \frac{\rho_{water} g \Delta z_{water}}{g \Delta z_{oil}} = \frac{\rho_{water} \Delta z_{water}}{\Delta z_{oil}}$$

$$= \frac{(997)(100)}{(125)} = 797.6 \text{ kg m}^{-3}$$

This is an easy way to determine the density of an unknown liquid.   **Ans:** $\boxed{798 \text{ kg m}^{-3}}$

Exercise 2–1

During a blood transfusion, the needle is inserted in a vein where the blood pressure (gauge pressure) is 2000 Pa. At what height must the blood container be placed so that the blood just enters the vein? The density of blood is 1.0595 g cm^{-3} and g = 9.80 m s^{-2}. ☺

Exercise 2-2

What pressure in atm would be exerted on a diver at a depth of 100 m below the surface of the ocean? The density of sea water may be taken as 1.03 g cm^{-3}. ☺

Exercise 2-3

When you are swimming underwater and holding your breath, the external increased pressure causes your body to contract in order to re-establish pressure equilibrium. As a result, the pressure of the air in your lungs will increase to match the pressure of the external surroundings. On the other hand, if you breathe through a snorkel tube, the air in your lungs will remain at the atmospheric surface pressure. Your lungs are able to withstand a pressure differential ΔP of up to about 1/20 atm and still function; when ΔP is greater than this you will not be able to expand your lungs to inhale. Not only will your lungs collapse, but also, pressurized blood will be forced into them, with death soon following. Calculate how far below the surface of a lake you can swim using a snorkel tube if the lake water has a density of 1000 kg m^{-3} and the acceleration due to gravity is 9.807 m s^{-2}. ☹

2.3 ARCHIMEDES' PRINCIPLE

Archimedes' Principle states that a body fully or partially immersed in a fluid is buoyed up by a force equal to the weight of the fluid that the body replaces. The following examples illustrate the application of this principle:

Example 2-4

A spherical helium-filled research balloon with a diameter of 10.0 m floats 30 km above the surface of the Earth. The total payload m (the instrument package plus the weight of the empty balloon) is $m = 8.41$ kg. If the helium were replaced with hydrogen, by what percentage would the payload increase? At 30 km above the surface of the Earth the densities of helium gas and hydrogen gas are 0.00259 and 0.001306 kg m^{-3}, respectively. The balloon may be assumed to be spherical.

Solution

According to Archimedes' Principle, the upward buoyancy force $F\uparrow$ is equal to the weight of the air displaced by the balloon. (Note that the "weight" of a body is the force exerted by gravity on its mass; therefore, *weight* is a force, and is expressed in *newtons*, not in *kilograms*. In outer space, for example, a person still would have a mass of 75 kg but her *weight* would be zero because of the absence of any significant gravitational field.)

The mass of air displaced by the balloon is $V\rho_{air}$

where V is the volume of the balloon and ρ_{air} is the density of the air in the surrounding atmosphere; therefore, from Newton's Law $(F = ma)$,

we get
$$F\uparrow = ma$$
$$= (V\rho_{air})g$$

where g is the acceleration due to gravity. The total downward force is the force of gravity acting on the mass m of the payload and on the mass m_{He} of the helium gas in the balloon:

Thus,
$$F\downarrow = ma$$
$$= (m + m_{He})g$$
$$= (m + V\rho_{He})g$$

From Archimedes' Principle, at mechanical equilibrium $\quad F\uparrow = F\downarrow$

that is,
$$V\rho_{air}g = (m + V\rho_{He})g$$

Eliminating g from each side: $\quad V\rho_{air} = m + V\rho_{He}$

from which
$$m = V(\rho_{air} - \rho_{He})$$

If we replace the helium with hydrogen the force balance becomes
$$V\rho_{air} = m' + V\rho_{H_2}$$

where m' is the mass of the new payload.

Thus,
$$m' = V(\rho_{air} - \rho_{H_2})$$

The increase in the payload is:
$$\Delta m = m' - m$$
$$= V(\rho_{air} - \rho_{H_2}) - V(\rho_{air} - \rho_{He})$$
$$= V(\rho_{He} - \rho_{H_2})$$

Therefore the fractional increase in the payload is

$$\frac{\Delta m}{m} = \frac{V(\rho_{He} - \rho_{H_2})}{V(\rho_{air} - \rho_{He})} = \frac{\rho_{He} - \rho_{H_2}}{\rho_{air} - \rho_{He}}$$

We can obtain the density of the air from the initial force balance; namely,

$$V\rho_{air} = m + V\rho_{He}$$

from which
$$\rho_{air} = \frac{m + V\rho_{He}}{V}$$

The volume of the gas in the balloon is

$$V = \tfrac{4}{3}\pi r^3$$

$$= \tfrac{4}{3}\pi\left(\frac{10.0}{2}\right)^3$$

$$= 523.6 \text{ m}^3$$

therefore:

$$\rho_{air} = \frac{m + V\rho_{He}}{V}$$

$$= \frac{8.41 + (523.6)(0.00259)}{523.6}$$

$$= 0.01865 \text{ kg m}^{-3}$$

and

$$\frac{\Delta m}{m} = \frac{\rho_{He} - \rho_{H_2}}{\rho_{air} - \rho_{He}}$$

$$= \frac{0.00259 - 0.001306}{0.01865 - 0.00259}$$

$$= 0.0800$$

or, expressed as a percentage, 8.00%.

Therefore the payload would increase by 8.00%.

Ans: $\boxed{8.00\%}$

It might be noted that at an elevation of 30 km above the surface of the Earth, the atmospheric pressure is only 0.0125 atm, the ambient temperature is –38°C, and the acceleration due to gravity is about 9.7 m s^{-2}.

Exercise 2-4

What volume fraction of ice floats beneath the surface of water at 0°C? At this temperature, the densities of water and ice are 1000 kg m^{-3} and 917 kg m^{-3}, respectively. ☺

Exercise 2-5

A hollow iron sphere weighs 6000 N in air and 4000 N in water. What is the volume of the cavity inside the sphere? The density of air and of whatever gas (if any) is inside the cavity of the sphere may be assumed to be negligible with respect to the density of iron and of water, which are, respectively, 7870 and 1000 kg m^{-3}. The acceleration due to gravity may be taken as 9.80 m s^{-2}. ☹

2.4 TEMPERATURE (T)

Heat flows spontaneously from a body at a higher temperature to a body at a lower temperature. When no heat flows between two bodies in thermal contact we have *thermal equilibrium* (i.e., the temperature of each body is the same).

Celsius scale (θ): 0.00°C is defined as the *ice point for water*, which is the temperature at which water saturated with air at one atm pressure is in equilibrium with ice. We use the symbol θ for degrees Celsius.

Kelvin scale (T): This is an absolute temperature scale. Absolute zero is 0 K. The relationship between the Celsius scale and the Kelvin scale is

$$T = \theta + 273.15$$

where T is the temperature in kelvin. Note that the unit of absolute temperature is "kelvin," not "degrees kelvin." We use the symbol T to represent the absolute temperature of a system. Absolute zero (*0 K*) corresponds to –273.15°C.

Two older scales still sometimes encountered are the *Fahrenheit* scale and its corresponding absolute scale, the *Rankine* scale. The formulas for these two scales are

$$°F = \tfrac{9}{5}\theta + 32$$

and
$$°R = °F + 459.69$$

Example 2-5

At what temperature is the Fahrenheit scale reading (a) three times the Celsius reading? (b) one-third the Celsius reading?

Solution

(a) Let C denote the temperature in °C, and F denote the corresponding temperature in °F:

$$F = \tfrac{9}{5}C + 32$$

if $F = 3C$, then
$$3C = \tfrac{9}{5}C + 32$$

i.e.,
$$3C - 1.8C = 32$$
$$1.2C = 32$$

Solving:
$$C = 26.67°C$$

Check: $F = \tfrac{9}{5}C + 32 = \tfrac{9}{5}(26.67) + 32 = 80.0°F$ [$= 3 \times 26.67$].   OK!

Therefore at 26.67°C the corresponding temperature in Fahrenheit (80.0°F) is three times the Celsius value.

Ans: $\boxed{26.67°C}$

(b) $F = \frac{9}{5}C + 32$

if $F = \frac{1}{3}C$, then $\frac{1}{3}C = \frac{9}{5}C + 32$

$$\frac{9}{5}C - \frac{1}{3}C = -32$$

i.e., $\frac{27}{15}C - \frac{5}{15}C = -32$

$$\frac{22}{15}C = -32$$

$$C = \frac{(-32)(15)}{22} = -21.82°C$$

Check: $F = \frac{9}{5}C + 32 = \frac{9}{5}(-21.82) + 32 = -7.28°F$ $[= \frac{1}{3} \times (-21.82)]$ Yes!

Therefore at –21.82°C the temperature in degrees Fahrenheit (–7.28°F) is one-third the Celsius value.

Ans: $\boxed{-21.82°C}$

Exercise 2-6

If the temperature of the human body is 98.6°F, what is the value in °C and in K? ☺

Exercise 2-7

Convert the following Fahrenheit temperatures to Celsius temperatures:

(a) –95.8°F, which was recorded in the Siberian village of Oymyakon in 1964.

(b) –70.0°F, which is the lowest officially recorded temperature in the continental United States (measured at Rogers Pass, in Montana).

(c) +134°F, which is the highest officially recorded temperature in the continental United States (measured in Death Valley, California). ☺

KEY POINTS FOR CHAPTER TWO

1. **The State of a System**: The *state* of a system refers to its condition as described by its pressure (P), volume (V), temperature (T), and amount of substance (n) or composition (x). The "state" of a simple, pure, homogeneous, chemical system is completely defined by its P, V, and T. Two samples of the same substance are in the same "state" if each has the same P, V, T, and *mass*.

2. **Static pressure**: $P = F/A$. P = pascals (Pa), F = newtons (N), A = square meters (m^2).

    $$P = N/m^2 = Pa = (kg)(m\ s^{-2})/m^2 = kg\ m^{-1}\ s^{-2}$$

    Pressure of a Gas: A measure of the average (macroscopic) rate of change of momentum of the molecules or atoms per unit area of wall:

    $$P \equiv \frac{d(m\upsilon)}{dt} \bigg/ A = \frac{(kg)(m/s)}{(s)} \bigg/ m^2 = kg\ m^{-1}\ s^{-2} = Pa$$

    Pressure Units: 1 standard atmosphere $\equiv$ 101 325 Pa = 1.01325 bar = 101.325 kPa
    = 14.6959 psi (pounds per square inch) = 760 Torr (*1 torr = 1 mm Hg*).

3. **Pascal's Principle**: At any given point in a fluid, the pressure is the same in all directions. Also, a *change* ΔP in pressure applied to an enclosed incompressible fluid is transmitted undiminished to every portion of the fluid and to the walls of the containing vessel.

4. The pressure exerted by a column of fluid of density ρ kg m^{-3} and height Δz meters is $P = \rho g \Delta z$ where g is the acceleration due to gravity. Standard g = 9.80665 m s^{-2}.

5. **Archimedes' Principle**: A body fully or partially immersed in a fluid is buoyed up by a force equal to the weight of the fluid that the body replaces.

6. **Temperature: Celsius scale** (θ): 0.00°C is defined as the *ice point for water*, which is the temperature at which water saturated with air at one atm pressure is in equilibrium with ice. **Kelvin scale** (T): Absolute temperature scale. Absolute zero is 0 K. The relationship between the Celsius scale and the Kelvin scale is $T = \theta + 273.15$.

PROBLEMS

1. The beam balance shown at the right is used to weigh out an amount of NaCl crystals. Exactly enough NaCl is added to the right-hand pan to counterbalance a brass weight placed on the left-hand pan. The brass weight on the left-hand pan has a true weight in vacuum of 100.0000 g. What is the 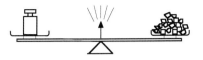 true weight of the NaCl crystals on the right-hand pan? The densities of air, brass, and crystalline NaCl are 1.174, 8400, and 2100 kg m^{-3}, respectively. ☺

2. The mass and diameter of a penny are 2.2785 g and 1.90 cm, respectively.
   (a) What would be the height of a stack of pennies that generates a static pressure of one atm? (b) How much would this stack of pennies be worth? You may assume that pennies are made of pure copper and that the acceleration due to gravity is constant at 9.806 m s^{-2}. ☺

3. An 80.0 kg man just floats in fresh water with virtually all his body just below the surface. What is his volume if the density of fresh water is 997 kg m^{-3}? ☺

4. An airtight partially evacuated cylindrical container of 10.0 cm inside diameter and inside length 15.0 cm has a lid on the bottom that is held in place by suction. If a mass m of 50.0 kg hanging from the lid is required to break the suction to remove the lid, what is the pressure inside the container? The surroundings are at 25°C and one atm pressure. The mass of the lid may be considered negligible. ☹

5. Estimate the pressure of the atmosphere at an elevation of 5.00 km. For this calculation you may assume that the average temperature of the air is 0°C, the average molar mass of air is $M = 28.8$ g mol^{-1}, the acceleration due to gravity is constant at $g = 9.806$ m s^{-2}, and the pressure at ground level ($z = 0$) is $P = P° = 1.01325$ bar. ☹*

6. How many moles of air is contained in a column of air of cross-sectional area 1.00 m^2 extending from the surface of the earth to a height of 1.00 km? For this calculation you may assume that the average temperature of the air is 0°C, the average molar mass of air is $M = 28.8$ g mol^{-1}, the acceleration due to gravity is constant at $g = 9.806$ m s^{-2}, and the pressure at ground level ($z = 0$) is $P = P° = 101\ 325$ Pa. ☹*

7. *Pascal's Principle* states that any change in pressure applied to an enclosed incompressible fluid is transmitted undiminished to every part of the fluid as well as to the walls of the container. This is observed when you squeeze a tube of toothpaste. The most common application of Pascal's principle is the hydraulic jack, in which a small force applied to a fluid can be used to lift a heavy weight. Thus, consider the following arrangement in which an incompressible fluid such as hydraulic oil is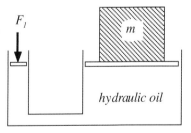

contained in a rigid system containing two pistons. What force F_1 is required to lift a mass of $m = 250$ kg if the area of the small piston face is $A_1 = 1.00$ cm^2 and that of the large piston face is $A_2 = 300$ cm^2? ☺

8. When you are standing, the blood pressure in your upper body is quite different from that in your lower body. Fortunately, the human circulatory system has evolved to solve the problems associated with quickly moving blood through relatively large distances against the force of gravity. Animals such as snakes, eels, and even rabbits, whose circulatory systems have not so adapted, will actually die if held head upward, because their blood pools in their extremities and is unable to reach their brains. When a certain woman stands, her brain is 40 cm higher than her heart; when she bends, it is 35 cm below her heart.

    (a) What is the blood pressure in her brain when she is standing?

    (b) What is the blood pressure in her brain when she is bending over?

    Her normal systolic blood pressure—the peak value upon heart contraction—is 120 Torr, while her normal diastolic blood pressure—the value in between beats when the heart relaxes and blood flows from the veins into the heart—is 80 Torr. (This is why the normal blood pressure for a healthy adult is reported as 120/80.) For the purpose of the calculation you may use an average blood pressure; also, any small frictional and velocity effects on blood pressure may be ignored. The density of blood may be taken as 1.0595 g cm^{-3}. ☺

9. The rigid L-shaped tank shown at the right is open at the top and filled with water of density 1000 kg m^{-3}. If the atmospheric pressure is 1.00 bar, what is the net force exerted by the water on the cross-hatched face of the tank? ☺

10. A glass of water that is filled to the brim contains an ice cube floating at the surface. Taking the buoyancy effects of both water and air into account, what would happen to the water level when the ice cube melts if no counteracting surface tension effects were present? ☹

11. A boat of mass m' containing a large heavy rock of mass m_R, volume V_R, and density ρ_R floats on a pond containing a volume of water V_W of density ρ_W. If the rock is thrown overboard, does the level of the pond rise, fall, or remain the same? Neglect any buoyancy contributions from the air. ☹

12. A blimp is filled with 5000 m^3 of helium at 20°C and 1.05 bar pressure. The volume of the cabin is 30 m^3 and the mass of the empty blimp structure (no helium) is 4200 kg. What is the maximum additional mass m' that can be loaded into the cabin and still permit lift off? The surrounding air is at 1.00 bar pressure and 20°C. The molar mass of air is 28.8 g mol^{-1}. ☺

13. What is the maximum height that water can rise in the pipes of a building if the water pressure at the ground floor is 50 psig? Assume the density of water is 0.997 g cm^{-3}. ☺

14. An object of mass m hangs from a spring balance, as shown in the figure at the right. In air, the balance reads 4.00 kg. When the object is immersed in water the balance reads 3.65 kg, and when immersed in an unknown liquid it reads 3.72 kg. What is the density of the unknown liquid? The air (molar mass = 28.8 g mol^{-1}) is at one bar pressure and 25°C. The density of water at 25°C is 0.997 g cm^{-3}. ☹

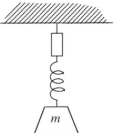

15. A vertical glass U-tube, open at each end, contains mercury. If 15.0 cm of water is poured into one arm of the tube, by how much does the level in the other arm rise above its initial level? The densities of water and of mercury can be taken as 0.997 g cm^{-3} and 13.55 g cm^{-3}, respectively. The atmospheric pressure is 1.024 bar. ☺

16. A piece of wood is floating in a pool of oil with 40.0% of the wood above the surface of the oil. If the oil has a density of 1.25 g cm^{-3}, what is the density of the wood in kg m^{-3}? The buoyancy contribution from the air may be neglected. The acceleration due to gravity is $g = 9.80$ m s^{-2}. ☺

THE IDEAL GAS

Equations of state are mathematical equations relating P, V, T, and n for a sample of substance. These equations usually are very complicated for solids and liquids, but are simpler for gases. P, V, T, and n are *not* independent; for the types of systems *we* deal with, specifying the values for any three automatically determines the fourth.

3.1 THE IDEAL GAS EQUATION OF STATE

$PV = nRT$ where P is the pressure in Pa, V is the volume in m^3, n is the number of moles, and T is the absolute temperature, in kelvin. This is the most important equation in physical chemistry— and you already know it! R is the universal gas constant, and has a value of $R = 8.314$ J mol^{-1} K^{-1} for all gases (this is why it's called a "universal" constant). Note the units of R:

$$R = \frac{PV}{nT} = \frac{Pa \cdot m^3}{mol \cdot K} = \frac{(N/m^2) \cdot m^3}{mol \cdot K} = \frac{N \cdot m}{mol \cdot K} = \frac{J}{mol \cdot K}$$

Since 1 Pa = 0.001 kPa and 1 m^3 = 1000 L, then 1 Pa·m^3 = (0.001 kPa)(1000 L) = 1 kPa·L; therefore we can use the same numerical value of R with both kPa·L and Pa·m^3. Thus,

$R = 8.314$ Pa m^3 mol^{-1} K^{-1} = 8.314 J mol^{-1} K^{-1} = 8.314 kPa L mol^{-1} K^{-1}.

A gas that obeys $PV = nRT$ at *all* pressures is called an *ideal gas* (or *perfect gas*). At common temperatures and pressures (say 25°C and 1 bar) most gases obey the ideal gas law to within a few percent. In this course, *unless otherwise stated, we shall assume all gases are ideal gases.*

The ideal gas equation becomes more accurate as $P \to 0$, and in the *limit* of $P = 0$ all gases behave ideally—this makes $PV = nRT$ a so-called "limiting law." At high pressures and low temperatures the ideal gas law becomes increasingly less accurate and it becomes necessary to use fancier, more complicated equations of state to describe the relationships between P, V, T, and n. $PV = nRT$ works best at high temperatures and low pressures, i.e., in a "hot vacuum."

The equation $PV = nRT$ incorporates three sets of observations:

(1) $P = \dfrac{nRT}{V}$; i.e., $(PV)_{n,T} = constant.$ (Fig. 1(a)) or $P_{n,T} \propto \dfrac{1}{V}$ (Fig. 1(b)).

This is **Boyle's Law**:

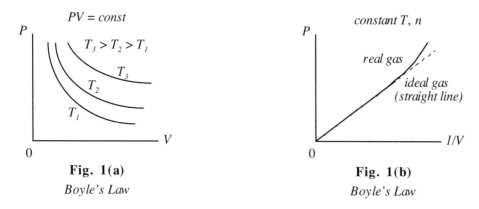

Fig. 1(a)

Boyle's Law

Fig. 1(b)

Boyle's Law

(2) $V = \dfrac{nRT}{P}$; i.e., $V_{n,P} \propto T$ (Fig. 2(a)). This is **Charles' Law**.

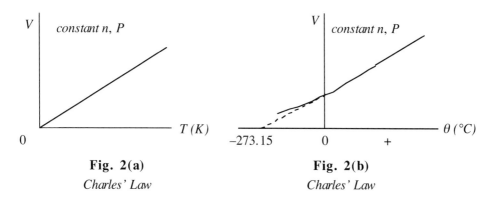

Fig. 2(a)

Charles' Law

Fig. 2(b)

Charles' Law

As indicated in Fig. 2(b), extrapolation to $V = 0$ on the Celsius scale intersects the x-axis at –273.15°C. Since V can't be less than zero, then –273.15°C must be the lowest temperature possible—"absolute zero."

(3)   **Avogadro's Principle**:   Consider two ideal gases, A and B:

From the ideal gas law: $V_A = \dfrac{n_A R T_A}{P_A}$ and $V_B = \dfrac{n_B R T_B}{P_B}$

Dividing V_A by V_B: $\dfrac{V_A}{V_B} = \dfrac{n_A T_A P_B}{n_B T_B P_A}$ . . . [1]

If both gases are at the same temperature and pressure, then $T_A = T_B$ and $P_A = P_B$, and Eqn [1] becomes

$$\frac{V_A}{V_B} = \frac{n_A}{n_B} \qquad\qquad \ldots [2]$$

If we take *equal volumes* of each gas (both at the same T and P), then, from Eqn [2],

$$\frac{V_A}{V_B} = \frac{n_A}{n_B} = 1 \qquad\qquad \ldots [3]$$

In other words, $\qquad\qquad\qquad\qquad\qquad n_A = n_B$

Therefore, *equal volumes of any gas at the same temperature and pressure contain the same number of moles (or molecules, since 1 mole = Avogadro's number N_A of molecules).* This is known as **Avogadro's Principle**.

Example 3-1

Calculate the density at 25°C and one bar pressure of a gas mixture consisting of 3.00 mol of H_2 and 5.00 mol of N_2. Relative molar masses: $H = 1.008$; $N = 14.01$.

Solution

Total mass of mixture = mass of 3 mol H_2 + mass of 5 mol N_2

$$= (3 \text{ mol}) \times \left(0.002016 \; \frac{\text{kg}}{\text{mol}}\right) + (5 \text{ mol}) \times \left(0.02802 \; \frac{\text{kg}}{\text{mol}}\right)$$

$$= 0.1461 \text{ kg}$$

The total volume of the gas mixture at 25°C and 1.00 bar pressure is

$$V = \frac{nRT}{P}$$

$$= \frac{(3+5)(8.314)(25+273.15)}{100\,000}$$

$$= 0.1983 \text{ m}^3$$

Therefore, the density is: $\qquad \rho = \dfrac{\text{mass}}{\text{volume}} = \dfrac{0.1461 \text{ kg}}{0.1983 \text{ m}^3}$

$$= 0.7368 \text{ kg m}^{-3}$$

$$= 0.7368 \text{ g L}^{-1}$$

(Note that $1 \text{ kg/m}^3 = 1000 \text{ g}/1000 \text{ L} = 1 \text{ g/L}$; thus the numerical value of density expressed in kg m^{-3} is the same as the value expressed in g L^{-1}.)

Ans: $\boxed{0.737 \text{ g L}^{-1}}$

Example 3-2

At sufficiently high temperatures, molecular hydrogen gas will start to dissociate into gaseous hydrogen atoms according to:

$$H_{2(g)} \rightleftharpoons 2H_{(g)}$$

Calculate the density of hydrogen gas at 2000°C and one atm pressure if molecular hydrogen is 33% dissociated at this temperature. Molecular hydrogen and atomic hydrogen both behave as ideal gases under these conditions.

Solution

Basis: Start initially with 100 moles of H_2 and heat to 2000°C at one atm:

		H_2	$\rightleftharpoons$	$2H$
Moles:	Initial:	100		0
	Dissociation:	-33		$+66$
	Equilibrium:	$100 - 33 = 67$		66

Total moles: $n = 67 + 66 = 133$

Note that 2 moles of H are produced for each mole of H_2 dissociated.

At equilibrium there is a total of 133 moles present.

Since mass is conserved,

$$\begin{aligned}
\text{Total mass of the mixture} &= \text{Mass of the initial } H_2 \text{ gas} \\
&= \left(100 \text{ mol}\right) \times \left(0.002016 \frac{\text{kg}}{\text{mol}}\right) \\
&= 0.2016 \text{ kg}
\end{aligned}$$

The total volume of gas mixture at 2000°C and 1.00 atm pressure is

$$\begin{aligned}
V &= \frac{nRT}{P} \\
&= \frac{(133)(8.314)(2000 + 273.15)}{101\,325} \\
&= 24.807 \text{ m}^3
\end{aligned}$$

Therefore the density is:

$$\begin{aligned}
\rho &= \frac{\text{mass}}{\text{volume}} = \frac{0.2016 \text{ kg}}{24.807 \text{ m}^3} \\
&= 0.008127 \text{ kg m}^{-3} \\
&= 8.127 \text{ g m}^{-3}
\end{aligned}$$

Ans: 8.13 g m^{-3}

Exercise 3-1

A two-liter vessel contains n moles of N_2 at 0.5 atm pressure and T K. When 0.01 mole of O_2 is added, it is necessary to cool the vessel to a temperature of 10°C in order to maintain the pressure at 0.5 atm. Calculate n and T. ☺

3.2 MOLAR VOLUME (v)

The molar volume v of any pure substance (not just gases) is just the volume occupied by one mole of the substance:

$$v = \frac{V}{n} \qquad \ldots [4]$$

where v is the molar volume in $m^3 \, mol^{-1}$, V is the total volume in m^3, and n is the number of moles of the substance.

For an ideal gas, $PV = nRT$

from which $v = \dfrac{V}{n} = \dfrac{RT}{P}$

Thus the molar volume of an ideal gas is just RT/P. At STP (Standard Temperature and Pressure, 0°C, 1 atm) this gives a value of

$$v = \frac{RT}{P} = \frac{(8.31451)(273.15)}{(101\ 325)}$$
$$= 0.022414 \text{ m}^3$$
$$= 22.414 \text{ L}$$

Thus, *one mole of an ideal gas occupies 22.414 L at STP.*

Often we are more interested in 25°C—which is closer to room (ambient) temperature—and one *bar* pressure (= 10^5 Pa)—a more convenient metric pressure than one atm. Thus, at SATP (Standard Ambient Temperature and Pressure):

$$v = \frac{RT}{P} = \frac{(8.31451)(298.15)}{(100\ 000)}$$
$$= 0.02479 \text{ m}^3$$
$$= 24.79 \text{ L}$$

3.3 COMBINED GAS EQUATION

For a fixed amount (n moles) of a given ideal gas,

$$\frac{PV}{T} = nR = \text{constant} \qquad \dots [5]$$

Since nR is constant, then all combinations of P, V, and T for this sample of gas must obey Eqn [5].

That is,
$$\frac{P_1V_1}{T_1} = \frac{P_2V_2}{T_2} = \frac{P_3V_3}{T_3} = \dots \dots = nR \qquad \dots [6]$$

Therefore, for a constant number of moles of gas,
$$\boxed{\frac{P_1V_1}{T_1} = \frac{P_2V_2}{T_2}} \qquad \textit{Combined Gas Equation} \qquad \dots [7]$$

You will find Eqn [7] to be a very useful equation.

3.4 DALTON'S LAW OF PARTIAL PRESSURES

Dalton's Law: *"The total pressure P exerted by a mixture of ideal gases is the sum of the pressures that would be exerted if each gas occupied the same volume alone."*

Expressed mathematically:
$$\boxed{P = P_A + P_B + P_C + \dots} \qquad \textit{Dalton's Law} \qquad \dots [8]$$

Dalton's Law follows naturally from the ideal gas law:

$$P = \frac{nRT}{V}$$

$$= (n_A + n_B + n_c + \dots)\frac{RT}{V}$$

$$= \frac{n_ART}{V} + \frac{n_BRT}{V} + \frac{n_CRT}{V} + \dots$$

$$= P_A + P_B + P_C + \dots$$

Note that V is the total volume of the gas and n is the total moles of gas. P_A, P_B, and P_C are called the **partial pressures** of A, B, and C, respectively.

3.5 MOLE FRACTION (x)

The mole fraction x_i of component i in a mixture is defined by

$$\boxed{x_i \equiv \frac{n_i}{n}}$$ *Mole Fraction Defined*   . . . [9]

where n_i is the number of moles of component i and n is the *total* number of moles of all components in the mixture. This formula can be applied to solids, liquids, or gases. For example, consider a mixture of three gases, A, B, and C:

$$x_A = \frac{n_A}{n_A + n_B + n_C}, \quad x_B = \frac{n_B}{n_A + n_B + n_C}, \quad \text{and} \quad x_C = \frac{n_C}{n_A + n_B + n_C}.$$

It is obvious that

$$x_A + x_B + x_C = 1$$

Therefore, in general,

$$\boxed{\sum_i x_i = 1}$$ . . . [10]

Now consider the same mixture of the same three ideal gases:

From Dalton's Law,
$$P = P_A + P_B + P_C$$
$$= \frac{n_A RT}{V} + \frac{n_B RT}{V} + \frac{n_C RT}{V}$$ . . . [11]

But for the total gas mixture,
$$PV = nRT$$

where n is the total number of moles of gas in the mixture.

Thus,
$$\frac{RT}{V} = \frac{P}{n}$$ . . . [12]

Substituting Eqn [12] into Eqn [11] gives

$$P = n_A \left(\frac{P}{n}\right) + n_B \left(\frac{P}{n}\right) + n_C \left(\frac{P}{n}\right)$$

$$= \frac{n_A}{n} \cdot P + \frac{n_B}{n} \cdot P + \frac{n_C}{n} \cdot P$$

$$= x_A P + x_B P + x_C P$$ . . . [13]

In other words, the partial pressure of any given gas in a gas mixture is just the mole fraction of that gas times the total pressure of the mixture:

$$\boxed{P_i = x_i P}$$ *Partial Pressure Defined*   . . . [14]

Sometimes if we want to distinguish between the mole fraction in the gas phase and that in the liquid or solid phase we use the notation x_i for the mole fraction in the liquid or solid phase and the notation y_i for the mole fraction in the gas or vapor phase. Thus, for a gas mixture, we may choose to write $P_i = y_i P$ instead of $P_i = x_i P$.

Example 3-3

If the air pressure drops by about 0.078 atm for every kilometer above sea level, and air may be assumed to be 21% by volume oxygen, at what altitude would breathing even pure oxygen result in a below-normal oxygen intake?

Solution

From Dalton's Law, the partial pressure of oxygen in normal air at one atm pressure is

$$P_{O_2} = y_{O_2} P = (0.21)(101\ 325)$$
$$= 21\ 278 \text{ Pa}$$

Let the altitude at which pure oxygen has $P_{O_2} = 21\ 278$ Pa be z' meters.

We are told that the pressure of the atmosphere varies with altitude z according to

$$P_z = 1.000 - \left(\frac{0.078 \text{ atm}}{1000 \text{ m}}\right)(z \text{ m})$$

$$= \left(1.000 - 7.8 \times 10^{-5} z \text{ atm}\right)\left(101\ 325\ \tfrac{\text{Pa}}{\text{atm}}\right)$$

$$= 101\ 325 - 7.90335z \text{ Pa}$$

Therefore,

$$P_{z'} = 21\ 278 = 101\ 325 - 7.90335\,z'$$

$$7.90335\,z' = 101\ 325 - 21\ 278$$
$$= 80\ 047$$

$$z' = 10\ 128 \text{ m} = 10.128 \text{ km}$$

Ans: $\boxed{\sim 10 \text{ km}}$

Exercise 3-2

When two grams of a gaseous substance A is introduced into an initially evacuated flask maintained at a constant temperature T, the pressure is found to be one bar. When three grams of a second gaseous substance B is added at the same temperature to the two grams of A, the pressure increases to 1.5 bar. Calculate the ratio of the molar mass of B to that of A. ☺

3.6 PARTIAL VOLUME (V_i)

The partial volume V_i of any gas i in a mixture of gases is defined as the volume that would be occupied by that gas if it were at the total pressure P of the mixture (Fig. 3). Of course in actual fact each gas occupies the *total* volume and has a partial *pressure*, not a partial *volume*.

The concept of partial volume is useful for describing the composition of a gas mixture, because for gases, the volume % is the same as the mole %:

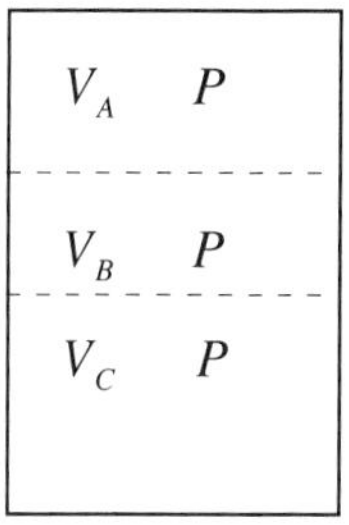

V_A	P
V_B	P
V_C	P

Fig. 3

Thus,
$$V_A = \frac{n_A RT}{P} \quad \text{and} \quad V = \frac{nRT}{P}$$

so that
$$\frac{V_A}{V} = \frac{n_A}{n}$$

and
$$\frac{V_A}{V} \times 100\% = \frac{n_A}{n} \times 100\% \qquad \qquad \dots [15]$$

We often report gas compositions as volume percents. Thus, when we say that a certain gas mixture is 25% by volume oxygen and 75% by volume argon, we mean that the mole fraction of oxygen in the mixture is 0.25 and the mole fraction of argon is 0.75.

Example 3-4

If air is 21% by volume oxygen and 79% by volume nitrogen, what is the molar mass of dry air? Relative molar masses: O = 16.00; N = 14.01.

Solution

1.00 mol of air contains 0.21 mol of O_2 + 0.79 mol of N_2

Therefore the mass of 1.00 mol of air = (mass of 0.21 mol O_2) + (mass of 0.79 mol N_2)

$$= (0.21 \text{ mol}) \times \left(32.00 \tfrac{g}{mol}\right) + (0.79 \text{ mol}) \times \left(28.02 \tfrac{g}{mol}\right)$$
$$= 6.72 + 22.14$$
$$= 28.86 \text{ g}$$

The molar mass of dry air is 28.9 g mol^{-1}.

Ans: $\boxed{28.9 \text{ g mol}^{-1}}$

The actual value will vary because of the presence of traces of other gases and of water vapor, all of which will contribute to the mass of one mole of air. Excluding water vapor and industrial pollutants, the standard composition of atmospheric air is that shown in Table 1:

| Table 1 | Components of Atmospheric Air | | |
| | (Excluding Water Vapor) | | |
Component	*Content* (% by volume)	*Component*	*Content* (ppm)
N_2	78.084 ± 0.004	Ne	18.18 ± 0.04
O_2	20.946 ± 0.002	He	5.24 ± 0.004
CO_2	0.033 ± 0.001	CH_4	2
Ar	0.934 ± 0.001	Kr	1.14 ± 0.01
		N_2O	0.5 ± 0.1
		H_2	0.5
		Xe	0.087 ± 0.001

Exercise 3-3

A polyethylene bag is filled with air at one atm pressure and taken underwater to provide oxygen for breathing. Oxygen becomes toxic to humans when its partial pressure reaches about 0.8 atm. At what depth will breathing the air from this bag lead to oxygen toxicity? ☺

KEY POINTS FOR CHAPTER THREE

1. **Equation of State**: A mathematical equation relating *P*, *V*, *T* and *n* for a sample of substance.

2. **Ideal Gas Equation of State**: $PV = nRT$, where *P* is the pressure in Pa, *V* is the volume in m^3, *n* is the number of moles, *T* is the absolute temperature, in kelvin, and *R* is the universal gas constant. $R = 8.314$ J mol^{-1} K^{-1}.

   Units: $$R = \frac{PV}{nT} = \frac{\text{Pa} \cdot \text{m}^3}{\text{mol} \cdot \text{K}} = \frac{(\text{N}/\text{m}^2) \cdot \text{m}^3}{\text{mol} \cdot \text{K}} = \frac{\text{N} \cdot \text{m}}{\text{mol} \cdot \text{K}} = \frac{\text{J}}{\text{mol} \cdot \text{K}}$$

   Note: $R = 8.314$ Pa m^3 mol^{-1} K^{-1} = 8.314 J mol^{-1} K^{-1} = 8.314 kPa L mol^{-1} K^{-1}

3. A gas which obeys $PV = nRT$ at all pressures is called an ideal gas. At common temperatures and pressures (say 25°C and 1 bar) most gases obey the ideal gas law to within a few per cent. As the pressure decreases, all gases increasingly obey the ideal gas law; in the limit of zero pressure, all gases behave ideally. $PV = nRT$ holds best for conditions approximating a "hot vacuum."

4. **Boyle's Law**: $(PV)_{n,T} = constant.$ or $P_{n,T} \propto \dfrac{1}{V}$

Charles' Law: $V_{n,P} \propto T$

Avogadro's Principle: Equal volumes of any gas at the same temperature and pressure contain the same number of moles (or molecules, since 1 mole = Avogadro's number N_A of molecules).

Combined Gas Equation: $\dfrac{P_1 V_1}{T_1} = \dfrac{P_2 V_2}{T_2}$ or $\dfrac{PV}{T} = constant$

5. **Molar Volume v**: The volume occupied by one mole of substance: $v = V/n$
The molar volume for an ideal gas is $v = V/n = RT/P$.

6. **Dalton's Law of Partial Pressures**: The total pressure P exerted by a mixture of ideal gases is the sum of the pressures that would be exerted if each gas occupied the same volume alone. $P = P_A + P_B + P_C + \ldots$

7. **Mole Fraction x**: $x_i \equiv \dfrac{n_i}{n}$ and $\sum_i x_i = 1$

where n_i is the number of moles of component i and n is the total number of moles of all components in the mixture. This formula can be applied to solids, liquids, or gases.

8. **Partial Pressure**: $P_i = x_i P,$ where P_i is the partial pressure of gas component i and P is the total pressure of the gas mixture. For a gas mixture, we may choose to write $P_i = y_i P$ instead of $P_i = x_i P$.

9. **Partial Volume V_i**: The partial volume V_i of any gas i in a mixture of gases is defined as the volume that would be occupied by that gas if it were at the total pressure P of the mixture. For a gas mixture, the volume % of any component is the same as its mole %:

$$\frac{V_i}{V} \times 100\% = \frac{n_i}{n} \times 100\%$$

10. **Dry Atmospheric Air** $\approx$ 78 % by vol. N_2 + 21 % by vol. O_2 + 1 % by vol. Ar

PROBLEMS

1. 10.0 L of a gaseous mixture contains 5.00 g of O_2, 3.00 g of N_2, and 1.00 g of He. Calculate the partial pressure of O_2 and the total pressure if the temperature is 100°C. ☺

2. The density of a mixture of methane (CH_4) and He at 127°C and 53.3 kPa is 0.141 g L^{-1}. Calculate the mole percent of CH_4 in the mixture. ☺

3. Two flasks of equal volume, which are filled with nitrogen initially at 20.0°C and 100.0 kPa pressure, are joined by a narrow tube. One of the flasks is then immersed in an ice bath at 0.00°C while the other is held in a steam bath at 100.0°C. Calculate the final pressure in the system. ☹

4. A piece of sodium metal undergoes complete reaction with water as follows:

$$2Na_{(s)} + 2H_2O_{(liq)} \rightarrow 2NaOH_{(aq)} + H_{2(g)}$$

The hydrogen gas generated is collected over water at 25.0°C. The volume of the gas is 246 mL measured at 1.00 atm. Calculate the number of grams of sodium used in the reaction. The vapor pressure of water at 25°C is 0.0313 atm. ☺

5. Nitric oxide (NO) reacts with molecular oxygen as follows:

$$2NO_{(g)} + O_{2(g)} \rightarrow 2NO_{2(g)}$$

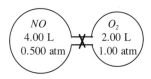

Initially NO and O_2 are separated as shown in the diagram at the right. When the valve is opened, the reaction quickly goes to completion. After the reaction has finished: (a) what is the final mole ratio of NO_2 to O_2? (b) what is the partial pressure of oxygen in the system? Assume that the temperature remains constant at 25°C. ☹

6. Phosphine is partially decomposed in a continuous flow reactor according to

$$4PH_{3(g)} \rightarrow P_{4(g)} + 6H_2$$

The mole fraction of the hydrogen in the product stream is 0.400. Assuming steady state operation of the reactor and a feed stream of pure phosphine, calculate: (a) the percent decomposition of the phosphine, (b) the mass of P_4 (molar mass = 123.9) produced per 100 m^3 of feed. The feed stream is at 900 K and 820 Torr. ☹

7. The reaction $2CO + O_2 \rightarrow 2CO_2$

proceeds to completion in a rigid reactor. The initial temperature and pressure were 21°C and 100 kPa, respectively. The final temperature and pressure were 204°C and 116 kPa. Calculate the composition of the initial mixture assuming that it consisted of CO in the presence of an excess of O_2. ☹

8. A closed container of humid air at a total pressure of 99.00 kPa and a temperature of 60.0°C has a pool of liquid water in the bottom that causes the air to be saturated with water vapor. (a) What is the partial pressure of the oxygen in this air? (b) How many kilograms of water is contained in one cubic meter of this air? (c) What is the density of this air?

Dry air may be considered to consist of 80.0% by volume nitrogen gas and 20.0% by volume oxygen gas. The equilibrium vapor pressure of water at 60°C is 149.38 Torr. ☺

9. When released at sea level at 25°C and one atm pressure, a certain weather balloon containing helium has a radius of 1.5 m. At its maximum altitude, where the temperature is −20°C, the radius increases to 4.0 m. What is the pressure inside the balloon at its maximum altitude? ☺

10. Calcium hypochlorite—$Ca(ClO)_2$—which is used in bleaching powder, is prepared by absorbing chlorine gas in a $Ca(OH)_2$ (milk of lime) solution:

$$2Cl_2 + 2Ca(OH)_2 \rightarrow Ca(ClO)_2 + CaCl_2 + 2H_2O$$

A gas mixture consisting of chlorine and inert gases enters an absorption apparatus at 25°C and a total pressure of 95.0 kPa. The partial pressure of the chlorine gas in this mixture is 8.0 kPa. The gas leaves the apparatus at a temperature of 30°C and a total pressure of 92.0 kPa with a Cl_2 partial pressure of 70 Pa. The inert gases are not absorbed in the apparatus. Calculate: (a) The volume of the gases leaving the absorber for every 100 m³ of inlet gas. (b) The mass of Cl_2 absorbed for every 100 m³ of inlet gas. ☹

11. An inverted bell jar contains 1.00 m³ of water-saturated air at 25°C and a total pressure of one atm; the surrounding atmosphere also is at one atm pressure. If the bell jar is lowered to a depth of 60.0 m in Lake Ontario, what will be the volume of the water-saturated air in the bell jar? At a depth of 60.0 m, it may be assumed that the water temperature is 5.0°C. The equilibrium vapor pressures of water at 25°C and at 5.0°C are 23.756 Torr and 6.543 Torr, respectively. Assume none of the air dissolves in the liquid water. $P_{atm} = 101$ 325 Pa. $g = 9.80$ m s^{-2}. The density of the water may be taken as constant at 1000 kg m^{-3}. ☹

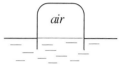

12. One mole of dry air (79% N_2 and 21% O_2 by volume) at 25°C is brought into contact with liquid water at the same temperature. (a) What is the volume of the dry air if the pressure is one bar? (b) What is the final volume of the air after it is saturated with water vapor if the total pressure is maintained at one bar? (c) What is the partial pressure of the O_2 in the moist air? The equilibrium vapor pressure of water at 25°C is $P_W^{\bullet} = 3168$ Pa. Assume the gases behave ideally, and that no oxygen or nitrogen dissolves in the water. ☺

13. Can a gas have both a concentration of 1.00 mol m^{-3} and a pressure of 1.00 Pa at the same time? ☺

14. A sample of dry air has the following composition in mole %: $N_2 = 78.0\%$, $O_2 = 21.0\%$, $Ar = 1.0\%$. Calculate its density at 21°C and 98.87 kPa.
Molar masses: $N_2 = 28.0$, $O_2 = 32.0$, $Ar = 39.9$ ☺

15. The density of a mixture of methane and helium at 127°C and 53.3 kPa is 0.141 g L^{-1}. Calculate the mole fraction of methane in the mixture. Molar masses: $CH_4 = 16.0$, $He = 4.00$ ☺

16. 100 g of water is placed into a one-liter container at 23°C and then sealed. The container also contains dry air at one atm pressure. The sealed container then is heated to 100°C. Note that at 100°C the partial pressure of water vapor must equal the value of its equilibrium vapor pressure at the same temperature (to be discussed later), which is 1.00 atm.

   (a) Calculate the equilibrium pressure inside the container assuming negligible solubility of air in the liquid water and that the volume of the container remains constant.

   (b) What mass of water is contained in the vapor at 100°C? The density of water at 23°C and 100°C is 0.998 g cm^{-3} and 0.958 g cm^{-3}, respectively. ☹

17. A 15-liter tank contains gaseous C_2H_6 at 68 atm and 60°C. Calculate the mass of C_2H_6 in the tank assuming ideal behavior. The molar mass of C_2H_6 is 30.1 g mol^{-1}. ☺

18. Assuming ideal gas behavior, calculate the density of air saturated with water vapor at 95°C if the total pressure is 1.00 bar. The vapor pressure of water at 95°C is 0.0845 MPa. The molar masses of water and dry air are, respectively, 18.015 and 28.965 g mol^{-1}. $R = 8.3145$ J mol^{-1} K^{-1}. ☹

19. A company sells dry oxygen-enriched air, which they claim contains at least 50% by volume pure oxygen. To test their claim, a 1.00 liter glass bulb was filled with the gas to a total pressure of 750 Torr at 20.0°C. The weight of the gas sample was found to be 1.236 grams. What is the volume percent of oxygen gas in the mixture? Assume ordinary air consists of 20.0% by volume O_2 and 80.0% by volume N_2. Molar masses: O = 16.00, N = 14.01. ☹

FOUR

THE KINETIC THEORY OF GASES

The theory is called a "kinetic" theory because it assumes that all the macroscopic properties of a gas result from molecular *motion*. To be considered "macroscopic" requires $\geq 10^6$ particles, which corresponds to only about 10^{-17} L of gas!

4.1 POSTULATES

(1) *A gas consists of a large number of small, identical particles, which are relatively far apart.* The particles must be identical because gases form uniform solutions with uniform properties. The particles must be small and relatively far apart because gases are transparent, easily compressed, and quickly diffuse through each other.

(2) *The molecules are in continuous, random, very rapid motion, colliding frequently with each other and with the retaining walls of the container.* The molecules must be in continuous motion because they don't settle with time. The motion must be random because pressure is exerted in all directions; furthermore, the molecules move in all directions to occupy the container. The motion must be rapid because gases quickly move in straight line motion to very quickly fill a vacuum. There must be frequent collisions between particles because diffusion in other gases is much slower than in a vacuum. Finally, there must be frequent bombardment with the walls in order to account for the pressure of the gas.

(3) *The collisions are perfectly elastic, with no loss in total energy.* If energy were gradually dissipated, there would be less bombardment with time, resulting in a gradual decrease in the pressure of the gas. This is not observed.

(4) *The only type of energy is kinetic energy (KE), and the average KE of translation is directly proportional to the absolute temperature; the same constant of proportionality holds for all gases.* For a constant volume of gas, increasing the temperature increases the pressure, which means increased bombardment. Since increased bombardment means increased kinetic energy, it follows that the KE is proportional to the temperature. The constant of proportionality must be the same for all gases because for all gases the same change in pressure (ΔP) is observed for the same change in temperature (ΔT).

4.2 SIMPLIFIED DERIVATION OF THE IDEAL GAS LAW

Consider W kilograms of an ideal gas at a temperature of T kelvin in a volume of V m^3, contained in a cubical vessel of sides L meters. Therefore, $V = L^3$. Let the total number of molecules $= N$, each of mass m. Therefore, the total mass of the gas is given by $W = Nm$. Because of frequent random collisions, different velocities will be present. Also, because the motion is random, no single direction of movement is favored; thus at any instant, $\frac{1}{3}$ of the molecules are moving in the x-direction, $\frac{1}{3}$ of the molecules are moving in the y-direction, and $\frac{1}{3}$ of the molecules are moving in the z-direction.

Consider the motion along the x-axis: of the various velocities present, consider a *representative* molecule, the velocity of which is v_x m s^{-1}. Earlier we saw that the pressure of a gas is defined as the rate of change of momentum (mv) per unit time per unit area of the walls of the container:

$$P = \frac{d(mv)}{dt}\bigg/A$$

But we also know that the pressure is the total force acting on the walls of the container divided by the area of the walls of the container:

$$P = F/A$$

Comparing the two definitions of pressure shows that the force F_x exerted by the molecules traveling in the x-direction on the walls of the container is

$$F_x = \frac{d(mv_x)}{dt} = \text{change in momentum per second}$$

Now, $\left(\dfrac{\text{change in momentum}}{\text{s}}\right) = \left(\dfrac{\text{change in momentum}}{\text{impact}}\right) \times \left(\dfrac{\text{no. of impacts}}{\text{s}}\right)$

Referring to Fig. 1, it can be seen that a molecule moving in the x-direction makes two impacts (one at A and one at B) while traveling a distance of $2L$ meters. That is, the molecule undergoes

$$\frac{2}{2L} = \frac{1}{L} \quad \frac{\text{impacts}}{\text{m}}$$

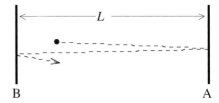

Fig. 1

Therefore, at a velocity of v_x m s^{-1} it makes

$$\left\{ v_x \; \frac{m}{s} \right\}\left\{ \frac{1}{L} \; \frac{impacts}{m} \right\} = \frac{v_x}{L} \quad \frac{\text{impacts}}{\text{s}}$$

Since the collisions are assumed to be perfectly elastic, the change in momentum per impact is from $+mv_x$ to $-mv_x$; that is, a total change in momentum of $2mv_x$ kg m s^{-1} per impact. Thus the force due to the impact of *one* molecule on the walls of the container is

$$F_x = \left\{ 2mv_x \; \frac{\text{kg m s}^{-1}}{\text{impact}} \right\}\left\{ \frac{v_x}{L} \; \frac{\text{impacts}}{\text{s}} \right\} = \frac{2mv_x^2}{L} \quad \text{newtons (per molecule)}$$

Since $\frac{1}{3}$ of the N molecules are traveling in the x-direction, the *total* force acting in the x-direction is

$$\left\{ \frac{N}{3} \right\}\left\{ \frac{2mv_x^2}{L} \right\} \text{ newtons}$$

The total force acting on the walls *in all 3 directions* is just three times the force in the x-direction:

$$F = 3\left\{\frac{N}{3}\right\}\left\{\frac{2mv_x^2}{L}\right\} = \frac{2Nmv_x^2}{L} \text{ newtons}$$

Since the total area of the walls of the cube is $6L^2$, then the *pressure* exerted by the gas is

$$P = \frac{F}{A} = \left\{\frac{2Nmv_x^2/L}{6L^2}\right\} = \frac{Nmv_x^2}{3V} \quad \text{pascals} \qquad \dots [1]$$

Cross-multiplying gives
$$PV = \tfrac{1}{3}Nmv_x^2 \qquad \dots [2]$$

According to postulate 4, the average translational *KE* of a molecule is proportional to *T*, i.e.,

$$\bar{e}_{transl} = bT \qquad \dots [3]$$

where b is a constant, having the same value for all gases. Since we know that $e_{transl} = \tfrac{1}{2}mv^2$,

then
$$\bar{e}_{transl} = \overline{\tfrac{1}{2}mv^2} = \tfrac{1}{2}m\overline{v^2} = \tfrac{1}{2}mc^2 \qquad \dots [4]$$

where $c^2 = \overline{v^2}$ is known as the *mean square velocity*. Note that $\overline{v^2}$ is *not* the same as $(\bar{v})^2$. Furthermore, since the molecule chosen was "representative," let it have as its *KE* the *average KE*,

i.e.,
$$\tfrac{1}{2}m\overline{v^2} = \tfrac{1}{2}mc^2 = \bar{e}_{transl} = bT \qquad \dots [5]$$

Assuming the velocity distribution is the same in all three coordinate directions, we now can replace v_x^2 in Eqn [2] with c^2 to obtain:

$$PV = \tfrac{1}{3}Nmc^2 = \tfrac{2}{3}N\left(\tfrac{1}{2}mc^2\right) = \tfrac{2}{3}N\bar{e}_{transl} = \tfrac{2}{3}NbT \qquad \dots [6]$$

Therefore,
$$PV = \tfrac{2}{3}NbT = \tfrac{2}{3}\left(\frac{N}{N_A}\right)(N_AbT) = \tfrac{2}{3}(n)(N_AbT) \qquad \dots [7]$$

where N_A is Avogadro's number and $N/N_A = n =$ the total number of moles of gas.

That is,
$$\boxed{PV = n\left(\tfrac{2}{3}N_Ab\right)T} \quad \text{*Ideal Gas Law*} \qquad \dots [8]$$

Eqn [8] is the *Ideal Gas Law*, $PV = nRT$, with $R = \tfrac{2}{3}N_Ab$.

4.3 THE MEANING OF PRESSURE

From Eqn [6],
$$PV = \tfrac{1}{3}Nmc^2 \qquad \dots [9]$$

where $P =$ pressure of the gas, $V =$ volume of the gas, $N =$ total number of molecules, $m =$ mass of each molecule, and

$$c^2 = \text{the mean square speed of the molecules} = \frac{s_1^2 + s_2^2 + \ldots + s_N^2}{N} \qquad \ldots [10]$$

In Eqn [10] s_i is the *speed* of a given molecule i.[1]

Furthermore,

$$Nm = \text{(no. molecules)(mass/molecule)}$$
$$= \text{total mass of gas}$$
$$= nM \qquad \ldots [11]$$

where n is the total number of moles of the gas and M is the mass per mole.

Substituting Eqn [11] into Eqn [9]:

$$PV = \tfrac{1}{3} nMc^2 \qquad \ldots [12]$$

from which

$$\boxed{P = \frac{nMc^2}{3V}} \quad \begin{array}{l}\textit{Meaning of}\\ \textit{Pressure}\end{array} \qquad \ldots [13]$$

Eqn [13] gives the pressure in terms of *fundamental* properties of the gas. Our derivation also showed—Eqn [8]—that $PV = n\left(\tfrac{2}{3} N_A b\right)T$, which we recognized as the *Ideal Gas Law*, with

$$\boxed{R = \tfrac{2}{3} N_A b} \quad \begin{array}{l}\textit{Meaning of the}\\ \textit{Gas Constant}\end{array} \qquad \ldots [14]$$

4.4 THE MEANING OF TEMPERATURE

Eqn [14] rearranges to

$$b = \frac{3}{2} \frac{R}{N_A} \qquad \ldots [15]$$

One of our postulates was that the kinetic energy of the representative gas molecule was directly proportional to the temperature, with b being the constant of proportionality (Eqn [3]):

$$\bar{e}_{transl} = bT \qquad \ldots [3]$$

Substituting the expression for b given by Eqn [15] into Eqn [3] gives

$$\bar{e}_{transl} = \frac{3}{2}\left(\frac{R}{N_A}\right)T \qquad \ldots [16]$$

where R/N_A is the gas constant *per molecule*. This ratio is so important that it has its own special name and symbol: **Boltzmann's Constant**, k. The numerical value of Boltzmann's constant is

---

[1] The speed s is not the same as the velocity v, the former being a *scalar* quantity (independent of direction of motion) while the latter is a *vector* quantity (depends on direction of motion). You may recall that c—defined in Eqn [5]—was chosen to be *representative* speed; that is, one connected with velocity v and not with speed s. However, when the velocity v of a molecule is squared, it loses its sign and the value becomes the same as its speed s squared, justifying Eqn [10].

$$k = \frac{R}{N_A} = \frac{8.314\,\text{J mol}^{-1}\,\text{K}^{-1}}{6.02 \times 10^{23}\,\text{mol}^{-1}} = 1.38 \times 10^{-23}\,\text{J K}^{-1} \qquad \ldots [17]$$

Therefore, the average KE per molecule is $\boxed{\bar{e}_{transl} = \tfrac{3}{2}kT}$ $\qquad \ldots [18]$

from which $\boxed{T = \left(\frac{2}{3k}\right)\bar{e}_{trans}}$ *Meaning of Temperature* $\qquad \ldots [19]$

Eqn [19] shows that *T is a measure of the average KE of the gas molecules.*

For one molecule, $\bar{e}_{transl} = \tfrac{3}{2}kT$, therefore the average translational KE per *mole* of gas is

$$\left(N_A \frac{\text{molecules}}{\text{mol}}\right)\left(\bar{e}_{trans} \frac{\text{joules}}{\text{molecule}}\right) = (N_A)\left(\tfrac{3}{2}kT\right) = N_A \frac{3}{2}\left(\frac{R}{N_A}\right)T = \tfrac{3}{2}RT$$

That is, $\boxed{\begin{array}{c} \textit{Average KE per} \\ \textit{Mole of Monatomic} \quad = \tfrac{3}{2}RT \\ \textit{Ideal Gas} \end{array}}$ $\qquad \ldots [20]$

Eqn [20] is a very important relationship often used in thermodynamics. Note that in the above derivations the gas molecules have been assumed to consist of "point" molecules (sort of like little ping pong balls with zero radius). That is, the gas has been assumed to be a *monatomic* gas, such as Ar, He, or Ne. For monatomic gases, we may assume that the kinetic energy of the molecules consists only of translational kinetic energy. For more complicated gases, other forms of kinetic energy also must be taken into consideration. For example, for diatomic gases such as H_2 or N_2, in addition to the straight line motion of translation, the molecules also can *vibrate* and *rotate*. Vibration and rotation also are modes of kinetic energy; thus a mole of a diatomic gas can store *more* kinetic energy than just $\tfrac{3}{2}kT$. Therefore it follows that Eqn [20] only gives the average kinetic energy for one mole of an *ideal monatomic gas.*

4.5 DIFFUSION AND EFFUSION

Gases expand very rapidly to uniformly occupy the space available; this is called *diffusion*. Light gases such as H_2 diffuse much faster than heavier gases such as Cl_2. The escape of a gas through a very small hole is called *effusion*. From the kinetic theory, we showed that

$$PV = \frac{Nmc^2}{3}$$

therefore, $\qquad c^2 = \frac{3PV}{Nm} = \frac{3PV}{W} = \frac{3P}{W/V} = \frac{3P}{\rho} \qquad \ldots [21]$

where c^2 is the *mean square speed*, P is the pressure of the gas, and ρ is the density of the gas. The **R**oot **M**ean **S**quare speed (*RMS*) is defined as

$$c_{RMS} \equiv \sqrt{\frac{s_1^2 + s_2^2 + \ldots + s_N^2}{N}} \qquad \ldots [22]$$

From the statistical theory of systems containing large numbers of particles, it can be shown that the *average* speed, $\bar{s}$, is given by

$$\bar{s} = \frac{s_1 + s_2 + \ldots + s_N}{N} = \frac{c_{RMS}}{\sqrt{3\pi/8}} \qquad \ldots [23]$$

Now, the rate of diffusion or effusion is proportional to the average speed of the molecules; but the average speed of the molecules is proportional to the root mean speed of the molecules;

therefore, $$RATE \propto c_{RMS} \qquad \ldots [24]$$

However, from Eqn [21], the mean square speed c^2 is equal to $3P/\rho$. Substituting this into Eqn [24] gives

$$RATE \propto \sqrt{\frac{3P}{\rho}} \qquad \ldots [25]$$

If we compare two different ideal gases A and B, both at the same temperature and pressure, then, from Eqn [25] it follows that

$$\frac{RATE\ A}{RATE\ B} = \sqrt{\frac{3P/\rho_A}{3P/\rho_B}} \qquad \ldots [26]$$

That is, $$\boxed{\frac{RATE\ A}{RATE\ B} = \sqrt{\frac{\rho_B}{\rho_A}}} \qquad \ldots [27]$$

Eqn [27] is known as **Graham's Law**, which states that

> *"At a given T and P, the rate of effusion of a gas is inversely proportional to the square root of its density."*

From $$PV = nRT = \frac{W}{M}RT$$

we get $$\frac{W}{V} = \rho = \left(\frac{P}{RT}\right)M$$

where W is the mass of the gas, V is its volume, and ρ is its density. Therefore, at constant pressure and temperature the density of a gas is directly proportional to its molar mass, so that

Eqn [27] also can be written $$\boxed{\frac{RATE\ A}{RATE\ B} = \sqrt{\frac{M_B}{M_A}}} \qquad Constant\ T,\ P \qquad \ldots [28]$$

Eqn [28] shows that the molar mass of an unknown gas can be determined by measuring its rate of effusion and comparing it with the rate for a gas of known molar mass. For example, H_2 has a molar mass of 2.02, while the molar mass of Cl_2 is 70.90; therefore H_2 effuses $\sqrt{70.90/2.02} = 5.9$ times faster than Cl_2 if both gases are at the same temperature and pressure.

Exercise 4-1

Calculate the mean molecular speed and the root mean speed for the following set of gas molecules: ☺

Number of molecules	1	6	9	4	1
Speed, m s^{-1}	100	500	1000	1500	2000

Example 4-1

A 5.00 L bulb containing nitrogen gas at 30°C and 5.0 bar pressure contains a small pinhole of area 0.036 mm^2 through which the gas leaks to the surroundings. The pressure in the bulb drops 24.0 torr when the gas is permitted to effuse for 15.0 seconds. When the process is repeated under the same conditions using a gaseous refrigerant instead of nitrogen, the same drop in pressure requires 28.3 seconds. Estimate the molar mass of the refrigerant.

Solution

$N_2 = 2 \times 14.0067 = 28.0134$ g mol^{-1} = 0.0280134 kg mol^{-1}

The drop in pressure in the bulb is given by $\Delta P = \dfrac{\Delta nRT}{V}$

therefore the same drop in pressure with the second gas results when the same number of moles of gas Δn escapes.

According go Graham's Law: $\dfrac{r_{refrig}}{r_{N_2}} = \sqrt{\dfrac{M_{N_2}}{M_{refrig}}}$

Since the effusion of the refrigerant takes longer than that of the N_2 for the same total moles to effuse, the rate of effusion of the refrigerant is slower that that of the N_2 for the same Δn moles:

Therefore, $\Delta n = \left(r_{refrig} \; \frac{mol}{s}\right)(28.3 \; s) = \left(r_{N_2} \; \frac{mol}{s}\right)(15.0 \; s)$

from which $\dfrac{r_{refrig}}{r_{N_2}} = \dfrac{15.0}{28.3}$

Therefore $\dfrac{r_{refrig}}{r_{N_2}} = \dfrac{15.0}{28.3} = \sqrt{\dfrac{M_{N_2}}{M_{refrig}}} = \sqrt{\dfrac{0.0280134}{M_{refrig}}}$

Squaring:
$$\left(\frac{15.0}{28.3}\right)^2 = 0.28094 = \frac{0.0280134}{M_{refrig}}$$

from which
$$M_{refrig} = \frac{0.0280134}{0.28094} = 0.0997 \text{ kg mol}^{-1} = 99.7 \text{ g mol}^{-1}$$

Ans: $\boxed{99.7 \text{ g mol}^{-1}}$

(The refrigerant dichloroethane ($C_2H_4Cl_2$) has a molar mass of 98.96 g mol^{-1}.)

Example 4-2

Naturally occurring hydrogen consists of 99.985 mol % $^1H^1$ isotope and 0.015 mol % $^1H^2$ isotope (known as deuterium, designated $^1D^2$). These two isotopes gradually can be separated using a large series of effusion cells, each containing very small pinholes. Calculate the composition of the gas passing from the first effusion cell if the cell contains natural hydrogen gas maintained at 25°C and 15 bar pressure. Relative molar masses: $^1H^1 = 1.007947$; $^1D^2 = 2.014102$

Solution

Using "dot" notation to indicate the rate dn_i/dt in moles per second of gas i effusing from the cell through the pinhole, it can be seen that

$$\dot{n}_i = \frac{dn_i}{dt} = \left(\begin{array}{c}\text{no. of moles} \\ \text{per cubic meter}\end{array}\right) \times \left(\begin{array}{c}\text{mean speed of the molecules} \\ \text{in meters per second}\end{array}\right) \times \left(\begin{array}{c}\text{cross - sectional area} \\ \text{of the pinhole in m}^2\end{array}\right)$$

i.e.,
$$\dot{n}_i = \left(\frac{n_i}{V} \frac{\text{mol}}{\text{m}^3}\right)\left(\bar{s}_i \frac{\text{m}}{\text{s}}\right)\left(A \text{ m}^2\right) = \frac{n_i \bar{s}_i A}{V} \text{ mol s}^{-1} \qquad \ldots \text{[a]}$$

From $P_iV = n_iRT$
$$\frac{n_i}{V} = \frac{P_i}{RT}$$

Substituting this into Eqn [a]:
$$\dot{n}_i = \frac{P_i}{RT}\bar{s}_i A \text{ mol s}^{-1} \qquad \ldots \text{[b]}$$

and it follows that the ratio of the moles of H_2 effusing through the pinhole to the moles of D_2 effusing through the pinhole is given by

$$\frac{\dot{n}_{H_2}}{\dot{n}_{D_2}} = \frac{P_{H_2}\bar{s}_{H_2}A/RT}{P_{D_2}\bar{s}_{D_2}A/RT} = \frac{P_{H_2}\bar{s}_{H_2}}{P_{D_2}\bar{s}_{D_2}} \qquad \ldots \text{[c]}$$

But the mean speed $\bar{s}$ is proportional to the *RMS* velocity,

$$\bar{s}_i \propto \left\{(c_{RMS})_i = \sqrt{\frac{3RT}{M_i}}\right\} \qquad \ldots \text{[d]}$$

Putting [d] into [c]:
$$\frac{\dot{n}_{H_2}}{\dot{n}_{D_2}} = \frac{P_{H_2}\sqrt{3RT/M_{H_2}}}{P_{D_2}\sqrt{3RT/M_{D_2}}} = \frac{P_{H_2}}{P_{D_2}}\sqrt{\frac{M_{D_2}}{M_{H_2}}} \qquad \ldots [e]$$

Inserting numerical values gives

$$\frac{\dot{n}_{H_2}}{\dot{n}_{D_2}} = \frac{99.985}{0.015}\sqrt{\frac{2 \times 2.014402}{2 \times 1.007947}} = 9422.5$$

Therefore, for every mole of deuterium effusing from the cell there are 9422.5 moles of hydrogen, and the mole % of deuterium in the effusing gas is

$$mole\ \%\ D_2 = \left[\frac{moles\ D_2}{total\ moles\ of\ gas}\right] \times 100\%$$

$$= \left[\frac{1\ mol\ D_2}{(1 + 9422.5)\ mol\ (D_2 + H_2)}\right] \times 100\%$$

$$= 0.01061\ \%$$

and
$$mole\ \%\ H_2 = 100 - 0.01061 = 99.98939\ \%$$

The concentration of deuterium in the effusing gas has changed by

$$\left(\frac{0.01061 - 0.015}{0.015}\right) \times 100\% = -29.3\%$$

$$\boxed{\begin{array}{c} 99.989\ mol\ \%\ H_2 \\ 0.011\ mol\ \%\ D_2 \end{array}}$$

Ans:

$$\boxed{\textbf{Exercise\ \ 4-2}}$$

Inspection of a natural gas pipeline operating at 60 psig reveals a tiny pinhole. A leak test using nitrogen under the same conditions shows a gas leakage rate of 42 liters per hour. If natural gas is worth 25¢ per cubic meter, what is the value of the gas lost in one year? Natural gas may be considered to be pure methane. ☺

4.6 THE SPEEDS OF GAS MOLECULES

The average *KE* of a single gas molecule is $\frac{1}{2}mc^2 = \frac{3}{2}kT$;

therefore,
$$c^2 = \frac{3kT}{m} = \frac{3(R/N_A)T}{m} = \frac{3RT}{N_A m} = \frac{3RT}{M}$$

and
$$\boxed{c_{RMS} = \sqrt{c^2} = \sqrt{\frac{3RT}{M}}} \qquad \ldots [29]$$

It can be shown from the kinetic theory of gases that the mean speed of a gas molecule is given by

$$\bar{s} = \frac{c_{RMS}}{\sqrt{3\pi/8}} \qquad \dots [30]$$

Putting Eqn [29] into Eqn [30]:

$$\bar{s} = \sqrt{\frac{3RT}{M}} \Big/ \sqrt{\frac{3\pi}{8}} = \sqrt{\frac{3RT}{M} \times \frac{8}{3\pi}} = \sqrt{\frac{8RT}{\pi M}}$$

Therefore:

$$\boxed{\bar{s} = \sqrt{\frac{8RT}{\pi M}}} \quad \begin{array}{l} \textit{Mean Speed of} \\ \textit{a Gas Molecule} \end{array} \qquad \dots [31]$$

Eqn [31] shows that the mean speed of a gas molecule increases with increasing temperature and decreases with increasing molar mass. Note that this speed is independent of the pressure.

Example 4-3

The kinetic theory of gases is amazingly versatile, and can be applied not only to the molecular scale, but also to the cosmic scale! This owes to the fact that cosmic bodies are so far apart as to behave almost like particles of ideal gas with very low densities. According to the book, *Astrophysical Concepts*,[2] galaxies, which, on the average have a mass of about 3×10^{41} kg and a random mean velocity of about 100 km s^{-1}, exert a cosmic pressure that affects the dynamics of the universe. The number density of galaxies is about 0.1 galaxy per cubic Mpc (megaparsec), where 1 pc = 3×10^{16} m.

(a)   Use the kinetic theory of gases to estimate the cosmic pressure exerted by galaxies.
(b)   Estimate the mean temperature within a galaxy.

Solution

(a)   From the kinetic theory of gases,

$$P = \frac{nMc_{RMS}^2}{3V} = \frac{n}{V} \cdot \frac{Mc_{RMS}^2}{3} \qquad \dots [a]$$

where P is the pressure in Pa, n is the number of moles of gas, M is the molar mass in kg, V is the volume in m^3, and c_{RMS} is the root mean square velocity of the molecules. We can treat a galaxy as a rather heavy molecule, with $M = 3 \times 10^{41}$ kg mol^{-1}.

Further, $1 \text{ Mpc} = 10^6 \times (3 \times 10^{16}) = 3 \times 10^{22}$ m

and $1 \text{ Mpc}^3 = (3 \times 10^{22})^3 = 27 \times 10^{66}$ m^3

Therefore $\dfrac{n}{V} = \dfrac{0.1 \text{ mol}}{27 \times 10^{66} \text{ m}^3} = 3.704 \times 10^{-69}$ mol m^{-3}

The root mean square velocity and the mean velocity are given, respectively, by

_____

[2] M. Harwit, *Astrophysical Concepts*, 3rd edition, Springer-Verlag, NY (1998).

$$c_{RMS} = \sqrt{\frac{3RT}{M}} \quad \text{and} \quad \bar{s} = \sqrt{\frac{8RT}{\pi M}},$$

therefore

$$\frac{c_{RMS}^2}{\bar{s}^2} = \frac{3RT/M}{8RT/\pi M} = \frac{3\pi}{8}$$

and

$$c_{RMS}^2 = \frac{3\pi}{8} \times \bar{s}^2 = \frac{3\pi}{8} \times \left(100 \times 10^3\right)^2 = 1.178 \times 10^{10} \text{ m}^2 \text{ s}^{-2}$$

Substituting numerical values into Eqn [a] gives

$$P = \frac{n}{V} \cdot \frac{Mc_{RMS}^2}{3} = \frac{(3.704 \times 10^{-69})(3 \times 10^{41})(1.178 \times 10^{10})}{3} = 4.36 \times 10^{-18} \text{ Pa}$$

Ans: $\boxed{4 \times 10^{-18} \text{ Pa}}$

(b) $c_{RMS} = \sqrt{\frac{3RT}{M}}$, from which $T = \frac{Mc_{RMS}^2}{3R} = \frac{(3 \times 10^{41})(1.178 \times 10^{10})}{(3)(8.3145)} = 1.42 \text{ K}$

This is consistent with the value of about 2.7 K for intergalactic space.

Ans: $\boxed{\sim 1.4 \text{ K}}$

Exercise 4-3

A rigid tank having a volume of 1.00 m^3 contains argon gas at 3.00 bar pressure. If the mean molecular speed of the argon atoms is 400 m s^{-1}, what would the pressure be if the mean molecular speed were increased to 600 m s^{-1}? ☺

Example 4-4

Calculate the density of nitrogen gas at 10 bar pressure if its root mean square speed is 2500 km h^{-1}.

Solution

Density: $\rho = \frac{\text{kg}}{\text{m}^3} = \frac{(n \text{ mol})(M \text{ kg/mol})}{V \text{ m}^3} = \left(\frac{n}{V}\right)M$ . . . [a]

From $PV = nRT$, $\frac{n}{V} = \frac{P}{RT}$ . . . [b]

Putting [b] into [a]: $$\rho = \left(\frac{P}{RT}\right) M = P\left(\frac{M}{RT}\right) \qquad \dots [c]$$

Given $$c_{RMS} = \left(2500 \ \frac{km}{h}\right)\left(\frac{1 \ h}{3600 \ s}\right)\left(1000 \ \frac{m}{km}\right) = 694.44 \text{ m s}^{-1}$$

But $c_{RMS} = \sqrt{\dfrac{3RT}{M}}$, therefore $\dfrac{3RT}{M} = c^2_{RMS}$ and $\dfrac{M}{RT} = \dfrac{3}{c^2_{RMS}}$ $\qquad \dots [d]$

Substituting [d] into [c]: $\rho = P\left(\dfrac{M}{RT}\right) = P\left(\dfrac{3}{c^2_{RMS}}\right) = \dfrac{(10 \times 10^5)(3)}{(694.44)^2} = 6.2209 \text{ kg m}^{-3}$

Ans: $\boxed{6.22 \text{ kg m}^{-3}}$

Exercise 4-4

Calculate the mean speed of the molecules in (a) hydrogen gas at 25°C and 10 bar pressure, (b) hydrogen gas at 1000°C and one bar pressure, and (c) nitrogen gas at 250°C and one bar pressure. ☺

4.7 EFFECT OF PRESSURE ON SPEED

We showed earlier that $\bar{e}_{transl} = \frac{1}{2}mc^2 = \frac{3}{2}kT$. This equation states that the *KE* of a gas molecule is determined only by the temperature. Increasing the pressure merely means an increased bombardment per unit area, *not* an increase in the speed of the molecules. Therefore *the KE is independent of pressure.*

4.8 DISTRIBUTION OF MOLECULAR SPEEDS

Probability and Distribution Functions

The frequent collisions and interchange of momentum among the gas molecules leads to a variation in their speeds. All the collisions are *elastic*, which means that there is conservation of momentum ($m \times s$) and conservation of kinetic energy ($\frac{1}{2}ms^2$) in the system.

Suppose we have a gas (the system) consisting of N molecules, of which N_s is the number of molecules having speed s. The mean speed $\bar{s}$ of the N molecules is defined as

$$\bar{s} = \frac{1}{N}\sum_{i=1}^{N} s_i \qquad \dots [32]$$

If the possible values of speed s are *discrete*,[3] such as the number of peas in a pod, more than one molecule may, in general, have exactly the same speed. Thus, if N_j molecules have velocity s_j, then

---

[3] That is, 1, 2, 3, 4, etc. as opposed to 1.324, 2.769, etc.

Eqn [32] can be written as

$$\bar{s} = \frac{1}{N} \sum_j N_j s_j \qquad \ldots [33]$$

where the index j now labels the possible values of s rather than the molecules $1 \ldots N$.

Note that
$$\sum_j N_j = N \qquad \ldots [34]$$

The probability[4] P_j of a molecule having a speed s_j is *defined* as

$$P_j \equiv \frac{N_j}{N} \qquad \ldots [35]$$

Therefore, we could express Eqn [33] alternatively as

$$\bar{s} = \sum_j P_j s_j \qquad \ldots [36]$$

The probabilities must, of course, add up to unity:

$$\sum_j P_j = 1 \qquad \ldots [37]$$

Now, if the possible values of speed s are *continuously* (as opposed to *discretely*) distributed, as is the case for the molecules of a gas, we must modify the above definitions. Instead of P_j, we now define what is known as a *distribution function* $\rho(s)$, such that

$$\rho(s')ds'$$

represents the (infinitesimal) probability that s has a value between s' and $s' + ds'$. The distribution function $\rho(s)$ represents not a probability, but what is called a *probability density*. The units of $\rho(s)$ are probability per unit speed.

The (infinitesimal) number of molecules dN_s having speed between s and $s + ds$ is given by:

$$dN_s = \{\rho(s)ds\}N \qquad \ldots [38]$$

Note the units in Eqn [38]:

$$\begin{Bmatrix} \text{Number of} \\ \text{molecules} \end{Bmatrix} = \left\{ \left(\text{probability density}, \frac{\text{fraction between 0 and 1}}{\text{per unit speed, m/s}} \right) \times \left(\text{speed, m/s} \right) \right\} \begin{Bmatrix} \text{Total number} \\ \text{of molecules} \end{Bmatrix}$$

From Eqn [38] it follows that the (infinitesimal) *fraction* f_s of the molecules having speed between s and $s + ds$ is just

$$f_s = \frac{dN_s}{N} = \rho(s)ds \qquad \ldots [39]$$

An integrated quantity that *does* represent a probability is

$$P(s_1 \le s \le s_2) = \int_{s_1}^{s_2} \rho(s)ds \qquad \ldots [40]$$

---

[4] Don't confuse probability P with pressure P.

where $P(s_1 \leq s \leq s_2)$ is the probability (having a value somewhere between *0* and *1*) that *s* has a value between s_1 and s_2.

When the integration extends over the complete range of values of *s*, i.e., from zero to infinity, then all the molecules in the system are taken account of, and this probability must equal unity (i.e., 100% probability):

$$P(0 \leq s \leq \infty) = \int_0^\infty \rho(s)ds = 1 \qquad \ldots [41]$$

Because it adds up to a value of unity, the distribution function given by Eqn [41] is said to be *normalized*. Using the normalized distribution function, the average speed $\bar{s}$ is given by

$$\bar{s} = \int_0^\infty s \cdot \rho(s)ds \qquad \ldots [42]$$

Maxwell-Boltzmann Speed Distribution

In 1860 the Scottish physicist James Clerk Maxwell derived the following normalized distribution function [5] for the molecular speeds in an equilibrium gas of molar mass *M*:

Maxwell-Boltzmann
Speed Distribution
$$\rho(s) = 4\pi \left(\frac{M}{2\pi RT}\right)^{3/2} s^2 \cdot exp\left[-\frac{Ms^2}{2RT}\right] \qquad \ldots [43]$$

Fig. 2 shows a plot of Eqn [43] for nitrogen gas at three different temperatures. The plot shows that the distribution of molecular speeds in a gas is *not* symmetrical. There are more molecules with very high speeds than with very low speeds. The lowest speed is limited to *zero*, whereas there is no theoretical limit to the highest speed. It also is seen that at higher temperatures there are more molecules with higher speeds than there are at lower temperatures. At higher temperatures there are more speeds among which to distribute the *N* molecules; therefore the curve is broader and flatter, with a lower peak. This explains the sensitivity of chemical reaction rates to temperature. Only molecules with "high energy" (i.e., those that have attained a certain minimum speed) collide to react.

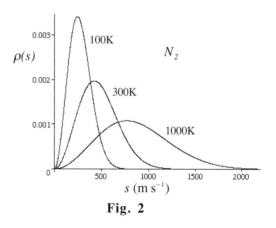

Fig. 2

At higher temperatures more molecules attain this necessary "activation" energy than at lower temperatures.

---

[5] Actually, Maxwell's equation gives the distribution of molecular *velocities*, not molecular *speeds*. As mentioned above, *velocity* is a vector quantity dependent on the direction (*x*-, *y*-, or *z*-) of motion, whereas *speed* is a scalar quantity, independent of the direction of motion. For many applications the distribution of molecular speeds, without regard to the direction of motion, is all that is required. Eqn [43], giving the distribution of *speeds*, is more properly called the Maxwell-Boltzmann distribution. The mathematics required to derive this equation are beyond the level of this course. You'll just have to take my word that these equations are correct. Wouldn't you buy a used car from me?

Substitution of Eqn [43] into Eqn [39] gives the (infinitesimal) number of molecules having speeds between s and $s + ds$ as

$$dN_s = N \cdot \rho(s)ds = 4\pi N \left(\frac{M}{2\pi RT}\right)^{3/2} s^2 \cdot exp\left[-\frac{Ms^2}{2RT}\right] \cdot ds \qquad \ldots [44]$$

The number of molecules ΔN_s having speeds in the *finite* range s to $s + \Delta s$ can be estimated by replacing ds in Eqn [44] with Δs, providing that Δs is not too wide.[6]

Thus,
$$\Delta N_s \approx 4\pi N \left(\frac{M}{2\pi RT}\right)^{3/2} s^2 \cdot exp\left[-\frac{Ms^2}{2RT}\right] \cdot \Delta s \qquad \ldots [45]$$

From Eqn [44] the (infinitesimal) *fraction* f_s of the total N molecules having speeds in the (infinitesimal) range s to $s + ds$ is

$$\boxed{f_s = \frac{dN_s}{N} = 4\pi \left(\frac{M}{2\pi RT}\right)^{3/2} s^2 \cdot exp\left[-\frac{Ms^2}{2RT}\right] \cdot ds} \qquad \ldots [46]$$

Again, providing Δs is not too great, Eqn [46] can be used to estimate the fraction of molecules having speeds in the *finite* range of s to $s + \Delta s$:

$$f_s \approx \frac{\Delta N_s}{N} \approx 4\pi \left(\frac{M}{2\pi RT}\right)^{3/2} s^2 \cdot exp\left[-\frac{Ms^2}{2RT}\right] \cdot \Delta s \qquad \ldots [47]$$

Example 4-5

Estimate the fraction of the nitrogen molecules that have speeds at 500 K in the range 290 m s^{-1} to 300 m s^{-1}.

Solution

It is clear that we have to use Eqn [47]; but what value do we use for s? The usual procedure is to take the *average* value in the range of interest, which, in this case, is $290 \sim 300$ m s^{-1}; i.e., use a value of $s = 295$ m s^{-1}.

The molar mass of N_2 gas is 28.02 g mol^{-1}, which must be expressed in base SI units in order to get the answer also to come out in base SI units (m s^{-1}). Therefore, we must use $M = 0.02802$ kg mol^{-1}. Substituting values into Eqn [47] with $T = 500$ K gives

$$f_s = 4\pi \left(\frac{0.02802}{2\pi(8.3145)(500)}\right)^{3/2} (295)^2 \cdot exp\left[-\frac{(0.02802)(295)^2}{2(8.3145)(500)}\right] \cdot (10)$$

$$= 4\pi \, (1.111 \times 10^{-9})(8.7025 \times 10^4) \, exp\,[-0.29329] \cdot (10)$$

$$= 9.062 \times 10^{-3} = \frac{0.91}{100} = 0.91\%$$

---

[6] No greater, say, than $\Delta s = 10$ m s^{-1}.

Ans: 0.91 %

Example 4-6

The Maxwell-Boltzmann equation for the distribution of molecular speeds has no upper limit set on the permissible speeds. How many molecules can we expect to find at velocities approaching the speed of light (2.99792458×10^8 m s^{-1})? To examine this, neglecting any changes in the mass of a molecule at such high speeds, calculate the fraction of the nitrogen molecules that have speeds at 500 K in the range 2.99000000×10^8 m s^{-1} to 2.99000001×10^8 m s^{-1}. (This is 99.74% the speed of light).

Solution

$$f \approx 4\pi \left(\frac{M}{2\pi RT} \right)^{3/2} s^2 \cdot exp\left[-\frac{Ms^2}{2RT} \right] \cdot \Delta s,$$

We choose a value of

$$s = \tfrac{1}{2}(2.99000000 \times 10^8 + 2.99000001 \times 10^8) = 2.990000005 \times 10^8 \text{ m s}^{-1}$$

with $\Delta s = 2.99000001 \times 10^8 - 2.99000000 \times 10^8 = 1.0$ m s^{-1}

The molar mass of N_2 gas is $M = 0.02802$ kg mol^{-1}.
Substituting values in the equation with $T = 500$ K gives

$$f = 4\pi \left(\frac{0.02802}{2\pi(8.3145)(500)} \right)^{3/2} (2.990000005 \times 10^8)^2 \cdot exp\left[-\frac{(0.02802)(2.990000005 \times 10^8)^2}{2(8.3145)(500)} \right] \cdot (1)$$

$$= 4\pi \, (1.111 \times 10^{-9})(8.9401 \times 10^{16}) \, exp\,[-3.0128 \times 10^{11}] \cdot (1)$$

$$= (1.2481 \times 10^9) \, exp\,[-3.0128 \times 10^{11}]$$

Most calculators can't evaluate $exp\,[-3.0128 \times 10^{11}]$; therefore we must break this quantity down into easier-to-handle quantities:

$$e^{-3.0128 \times 10^{11}} = e^{-301.28 \times 10^9} = \left(e^{-301.28} \right)^{10^9} = \left(1.4314 \times 10^{-131} \right)^{1,000,000,000} = 1.4314 \times 10^{-131,000,000,000}$$

Therefore, $f = (1.2481 \times 10^9)(1.4314 \times 10^{-131,000,000,000})$

$$= 1.7865 \times 10^{-130,999,999,991}$$

I guess we can round this off to zero!   **Ans:** 0%

Exercise 4-5

Calculate the fraction of the nitrogen molecules that have speeds at 25°C in the range 1999 m s^{-1} to 2001 m s^{-1}. ☺

4.9 THE MAXWELL–BOLTZMANN DISTRIBUTION AS AN ENERGY DISTRIBUTION[7]

In the previous section it was mentioned that for many chemical reactions in the gas phase, only those molecules that have a speed—i.e., a kinetic energy—equal to or greater than a threshold value will undergo chemical reaction. Thus, often we are more interested in the *energy* of the molecules as opposed to their *speed*. For this purpose we shall see how to convert Eqn [43] from a distribution of speeds to one of energies. As before, in all cases we shall assume that all the energy is purely kinetic energy.

We start with Eqn [46] for the *fraction f_s* of the N molecules having speeds in the range s to $s + ds$:

$$f_s = \frac{dN_s}{N} = 4\pi \left(\frac{M}{2\pi RT} \right)^{3/2} s^2 \cdot exp\left[-\frac{Ms^2}{2RT} \right] \cdot ds \qquad \ldots [46]$$

The kinetic energy ε of a single molecule of mass m is

$$\varepsilon = \tfrac{1}{2}ms^2 \qquad \ldots [48]$$

Therefore, the kinetic energy E of one *mole* of molecules, if each had the same speed s, would be

$$E = \left(\tfrac{1}{2}ms^2 \ \frac{J}{molecule} \right)\left(N_A \ \frac{molecules}{mol} \right) = \tfrac{1}{2}mN_A s^2 \ \frac{J}{mol} \qquad \ldots [49]$$

But

$$\left(m \ \frac{kg}{molecule} \right)\left(N_A \ \frac{molecules}{mol} \right) = mN_A \ \frac{kg}{mol} = M \qquad \ldots [50]$$

where M is just the molar mass of the gas, in kg mol^{-1}.

Putting Eqn [50] into Eqn [49] gives the energy of the gas per mole as

$$E = \tfrac{1}{2}Ms^2 \qquad \ldots [51]$$

Therefore it can be seen that the quantity $Ms^2/2$ in the exponential term in Eqn [46] is just the kinetic energy of one mole of gas if all the molecules had the same speed s.

Differentiating Eqn [51]:
$$\frac{dE}{ds} = 2\left(\tfrac{1}{2}Ms \right) = Ms$$

from which
$$ds = \frac{dE}{Ms} \qquad \ldots [52]$$

Also, from Eqn [51],
$$\frac{M}{2} = \frac{E}{s^2} \qquad \ldots [53]$$

Substituting Eqns [53], [51], and [52] into Eqn [46] gives f in terms of energy (f_E) instead of in terms of speed (f_s):

---

[7] To understand this section properly you need to know some integral calculus; otherwise, you'll just have to take my word for it. Sorry.

$$f_E = \frac{dN_E}{N} = 4\pi \left(\frac{E/s^2}{\pi RT}\right)^{3/2} s^2 \cdot exp\left[-\frac{E}{RT}\right] \cdot \frac{dE}{Ms}$$

$$= 4\pi \left(\frac{E}{\pi RT}\right)^{3/2} \frac{s^2}{s^3} \cdot exp\left[-\frac{E}{RT}\right] \cdot \frac{dE}{Ms}$$

$$= 4\pi \left(\frac{E}{\pi RT}\right)^{3/2} \frac{1}{Ms^2} \cdot exp\left[-\frac{E}{RT}\right] \cdot dE$$

$$= 4\pi \left(\frac{E}{\pi RT}\right)^{3/2} \frac{1}{2E} \cdot exp\left[-\frac{E}{RT}\right] \cdot dE$$

$$= 2\pi \left(\frac{1}{\pi RT}\right)^{3/2} \frac{E^{3/2}}{E} \cdot exp\left[-\frac{E}{RT}\right] \cdot dE$$

$$= 2\pi \left(\frac{1}{\pi RT}\right)^{3/2} \sqrt{E} \cdot exp\left[-\frac{E}{RT}\right] \cdot dE \qquad \ldots [54]$$

$$\boxed{f_E = \frac{dN_E}{N} = 2\pi \left(\frac{1}{\pi RT}\right)^{3/2} \sqrt{E} \cdot exp\left[-\frac{E}{RT}\right] \cdot dE} \qquad \ldots [55]$$

where f_E now is the fraction of the molecules having energy in the range of E to $E + dE$.

The normalized energy distribution function is thus

Maxwell-Boltzmann
Energy Distribution $$\boxed{\rho(E) = \frac{1}{N}\frac{dN_E}{dE} = 2\pi \left(\frac{1}{\pi RT}\right)^{3/2} \sqrt{E} \cdot exp\left[-\frac{E}{RT}\right]} \qquad \ldots [56]$$

Eqn [56] is plotted in Fig. 3 for a gas at three different temperatures.

It should be noted that the *energy* distribution has a different shape than the speed distribution. In particular, there is a *vertical* tangent at the origin, which means that the function rises much more quickly than does the speed distribution, which has a *horizontal* tangent at the origin. Also, after passing through a maximum, the energy distribution falls off more gently than does the speed distribution. However, in common with the speed distribution, the energy distribution broadens at higher

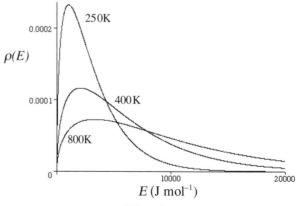

Fig. 3

temperatures, reflecting the fact that at higher temperatures a greater fraction of the molecules have high energies.

Example 4-7

Suppose it were possible to fill a container with gas molecules such that 100% of the molecules each had the same kinetic energy of 5.0×10^{-21} J. After a given length of time, owing to random collisions the initially uniform motion of the molecules would become chaotic, eventually falling into a Maxwell-Boltzmann distribution of energies.

(a) What would be the temperature of the system?

(b) What percent of the molecules finally would be in the range of 4.98×10^{-21} to 5.02×10^{-21} J?

Solution

(a) Initially, the kinetic energy of each molecule (of mass m) is 5.0×10^{-21} J. If s_o is the initial uniform speed of each molecule,

then, $\quad\quad \tfrac{1}{2} m s_o^2 = 5.0 \times 10^{-21} \quad\quad$ from which $\quad\quad s_o^2 = \dfrac{1.0 \times 10^{-20}}{m} \quad\quad$ . . . [a]

Since all the molecules have the same initial speed, then s_o^2 also must be the initial mean square speed. Therefore,

$$s_o^2 = \overline{c_o^2} = \frac{3RT}{M} = \frac{3(N_A k)T}{N_A m} = \frac{3kT}{m} \quad\quad\quad \text{. . . [b]}$$

Substituting Eqn [a] into Eqn [b]: $\quad\quad \dfrac{1.0 \times 10^{-20}}{m} = \dfrac{3kT}{m}$

from which $\quad\quad\quad T = \dfrac{1.0 \times 10^{-20}}{3k} = \dfrac{1.0 \times 10^{-20}}{3(1.381 \times 10^{-23})} = 241.37 \text{ K}$

Ans: $\boxed{241 \text{ K}}$

(b) Writing Eqn [55] for a finite energy range:

$$\frac{\Delta N_E}{N} \approx 2\pi \left(\frac{1}{\pi RT} \right)^{3/2} \sqrt{E} \cdot exp\left[-\frac{E}{RT} \right] \cdot \Delta E \quad\quad\quad \text{. . . [a]}$$

where $\Delta N_E / N$, the fraction of the molecules having energy in the range of E to $E + \Delta E$ J mol^{-1}.

To convert Eqn [a] to values per molecule rather than per mole, we note that the energy ε of a single molecule (in joules) is related to the energy E per mole of molecules by

$$E \tfrac{J}{mol} = \left(\varepsilon \tfrac{J}{molecule} \right) \left(N_A \tfrac{molecules}{mol} \right) \quad\quad\quad \text{. . . [b]}$$

Also, recall that Boltzmann's constant k is just the gas constant per molecule, that is, $k = R/N_A$, from which $\qquad\qquad R = N_A k$ $\qquad\qquad\qquad$ . . . [c]

Substituting Eqns [b] and [c] into Eqn [a] gives the expression in terms of single molecular properties:

$$\frac{\Delta N_E}{N} \approx 2\pi \left(\frac{1}{\pi(N_A k)T}\right)^{3/2} \sqrt{N_A e} \cdot exp\left[-\frac{N_A e}{(N_A k)T}\right] \cdot \Delta(N_A e)$$

$$\approx 2\pi \left(\frac{1}{\pi kT}\right)^{3/2} \frac{N_A^{1/2} \cdot N_A}{N_A^{3/2}} \sqrt{e} \cdot exp\left[-\frac{e}{kT}\right] \cdot \Delta e$$

$$\approx 2\pi \left(\frac{1}{\pi kT}\right)^{3/2} \sqrt{e} \cdot exp\left[-\frac{e}{kT}\right] \cdot \Delta e \qquad\qquad \text{. . . [d]}$$

Inserting values into Eqn [d]:

$$\frac{\Delta N_E}{N} \approx 2\pi \left(\frac{1}{\pi(1.381 \times 10^{-23})(241.37)}\right)^{3/2} \sqrt{5.0 \times 10^{-21}} \cdot exp\left[\frac{-5.0 \times 10^{-21}}{(1.381 \times 10^{-23})(241.37)}\right] \cdot \left\{0.04 \times 10^{-21}\right\}$$

$$= \left(5.8633 \times 10^{30}\right)\left(7.0711 \times 10^{-11}\right)\left(0.22313\right)\left(0.04 \times 10^{-21}\right)$$

$$= 0.00370 = 0.370\% \qquad\qquad \text{Less than 1\% would be in this speed range.}$$

Ans: $\boxed{0.37\%}$

4.10   FRACTION OF MOLECULES HAVING $E > E'$ [8]

In the previous section it was mentioned that for many chemical reactions in the gas phase only those molecules having energy equal to or greater than some minimum value E' J mol^{-1} will undergo reaction. In this section we shall learn how to calculate this quantity.

We start with Eqn [55] for the infinitesimal fraction of molecules having molar energies in the range of E to $E + dE$:

$$\frac{dN_E}{N} = 2\pi \left(\frac{1}{\pi RT}\right)^{3/2} \sqrt{E} \cdot exp\left[-\frac{E}{RT}\right] \cdot dE \qquad\qquad \text{. . . [55]}$$

Multiplying by N gives $\qquad dN_E = 2\pi N \left(\frac{1}{\pi RT}\right)^{3/2} \sqrt{E} \cdot exp\left[-\frac{E}{RT}\right] \cdot dE \qquad\qquad \text{. . . [56]}$

where dN_E is the number of molecules having energy in the range E to $E + dE$.

The total number $N_{E \geq E'}$ of molecules having energy *equal to or greater than* E' is given by

$$N_{E \geq E'} = \int_{E=E'}^{E=\infty} dN_E \qquad\qquad \text{. . . [57]}$$

---

[8] Again, to understand this section properly you need to know some integral calculus.

Substituting Eqn [56] into Eqn [57] and taking the constants outside the integral sign:

$$N_{E \geq E'} = 2\pi N \left(\frac{1}{\pi RT}\right)^{3/2} \cdot \int_{E=E'}^{E=\infty} \sqrt{E} \cdot exp\left[-\frac{E}{RT}\right] \cdot dE \qquad \ldots [58]$$

Dividing Eqn [58] by N gives the *fraction* of the total molecules having energy $\geq E'$:

$$\frac{N_{E \geq E'}}{N} = 2\pi \left(\frac{1}{\pi RT}\right)^{3/2} \cdot \int_{E=E'}^{E=\infty} \sqrt{E} \cdot exp\left[-\frac{E}{RT}\right] \cdot dE \qquad \ldots [59]$$

We now have to evaluate the integral in Eqn [59]. Get ready for a thrill ride; otherwise, proceed directly to Eqn [88]!

Designate the integral as *(I)*: $(I) = \int_{E=E'}^{E=\infty} \sqrt{E} \cdot exp\left[-\frac{E}{RT}\right] \cdot dE$ $\ldots [60]$

Integral *(I)* can be expressed as the difference between *two* integrals, Integral *(A)*, integrating from $x = 0$ to $x = \infty$, and Integral *(B)*, integrating from $x = 0$ to $x = E'$:

$$(I) = \int_{E=E'}^{E=\infty} \sqrt{E} \cdot exp\left[-\frac{E}{RT}\right] \cdot dE$$

$$= \underbrace{\int_{0}^{\infty} \sqrt{E} \cdot exp\left[-\frac{E}{RT}\right] \cdot dE}_{(A)} - \underbrace{\int_{0}^{E'} \sqrt{E} \cdot exp\left[-\frac{E}{RT}\right] \cdot dE}_{(B)} \qquad \ldots [61]$$

Integral *(A)*

First we evaluate Integral *(A)*: $(A) = \int_{0}^{\infty} \sqrt{E} \cdot exp\left[-\frac{E}{RT}\right] \cdot dE$ $\ldots [62]$

From tables of integrals,[9] we find that some mathematician has been able to show that

$$\int_{0}^{\infty} \sqrt{x} \cdot exp[-ax] \cdot dx = \frac{1}{2a}\sqrt{\frac{\pi}{a}} \qquad \ldots [63]$$

Comparing Integral *(A)* with the integral in Eqn [63] shows that our integral is the same, with $a = 1/(RT)$. Therefore, the solution to Integral *(A)* is

$$\frac{1}{2a}\sqrt{\frac{\pi}{a}} = \frac{1}{2(1/RT)}\sqrt{\frac{\pi}{(1/RT)}} = \frac{RT}{2}\sqrt{\pi RT} = (RT)^{3/2}\frac{\sqrt{\pi}}{2}$$

and $(A) = \int_{0}^{\infty} \sqrt{E} \cdot exp\left[-\frac{E}{RT}\right] \cdot dE = (RT)^{3/2}\frac{\sqrt{\pi}}{2}$ $\ldots [64]$

---

[9] I used the *Handbook of Chemistry and Physics*, 84th ed., D.R. Lide, editor, CRC Press, Boca Raton (2003).

Integral *(B)*

Now we evaluate Integral *(B)*: $(B) = \int_0^{E'} \sqrt{E} \cdot exp\left[-\dfrac{E}{RT}\right] \cdot dE$ . . . [65]

This integral is more difficult than Integral *(A)*, but can be evaluated by considering a change of variable:

Thus, let $\dfrac{E}{RT} = x^2$ . . . [66]

Therefore,[10] $x = \sqrt{\dfrac{E}{RT}}$ . . . [67]

and $E = RTx^2$ . . . [68]

from which $\dfrac{dE}{dx} = 2RTx$ or $dE = 2RTx\,dx$ . . . [69]

Also, from Eqn [68], $\sqrt{E} = \sqrt{RT}\,x$ . . . [70]

Substituting Eqns [70], [66], and [69] into the integral in Eqn [65] gives:

$$(B) = \int \left(\sqrt{RT}\,x\right) \cdot \left(exp\left[-x^2\right]\right) \cdot \left(2RTx\,dx\right)$$ . . . [71]

We need to know the limits of the integration when we switch the independent variable from E to x: Thus, when

$$E = 0, \quad x = \sqrt{\dfrac{E}{RT}} = \sqrt{\dfrac{0}{RT}} = 0$$ . . . [72]

and when $E = E', \quad x = \sqrt{\dfrac{E'}{RT}}$ . . . [73]

Therefore, with the new integration limits Eqn [71] becomes

$$(B) = \int_0^{\sqrt{E'/RT}} \left(\sqrt{RT}\,x\right) \cdot \left(exp\left[-x^2\right]\right) \cdot \left(2RTx\,dx\right)$$

$$= 2(RT)^{3/2} \int_0^{\sqrt{E'/RT}} x^2 \cdot exp\left[-x^2\right] \cdot dx$$ . . . [74]

Now we are stuck with a new integral, which we will call Integral *(C)*:

$$(C) = \int x^2 \cdot exp\left[-x^2\right] \cdot dx$$ . . . [75]

---

[10] Only the positive roots are required since E is always greater than zero.

This one we can solve using *integration by parts*, according to which

$$\int u\,dv = uv - \int v\,du \qquad \ldots [76]$$

If we let

$$u = exp\left[-x^2\right] \qquad \ldots [77]$$

then,

$$\frac{du}{dx} = exp\left[-x^2\right]\cdot(-2x) \quad \text{and} \quad du = -2x\cdot exp\left[-x^2\right]\cdot dx \qquad \ldots [78]$$

Also, if we let

$$v = x \qquad \ldots [79]$$

then,

$$dv = dx \qquad \ldots [80]$$

Substituting Eqns [77], [80], [79], and [78] into Eqn [76]:

$$\overset{u}{\overleftrightarrow{}}\ \overset{dv}{\overleftrightarrow{}}\qquad\overset{u}{\overleftrightarrow{}}\ \overset{v}{\overleftrightarrow{}}\quad\overset{v}{\overleftrightarrow{}}\ \overset{du}{\overleftrightarrow{}}$$

$$\int\left(exp\left[-x^2\right]\right)(dx) = \left(exp\left[-x^2\right]\right)(x) - \int(x)\left(-2x\cdot exp\left[-x^2\right]\cdot dx\right)$$

$$= x\cdot exp\left[-x^2\right] + 2\int x^2\cdot exp\left[-x^2\right]\cdot dx \qquad \ldots [81]$$

Rearranging:

$$\int x^2\cdot exp\left[-x^2\right]\cdot dx = -\frac{x}{2}\cdot exp\left[-x^2\right] + \frac{1}{2}\int exp\left[-x^2\right]\cdot dx \qquad \ldots [82]$$

Substituting Eqn [82] into Eqn [74]:

$$(B) = 2(RT)^{3/2}\left\{ -\left[\frac{x}{2}\cdot exp\left[-x^2\right]\right]_0^{\sqrt{E'/RT}} + \frac{1}{2}\int_0^{\sqrt{E'/RT}} x^2\cdot exp\left[-x^2\right]\cdot dx \right\}$$

$$= 2(RT)^{3/2}\left\{ -\left[\frac{\sqrt{E'/RT}}{2}\cdot exp\left[-\frac{E'}{RT}\right] - \frac{0}{2}\cdot exp\left[-0\right]\right] + \frac{1}{2}\int_0^{\sqrt{E'/RT}} exp\left[-x^2\right]\cdot dx \right\}$$

$$= 2(RT)^{3/2}\left\{ -\frac{\sqrt{E'/RT}}{2}\cdot exp\left[-\frac{E'}{RT}\right] + \frac{1}{2}\int_0^{\sqrt{E'/RT}} exp\left[-x^2\right]\cdot dx \right\}$$

$$= -(RT)^{3/2}\sqrt{\frac{E'}{RT}}\cdot exp\left[-\frac{E'}{RT}\right] + (RT)^{3/2}\int_0^{\sqrt{E'/RT}} exp\left[-x^2\right]\cdot dx \qquad .$$

Therefore

$$(B) = -(RT)^{3/2}\sqrt{\frac{E'}{RT}}\cdot exp\left[-\frac{E'}{RT}\right] + (RT)^{3/2}\int_0^{\sqrt{E'/RT}} exp\left[-x^2\right]\cdot dx \qquad \ldots [83]$$

Now we substitute Eqns [64] and [83] into Eqn [61] to give

$$(I) = (A) - (B) = \left\{ (RT)^{3/2}\frac{\sqrt{\pi}}{2} \right\} - \left\{ -(RT)^{3/2}\sqrt{\frac{E'}{RT}}\cdot exp\left[-\frac{E'}{RT}\right] + (RT)^{3/2}\int_0^{\sqrt{E'/RT}} exp\left[-x^2\right]\cdot dx \right\}$$

$$(I) = (RT)^{3/2} \frac{\sqrt{\pi}}{2} + (RT)^{3/2} \sqrt{\frac{E'}{RT}} \cdot exp\left[-\frac{E'}{RT}\right] - (RT)^{3/2} \int_0^{\sqrt{E'/RT}} exp\left[-x^2\right] \cdot dx \quad \ldots [84]$$

The **Error Function** (*erf*) of *y* is *defined* as

$$erf(y) \equiv \frac{2}{\sqrt{\pi}} \int_0^y exp\left[-x^2\right] \cdot dx \qquad \ldots [85]$$

This is an integral that is encountered frequently in the physical sciences. There is no analytical solution for this integral, but values can be evaluated numerically to high accuracy. A table of these values is presented below in Table 1.

Comparison of the integral in Eqn [84] with Eqn [85] shows that

$$\int_0^{\sqrt{E'/RT}} exp\left[-x^2\right] \cdot dx = \frac{\sqrt{\pi}}{2} erf\left(\sqrt{\frac{E'}{RT}}\right) \qquad \ldots [86]$$

Substituting Eqn [86] into Eqn [84]:

$$(I) = \int_{E=E'}^{E=\infty} \sqrt{E} \cdot exp\left[-\frac{E}{RT}\right] \cdot dE$$

$$= (RT)^{3/2} \frac{\sqrt{\pi}}{2} + (RT)^{3/2} \sqrt{\frac{E'}{RT}} \cdot exp\left[-\frac{E'}{RT}\right] - (RT)^{3/2} \frac{\sqrt{\pi}}{2} erf\left(\sqrt{\frac{E'}{RT}}\right) \quad \ldots [87]$$

Finally, we substitute Eqn [87] into Eqn [59] to give

$$\frac{N_{E \geq E'}}{N} = 2\pi \left(\frac{1}{\pi RT}\right)^{3/2} \cdot \int_{E=E'}^{E=\infty} \sqrt{E} \cdot exp\left[-\frac{E}{RT}\right] \cdot dE$$

$$= 2\pi \left(\frac{1}{\pi RT}\right)^{3/2} \left\{ (RT)^{3/2} \frac{\sqrt{\pi}}{2} + (RT)^{3/2} \sqrt{\frac{E'}{RT}} \cdot exp\left[-\frac{E'}{RT}\right] - (RT)^{3/2} \frac{\sqrt{\pi}}{2} erf\left(\sqrt{\frac{E'}{RT}}\right) \right\}$$

$$= \pi\sqrt{\pi} \left(\frac{RT}{\pi RT}\right)^{3/2} + 2\pi \left(\frac{1}{\pi RT}\right)^{3/2} \frac{(RT)^{3/2}}{(RT)^{1/2}} \sqrt{E'} \cdot exp\left[-\frac{E'}{RT}\right] - \pi\sqrt{\pi} \left(\frac{RT}{\pi RT}\right)^{3/2} erf\left(\sqrt{\frac{E'}{RT}}\right)$$

$$= \left(\frac{\pi RT}{\pi RT}\right)^{3/2} + 2\pi(\pi)^{-3/2} \left(\frac{RT}{RT}\right)^{3/2} \sqrt{\frac{E'}{RT}} \cdot exp\left[-\frac{E'}{RT}\right] - \left(\frac{\pi RT}{\pi RT}\right)^{3/2} erf\left(\sqrt{\frac{E'}{RT}}\right)$$

$$= 1 + 2\sqrt{\frac{E'}{\pi RT}} \cdot exp\left[-\frac{E'}{RT}\right] - erf\left(\sqrt{\frac{E'}{RT}}\right)$$

Fraction of Molecules
Having E ≥ E'

$$\boxed{\frac{N_{E \geq E'}}{N} = 1 + 2\sqrt{\frac{E'}{\pi RT}} \cdot exp\left[-\frac{E'}{RT}\right] - erf\left(\sqrt{\frac{E'}{RT}}\right)} \qquad \ldots [88]$$

There now, wasn't that fun?

Table 1. *Values of the Error Function* $\quad erf\,[y] = \dfrac{2}{\sqrt{\pi}}\displaystyle\int_{0}^{y}exp\,[-x^2\,]dx$

y	0	1	2	3	4	5	6	7	8	9
0.0	0.0000	0.0113	0.0226	0.0338	0.0451	0.0564	0.0676	0.0789	0.0901	0.1013
0.1	0.1125	0.1236	0.1348	0.1459	0.1569	0.1680	0.1790	0.1900	0.2009	0.2118
0.2	0.2227	0.2335	0.2443	0.2550	0.2657	0.2763	0.2869	0.2974	0.3079	0.3183
0.3	0.3286	0.3389	0.3491	0.3593	0.3694	0.3794	0.3893	0.3992	0.4090	0.4187
0.4	0.4284	0.4380	0.4475	0.4569	0.4662	0.4755	0.4847	0.4937	0.5027	0.5117
0.5	0.5205	0.5292	0.5379	0.5465	0.5549	0.5633	0.5716	0.5798	0.5979	0.5959
0.6	0.6039	0.6117	0.6194	0.6270	0.6346	0.6420	0.6494	0.6566	0.6638	0.6708
0.7	0.6778	0.6847	0.6914	0.6981	0.7047	0.7112	0.7175	0.7238	0.7300	0.7361
0.8	0.7421	0.7480	0.7538	0.7595	0.7651	0.7707	0.7761	0.7814	0.7867	0.7918
0.9	0.7969	0.8019	0.8068	0.8116	0.8163	0.8209	0.8254	0.8299	0.8342	0.8385
1.0	0.8427	0.8468	0.8508	0.8548	0.8586	0.8624	0.8661	0.8698	0.8733	0.8768
1.1	0.8802	0.8835	0.8868	0.8900	0.8931	0.8961	0.8991	0.9020	0.9048	0.9076
1.2	0.9103	0.9130	0.9155	0.9181	0.9205	0.9229	0.9252	0.9275	0.9297	0.9319
1.3	0.9340	0.9361	0.9381	0.9400	0.9419	0.9438	0.9456	0.9473	0.9490	0.9507
1.4	0.9523	0.9539	0.9554	0.9569	0.9583	0.9597	0.9611	0.9624	0.9637	0.9649
1.5	0.9661	0.9673	0.9684	0.9695	0.9706	0.9716	0.9726	0.9736	0.9745	0.9755
1.6	0.9763	0.9772	0.9780	0.9788	0.9796	0.9804	0.9811	0.9818	0.9825	0.9832
1.7	0.9838	0.9844	0.9850	0.9856	0.9861	0.9867	0.9872	0.9877	0.9882	0.9886
1.8	0.9891	0.9895	0.9899	0.9903	0.9907	0.9911	0.9915	0.9918	0.9922	0.9925
1.9	0.9928	0.9931	0.9934	0.9937	0.9939	0.9942	0.9944	0.9947	0.9949	0.9951
2.0	0.99532	0.99552	0.99572	0.99591	0.99609	0.99626	0.99642	0.99658	0.99673	0.99688
2.1	0.99702	0.99715	0.99728	0.99741	0.99753	0.99764	0.99775	0.99785	0.99795	0.99805
2.2	0.99814	0.99822	0.99831	0.99839	0.99841	0.99854	0.99861	0.99867	0.99874	0.99880
2.3	0.99886	0.99891	0.99897	0.99902	0.99906	0.99911	0.99915	0.99920	0.99924	0.99928
2.4	0.99931	0.99935	0.99938	0.99941	0.99944	0.99947	0.99950	0.99952	0.99955	0.99957
2.5	0.99959	0.99961	0.99963	0.99965	0.99967	0.99969	0.99971	0.99972	0.99974	0.99975
2.6	0.99976	0.99978	0.99979	0.99980	0.99981	0.99982	0.99983	0.99984	0.99985	0.99986
2.7	0.99987	0.99987	0.99988	0.99989	0.99989	0.99990	0.99991	0.99991	0.99992	0.99992
2.8	0.999925	0.999929	0.999933	0.999937	0.999941	0.999944	0.999948	0.999951	0.999954	0.999956
2.9	0.999959	0.999961	0.999964	0.999966	0.999968	0.999970	0.999972	0.999973	0.999975	0.999977

Example 4-8

When the gas of Example 4-7 reaches equilibrium, what fraction of the molecules will have energies $\geq 5.0 \times 10^{-21}$J?

Solution

The fraction of the molecules having energy $\varepsilon \geq \varepsilon'$ is given by

$$\frac{N_{\varepsilon \ge \varepsilon'}}{N} = 1 + 2\sqrt{\frac{\varepsilon'}{\pi kT}} \cdot exp\left(-\frac{\varepsilon'}{kT}\right) - erf\left(\sqrt{\frac{\varepsilon'}{kT}}\right) \qquad \ldots \text{[a]}$$

where $\varepsilon' = 5.0 \times 10^{-21}$J   and $T = 241.37$ K

Therefore,
$$\sqrt{\frac{\varepsilon'}{kT}} = \sqrt{\frac{5.0 \times 10^{-21}}{(1.381 \times 10^{-23})(241.37)}} = 1.2247$$

From the Error Function table, using linear interpolation, $erf[1.2247] = 0.91672$

Therefore, using Eqn [a] the fraction of the molecules having energy $\varepsilon \ge 5.0 \times 10^{-21}$ J is

$$\frac{N_{\varepsilon \ge \varepsilon'}}{N} = 1 + 2\sqrt{\frac{\varepsilon'}{\pi kT}} \cdot exp\left(-\frac{\varepsilon'}{kT}\right) - erf\left(\sqrt{\frac{\varepsilon'}{kT}}\right)$$

$$= 1 + 2\sqrt{\frac{5.0 \times 10^{-21}}{\pi(1.381 \times 10^{-23})(241.37)}} \cdot exp\left(-\frac{5.0 \times 10^{-21}}{(1.381 \times 10^{-23})(241.37)}\right) - erf(1.2247)$$

$$= 1 + (1.38198)(0.223129) - 0.91672$$

$$= 0.3916$$

39.2% of the molecules have this much energy or more.

Ans: | 39.2% |

4.11 CONCLUDING REMARKS

The above, relatively simple treatment of the Kinetic Theory of Gases shows how, by starting with only a few perfectly reasonable postulates based on our observations, it is possible to derive, quantitatively, many properties of gases, as well as to clarify the meaning of important concepts such as temperature and pressure. This illustrates the beauty and the power of inductive reasoning.

KEY POINTS FOR CHAPTER FOUR

The following relationships emerge from the derivation of the kinetic theory of gases:

1. **Average *KE* of a Molecule:** $\bar{e}_{transl} = \frac{1}{2}m\overline{\upsilon^2} = \frac{1}{2}m\overline{\upsilon}^2 = \frac{1}{2}mc^2$

    where m is the mass of the molecule in kg, υ is the velocity of the molecule in m s^{-1}, and $\overline{\upsilon^2}$ is the mean square velocity of the molecule, in m^2 s^{-2}.

2. **Ideal Gas Law:** $PV = n\left(\frac{2}{3}N_A b\right)T$, where n is the number of moles of gas, N_A is

Avogadro's number, T is the absolute temperature, and b is $\bar{e}_{transl}/T$. This equation, derived from the kinetic theory of gases, is the ideal gas law, with $R = \frac{2}{3}N_A b$. This latter equation defines the gas constant in terms of fundamental properties of the gas.

2. **Pressure**: $P = \dfrac{nMc^2}{3V}$, where P is the pressure in Pa, n is the number of moles of gas, M is the molar mass of the gas in kg mol^{-1}, V is the volume of the gas in m^3, 3 is the number that follows 2, and c^2 is the mean square speed of the molecules, defined as

   $c^2 = \dfrac{s_1^2 + s_2^2 + \ldots + s_N^2}{N}$, where s_i is the velocity of any given molecule in m s^{-1}. This equation defines the pressure in terms of fundamental properties of the gas.

3. **Temperature**: The theory shows that the average kinetic energy per molecule of gas is given by $\bar{e}_{transl} = \frac{3}{2}kT$, where k is **Boltzmann's constant**, which is just the gas constant per molecule: $k = R/N_A$. Rearrangement gives $T = \left(\dfrac{2}{3k}\right)\bar{e}_{trans}$. This equation shows that *temperature is a measure of the average KE of the gas molecules.*

4. The average kinetic energy *per mole of a monatomic ideal gas* is shown to be $KE = \frac{3}{2}RT$. This is an important relationship that often is made use of in thermodynamic derivations. In the derivation of this equation it has been assumed that the kinetic energy of the molecules consists only of *translational* kinetic energy. For more complicated gases, other forms of kinetic energy also must be taken into consideration, such as vibration and rotation. Therefore, a mole of a diatomic gas such as N_2 can store *more* kinetic energy than just $\frac{3}{2}RT$.

5. **Diffusion and Effusion**: The root mean square speed (*RMS*) of the molecules is defined as $c_{RMS} = \sqrt{\dfrac{s_1^2 + s_2^2 + \ldots + s_N^2}{N}}$, where N is the total number of molecules in the sample of gas. From statistical mechanics it can be shown that

   $\bar{s} = \dfrac{s_1^2 + s_2^2 + \ldots + s_N^2}{N} = \dfrac{c_{RMS}}{\sqrt{3\pi/8}}$. This equation can be used to derive **Graham's Law**, which states that the rate of effusion (leakage) of a gas through a small pinhole in a container is inversely proportional to the square root of its density.

   Expressed mathematically, $\dfrac{RATE\ A}{RATE\ B} = \sqrt{\dfrac{\rho_B}{\rho_A}} = \sqrt{\dfrac{M_B}{M_A}}$, where A and B are two different gases having densities ρ_A and ρ_B and molar masses M_A and M_B, respectively. By comparing the rates of leakage of two different gases through the same pinhole, if the molar mass of one of the gases is known, Graham's law can be used to evaluate the molar mass of the other (unknown) gas.

6.   **Speeds of Molecules**: From statistical theory it can be shown that the mean speed of a gas molecule is given by $\bar{s} = \dfrac{c_{RMS}}{\sqrt{3\pi/8}}$. Also, we showed in this chapter that the root mean speed of a molecule is given by $c_{RMS} = \sqrt{c^2} = \sqrt{\dfrac{3RT}{M}}$. Combining these two equations gives the following expression for the mean speed of a gas molecule: $\bar{s} = \sqrt{\dfrac{8RT}{\pi M}}$, where $\bar{s}$ is the mean speed in m s^{-1}, R is the gas constant (8.314 J mol^{-1} K^{-1}), T is the absolute temperature of the gas in kelvin, and M is the molar mass of the gas in kg mol^{-1}.

7.   **Effect of Pressure on Speed**: Increasing the pressure merely means an increased bombardment per unit area, *not* an increase in the speed of the molecules. Therefore *the KE is independent of pressure.*

8.   **Probability**: The probability P_j of a molecule having a speed s_j is *defined* as $P_j \equiv N_j/N$, which is just the fraction of the total molecules having a speed of s_j.

   Distribution function: The quantity $\rho(s')ds'$, where $\rho(s)$ is the speed distribution function, represents the (infinitesimal) probability that s has a value between s' and $s' + ds'$. The distribution function $\rho(s)$ is also called a *probability density*. The units of $\rho(s)$ are probability (a unitless number between *0* and *1*) divided by speed.

   The (infinitesimal) number of molecules dN_s having speed between s and $s + ds$ is given by
   $$dN_s = \{\rho(s)ds\}N$$

   The (infinitesimal) *fraction* f_s of the molecules having speed between s and $s + ds$ is just
   $$f_s = \frac{dN_s}{N} = \rho(s)ds$$

   The probability (having a value somewhere between *0* and *1*) that s has a value between s_1 and s_2 is given by
   $$P(s_1 \leq s \leq s_2) = \int_{s_1}^{s_2} \rho(s)ds$$

   When the integration extends over the complete range of values of s, i.e., from zero to infinity, then all the molecules in the system are taken account of, and this probability must equal unity (i.e., 100% probability):
   $$P(0 \leq s \leq \infty) = \int_{0}^{\infty} \rho(s)ds = 1$$

9.   The **Speed Distribution Function** giving the distribution of molecular speeds in a gas of molar mass M is given by the **Maxwell-Boltzmann** equation:

$$\rho(s) = 4\pi \left(\frac{M}{2\pi RT}\right)^{3/2} s^2 \cdot exp\left[-\frac{Ms^2}{2RT}\right]$$

There are more molecules with very high speeds than with very low speeds. The lowest speed is limited to *zero*, while there is no theoretical limit to the highest speed. At higher temperatures there are more molecules with higher speeds than there are at lower temperatures. At higher temperatures there are more speeds among which to distribute the N molecules; therefore the curve is broader and flatter, with a lower peak.

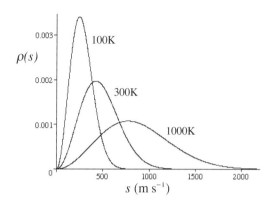

This explains the sensitivity of chemical reaction rates to temperature. Only molecules with "high energy" (i.e., those that have attained a certain minimum speed) collide to react. At higher temperatures more molecules attain the necessary "**activation**" energy than at lower temperatures.

10. The (infinitesimal) fraction f_s of the total N molecules having speeds in the (infinitesimal) range s to $s + ds$ is

$$f_s = \frac{dN_s}{N} = 4\pi \left(\frac{M}{2\pi RT}\right)^{3/2} s^2 \cdot exp\left[-\frac{Ms^2}{2RT}\right] \cdot ds$$

Providing Δs is not too great (≤ 10 m s^{-1}), this equation can be used to estimate the fraction of molecules having speeds in the *finite* range of s to $s + \Delta s$:

$$\frac{\Delta N_s}{N} \approx 4\pi \left(\frac{M}{2\pi RT}\right)^{3/2} s^2 \cdot exp\left[-\frac{Ms^2}{2RT}\right] \cdot \Delta s$$

Choose s as the value in the middle of the speed range.

11. The **Energy Distribution Function** is given by

$$\rho(E) = \frac{1}{N}\frac{dN_E}{dE} = 2\pi \left(\frac{1}{\pi RT}\right)^{3/2} \sqrt{E} \cdot exp\left[-\frac{E}{RT}\right]$$

The *energy* distribution has a different shape than the speed distribution. In particular, there is a *vertical* tangent at the origin, which means that the function rises much more quickly than does the speed distribution, which has a *horizontal* tangent at the origin. Also, after passing through a maximum, the energy distribution falls off more gently than does the speed distribution. However, in common with the speed distribution, the energy distribution broadens at higher temperatures, reflecting the fact that at higher temperatures a greater fraction of the molecules have high energies.

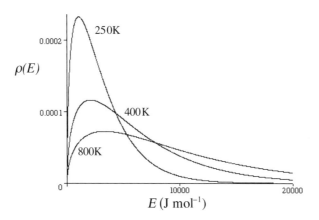

12.   The (infinitesimal) fraction f_E of the molecules having energy in the range of E to $E + dE$

is
$$f_E = \frac{dN_E}{N} = 2\pi \left(\frac{1}{\pi RT}\right)^{3/2} \sqrt{E} \cdot exp\left[-\frac{E}{RT}\right] \cdot dE$$

This equation also can be used to estimate the fraction of molecules having energies in the *finite* range of E to $E + \Delta E$, providing ΔE is not too large:

$$f_E \approx \frac{\Delta N_E}{N} = 2\pi \left(\frac{1}{\pi RT}\right)^{3/2} \sqrt{E} \cdot exp\left[-\frac{E}{RT}\right] \cdot \Delta E$$

13.   The fraction of molecules having energy $\geq E'$ is given by

$$\frac{N_{E \geq E'}}{N} = 1 + 2\sqrt{\frac{E'}{\pi RT}} \cdot exp\left[-\frac{E'}{RT}\right] - erf\left(\sqrt{\frac{E'}{RT}}\right)$$

where $erf(y)$ is the **Error Function**, defined by

$$erf(y) \equiv \frac{2}{\sqrt{\pi}} \int_0^y exp\left[-x^2\right] \cdot dx$$

Values of the error function are obtained from tables.

PROBLEMS

1.   A gas maintained at 25°C and 2.0 bar pressure effuses through a small pinhole at the rate of 7.53×10^{11} molecules per second. If the gas has a molecular weight of 125, how many grams of gas will escape from the same container in one week if the temperature and pressure are increased to 100°C and 5.0 bar, respectively? ☺

2. A mixture of helium and argon containing 70.0 mole per cent helium effuses through a small hole into an evacuated tank. What is the composition of the first mixture that passes into the evacuated tank? ☺

3. Calculate the root mean square speed and the average speed of CO_2 molecules at 300 K and 3 bar pressure. ☺

4. The speed of sound in a gas is related to the *RMS* speed of the molecules in the gas. A sound wave passing through a gas is transmitted by collisions from molecule to molecule; since not all the gas molecules are traveling in the same direction as the sound wave, it follows that a sound wave propagating through a gas must travel *slower* than the average speed of the gas molecules. In an ideal gas it can be shown that the speed of sound c_s is given by

$$c_S = \sqrt{\frac{\gamma RT}{M}}$$

where γ is the dimensionless ratio of the heat capacity at constant pressure to that at constant volume for the gas. (These quantities are discussed in Chapter 7.) At a certain temperature the *RMS* speed of the molecules in dry air is 547.6 m s^{-1}.

(a) What is the speed of sound in air under these conditions?

(b) What is the mean molecular speed?

The molar mass of air is 28.96 g mol^{-1} and the heat capacities at constant pressure and at constant volume are 29.19 J mol^{-1} K^{-1} and 0.721 J g^{-1} K^{-1}, respectively. ☺

5. For a certain reaction involving nitrogen gas to take place the nitrogen molecules must have a kinetic energy of at least 5.00 kJ mol^{-1}. What percentage of the N_2 molecules have this or greater energy (a) at 25°C, (b) at 500°C, and (c) at 1000°C? ☺

6. In Exercise 4-5 you were asked to calculate the fraction of the molecules in nitrogen gas at 25°C that had speeds between 1999 m s^{-1} and 2001 m s^{-1}. What percentage of the molecules in this gas would have energies equal to or greater than the kinetic energy of those molecules traveling at 2000 m s^{-1}? ☺

7. A one-dimensional gas is one in which the speeds are restricted to only one direction, such as the *x*-direction. Although such a gas does not in reality exist, the concepts involved in its analysis are similar to those for the three-dimensional case, and provide a good test of your understanding. The speed distribution function for a *one-dimensional* gas is

$$\rho(s) = \frac{1}{N}\frac{dN_s}{ds} = 2\sqrt{\frac{M}{2\pi RT}} \cdot exp\left[-\frac{Ms^2}{2RT}\right]$$

(a) Show that the equation for the *energy* distribution in a one-dimensional gas is given by

$$\rho(E) = \frac{1}{N}\frac{dN_E}{dE} = \sqrt{\frac{1}{\pi RT}} \cdot \frac{1}{\sqrt{E}} \cdot exp\left[-\frac{E}{RT}\right]$$

(b) Show that for a one-dimensional gas the equation giving the fraction of the molecules having energy $\geq E'$ J mol^{-1} is

$$\frac{N_{E \geq E'}}{N} = 1 - erf \sqrt{E'/RT}$$

Hint: you need to know that $\int_0^\infty exp\left[-a^2 x^2\right]dx = \frac{\sqrt{\pi}}{2a}$ (for all $a > 0$) ☠*

8. The speed distribution function for a *one-dimensional* gas, in which the speeds are restricted to only one direction, such as the x-direction, is

$$\rho(s) = \frac{1}{N}\frac{dN_s}{ds} = 2\sqrt{\frac{M}{2\pi RT}} \cdot exp\left[-\frac{Ms^2}{2RT}\right]$$

Show that for such a gas the root mean square speed is $\sqrt{RT/M}$. (In other words, the mean square speed is RT/M, which is exactly $\frac{1}{3}$ the value for a three-dimensional gas.) ☠*

Hint: you need to know that the mathematicians can show that $\int_0^\infty x^{1/2} exp\left[-x\right]dx = \sqrt{\pi}/2$

9. A two-dimensional gas is one in which the speeds are distributed only in two directions, such as in the x- and y-directions. The Maxwell-Boltzmann speed distribution function for a *two-dimensional* gas is

$$\rho(s) = \frac{1}{N}\frac{dN_s}{ds} = \left(\frac{M}{RT}\right) s \cdot exp\left[-\frac{Ms^2}{2RT}\right]$$

For a two-dimensional gas, show that the fraction of the molecules that have molar energies in the range of E to $E + dE$ is given by

$$f_E = \frac{dN_E}{N} = \left(\frac{1}{RT}\right) \cdot exp\left[-\frac{E}{RT}\right] \cdot dE \; ⊗$$

10. Calculate the percentage of molecules in a two-dimensional gas at 25°C that have energies in the range of 9.95 kJ mol⁻¹ to 10.05 kJ mol⁻¹. ☺

11. The speed distribution function for a *two-dimensional* gas in which the speeds are distributed only in two directions, such as in the x- and y-direction, is

$$(s) = \frac{1}{N}\frac{dN_s}{ds} = \left(\frac{M}{RT}\right) s \cdot exp\left[-\frac{Ms^2}{2RT}\right]$$

For a two-dimensional gas show that the fraction of the molecules that have a kinetic energy ($= \frac{1}{2}ms^2$) equal to or greater than some value E' J mol⁻¹ is

$$\frac{N_{E \geq E'}}{N} = exp\left(-\frac{E'}{RT}\right)$$

Hint: $\int exp[ax]dx = \frac{1}{a} \cdot exp[ax]$ ☹*

12. Using the results of Problem 11, calculate the percentage of the molecules in a two-dimensional gas that have an energy equal to or greater than 10 kJ mol^{-1} if the temperature of the gas is (a) 25°C, or (b) 500°C. ☺

REAL GASES

5.1 REAL GASES

The total energy of a gas is equal to the sum of its kinetic energy KE plus its potential energy PE. The PE of a gas is energy due to molecular forces of interaction between the molecules; these forces fall off very quickly as the distance of separation between the molecules increases, and usually can be ignored at separation distances greater than about 100 molecular diameters. Under these conditions, the total energy of the gas then becomes equal to its *kinetic energy*. An ideal gas is one which, by definition, has no interactive forces between its particles; furthermore, the particles of an ideal gas are *point particles*, which means they occupy no volume. Thus the total energy of an ideal gas is just equal to its KE.

$$(Total\ Energy)_{ideal\ gas} = KE$$

Real gases behave as ideal gases at low pressures because, under these conditions, the molecules are relatively far apart. However, as the pressure of a real gas is increased, the particles become closer together and start to feel forces of interaction. At distances of separation of 2-3 molecular diameters, the molecules of most real gases start to attract each other; attractive forces are considered to be *negative* forces, and *lower* the potential energy of the gas (Fig. 1). At these close distances of separation, if the temperature is made sufficiently low, the kinetic motion of the particles becomes insufficient to keep them apart and they *condense* to form a *liquid*. If the molecules get even closer together the attractive forces start to become *repulsive* forces, which are *positive*

forces. Repulsive forces *increase* the potential energy of the system. If it is attempted to push the molecules so close together that they actually start to *penetrate* each other, the positive potential energy jumps through the ceiling. This is similar to trying to push two tennis balls into each other, a very strong positive force is required to hold them in such a squashed condition. The work (energy) required to do this is stored in the tennis balls as increased positive energy. When the force pushing them together is released, the balls spring apart, releasing this increased en-

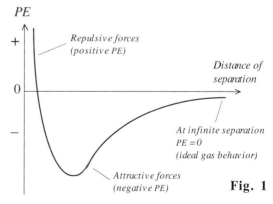

Fig. 1

ergy. The reason it is so difficult to push two gas molecules together is because the outer orbitals (negatively charged electron clouds) strongly *repel* each other. It thus takes a tremendous amount of energy to push them together against such repulsive forces. Fig. 1 shows the typical variation of the potential energy of two gas molecules as a function of their distance of separation.

Fig. 2 shows a plot of *P vs. v* for one mole of an ideal gas at different temperatures (*v* is the molar volume). A line at constant temperature is called an *isotherm*. At any given fixed temperature, $Pv = RT = $ constant; i.e., $Pv = $ const. (This is Boyle's law). The value of the constant, of course, increases as *T* increases; this has the effect of lifting the curves upward at higher temperatures.

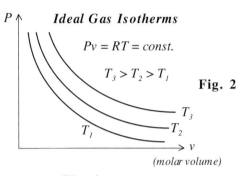

Fig. 2

5.2 ISOTHERMS FOR REAL GASES

The interactive forces present in a *real* gas cause deviations from the type of ideal gas law behavior shown in Fig. 2. Fig. 3 shows a plot of *P vs. V* for a typical real gas. To understand this figure, suppose we start with a gas at some temperature T_1 (point A in Fig. 3) and use a piston to compress the gas from point A to point B (see Fig. 4a). As V decreases, P increases. At point B, a further decrease in V does not result in a further increase in P; instead, P stays constant until point D is reached. Over the region from B to D the gas molecules become pushed sufficiently close together that they start to attract, condensing to form liquid (Fig. 4b).

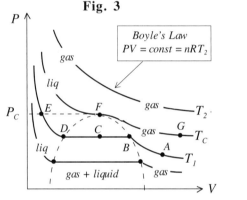

Fig. 3

At point D all the gas has been converted to liquid, and any further reduction of the volume requires great pressure because of the difficulty of compressing a liquid (see Fig. 4c).

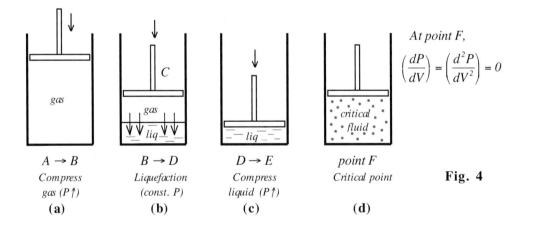

At point F,

$$\left(\frac{dP}{dV}\right) = \left(\frac{d^2P}{dV^2}\right) = 0$$

$A \rightarrow B$	$B \rightarrow D$	$D \rightarrow E$	point F	
Compress	Liquefaction	Compress	Critical point	Fig. 4
gas (P↑)	(const. P)	liquid (P↑)		
(a)	**(b)**	**(c)**	**(d)**	

At T_1, the point C represents a region in which both liquid and gas are present (in equilibrium) at the same temperature and pressure. There will be a definite interface separating the gas phase from the liquid phase. If, starting at point C we slowly raise the temperature to T_c (point F) while maintaining the same volume, we find that more liquid vaporizes, increasing both the pressure exerted by the vapor and the density of the vapor phase; at the same time, the density of the liquid phase decreases owing to the thermal expansion of the liquid.

When the system has reached equilibrium at T_c (point F) it is observed that the density of the vapor has increased to the point where it has the same value as that of the liquid, and the two phases in effect become indistinguishable (see Fig. 4d). That is, there now is only one phase—best described as a very dense vapor—and there will be no distinguishable interface present. On the isotherm for T_c, at any point to the right of point F the system exists as a very dense vapor, while at any point to the left of point F it exists as a liquid of rather low density.

At temperature T_c, known as the **critical temperature**, as the volume of the system decreases from point G, the system passes through point F from the gaseous state to the liquid state with no visible transition from the gas phase to the liquid phase! Point F is called the **critical point** of the gas, and the corresponding pressure is called the critical pressure P_c. At any temperature greater than T_c—for example T_2—the gas cannot be condensed to a liquid no matter how high a pressure is applied.

Thus, it is seen that at low pressure the isotherms for a real gas approach ideal behavior only at temperatures $>T_c$, the *critical temperature* of the gas. Also, as stated above, at any temperature greater than T_c a gas cannot be liquefied, no matter how high a pressure is applied. For this reason, gases that have low critical temperatures, such as He ($T_c = -268°C$), H_2 ($T_c = -240°C$), and N_2 ($T_c = -147°C$), are often called "permanent gases."

5.3 EQUATIONS OF STATE FOR REAL GASES

For one mole of an ideal gas,
$$Pv = RT$$

it follows therefore that
$$\frac{Pv}{RT} = 1$$

at *any* pressure. For a *real* gas this is not necessarily true, and we let

$$\boxed{\frac{Pv}{RT} = Z} \quad \begin{array}{l} Compressibility \\ Factor \end{array}$$

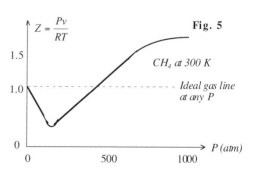

where Z is called the **compressibility factor** for the gas. Thus, the compressibility factor for an ideal gas is $Z = 1$ at all pressures. Fig. 5 shows a plot of the compressibility factor for a typical real gas (in this case methane, CH_4). As the pressure is increased, the molecules get closer together and start to experience attractive forces; these forces tend to make the volume *smaller*, so that $Pv < RT$, and therefore $Z < 1$. At pressures of around 200 atm (in the case of

CH_4), the molecules begin to get sufficiently close together such that repulsive forces start to dominate; this tends to make v bigger than predicted by the ideal gas law, so that eventually, at sufficiently high pressures (about 500 atm in the case of methane) $Pv > RT$, and $Z > 1$. Note that as the pressure approaches *zero*, all gases approach ideal behavior (i.e., $Z \rightarrow 1$, and $Pv \rightarrow RT$ as $P \rightarrow 0$).

5.4 THE VIRIAL EQUATION

In order to better describe the P–V–T relationships of real gases, especially at high pressures and low temperatures, equations of state fancier than $Pv = RT$ are required. One of the most accurate of the various equations that have been developed is the *virial equation:*

$$\frac{Pv}{RT} = 1 + \frac{B(T)}{v} + \frac{C(T)}{v^2} + \dots \qquad \begin{array}{l} \textit{Virial Equation} \\ \textit{for One Mole of Gas} \end{array}$$

In the virial equation, 1 is called the *first* virial coefficient, $B(T)$ is the *second* virial coefficient, $C(T)$ is the *third* virial coefficient, etc. Note that these coefficients usually are determined *experimentally* as functions of temperature for each gas. Thus, except for the first coefficient, the others have different values at different temperatures ($B(T)$ means that B is a function of T).

The virial equation is a very accurate equation of state; but the accuracy is gained at the expense of simplicity. An equation of state such as the virial equation, for which the various coefficients must be determined experimentally in the laboratory by actually measuring the volumes that result when a predetermined pressure is applied to a sample of gas in a piston/cylinder arrangement, is known as an *empirical equation of state* (i.e., an *observation-based* equation of state). Since, for any gas, as $P \rightarrow 0$, $v \rightarrow \infty$, it can be seen that for the virial equation, as $P \rightarrow 0$, both $B(T)/v$ and $C(T)/v^2$ approach *zero*, and, in the limit of zero pressure, the equation just becomes $Pv = RT$. This is true for *all* equations of state for gases: in the limit of vanishing pressure, they all must turn into the ideal gas equation.

Example 5-1

Calculate the pressure using (a) the ideal gas law, and (b) the virial equation, if one mole of water vapor is constrained in a volume of 22.414 L at 723.19 K. The second and third virial coefficients for water vapor at 723.19 K are -59.25×10^{-6} m^3 mol^{-1} and $+840 \times 10^{-12}$ m^6 mol^{-2}, respectively.

Solution

(a) $P = \dfrac{RT}{v} = \dfrac{(8.314)(723.19)}{0.022414} = 268.25 \times 10^3$ Pa $= 268.3$ kPa.

Ans: 268.3 kPa

(b)   The virial equation truncated after the third term is: $\dfrac{Pv}{RT} \approx 1 + \dfrac{B}{v} + \dfrac{C}{v^2}$

(Higher terms are almost never encountered.)

Multiplying both sides by $\dfrac{RT}{v}$ gives: $P \approx \dfrac{RT}{v} + \dfrac{RTB}{v^2} + \dfrac{RTC}{v^3}$

Putting in values:

$P = \dfrac{(8.314)(723.19)}{0.022414} + \dfrac{(8.314)(723.19)(-59.25 \times 10^{-6})}{(0.022414)^2} + \dfrac{(8.314)(723.19)(840 \times 10^{-12})}{(0.022414)^3}$

$= 268\ 252.059 - 709.107 + 0.449$

$= 267\ 543$ Pa

$= 267.5$ kPa

Ans: 267.5 kPa

The ideal gas law works best for high temperatures and low pressures ("hot vacuum"). Here, the pressure is fairly low (less than 3 atm) and the temperature is fairly high; accordingly, we would expect that the ideal gas equation gives reasonably accurate results. Comparing the answers to (a) and (b) shows that the pressure calculated using the ideal gas law is only 0.3% higher than that obtained using the more accurate virial equation. In fact, in such cases for which the pressure is fairly low and the temperature fairly high, sometimes it is only necessary to use the first *two* terms of the virial equation; thus:

$$P \approx \frac{(8.314)(723.19)}{0.022414} + \frac{(8.314)(723.19)(-59.25 \times 10^{-6})}{(0.022414)^2}$$

$$= 267\ 542.952 \text{ Pa}$$

$$= 267.5 \text{ kPa}$$

This value is essentially identical with the value obtained using three terms.

Example 5-2

Using the virial equation, calculate the molar volume of argon at 0°C and 100 atm pressure. The second and third virial coefficients for argon at 0°C are -21.7×10^{-6} m^3 mol^{-1} and 1.200×10^{-9} m^6 mol^{-2}, respectively.

Solution

$\dfrac{Pv}{RT} \approx 1 + \dfrac{B}{v} + \dfrac{C}{v^2}$. In this problem we are given T and P and asked to solve for v.

Multiplying both sides by $\dfrac{RT}{P}$ gives: $v = \dfrac{RT}{P} + \dfrac{RTB}{Pv} + \dfrac{RTC}{Pv^2}$

Multiplying both sides by v^2 gives: $v^3 = \left(\dfrac{RT}{P}\right)v^2 + \left(\dfrac{RTB}{P}\right)v + \left(\dfrac{RTC}{P}\right)$

Rearranging: $v^3 - \left(\dfrac{RT}{P}\right)v^2 - \left(\dfrac{RTB}{P}\right)v - \left(\dfrac{RTC}{P}\right) = 0$ . . . [a]

Unlike P, which can be solved for explicitly using the virial equation—as was done in the previous example—solving the virial equation for v is not so straightforward. Eqn [a] is a cubic equation in v, which is not readily solved without a calculator. One handy "trial and error" method of solving for v is to recognize that the second and third terms of the virial equation are minor contributors to the value of Pv/RT, which is recognized as the compressibility factor Z for the gas;

thus: $$Z = 1 + \frac{B}{v} + \frac{C}{v^2} \qquad \text{. . . [b]}$$

As a first approximation to v we can assume the value from the ideal gas equation; namely:

$$v_1 \approx \frac{RT}{P} = \frac{(8.314)(273.15)}{(100 \times 101\,325)}$$

$$= 2.2413 \times 10^{-4} \text{ m}^3 \text{ mol}^{-1}$$

This now can be substituted into Eqn [b] to give a first approximation for Z:

$$Z_1 \approx 1 + \frac{B}{v_1} + \frac{C}{v_1^2} \qquad \text{. . . [c]}$$

$$= 1 + \frac{-21.7 \times 10^{-6}}{2.2413 \times 10^{-4}} + \frac{1.200 \times 10^{-9}}{(2.2413 \times 10^{-4})^2}$$

$$= 0.92707$$

This value of Z now is used to obtain v_2—a more accurate approximation of v:

Since $Z = Pv/RT$ then $v_2 = \left(\dfrac{RT}{P}\right)Z_1$

$$= (2.2413 \times 10^{-4})(0.92707)$$

$$= 2.0778 \times 10^{-4} \text{ m}^3 \text{ mol}^{-1}.$$

This new value $v_2 = 2.0778 \times 10^{-4}$ is a better approximation of v than the first guess of $v_1 = 2.2413 \times 10^{-4}$. We now put this better value of v into Eqn [c] to obtain a *second* estimate of Z:

$$Z_2 \approx 1 + \frac{B}{v_2} + \frac{C}{v_2^2}$$

$$= 1 + \frac{-21.7 \times 10^{-6}}{2.0778 \times 10^{-4}} + \frac{1.200 \times 10^{-9}}{(2.0778 \times 10^{-4})^2}$$

$$= 0.92336$$

Z_2 is used to obtain a third (even better) approximation of v:

$$v_3 = \left(\frac{RT}{P}\right) Z_2$$

$$= (2.2413 \times 10^{-4})(0.92336)$$

$$= 2.0695 \times 10^{-4}$$

This is continued until a constant value of v is obtained; thus:

$$Z_3 = 1 + \frac{B}{v_3} + \frac{C}{v_3^2}$$

$$= 1 + \frac{-21.7 \times 10^{-6}}{2.0695 \times 10^{-4}} + \frac{1.200 \times 10^{-9}}{(2.0695 \times 10^{-4})^2}$$

$$= 0.92316$$

and
$$v_4 = \left(\frac{RT}{P}\right) Z_3$$

$$= (2.2413 \times 10^{-4})(0.92316)$$

$$= 2.0691 \times 10^{-4}$$

$$Z_4 = 1 + \frac{B}{v_4} + \frac{C}{v_4^2}$$

$$= 1 + \frac{-21.7 \times 10^{-6}}{2.0691 \times 10^{-4}} + \frac{1.200 \times 10^{-9}}{(2.0691 \times 10^{-4})^2}$$

$$= 0.92315$$

and
$$v_5 = \left(\frac{RT}{P}\right) Z_4$$

$$= (2.2413 \times 10^{-4})(0.92315)$$

$$= 2.0691 \times 10^{-4}$$

Since $v_5 = v_4$, the equation has "converged" and all future values of v will have this value, which is the solution to the cubic equation that we have been seeking. Thus, the molar volume is

2.0691 × 10^{-4} m^3 mol^{-1}. Actually, owing to the accuracy of the supplied data, we only need to report the answer to three significant figures; therefore we could have stopped at $v_4 = 2.07 \times 10^{-4}$ m^3 mol^{-1} = 207 mL mol^{-1}.

Ans: | 207 mL mol^{-1} |

Exercise 5-1

At 300 K and 20.0 atm pressure the compressibility factor Z for a certain gas is 0.86. Estimate the second virial coefficient B for this gas at 300 K. ☺

5.5 THE VAN DER WAALS EQUATION

This equation of state for a gas was derived in 1873 by the Dutch physicist Johannes van der Waals. Van der Waals reasoned that the attractive forces which operate at intermediate pressures tend to pull the gas molecules closer together, thereby effectively *lowering* the pressure from the value that would be expected based on the ideal gas law. In order to compensate for this effect, a term must be added to the pressure P to get it back to the "expected" value. Thus, the pressure term in the ideal gas law is replaced with $P + a(n/V)^2$, where a is a constant having the units of Pa m^6 mol^{-2}.

Similarly, the repulsive forces that prevent the molecules from getting too close together at very high pressures effectively exclude part of the volume available for molecular motion. Therefore the effective volume must be modified from just V, as given by the ideal gas law, to a smaller value, $(V - nb)$, where n is the number of moles of gas, and b is a constant having the units of m^3 mol^{-1}. The term nb can be viewed as the actual volume occupied by the molecules themselves; since this volume cannot be compressed, it must be subtracted from the total compressible volume V. Accordingly, van der Waals modified the ideal gas equation from $PV = nRT$ to

$$\left[P + a\left(\frac{n}{V}\right)^2 \right](V - nb) = nRT \qquad \text{\textit{van der Waals Equation}}$$

The van der Waals equation is more accurate than the ideal gas equation, but less accurate than the virial equation. For one mole of gas, $n = 1$ and $V = v$, and the van der Waals equation becomes

$$\left[P + \frac{a}{v^2} \right](v - b) = RT \qquad \begin{array}{l} \textit{Van der Waals Equation} \\ \textit{for One Mole of Gas} \end{array}$$

Example 5-3

Use the van der Waals equation to calculate the pressure of 92.4 kg of nitrogen gas contained in a 1.000 m^3 vessel at 500 K. For nitrogen, $a = 1.408$ L^2 atm mol^{-2} and $b = 0.0391$ L mol^{-1}. The molar mass of nitrogen is N = 14.01 g mol^{-1}.

Solution

The van der Waals equation for n moles of gas can be rearranged to: $P = \dfrac{nRT}{V - nb} - a\left(\dfrac{n}{V}\right)^2$

In order to avoid memorizing different values for the gas constant, we convert the values of a and b to SI units; thus, in SI units:

$$a = 1.408 \, \frac{L^2 \, atm}{mol^2} \times \frac{101\,325 \, Pa}{atm} \times \left(\frac{1 \, m^3}{1000 \, L}\right)^2$$

$$= 0.1427 \, \frac{m^6 \, Pa}{mol^2}$$

$$b = 0.0391 \, \frac{L}{mol} \times \frac{1 \, m^3}{1000 \, L}$$

$$= 39.1 \times 10^{-6} \, \frac{m^3}{mol}$$

No. of moles of N_2 gas:[1] $\qquad n = \dfrac{92.4 \times 10^3 \, g}{28.02 \, g/mol} = 3297.6$

Substituting values into the van der Waals equation:

$$P = \frac{(3297.6)(8.314)(500)}{1.000 - (3297.6)(39.1 \times 10^{-6})} - 0.1427\left(\frac{3297.6}{1.000}\right)^2$$

$$= 1.5737 \times 10^7 - 0.15517 \times 10^7$$

$$= 1.4185 \times 10^7 \, Pa$$

$$= 14.185 \, MPa$$

Ans: $\boxed{14.2 \, MPa}$

For purposes of comparison, the value given by the ideal gas equation is

---

[1] Note that $N_2 = 2(14.01) = 28.02$ g mol^{-1}.

$$P = \frac{nRT}{V} = \frac{(3297.6)(8.314)(500)}{1.000}$$

$$= 13.71 \times 10^6$$

$$= 13.7 \text{ MPa}$$

This value is 3.5% smaller than the value given by the more accurate van der Waals equation.

Exercise 5-2

Calculate the pressure in a 100 cm^3 flask containing 1.000 mol of H$_2$ at 0°C using (a) the ideal gas equation, (b) the van der Waals equation, and (c) the virial equation. The van der Waals constants for hydrogen are $a = 0.0259$ m^6 Pa mol^{-2} and $b = 2.661 \times 10^{-5}$ m^3 mol^{-1}. The virial coefficients for hydrogen at 0°C are $B = 14.00 \times 10^{-6}$ m^3 mol^{-1} and $C = 304.6 \times 10^{-12}$ m^6 mol^{-2}. ☺

Example 5-4

We saw earlier that on a plot of *P vs. V* for a real gas (Fig. 3, p. 5-2), it is observed experimentally that at the critical point the isotherm (i) has a slope of zero, and (ii) encounters a point of inflection. These two observations can be expressed mathematically by:

$$\left(\frac{\partial P}{\partial v} \right)_{T_c} = 0$$

and

$$\left(\frac{\partial^2 P}{\partial v^2} \right)_{T_c} = \left[\frac{\partial}{\partial v} \left(\frac{\partial P}{\partial v} \right) \right]_{T_c} = 0$$

Using this information, show that for a gas obeying the van der Waals equation of state, the values of the van der Waals constants are given by:

$$a = \frac{27 R^2 T_c^2}{64 P_c} \quad \text{and} \quad b = \frac{RT_c}{8P_c}$$

and that the molar volume at the critical point is given by $v_c = \dfrac{3RT_c}{8P_c}$. [This is a ☠ problem.]

Solution

For one mol of gas, the van der Waals equation $\left(P + \dfrac{a}{v^2} \right)(v - b) = RT$

can be rearranged to
$$P = \frac{RT}{v - b} - \frac{a}{v^2}$$
. . . [a]

At the critical point, Eqn [a] becomes $\quad P_c = \dfrac{RT_c}{v_c - b} - \dfrac{a}{v_c^2}$ $\qquad \ldots$ [b]

where P, v, and T have been replaced by the corresponding values at the critical point. Differentiating Eqn [a] with respect to volume at constant temperature gives

$$\left(\frac{\partial P}{\partial v}\right)_T = -\frac{RT}{(v-b)^2} + \frac{2a}{v^3} \qquad \ldots \text{[c]}$$

Taking the second differential: $\qquad \left(\dfrac{\partial^2 P}{\partial v^2}\right)_T = \dfrac{2RT}{(v-b)^3} - \dfrac{6a}{v^4}$ $\qquad \ldots$ [d]

At the critical point, $\qquad \left(\dfrac{\partial P}{\partial v}\right)_T = \left(\dfrac{\partial^2 P}{\partial v^2}\right)_T = 0$

so that Eqn [c] becomes $\qquad 0 = -\dfrac{RT_c}{(v_c - b)^2} + \dfrac{2a}{v_c^3}$

which rearranges to $\qquad \dfrac{2a}{v_c^3} = \dfrac{RT_c}{(v_c - b)^2}$ $\qquad \ldots$ [e]

Similarly, at the critical point, $\left(\partial^2 P / \partial v^2\right)_T = 0$, and Eqn [d] rearranges to

$$\frac{6a}{v_c^4} = \frac{2RT_c}{(v_c - b)^3} \qquad \ldots \text{[f]}$$

Dividing Eqn [e] by Eqn [f]: $\qquad \dfrac{2a/v_c^3}{6a/v_c^4} = \dfrac{RT_c/(v_c - b)^2}{2RT_c/(v_c - b)^3}$

which simplifies to $\qquad \dfrac{v_c}{3} = \dfrac{(v_c - b)}{2}$

Cross-multiplying and rearranging: $\qquad 2v_c = 3v_c - 3b$

from which $\qquad v_c = 3b$ $\qquad \ldots$ [g]

Substituting [g] into [e]: $\qquad \dfrac{2a}{(3b)^3} = \dfrac{RT_c}{(3b - b)^2}$

$$\frac{2a}{27b^3} = \frac{RT_c}{4b^2}$$

from which
$$a = \frac{27RT_cb}{8} \qquad \ldots [h]$$

Putting Eqns [g] and [h] into Eqn [b]:

$$P_c = \frac{RT_c}{3b - b} - \frac{27RT_cb/8}{(3b)^2}$$

$$= \frac{RT_c}{2b} - \frac{27RT_c}{72b}$$

$$= \frac{36RT_c - 27RT_c}{72b}$$

$$= \frac{RT_c}{8b} \qquad \ldots [i]$$

Rearranging [i] gives:
$$b = \frac{RT_c}{8P_c} \qquad \ldots [j]$$

Putting [j] into [h]:
$$a = \frac{27RT_c}{8}\left(\frac{RT_c}{8P_c}\right)$$

that is,
$$a = \frac{27R^2T_c^2}{64P_c} \qquad \ldots [k]$$

Putting [j] into [g]:
$$v_c = 3b = 3\left(\frac{RT_c}{8P_c}\right) \qquad \ldots [l]$$

Therefore, from Eqns [k], [j], and [l], we see that

Ans: $\boxed{a = \dfrac{27R^2T_c^2}{64P_c}}$ $\boxed{b = \dfrac{RT_c}{8P_c}}$ $\boxed{v_c = \dfrac{3RT_c}{8P_c}}$

5.6 LIQUEFACTION OF GASES

In most gases, attractive interactions dominate over repulsive interactions. Such gases can be *cooled* by first pressurizing to a high pressure P_{Hi}, followed by a *rapid expansion* to a low pressure P_{Lo} within an insulated system. The insulation prevents heat from flowing into or out of the system during the expansion, so that the total energy of the gas before the expansion must be the *same* as after the expansion. Depending on the conditions, the cooling effect can be so great as to actually *liquefy* the gas. This effect is known as the *Joule-Thompson effect* and is used industrially to liquefy many gases. The Joule-Thompson effect can be understood by referring to the potential energy diagram shown below (Fig. 6) for the gas molecules.

Thus, $$E_{total} = KE_1 + PE_1 = KE_2 + PE_2 = constant$$

It is seen from Fig. 6 that the initial *PE* of the gas molecules at the high pressure where the molecules are close together, PE_1, is

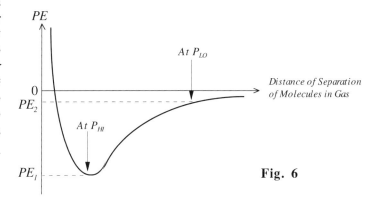

lower than the final *PE* after expansion of the gas to the low pressure where the molecules are far apart, PE_2. Thus, $PE_2 > PE_1$, and it follows that for the total energy to remain constant, KE_2 must be *less than* KE_1. But, since $KE = \frac{3}{2}RT$, it follows that $T_2 < T_1$, thereby explaining the cooling effect.

Fig. 6

KEY POINTS FOR CHAPTER FIVE

1. Total energy of a gas = *KE* (energy of motion) plus *PE* (energy of interaction between molecules). At more than 100 molecular diameters separation *PE* can be neglected, and total energy becomes equal to *kinetic energy*. Ideal gases are point particles with no interactive forces; therefore the total energy of an ideal gas is just equal to its *KE*.

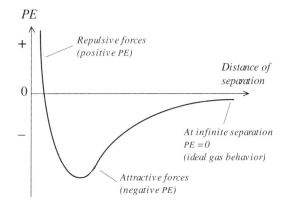

2.

Ideal Gas Isotherms

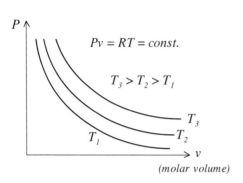

$Pv = RT = const.$

$T_3 > T_2 > T_1$

$T_1 \quad T_2 \quad T_3$

(molar volume)

Real Gas Isotherms

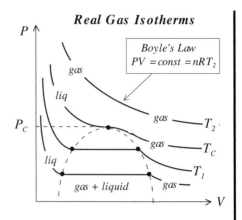

3. **The Critical Point**: The temperature T_c and pressure P_c at which the density of the liquid equals the density of the gas and the two phases can no longer be distinguished. At any temperature greater than T_c the gas cannot be condensed to a liquid no matter how great the applied pressure.

4. **Ideal Gas**: $\dfrac{Pv}{RT} = 1$

   Real Gas: $\dfrac{Pv}{RT} = Z$

   Z = the *compressibility factor*

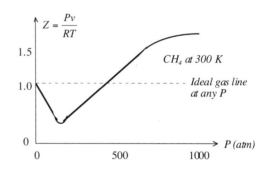

5. **The Virial Equation**: $\dfrac{Pv}{RT} = 1 + \dfrac{B(T)}{v} + \dfrac{C(T)}{v} + \dots$ (Very accurate)

   1 = 1st virial coefficient, $B(T)$ = 2nd virial coefficient, $C(T)$ = 3rd virial coefficient

   B and C are different for different gases, and both are functions of temperature.

   Since $v \to \infty$ as $P \to 0$ for all gases, it can be seen that for the virial equation, as

   $P \to 0$, both $B(T)/v$ and $C(T)/v^2$ approach *zero*, and, in the limit of zero pressure, the equation just becomes $Pv = RT$. This is true for *all* equations of state for gases: in the limit of vanishing pressure, they all must turn into the ideal gas equation. When the pressure is fairly low and the temperature fairly high, usually it is only necessary to use the first *two* terms of the virial equation.

6. **The van der Waals Equation**: More accurate than $PV = nRT$ but less accurate than the virial equation.

$$\left[P + \frac{a}{v^2}\right](v - b) = RT \quad \text{(for 1 mole])} \qquad \left[P + a\left(\frac{n}{V}\right)^2\right](V - nb) = nRT \quad \text{(for } n \text{ moles)}$$

Each gas has two van der Waals constants, a and b.

Units: $a = $ Pa m^6 mol^{-2} (accounts for attractive forces)
$b = $ m^3 mol^{-1} (accounts for volume of molecule)

PROBLEMS

1. A cylindrical tank of known dimensions containing 4.00 kg of carbon monoxide gas (molar mass = 28.01 g mol^{-1}) at –50.0°C, has an inner diameter of 0.200 m and a length of 1.00 m. Calculate the pressure, in bar, exerted by the gas using (a) the ideal gas law, and (b) the van der Waals equation of state. The actual value is about 76 bar. For CO, $T_c = 133$ K and $P_c = 35.0$ bar. ☺

2. For ammonia gas (NH_3, 17.03 g mol^{-1}) at 100°C and 1.00 MPa pressure, calculate the molar volume using (a) the ideal gas law, and (b) the van der Waals equation. [For part (b), use your own calculated values of the van der Waals constants for ammonia.] For ammonia, $T_c = 405.5$ K and $P_c = 11.28$ MPa. The actual volume is about 2.98 L mol^{-1}.

   Part (b) of the problem requires you to solve a cubic equation; you can do this using the Newton-Raphson method explained in Appendix 7. ☹

3. A 30.0 m^3 tank contains 14.0 m^3 of liquid n-butane in equilibrium with its vapor at 25°C. Calculate the values of the van der Waals constants for this vapor and, using the van der Waals equation and the Newton-Raphson method, estimate the mass of n-butane vapor in the tank. The vapor pressure of n-butane at 25°C is $P^{\bullet} = 2.43$ bar. For n-butane, $T_c = 425$ K and $P_c = 38.0$ bar. The molar mass of n-C$_4$H$_{10}$ is 58.12 g mol^{-1}. ☹

4. Show that at the critical point the compressibility factor for a van der Waals gas is given by

$$Z_c = \frac{P_c v_c}{RT_c} = 0.375 \quad ☺$$

5. For carbon monoxide gas (CO), the constant b in the van der Waals equation of state is $b = 0.0395$ L mol^{-1}. When CO gas is compressed at -50°C to a molar volume of 0.220 L mol^{-1}, its pressure is 76.0 bar. (a) Find the van der Waals constant a. (b) Find the new pressure if the molar volume is kept constant while the CO is cooled to –100°C. ☺

6. A strong, rigid, thermally insulated tank has a fixed volume of 6.00 L. Initially it contains 70.0 moles of gaseous ethylene (C_2H_4) at 30.0°C. Then a valve is opened and 10.0 moles of ethylene flows out of the tank, leaving behind 60.0 moles of the gas. The final uniform temperature of the ethylene remaining in the tank is 13.0°C. (a) Assuming the ethylene behaves as an ideal gas, calculate ΔP, the change in the pressure of the gas in the tank as a result of this process. (b) What would the change in pressure be if, instead of obeying the ideal gas law, the gas obeyed the van der Waals equation of state? For ethylene gas, $a = 0.462$ Pa m^6 mol^{-2} and $b = 5.83 \times 10^{-5}$ m^3 mol^{-1}. ☹

7. Assuming that water vapor behaves as a van der Waals gas, calculate the molar volume of water vapor at 330°C and 9.70 MPa pressure. For water vapor, $a = 0.5531$ Pa m^6 mol^{-2} and $b = 3.05 \times 10^{-5}$ m^3 mol^{-1}. $H_2O = 18.015$ g mol^{-1} ☺

8. For a van der Waals gas, providing the pressure does not become inordinately high—say not any higher than about 100 bar—the term $(b/v) < 1$; under these conditions show that for a van der Waals gas the compressibility factor is given by

$$Z \approx 1 + \frac{bP}{RT} - \frac{aP}{(RT)^2}$$

    Hint: Make use of the Maclaurin series $\dfrac{1}{1-x} = 1 + x + x^2 + x^3 + \dots$ (for $x < 1$) ☹

9. The specific volume of a sample of steam at 400°C is 0.00279 m^3 kg^{-1}. Calculate the pressure of the steam using (a) the ideal gas law, and (b) the van der Waals equation. For steam the van der Waals constants are $a = 0.5536$ Pa m^6 mol^{-2} and $b = 30.49 \times 10^{-6}$ m^3 mol^{-1}. ☺

10. A 1.00 L rigid cylinder contains one mole of neon gas at 25°C. Calculate the pressure using (a) the ideal gas law, and (b) the van der Waals equation.

    (c) A much more accurate equation is the virial equation, $\dfrac{Pv}{RT} \approx 1 + \dfrac{B(T)}{v} + \dfrac{C(T)}{v^2}$

    where $B(T)$ and $C(T)$ are functions of temperature. For neon at 25°C, the values of $B(T)$ and $C(T)$ are 11.42 cm^3 mol^{-1} and 221 cm^6 mol^{-2}, respectively. Assuming the pressure calculated using the virial equation is the "true" value, determine the % error in the pressure by using the ideal gas law and the van der Waals equation. ☺

11. When one mole of a certain van der Waals gas at 300 K is confined to a volume of 5.00 L its pressure is 497 kPa. At 500 K the pressure increases to 832 kPa. Identify the gas. ☺

12. A rigid 2.10 L steel cylinder contains oxygen at 0°C and 14.9 bar pressure. Calculate the number of moles of O_2 in the cylinder using (a) the ideal gas law, and (b) the van der Waals equation of state. The van der Waals constants for oxygen are $a = 0.1378$ Pa m^6 mol^{-2} and $b = 31.83 \times 10^{-6}$ m^3 mol^{-1}. ☹

13. If one mole of *n*-pentane occupies 25.0 L at 25°C, what is its pressure according to (a) the ideal gas law, and (b) the van der Waals equation? The actual pressure is about 94.9 kPa. ☺

THERMODYNAMICS (I)

6.1 THERMODYNAMICS

Thermodynamics is the study of energy transformation, especially the transformation of heat (Q) to work (W) and its reverse, the transformation of work into heat.

Classical Thermodynamics—deals with the observable properties of bulk materials, and is independent of any models of atomic structure.

Statistical Thermodynamics—requires a model of atomic structure, and relates properties to molecular behavior. In this course we deal primarily with classical thermodynamics.

Energy—We learned in high school that energy is "the ability to do work."

Conservation of Energy—"The energy of the universe is constant." This is one way to state the First Law of Thermodynamics, and is based on long observation and experience. Question: If energy is the ability to do work, and the energy of the universe is constant, then doesn't it follow that "the ability to do work of the universe is constant"? If this is true, and energy is constant no matter what we do, then why all the fuss about saving energy?

6.2 DEFINITIONS USED IN THERMODYNAMICS

System—That part of the universe under study, delineated by a **boundary**. The boundary can be real or imaginary, depending on which is more convenient. If the system changes size, such as an expanding gas, the boundary also may change size.

Surroundings—Everything else.

Universe—The system plus the surroundings.[1]

Closed System—No *matter* enters or leaves the system, but the system can exchange *energy* (in the form of work or heat) with the surroundings.

---

[1] However, in thermodynamics, *unless otherwise stated*, "the surroundings" usually means the immediate or local surroundings adjacent to the system; therefore, the "universe" is the "local" universe, not the actual cosmological universe that includes Jupiter, Saturn, and all the galaxies.

Open System—Matter as well as energy can be exchanged with the surroundings.

Isolated System—Neither matter nor energy is exchanged with the surroundings.

Heat (Q)—A flow of energy that results from a difference in temperature (ΔT).

Adiabatic System—A thermally insulated system (no heat Q can flow into or out of the system).

Work (W)—A flow of energy that can result in a weight being lifted or lowered in the surroundings. In closed systems energy transfers take place in the form of work and/or heat flow.

Exothermic Process in the system: Heat flows *from* the system *to* the surroundings; the system is hotter than the surroundings ($T_{syst} > T_{surr}$).

Endothermic Process in the system: Heat flows *from* the surroundings *into* the system; the surroundings are hotter than the system ($T_{surr} > T_{syst}$).

6.3 WORK

In the most fundamental definition, work is defined as a force acting through a displacement:

$$\text{Work} \equiv \begin{array}{l} \textit{A force acting} \\ \textit{through a dis-} \\ \textit{placement} \end{array} \qquad \textit{Work Defined}$$

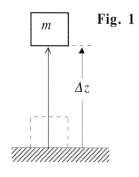

Fig. 1

The lifting of a body of mass m through a vertical displacement of Δz in the gravitational field is known as *mechanical* work (Fig. 1). The gravitational downward force acting on the body is given by Newton's second law:

$$F = ma = mg$$

where a is the acceleration of the body and g is the acceleration due to gravity. Therefore, the mechanical work done on overcoming this force and lifting the body is

$$W = F\Delta z = (mg)\,\Delta z$$

$$\boxed{W = mg\Delta z}$$

Note the units of work: $(\text{kg})(\text{m s}^{-2})(\text{m}) = (\text{N})(\text{m}) = \text{J}$

Example 6-1

A human leg will break if the force of compression exceeds about 2×10^5 N. If a person of 75 kg mass lands on one leg, what acceleration is required for the femur to fracture? How many times the acceleration of gravity is this? The acceleration due to gravity can be taken as $g = 9.8$ m s^{-2}.

Solution

The maximum force sustainable without fracture is $F_{max} \approx 2 \times 10^5$ N.

From Newton's second law, $F = ma$, the acceleration a required for a 75 kg person to generate

this force is
$$a = \frac{F_{max}}{m} = \frac{2 \times 10^5}{75} = 2.7 \times 10^3 \text{ m s}^{-2}$$

which is $\dfrac{2.7 \times 10^3}{9.8} = 275$ times the acceleration due to gravity.

Ans: $\boxed{2.7 \times 10^3 \text{ m s}^{-2} = 275 \times g}$

Example 6-2

Niagara Falls is about 50 m high and 800 m wide. As it goes over the falls the water has a depth of about 1.0 m and flows with a velocity of 10 m s^{-1}. If the potential energy of this water could be completely harnessed and converted to electrical energy, what would be the power output of Niagara Falls?

Solution

The volume of water flowing over the falls per second is
$$(800 \text{ m})(1.0 \text{ m})(10 \tfrac{\text{m}}{\text{s}}) = 8000 \text{ m}^3 \text{ s}^{-1}$$

Since water has a density of 1000 kg m^{-3}, the mass flow rate of the water is
$$\dot{m} = \left(8000 \tfrac{\text{m}^3}{\text{s}}\right)\left(1000 \tfrac{\text{kg}}{\text{m}^3}\right) = 8.0 \times 10^6 \text{ kg s}^{-1}$$

(The "dot" notation is used to express a *rate*; thus $m = $ kg, and $\dot{m} = $ kg s^{-1}.)

Power P is defined as *the rate of doing work*; thus,
$$
\begin{aligned}
P = \dot{W} &= \dot{m}g\Delta z \\
&= (8.0 \times 10^6)(9.8)(50) \\
&= 3.92 \times 10^9 \text{ J s}^{-1} \\
&= 3.92 \times 10^9 \text{ W} \\
&= 3.92 \text{ GW}
\end{aligned}
$$

This is almost four billion watts!   **Ans:** $\boxed{3.92 \text{ GW}}$

Exercise 6-1

Lake Ontario is 75 m above sea level. As it leaves Lake Ontario the St. Lawrence River has a flow rate of 6800 m^3 s^{-1}. Ignoring any water entering the river further downstream, what is the maximum electrical power production that could, in principle, be generated by the potential energy of the St. Lawrence River after it leaves Lake Ontario? ☺

Exercise 6-2

A bodybuilder lifts a 20-kg barbell a distance of 0.50 m 1000 times. If the energy required for this task is provided by burning fat, how much fat will the bodybuilder consume? (The potential energy lost each time she lowers the weight is dissipated as heat.) Fat supplies 38 MJ of energy per kilogram; this chemical energy is converted to mechanical energy with a conversion efficiency of about 20%. ☺

6.4 PV-WORK

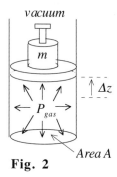

Consider the piston/cylinder arrangement shown in Fig. 2, in which a gas at a high pressure P_{gas} acts on the face of a piston of area A m^2 and lifts a mass of m kg through a vertical displacement of Δz meters in the gravitational field. The m-kg mass exerts a downward external force of $F_{ext} = mg$ on the piston. The expansion of the gas overcomes this force and causes the weight to lift, thereby performing work.

$$\text{Work done by gas} = (\text{force})(\text{displacement})$$
$$= (mg)(\Delta z)$$
$$= (F_{ext})(\Delta z)$$

Fig. 2

Multiplying each side by $A/A = 1$ gives

$$\text{Work done by gas} = \left(\frac{A}{A}\right)(F_{ext})(\Delta z)$$

$$= \left(\frac{F_{ext}}{A}\right)(A\,\Delta z) \qquad \ldots [1]$$

But a force acting on an area $\left(F_{ext}/A\right)$ defines a *pressure* P_{ext}, while an area A moving through a displacement Δz defines a *change in volume* ΔV. Therefore, Eqn [1] can be written

$$\text{Work done by gas} = P_{ext}\Delta V_{gas}$$

where P_{ext} is the external pressure against which the expanding gas is doing work, and ΔV_{gas} is the change in the volume of the gas.

Consider the units of $P\Delta V$: $\qquad \text{Pa m}^3 = \left(\frac{\text{N}}{\text{m}^2}\right)\left(\text{m}^3\right) = \text{N m} = \text{J}$

Therefore $P\Delta V$ has the units of *work*, and the work done by the system (the gas) is

$$\boxed{W_{PV} \equiv -P_{ext}\,\Delta V_{syst}} \quad \textit{PV-Work Defined}$$

This kind of work is called *PV-work*, or *expansion work* or *boundary work*. For reasons that we will discuss in the next chapter, when calculating *PV*-work we must, as indicated in the definition, put a *negative* sign in front of the expression.

6.5 MAXIMUM WORK OBTAINABLE FROM THE ISOTHERMAL EXPANSION OF AN IDEAL GAS

As indicated in Fig. 2, when a gas under pressure is allowed to expand, it can lift a weight, and thereby deliver work. Under isothermal (constant temperature) conditions the maximum work is done by the expanding gas when the external pressure P_{ext} against which the gas must do work has its maximum possible value. Assuming the piston itself is weightless and frictionless, in order for the gas to expand and lift the weight it is required that $P_{gas} > P_{ext}$. Therefore, to obtain the maximum work from the expanding gas, P_{ext} must be *just slightly* smaller than P_{gas}, and, in the *limit*, for the *maximum* work to be delivered, the external pressure must be so close to the pressure of the gas that it is, for all intents and purposes, the same as the pressure of the gas. Thus, to extract the maximum work from the expanding gas requires that $P_{ext} \rightarrow P_{gas}$, so that $W_{max} = -(P_{ext})_{max} \Delta V$ $\rightarrow -P_{gas} \Delta V$.

In the example of Fig. 2, it was implied that the gas had a very high pressure that was *more* than sufficient to lift the weight *m*; and the work done by the gas was defined in terms of the work done against the constant external pressure P_{ext}. Under these conditions, the *actual* work done by the gas in lifting the weight was *less* than the maximum work available from the gas, because the gas was capable of lifting a heavier weight than the one actually lifted.

Now, however, we are interested in evaluating the *maximum* work that the expanding gas could, in the limit, deliver; and, as mentioned, in order to do this the external pressure must be only infinitesimally less than the pressure of the expanding gas itself. However, as the gas expands at constant temperature, its pressure *continuously decreases*. Thus, we can't write $W = -P_{gas} \Delta V_{gas}$ because P_{gas} changes as V_{gas} changes. But over a small change dV in the gas volume, P_{gas} may be considered to be constant at a value determined by $P_{gas} = nRT/V$. Thus, for an infinitesimal increment δ of work, we write $\delta W = -P_{gas} dV$, and integrate[2] to get the total maximum work:

$$W_{max} = -\int_{V_1}^{V_2} (P_{ext})_{max} dV_{syst}$$

$$= -\int_{V_1}^{V_2} P_{gas} dV_{gas}$$

$$= -\int_{V_1}^{V_2} \frac{nRT}{V} dV$$

$$= -nRT \int_{V_1}^{V_2} \frac{dV}{V}$$

---

[2] If you haven't yet studied integral calculus you had better carefully read Appendices 8 and 9 at the end of the book, where I provide a crash course in integration. Even if you have studied integration, read them anyway; in my unbiased opinion they explain basic integration and natural logarithms better than most calculus textbooks. All problems that require integration are marked beside the face icons with an asterisk; e.g., ☺ *

$$= -nRT \, ln\left(\frac{V_2}{V_1}\right).^3$$

$$\boxed{W_{max} = -nRT \, ln\left(\frac{V_2}{V_1}\right)} \quad \begin{array}{l} \textit{Maximum PV-Work Output} \\ \textit{from the Isothermal Expansion} \\ \textit{of an Ideal Gas} \end{array}$$

Thus, the *maximum* work is delivered when the expansion takes place under the condition that $P_{ext} \approx P_{gas}$; i.e., under the condition of almost mechanical (pressure) equilibrium at all times.

6.6 REVERSIBLE PROCESSES

Consider the three cases shown in Fig. 3: In Case (*a*), we have mechanical equilibrium, in which $P_{gas} = P_{ext}$. In this case there is no movement of the piston, $\Delta V = 0$, and therefore no work is done.

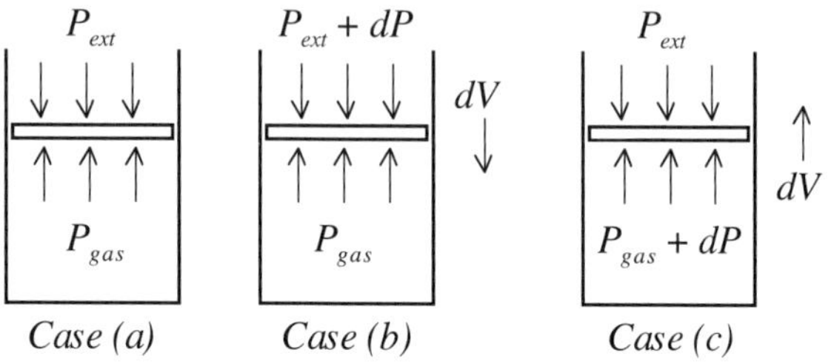

In Case (*b*), $(P_{ext} + dP) > P_{gas}$, which causes an infinitesimal *decrease* dV in the volume of the gas; work is done *by* the surroundings *on* the gas.

Fig. 3

In Case (*c*), $(P_{gas} + dP) > P_{ext}$. Now the small imbalance in P_{gas} causes an infinitesimal *increase* dV in the volume of the gas, and work is done *by* the gas *on* the surroundings.

An infinitesimal change in P in either direction from pressure equilibrium can cause an infinitesimal *reversal* in the *direction* of the movement of the piston. Any process that can be reversed by an infinitesimal change in a parameter (variable) such as P is called a **reversible process**,[4] and the work output under these conditions will be the maximum work, which is called the **reversible work**, w_{rev}.[5]

$$\boxed{w_{max} = w_{rev}}$$

Example 6-3

A piston that applies a constant pressure of 5.00 bar is used to compress three moles of an ideal gas isothermally at 25°C from an initial pressure of 0.50 bar to a final pressure of 3.00 bar.

---

[3] *ln x* is the natural logarithm of *x*, defined by: $ln \, x = \int\left(\frac{1}{x}\right) dx$. Thus, $\int_{x=a}^{x=b}\left(\frac{1}{x}\right) dx = \left[ln \, x\right]_{x=a}^{x=b} = ln \, b - ln \, a = ln\left(\frac{b}{a}\right)$.

[4] Strictly speaking, a reversible process is one that does not result in an increase in the entropy of the universe.

[5] If the gas were being compressed, then the surroundings would be doing work on the system (instead of the other way around). In this case, the reversible work would be the *minimum* work input required for the compression. Thus, the reversible work can be the maximum work or the minimum work, depending on the direction of work flow.

(a) How much work is done to compress the gas? (b) What is the *minimum* work required to carry out this compression?

Solution

(a) The *PV*-work for a system is defined as $W_{syst} = -P_{ext} \Delta V_{syst}$, where P_{ext} is the external pressure applied to the system (i.e., to the gas), and ΔV_{syst} is the change in volume of the system.

Thus:
$$W_{syst} = -P_{ext} \Delta V_{syst}$$
$$= -P_{ext}(V_2 - V_1)_{gas}$$
$$= -P_{ext}\left(\frac{nRT_2}{P_2} - \frac{nRT_1}{P_1}\right)$$

Since $T_1 = T_2 = T = 298.15$ K, and n and R are constant, we can write
$$W_{syst} = -P_{ext}\, nRT\left(\frac{1}{P_2} - \frac{1}{P_1}\right)$$
$$= -(5.0 \times 10^5)(3.0)(8.314)(298.15)\left(\frac{1}{3.0 \times 10^5} - \frac{1}{0.5 \times 10^5}\right)$$
$$= +61\ 970 \text{ J}$$
$$= +61.970 \text{ kJ}$$

The positive sign indicates that work is done *on* the system (the gas) *by* the surroundings (the piston).

Ans: $\boxed{+61.9 \text{ kJ}}$

(b) The minimum (reversible) work is done when the applied external pressure is just infinitesimally greater than the pressure of the gas. Under these conditions—which can never be attained in practice—the compression is carried out *reversibly* and we can say $P_{ext} \approx P_{gas} = nRT/V$. However, since the pressure of the gas is increasing at all times during the compression, as discussed above, we must write the work in a differential form δW and integrate it over the complete process.

Thus,
$$\delta W_{syst}^{min} = \delta W_{syst}^{rev} = -P_{ext}^{min}\, dV_{syst}$$
$$= -P_{gas}\, dV_{gas}$$
$$= -\frac{nRT}{V}\, dV$$
$$= -nRT\frac{dV}{V}$$

Integrating:
$$W_{syst}^{min} = \int_{V_1}^{V_2}\left(-nRT\frac{dV}{V}\right)$$

$$= -nRT \int_{V_1}^{V_2} \frac{dV}{V}$$

$$= -nRT \ ln\left(\frac{V_2}{V_1}\right) \qquad \qquad \ldots \text{[a]}$$

At constant temperature, $PV = nRT = const$

therefore, $P_1V_1 = P_2V_2$

from which $\dfrac{V_2}{V_1} = \dfrac{P_1}{P_2}$

and Eqn [a] becomes

$$W_{syst}^{min} = -nRT \ ln\left(\frac{P_1}{P_2}\right)$$

$$= -(3.0)(8.314)(298.15) \ ln\left(\frac{0.5}{3.0}\right)$$

$$= +13.32 \times 10^3 \ J$$

Ans: $\boxed{+13.3 \ kJ}$

It should be noted that the reversible work (+13.3 kJ) is *less than* the actual work (+61.9 kJ). Any real compression always requires a greater work input than the reversible work input. Similarly, for an expansion, in which the expanding gas *delivers* work, any real expansion always delivers *less* work than the amount of work that would be delivered by a reversible expansion. This is one of those facts of life:

> *You always have to put more in than the minimum,*
> *and you always get less out than the maximum.*

Example 6-4

How much work is done when one mole of metallic sodium (23 g) is dropped into a beaker of water at 35°C and allowed to react isothermally at this temperature?

$$Na_{(s)} + nH_2O \rightarrow NaOH_{(aq)} + (n-1)H_2O_{(liq)} + \tfrac{1}{2}H_{2(g)} \uparrow$$

Solution

We take the system as the contents of the beaker. When the sodium reacts it generates hydrogen gas, which—neglecting any small amount that may dissolve in the water—is evolved into the atmosphere. Since there are no restraining walls permitting the pressure to build up, the pressure of the evolved gas is essentially that of the surrounding atmosphere. However, in escaping from the beaker, the evolved hydrogen gas must *push back* the atmosphere. In so doing, it does work against the external pressure of the atmosphere.

Applying the usual equation for *PV*-work:

$$W_{syst} = -P_{ext}\,\Delta V_{syst}$$

The volume change for the system will be the total volume *after* the reaction less the total volume *before* the reaction; namely:

$$\Delta V_{syst} = \left[V_{NaOH_{(aq)}} + V_{(n-1)H_2O} + V_{\frac{1}{2}H_{2(g)}}\right] - \left[V_{Na_{(s)}} + V_{nH_2O}\right]$$

Since the volume of a gas is much greater than that of a solid or of a liquid,[6] we can ignore the volumes of the liquid and solid components in comparison with that of the gas so that

$$\Delta V_{syst} \approx V_{\frac{1}{2}H_{2(g)}} = \frac{n_{H_2}RT}{P_{H_2}}$$

As explained above, the pressure of the generated hydrogen gas is essentially that of the surrounding atmosphere; i.e., $P_{H_2} = P_{ext}$; thus:

$$\begin{aligned} W_{syst} &= -P_{ext}\,\Delta V_{syst} \\ &\approx -P_{ext}\left(\frac{n_{H_2}RT}{P_{H_2}}\right) \\ &= -P_{ext}\left(\frac{n_{H_2}RT}{P_{ext}}\right) \\ &= -n_{H_2}RT \end{aligned}$$

Inserting numerical values:

$$\begin{aligned} W_{syst} &= -(0.5)(8.314)(35 + 273.15) \\ &= -1281\text{ J} = -1.28\text{ kJ} \end{aligned}$$

(The negative sign indicates that the system does work *on* the surroundings. We will discuss this further in the next chapter.)

Ans: 1.28 kJ

Exercise 6-3

An airlock on a spaceship normally contains 10 m^3 of air at 20°C and 1.1 atm pressure. The airlock was accidentally opened, during which time half the air escaped into outer space. When the airlock was finally closed again the temperature of the remaining air had dropped to 5°C. How much work was done during this process? ☺

---

[6] For example, 1 mole of liquid water has a volume of about 18 cm^3 while 1 mole of water vapor (at *STP*) has a volume of 22 414 cm^3.

KEY POINTS FOR CHAPTER SIX

1. **Thermodynamics**: The study of energy transformation, especially the transformation of heat (Q) to work (W) and its reverse, the transformation of work into heat.

2. **System**—That part of the universe under study, delineated by a **boundary**. The boundary can be real or imaginary, depending on which is more convenient. If the system changes size, such as an expanding gas, the boundary also may change size.

    Surroundings—Everything else outside the system, but usually the area adjacent to the system—the "local" surroundings—as opposed to the other side of Mars.

    Universe—The system plus the surroundings.

3. **Closed System**—No *matter* enters or leaves the system, but the system can exchange *energy* with the surroundings in the form of work or heat.

    Open System—Both matter and energy can be exchanged with the surroundings.

    Isolated System—Neither matter nor energy is exchanged with the surroundings.

    Heat (Q)—A flow of energy that results from a difference in temperature (ΔT).

    Adiabatic System—A thermally insulated system (no heat Q can flow into or out of the system).

    Work (W)—A flow of energy that can result in a weight being lifted or lowered in the surroundings. In closed systems energy transfers take place in the form of work and/or heat flows.

    Exothermic Process in system: Heat flows *from* the system *to* the surroundings; the system is hotter than the surroundings ($T_{syst} > T_{surr}$).

    Endothermic Process in system: Heat flows *from* the surroundings *into* the system; the surroundings are hotter than the system ($T_{surr} > T_{syst}$).

4. **Isothermal Process**: A constant temperature process.

    Adiabatic Process: No heat is transferred (i.e., the system is thermally insulated).

    Isobaric Process: A constant pressure process.

    Isochoric Process: A constant volume process.

5. **Work**: A force moving through a displacement.

    Mechanical work: The lifting of a body of mass m through a vertical displacement Δz in the gravitational field g: $W_{mech} = mg\Delta z$

    ***PV*-Work**: The work done on or by a system when its volume changes.

    For constant external pressure: $W_{PV} = -P_{ext}\,\Delta V_{syst}$ (Note the negative sign.)

    where W_{PV} is the *PV*-work in joules, P_{ext} is the external pressure acting on the system in Pa, and ΔV_{syst} is the volume change of the system in m³.

    If work is done *on* the system *by* the surroundings, then $\Delta V < 0$ and the work is *positive*; if work is done *by* the system *on* the surroundings, then $\Delta V > 0$ and the work is *negative*.

6. **Reversible Process**: Any process that can be reversed by an infinitesimal change in a parameter (variable) such as P is called a **reversible process**, and the work output under

these conditions will be the maximum work, which is called the **reversible work**, W_{rev}.

A reversible process is one in which there is no friction of any sort and which takes place so slowly that there are no gradients of temperature, pressure, or concentration within the system during the process. For a reversible process the entropy of the universe does not change.

7. The **Maximum (= reversible) Work available from the Isothermal Expansion of an Ideal Gas** is given by

$$W_{max} = -\int_{V_1}^{V_2} P_{ext}^{max} dV_{syst} = -\int_{V_1}^{V_2} P_{gas} dV_{syst} = -\int_{V_1}^{V_2} \frac{nRT}{V} dV = -nRT \int_{V_1}^{V_2} \frac{dV}{V} = -nRT \ln\left(\frac{V_2}{V_1}\right)$$

When the gas expands, since $V_2 > V_1$, the reversible work is negative and is the *maximum* work that can be delivered by the gas; conversely, when the gas is compressed, $V_2 < V_1$, and the reversible work is positive and is the *minimum* work required to compress the gas. The same formula is used in each case.

PROBLEMS

1. The work done by an engine may depend on its orientation in a gravitational field, because the mass of the piston is relevant when the expansion is vertical. A chemical reaction takes place in a container of cross-sectional area 55.0 cm²; the container has a piston of mass 250 g at one end. As a result of the reaction, the piston can be pushed out (a) horizontally, or (b) vertically through 155 cm against an external pressure of 105 kPa. Neglecting friction, calculate the work done by the system in each case. $g = 9.807$ m s⁻² ☹

2. A sample of methane gas (CH_4)—the system—of mass 4.50 g occupies 12.7 L at 310 K.

   (a) Calculate W_{syst} when the gas expands isothermally against a constant external pressure of 200 Torr until its volume has increased by 3.30 L.

   (b) Calculate W_{syst} if the same expansion occurred isothermally and reversibly. ☹

3. A piece of pure tin (Sn) weighing 8.4210 g was placed into an evacuated rigid vessel of constant volume. Pure oxygen then was introduced into the vessel until the pressure was 2.000 bar and the temperature was 16.85°C, which was the same temperature as that of the surroundings. The volume of the pressurized oxygen in the vessel was 3.108 L. Using an accurate balance, the sealed vessel and its contents were found to weigh x grams. Next, the vessel and its contents were heated until all the tin was irreversibly oxidized to a white, stable powder consisting of SnO_2; then the vessel and its contents were allowed once more to cool down to the temperature of the room, which was still at 16.85°C. The barometric pressure of the surrounding air in the room was 102.14 kPa. After the vessel had cooled back down to room temperature, the valve in the vessel was opened slightly, and an amount of unreacted oxygen permitted to slowly leak out of the vessel until the pressure inside the vessel was the same as that of the air in the room. At this point the valve was shut again and the vessel and its con-

tents re-weighed using the same accurate balance as before. It was found that the vessel and its contents now weighed $(x - 1.7666)$ grams. ☹

(a) Neglecting any small differences in the densities of Sn and SnO_2, and assuming that the ideal gas law holds, use the above data to determine the molar mass of Sn. Report the answer in $g\ mol^{-1}$, to two decimal places.

(b) Taking the system as the vessel and its contents, calculate W_{syst} during the period when the valve was opened and some of the oxygen was permitted to escape from the vessel.

4.   (a) Calculate the minimum cost of the energy required to lift 50,000 tons of seawater per day through a vertical displacement of 4.00 meters.

   • acceleration due to gravity = $9.80\ m\ s^{-2}$ • cost of electricity = 5.00¢ per kWh
   • 1 ton = 2000 pounds   • 1 kg = 2.20 lb   • density of seawater = $1.030\ g\ cm^{-3}$

   (b) Give your opinion as a four or five word comment for each of the following:
   (i)   The electric charges are $14.85 per day
   (ii) The electric charges are $24.75 per day
   (iii) The electric charges are $33.33 per day
   (iv) The electric charges are $248.00 per day ☺

5.   The average energy consumption of an adult human being at rest is approximately equivalent to the output from a 100-watt light bulb. What minimum daily caloric food intake is required by a manual laborer who delivers one-tenth of a horsepower of mechanical work output for eight hours per day? Report the answer in food Calories per day.

   Data: 1 hp = 746 watts;   1 food Calorie = 1 kcal;   1 cal = 4.184 J ☺

6.   A manual laborer with a shovel lifts five kg of earth two meters through the gravitational field every ten seconds.

   (a) How many kJ of useful work can he deliver in an eight-hour shift if he works continuously?

   (b) If you must pay him a minimum wage of $8.50 per hour, how many kJ of useful work can you buy from him for $1.00?

   (c) Electricity costs 10.0¢ per kWh, and can be converted to mechanical energy at 95% efficiency in an electric motor. How many kJ of useful work can you get for one dollar's worth of electricity using an electrically driven ditch digging machine if 35% of the mechanical energy output from the electric motor ends up doing useful mechanical work (i.e., lifting earth)?

   Acceleration due to gravity = $9.807\ m\ s^{-2}$ ☹

7.   The first Super Titan rocket burned 500 tons of solid rocket propellant in 90 s in the first stage boosters. The heat of combustion of rocket fuel is approximately 500 kcal lb^{-1}. Calculate the average horsepower of the booster rockets if all this energy could be converted to work.

   1 hp = 746 W;   1 ton = 2000 lb ☺

8.   Calculate the work required to exclude one atmosphere from one cubic centimeter, using the information that one atmosphere is just balanced by a column of mercury 760 mm in height and that the density of the mercury is 13.5955 g cm^{-3}. The acceleration due to gravity is 9.80665 m s^{-2}. ☺

9.   Calculate the net work for one cycle consisting of the isothermal expansion ($A \rightarrow B$) and compression ($B \rightarrow A$) of one mole of ideal gas at 273.15 K between one atm and 0.500 atm. The expansion involves lifting 5.0 kg through 22 m. The compression is accomplished by a mass of 40 kg falling through 22 m. Friction may be neglected. The acceleration due to gravity is 9.807 m s^{-2}. ☺

10. A solid metal cylinder of the dimensions shown weighs 300 kg and fits like a piston into a gas-tight frictionless sleeve attached to a rigid metal sphere. The piston is prevented from lowering into the sphere by a retaining pin, as shown. The metal sphere contains an ideal gas initially at a pressure of 150 kPa. When the retaining pin is removed, the gas in the sphere undergoes an isothermal compression as the cylinder slowly lowers into the sphere. The atmospheric pressure is 100 kPa, and the acceleration due to gravity is $g = 9.80$ m s^{-2}. The formula for the volume of a sphere is $V = \frac{4}{3}\pi r^3$.

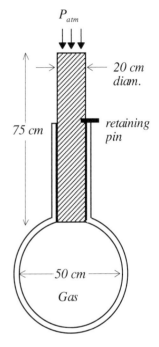

     (a) What is the final pressure of the gas?
     (b) How far does the cylinder descend into the sphere?
     (c) How much work is done on the gas? ☹

11. The barometric reading at the top of a building is 720 mm Hg, while the reading at the bottom is 760 mm Hg. Estimate the height of the building. The acceleration due to gravity is 9.807 m s^{-2}; the density of the air may be taken as constant at 1.18 kg m^{-3}. ☺

12. 1000 m^3 of nitrogen gas (N_2) is to be compressed isothermally at 32.0°F from an initial pressure of 1.00 atm to a final pressure of 100 atm. Calculate the minimum work required to carry out the compression. Nitrogen can be assumed to behave as an ideal gas. N = 14.0067 g mol^{-1}. ☺

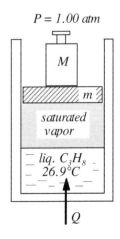

13. The piston and cylinder arrangement shown at the right contains liquid propane in equilibrium with its vapor at 26.9°C. The apparatus is well insulated to prevent any heat loss and initially is in mechanical equilibrium. The system (the propane) is slowly heated to evaporate exactly 1.00 mol of liquid. Calculate the maximum work done during this process *other* than the work of pushing back the atmosphere. *Data:* CH$_4$ = 44.11 g mol^{-1}; vapor pressure of propane at 26.9°C = 10.00 atm ☹

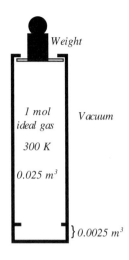

14. The frictionless piston/cylinder assembly shown at the right is located vertically in the gravitational field. The mass of the piston face may be assumed to be negligible, and the heat transfer with the surroundings may be taken as reversible, with both system and surroundings always at 300 K. The area of the piston face is 0.0300 m^2. What is the minimum weight required to compress the gas isothermally in one step to 0.00250 m^3?

The acceleration due to gravity is 9.807 m s^{-2}. ☹

15. 500 moles of oxygen gas (O_2) is to be expanded isothermally at 250°C from an initial pressure of 100 bar to a final pressure of 1.00 bar. Calculate the limiting maximum work that you will never be able to get from this process. Oxygen may be assumed to behave as an ideal gas. O = 15.9994 g mol^{-1}. ☹

16. One mole of nitrogen gas at 301 K occupies a volume of 12.5 L in the frictionless piston/cylinder arrangement shown. The mass of the piston may be assumed to be negligible. The cross-sectional area of the piston face is 300 cm^2, and the pressure of the external atmosphere is 100 kPa. When a 600 kg weight is placed on the piston, the piston lowers by 13.49 cm. The process may be assumed to be isothermal, and nitrogen may be assumed to behave as an ideal gas. The acceleration due to gravity is 9.80 m s^{-2}.

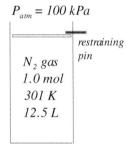

(a) How much work is done on the gas?

(b) What is the minimum work required to carry out this compression? ☹

17. What is (a) the minimum constant external pressure, and (b) the corresponding work, required to compress 2.00 moles of gaseous ammonia (NH_3) isothermally at 300 K from an initial volume of 2.00 L to a final volume of 0.50 L if the ammonia is assumed to behave as an ideal gas? ☹

18. What is (a) the minimum constant external pressure, and (b) the corresponding work, required to compress 2.00 moles of gaseous ammonia (NH_3) isothermally at 300 K from an initial volume of 2.00 L to a final volume of 0.50 L if the ammonia is assumed to behave as a van der Waals gas? For ammonia, the van der Waals constants are a = 0.4225 Pa m^6 mol^{-2} and b = 37.07 × 10^{-6} m^3 mol^{-1}. ☹

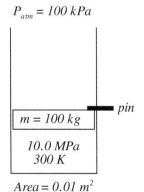

19. 1.00 L of compressed nitrogen gas at 10.0 MPa pressure and 300 K is located in the cylinder/piston assembly shown at the right. The piston weighs 100 kg and the surrounding atmosphere is at a pressure of 100 kPa. The cross-sectional area of the piston face is 100 cm^2. For the following calculations it may be

assumed that nitrogen behaves as an ideal gas and that any friction may be neglected. The acceleration due to gravity is 9.8 m s^{-2}.

(a) The retaining pin is removed and the gas is allowed to expand until the piston comes to rest and the system re-attains thermal equilibrium at 300 K. How much work will have been done by the gas?

(b) If the gas were able to expand isothermally and reversibly between the same initial and final states as in (a), how much work would the gas have delivered? 😐

20. 4.00 L of nitrogen gas at 100 kPa pressure and 300 K is located in the cylinder/piston assembly shown at the right. The piston weighs 100 kg and the surrounding atmosphere is at a pressure of 100 kPa. The cross-sectional area of the piston face is 100 cm^2. For the following calculations, it may be assumed that nitrogen behaves as an ideal gas and that any friction may be neglected. The acceleration due to gravity is 9.8 m s^{-2}.

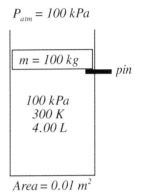

(a) The retaining pin is removed and the gas is spontaneously compressed until the piston comes to rest and the system re-attains thermal equilibrium at 300 K. How much work will have been done on the gas?

(b) If the gas were compressed isothermally and reversibly between the same initial and final states as in (a), how much work would have to be done on the gas? 😐

21. 400.92 moles of nitrogen (which may be assumed to behave as an ideal gas) is to be compressed isothermally at 300.0 K from an initial pressure of 1.000 atm to a final pressure of 100.0 atm. Calculate the minimum (no friction) work required for this process if:

(a) the external pressure is set at 100$^+$ atm (i.e., just slightly greater than 100 atm) and the compression is carried out in one fun, fast, and easy step.

(b) the compression is carried out in two easy steps by first compressing to V_b = 0.9869 m^3 using a constant external pressure of P_b throughout, followed by compressing to V_2 using a constant external pressure of P_2 = 100+ atm throughout.

(c) the compression is carried out in *four* steps: to V_a = 3.1209 m^3 using P_a^+, then to V_b = 0.9869 m^3 using P_b^+, next to V_c = 0.31209 m^3 using P_c^+, and finally to V_2 using P_2 = 100$^+$ atm.

(d) the compression is carried out using a *very large* number of "quasi-static" steps that are not fun, infinitely slow, and definitely not easy to carry out. 🙁

22. What is the minimum work required to compress one mole of nitrogen gas isothermally at 300 K from an initial volume of 5.00 L to a final volume of 1.00 L if

(a) the gas is an ideal gas? and

(b) the gas is a van der Waals gas?
The van der Waals constants for N$_2$ are a = 0.1408 Pa m^6 mol^{-2} and b = 3.913 × 10^{-5} m^3 mol^{-1}. 😊*

23. What is the minimum *one-step* work required to compress one mole of nitrogen gas isothermally at 300 K from an initial volume of 5.00 L to a final volume of 1.00 L if

    (a)   the gas is an ideal gas? and

    (b)   the gas is a van der Waals gas?

        The van der Waals constants for N_2 are $a = 0.1408$ Pa m^6 mol^{-2} and $b = 3.913 \times 10^{-5}$ m^3 mol^{-1}. ☹

24. 0.8019 moles of an ideal gas is contained in a piston/cylinder assembly having a weightless, frictionless piston of internal cross-sectional area 0.01656 m^2. The gas is originally at 300 K and 20.0 bar pressure. The pressure of the surrounding atmosphere is 1.00 bar. A weight of mass *m* kilograms is placed on the top of the piston. The retaining pin is removed, permitting the gas to expand isothermally at 300 K, lifting the weight until the final volume of the gas is 2.00 L. What is the mass *m* of the weight in kilograms if the total work done by the gas is exactly 50% of the maximum reversible work that could, in the limit, be delivered by the gas in passing from the same initial and final states? The acceleration due to gravity is 9.7875 m s^{-2}. ☹

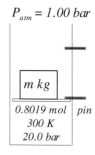

SEVEN

THERMODYNAMICS (II)

7.1 INTERNAL ENERGY (U) AND THE FIRST LAW OF THERMODYNAMICS

The total chemical energy contained in a chemical system is called the *internal energy* of the system, and is given the symbol U. This internal energy consists of the sum of all the various kinetic and potential energy contributions made to the system by its constituent atoms and molecules. This includes contributions from electrons, chemical bonds, subatomic particles contained in the atomic nuclei, etc., as well as from any other energy contributions not yet discovered. Because of our lack of knowledge about all the types of energies that may be present, it is impossible to measure absolute values of U.

It is found, however, that the internal energy of a system can be increased both by *heating* the system and by *doing work* on it. Thus, consider a closed insulated system (Fig. 1) containing liquid water at 25°C and one atm pressure (the system is insulated in order to ensure that any heat flowing into the system cannot escape). If we now cause 1000 J of heat Q to flow into the system, we observe that the temperature of the system increases by some amount ΔT, indicating that there has been a change in the *state* of the system, and of its *internal energy*.

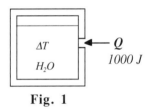

Fig. 1

We find, however, that we also can bring about the *same* change in ΔT —and therefore in ΔU_{system}—by letting a weight of 102 kg fall through a displacement Δz of one meter in the surroundings, causing a stirring paddle immersed in the water to rotate (Fig. 2). The weight falling in the surroundings clearly is a force moving through a displacement, and therefore is a *work* term. The magnitude of this work is readily calculated to be

$$W = mg\Delta z = (102)(9.8)(1.0) = 1000 \text{ J.}$$

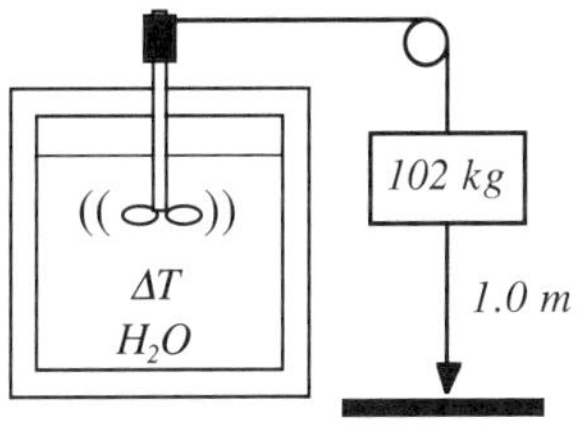

Fig. 2

Neglecting any friction at the pulley, this 1000 J of work is being transferred to the water from the falling weight *via* the mechanical action of the pulley/stirrer assembly. It is thus seen that the same change in internal energy of the system can be brought about either by "adding" 1000 J of *heat to* the system or by "doing" 1000 J of *work on* the system. We know that each process has brought about the same ΔU for the system because the ΔT for the system is the same in each case, and we know that an increase in temperature means an increase in the energy of the system.

Therefore both Q and W can bring about the same change in the state of the system. We also find that the same change in the state of the system can be brought about by using a *combination* of Q and W. For example, any of the following combinations would bring about the same ΔT and therefore the same ΔU for the system:

                    1000 J of heat + no work,
        or   500 J of heat + a falling mass of 51.0 kg,
        or   250 J of heat + a falling mass of 76.5 kg,
        or   no heat + a falling mass of 102 kg,   etc.

We therefore conclude that *work and heat are equivalent ways of changing the internal energy of a system*. Even though work is not the same thing as heat—lifting a weight isn't the same as feeling the sun shining on your face—both work and heat can bring about the same change in the internal energy of a system; there is some kind of "equivalency" between the two. This "equivalency" can be expressed by the equation

$$\Delta U = Q + W \qquad \begin{array}{l} \textit{First Law of} \\ \textit{Thermodynamics} \end{array} \qquad \dots [1]$$

Eqn [1], which holds for a *closed system*, is a statement of the *First Law of Thermodynamics*.

Exercise 7-1

The state of a certain system is changed when 500 J of heat flows into the system. How much heat must flow into the system to bring about the same change in state if mechanical work is done on the system equivalent to a mass of 25 kg falling through a displacement of (a) 1.00 m (b) 2.04 m (c) 3.00 m? The acceleration due to gravity is 9.8 m s^{-2}. ☺

7.2 STATE FUNCTIONS

A property of a system that depends only on the current *state* of the system (i.e., on its P, V, T, and n), and is independent of the path taken to reach that state, is called a *state function*.

From experiment, U is found to be a state function. Suppose we have a system in some initial state, state A, defined by (P_1, V_1, T_1, U_1) and some final state, state B, having (P_2, V_2, T_2, U_2); and that the system undergoes some process that takes it directly from state A to state B in one step, as indicated in Fig. 3 by path a. There is more than one way, however, for the system to go

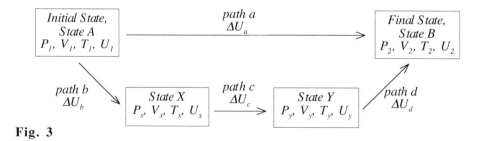

Fig. 3

from state A to State B: For example, it could pass through two *intermediate* states, labeled state X and state Y, in which the pressure, temperature, volume, and internal energy of the system have values that are different from those of state A and state B. Thus, the system can go from state A to state B either directly by path a, or by following the three intermediate paths b, c, and d, each of which has its own value of ΔU. When we talk about the "path" between two states, we merely mean the *process* or *processes* that occur for the system to pass from one state to the other. Because U has been found to be a state function, we can say that, for the processes of Fig. 3,

$$\Delta U_{AB} = \Delta U_a = \Delta U_b + \Delta U_c + \Delta U_d$$

where $\Delta U_{AB} \equiv U_B - U_A$. In other words, it doesn't matter which path the system takes in passing from state A to state B, ΔU_{AB} depends only on (P_1, V_1, and T_1) and on (P_2, V_2, and T_2), not on any intermediate states.

7.3 WORK AND HEAT ARE NOT STATE FUNCTIONS

We have shown that internal energy U is a state function. Since $\Delta U = Q + W$, does this mean that Q and W also are state functions? NO! Why not? Because, as we showed above when discussing the paddle wheel experiment with the water, we find experimentally that we can use *different* combinations of Q and W to bring about the *same* change in U for the water. This means that, unlike U, we cannot associate only one specific value of Q or of W with the state of the system.

For example, suppose we start with one kg of water at 20°C and 0.5 bar pressure and subject it to some process by which its state is changed to 50°C and 1.0 bar pressure. The internal energy U of the one kg of water at 50°C and one bar pressure has only one, fixed value. However, as we saw above, to put the water into this state from its initial state, we can use any number of different combinations of Q and W. The only restriction imposed by the First Law—Eqn [1]—is that the *sum* ($Q + W$) has to add up to the same value of ΔU. Thus, we cannot associate only one value of Q or one value of W with a specific state of the water, and it follows therefore that neither Q nor W can be a state function.

Any property, such as Q or W, that is not independent of the path; i.e., that is not a state function, is called a **path function**.

Thus, for any state property B of a system, the change in the value of the property as the system passes from State *1* to State *2* is independent of the actual nature of the path taken. Rather, it depends only on the initial and final values of B, and we can write:

$$\int_{State\ 1}^{State\ 2} dB = B_2 - B_1 = \Delta B \qquad\qquad State\ Function$$

For *path* functions such as Q and W, however, we can only write

$$\int_{State\ 1}^{State\ 2} \delta Q = Q_{12} \quad \text{and} \quad \int_{State\ 1}^{State\ 2} \delta W = W_{12} \qquad\qquad Path\ Functions$$

Notice that we don't write dQ or dW; but, instead, δQ and δW. This is done to remind us that Q and W are path functions, and not state functions.

Thus we CANNOT write $\displaystyle\int_{State\,1}^{State\,2}\delta Q = Q_2 - Q_1 = \Delta Q$ or $\displaystyle\int_{State\,1}^{State\,2}\delta W = W_2 - W_1 = \Delta W$ No! No!

The integrals of δQ and δW are just Q and W, not ΔQ and ΔW. You can *never* write ΔQ or ΔW!

7.4 Q AND W HAVE ALGEBRAIC SIGNS

In order to be consistent and to avoid confusion, in thermodynamics we use the following sign convention:

- W_{syst} is **positive** if work is done **on** the system (work flows *into* the system);
- W_{syst} is **negative** if work is done **by** the system (work flows *out of* the system);
- Q_{syst} is **positive** if heat flows **into** the system;
- Q_{syst} is **negative** if heat flows **out of** the system.

The sign is very important. It can be like the difference between having +100,000 dollars in your bank account or –100,000 dollars! It can be the difference between designing a 10 megawatt cooling system for an industrial process or a 10 megawatt heating system. So, be very, very careful about signs.

7.5 ANOTHER LOOK AT PV-WORK

Earlier we showed that the work done by an expanding gas (Fig. 4) against an external pressure P_{ext} was

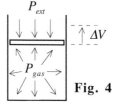

Fig. 4

$$W_{syst} = P_{ext}\,\Delta V_{gas} = P_{ext}(V_2 - V_1)_{gas}$$

This value of W_{syst} will be *positive* because P_{ext} is always positive (there is no such thing as a negative pressure in a gas—a vacuum is just a pressure that is *less than* the surrounding pressure, while a perfect vacuum is just $P = 0$) and, clearly, if the gas expands, ΔV also is positive. However, since the system is delivering work to the surroundings, it also is clear (from the First Law) that the system is *losing energy*; in other words, W_{syst} must be *negative*. Accordingly, our sign convention requires that we have to insert a negative sign and *define PV-work* as

$$W_{PV}^{syst} \equiv -P_{ext}\,\Delta V_{syst} \qquad\qquad \text{. . . [2]}$$

Similarly, for an infinitesimally small increment of work:

$$\delta W_{PV}^{syst} \equiv -P_{ext}\,dV_{syst} \qquad\qquad \text{. . . [3]}$$

Also, for the same reasons, the maximum work that can be delivered by the isothermal expansion of an ideal gas must be written

$$W_{PV}^{max} = -nRT \, ln\left(\frac{V_2}{V_1}\right) \quad \begin{array}{l} \textit{Isothermal} \\ \textit{Expansion} \\ \textit{of an Ideal Gas} \end{array} \quad \ldots [4]$$

Why? Because for an expanding gas, $V_2 > V_1$, therefore the logarithmic term is positive; however, according to our convention, since the gas is losing internal energy by delivering work to the surroundings, the work must be negative, so we have to insert a negative sign in front of the expression.

Note that Q and W both are *energy in transit*. This means that they exist *only* when in the act of crossing the boundary of a system. A system doesn't "contain" heat or "contain" work; it only contains internal energy. We sometimes say that a system has a lot of "heat" in it, but this is sloppy speech; we should say that it has a lot of *thermal energy*.

Although we can't measure the absolute value of U_{syst}, we *can* measure *changes* in U_{syst}; i.e., ΔU_{syst}, because we know that $\Delta U = Q + W$, and we *can* measure Q and W. Furthermore, the measurement of Q and W is made in the *surroundings*, which are readily accessible (heat flows to or from the *surroundings*; weights are raised or lowered *in the surroundings*).

You may recall that an *isolated* system is one that is insulated both thermally and mechanically from its surroundings; i.e., $Q = 0$ and $W = 0$. Therefore, for an isolated system $\Delta U = Q + W = 0$. But if $\Delta U = 0$, then U doesn't change; therefore:

> *The internal energy of an isolated system is constant.*

Since the *Universe* is an isolated system (it is assumed that nothing exists outside the Universe), then it follows that "the internal energy of the universe must be . . ."

To reiterate, a mathematical statement of *the First Law of Thermodynamics* is

$$\Delta U = Q + W \quad \begin{array}{l} \textit{First Law of} \\ \textit{Thermodynamics} \\ \textit{for a Closed System} \end{array} \quad \ldots [5]$$

For an infinitesimally small change dU in the internal energy of a system Eqn [5] must be written in *differential* form

$$dU = \delta Q + \delta W \quad \begin{array}{l} \textit{First Law of} \\ \textit{Thermodynamics} \\ \textit{for a Closed System} \end{array} \quad \ldots [6]$$

Note once again that the differential forms of Q and W are δQ and δW, not dQ and dW.

7.6 CHEMICAL REACTION IN A
CONSTANT VOLUME SYSTEM

Suppose we have a chemical reaction $A + B \rightarrow C$ taking place in a rigid container. Since the container is rigid, its volume doesn't change, so for this process we can say that $\Delta V = 0$. For chemically reacting systems, usually the only kind of work involved is PV-work that is done when the system expands (or contracts). If the chemical reaction is carried out in a rigid container (i.e., at constant volume), then

$$W_{PV} = -P_{ext} \Delta V_{syst}$$

$$= -P_{ext}(0) \; = \; 0$$

and

$$\Delta U_V = Q + W$$

$$= Q + 0 \; = \; Q$$

Therefore, for a closed system constant volume process,

$$\boxed{\Delta U_V = Q} \quad \begin{array}{l} \textit{Constant Volume} \\ \textit{Process, Closed System} \end{array} \quad \cdots \; [7]$$

For a constant volume process, measuring Q gives us ΔU! [1]

7.7 THE HEAT CAPACITY
OF A SINGLE PHASE SYSTEM

When a homogeneous chemical system is heated with no phase changes or chemical reactions taking place, its temperature increases: the bigger the Q, the bigger the increase in T:

therefore, $\Delta T \propto Q = (const) \cdot Q$

We let $(1/C)$ be the constant of proportionality between ΔT and Q, where C is known as the **heat capacity** of the system.

Thus, $\Delta T = \left(\dfrac{1}{C}\right) Q$

and it follows directly that $\boxed{C \equiv \dfrac{Q}{\Delta T}} \quad \begin{array}{l} \textit{Heat Capacity} \\ \textit{Defined} \end{array} \quad \cdots \; [8]$

There are different ways to represent the heat capacity of a substance:

(i)   We use an upper case "C" to represent the *total heat capacity* of the substance; C has the units of J K^{-1}.

---

[1] Eqn [7] holds both for processes involving chemical reactions and for those in which no chemical reactions take place.

(ii)   A lower case "*c*" with a macron—$\bar{c}$—is used to designate the heat capacity per unit mass of the substance, and is called the *specific heat capacity*; $\bar{c}$ has the units of J kg^{-1} K^{-1}.

(iii) A lower case "*c*" is used for the *molar heat capacity*, which is the heat capacity of one mole of the substance; c has the units of J mol^{-1} K^{-1}.

If we know the heat capacity of a system, the heat is easily calculated by measuring the change in temperature of the system when we heat it:

Thus, $$Q = C\Delta T = m\bar{c}\,\Delta T = nc\Delta T$$ . . . [9]

where *m* and *n* are, respectively, the mass and the number of moles of the substance being heated or cooled.

There are two common ways to heat a system:

(1) **Constant Volume Heating**—in which the system is enclosed in a rigid container at constant volume ($\Delta V_{syst} = 0$). For a constant volume heating process the *pressure* of the system is allowed to change. Thus, C_V, $\bar{c}_V$, and c_V are, respectively, the *total heat capacity at constant volume*, the *specific heat capacity at constant volume*, and the *molar heat capacity at constant volume*.

(2) **Constant Pressure Heating**—in which the pressure is kept constant and the *volume* of the system is allowed to change. In this case we use the symbols C_P, $\bar{c}_P$, and c_P to represent the *total heat capacity at constant pressure*, the *specific heat capacity at constant pressure*, and the *molar heat capacity at constant pressure*. For **solids** and **liquids** it is found that the heat capacity at constant volume is approximately the same as the heat capacity at constant pressure. (This is a consequence of the fact that the volume of a solid or liquid doesn't change appreciably on heating, so that negligible *PV*-work is done when a solid or liquid is heated at constant pressure.)

From thermodynamics it can be shown that for an ideal gas,

$$c_P = c_V + R \qquad \textit{Ideal Gas}$$ . . . [10]

where *R* is the gas constant. Eqn [10] holds for diatomic and polyatomic ideal gases as well as for monatomic ideal gases. For most *real* gases, too, the relationship is approximately valid, and can be used in the absence of more detailed information.

We showed that for any constant volume process in a closed system in which no work of any kind is done, $\Delta U_V = Q_V$. Furthermore, *if* the process is merely the heating or cooling of a homogeneous chemical system (i.e., no phase changes or chemical reactions take place), then we also can say that $Q_V = C_V\Delta T$. Combining the two equations gives

$$\Delta U_V = Q_V = C_V\Delta T \qquad \begin{array}{l} \textit{W = 0; Closed System,} \\ \textit{No Chemical Reactions} \\ \textit{or Phase Changes} \end{array}$$ . . . [11]

Thus, for a closed system, when $W = 0$, $\Delta U_V = Q_V$ *always*; but $\Delta U_V = C_V\Delta T$ *only* when the constant volume process taking place is just a simple heating or cooling process with no phase changes or chemical reactions occurring. When heating or cooling m kilograms, of course, $\Delta U_V = m\bar{c}_V \Delta T$, and when heating or cooling n moles, $\Delta U_V = nc_V \Delta T$.

Example 7-1

A candle is burning in a sealed insulated room—the "system"—releasing 100 J s^{-1} of thermal energy into the room, which is at one atm pressure. The total volume of the room is 100 m^3, and the initial room temperature is 20.0°C. For air, $c_V = 20.47$ J mol^{-1} K^{-1}. The molar mass of air is 28.97 g mol^{-1}.

(a) Neglecting any compositional changes in the air resulting from the combustion products of the burning candle, and assuming uniform mixing, what is the temperature of the air in the room after one hour? The heat capacity of the walls can be neglected.

(b) What is Q_{syst} as a result of the candle burning for one hour?

(c) What is ΔU_{syst} as a result of this process?

Solution

(a) Let the final temperature of the air be θ_f °C.

The heating of the air is a constant volume process, for which $Q_V = nc_V \Delta T$.

Number of moles of air in the room: $n = \dfrac{PV}{RT} = \dfrac{(101\,325)(100)}{(8.314)(20 + 273.15)} = 4157$ mol

The heat released by the burning candle in one hour is $(100$ J s$^{-1})(3600$ s$) = 3.60 \times 10^5$ J.

This heat is transferred to the air in the room, which becomes warmer. Thus,

$$(Q_V)_{air} = nc_V \Delta T$$

$$3.60 \times 10^5 = (4157)(20.47)\Delta T$$

$$\Delta T = 4.23 \text{ K} = \theta_f - 20$$

$$\theta_f = 20 + 4.23 = 24.23°C$$

Ans: $\boxed{24.2°C}$

(b) The system is the room itself. Since the room is insulated, no heat can cross its boundaries; therefore $Q_{syst} = 0$. There is, of course, heat exchange between the *subsystems* that make up the contents of the room; namely, the candle and the air. The heat *evolved* by the candle ($Q_{candle} = -3.60 \times 10^5$ J) is equal to the heat *absorbed* by the air ($Q_{air} = +3.60 \times 10^5$ J).

Ans: $\boxed{Q_{syst} = 0}$

(c) Similarly, no work crosses the boundary of the room; therefore, $W_{syst} = 0$. Thus:

$$\Delta U_{syst} = Q_{syst} + W_{syst} = 0 + 0 = 0$$

Since nothing enters or leaves the room, its internal energy *must* remain the same; i.e., $\Delta U = 0$. All that happens *inside* the room is that one form of concentrated energy (the chemical energy contained in the wax of the candle) is converted into another, lower grade, form of energy (dispersed thermal energy of the air molecules).

Ans: $\boxed{\Delta U_{syst} = 0}$

Example 7-2

An insulated container made of a certain metal has a mass of 3.6 kg and contains 14.0 kg of water. Both the container and the water are initially at a temperature of 16.0°C. A 10.0 kg piece of the same type of metal initially at a temperature of 50°C is dropped into the water. After reaching thermal equilibrium, the entire system is at 18.0°C. What is the specific heat capacity of the metal? The specific heat capacity of water is $(\bar{c}_P)_W = 4.18$ kJ K^{-1} kg^{-1}. The process takes place at constant pressure.

Solution

Let the specific heat capacity of the metal be $\bar{c}_P$ J K^{-1} kg^{-1}.

The heat gained by the container is
$$\begin{aligned}
Q_1 &= m_1 \bar{c}_P \Delta T_1 \\
&= (3.6)\bar{c}_P(18.0 - 16.0) \\
&= +7.2\,\bar{c}_P \text{ J}
\end{aligned}$$

The heat gained by the water is
$$\begin{aligned}
Q_2 &= m_2(\bar{c}_P)_w \Delta T_2 \\
&= (14.0)(4180)(18.0 - 16.0) \\
&= +117\,040 \text{ J}
\end{aligned}$$

The heat lost by the piece of metal is
$$\begin{aligned}
Q_3 &= m_3 \bar{c}_P \Delta T_3 \\
&= (10.0)\bar{c}_P(18.0 - 50.0) \\
&= -320.0\,\bar{c}_P \text{ J}
\end{aligned}$$

Since the system is insulated, no heat escapes, so that for the overall system $Q_{syst} = 0$. Thus the sum of the various subsystem internal heat transfers must equal zero:

$$Q_1 + Q_2 + Q_3 = 0$$
$$+7.2\,\bar{c}_P + 117\,040 - 320.0\,\bar{c}_P = 0$$
$$312.8\,\bar{c}_P = 117\,040 \quad \text{from which} \quad \bar{c}_P = 374.2 \text{ J K}^{-1} \text{ kg}^{-1} = 0.3742 \text{ kJ K}^{-1} \text{ kg}^{-1}$$

Ans: $\boxed{0.37 \text{ kJ K}^{-1} \text{ kg}^{-1}}$

Exercise 7-2

A 75 g piece of copper metal is heated to 312°C and dropped into an insulated glass beaker containing 220 g of water at 12°C. What is the final temperature after thermal equilibrium has been attained? The heat capacity of the glass beaker is $C_b = 188$ J K^{-1}. The specific heat capacities of water and copper are 4.18 kJ K^{-1} kg^{-1} and 386 J K^{-1} kg^{-1}, respectively. ☺

Example 7-3 [2]

The molar heat capacity at constant pressure for water vapor varies with temperature according to

$$c_P = 30.54 + 0.0103T \qquad \qquad \dots \text{[a]}$$

where c_P is in J mol^{-1} K^{-1} and T is the temperature in kelvin.

Calculate: (a) W (b) Q (c) ΔU when one mole of water vapor is heated at constant volume from 25°C to 200°C. Assume ideal gas behavior.

Solution

(a) Since this is a constant volume process, $\Delta V_{syst} = 0$; therefore, the PV-work, defined as $W_{PV} = -P_{ext} \Delta V_{syst}$, also must be zero.

Ans: $\boxed{W = 0}$

(b) Q_V, the heat at constant volume, is given by $Q_V = n c_V \Delta T$ where n is the number of moles of gas being heated, c_V is the molar heat capacity at constant volume, and $\Delta T = T_2 - T_1$ is the temperature difference through which the gas is heated.

We are given c_P, the molar heat capacity at constant *pressure*, but we need to know c_V, the molar heat capacity at constant *volume*, because this is a constant volume process.

For an ideal gas, $c_P = c_V + R$;

therefore, $$c_V = c_P - R$$
$$= (30.54 + 0.0103T) - 8.314$$
$$= 22.226 + 0.0103T$$

We can use the formula $Q_V = n c_V \Delta T$ only when c_V is constant over the temperature range $T_1 \rightarrow T_2$. In this problem, however, Eqn [a] tells us that c_V has a different value at each temperature. However, over an infinitesimally small temperature range from T to $T + dT$ the value of c_V will be constant at the value for T; therefore, we can calculate the incremental bit of heat δQ over this small temperature range dT using the expression[3] $\delta Q_V = c_V dT$ and integrate (add up) all

---

[2] This is an important type of calculation often encountered. If you're deficient in integration, read Appendix 8.
[3] In this problem, $n = 1$ so that $\delta Q_V = c_V dT$.

the small increments of δQ to obtain the *total* heat Q over the complete temperature range ΔT.

Thus:
$$Q_V = \int_{T_1}^{T_2} c_V \, dT$$

$$= \int_{298.15}^{473.15} (22.226 + 0.0103T) \, dT$$

$$= \left[22.226T \right]_{298.15}^{473.15} + \left[0.0103 \left(\frac{T^2}{2} \right) \right]_{298.15}^{473.15}$$

$$= (22.226)(473.15 - 298.15) + (0.00515)\left(473.15^2 - 298.15^2\right)$$

$$= 3889.5 + 695.1 = 4584.2 \text{ J} = +4.58 \text{ kJ}$$

Ans: $\boxed{Q_V = +4.58 \text{ kJ}}$

Three important points can be made here:

First, when calculating a ΔT it doesn't matter whether you use °C or kelvin, since the *difference* between the two numbers will always be the same. However, when carrying out integrations we often end up with expressions involving T^2 or even higher powers of T, as well as with the logarithm of T. For these cases you **must** use the absolute temperature, otherwise you will get the wrong answer. For this reason, when in doubt, it is always safe to express the temperature as the absolute temperature.

Second, be careful to remember that $\left(T_2^2 - T_1^2\right)$ is *not* the same as $\left(T_2 - T_1\right)^2$.

Third, note that although $(\ln a - \ln b) = \ln\left(\frac{a}{b}\right)$, $\dfrac{\ln a}{\ln b}$ does *not* equal $\ln\left(\frac{a}{b}\right)$.

(c) For a constant volume process $\Delta U_V = Q_V$, therefore $\Delta U = +4.58$ kJ.

Ans: $\boxed{\Delta U = +4.58 \text{ kJ}}$

Exercise 7-3

When water is shaken, the mechanical kinetic energy input by each shake dissipates 100% through friction to thermal energy. How long would it take to raise the temperature of 500 mL of water initially at 20°C to its normal boiling point by putting the water into a completely insulated vacuum flask and shaking it? Assume that you can do 40 shakes per minute, and that for each shake the water falls through a distance of 20 cm. The heat capacity of water is 4.18 kJ K^{-1} kg^{-1} and the acceleration due to gravity is 9.8 m s^{-2}. The heat capacity of the vacuum flask can be ignored. ☺

7.8 ΔU FOR THE ISOTHERMAL EXPANSION OF AN IDEAL GAS

In Chapter 4 it was shown that the internal energy of an ideal gas is proportional to its temperature; therefore, it follows that if the temperature of an ideal gas is *constant*, then its internal energy U also must be constant.

Thus: $\Delta U = 0$ *for an isothermal expansion of an ideal gas.*

But $\Delta U = Q + W$

therefore, for the isothermal expansion of an ideal gas,

$$\Delta U = Q + W = 0$$

and it follows immediately that

$$\boxed{Q = -W} \quad \begin{array}{l} \textit{Isothermal Expansion} \\ \textit{of an Ideal Gas} \end{array} \quad \cdots [12]$$

7.9 THE INTERNAL ENERGY
OF A MONATOMIC IDEAL GAS

A monatomic ideal gas can't store energy internally by vibration or rotation because it consists of single atoms. Furthermore, ideal gases don't have *PE* between the particles because, by definition, there are no interactions. Thus, for a *monatomic ideal gas* the internal energy consists only of the kinetic energy of translation: $U = \frac{3}{2}kT$ per *molecule* or $u = \frac{3}{2}RT$ per *mole*.[4] Under normal conditions most real monatomic gases behave like ideal monatomic gases. Thus, for the special case of an ideal monatomic gas we can assign an absolute value for U.

KEY POINTS FOR CHAPTER SEVEN

1. **Internal Energy** U: The total chemical energy contained in a chemical system. Can be increased both by *heating* the system and by *doing work* on it. Work and heat are equivalent ways of changing the internal energy of a system. Even though work is not the same thing as heat, both work and heat can bring about the same change in the internal energy of a system. For a closed system the "equivalency" between the two is given by the **First Law of Thermodynamics**:

    $$\Delta U = Q + W$$

    where ΔU is the change in the internal energy of the system in joules, Q is the heat transfer *to* (positive) or *from* (negative) the system in joules, and W is the work in joules done *on* the system (positive) or done *by* the system (negative).

---

[4] See Chapter 4 for the derivation of this relationship.

2.   **State Functions**: A property of a system that depends only on the current *state* of the system (i.e., on its P, V, T, and n), and is independent of the path taken to reach that state. From experiment, the internal energy U is found to be a state function.

When a system passes from an initial state *1* having P_1, V_1, T_1, and U_1 to a final state *2* having P_2, V_2, T_2, and U_2, the change ("process") may be brought about by any number of ways or sub-processes ("paths"). The most important characteristic of any state function X is that the change ΔX_{12} in the state function as a result of the process $1 \rightarrow 2$ depends *only* on the initial and final state of the system, and is *independent of the path*. Thus, since U is a state function, we can write

$$\Delta U_{12} \equiv U_2 - U_1 \quad \text{(State function)}$$

3.   **Work and Heat are NOT State Functions**: Since the same overall change in state for a system can be brought about by using any number of different combinations of Q and W, in general there is no single value of Q or of W associated with any given process. Therefore the values of Q and of W depend on the particular path taken in going from state *1* to state 2. Such functions are called **path functions**. For *path* functions such as Q and W we can only write

$$\int_{State\,1}^{State\,2} \delta Q = Q_{12} \quad \text{and} \quad \int_{State\,1}^{State\,2} \delta W = W_{12} \qquad \text{(Path functions)}$$

We write δQ and δW instead of dQ and dW to remind ourselves that Q and W are path functions, and not state functions.

We CANNOT write $\quad \int_{State\,1}^{State\,2} \delta Q = Q_2 - Q_1 = \Delta Q \quad$ or $\quad \int_{State\,1}^{State\,2} \delta W = W_2 - W_1 = \Delta W$

The integrals of δQ and δW are just Q and W, not ΔQ and ΔW. You can *never* write ΔQ or ΔW!

4.   *Q and W both are energy in transit*: they exist *only* during the act of crossing the boundary of a system. A system doesn't "contain" heat or "contain" work; it only contains internal energy. We sometimes say that a system has a lot of "heat" in it, but this is sloppy speech; we should say that it has a lot of *thermal energy*.

5.   **Sign Convention**:

   - W_{syst} is **positive** if work is done **on** the system (work flows *into* the system)
   - W_{syst} is **negative** if work is done **by** the system (work flows *out of* the system)
   - Q_{syst} is **positive** if heat flows **into** the system
   - Q_{syst} is **negative** if heat flows **out of** the system

6.   **Constant Volume Process**: For a chemical reaction $A + B \rightarrow C$ taking place in a rigid container, the volume of the system doesn't change; therefore, $\Delta V = 0$. For chemically reacting systems, usually the only kind of work involved is PV-work, which is done when the system expands (or contracts). Therefore, for a chemical reaction in which the only kind of work done is PV-work, $\quad W = W_{PV} = -P_{ext}\,\Delta V = -P_{ext}(0) = 0$

and
$$\Delta U_V = (Q + W)_V = Q + 0 = Q$$

For a constant volume process in a closed system, $\Delta U = Q$, and measuring the heat for the process gives us the change in internal energy for the system as a result of the process.

7. **Heat Capacity of a Single Phase System**: C, the heat capacity of the system in J K^{-1}, is the number of joules of heat Q required to raise the temperature of the system by one kelvin (or by one °C):
$$C = Q/\Delta T$$

C = *total* heat capacity of system, in J K^{-1}; c = *molar* heat capacity of system, in J K^{-1} mol^{-1}; $\bar{c}$ = *specific* heat capacity of system, in J kg^{-1} K^{-1}.

The heat required to change the temperature by ΔT is
$$Q = C\Delta T = m\bar{c}\Delta T = nc\Delta T$$

where m and n are the number of kilograms or moles, respectively, of the system.

8. For **constant volume heating** the *pressure* of the system is allowed to change. Thus, C_V, $\bar{c}_V$, and c_V are the *total heat capacity at constant volume*, the *specific heat capacity at constant volume*, and the *molar heat capacity at constant volume*, respectively.

For **constant pressure heating** the *volume* of the system is allowed to change. In this case we use the symbols C_P, $\bar{c}_P$, and c_P to represent the *total heat capacity at constant pressure*, the *specific heat capacity at constant pressure*, and the *molar heat capacity at constant pressure*, respectively. For **solids** and **liquids** it is found that the heat capacity at constant volume is approximately the same as the heat capacity at constant pressure.

9. For *any* **Ideal Gas**—monatomic, diatomic, triatomic, etc.—the molar heat capacities are related by
$$c_P = c_V + R$$

where R is the gas constant. This relation also is approximately valid for most real gases.

10. For simple heating or cooling of a homogeneous chemical system at constant volume with no phase changes or chemical reactions taking place and no work of any kind being done:
$$\Delta U_V = Q_V = C_V\Delta T = nc_V\Delta T = m\bar{c}_V\Delta T$$

11. Since the internal energy of an ideal gas is proportional to its temperature, if the temperature of an ideal gas doesn't change, then its internal energy also doesn't change. Therefore, for the **isothermal expansion or compression of an ideal gas**, $\Delta U = 0$, from which $\Delta U = 0 = Q + W$ and $Q = -W$.

12. A monatomic ideal gas can't store energy internally by vibration or rotation because it consists of single atoms. Furthermore, ideal gases don't have *PE* between the particles because, by definition, there are no interactions. Thus, **for a monatomic ideal gas** the internal energy consists only of the kinetic energy of translation:
$$U = \tfrac{3}{2} kT \qquad \text{per } molecule$$

or
$$u = \tfrac{3}{2} RT \qquad \text{per } mole$$

PROBLEMS

1. 100.0 g of liquid water is heated from 4°C to 85°C at a constant pressure of one bar. For this process, calculate (a) Q, (b) W, (c) ΔU. For liquid water, $\bar{c}_P = 4.18$ J g^{-1} K^{-1}. The densities of water at 4°C and at 85°C are, respectively, 1.0000 g cm^{-3} and 0.96865 g cm^{-3}. ☺

2. 100.0 g of nitrogen gas is heated reversibly from 4°C to 85°C at a constant pressure of one bar. For this process, calculate (a) Q, (b) W, (c) ΔU. Nitrogen may be assumed to behave as an ideal gas. For nitrogen gas, $c_P = 29.124$ J mol^{-1} K^{-1}. ☺

3. 2.00 moles of an ideal gas occupies a volume of 50.0 L at a temperature of 30.0°C and a pressure of P_1. 1000 J of work is used to compress the gas at constant pressure. The heat capacity of the gas at constant volume is $\bar{c}_V = \frac{3}{2}R$ J mol^{-1} K^{-1}. Calculate: (a) V_2, the final volume of the gas, (b) T_2, the final temperature of the gas, (c) ΔU for the gas, and (d) Q_{gas}. ☺

4. Complete the following table of market values of various forms of energy in ¢/MJ:

   (a) 1 MJ (electric) at 10.0¢ per kWh =   ¢/MJ
   (b) 1 MJ (oil) at $30.00 per barrel   =   ¢/MJ
   (c) 1 MJ (oats) at $1.50 per kg   =   ¢/MJ
   (d) 1 MJ (firewood) at $350 a cord   =   ¢/MJ

   The following conversion factors are given to save your time, but you should try to find them in handbooks and textbooks:

   • 1 barrel of oil = 42 U.S. gallons
   • 1 U.S. gallon of water (density 1 g cm^{-3}) weighs 3.82 kg
   • The density of most long chain liquid hydrocarbons is about 0.8 g cm^{-3}
   • The heat of combustion of petroleum oil is about 11,000 cal g^{-1}
   • The food value of oats is given as 1700 food calories per pound
   • A cord is $4 \times 4 \times 8$ feet of loosely piled split wood
   • 1 inch = 2.54 cm
   • 1 lb = 453.59 g
   • The heat of combustion of dry hardwood is about 5,000 cal g^{-1} ☺

5. Use the data given in the following table to calculate the food value of protein in MJ kg^{-1}.

Food	% Protein	% Fat	% Carbohydrate	% Water	Food Value (kcal lb^{-1})
lard	0	100	0	0	4080
maple sugar	0	0	82.8	17.2	1540
corn bread	7.90	4.70	46.3	41.1	1205	☺

6. One mole of oxygen at 25°C and 1.00 bar (state A) is allowed to expand isothermally to 0.100 bar (state B). What is the value of Q if:

(a) The maximum weights are lifted for two equal volume step expansions?

(b) The maximum weights are lifted for a many (i.e., infinite) step expansion? ☺

7. Strictly speaking, the "Law of Conservation of Energy" should be called the "Law of Conservation of Mass-Energy." Thermodynamic and chemical process calculations make extensive use of the conservation of elemental mass in chemical reactions. However, whenever energy is released (or absorbed) in a chemical reaction there will be a small corresponding loss (or gain) of mass through the mass-energy relationship $E = mc^2$ derived by Albert Einstein. The combustion of one liter of gasoline releases 33.7 MJ of heat. Calculate the mass of the system that is lost on account of this energy release, and express it as a percentage of the initial mass of the one liter of gasoline.

Density of gasoline = 0.739 g cm^{-3}; velocity of light = 3.00×10^8 m s^{-1}. ☺

8. The energy of the sun originates in the "nuclear burning" of hydrogen, evolving about 4×10^{26} J s^{-1} of heat.

(a) How much mass is burned in one second?

(b) If the present solar mass is about 2×10^{30} kg, how many years will it take to burn 1% of the sun's mass? (1 year = 365.25 days) ☺

9. One kilogram of solid copper cools at one atm pressure from 100°C to 25°C. By how much, if at all, does its mass change as a result of this process? The molar heat capacity of copper may be taken as constant at $c_p = 24.5$ J K^{-1} mol^{-1}. ☺

10. A house is heated by a thermostatted electric resistance heater that maintains the temperature inside the house at 27°C by controlling the flow of electricity to the heater. This heater operates at an average power consumption of 4.00 kW. An electric freezer that draws an average current of 10.0 A at 100 V is to be installed. Electricity sells for 5.00¢ per kWh. The owner of the house has the option of installing the freezer outside the house in an unheated building. Because of lower temperatures outside the freezer, only half the current (i.e., 5.00 A) will be required to maintain the desired freezer temperature if the owner selects this option. How much will the daily total electric bill be

(a) With the freezer installed inside the house?

(b) With the freezer installed outside the house? ☺

11. A 1.00-hp motor is used to turn a stirring paddle in a tank of water containing 25.0 kg of water, which is initially in thermal equilibrium with the surroundings at 25.0°C and one atm pressure. The specific heat of the water can be assumed constant at $\bar{c}_V = 4.18$ J g^{-1} K^{-1}.

(a) If the tank is well insulated and the stirring action is applied for 1.00 hour, assuming the process takes place at constant volume, calculate:

(i) the change in internal energy of the water

(ii) the final temperature of the water

(b) Calculate the final temperature of the water if the tank loses heat at an average rate of 200 J s^{-1} while the stirring action is applied.

Data: 1 horsepower = 746 W ☺

12. A house is heated by a thermostatted electric resistance heater that maintains the temperature inside the house at 25°C by controlling the flow of electricity to the heater. The heater operates at an average power consumption of 10.0 kW. You may assume that the air in the house is well circulated to prevent significant temperature differences.

    (a) Calculate the daily cost of electricity at 10.0¢ per kWh.
    (b) Calculate the daily cost of the electricity required to operate an electric freezer that operates inside the house if the device draws an average current of 10.0 amperes at 100 V.
    (c) After the freezer has been installed in the house, what will be the *total* daily cost of electricity (i.e., the *sum* of the cost of the electricity to heat the house plus that to operate the freezer)?
    (d) What is the net additional cost per day (i.e., the *increase* in the total electricity bill) of installing and operating the freezer? ☹

13. The owner of an electrically heated house has a very large freezer that is not being used. She is told by her friend *A* that if she connects the very low temperature region in the freezer (at −20°C) to the outside temperature (0°C) using a heat exchanger, it will automatically reduce her electric bill, by an amount greater than the extra electricity required to operate the freezer. Another friend *B* tells her that she will just wear out her freezer for nothing, since *x* joules of w_{elec} used will provide exactly *x* joules of heat according to the law of energy conservation. Which advice is correct? ☹

14. Heat is leaking through imperfect insulation into a controlled cold region at the rate of 3000 BTU h^{-1}. A thermoelectric system maintains the region at a steady state temperature of 5.0°C and operates on a current of 4.0 amperes at a voltage of 240 V. Find the heat rejected by the thermoelectric system. *Data:* 1 BTU = 252 cal ☺

15. A house is heated by an electric resistance heater controlled by a thermostat that maintains a constant temperature inside the house. The air is well circulated to prevent temperature differences in the house. The heater draws an average current of 20.0 amperes at 110 volts. The cost of electricity is 6.0 cents per kWh. An electric freezer is available that draws an average current of 3.00 amperes at 110 volts.

    (a) Calculate the daily cost of the electricity to operate the freezer if it is located *outside* the house.
    (b) If the freezer described in (a) is installed *inside* the house, calculate the total combined cost of electricity per day for operating the heater and the freezer. ☺

16. The owner of an electrically heated house fitted with heaters that consume 200 kWh per day has in the house a very large freezer that is not being used. He is testing an idea by thermally connecting the low temperature region in the freezer (−20.0°C when the freezer is operating) to the environment outside the house (at 0.00°C) by means of a heat exchanger (the freezer remains inside the house). This heat exchanger transfers 1000 kcal h^{-1} between the cold box of the freezer and the outside environment. The electrical consumption of the freezer is 50.0 kWh of electricity per day. By using the freezer in this manner, will the total electricity used to keep the house warm *increase*, *decrease*, or *remain the same*? and by how many kWh per day? ☺

17. The isothermal compression of 100 moles of nitrogen gas at 300 K from 1.00 atm to 2.00 atm is accomplished using two different machines: (i) an electrically-driven compressor and (ii) a mechanically-driven compressor powered by a falling weight.

   (a) What is the change of the internal energy of the gas during the compression? The gas may be assumed to behave ideally.

   (b) What is the value of Q calculated for this process if the oversimplified, imaginary, textbook model process called "reversible" is used for the calculation?

   (c) What is the value of Q if compressor (i) uses 0.685 ampere-hours of electricity at 100 V for the job?

   (d) What is the value of Q if compressor (ii) uses 200 000 ft-lb for the job?

      Data: 1 in = 2.54 cm   1 ft = 12 in   1 kg = 2.20 lb   g = 9.807 m s^{-2} ☺

18. A fully-charged 60 ampere-hour car battery was continuously discharged at a current of 30.0 amperes to run an electrical device. During the discharge the battery delivered current at an average voltage of 8.0 V. The discharge process used up 25.0% of the battery's capacity. Afterwards, the battery was recharged to its initial state using a battery charger that delivered an average charging current of 5.0 amperes at 13.0 V.

   (a) What is Q_{net} for the battery for the combined discharge-charging cycle?

   (b) Is this heat absorbed or evolved by the battery? ☹

19. A mass of 20.4 kg falls 5.00 m through the gravitational field, and in so doing causes a paddle wheel to turn inside a rigid, perfectly insulated tank containing 1.00 mole of an ideal gas (the system). The initial temperature of the gas is 300.0 K, and its molar heat capacity is constant, at $c_P = 28.3143$ J mol^{-1} K^{-1}. It may be assumed that no frictional or heat losses occur outside the system, and that the temperature of the surroundings is 300 K. The local value of the acceleration due to gravity is 9.804 m s^{-2}.

   (a) What is Q_{syst} as a result of this process?

   (b) What is the final temperature of the system? ☺

20. A small cabin in northern Ontario is heated in the winter by means of an electric resistance heater. It costs the owner $50 per month for electricity to maintain the inside of the cabin at a constant temperature of 15°C. What will she have to pay for electricity per month if she wants to increase the room temperature to 25°C? The outside temperature is constant at –5°C.

   Hint: The rate of heat loss from the cabin may be calculated using Newton's cooling law:

   $$\dot{q}_{loss} = kA\Delta T$$

   where $\dot{q}_{loss}$ is the rate of heat loss, k is a constant (the heat transfer coefficient), A is the area through which heat transfers, and ΔT is the temperature difference between the inside and the outside of the cabin. ☹

21. 100 moles of gaseous CO_2 is taken from 1.000 bar pressure and 300 K to 1.167 bar pressure and 350 K by two reversible paths:

   Path *A*: The volume is held constant and the gas heated directly to 350 K.

   Path *B*: The gas is expanded isothermally to twice its original volume, then heated at constant volume to 350 K, and finally compressed isothermally to the final volume.

(a) Is this a constant volume process?

(b) For Path A, evaluate ΔU, Q, and W.

(c) For Path B, evaluate ΔU, Q, and W.

For this calculation, carbon dioxide may be considered to behave as an ideal gas with a constant heat capacity of $c_p = 37.11$ J mol^{-1} K^{-1}. ☹

22. For the isothermal expansion ($A{\rightarrow}B$) and compression ($B{\rightarrow}A$) of one mole of ideal gas from one atm (state A) to one-half atm (state B) at 0°C, calculate:

(a) The maximum work obtainable from the system in a 3-step (each step of equal volume) expansion of the gas $A{\rightarrow}B$.

(b) The minimum work that must be done on the gas to restore the system ($B{\rightarrow}A$) in a 3-step (each step of equal volume) compression.

(c) The value of Q for the above six-step cycle $A{\rightarrow}B{\rightarrow}A$.

The local value of the acceleration due to gravity is $g = 9.806$ m s^{-2}. ☹

23. A piston/cylinder arrangement contains one mole of an ideal gas (the system) initially at 10.0 atm pressure and 300 K, as shown in the accompanying illustration. Neglecting the mass of the piston, neglecting friction, and assuming isothermal conditions throughout, the pin restraining the piston is removed. For the resulting process:

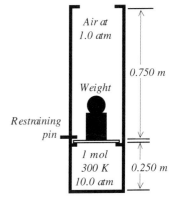

(a) What is Q_{syst} if the mass of the weight is zero?

(b) What is Q_{syst} if the mass of the weight is 100 kg?

(c) What is Q_{syst} if the mass of the weight is 254.35 kg?

(d) What is Q_{syst} if the mass of the weight is 1017.4 kg?

Data: $g = 9.807$ m s^{-2} ☹

24. One mole of a combustible ideal gas, having constant $c_p = 29.0$ J mol^{-1} K^{-1}, is initially at 300 K and 40.0 atm pressure. Calculate the *maximum* work output from the gas for each of the following processes:

(a) Adiabatic expansion against a constant external pressure of 10.0 atm to a final gas temperature of 250 K.

(b) Isothermal expansion into a vacuum until the final gas pressure is 10.0 atm.

(c) Isothermal expansion to a final volume of 2.462 L.

(d) Combustion of the gas in a heat engine at 2500 K and one atm pressure, with exhaustion of the product gases to the surrounding atmosphere at 300 K. The heat of combustion of the gas at 2500 K and one atm pressure is 795.5 kJ mol^{-1}.

Note: A heat engine takes in heat at a high temperature, converts some of it to work output, and exhausts the remaining heat to the surroundings, which is at a lower temperature. The maximum (reversible) efficiency η_{rev} of such an engine is defined as the maximum work output w_{max} obtainable from a given amount of high temperature heat input q_{Hi}, and can be shown to be given by

$$\eta_{rev} = \frac{w_{max}}{q_{Hi}} = \left(\frac{T_{hi} - T_{lo}}{T_{hi}} \right) \ \odot$$

25. Consider the isothermal expansion of 1.00 mole of oxygen at 273.15 K from an initial volume of 22.414 L and pressure of 1.00 atm to a final volume of 44.828 L and pressure of 0.500 atm in the apparatus shown, which is located in a vacuum chamber. For this calculation assume that the mass of the piston itself is zero, and that the acceleration due to gravity is 9.807 m s^{-2}. Oxygen gas may be assumed to behave as an ideal gas. Neglect any friction.

For this process:

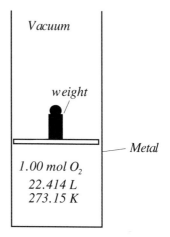

(a) Calculate the heat Q if no weight is lifted during the expansion.

(b) Calculate Q if a weight of 1.00 kg is lifted through 22.414 m.

(c) Calculate Q if a weight of 3.00 kg is lifted through 22.414 m.

(d) Calculate Q if the maximum single weight that could be lifted in one step through 22.414 m is lifted.

(e) What is the weight lifted in (d)?

(f) Calculate Q if the maximum weights that can be lifted through two equal volume steps of 11.207 L each are lifted.

(g) Calculate the limiting value of Q if the maximum weights are lifted in a many-step expansion. $\otimes$

26. One mole of oxygen gas at 273 K and 1.00 atm pressure (state 1) is expanded isothermally to 0.500 atm pressure (state 2). What is the value of Q if:

(a) no weight is lifted by the expanding gas?

(b) the maximum weight that could be lifted to 0.500 atm is lifted in one step? $\odot$

27. Consider the isothermal expansion ($A{\rightarrow}B$) and compression ($B{\rightarrow}A$) of one mole of an ideal gas from 1.0 atm (state A) to 0.50 atm (state B) at 273 K. Calculate the following maximum-minimum quantities assuming weightless, frictionless pistons for the following machines designed to carry out the process. Do the calculations making use of weights lifted through displacements to evaluate the work terms. At 273 K and one atm pressure the volume occupied by the gas can be taken as 22.4 L. For this problem, $g = 9.8$ m s^{-2}.

(a) The maximum work obtainable in a 1-step expansion of the gas $A{\rightarrow}B$.

(b) The minimum work that must be done on the gas to restore the system ($B{\rightarrow}A$) in a 1-step compression.

(c) The maximum work obtainable in an expansion of two equal volume steps $A{\rightarrow} A'{\rightarrow}B$.

(d) The minimum work that must be done on the gas to restore the system ($B \rightarrow A' \rightarrow A$) in two equal volume steps.

(e) The value of Q for the four-step cycle (2-step expansion + 2-step compression) $A \rightarrow A' \rightarrow B \rightarrow A' \rightarrow A$. ☹

28. One kilogram of O_2 is heated in a frictionless cylinder-piston apparatus at a constant pressure of one atm from 20°C to 30°C. Assuming ideal gas behavior, calculate: (a) ΔH for the process, (b) ΔU for the process, and (c) c_V for oxygen gas. The molar mass of O_2 is 32.00 g mol^{-1}. ☺

29. An electric motor rated at 1.50 horsepower is turned on and operated for 30 minutes, during which time it delivers 1.85 MN·m of mechanical work. How much energy, in kilojoules, is dissipated as friction and in the windings of the motor as a result of this process? ☹

30. The molar heat capacity at constant pressure for ammonia vapor is $c_p = 29.75 + 0.0251T$, where c_p is in J K^{-1} mol^{-1} and T is in kelvin. Assuming ammonia behaves as an ideal gas, calculate the change in internal energy ΔU when 10.0 moles of ammonia vapor is heated at a constant pressure of 0.100 bar from 25.0°C to 75.0°C. ☺

31. The total electrical power consumed by France is about 81 GW, of which 78% is generated by nuclear reactors, each, on the average, producing about a GW of electricity. It has been suggested that these nuclear power plants could be replaced with wind turbines along the coast of Brittany, where the wind is abundant and relatively stable. How feasible is this suggestion? For your calculation, you may assume the following: The diameter of the blades of a giant wind turbine is about 100 m across. Under optimum conditions such devices are capable of generating about 2 MW of electrical energy, but in effect optimal conditions prevail only about 25% of the time. For optimal operation the turbines must be spaced about 400 m apart (center-to-center). The coastline of Brittany is about 2700 km long, of which you may assume that 50% could be used for wind turbine installation. ☺

THERMODYNAMICS (III)

8.1 ENTHALPY (*H*)

Many processes are carried out at constant pressure, usually that of the atmosphere. For most chemical reactions, the only kind of work done by or on the system is the *PV*-work that is associated with the volume changes that occur during the process. We saw earlier that we can't use the expression $-P\Delta V$ if the pressure does not stay constant during the process; instead, we must use $-PdV$, which indicates that P can be considered constant only over the small volume change dV of the system. To obtain the *total PV*-work done during the process we must *integrate*, to get

$$W_{PV} = -\int_{V_1}^{V_2} P_{ext}\, dV_{syst}$$

... [1]

On the other hand, if P_{ext} *is* constant during the process, then Eqn [1] reduces to

$$W_{PV} = -\int_{V_1}^{V_2} P_{ext} dV_{syst} = -P_{ext}\int_{V_1}^{V_2} dV_{syst} = -P_{ext}(V_2 - V_1) = -P_{ext}\Delta V_{syst}$$

... [2]

which is just the same expression we had before; however, if the external pressure varies during the process,[1] then the value of the integral must be determined in order to evaluate *W*.

For a closed system, we showed that $\Delta U = Q + W$. Furthermore, for a constant pressure process in which the only work is *PV*-work, we saw in Chapter 6 that in the limiting case of a *reversible* process, the external pressure becomes identical with the pressure of the gas itself, so that $W_{syst} = -P_{syst}\Delta V_{syst}$. Substituting this into the above expression for ΔU gives

$$\Delta U_P = Q - P\Delta V \qquad \begin{array}{l}\textit{Constant P,}\\ \textit{Only PV-Work}\end{array} \qquad \text{... [3]}$$

---

[1] As, for example, in the case of a *reversible* compression, during which the external pressure is continuously increased so that it is always just infinitesimally greater than the pressure of the gas being compressed. The reversible work of compression is needed in order to evaluate the *minimum* work required to carry out the compression. Similarly, the work delivered by the reversible expansion of the gas gives the maximum work output potentially available from the gas expansion. In this latter case, the external pressure is continuously adjusted so that it is always just infinitesimally *less* than the pressure of the expanding gas at any moment during the expansion.

where ΔU_P is the change at constant pressure in the internal energy of the system, Q is the heat transferred to or from the system, P is the pressure of the system, and ΔV is the volume change for the system during the process. This equation shows that at constant pressure, ΔU is just the *heat* for the process plus a *work term correction*. *PV*-work is so common that it is convenient to define a new property H called the **enthalpy**, which takes into account this work term correction:

$$\boxed{H \equiv U + PV} \quad \textit{Enthalpy Defined} \quad \ldots [4]$$

Since each of its components is a state function, then H also must be a state function.

8.2 CONSTANT PRESSURE PROCESSES

For any closed system process in which the system changes from State *1* to State *2*,

$$\Delta H = H_2 - H_1$$
$$= (U_2 + P_2 V_2) - (U_1 + P_1 V_1)$$
$$= (U_2 - U_1) + (P_2 V_2 - P_1 V_1)$$
$$= \Delta U + \Delta(PV)$$

$$\boxed{\Delta H = \Delta U + \Delta(PV)} \quad \ldots [5]$$

But $\Delta U = Q + W$; therefore $\qquad \Delta H = Q + W + \Delta(PV)$

Suppose the pressure of the system is kept constant and is the same as the pressure of the surroundings; furthermore, if the only type of work done is *PV*-work, then,

$$W = -P_{surr} \Delta V_{syst} = -P_{syst} \Delta V_{syst}$$

and

$$\Delta H = Q + W + P\Delta V$$
$$= Q - P\Delta V + P\Delta V$$
$$= Q$$

Thus, for a constant pressure process in which $P_{syst} = P_{surr}$:

$$\boxed{\Delta H_P = Q_P} \quad \begin{array}{l} \textit{Constant Pressure} \\ \textit{Process, Only PV-Work} \end{array} \quad \ldots [6]$$

Q_P is the *heat* for the process; we use the subscript P to remind ourselves that the process is a constant pressure process. Therefore, for a constant pressure process where $P_{syst} = P_{surr}$ and only *PV*-work is done against the surroundings, the change in enthalpy for the process is just the heat!

If the constant pressure process is just a heating or cooling process (no chemical reaction or phase changes in the system), then we know that the heat is just given by the familiar $Q_P = C_P \Delta T$, and therefore

*Const P, only PV-Work, Simple
Heating or Cooling Process*

$$\Delta H_P = Q_P = C_P\,\Delta T = nc_P\,\Delta T = m\bar{c}_P\,\Delta T \qquad \dots [7]$$

For constant pressure systems in which only solids or liquids are involved, usually $\Delta(PV) = P\Delta V$ is very small because such systems are more or less incompressible ($\Delta V \approx 0$), so that

$$\Delta H = \Delta U + \Delta(PV) \approx \Delta U$$

Therefore, for processes involving small volume changes,

$$\Delta H \approx \Delta U \qquad \text{\textit{Solid/Liquid Systems,}} \atop \text{\textit{Constant Pressure}} \qquad \dots [8]$$

On the other hand, for reactions in which *gases* are involved, ΔV may be very large. For example, the volume at *STP* of one mole of *liquid* water is only about 18 cm^3, while the volume at *STP* of one mole of water *vapor* is about 22 400 cm^3. Since the former volume is negligible when compared with the latter, we can say that $22\,400 + 18 \approx 22\,400$ without introducing any significant error. Therefore, for reactions involving gases, it is usually accurate to say

$$\Delta V_{reaction} \approx \Delta V_{gases} \qquad \dots [9]$$

For a real gas, $\Delta H = C_P\,\Delta T$ only when the gas is heated or cooled at *constant pressure*. However, when an *ideal* gas is heated or cooled, we still use $\Delta H = C_P\,\Delta T$, *even though the pressure may not be constant during the process*. This is a consequence of the fact that for an ideal gas the enthalpy is determined only by the *temperature*.[2] Similarly, for the same reason, for the heating or cooling of an *ideal* gas $\Delta U = C_V\,\Delta T$ *even though the volume may not be constant during the process*. For most real gases as well, to a good approximation, when we heat or cool a real gas we also can say that $\Delta H \approx C_P\,\Delta T$ and $\Delta U \approx C_V\,\Delta T$, even though the pressure and volume may change during the process.

Example 8-1

One mole of nitrogen gas is compressed from an initial state of 1.00 bar and 25°C to 10.0 bar and 100°C. Assuming the gas behaves as an ideal gas, what is ΔH for this process? For N_2 gas, $c_V = 20.81$ J mol^{-1} K^{-1}.

Solution

For one mole of an ideal gas, $\Delta H = c_P\,\Delta T$ and $c_P = c_V + R$;

therefore:
$$\Delta H = (c_V + R)\Delta T$$
$$= (20.81 + 8.314)(100 - 25)$$
$$= 2184 \text{ J} = 2.184 \text{ kJ.}$$

Ans: | 2.18 kJ |

---

[2] This will be proved in your thermodynamics course.

This problem also can be solved as follows:

$$\Delta H = \Delta U + \Delta(PV)$$
$$= \Delta U + P_2V_2 - P_1V_1$$
$$= \Delta U + nRT_2 - nRT_1$$
$$= \Delta U + nR\Delta T$$

Note how, for an ideal gas, ΔH depends only on temperature and not on pressure!

For one mole of an ideal gas, $\Delta U = c_V \Delta T$;

so that $\qquad\qquad\qquad\qquad \Delta U = (20.81)(100 - 25) = 1560.8 \text{ J}$

Therefore:
$$\Delta H = \Delta U + nR\Delta T$$
$$= 1560.8 + (1)(8.314)(100 - 25)$$
$$= 1560.8 + 623.55$$
$$= 2184 \text{ J} = 2.184 \text{ kJ} \quad \text{(Same answer as before.)}[3]$$

Example 8-2

One mole of gaseous A is injected at 25.0°C into a rigid reactor having a total constant volume of $V = 1.00$ L, and initially containing 2.00 moles of liquid B. The reactor is sealed and the following chemical reaction proceeds to completion, with the evolution of 285.0 kJ of heat:

$$A_{(g)} + 2B_{(liq)} \rightarrow 3C_{(s)} + 2D_{(g)}$$

At the completion of the reaction the temperature in the reactor is 500 K. The molar volumes of liquid B at 25.0°C and of solid C at 500 K are 24.0 mL and 14.2 mL, respectively. What is ΔH for this process? The vapor pressures of liquid B and solid C may be assumed to be negligible.

Solution

The volume V_B of liquid B initially present is

$$V_B = n_B v_B = (2 \text{ mol})(24.0 \text{ mL mol}^{-1}) = 48.0 \text{ mL}$$

Therefore the initial volume occupied by gas A is

$$V_A = V - V_B = 1000 - 48.0 = 952 \text{ mL} = 952 \times 10^{-6} \text{ m}^3$$

and the initial pressure in the reactor is

$$P_1 = P_A = \frac{n_A RT_1}{V_A} = \frac{(1.00)(8.314)(298.15)}{952 \times 10^{-6}} = 2.604 \times 10^6 \text{ Pa}$$

Since the reaction proceeds to completion, from the reaction stoichiometry 3.00 moles of solid C and 2.00 moles of gaseous D will be produced.

---

[3] Why? Because for an ideal gas (ig), $\Delta U^{ig} + nR\Delta T = nc_V \Delta T + nR\Delta T = n(c_V + R)\Delta T = nc_P \Delta T = \Delta H^{ig}$.

The final volume of gaseous D in the reactor will be

$$V_D = V - V_C = 1000 - 3 \times 14.2 = 957.4 \text{ mL} = 957.4 \times 10^{-6} \text{ m}^3$$

and the final pressure P_2 will be

$$P_2 = P_D = \frac{n_D R T_2}{V_D} = \frac{(2.00)(8.314)(500)}{957.4 \times 10^{-6}} = 8.684 \times 10^6 \text{ Pa}$$

Since the volume of the system is constant, $\Delta V = 0$ and therefore $W = W_{PV} = 0$.

So, $\qquad \Delta U = Q + W = Q = -285.0 \text{ kJ}$

and $\qquad \Delta H = \Delta U + \Delta(PV) = \Delta U + P_2 V_2 - P_1 V_1$

But $\qquad V_2 = V_1 = V = constant,$

therefore, $\qquad \Delta H = \Delta U + (P_2 - P_1)V$

$$= -285\ 000 + (8.684 \times 10^6 - 2.604 \times 10^6)(0.00100)$$

$$= -285\ 000 + 6084$$

$$= -278\ 916 \text{ J} = -278.9 \text{ kJ}$$

Ans: $\boxed{-278.9 \text{ kJ}}$

Exercise 8-1

One mole of a real gas initially at a pressure of 1.98 atm and contained in a flask having a volume of 8.27 L is suddenly expanded by opening a valve connecting the flask to a second flask, the latter having a volume of 7.64 L and containing a vacuum. During the process, 0.71 J of heat is absorbed by the gas from the surroundings. The final pressure of the gas is 1.033 atm. (a) What is ΔU for this process? (b) What is ΔH? ☺

8.3 THERMOCHEMISTRY

Thermochemistry focuses on the *heats* associated with chemical reactions. Using as a basis the formulas $\Delta U = Q + W$, $\Delta H = \Delta U + \Delta(PV)$, $\Delta U_V = Q_V$, and $\Delta H_P = Q_P$, it is possible to find ΔH for any process, including chemical reactions.

For now, we shall deal with the enthalpy changes that occur during *phase changes* taking place at constant pressure. Commonly encountered phase changes include:

$$liquid\ (liq) \rightarrow vapor\ (V)$$

$$solid\ (S) \rightarrow liquid\ (liq)$$

$$solid\ (S) \rightarrow vapor\ (V)$$

$$solid\ A \rightarrow solid\ B\ ^4$$

---

[4] A typical example of *solid A → solid B* would be the transition at some temperature of one crystalline structure to another, such as *monoclinic sulfur → rhombic sulfur*.

8.4 ΔH FOR FUSION (MELTING) AND FREEZING

6.01 kJ of heat is absorbed when one mole of ice at 0°C and one atm pressure melts to form one mole of liquid water at 0°C and one atm pressure. This heat is known as the *molar enthalpy change of fusion*, and is given the symbol Δh_{fus}:[5]

$$1 \ mol \ H_2O_{(s)} \ (State \ 1) \ \xrightarrow{\ 0°C, \ 1 \, atm \ } \ 1 \ mol \ H_2O_{(liq)} \ (State \ 2)$$

$$\Delta H_P = H_2 - H_1 = \Delta h_{fus, \ 273.15} = +6.01 \ kJ$$

Note that heat is *absorbed* when ice melts, so that the *sign* of Δh_{fus} is *positive*. Also, note that we use the lower case h to denote *molar* enthalpy. Because the enthalpy change at constant pressure is just the heat, we often call Δh_{fus} the molar *heat* of fusion. When 1 kg of ice melts the enthalpy change is

$$\left(\frac{1000 \ g}{18.015 \ g/mol} \right)\left(+6.01 \ \frac{kJ}{mol} \right) = +333.6 \ kJ$$

Thus, the *specific enthalpy change of fusion* is $\Delta \overline{h}_{fus} = +336.6 \ kJ \ kg^{-1}$, which also is often referred to as the *specific heat of fusion*.

The process of *freezing* (solidification) is just the opposite of melting (fusion):

$$1 \ mol \ H_2O_{(liq)} \ (State \ 1) \ \xrightarrow{\ 0°C, \ 1 \, atm \ } \ 1 \ mol \ H_2O_{(s)} \ (State \ 2)$$

Because enthalpy is a state function, its value is independent of path; hence the change in enthalpy is always just the value in State *2* minus the value in State *1*; thus, for the freezing of one mole of water,

$$\Delta H_P = H_2 - H_1 = \Delta h_{freezing, \ 273.15} = -6.01 \ kJ$$

The heat *absorbed* when ice melts, is *evolved* when water freezes; hence, when the direction of the process is reversed, ΔH_P has the same magnitude as before, but the *opposite sign*. This is true for the changes in all state functions: If ΔH (or ΔU, or ΔS, or ΔV, etc.) for $A \rightarrow B$ has a value of $+x$ joules, then ΔH (or ΔU, or ΔS, or ΔV, etc.) for $B \rightarrow A$ has a value of $-x$ joules. This is one of the fundamental properties of state functions that makes them so useful.

8.5 ΔH FOR VAPORIZATION AND CONDENSATION

The same ideas apply to the opposite processes of vaporization and condensation. Thus, for the vaporization of one mole of liquid water to water vapor at one atm pressure and 100°C (the normal boiling point):

$$1 \ mol \ H_2O_{(liq)} \ \xrightarrow{\ 100°C, \ 1 \, atm \ } 1 \ mol \ H_2O_{(v)} \quad \Delta h_{vap, \ 373.15} = +40.7 \ kJ \ mol^{-1}$$

and for

$$1 \ mol \ H_2O_{(v)} \ \xrightarrow{\ 100°C, \ 1 \, atm \ } 1 \ mol \ H_2O_{(liq)} \quad \Delta h_{cond, \ 373.15} = -40.7 \ kJ \ mol^{-1}$$

---

[5] The enthalpy change for a phase change is often called a *latent heat*; for example, the *latent heat of fusion*.

Note that heat is *absorbed* when a liquid is vaporized, but is *released* when a vapor condenses.

8.6 ΔH FOR SUBLIMATION

Usually when a solid is heated at constant pressure, it eventually melts and is converted to a liquid; if the liquid is further heated, it eventually vaporizes and is converted to a vapor. Sometimes, depending on the conditions (we will examine this later), a solid transforms *directly* to the vapor state, bypassing the liquid state altogether. This process is called *sublimation*. (The opposite process, in which a vapor changes directly to a solid, is called *vapor deposition*). Since H is a state function, it is independent of the path followed from State *1* to State *2*. Thus, referring to Fig. 1, we can see that

$$\boxed{\Delta h_{sub} = \Delta h_{fus} + \Delta h_{vap}} \quad \ldots [10]$$

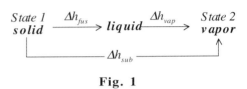

Fig. 1

For example, at 0°C and one atm pressure, the molar heat of fusion of ice is 6.01 kJ mol^{-1}, while the molar heat of vaporization is 45.07 kJ mol^{-1}. Thus the molar enthalpy of sublimation is $+6.01 + 45.07 = +51.08$ kJ mol^{-1}, and the molar enthalpy of vapor deposition is -51.08 kJ mol^{-1}.

Note from the above values that $\Delta h_{vap} > \Delta h_{fus}$. This can be understood if you refer to the potential energy diagram shown in Fig. 2. As a consequence of the fact that the intermolecular distance between the solid state and the liquid state is much smaller than the intermolecular distance between the liquid state and the vapor state, the potential energy difference between the solid state and the liquid state (Δh_{fus}) also must be smaller than the potential energy difference between the liquid state and the vapor state (Δh_{vap}).

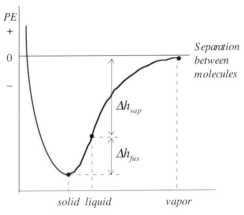

Fig. 2

Note that any value of ΔH must have an *algebraic sign* to indicate whether heat is absorbed or evolved. Thus, if we say the *enthalpy* change for a process is 100 kJ, this means that $\Delta H = +100$ kJ. This is not necessarily true, however, when we use the word "heat." For example, if we say that the molar *heat* of combustion of methane gas is 890 kJ mol^{-1}, we mean that

$$\Delta h_{combustion} = -890 \text{ kJ mol}^{-1}$$

In other words, you are supposed to know that a combustion is an exothermic process, and that the sign of ΔH is negative. Similarly, if it is stated that the molar *heat* of combustion of hydrogen gas to form liquid water is 286 kJ mol^{-1}, this means that for the reaction

$$H_{2(g)} + \frac{1}{2}O_{2(g)} \rightarrow H_2O_{(liq)} \qquad \Delta H = -286 \text{ kJ}$$

On the other hand, when we use the phrase "enthalpy change," we must always state the sign. Thus, although it is all right to say that the molar *heat* of combustion of H_2 is 286 kJ mol^{-1}, we must say that the molar *enthalpy*[6] of combustion of H_2 is −286 kJ mol^{-1}. Note also that, since the *combustion* process for hydrogen also can be viewed as the *formation* process for water, we also can say that the *heat of formation* of liquid water is 286 kJ mol^{-1}; again, you are expected to know that the formation of water from gaseous hydrogen and oxygen is a combustion process, and therefore is exothermic.

Hence, for liquid water, $\Delta h_{formation} = -286$ kJ mol^{-1}. Similarly, the heat of formation of CO_2 gas is 394 kJ mol^{-1}, which means that for this gas $\Delta h_{form} = -394$ kJ mol^{-1}; again, you are supposed to recognize that the formation of CO_2 from its elements is a combustion process, and therefore has a negative enthalpy change:

$$C_{(s)} + O_{2\,(g)} \rightarrow CO_{2\,(g)} \qquad \Delta H = -394 \text{ kJ}$$

8.7 IONIZATION ENTHALPIES

The ionization enthalpy, Δh_{ion}, is the energy required to remove electrons from atoms or ions in the *gaseous* state. Ionization enthalpies are always positive (endothermic), because energy must be pumped into an atom or molecule to dislodge an electron.

$Al_{(g)} \rightarrow Al^+_{(g)} + e^-_{(g)}$ $\Delta h_{ion} = +577$ kJ mol^{-1} (first ionization enthalpy for aluminum)

$Al^+_{(g)} \rightarrow Al^{2+}_{(g)} + e^-_{(g)}$ $\Delta h_{ion} = +1820$ kJ mol^{-1} (second ionization enthalpy for aluminum)

$Al^{2+}_{(g)} \rightarrow Al^{3+}_{(g)} + e^-_{(g)}$ $\Delta h_{ion} = +2740$ kJ mol^{-1} (third ionization enthalpy for aluminum)

The first electron is fairly easy to remove from the outer orbital, but the second and third are increasingly difficult to remove; therefore each successive ionization enthalpy is greater than the previous ionization enthalpy.

8.8 ELECTRON AFFINITIES

The electron *affinity*—also called the *electron gain enthalpy*—of a gaseous species is a measure of how strongly the species attracts an electron. The process involved is basically the opposite process to ionization.

8.9 BOND ENTHALPIES

The bond enthalpy—also called the bond dissociation energy—is the enthalpy input required to break open a chemical bond. Bond dissociation energies are always positive because an energy input is required to rip open a chemical bond. The more stable the bond, the greater the amount of energy that must be pumped into the bond to break it open. Table 1 lists some selected values: It should be noted that the enthalpy required to break open a given bond depends on the specific molecule in which the bond is found. If the specific bond dissociation energy for a given bond *AB* in a certain molecule is unknown, as an *approximation*, we can use a *mean bond enthalpy* (Table 2), which is a value determined by taking an average over a large number of compounds containing the bond *AB*. For example, the bond energy for the *C–H* bond in methane, ethane, benzene, and chloroform is 439, 423, 473, and 393 kJ mol^{-1}, respectively; while the mean *C–H* bond energy taken over many compounds containing this bond is 412 kJ mol^{-1}.

---

[6] Strictly speaking, the molar enthalpy *change* of combustion.

Table 1. *Bond Enthalpies for Selected Gases (kJ mol^{-1})*

Diatomic Molecules		Polyatomic Molecules							
Br–Br	192.81	Cl–CCl$_3$	305.9	H–CH$_2$Cl	419.0	HO–OH	213.0		
C–O	1076.50	H–CH	424.0	H–CHO	368.5	HO-CH$_3$	377.0		
Cl–Cl	242.58	H–CH$_3$	438.9	H–CH$_2$OH	401.8	H–SH	381.6		
F–F	158.78	H–C$_2$H$_5$	423.0	H–CN	527.6	H$_2$C=CH$_2$	728.3		
H–Cl	431.62	H–n-C$_3$H$_7$	423.3	H–NH$_2$	452.7	H$_3$C–CH$_3$	376.0		
H–H	435.99	H–i-C$_3$H$_7$	409.1	H–OCH$_3$	436.0	H$_2$N–NH$_2$	275.3		
H–D	439.43	H–n-C$_4$H$_9$	425.4	H–OC$_2$H$_5$	437.7	O=CO	532.2		
D–D	443.53	H–i-C$_4$H$_9$	425.2	H–OCOCH$_3$	442.7	O–ClO	247.0		
N–N	945.33	H–C$_6$H$_5$	473.1	H–OH	498.0	O–N$_2$	167.0		
N–O	630.57	HC≡CH	962.0	H–O$_2$H	369.0	O–NO	305.0		
O–O	498.36	H–CHCH$_2$	465.3	H–ONO$_2$	423.4	O–SO	552.0		

Table 2. *Mean Bond Enthalpies (kJ mol^{-1})*

	H	C	N	O	Cl	Br	I	F	S	P	Si
H	436										
C	412	348 (1)									
		612 (2)									
		838 (3)									
		518 (ar)									
N	388	305 (1)	163 (1)								
		613 (2)	409 (2)								
		890 (3)	945 (3)								
O	463	360 (1)	157	146 (1)							
		743 (2)		498 (2)							
Cl	432	338	200	203	243			254			
Br	366	276			219	193					
I	299	238			210	178	151				
F	565	484	270	185				159			
S	338	259			250	212		496	264		
P	322									201	
Si	318		374	466							226

(1) = single bond; (2) = double bond; (3) = triple bond; (ar) = aromatic bond

KEY POINTS FOR CHAPTER EIGHT

1. *PV*-**Work When** *P* **Varies During the Process**: If the external pressure is constant, the *PV*-work done on or by the system during the process is given by $W_{PV} = -P_{ext} \Delta V_{syst}$. If the external pressure changes during the process, then we must use $\delta W_{PV} = -P_{ext} dV_{syst}$, which indicates that P_{ext} can be considered constant only over the small volume change dV of the system. To obtain the *total PV*-work done during the process we must *integrate*:

$$W_{PV} = -\int_{V_1}^{V_2} P_{ext} dV_{syst}$$

2. The **Enthalpy** of a system is defined as $H \equiv U + PV$,

   where H is the enthalpy of the system in J, U is the internal energy of the system in J, P is the pressure of the system in Pa, and V is the volume of the system in m^3. Since U, P, and V all are state functions, it follows that enthalpy also is a state function.

3. **Constant Pressure Process**: The change in the enthalpy ΔH of the system for a process during which the pressure *stays constant throughout* the process—a "constant pressure" process—is given by

$$\Delta H = \Delta(U + PV) = \Delta U + P\Delta V$$

   where ΔU is the change in the internal energy of the system, P is the pressure of the system, and ΔV is the volume change of the system. If $P_{syst} = P_{surr}$ and the only type of work done is the *PV*-work in pushing back the surroundings, then, replacing ΔU with $Q + W$, and W with $-P\Delta V$ gives

$$\Delta H = \Delta U + P\Delta V = (Q - P\Delta V) + P\Delta V = Q$$

   That is, $$\Delta H_P = Q_P$$

   where ΔH_P is the change in the enthalpy of the system for the constant pressure process and Q_P is the heat transferred at constant pressure to or from the system. To evaluate the enthalpy at constant pressure all we have to do is measure the heat for the process!

4. For a simple heating or cooling process at constant pressure (no chemical reactions or phase changes), then

$$\Delta H_P = Q_P = C_P \Delta T = nc_P \Delta T = m\bar{c}_P \Delta T$$

   where C_P is the heat capacity at constant pressure of the substance being heated or cooled in J K^{-1}, n is the number of moles of the substance, c_P is its molar heat capacity at constant pressure in J mol^{-1} K^{-1}, m is its mass in kg, and $\bar{c}_P$ is its specific heat capacity at constant pressure in J kg^{-1} K^{-1}.

5. **For constant pressure systems in which only solids or liquids are involved**, usually $P\Delta V$ is very small because such systems are essentially incompressible ($\Delta V \approx 0$), so that

$$\Delta H_P = \Delta U + P\Delta V \approx \Delta U$$

6. For chemical reactions in which *gases* are involved (either as reactants or products), the contribution by the gases to the volume change for the process is usually much greater than

that for any solids or liquids involved, since the molar volumes of solids and liquids are much smaller than those for gases. (Solids and liquids have much higher densities than gases.) Accordingly, the volume change for the system usually can be assumed to be the same as the volume change for the gaseous species:

$$\Delta V_{reaction} \approx \Delta V_{gases}$$

7.   For heating or cooling an **ideal gas**, $\Delta H^{ig} = C_p \Delta T = nc_p \Delta T = m\bar{c}_p \Delta T$

That is, even when the pressure is *not* constant we still use c_p to evaluate ΔH^{ig}.

Similarly, for heating or cooling an **ideal gas**, $\Delta U^{ig} = C_V \Delta T = nc_V \Delta T = m\bar{c}_V \Delta T$

Even when the volume is *not* constant, we still use c_V to evaluate ΔU.

To calculate ΔU and ΔH for ideal gases we can use the heat capacities at constant pressure and at constant volume even when the pressure and volume change because the internal energy U and the enthalpy H for an *ideal* gas depend only on the *temperature* of the gas, and not on its pressure or volume. For real gases, this still is approximately true.

8.   For an **ideal gas**, $\Delta H^{ig} = \Delta U + \Delta(PV) = \Delta U + nR\Delta T$

(Again, ΔH^{ig} depends only on the *temperature*, not on the pressure or volume.)

9.   For constant pressure **fusion** (melting): $\Delta H_{fus} = n\Delta h_{fus} = m\Delta \bar{h}_{fus}$

For constant pressure **freezing**: $\Delta H_{freezing} = n\Delta h_{freezing} = m\Delta \bar{h}_{freezing}$

For constant pressure **vaporization**: $\Delta H_{vap} = n\Delta h_{vap} = m\Delta \bar{h}_{vap}$

For constant pressure **condensation**: $\Delta H_{cond} = n\Delta h_{cond} = m\Delta \bar{h}_{cond}$

where n and m are the number of moles and kg of the substance, Δh_{fus}, $\Delta h_{freezing}$, Δh_{vap}, and Δh_{cond} are its *molar* enthalpy changes for fusion, freezing, vaporization, and condensation, respectively, and $\Delta \bar{h}_{fus}$, $\Delta \bar{h}_{freezing}$, $\Delta \bar{h}_{vap}$, and $\Delta \bar{h}_{cond}$ are its *specific* enthalpy changes for fusion, freezing, vaporization, and condensation.

Note that $\Delta h_{fus} = -\Delta h_{freezing}$, $\Delta \bar{h}_{fus} = -\Delta \bar{h}_{freezing}$, $\Delta h_{vap} = -\Delta h_{cond}$, and $\Delta \bar{h}_{vap} = -\Delta \bar{h}_{cond}$.

10.   For **sublimation**: $\Delta H_{sub} = \Delta H_{fus} + \Delta H_{vap}$

11.   The **ionization enthalpy**, Δh_{ion}, is the energy required to remove electrons from atoms or ions in the *gaseous* state. Ionization enthalpies are always *positive* (endothermic), because energy must be pumped into an atom or molecule to dislodge an electron.

The **electron affinity**—electron gain enthalpy—of a gaseous species is a measure of how strongly the species attracts an electron. The process involved is basically the opposite of ionization

12.   The **bond enthalpy**—bond dissociation energy—is the enthalpy input required to break open a chemical bond. Bond dissociation energies are always positive because an energy

input is required to rip open a chemical bond. The bond enthalpy for the same type of bond can vary slightly, depending on the actual compound containing the bond.

The **mean bond enthalpy** of a given type of bond is an approximate value determined by taking an average over a large number of compounds containing the same type of bond. Mean bond enthalpies are useful when the actual value of the bond enthalpy in a particular compound is unknown.

PROBLEMS

1. When 3.00 moles of $O_{2(g)}$ is heated at a constant pressure of 3.25 atm, its temperature increases from 260 K to 285 K. Given that the molar heat capacity of oxygen at constant pressure is 29.4 J K^{-1} mol^{-1}, calculate (a) Q, (b) ΔH, and (c) ΔU. ☺

2. Refrigerators make use of the heat absorption required to vaporize a volatile liquid. A fluorocarbon liquid being investigated to replace a chlorofluorocarbon has a molar heat of vaporization of 26.0 kJ mol^{-1} at 250 K and 750 Torr. Calculate (a) Q, (b) W, (c) ΔH, and (d) ΔU when 1.50 mol is vaporized at this temperature and pressure. ☹

3. A rigid steel tank contains 100 moles of nitrogen gas at 320 K and 1.00 bar pressure. The tank is heated until the pressure is 1.50 bar. Assuming the gas behaves ideally, calculate: (a) ΔH_{syst}, and (b) Q_{syst}. For nitrogen gas, $c_V = 20.811$ J mol^{-1} K^{-1}. ☹

4. An insulated piston/cylinder arrangement contains a 1.000-gram layer of ice at 0°C on the bottom of the cylinder. (The insulation prevents any heat input to the system from escaping.) As indicated, the piston lies on top of the ice, with no vapor space initially present. The inside cross-sectional area of the cylinder is 100 cm^2, and the piston may be considered to have no mass and be frictionless. The pressure of the surroundings is one atmosphere. Sufficient heat is passed into the system to increase the temperature of the contents of the cylinder to 200°C. Taking the water inside the cylinder as the system, and neglecting the heat capacity of the piston and cylinder, for the above process calculate:

    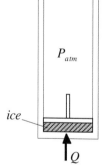

    (a) W_{syst} (b) ΔH_{syst} (c) Q_{syst} (d) ΔU_{syst}.

    H_2O = 18.02 g mol^{-1}. Heat capacities (in J mol^{-1} K^{-1}) are as follows: c_P [water] =75.3; c_P [steam] = 30.54 + 0.0103T where T is the temperature in kelvin. For ice and water, the molar heats of fusion and vaporization are, respectively, 6.01 kJ mol^{-1} and 40.66 kJ mol^{-1}. The density of ice is 0.917 g cm^{-3}. ☹*

5. Calculate the change in enthalpy for the following process carried out at a constant pressure of one atmosphere:

$$\begin{pmatrix} 1.00 \ mole \ liquid \\ water \ at \ 25°C \end{pmatrix} \rightarrow \begin{pmatrix} 1.00 \ mole \ steam \\ at \ 726.85°C \end{pmatrix}$$

H_2O = 18.02 g mol^{-1}. For liquid water, $\bar{c}_p$ = 4.18 J g^{-1} K^{-1}. The heat capacity of steam at one atm pressure is c_p = 30.54 + 0.0103T where c_p is in J mol^{-1} K^{-1} and T is the temperature in kelvin. The heat of condensation of saturated water vapor at 100°C is 2.257 kJ g^{-1}. ☺*

6. The enthalpy of vaporization of water at 200°C is 833 BTU lb^{-1}. What is the value expressed in kJ mol^{-1}? One BTU (British Thermal Unit) is the amount of heat required to raise the temperature of 1 lb of liquid water by 1°F.

    Data: $\bar{c}_p$ [H_2O] = 4.184 J g^{-1} K^{-1}; 1 kg = 2.20462 lb; 1 K = 1.8 F°; H_2O = 18.015 g mol^{-1}. ☺

7. Calculate the time required to heat the water in a forty gallon electric hot water heater from 22.0°C to 70.0°C using a 3000 watt electric immersion heater when (a) the tank is perfectly insulated, and (b) some heat escapes to the surroundings through the walls of the tank.

    Data: 1 gallon = 4.55 L; $\bar{c}_p$ [water] = 4184 J kg^{-1} K^{-1}; density of water = 1.00 kg L^{-1}; room temperature = 22.0°C; area of tank walls = 2.27 m^2; k = coefficient of thermal resistance of the tank walls = 0.180 s m^2 K J^{-1}. The rate of heat loss from the hot water tank is given by the following equation:

$$\dot{q}_{loss} = \frac{dq_{out}}{dt} = \frac{A}{k}\left(T - T_{air}\right)$$

where $\dot{q}_{loss}$ is the rate of heat loss from the tank in J s^{-1}, A is the area of the tank through which heat flows, k is the coefficient of thermal resistance of the tank walls, T is the temperature of the water inside the tank (assumed to be uniform), and T_{air} is the temperature of the air surrounding the tank. The following integral will be useful:

$$\int \frac{dT}{a - bT} = -\frac{1}{b}ln(a - bT) \quad ☹*$$

8. Using mean bond enthalpy values from Table 1, estimate the standard molar enthalpy of combustion of α-D-glucose at 25°C according to

$$C_6H_{12}O_{6\,(s)} + 6O_{2(g)} \rightarrow 6CO_{2(g)} + H_2O_{(liq)}$$

    Remember that heat is absorbed in breaking bonds and evolved on making bonds. At 25°C the standard enthalpy of sublimation of glucose is 144 kJ mol^{-1}, and the enthalpy of vaporization of water is 44.01 kJ mol^{-1}. The enthalpies of complete atomization of gaseous CO_2 and gaseous H_2O are 1609 kJ mol^{-1} and 920 kJ mol^{-1}, respectively. ☺

Structure of α-D-Glucose

9. Using bond enthalpy values from Tables 1 and 2, estimate the standard molar enthalpy of combustion of ethanol at 25°C according to

$$C_2H_5OH_{(liq)} + 3O_{2(g)} \rightarrow 2CO_{2(g)} + 3H_2O_{(liq)}$$

At 25°C the standard enthalpies of vaporization of liquid water and of liquid ethanol are 44.01 kJ mol^{-1} and 42.59 kJ mol^{-1}, respectively. The enthalpies of complete atomization of gaseous CO_2 and gaseous H_2O are 1609 kJ mol^{-1} and 920 kJ mol^{-1}, respectively. ☺

10. The latent heat of fusion of ice is 333.5 J g^{-1} at 0°C and one atm pressure.

    (a) What is the contribution of the work term on melting one mole of ice under these conditions?

    (b) Calculate ΔH and ΔU for this process.

        Data: density of ice at 0°C = 0.917 g cm^{-3}; density of liquid water at 0°C = 1.000 g cm^{-3}

           Relative molar masses: H = 1.0079; O = 15.9994 ☺

11. A house is heated by an electric resistance heater controlled by a thermostat that maintains a constant temperature in the house. The air is well circulated to prevent temperature differences in the house. The heater draws an average current of 25.0 amperes at 110 volts. The cost of electricity is 8.00 cents per kWh. The owner of the house, having discovered that her costs for operating a freezer are very low, offers to provide a non-profit fish freezing service for her friends. The only charge for the service is to be the *change* in the cost of her electricity bill. The average heat capacity of a fish and its heat of freezing may be taken as a water equivalent of 0.75 kg water/kg fish. The fish come into the house at + 20°C unfrozen, and leave the house at −20°C frozen.

How much should the owner of the house charge her friends per 100 kg of fish frozen?

Data: Heat capacity of liquid water = 4.18 J g^{-1} K^{-1}; Heat capacity of ice = 1.88 J g^{-1} K^{-1};
       Heat of fusion for ice = 335 J g^{-1} ☺

12. 1000 moles of carbon monoxide at 2.758 MPa and 700K is subjected to the following series of steps:

        (1) expansion isothermally to 0.552 MPa, followed by

        (2) cooling at constant volume to 437.5K, followed by

        (3) cooling at constant pressure to 350K, followed by

        (4) adiabatic compression to 2.758 Mpa and 631.365 K, followed by

        (5) heating at constant pressure to 700K.

    (a) For each step of the above process, calculate:

       (i) W_{rev} (ii) Q_{rev} (iii) ΔH (iv) ΔU

    (b) Calculate $\sum W_{rev}$, $\sum Q_{rev}$, $\sum \Delta H$, and $\sum \Delta U$ for the complete cycle.

    (c) Sketch the cycle on a *PV* diagram.

        Data: c_P for carbon monoxide can be taken as constant at 29.3 J mol^{-1} K^{-1}. ☹

13. Ten moles of methane (CH_4) initially at one bar pressure and 300 K is heated at *constant pressure* to 1000 K. Assuming methane behaves as an ideal gas, calculate Q, W, ΔU, and ΔH for this process. The heat capacity of methane may be taken as $c_p = 23.64 + 0.0479T$, where c_p is in J mol^{-1} K^{-1} and T is in kelvin. ☺*

14. Ten moles of methane (CH_4) initially at one bar pressure and 300 K is heated at *constant volume* to 1000 K. Assuming methane behaves as an ideal gas, calculate Q, W, ΔU, and ΔH for this process. The heat capacity of methane may be taken as $c_p = 23.64 + 0.0479T$, where c_p is in J mol^{-1} K^{-1} and T is in kelvin. ☺*

15. One mole of nitrogen (N_2) is cooled from an initial temperature and pressure of 700 K and 10.0 bar to a final temperature of 300 K. The heat capacity of nitrogen may be taken as $c_p = 28.58 + 0.00377T$, where c_p is in J mol^{-1} K^{-1} and T is in kelvin. Assuming nitrogen behaves as an ideal gas, calculate Q, W, ΔU, and ΔH for this process when it is carried out

   (a) at constant pressure, and

   (b) at constant volume.

   Compare the values obtained for the two cases. ☹*

16. A 350 g metal bar initially at 1000 K is removed from a furnace and quenched by quick immersion into a well-insulated closed tank containing 10.00 kg of water initially at 300 K. The heat capacities of the metal and the water are constant at 0.45 J g^{-1} K^{-1} and 4.18 J g^{-1} K^{-1}, respectively. Neglecting any heat loss from the tank or water lost by vaporization, what is the final equilibrium temperature in the tank? ☺

17. A closed system undergoes a cycle consisting of two processes. During the first process, which starts at an initial system pressure of 1.00 atm and system temperature of 300 K, 60 kJ of heat is transferred from the system while 25 kJ of work is done on the system. At the end of the first process, the system pressure is 0.50 atm. During the second process, 15 kJ of work is done by the system. The temperature and pressure of the surroundings are constant throughout at 300 K and one atm, respectively.

   (a) Determine Q_{syst} for the second process.

   (b) Determine Q_{syst} for the cycle.

   (c) Determine W_{syst} for the cycle.

   (d) Determine ΔU_{syst} for the cycle.

   (e) Determine ΔH_{syst} for the cycle. ☺

18. You are given three identical 2-liter Dewar flasks (vacuum flasks), A, B, and C, and an empty, smaller 600 mL flask D, which has thermally conducting walls. Flask D easily can be placed inside a Dewar flask. 1000 g of hot water at 80.0°C is poured into flask A, and 1000 g of cold water at 20.0°C is poured into flask B. Now, using all four containers, and without

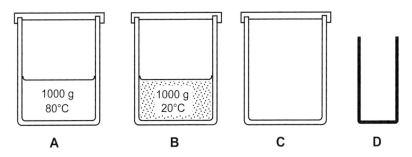

A   **B**   **C**   **D**

mixing the hot water with the cold water, is it possible to heat the cold water with the aid of the hot water so that the final temperature of the cold water will be higher than the final temperature of the hot water? No heat exchange with the surrounding air is permitted. The specific heat capacity of water can be taken as constant at 4.184 J g^{-1} K^{-1}. 🔬

19. One mole of nitrogen gas at 250 K occupies a volume of 10.0 L in the frictionless piston/cylinder arrangement shown. The mass of the piston may be assumed to be negligible. The cross-sectional area of the piston face is 300 cm², and the pressure of the external atmosphere is 100 kPa. When a 500 kg weight is placed on the piston, the piston lowers, compressing the gas. The process may be assumed to be isothermal, and nitrogen may be assumed to behave as an ideal gas. The acceleration due to gravity is 9.80 m s^{-2}.

    (a) How much work is done on the gas?
    (b) What is the minimum work required to carry out this compression?
    (c) What is ΔU for the gas compression of part (a)?
    (d) What is Q for the gas compression of part (a)?
    (e) What is ΔH for the gas? Confirm your answer with a numerical calculation. ☹

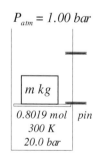

$P_{atm} = 100\,kPa$

restraining pin

N_2 gas
1.0 mol
250 K
10.0 L

20. Consider one mole of ideal gas:

    (a) Knowing that $dH^{ig} = nc_p\,dT$ and $dU^{ig} = nc_v\,dT$, show that $c_p - c_v = R$.
    (b) When expanded isothermally at temperature T from v_1 to v_2, show that

    (i) $-W = RT\ln(v_2/v_1) = q$, when the expansion is carried out reversibly, and

    (ii) $-W = \left[RT/v_2\right](v_2 - v_1)$ when the external pressure is set at the value corresponding to v_2. ☹

21. 0.8019 moles of an ideal gas is contained in a piston/cylinder assembly having a weightless, frictionless piston of internal cross-sectional area 0.01656 m². The gas is originally at 300 K and 20.0 bar pressure. The pressure of the surrounding atmosphere is 1.00 bar. A weight of mass m kilograms is placed on top of the piston. The lower retaining pin is removed and the gas expands isothermally at 300 K, lifting the weight, until the final volume of the gas is 2.00 L. What is the mass m of the weight in kilograms if the total work done by the gas is exactly 50% of the maximum reversible work that could, in the limit, be delivered by the gas in passing from the same initial to final state? The acceleration due to gravity is 9.7875 m s^{-2}. ☹

$P_{atm} = 1.00\,bar$

$m\,kg$

0.8019 mol   pin
300 K
20.0 bar

22. Assuming ammonia (NH_3) behaves as an ideal gas, calculate the change in internal energy ΔU when 10.0 moles of ammonia vapor is heated at a constant pressure of 0.100 bar from 25.0°C to 75.0°C. The molar heat capacity at constant pressure for ammonia vapor is $c_p = 29.75 + 0.0251T$, where c_p is in J mol^{-1} K^{-1} and T is in kelvin. ☹

THERMODYNAMICS (IV)

9.1 THE STANDARD STATE FOR CHEMICAL REACTIONS

The enthalpy change ΔH for any chemical reaction $aA + bB \rightarrow cC + dD$ depends on the temperature, the pressure, and the purity of the reactants. For convenience, thermodynamic data usually are reported under a standard set of conditions called the *Standard State*. The standard state at any given temperature T is taken as the pure substance at one bar pressure[1] at the temperature T. Thus, ΔH_{298}^{o} is the standard enthalpy change at one bar pressure and 25°C. The superscript " ° " indicates that the pressure is at *one bar*. Although most thermodynamic data are reported at 25°C, strictly speaking, the standard state is not confined to this temperature. Therefore, ΔH_{500}^{o} would be the standard enthalpy change at one bar and 500 K.

9.2 HESS'S LAW OF CONSTANT HEAT SUMMATION

Hess's Law states that the standard enthalpy change of a chemical reaction is the sum of the standard enthalpy changes of the individual component reactions into which the overall reaction may be divided. Thus for the consecutive processes

$$A \rightarrow B \rightarrow C \rightarrow D$$

shown in Fig. 1, Hess's Law states that

$$\Delta H_4 = \Delta H_1 + H_2 + \Delta H_3$$

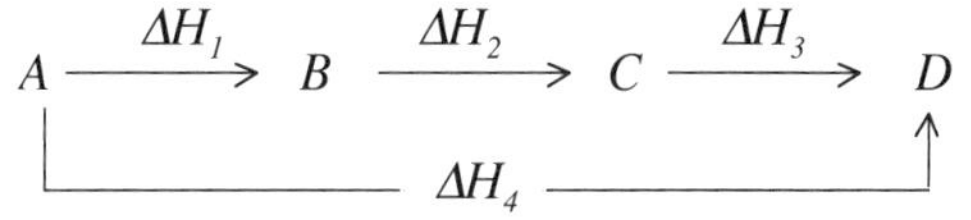

$$A \xrightarrow{\ \Delta H_1\ } B \xrightarrow{\ \Delta H_2\ } C \xrightarrow{\ \Delta H_3\ } D$$

with ΔH_4 connecting A to D.

Fig. 1

Strictly speaking, Hess's Law actually deals with the sum of *enthalpy* changes, and should therefore be called *Hess's Law of Constant Enthalpy Summation*; but, since ΔH at constant pressure is just the heat, it is usually called the Law of Constant Heat Summation. You should note that Hess's Law is valid only if the only type of work involved is *PV*-work.[2]

---

[1] The older literature uses one *atm* as the standard pressure. However, one atm is (exactly, by definition) equal to 101 325 Pa or 1.01325 bar, which is not a very convenient number in SI units. Accordingly, the trend is towards the use of one *bar* as the standard pressure. Fortunately, both pressures give almost identical results for the purpose of calculation.

[2] It can be shown that for a process taking place in a closed system (or in a steady state flow system) the enthalpy change at constant pressure is given by $\Delta H_P = Q + W^{\circ}$, where W° is any work done by (or on) the system *other than* *PV*-work. For example, a battery delivers *electrical* work.

$$\boxed{\Delta H = \sum \Delta H_i} \quad \textit{Hess's Law} \qquad \ldots [1]$$

9.3 STANDARD ENTHALPIES OF FORMATION

Any chemical reaction can be broken down into two hypothetical steps:

      (1) decompose the reactants into their constituent *elements*, and then

      (2) reassemble the elements into *products*.

The *standard enthalpy of formation* of a substance, Δh_f^o, is *the standard enthalpy change per mole for its formation from its constituent elements in their reference states*. The **reference state** of an element is its *most stable form under the prevailing conditions, and at one bar pressure*.

For example, consider the gas reaction $\quad CO + \frac{1}{2}O_2 \rightarrow CO_2$. To obtain the *standard reaction enthalpy change* ΔH_R^o for this reaction, we first write the formation reaction for each of the substances from its elements:

Thus, for CO: $\qquad C_{(s)} + \frac{1}{2}O_{2\,(g)} = CO_{(g)} \quad \ldots [a] \qquad\qquad \Delta H_a^o = \Delta h_f^o[CO]$

And for CO_2: $\qquad C_{(s)} + O_{2\,(g)} = CO_{2\,(g)} \quad \ldots [b] \qquad\qquad \Delta H_b^o = \Delta h_f^o[CO_2]$

Note that there is no enthalpy of formation reaction for O_2 because it already *is* an element, and therefore there is no enthalpy change involved in forming it from its elements. Thus, *the enthalpies of formation are zero for all elements in their most stable states*. Since enthalpy is a state function, if reaction [a] proceeds in the *opposite* direction, then the ΔH has the same magnitude as, but the opposite sign from, the ΔH in the forward direction. Accordingly, reversing reaction [a] gives

$$CO_{(g)} = C_{(s)} + \frac{1}{2}O_{2\,(g)} \quad \ldots -[a] \qquad\qquad -\Delta h_f^o[CO]$$

As before: $\qquad C_{(s)} + O_{2\,(g)} = CO_{2\,(g)} \qquad \ldots [b] \qquad\qquad +\Delta h_f^o[CO_2]$

Adding $-[a] + [b]$:

$$CO_{(g)} + C_{(s)} + O_{2\,(g)} = C_{(s)} + \frac{1}{2}O_{2\,(g)} + CO_{2\,(g)} \qquad -\Delta h_f^o[CO] + \Delta h_f^o[CO_2]$$

Gathering terms: $\qquad CO_{(g)} + O_{2\,(g)} - \frac{1}{2}O_{2\,(g)} = CO_{2\,(g)}$

That is: $\qquad\qquad CO_{(g)} + \frac{1}{2}O_{2\,(g)} = CO_{2\,(g)} \qquad \Delta H_R^o = \Delta h_f^o[CO_2] - \Delta h_f^o[CO]$

Thus, the reactions can be added and subtracted just as if they were algebraic equations!

In general, the standard enthalpy change ΔH_R^o for the reaction $\quad aA + bB \rightarrow cC + dD$ is given by

$$\Delta H_R^o = c\Delta h_{f(C)}^o + d\Delta h_{f(D)}^o - a\Delta h_{f(A)}^o - b\Delta h_{f(B)}^o$$

That is, $\qquad\qquad \boxed{\Delta H_R^o = \sum_{Products} n_i\,\Delta h_{f(i)}^o - \sum_{Reactants} n_i\,\Delta h_{f(i)}^o} \qquad\qquad \ldots [2]$

Example 9-1

Liquid ethylcyclohexane (C_8H_{16}) can be formed from either ethylbenzene (C_8H_{10}) or from styrene (C_8H_8) according to the following reactions, respectively:

$$C_8H_{10(liq)} + 3H_{2(g)} \rightarrow C_8H_{16(liq)} \qquad \Delta H^o_{298.15} = -199.7 \text{ kJ}$$

$$C_8H_{8(liq)} + 4H_{2(g)} \rightarrow C_8H_{16(liq)} \qquad \Delta H^o_{298.15} = -315.6 \text{ kJ}$$

Given that at 298.15 K and one bar pressure the heat of combustion of C_8H_{16} to water vapor and CO_2 is 4870.5 kJ mol^{-1}, and that the heats of formation of water vapor and CO_2 are 241.82 and 393.51 kJ mol^{-1}, respectively, calculate:

(a) The enthalpy of hydrogenation of styrene to ethylbenzene.

(b) The enthalpy of formation of ethylbenzene.

Solution

The first step is to write down the reactions corresponding to the given data:

$$C_8H_{10} + 3H_2 \rightarrow C_8H_{16} \qquad \ldots \text{[a]} \qquad \Delta H^o_a = -199.7 \text{ kJ}$$

$$C_8H_8 + 4H_2 \rightarrow C_8H_{16} \qquad \ldots \text{[b]} \qquad \Delta H^o_b = -315.6 \text{ kJ}$$

$$C_8H_{16} + 12O_2 \rightarrow 8CO_2 + 8H_2O_{(g)} \quad \ldots \text{[c]} \qquad \Delta H^o_c = -4870.5 \text{ kJ}$$

$$H_2 + \tfrac{1}{2}O_2 \rightarrow H_2O_{(g)} \qquad \ldots \text{[d]} \qquad \Delta H^o_d = -241.82 \text{ kJ}$$

$$C + O_2 \rightarrow CO_2 \qquad \ldots \text{[e]} \qquad \Delta H^o_e = -393.51 \text{ kJ}$$

(Note the negative signs on the ΔH^o's for reactions [c], [d], and [e], which are all combustion reactions, and therefore evolve heat.)

(a) We want to find ΔH^o_R for $\quad \underset{styrene}{C_8H_8} + H_2 \rightarrow \underset{ethylbenzene}{C_8H_{10}}$

We do this by manipulating the above equations until we obtain the combination giving us the desired reaction. Thus, we want one C_8H_8 on the left and one C_8H_{10} on the right; therefore:

$$[b]: \qquad C_8H_8 + 4H_2 \rightarrow C_8H_{16} \qquad \Delta H^o_b = -315.6 \text{ kJ}$$

$$-[a]: \qquad C_8H_{16} \rightarrow C_8H_{10} + 3H_2 \qquad -\Delta H^o_a = +199.7 \text{ kJ}$$

$$\text{Adding:} \quad C_8H_8 + H_2 \rightarrow C_8H_{10} \ \ldots \text{[f]} \qquad \Delta H^o_f = \Delta H^o_b - \Delta H^o_a$$

$$\text{[This is the desired reaction.]} \qquad = -315.6 + 199.7 = -115.9 \text{ kJ}$$

Ans: $\boxed{-115.9 \text{ kJ mol}^{-1}}$

(b) Now we want to find ΔH_R^o for $8C + 5H_2 \rightarrow$ $\underset{\text{ethylbenzene}}{C_8H_{10}}$

Proceed in the same way. First, we want 8 carbons on the left and one C_8H_{10} on the right; therefore use Eqns [e] and [f]:

$8 \times$[e]: $8C + 8O_2 \rightarrow 8CO_2$ $8 \Delta H_e^o = 8(-393.51) = -3148.08$ kJ

[f]: $C_8H_8 + H_2 \rightarrow C_8H_{10}$ $\Delta H_f^o = -115.9$ kJ

Now we have to get rid of the unwanted C_8H_8 on the left; therefore use Eqn [b]:

–[b]: $C_8H_{16} \rightarrow C_8H_8 + 4H_2$ $-\Delta H_b^o = -(-315.6) = +315.6$ kJ

Now make use of Eqn [c] to eliminate the unwanted C_8H_{16} on the left:

–[c]: $8CO_2 + 8H_2O_{(g)} \rightarrow C_8H_{16} + 12O_2$ $-\Delta H_c^o = -(-4870.5) = +4870.5$ kJ

Finally, we have to add four O_2's to the left to get 12 O_2's on each side so we can eliminate the unwanted O_2's:

$8 \times$[d]: $8H_2 + 4O_2 \rightarrow 8H_2O_{(g)}$ $8 \Delta H_d^o = 8(-241.82) = -1934.56$ kJ

Adding: $8C + 5H_2 \rightarrow C_8H_{10}$ $\Delta H = 8 \Delta H_e^o + \Delta H_f^o - \Delta H_b^o - \Delta H_c^o + 8 \Delta H_d^o$

$$= -3148.08 - 115.9 + 315.6 + 4870.5 - 1934.56$$

$$= -12.44 \text{ kJ}$$

Ans: $\boxed{-12.4 \text{ kJ mol}^{-1}}$

Exercise 9-1

Calculate the enthalpy of fusion for LiCl at 883 K, given that the enthalpy of formation at 883 K of liquid LiCl is -92.347 kcal mol^{-1} and that the enthalpy of formation at 883 K of solid LiCl is -97.105 kcal mol^{-1}. One calorie = 4.184 J. ☺

Exercise 9-2

The standard heats of combustion at 25°C (to liquid water) of propylene (C_3H_6) and of hydrogen are 2058.44 kJ mol^{-1} and 141.78 kJ g^{-1}, respectively. The standard heat of formation of carbon dioxide at the same temperature is 393.5 kJ mol^{-1}.

(a) Calculate the standard molar enthalpy of formation at 25°C of propylene.

(b) Is the formation reaction endothermic or exothermic? ☺

Example 9-2

Exactly 1.00 mole of ethane is burned such that 80.0% of the carbon burns to CO_2 and the rest to CO. Determine the enthalpy of reaction, ΔH_R^o, for this process at 25°C and one bar pressure if liquid water is formed as product.

Solution

(Standard enthalpies of formation are obtained from the Appendices.)

Complete combustion (*cc*): $C_2H_6 + \frac{7}{2}O_2 \rightarrow 2CO_2 + 3H_2O$

$$\Delta H_{cc}^o = 3(-285.83) + 2(-393.51) - (-84.68) - \frac{7}{2}(0) = -1559.83 \text{ kJ}$$

Incomplete combustion (*ic*): $C_2H_6 + \frac{5}{2}O_2 \rightarrow 2CO + 3H_2O$

$$\Delta H_{ic}^o = 2(-110.53) + 3(-285.83) - (-84.68) - \frac{5}{2}(0) = -993.87 \text{ kJ}$$

80% of the enthalpy is released by complete combustion, and 20% by incomplete combustion:

$$\Delta H_{total}^o = 0.80 \times \Delta H_{cc}^o + 0.20 \times \Delta H_{ic}^o$$

$$= (0.80)(-1559.83) + (0.20)(-993.87)$$

$$= -1446.64 \text{ kJ mol}^{-1} = -1.446 \text{ MJ mol}^{-1}$$

Ans: $\boxed{-1.45 \text{ MJ mol}^{-1}}$

Exercise 9-3

Calculate the enthalpy change at 25°C and one bar pressure for the combustion of one mole of liquid ethanol (C_2H_5OH) to produce carbon dioxide and water vapor. ☺

9.4 VARIATION OF ΔH WITH TEMPERATURE AT CONSTANT PRESSURE

If ΔH_R is known at one temperature T_1, we can calculate it at any other temperature T_2 as follows:

Because H is a state function, we can take two different paths from State *1* to State *2*, as indicated in Fig. 2.

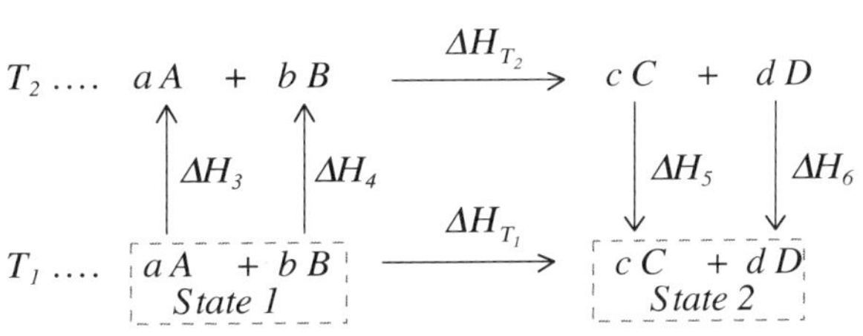

Fig. 2

From Hess's Law:

$$\Delta H_{T_I} = \Delta H_3 + \Delta H_4 + \Delta H_{T_2} + \Delta H_5 + \Delta H_6 \qquad \ldots [3]$$

It is recognized that the processes having enthalpy changes ΔH_3 and ΔH_4 are just ordinary *heating* processes at constant pressure in which the reactants (*a* moles of *A* and *b* moles of *B*) are heated from T_1 to T_2. Similarly, ΔH_5 and ΔH_6 are just the enthalpy changes for simple constant pressure *cooling* processes in which the products (*c* moles of *C* and *d* moles of *D*) are cooled from T_2 to T_1. These four processes are not chemical reactions—no chemical bonds are broken—and therefore ΔH_P for each of these processes is given by $\Delta H_P = C_P \Delta T = n c_P \Delta T$. Thus, Eqn [3] becomes:

$$\Delta H_{T_I} = a c_{P(A)}(T_2 - T_1) + b c_{P(B)}(T_2 - T_1) + \Delta H_{T_2} + c c_{P(c)}(T_1 - T_2) + d c_{P(D)}(T_1 - T_2)$$

$$= a c_{P(A)}(T_2 - T_1) + b c_{P(B)}(T_2 - T_1) + \Delta H_{T_2} - c c_{P(c)}(T_2 - T_1) - d c_{P(D)}(T_2 - T_1)$$

$$= a c_{P(A)} \Delta T + b c_{P(B)} \Delta T + \Delta H_{T_2} - c c_{P(c)} \Delta T - d c_{P(D)} \Delta T$$

Note in the above derivation that we have assumed the heat capacity of each component is not a function of temperature over the temperature range $T_1 \leftrightarrow T_2$. If the heat capacities *are* dependent on temperature, then integrations will have to be made for each heating and cooling term.

Rearranging:

$$\Delta H_{T_2} = \Delta H_{T_I} + \left\{ c c_{P(C)} + d c_{P(D)} - a c_{P(A)} - d c_{P(d)} \right\} \Delta T$$

Therefore,

$$\boxed{\Delta H_{T_2} = \Delta H_{T_I} + \Delta C_P \cdot \Delta T} \qquad \begin{array}{l} \textit{Constant Pressure,} \\ \textit{Constant Heat} \\ \textit{Capacities} \end{array} \qquad \ldots [4]$$

where

$$\boxed{\Delta C_P \equiv \sum_{Products} n_i (c_P)_i - \sum_{Reactants} n_i (c_P)_i} \qquad \ldots [5]$$

Example 9-3

Hydrogen gas for use in fuel cells can be made by reacting carbon monoxide gas with steam in the presence of a catalyst. One mole of carbon monoxide reacts stoichiometrically with water vapor to form carbon dioxide and hydrogen. How much heat is evolved by this reaction at 525.0°C and one bar pressure? The heat capacities provided are average values over the temperature range of 25°C to 525°C.

Enthalpies of Formation		$CO_{(g)}$	$H_2O_{(v)}$	$CO_{2(g)}$	$H_{2(g)}$
$\Delta h^o_{f(298.15)}$	(kJ mol^{-1})	−110.525	−241.818	−393.509	0
c_P	(J mol^{-1} K^{-1})	30.35	36.00	45.64	29.30

Solution

The reaction is
$$CO_{(g)} + H_2O_{(v)} \rightarrow CO_{2(g)} + H_{2(g)}$$

Since $\Delta H_P = Q$, in order to determine the heat at 525°C we need to determine ΔH_R^o at 525°C.

$$\Delta H_{T_2}^o = \Delta H_{T_1}^o + \Delta C_P \cdot \Delta T$$

$$\Delta H_{525°C}^o = \Delta H_{25°C}^o + \Delta C_P \cdot \Delta T$$

$$\Delta H_{25°C}^o = \left[\Delta h_f^o[CO_2] + \Delta h_f^o[H_2] - \Delta h_f^o[CO] - \Delta h_f^o[H_2O] \right]_{25°C}$$
$$= (-393.509) + (0) - (-110.525) - (-241.818)$$
$$= -41.166 \text{ kJ mol}^{-1}$$

$$\Delta C_P = c_P[CO_2] + c_P[H_2] - c_P[CO] - c_P[H_2O]$$
$$= 45.64 + 29.30 - 30.35 - 36.00 = +8.59 \text{ J mol}^{-1} \text{ K}^{-1}$$

Therefore: $\Delta H_{525°C}^o = \Delta H_{25°C}^o + \Delta C_P \cdot \Delta T$
$$= -41\ 166 + (8.59)(525 - 25)$$
$$= -41\ 166 + 4295 = -36\ 871 \text{ J mol}^{-1}$$
$$= -36.9 \text{ kJ mol}^{-1} = Q$$

Less heat is evolved at 525°C than at 25°C.   **Ans**: $\boxed{-36.9 \text{ kJ mol}^{-1}}$

Exercise 9-4

At 0°C and one bar pressure the specific heat of fusion of ice is 334.7 J g^{-1}. At one bar pressure the specific heat capacities of liquid water and of ice are 4.18 and 2.09 J g^{-1} K^{-1}, respectively. What is the standard molar enthalpy of fusion of ice at –15°C? ☹

KEY POINTS FOR CHAPTER NINE

1. The **Standard State** for a chemical reaction at any given temperature T is taken as the pure substance at one bar pressure at the temperature T. Thus, $\Delta H_{298.15}^o$ and ΔH_{500}^o are, respectively, the standard enthalpy changes for a reaction at one bar pressure and 25°C and at one bar pressure and 500 K.

2. **Hess's Law of Constant Heat Summation**: The standard enthalpy change of a chemical reaction at constant pressure during which the only kind of work done is PV-work is the sum of the standard enthalpy changes of the individual component reactions into which the overall reaction may be divided. Huh?

Thus for $A \rightarrow B \rightarrow C \rightarrow D$, $\Delta H_{AD} = \Delta H_{AB} + H_{BC} + \Delta H_{CD}$ (Hess's Law)

3. The **Standard Enthalpy of Formation** of a substance, Δh_f^o, is the standard enthalpy change per mole for its formation from its constituent elements in their reference states.

The **Reference State** of an element is its most stable form under the prevailing conditions, and at one bar pressure.

The standard enthalpy change ΔH_R^o for the reaction $aA + bB \rightarrow cC + dD$ is given by

$$\Delta H_R^o = c\Delta h_{f(C)}^o + d\Delta h_{f(D)}^o - a\Delta h_{f(A)}^o - b\Delta h_{f(B)}^o$$

i.e.,

$$\Delta H_R^o = \sum_{Products} n_i\,\Delta h_{f(i)}^o - \sum_{Reactants} n_i\,\Delta h_{f(i)}^o$$

4. **Variation of ΔH with Temperature at Constant Pressure**: At constant pressure—not necessarily the standard pressure—if the enthalpy change for the reaction

$aA + bB \rightarrow cC + dD$ at temperature T_1 is ΔH_{T_1}, then its value at T_2 is

$$\Delta H_{T_2} = \Delta H_{T_1} + \Delta C_P \cdot \Delta T$$

where

$$\Delta C_P \equiv \sum_{Products} n_i (c_P)_i - \sum_{Reactants} n_i (c_P)_i$$

In the above equations, n_i is the number of moles of species i and $(c_P)_i$ is the molar heat capacity of species i. The individual molar heat capacities have been assumed not to vary with temperature. If they do, integrations will have to be carried out.

PROBLEMS

1. Evaluate ΔH at one atm pressure and 19°C for the formation of one mole of gaseous hydrogen iodide (HI) from hydrogen gas and solid iodine (I_2). Report the answer in kJ mol^{-1}. The heats evolved at 19°C and one atm pressure for various reactions are listed below.[3]

$H_{2(g)} + Cl_{2(g)} \rightarrow 2HCl_{(g)}$	44.00	$KOH_{(aq)} + HCl_{(aq)} \rightarrow KCl_{(aq)}$ 13.74
$HCl_{(g)} + water \rightarrow HCl_{(aq)}$	17.31	$KOH_{(aq)} + HI_{(aq)} \rightarrow KI_{(aq)}$ 13.67
$HI_{(g)} + water \rightarrow HI_{(aq)}$	19.21	$Cl_{2(g)} + 2KI_{(aq)} \rightarrow 2KCl_{(aq)} + I_{2(s)}$ 52.42 ☹

2. Calculate the enthalpy change on the formation of methane gas from the elements, using the following data at 25°C and one bar pressure:

---

[3] 1 kcal = 1000 cal; 1 calorie = 4.184 J.

> • heat of combustion of carbon = 393.51 kJ mol^{-1}
> • heat of formation of water vapor = 241.82 kJ mol^{-1}
> • heat of combustion of methane gas to liquid water = 890.36 kJ mol^{-1}
> • heat of vaporization of water = 583.91 cal g^{-1} ☺

3. Liquid benzene (C_6H_6) and gaseous oxygen (O_2) at 25°C are fed into a reactor with a 10.0% by volume excess of oxygen. The reaction goes to completion with the gases CO_2, H_2O, and O_2 leaving the reactor at 125°C and one bar pressure. What is the heating or cooling requirement so that the exhaust gases remain at 125°C if the benzene feed rate is 100 kg h^{-1}? For this calculation you may assume that the heat capacities of the various reactants and products are independent of temperature. ☹

4. A gaseous equimolar (equal moles of each component) mixture of steam and oxygen at 225°C and one bar pressure is required as the feedstock for a certain process. This feedstock is produced by burning waste hydrogen gas with three times the stoichiometric amount of oxygen gas in a reactor. For each mole of steam produced, calculate: (a) Q, (b) ΔU, and (c) W. ☹

5. Estimate the standard molar internal energy of formation of liquid methyl acetate (CH_3COOCH_3) at 25°C, given that the standard molar enthalpy of formation is -442 kJ mol^{-1}. ☹

6. The expression $C_p = Q_p/\Delta T$ assumes that the heat capacity is constant over the temperature range ΔT. For small temperature ranges this is approximately true; however, over wider temperature ranges it is usually found that c_p varies with temperature. Thus, at any temperature T, c_p is more properly defined as $(c_p)_T = \delta Q_p/dT$. Noting this relationship, calculate ΔH for the following process carried out at a constant pressure of 1 atm:

$$\begin{pmatrix} 1 \; mol \; liquid \\ water \; at \; 25°C \end{pmatrix} \xrightarrow{\;1 \; atm\;} \begin{pmatrix} 1 \; mol \; steam \\ at \; 726.85°C \end{pmatrix}$$

The specific heat capacity of liquid water may be considered to be independent of temperature, with a value of $\bar{c}_p = 4.18$ J g^{-1} K^{-1}; the heat of condensation of saturated water vapor at 100°C is 2.257 kJ g^{-1}; for water vapor, the molar heat capacity varies with temperature according to $c_p = 34.39 + 0.00063T$ J mol^{-1} K^{-1}, where T is the absolute temperature, in kelvin. ☺*

7. Calculate the enthalpy of formation at one bar pressure and 25°C of methane gas from solid carbon and hydrogen gas at the same temperature and pressure given the following standard enthalpy changes at 25°C:

$$C_{(s)} + O_{2(g)} \rightarrow CO_{2(g)} \qquad\qquad -393.509 \; \text{kJ mol}^{-1}$$

$$H_{2(g)} + \tfrac{1}{2}O_{2(g)} \rightarrow H_2O_{(liq)} \qquad\qquad -285.830 \; \text{kJ mol}^{-1}$$

$$CH_{4(g)} + 2O_{2(g)} \rightarrow CO_{2(g)} + 2H_2O_{(liq)} \qquad -890.359 \; \text{kJ mol}^{-1} \quad ☺$$

8. Calculate the standard enthalpy of formation of acetylene gas (C_2H_2) at 25°C using the following standard heats of combustion at 25°C:

$C_2H_{2(g)} = 1299.578$ kJ mol^{-1}; $C_{(s)} = 393.509$ kJ mol^{-1}; $H_{2(g)} = 285.830$ kJ mol^{-1}

Is the formation reaction for acetylene endothermic or exothermic? ☺

9. The standard heat of combustion of liquid benzene (giving water vapor) is 40.145 kJ g^{-1} at 25°C. The heats of formation of carbon dioxide and water vapor are 393.509 and 241.818 kJ mol^{-1}, respectively, at 25°C and one bar pressure. Calculate the enthalpy change for the formation of benzene (C_6H_6) from the elements at 25°C and one bar. The molar mass of benzene is 78.11 g mol^{-1}. ☺

10. Very dilute solutions of strong electrolytes may be treated as solutions of mixtures of ions. Thus the enthalpy change (-74.859 kJ) on dissolving one mole of hydrogen chloride gas in a large quantity of water may be considered to be the sum of the molar heats of formation from $HCl_{(g)}$ of its two ions (in dilute solution). Similarly, the heat absorbed ($+65.488$ kJ) on dissolving one mole of solid silver chloride may be taken as the heat of dissolution and ionic dissociation in dilute solution of that compound $(AgCl_{(s)})$. By convention, the enthalpy of formation of aqueous hydrogen ion from elemental hydrogen in its standard state is taken to be zero. The standard enthalpies of formation from the elements in their standard states for $HCl_{(g)}$ and for $AgCl_{(s)}$ are -92.307 kJ mol^{-1} and -127.068 kJ mol^{-1}, respectively. All values are for 25°C and one bar pressure. Calculate the standard enthalpy of formation of the aqueous silver ion from elemental silver. ☹

11. The heat evolved at 25°C and one bar pressure on the formation of one mole of gaseous carbon dioxide from graphite (a form of carbon) is 393.509 kJ. The heat of combustion of one mole of carbon monoxide is 282.984 kJ. Calculate the standard enthalpy change at 25°C for the formation of gaseous carbon monoxide from graphite and carbon dioxide. Report the answer in GJ per metric tonne of graphite reacted. The molar mass of graphite is 12.011 g mol^{-1}. One metric tonne is 1000 kg. ☺

12. Methane (CH_4) is fed to a household furnace, where in the presence of 40 mole % excess air it is burned completely to $CO_{2(g)}$ and $H_2O_{(g)}$. The products of combustion (CO_2, H_2O, N_2, O_2) pass up the flue and leave the house at 200°C. If the methane and air are initially at 5°C, calculate the amount of heat supplied to the house per mole of methane fed. Air may be assumed to consist of 79% N_2 and 21% O_2 by volume.

Data:	c_P (J mol^{-1} K^{-1})	$\Delta h^o_{f\,(298.15)}$ (kJ mol^{-1})
$CO_{2(g)}$	37.11	-393.506
$N_{2(g)}$	29.125	0.000
$H_2O_{(g)}$	35.577	-241.818
$O_{2(g)}$	29.355	0.000
$CH_{4(g)}$	35.309	-74.81

☺

13. Using only the data given below, calculate the standard enthalpy change of reaction for

$$C_{(gr)} + 2H_2O_{(g)} \rightarrow CO_{2(g)} + 2H_{2(g)}$$

Data:	$\Delta h^o_{298.15}$ [combustion] (kJ mol^{-1})	$\Delta h^o_{f(298.15)}$ (kJ mol^{-1})
$CO_{(g)}$	-282.984	-110.525
$CH_{4(g)}$	-802.332	-74.81 ☺

14. (a) The standard enthalpy of combustion at 25°C of crystalline benzoic acid (C_6H_5COOH) to $CO_{2(g)}$ and $H_2O_{(liq)}$ is -3226.953 kJ mol^{-1}. Find the standard enthalpy of formation of crystalline benzoic acid at 25.0°C. The standard enthalpies of formation at 25.0°C of carbon dioxide and liquid water are -393.509 kJ mol^{-1} and -285.830 kJ mol^{-1}, respectively.

(b) Calculate the enthalpy of formation of crystalline benzoic acid at 200°C and one bar pressure if the molar heat capacities at constant pressure of crystalline benzoic acid, carbon solid, hydrogen gas, and oxygen gas, are constant at 146.8, 8.53, 28.8, and 29.4 J mol^{-1} K^{-1}, respectively. ☺

15. Assuming constant heat capacities, using the data provided calculate the molar change in enthalpy for the vaporization of liquid water to steam at 100°C and one bar pressure:

$$H_2O_{(liq)} \rightarrow H_2O_{(g)}$$ ☺

	$\Delta h^o_{f(298.15)}$ (kJ mol^{-1})	c_P (J mol^{-1} K^{-1})
$H_2O_{(liq)}$	-285.830	75.291
$H_2O_{(g)}$	-241.818	33.577

16. Fuel cells operate best on hydrogen fuel. Natural gas (methane), which is not very reactive in fuel cells, can be converted to hydrogen fuel by the following reaction, known as *steam reforming*:

$$CH_{4(g)} + H_2O_{(g)} \rightarrow 3H_{2(g)} + CO_{(g)}$$

	$\Delta h^o_{f(298.15)}$ (kJ mol^{-1})	c_P (J mol^{-1} K^{-1})*
$CO_{(g)}$	-110.525	30.35
$H_2O_{(g)}$	-241.818	36.00
$CO_{2(g)}$	-393.509	45.64
$H_{2(g)}$	0	29.30

* assume c_P is independent of temperature

Unfortunately, the carbon monoxide formed is a fuel cell catalyst poison and must be removed from the mixture. This can be achieved by the *water gas shift reaction*, which involves reaction with further steam in the presence of a catalyst to convert the carbon monoxide to even more hydrogen:

$$CO_{(g)} + H_2O_{(g)} \xrightarrow{catalyst} CO_{2(g)} + H_{2(g)}$$

We wish to design the cooling system needed for the shift reaction. The feed, which is to consist of an equimolar mixture of CO and steam, enters the reactor at 25.0°C and 1.00 bar pressure (initial state) and the products exit at 1.00 bar pressure and various temperatures as noted below.

(a) What is the enthalpy change for the shift reaction at 25.0°C and 1.00 bar pressure?

(b) What is the enthalpy change for the reaction at 325.0°C and 1.00 bar pressure?

(c) Determine the heat Q that must be added or removed in the reactor per mole of H_2 produced at 325.0°C. Report the answer giving the appropriate sign.

(d) If only 75.0% of the CO is converted to CO_2 at 325°C, what is Q per mole of H_2 produced? Report the answer giving the appropriate sign. ☹

17. Calculate ΔH for the following process at one atm pressure and –15°C:

$$1 \ mol \ liquid \ water \rightarrow 1 \ mol \ ice$$

Data: $\bar{c}_p[H_2O_{(liq)}] = 4.18 \ J \ g^{-1} \ K^{-1}$; $\bar{c}_p[ice] = 2.09 \ J \ g^{-1} \ K^{-1}$; $H_2O = 18.02 \ g \ mol^{-1}$

At 0°C and one atm pressure, $\Delta \bar{h}_{fus}[ice] = +80.0 \ cal \ g^{-1}$ ☺

18. Calculate ΔH for the following process at one atm pressure and +15°C:

$$1 \ mol \ ice \rightarrow 1 \ mol \ liquid \ water$$

Data: $\bar{c}_p[H_2O_{(liq)}] = 4.18 \ J \ g^{-1} \ K^{-1}$; $\bar{c}_p[ice] = 2.09 \ J \ g^{-1} \ K^{-1}$; $H_2O = 18.02 \ g \ mol^{-1}$

At 0°C and one atm pressure, $\Delta \bar{h}_{fus}[ice] = +80.0 \ cal \ g^{-1}$ ☺

19. If the absolute molar enthalpy of liquid water at one atm pressure is arbitrarily assigned a value of 0.00 kJ mol⁻¹ at 0°C, what is the value at 25°C and one atm pressure,

(a) assuming the specific heat of liquid water is constant, at 1.000 cal $g^{-1} \ K^{-1}$?

(b) if the variation of the specific heat at constant pressure (in J $g^{-1} \ K^{-1}$) varies as a function of the temperature, T (in kelvin) according to

$$\bar{c}_p = a + bT + cT^2 + dT^3 + eT^4 + fT^5$$

where the constants a, b, c, d, e, and f have the following values: $a = 186.98$; $b = -2.7072$; $c = 0.016021$; $d = -4.7342 \times 10^{-5}$; $e = 6.9834 \times 10^{-8}$; $f = -4.1112 \times 10^{-11}$.

(c) What *average* value of $\bar{c}_p$ should you use when heating water at one atm pressure from 0°C to 25°C? The relative molar mass of water is 18.015. ☺*

20. Calculate the enthalpy change ΔH_R^o of the following reaction at 400 K and one bar pressure:

$$A_{(liq)} + B_{(liq)} \rightarrow C$$

given the following information:

- A and B both are liquids at all temperatures between 290 K and 410 K
- The standard melting point of solid C is 350 K
- At 300 K, $\Delta H_R^o = 0 \ J \ mol^{-1}$
- The molar heat of fusion of solid C is 5750 J mol⁻¹
- The molar heat capacities of the various species are given in the accompanying table. ☺*

Species	c_p in J mol⁻¹ K⁻¹ T in kelvin
Liquid A	$c_p = 50$
Liquid B	$c_p = 50 + 0.1T$
Solid C	$c_p = 50$
Liquid C	$c_p = 100 + 3.55 \times 10^{-5}T^2$

21. At 298.15 K and one bar pressure, the enthalpy of combustion (to give water vapor) of natural gas (methane, CH_4) is –802.335 kJ mol⁻¹. Calculate the enthalpy of combustion at 1000 K. You may assume that, over the temperature range involved, the molar heat capacities of oxygen gas, carbon dioxide gas, and water vapor are constant at 33.0, 49.6, and 37.5

J mol^{-1} K^{-1}, respectively. The molar heat capacity of methane varies with temperature according to the equation

$$c_P = 19.24 + 5.21 \times 10^{-2} T + 1.20 \times 10^{-5} T^2 - 1.13 \times 10^{-8} T^3$$

where c_P is the heat capacity in J mol^{-1} K^{-1}, and T is the temperature in kelvin. ☺*

22. Assuming ideal gas behavior, calculate ΔH for one mole of oxygen gas undergoing the following process:

$$\begin{pmatrix} 220 \ K \\ 5.00 \ MPa \end{pmatrix} \rightarrow \begin{pmatrix} 300 \ K \\ 10.0 \ MPa \end{pmatrix}$$

The heat capacity for oxygen gas (with T in kelvin) is

$$c_P = 25.48 + 1.520 \times 10^{-2} T - 0.7155 \times 10^{-5} T^2 + 1.312 \times 10^{-9} T^3 \quad \text{J mol}^{-1} \text{K}^{-1} ☺*$$

23. The electrode reactions in a hydrogen–oxygen fuel cell are as follows:

$$\text{anode:} \quad H_{2(g)} \rightarrow 2 H^+_{(aq)} + 2e^-$$

$$\text{cathode:} \quad \tfrac{1}{2} O_{2(g)} + 2 H^+_{(aq)} + 2e^- \rightarrow H_2O_{(liq)}$$

$$\begin{array}{ll} \text{overall} \\ \text{cell:} \end{array} \quad H_{2(g)} + \tfrac{1}{2} O_{2(g)} \rightarrow H_2O_{(liq)} \qquad \Delta H^o_{298.15} = -285.830 \text{ kJ mol}^{-1}$$

At one bar pressure the molar heat capacities of hydrogen gas and oxygen gas vary with temperature (T in kelvin) according to $c_P[\text{H}_2] = 27.2 + 0.0038\,T$ J mol^{-1} K^{-1} and $c_P[\text{O}_2] = 27.2 + 0.0048\,T$ J mol^{-1} K^{-1}. The molar heat capacity of liquid water may be assumed to be independent of temperature, with a value of $c_P[\text{H}_2\text{O}_{(liq)}] = 75.48$ J mol^{-1} K^{-1}.

(a) Derive a numerical expression that gives ΔH^o_T for the overall cell reaction as a function of the absolute temperature T.

(b) What is the numerical value of ΔH^o_T at 80.0°C? ☹*

24. Below 100 K the specific heat capacity of diamond varies as the cube of the temperature: $\bar{c}_P = aT^3$, where $\bar{c}_P$ is in J g^{-1} K^{-1}, and T is in kelvin. A small diamond, of mass 100 mg, is cooled to 77.0 K by immersion in liquid nitrogen. It is then dropped into a bath of liquid helium at 4.2 K, which is the normal boiling point of helium. In cooling the diamond to 4.2 K, some of the helium is boiled off. The gas is collected and found to occupy a volume of 2.48×10^{-5} m^3 when measured at 0.00°C and 1.00 atm pressure. What is the value of a in the formula for the specific heat capacity of diamond? At 4.2 K and one atm pressure the latent heat of vaporization of helium is 84.0 J mol^{-1}. ☹*

25. At 25°C and one bar pressure the heats of combustion of hydrogen gas to water vapor and to liquid water are 241.818 and 285.830 kJ mol^{-1}, respectively. The heat of combustion of gaseous propane (C_3H_8) to water vapor at the same temperature and pressure is 46.35 kJ g^{-1}. Calculate the standard heat of combustion of 100 kg of propane to liquid water at 25°C. The molar mass of propane is 44.096 g mol^{-1}. ☺

26. Calculate the standard enthalpy of formation of acetylene gas (C_2H_2) from the elements at 25°C, using the following standard heats of combustion at 25°C:
    - heat of combustion of $C_2H_{2(g)}$ to carbon dioxide and liquid water = 1299.578 kJ mol^{-1}
    - heat of combustion of carbon (graphite) to carbon dioxide = 393.509 kJ mol^{-1}
    - heat of combustion of $H_{2(g)}$ to liquid water = 285.830 kJ mol^{-1} ☹

27. Calculate the standard enthalpy of formation of one mole of carbon monoxide from graphite and carbon dioxide at 25°C, using the following data at 25°C and one bar pressure:
    - heat of formation of carbon monoxide = 110.525 kJ mol^{-1}
    - heat of combustion of carbon (graphite) to carbon dioxide = 393.509 kJ mol^{-1} ☹

28. Calculate the standard enthalpy change at 25°C for the combustion of one mole of methanol vapor to carbon dioxide and water vapor given the following standard molar enthalpies of formation at 25°C (all values in kJ mol^{-1}): $CH_3OH_{(liq)}$ = −238.66; $CH_3OH_{(g)}$ = −200.66; $CO_{2(g)}$ = −393.509; $H_2O_{(liq)}$ = −285.830; $H_2O_{(g)}$ = −241.818 ☺

29. Calculate the standard enthalpy of formation of methane gas from the elements at 25°C, using the following data at 25°C and one bar pressure:
    - heat of combustion of carbon = 393.509 kJ mol^{-1}
    - heat of formation of water vapor = 241.818 kJ mol^{-1}
    - heat of combustion of methane gas to liquid water = 890.359 kJ mol^{-1}
    - heat of vaporization of water = 583.909 cal g^{-1}

    The molar mass of water is 18.015 g mol^{-1}. ☹

30. If burned with insufficient air, methane may produce $CO_{(g)}$ + $H_2O_{(g)}$ instead of $CO_{2(g)}$ + $H_2O_{(g)}$. If, during the combustion of 1.00 mole of methane, 0.50 mole each of CO and CO_2 are produced, calculate the reduction in the standard heat of combustion as a percent of the maximum value. Standard enthalpies of formation at the temperature of the combustion (kJ mol^{-1}): $H_2O_{(g)}$ = −241.818; $CO_{(g)}$ = −110.525; $CO_{2(g)}$ = −393.509; $CH_{4(g)}$ = −74.81. ☹

31. Exactly 1.00 mole of gaseous ethane (C_2H_6) is burned such that 80.0% of the carbon burns to $CO_{2(g)}$ and the rest to $CO_{(g)}$. Determine ΔH for this process at 25°C and one bar pressure if liquid water is formed as product. The following standard enthalpies of formation at 25°C (in kJ mol^{-1}) are given: $H_2O_{(liq)}$ = −285.830; $CO_{(g)}$ = −110.525; $CO_{2(g)}$ = −393.509; $C_2H_{6(g)}$ = −84.68. ☹

32. From the following standard enthalpies of combustion, calculate the enthalpy of formation of ethylene (C_2H_4) from its elements.

$$H_{2(g)} + \tfrac{1}{2}O_{2(g)} \rightarrow H_2O_{(liq)} \qquad \Delta H^o_{combust} = -285.830 \text{ kJ mol}^{-1}$$

$$C_2H_{4(g)} + 3O_{2(g)} \rightarrow 2CO_{2(g)} + 2H_2O_{(liq)} \qquad \Delta H^o_{combust} = -1410.94 \text{ kJ mol}^{-1}$$

$$C_{(s)} + O_{2(g)} \rightarrow CO_{2(g)} \qquad \Delta H^o_{combust} = -393.509 \text{ kJ mol}^{-1} ☹$$

33. If a compound is burned under conditions such that all the heat evolved is used in heating the product gases, then the maximum temperature reached is called *the adiabatic flame temperature*. Calculate the adiabatic flame temperature for the burning of carbon monoxide in air (79.0% N_2, 21.0% O_2 by volume) using 10% excess air to ensure complete combustion to carbon dioxide. The initial temperature of the gases is 25.0°C. The gases may be assumed to behave ideally with constant heat capacities.

Data:	c_P (J mol^{-1} K^{-1})	$\Delta h^o_{f(298.15)}$ (kJ mol^{-1})
$CO_{2(g)}$	37.11	-393.506
$N_{2(g)}$	29.125	0
$CO_{(g)}$	29.142	-110.525
$O_{2(g)}$	29.355	0

THERMODYNAMICS (V)

10.1 SPONTANEOUS PROCESSES

Some things happen, some things don't: (i) Niagara Falls falls down, but not up. (ii) When left on the kitchen table, a hot cup of coffee cools to room temperature, but a glass of water at room temperature doesn't heat up to 80°C. (iii) A gas under pressure expands to fill a vacuum, but a gas doesn't spontaneously compress and move to one side of the container. (iv) Salt and water mix to form a salt solution, but a salt solution doesn't separate into salt and pure water.

All changes can be classified as either spontaneous or non-spontaneous. *A spontaneous* process is one that has a natural tendency to occur; this kind of process can happen all by itself, and, in theory, is potentially capable of being harnessed to deliver work to the surroundings. A *non-spontaneous* process is one that has *no* natural tendency to occur; this kind of process requires some kind of work input from outside the system in order to proceed.

Some chemical reactions are spontaneous, e.g., the neutralization of an acid with a base to produce a salt plus water; others are not, e.g., $H_2O \rightarrow H_2 + \frac{1}{2}O_2$. A reaction mixture at **chemical equilibrium** has no tendency to move spontaneously in *either* direction—no net process takes place.

Note that when we say a process is a "spontaneous" process, we are referring to its *tendency* to occur; this doesn't necessarily tell us that it *will* occur, or, if it does, at what *rate* it will occur. For example, $H_2 + \frac{1}{2}O_2 \rightarrow H_2O$ is a spontaneous process, but if you mix some hydrogen gas together with some oxygen gas you don't suddenly get a tankful of water—all you get is a gaseous mixture of H_2 and O_2. The actual reaction *rate* is so slow that in practice it is zero. In order to cause this process to occur at a useful rate we need a catalyst or a source of ignition, such as a spark.

Why are some processes spontaneous and others not? Is it the tendency to move towards lower energy that decides? Consider a piece of hot copper metal and a piece of cold copper metal, each of the same mass, in contact with each other in an insulated container (Fig. 1). The hot copper spontaneously *cools*; but, the cold copper spontaneously *heats*—the cold copper seems to *spontaneously* move to a state of *higher* energy, not *lower* energy! Hmm.... so much for the lower energy hypothesis.

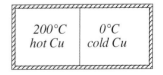

Fig. 1

The tendency for a system to move towards a state of lower energy is one factor that helps to determine whether or not a process will occur spontaneously, but another tendency also is important:

The tendency for energy and matter to become disordered.

Ordered matter tends to become disordered matter; e.g., dissolution of a crystalline salt in water, diffusion from a region of high concentration to a region of low concentration. Concentrated energy tends to become dispersed energy; e.g., the thermal energy in a cup of hot coffee tends to dissipate to the surroundings.

Returning to the two blocks of copper: If you consider the atomic vibrations in them, it is apparent that there is a natural tendency for the concentrated vibrational kinetic energy of the hot copper atoms to become evenly dispersed between the two blocks (Fig. 2).

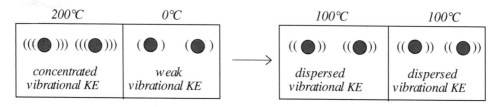

Fig. 2

Work (the movement of a force through a displacement) has *direction*, and therefore is an *ordered* type of energy flow. Heat, on the other hand, is non-directional (it disperses in all directions), and thus is a *disordered* type of energy flow. Because we know that ordered energy tends towards disordered energy, it is not surprising that work can be converted to heat very easily, but that heat is not so readily converted to work.

10.2 ENTROPY (S)

Entropy, designated by the symbol S, is *a measure of the disorder of matter and energy*. But we know from experience that disorder tends to increase;[1] therefore, we can generalize this observation by stating that *the entropy of the universe tends to increase*; this is a statement of **The Second Law of Thermodynamics**. A more accurate statement would be

> *The entropy of the universe never decreases:* $\Delta S_{univ} \geq 0$ *The Second Law of Thermodynamics*

The above statement of the Second Law is more useful for our purposes if we make use of the following mathematical formulation, which quantitatively defines entropy by its differential:

$$dS \equiv \frac{\delta Q_{rev}}{T}$$ *Entropy Defined*   . . . [1]

where Q_{rev} is the *reversible* heat for the process and T is the temperature at which the heat transfer

---

[1] After you finish tidying up your room, does it "spontaneously" become neater or messier?

takes place.

We mentioned earlier in Chapter 6 that a reversible process is one whose direction can be reversed by an infinitesimal change in one of its thermodynamic parameters. For example, to obtain the reversible (i.e., the *maximum*) work from an isothermally expanding gas, we saw that the external pressure against which the gas does work continuously has to be adjusted so that it is just infinitesimally lower than the pressure of the gas itself. If the gas pressure is 10000 bar and the external pressure is only one bar, when the pin restraining the piston is removed, will the expansion take place fast or slowly? Extremely fast, of course. The greater the ΔP between the gas and the surroundings, the faster the piston will shoot up.

So how fast do you think the piston will rise if the difference in pressure between the gas and the external pressure is only infinitesimally small, say $\Delta P = 0.0000001$ bar? Obviously very, very slowly. In fact, if the pressure difference is infinitely small, as is required for a truly reversible expansion, then the piston will rise infinitely slowly.

One of the characteristics of a reversible process is that it takes place extremely slowly; the greater the rate at which a process occurs, the more *irreversible* the process is. How do these ideas apply to heat transfer? Well, heat transfer is driven by a *difference in temperature* ΔT between the system and the surroundings. The greater the ΔT, the faster the heat transfer. So, in order to have *reversible* heat transfer, it is required that the heat must flow very, very slowly, which means that ΔT must be small, small, small. In the limit, for truly reversible heat transfer, $\Delta T \to 0$, which means that $T_{syst} \approx T_{surr}$.

When heat transfers very slowly, local "hot spots" are prevented from forming; hot spots add to the irreversibility because hot spots later disperse spontaneously, adding to the entropy increase. When the heat is the reversible heat δQ_{rev}, the entropy change dS is a *state function*.[2] It should be noted that the definition of entropy involves $\delta Q_{rev}/T$, not just δQ_{rev} itself. It turns out that Q_{rev} alone is not a state function.[2]

The total change in the entropy for the process is obtained by integrating the expression for dS:

$$\boxed{\Delta S = \int \frac{\delta Q_{rev}}{T}} \qquad \qquad \dots [2]$$

If the temperature is kept constant during the process, then the $1/T$ can be taken outside the integral sign to give

$$\Delta S = \int \frac{\delta Q_{rev}}{T} = \frac{1}{T} \int \delta Q_{rev}$$

i.e.,

$$\boxed{\Delta S = \frac{Q_{rev}}{T} \quad \begin{array}{l} \textit{Constant Temperature} \\ \textit{Process} \end{array}} \qquad \dots [3]$$

_____

[2] You'll have to wait for your thermodynamics course for the proof.

Since entropy is a state function it is independent of path. Thus, the entropy of 1.00 kg of hot water at 60°C and one bar pressure is a measure of the current state of disorder of the system. It doesn't matter how the water got to 60°C, the disorder will be the *same* as long as the water is in the *same state* (60°C, 1 bar pressure). In other words, 60°C water has "no idea" how it got that way; all it "knows" is that it is at 60°C, and is in a reproducible state of disorder.

Note the *units* of entropy:
$$\text{\textit{Total} entropy:} \quad S = \text{J K}^{-1}$$
$$\text{\textit{Molar} entropy:} \quad s = \text{J K}^{-1} \text{mol}^{-1}$$
$$\text{\textit{Specific} entropy:} \quad \bar{s} = \text{J K}^{-1} \text{kg}^{-1}$$

10.3 ΔS FOR THE ISOTHERMAL EXPANSION OF AN IDEAL GAS

In Chapter 7 it was shown that, because the internal energy of an ideal gas is a function only of temperature, ΔU for the isothermal expansion (or compression) of an ideal gas is zero.[3]

Therefore, for the isothermal expansion of an ideal gas,

$$\Delta U = Q + W = 0$$

and
$$Q = -W \qquad \qquad \ldots [4]$$

But if the expansion is carried out *reversibly*,[4] then

$$Q = Q_{rev}$$

and
$$W = W_{rev}$$

But we saw earlier that for the isothermal expansion of an ideal gas

$$W_{rev} = -nRT \, ln\left(\frac{V_2}{V_1}\right)$$

Therefore
$$Q_{rev} = -W_{rev} = -\left\{-nRT \, ln\left(\frac{V_2}{V_1}\right)\right\} = +nRT \, ln\left(\frac{V_2}{V_1}\right)$$

and
$$\Delta S = \frac{Q_{rev}}{T} = \frac{+nRT \, ln(V_2/V_1)}{T} = nR \, ln\left(\frac{V_2}{V_1}\right)$$

Therefore,
$$\boxed{\Delta S^{ig} = nR \, ln\left(\frac{V_2}{V_1}\right)} \quad \begin{array}{l}\textit{Isothermal} \\ \textit{Expansion of an} \\ \textit{Ideal Gas}\end{array} \qquad \ldots [5]$$

What if the expansion is *not* reversible? Is ΔS larger or smaller or the same as for the reversible

---

[3] This also is a very good approximation for most real gases.

[4] It should be pointed out that there can be more than one way to carry out a process reversibly, so there can be many different values of Q_{rev} and W_{rev}, depending on the path chosen for the process. Don't make the mistake of saying *the* reversible heat, or *the* reversible work for a process!

expansion?[5]

Example 10-1

Five moles of an ideal gas expands isothermally at 300 K from an initial volume of 100 L to a final volume of 500 L. Calculate: (a) the maximum potential work the gas can deliver, (b) the heat accompanying the process in (b), and (c) ΔS for the gas.

Solution

(a) In the limit, the maximum work that the gas could deliver is the reversible work W_{rev}, which, for an isothermal expansion, is

$$W_{syst} = W_{rev} = -nRT \ln\left(\frac{V_2}{V_1}\right)$$

$$= -(5.0)(8.314)(300)\ln\left(\frac{500}{100}\right)$$

$$= -20\ 071\ \text{J}$$

$$= -20.1\ \text{kJ}$$

The negative sign indicates that the gas *delivers* work.   **Ans:** $\boxed{20.1\ \text{kJ}}$

(b) Since the internal energy of an ideal gas depends only on the temperature, when an ideal gas expands at constant temperature, $\Delta U = 0$. Therefore, $\Delta U = Q + W = 0$ and

$$Q = -W = -(-20\ 071)$$

$$= +20\ 071\ \text{J}$$

$$= +20.1\ \text{kJ}$$

Ans: $\boxed{+20.1\ \text{kJ}}$

(c) Since the expansion is carried out reversibly (in order to obtain the maximum work), it follows that for this process the heat Q is the *reversible* heat Q_{rev}.

Therefore, $Q_{rev} = +20\ 071\ \text{J}$

and, from Eqn [3], $\Delta S = \dfrac{Q_{rev}}{T} = \dfrac{+20\ 071}{300} = +66.90\ \text{J K}^{-1}$

The direct application of Eqn [5] gives the same answer.

Ans: $\boxed{+66.9\ \text{J K}^{-1}}$

---

[5] The *same!* providing the initial and final states are the same; S is a state function, remember?

Exercise 10-1

Five moles of an ideal gas expands isothermally at 300K from an initial volume of 100 L to a final volume of 500 L. During the expansion the gas delivers an amount of work equal to the lifting of a mass of 100 kg a distance of 1.02 m through the gravitational field. Calculate: (a) the work delivered by the gas, (b) the heat for the process, and (c) ΔS for the gas. The acceleration due to gravity is 9.80 m s^{-2}. ☺

10.4 ΔS FOR A CONSTANT PRESSURE HEATING OR COOLING PROCESS

Suppose we heat a homogenous chemical system—such as a pure gas or liquid—from T_1 to T_2 at *constant pressure*. What is ΔS? We can't use the equation $\Delta S = Q_{rev}/T$ because T *changes* during the process, and we don't know which value of T to use. We do know, however, that for a constant pressure heating process in which the system is heated from T_1 to T_2, the heat is given by $Q = C_P \Delta T$, where $\Delta T = T_2 - T_1$. Now, as we discussed above, if we heat the system very very very slowly, continuously increasing the temperature of the surroundings by infinitesimal amounts to ensure at all times that it is only infinitesimally hotter than the temperature of the system, then the ΔT for this type of heat transfer becomes dT, and, instead of using the form $Q = C_P \Delta T$, we now write $\delta Q = C_P dT$. As discussed above, under these special conditions, just as the heat Q becomes the reversible heat Q_{rev}, so also δQ becomes δQ_{rev}. Therefore, for infinitely slow heating, $\delta Q_{rev} = C_P dT$, and it follows that for a constant pressure heating process,

$$dS_P = \frac{\delta Q_{rev}}{T} = \frac{C_P dT}{T} \qquad \cdots [6]$$

Integrating:
$$\Delta S_P = S_2 - S_1 = \int_{State\ 1}^{State\ 2} dS = \int_{T_1}^{T_2} \frac{C_P dT}{T}$$

$$\boxed{\Delta S_P = \int_{T_1}^{T_2} \frac{C_P dT}{T}} \qquad \begin{array}{l} \textit{Constant Pressure} \\ \textit{Heating (or Cooling)} \\ \textit{Process} \end{array} \qquad \cdots [7]$$

If C_P for the system doesn't change with temperature over the temperature range $T_1 \leftrightarrow T_2$, it can be taken outside the integral sign in Eqn [7] to give

$$\Delta S_P = C_P \int_{T_1}^{T_2} \frac{dT}{T} = C_P ln\left(\frac{T_2}{T_1}\right)$$

$$\boxed{\Delta S_P = C_P ln\left(\frac{T_2}{T_1}\right)} \qquad \begin{array}{l} \textit{Constant Pressure} \\ \textit{Heating or Cooling} \\ \textit{Process with} \\ C_P \neq f(T) \end{array} \qquad \cdots [8]$$

Make sure to keep in mind that Q is *not* a state function; even Q_{rev} is not a state function. There is no single value of Q for any state of a system. It is not possible to ask "what is Q for 100 g of water at 75°C?" We saw earlier that different values of Q can be used to take the system to this state, depending on the corresponding value of W. This is why we say δQ instead of dQ.

In thermodynamics, if we write dQ, it means that Q has what we call "an exact differential," meaning that it can be integrated to give $\int_{State\,1}^{State\,2} dQ = Q_2 - Q_1$. If this were true, it would mean that the value of the heat depends only on the initial and the final state of the system; i.e., heat is a state function, independent of the path. This, we know, is not true. BUT, although Q_{rev} is not a state function, Q_{rev}/T *is* a state function, and we *can* integrate Q_{rev}/T to obtain not, $\left(Q_{rev}/T\right)_2 - \left(Q_{rev}/T\right)_1$; but, rather, $S_2 - S_1 = \Delta S$![6]

10.5 ΔS FOR A CONSTANT VOLUME HEATING OR COOLING PROCESS

Constant volume heating or cooling is treated in exactly the same way as constant pressure heating or cooling, except that instead of using C_P, we use C_V, and obtain

$$\Delta S_V = \int_{T_1}^{T_2} \frac{C_V dT}{T}$$
Constant Volume Heating (or Cooling) Process   ... [9]

Again, if C_V is independent of temperature over the temperature range of interest, then

$$\Delta S_V = C_V ln\left(\frac{T_2}{T_1}\right)$$
Constant Volume Heating or Cooling Process with $C_V \neq f(T)$ ... [10]

Example 10-2

Two 1000 g copper blocks, one initially at a temperature of 200.0°C and the other initially at a temperature of 0.0°C, are put into a well-insulated box[7] and allowed to come to thermal equilibrium at one bar pressure (refer to Fig. 1). What is the change in the entropy of the universe as a result of this process? The molar mass and the molar heat capacity of copper are 63.54 g mol^{-1} and $c_P = 24.44$ J mol^{-1} K^{-1}, respectively.

Solution

From Eqn [7], for heating or cooling a substance at constant pressure, $\Delta S_P = \int_{T_1}^{T_2} \left(C_P/T\right)dT$

If the heat capacity of the substance is constant over the temperature range $T_1 \leftrightarrow T_2$, then,

---

[6] You're probably getting confused at this point, but hang in there!

[7] You may neglect the heat capacity of the insulation.

from Eqn [8], the expression simplifies to $\Delta S_p = C_p \ln(T_2/T_1)$. In this example, the hot copper block will cool down and the cold copper block will heat up, until, at thermal equilibrium, each will be at the same final temperature θ_f. Because each block has the same mass, at thermal equilibrium the thermal energy will be distributed evenly between the two blocks, and the final temperature θ_f will be exactly half-way between the two initial temperatures.

Therefore: $$\theta_f = \tfrac{1}{2}(200.0 + 0.0) = 100.0°C = 373.15 \text{ K } [8]$$

Thus, remembering that absolute temperatures must be used for T, for the *hot* copper block,

$$\Delta S_{hot} = n c_P \ln\left(\frac{T_2}{T_1}\right)$$

$$= \left(\frac{1000}{63.54}\right)(24.44)\ln\left(\frac{373.15}{200 + 273.15}\right)$$

$$= -91.33 \text{ J K}^{-1}$$

Similarly, for the *cold* block:

$$\Delta S_{cold} = n c_P \ln\left(\frac{T_2}{T_1}\right)$$

$$= \left(\frac{1000}{63.54}\right)(24.44)\ln\left(\frac{373.15}{0 + 273.15}\right)$$

$$= +119.99 \text{ J K}^{-1}$$

The total entropy change for the system is

$$\Delta S_{total} = \Delta S_{hot} + \Delta S_{cold}$$

$$= -91.33 + 119.99$$

$$= +28.66 \text{ J K}^{-1}$$

Since the box is insulated, no heat—reversible or otherwise—enters the surroundings, so that

$$\Delta S_{surr} = \frac{Q_{surr}^{rev}}{T_{surr}} = 0$$

and therefore $$\Delta S_{univ} = \Delta S_{syst} + \Delta S_{surr}$$

$$= +28.66 + 0$$

$$= +28.66 \text{ J K}^{-1}$$

Ans: $\boxed{+28.66 \text{ J K}^{-1}}$

This value is greater than zero, as demanded by the Second Law for a spontaneous process.

---

[8] We could have used absolute temperatures from the start to get $\theta_f = 0.5(473.15 + 273.15) = 373.15$ K.

Exercise 10-2

A hot 1.00 kg copper block initially at a temperature of 300°C is put into a well-insulated evacuated box containing a 500 g copper block initially at 20°C and allowed to come to thermal equilibrium. What is ΔS for this process? The molar mass and the molar heat capacity of copper are 63.54 g mol^{-1} and $c_p = 24.44$ J mol^{-1} K^{-1}, respectively. ☹

10.6 ΔS FOR A REVERSIBLE PHASE CHANGE

When ice melts reversibly to form liquid water, the ice must be *heated* reversibly. As we saw earlier, reversible heating would mean that $\Delta T = T_{water} - T_{ice} \rightarrow 0$; i.e., $T_{water} \approx T_{ice}$. If the liquid water and the solid ice are both at the same temperature (0°C at one atm pressure), then the ice and water are *in equilibrium*. Thus, *a reversible phase change means that the change takes place when the system essentially is in equilibrium*. Since at constant pressure $\Delta H_P = Q_P$, it follows that *for two phases in equilibrium* $\Delta H_P = Q_{rev}$. Since the temperature remains constant for this type of phase transition, ΔS for the process is easily calculated using the simple formula $\Delta S = Q_{rev}/T$.

Thus, for a melting (fusion) process:

$$\Delta S_{fus} = \frac{Q_{fus}^{rev}}{T} = \frac{\Delta H_{fus}}{T_{mp}}$$

. . . [11]

where T_{mp} is the *melting point*.[9]

For the reverse process of *freezing*, we use $\Delta H_{freezing} = -\Delta H_{fus}$

to give

$$\Delta S_{freezing} = -\Delta S_{fus}$$

. . . [12]

Similarly, for constant temperature *boiling* (vaporization):

$$\Delta S_{vap} = \frac{\Delta H_{vap}}{T_{bp}}$$

. . . [13]

where T_{bp} is the liquid *boiling point*.

For the reverse process of *condensation*, $\Delta H_{cond} = -\Delta H_{vap}$

and we get

$$\Delta S_{cond} = -\Delta S_{vap}$$

. . . [13]

Note that the melting point and the boiling point don't have to be just the *normal* melting point and the *normal* boiling point (at one atm pressure); these equations hold for *any* points located on the equilibrium melting point curve or vapor pressure curve, respectively.

---

[9] Don't forget to use the *absolute* temperature! If you try using $T = 0$ for the melting point of ice in Eqn [11] you'll see why.

Example 10-3

The standard molar enthalpy of vaporization of liquid carbon tetrachloride (CCl_4) at its normal boiling point (76.75°C) is 29.82 kJ mol^{-1}. What is the change in entropy when one mole of carbon tetrachloride vaporizes at its normal boiling point?

Solution

At constant pressure, $\Delta H_P = Q_P$. Since, at the normal boiling point, liquid carbon tetrachloride is in *equilibrium* with its vapor, under these conditions the heat Q_P will be the *reversible* heat Q_{rev}.

Thus, the entropy change of vaporization is

$$\Delta S_{vap} = \frac{Q_{rev}}{T} = \frac{\Delta H_{vap}}{T_{bp}}$$

$$= \frac{29\ 820}{76.75 + 273.15}$$

$$= +85.22 \text{ J K}^{-1} \text{ mol}^{-1}$$

Ans: $\boxed{+85.22 \text{ J K}^{-1} \text{ mol}^{-1}}$

Exercise 10-3

At one bar pressure, solid mercury melts at –38.85°C. The molar latent heat of fusion for the metal is 2292 J mol^{-1}. What is the change in entropy when 1.00 kg of mercury melts at its normal melting point? Hg = 200.59 g mol^{-1}. ☺

10.7 WHENEVER A REAL PROCESS TAKES PLACE THE ENTROPY OF THE UNIVERSE ALWAYS INCREASES

The first law of thermodynamics was introduced in Chapter 7; in this chapter (Section 10.2) we presented the second law of thermodynamics; and in Chapter 11 we introduce the third law of thermodynamics. Contrary to what you may think, the laws of thermodynamics cannot be proven; instead, they are *postulated*. They are based on the experience of humankind throughout the ages. As stated earlier, in the real world some things are observed to happen and others are not observed to happen. It is our accumulated experience with these observations (and non-observations) that has led us to postulate the laws of thermodynamics. In effect, these laws predict for us how we can expect the universe to behave in any given situation. Who knows? perhaps tomorrow Niagara Falls *will* run uphill on its own accord, or the glass of cold coffee at room temperature sitting on the kitchen table *will* spontaneously heat up to become hot again. But don't wager money on these events, because they have *never* been observed to occur. Countless experiments have been attempted to disprove or find exceptions to the laws of thermodynamics; but *all* these experiments have failed miserably. Consequently, with each failed experiment we become increasingly confident of the validity of the laws.

The second law statement that the entropy of the universe never decreases means that it either must increase or not change at all. In effect, what this means is that whenever something real happens in the universe, the entropy of the universe increases. The only way the entropy of the universe could remain constant would be if its entropy had "maxed out." This would mean that nothing could ever happen again; that is, all processes in the universe would have come to a standstill. The universe would be dead (as would you and I). This ultimate ending is referred to as the *Heat Death of the Universe*. But don't worry, it'll be a few more years before this happens.

The upshot of all this is that *whenever any real process takes place, the entropy of the universe must increase. If it does not increase, then the process can never occur*. This presents us with a powerful tool for being able to predict what is feasible and what is not.

An important key to understanding this aspect of the second law is to remember that the "universe" consists of the system *plus* the surroundings:

$$\Delta S_{univ} = \Delta S_{syst} + \Delta S_{surr}$$

The "surroundings" can mean one of two things: If we are talking about the actual cosmological universe; i.e., everything, including the sun moon, stars, galaxies, etc., then the "surroundings" literally means *everything* other than the system. When we use this meaning of the "surroundings" we are often talking in philosophical terms, trying to show how thermodynamics can yield insight into the very workings of nature itself. On the other hand, for most problems of practical interest, the word "surroundings" refers to the *local* surroundings adjacent to the system. For example, if the system is a chemical processing plant, then the surroundings might include the atmosphere surrounding the plant or an adjacent body of water. In such cases any heat or work flowing into or out of the system can be considered to be exchanged only with the local surroundings, which usually is at a fixed temperature such as 25°C. For these kinds of problems we don't care about the temperature of outer space or how much heat eventually might find its way to Jupiter, and the "universe" is a *local* universe.

For a process to be feasible, it is not just the change in the entropy of the system that must be positive; it is the change in the entropy of the *universe* that must be positive; i.e., the *sum* of the change for the system plus that for the surroundings. As it turns out, for many real processes the entropy change for the system actually can be *negative*; but, when combined with that for the surroundings the two must add up to a positive value for the process to be feasible.

A second key is the idea of "the minimum entropy change for the universe." Usually we have no control over the surroundings, which, as we have seen, could include almost any type of process imaginable. For example, suppose, as in the example below, our system were a cup of hot coffee initially at 60°C cooling down to the temperature of the room at 25°C. We know that this process actually takes place—we all have observed it many times. We know that heat will spontaneously transfer from the hot coffee to the surrounding air at 25°C. However, if we were to choose our surroundings to include the sun, for example, which is at 8000°C, then could we still say that heat will flow from our system into the surroundings? Will heat spontaneously flow from a body at 60°C into a body at 8000°C? Obviously not. So, in a nutshell, to evaluate whether or not the process in the system will take place we consider only the local surroundings adjacent to the system.

When heat transfers from the hot coffee into the surrounding air in the room, we know that the coffee gets colder and, consequently, the air must get warmer. However, the volume of the surrounding air, and consequently, its heat capacity, is so much greater than that of the system that, in practical terms, the temperature change of the air can be neglected. The air is considered to be what is called *an infinite heat reservoir* or *an infinite heat sink*. This just means that we always assume that its temperature never changes and that it is so well stirred that it contains no temperature gradients. As we saw earlier,[10] heat flow through a negligibly small temperature gradient constitutes *reversible* heat transfer. Thus, when carrying out our calculations, unless otherwise specified, we always assume that any heat entering or leaving the surroundings is *reversible* heat, Q_{rev}. This is important, because when calculating changes in entropy it is the *reversible* heat Q_{rev} that we must use, not the *actual* (observed) heat Q. What this all means is that, in effect, we are assuming that *all processes taking place in the surroundings are reversible processes*. Under this condition, the change in the entropy of the surroundings ΔS_{surr} will be the minimum entropy change for the surroundings; consequently, the resulting entropy change for the universe ΔS_{univ} will be the minimum entropy change for the universe. The calculations are done this way because our focus is almost always on the system, not the surroundings; therefore we minimize the effects of any processes taking place in the surroundings, which makes it easier to see what is going on in the system.

A few examples should help clarify these ideas:[11]

Example 10-4

A 300 mL cup of hot coffee initially at 60.0°C is placed on the table and allowed to cool to room temperature (25.0°C). Estimate the minimum change in the entropy of the universe as a result of this process. It may be assumed that $\bar{c}_P(\text{coffee}) = \bar{c}_P(\text{water}) = 4.18$ J g^{-1} K^{-1} and that the density of coffee is $\rho = 1.00$ g mL^{-1}. You may neglect the heat capacity of the cup itself.

Solution

Let the coffee be the system.

The minimum change in entropy for the universe results when the heat transfer in both the system and the surroundings is reversible; i.e., there are no temperature gradients in either system or surroundings. This means that the temperature of the surroundings, which is taken as an infinite heat reservoir (i.e., of infinite heat capacity), stays constant at 25°C as it receives heat from the system. The system, on the other hand, although its temperature is constantly changing as it cools down, is assumed always to be at the same uniform temperature throughout *at any given time* (no gradients). This type of a heating or cooling process is known as an "internally reversible" process.

Thus, we use Eqn [8] to evaluate the change in the entropy of the system (the cup of coffee):

$$\Delta S_{syst} = \int_{T_1}^{T_2} \frac{m\bar{c}_P}{T} \, dT$$

$$= m\bar{c}_P \ln\left(\frac{T_2}{T_1}\right)$$

$$= \left(300 \text{ mL} \times 1 \, \tfrac{g}{mL}\right)\left(4.18 \, \tfrac{J}{g \, K}\right) \ln\left(\frac{273.15 + 25}{273.15 + 60}\right)$$

---

[10] Section 10.2.

[11] But beware; students usually find these concepts to be extremely tricky to truly understand. So pay attention!

$$= -139.19 \text{ J K}^{-1}$$

The actual heat transferred from the system (at constant pressure) is

$$Q_{syst} = \left(m\bar{c}_P \Delta T\right)_{syst} = (300)(4.18)(25 - 60) = -43\ 890 \text{ J}$$

The heat flowing *out of* the system flows *into* the surroundings; therefore if Q_{syst} is negative, Q_{surr} is of the same magnitude, but of opposite sign:

$$Q_{surr} = -Q_{syst} = +43\ 890 \text{ J}$$

Since the process in the surroundings is considered to be reversible, then the heat that leaves the system becomes reversible heat when it enters the surroundings:

$$Q_{surr}^{rev} = Q_{surr} = +43\ 890 \text{ J}$$

Furthermore, since the surroundings is maintained at a constant temperature of 25°C = 298.15 K, then Eqn [3] can be used to evaluate ΔS_{surr}:

$$\Delta S_{surr} = \left(\frac{Q_{rev}}{T}\right)_{surr} = \frac{+43\ 890}{298.15} = +147.21 \text{ J K}^{-1}$$

Finally,

$$\begin{aligned} \Delta S_{univ} &= \Delta S_{syst} + \Delta S_{surr} \\ &= -139.19 + 147.21 \\ &= +8.02 \text{ J K}^{-1} \end{aligned}$$

A real process has taken place—coffee really does cool down spontaneously!—therefore, as expected, the entropy change for the universe is positive. Note that in this example of a real process, the entropy change ΔS_{syst} of the system is actually negative. It is ΔS_{univ}, not ΔS_{syst}, that determines whether or not the process is possible.

Ans: $\boxed{+8.02 \text{ J K}^{-1}}$

Read this next example very carefully, and more than once, if necessary:

Example 10-5

For a *one-step*[12] isothermal expansion of one mole of an ideal gas at 0°C from 1.00 atm to 0.500 atm in which all other irreversibilities may be taken to be negligible, calculate the minimum entropy change in the universe.

Solution

We must evaluate

$$\Delta S_{univ} = \Delta S_{syst} + \Delta S_{surr}$$

For an isothermal process involving an ideal gas (*ig*),

---

[12] A *one-step* expansion means that the gas expands against a *constant* external pressure, not an external pressure that is continuously adjusted to be just infinitesimally less than the pressure of the gas.

$$\Delta U^{ig} = Q + W = 0 \qquad \qquad \ldots [a]$$

We have seen that there are many possible combinations of Q's and W's that can bring about the same change of state of the gas, and that Eqn [a] must hold for all such combinations. One special subset of values consists of *reversible* values, Q^{rev}_{syst} and W^{rev}_{syst}.

Thus, we can write
$$\Delta U_{syst} = Q^{rev}_{syst} + W^{rev}_{syst} = 0$$

from which
$$Q^{rev}_{syst} = -W^{rev}_{syst} = -\left\{-nRT\ln\left(\frac{V_2}{V_1}\right)\right\} = +nRT\ln\left(\frac{V_2}{V_1}\right) \qquad \ldots [b]$$

$$= (1.00)(8.314)(273.15)\ln(2) = +1574 \text{ J}$$

For the isothermal expansion of an ideal gas, ΔS_{syst} can be calculated from either

$$\Delta S_{syst} = \frac{Q^{rev}_{syst}}{T_{syst}} \quad \text{or from} \quad \Delta S_{syst} = +nR\ln\left(\frac{V_2}{V_1}\right).$$

Thus,
$$\Delta S_{syst} = \frac{Q^{rev}_{syst}}{T_{syst}} = \frac{+1574}{273.15} = +5.762 \text{ J K}^{-1}$$

To evaluate ΔS_{surr} we need to know Q^{rev}_{surr}; and in order to evaluate Q^{rev}_{surr} we have to know Q_{surr}; and in order to evaluate Q_{surr} we need to know Q_{syst}.

Now comes the conceptually difficult part:

$$Q_{syst} \text{ is } \textbf{not} \text{ the same as } Q^{rev}_{syst} \text{!}$$

Q^{rev}_{syst} is what the heat for the system would be if the expansion in the system were carried out *reversibly*; i.e., infinitely slowly, with the external pressure continuously being adjusted so that it is just infinitesimally less than the pressure of the gas, etc. The *actual* heat Q_{syst} for the system depends through Eqn [a] on the value of the *actual work* W_{syst} for the system. In turn, W_{syst} depends on the value of the constant external pressure against which the gas has to do work when it expands. Since it is stated that the gas must expand to a final pressure of 0.50 atm, clearly the external pressure cannot be greater than 0.50 atm. If it were, the gas would be unable to expand to the required final pressure. If the external pressure were zero; i.e., a vacuum, then the gas would encounter no resistance and deliver no work in expanding to 0.50 atm. Therefore we see that the external pressure must be somewhere between zero atm and 0.50 atm. What value do we use for the calculation?

The problem asks for the minimum change in ΔS_{univ}. For the system (the gas), since S is a state function, ΔS_{syst} is determined only by the specified initial and final state of the gas, and has a value of $+5.762$ J K^{-1}, as calculated above. Therefore, since $\Delta S_{univ} = \Delta S_{syst} + \Delta S_{surr}$, in order to minimize the change in the entropy of the universe, it is necessary to have the *smallest possible* value for ΔS_{surr}, preferably a negative value, since this would offset the positive value of ΔS_{syst}. Since we can assume that any heat transfer into or out of the surroundings is reversible heat transfer, then the value of Q_{rev}^{surr} will be the same as Q_{surr} (the actual heat for the surroundings), which, of course, is just $-Q_{syst}$.

The value of Q_{syst} is equal to $-W_{syst}$ through Eqn [a]. Therefore, Q_{surr} will have its most negative value when Q_{syst} has its most positive value; and Q_{syst} will have its most positive value when W_{syst} has its most *negative* value; i.e., when the gas delivers the maximum amount of work possible. This, then, requires that the constant external pressure be the maximum value that the gas is capable of expanding against in one step, which is $P_{ext} = 0.50$ atm. There! Were you able to follow the above line of reasoning? If not, read it over a couple of times.

Since the heat *leaving* the system *enters* the surroundings[13]

$$Q_{surr} = -Q_{syst} \qquad \ldots \text{[c]}$$

But, for the minimum entropy change in the universe, we let all the irreversibilities be located in the system, and none in the surroundings. Thus, as stated earlier, the process in the surroundings (the heat transfer from the system) is taken to be reversible, so that

$$Q_{surr} = Q_{surr}^{rev}$$

In view of Eqn [c], it follows then that

$$Q_{surr}^{rev} = -Q_{syst} \qquad \ldots \text{[d]}$$

But, from Eqn [a]

$$-Q_{syst} = +W_{syst}$$

where

$$W_{syst} = -P_{ext} \Delta V_{syst} = -P_{ext}(V_2 - V_1)$$

Since this process is an isothermal ideal gas expansion,

$$V_2 = \frac{P_1 V_1}{P_2} = \frac{(1.00 \text{ atm})}{(0.50 \text{ atm})} \times V_1 = 2V_1$$

and

$$V_2 - V_1 = 2V_1 - V_1 = V_1$$

$$= \frac{nRT}{P_1} = \frac{(1.00)(8.314)(273.15)}{(101\ 325)}$$

$$= 0.02241 \text{ m}^3$$

Therefore

$$Q_{surr}^{rev} = W_{syst} = -P_{ext}(V_2 - V_1)$$

$$= -(0.50 \times 101\ 325)(0.02241)$$

$$= -1135 \text{ J}$$

Note that the process taking place in the surroundings—reversible heat transfer into the surroundings—is NOT just the opposite process to that taking place in the system—an isothermal expansion of an ideal gas. Therefore, although it is true that

---

[13] Where else *could* it go? Certainly it can't hide in the boundary!

$$Q_{surr} = -Q_{syst}$$

it is *not* true that $\qquad\qquad Q_{surr}^{rev} = -Q_{syst}^{rev}$ **No! Nein! Nyet!**

Now we can calculate ΔS_{surr}: $\qquad \Delta S_{surr} = \dfrac{Q_{surr}^{rev}}{T_{surr}} = \dfrac{-1135}{273.15} = -4.155 \text{ J K}^{-1}$

and, finally, $\qquad\qquad\qquad \Delta S_{univ}^{min} = \Delta S_{syst} + \Delta S_{surr}$

$$= +5.762 + (-4.155)$$

$$= +1.607 \text{ J K}^{-1}$$

The positive value indicates that this process is feasible.

Ans: $\boxed{+1.61 \text{ J K}^{-1}}$

Example 10-6

An electric current of 10.0 amperes flows through a resistor of 20.0 ohms, which, in the steady state, is maintained at a constant temperature of 10.0°C by immersion in a flowing stream of cold water at 10.0°C.

(a) What is the rate of change of the entropy of the resistor?

(b) For this process, what is the minimum rate of change of the entropy of the universe?

Solution

10.0°C = 283.15 K

(a) The resistor—the system—is in a steady state, at constant temperature and constant pressure. That is, its state doesn't change with time. If you look at it ten minutes from now it seems to be exactly the same as it is now.

Therefore the rate of change of the internal energy of the resistor (the system) is zero:

$$\dot{U} = \dot{Q} + \dot{W} = 0$$

Since the volume of the resistor is not changing, the system does not perform any *PV*-work. However, since there is a steady flow of electricity into (and out of) the resistor, *electrical work* is done on the resistor, as defined by

$$\dot{W}_{elec} = I^2 R = (10)^2(20.0) = +2000 \text{ W} = +2000 \text{ J s}^{-1}$$

Therefore, $\qquad\qquad\qquad \dot{Q}_{syst} = -\dot{W}_{syst} = -\dot{W}_{elec} = -2000 \text{ J s}^{-1}$

What is $\dot{S}$ for the resistor? Since S is a state function, and, at steady state the state of the resistor does not change, then $\dot{S}_{resistor} = 0$.

Ans: $\boxed{0}$

(b) To calculate the minimum $\dot{S}$ for the universe, we must assume the processes in the surroundings are reversible. That is, the heat transfer in the surroundings is reversible, and

$$\dot{Q}_{surr} = \dot{Q}_{surr}^{rev}$$

But

$$\dot{Q}_{surr} = -\dot{Q}_{syst} = -(-2000) = +2000 \text{ J s}^{-1}$$

Therefore,

$$\dot{Q}_{surr}^{rev} = +2000 \text{ J s}^{-1}$$

and

$$\Delta \dot{S}_{surr} = \frac{\dot{Q}_{surr}^{rev}}{T_{surr}} = \frac{+2000}{283.15} = +7.063 \text{ J K}^{-1} \text{ s}^{-1}$$

Finally

$$\dot{S}_{univ} = \dot{S}_{syst} + \dot{S}_{surr} = 0 + 7.063 = +7.063 \text{ J K}^{-1} \text{ s}^{-1}$$

Ans: $\boxed{+7.06 \text{ J K}^{-1} \text{ s}^{-1}}$

As expected, the rate of change of the entropy of the universe is positive.

Example 10-7

A piston/cylinder device contains water at 100°C and one atm pressure in equilibrium with its vapor. 1000 J of heat is transferred to the surrounding air, which is at 25.0°C. The heat is transferred sufficiently slowly that the temperature and pressure inside the cylinder remain constant during the process. As a result of this process, some of the water vapor inside the cylinder is caused to condense to liquid water.

(a) What is the entropy change of the water?
(b) What is the entropy change of the surroundings?
(c) What is the entropy change of the universe?
(d) Is the overall process reversible or irreversible?
(e) If the opposite process took place—i.e., 1000 J of heat transferred *into* the system, causing some of the liquid water to vaporize to water vapor—what would be the value of the entropy change of the universe?
(f) If the temperature of the surrounding air were 99.9999°C what would be the entropy change of the universe?

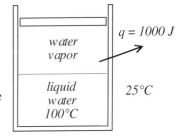

The heat of vaporization for water at 100°C and one atm pressure is 40.656 kJ mol⁻¹.

Solution

(a) $Q_{syst} = -1000 \text{ J}$

If the temperature of the water stays constant at 100°C, then the system remains at all times in equilibrium; i.e., the heat is removed from the system *reversibly*.

Therefore,
$$Q_{syst} = Q_{syst}^{rev} = -1000 \text{ J}$$

and
$$\Delta S_{syst} = \frac{Q_{syst}^{rev}}{T_{syst}} = \frac{-1000 \text{ J}}{373.15 \text{ K}} = -2.680 \text{ J K}^{-1}$$

Ans: $\boxed{-2.68 \text{ J K}^{-1}}$

(b) $Q_{surr} = -Q_{syst} = -(-1000) = +1000 \text{ J}$

It may be assumed that the surroundings is so large that its temperature remains constant at 25°C as it receives the heat from the system. That is, the heat transfer into the surroundings also takes place in a *reversible* manner (no temperature gradients in the surroundings).

Therefore,
$$Q_{surr} = Q_{surr}^{rev} = +1000 \text{ J}$$

and
$$\Delta S_{surr} = \frac{Q_{surr}^{rev}}{T_{surr}} = \frac{+1000 \text{ J}}{298.15 \text{ K}} = +3.354 \text{ J K}^{-1}$$

Ans: $\boxed{+3.35 \text{ J K}^{-1}}$

(c)
$$\begin{aligned}
\Delta S_{univ} &= \Delta S_{syst} + \Delta S_{surr} \\
&= -2.680 + 3.354 \\
&= +0.674 \text{ J K}^{-1}
\end{aligned}$$

Ans: $\boxed{+0.674 \text{ J K}^{-1}}$

(d)   Although the heat transfer in both the system and in the surroundings occurs in a reversible manner, the overall process is irreversible because the heat transfer takes place through a *finite* temperature difference, namely
$$T_{syst} - T_{surr} = 100 - 25 = 75 \text{ K}$$

The irreversibility of the overall process is confirmed from the fact that ΔS_{univ} is greater than zero; if the overall process were reversible, ΔS_{univ} would be zero.

Ans: $\boxed{\text{Irreversible}}$

(e)   If the opposite process were to take place, then

$$Q_{syst} = Q_{syst}^{rev} = +1000 \text{ J}$$

and
$$\Delta S_{syst} = \frac{Q_{syst}^{rev}}{T_{syst}} = \frac{+1000 \text{ J}}{373.15 \text{ K}} = +2.680 \text{ J K}^{-1}$$

Similarly,
$$Q_{surr} = Q_{surr}^{rev} = -1000 \text{ J}$$

and
$$\Delta S_{surr} = \frac{Q_{surr}^{rev}}{T_{surr}} = \frac{-1000 \text{ J}}{298.15 \text{ K}} = -3.354 \text{ J K}^{-1}$$

Under these conditions
$$\Delta S_{univ} = \Delta S_{syst} + \Delta S_{surr}$$
$$= +2.680 + (-3.354)$$
$$= -0.674 \text{ J K}^{-1}$$

Ans: $\boxed{-0.674 \text{ J K}^{-1}}$

The negative sign indicates that this process is impossible. We know from experience that heat will not transfer spontaneously from a lower temperature of 25°C to a higher temperature of 100°C.

(f) If the temperature of the surroundings were 99.9999°C, then

$$\Delta T = T_{syst} - T_{surr}$$
$$= 100 - 99.9999$$
$$\approx 0 \text{ K}$$

and there would be reversible heat transfer between the system and the surroundings. Under these conditions, as before,

$$\Delta S_{syst} = \frac{Q_{syst}^{rev}}{T_{syst}} = \frac{-1000 \text{ J}}{373.15 \text{ K}} = -2.680 \text{ J K}^{-1}$$

but now
$$\Delta S_{surr} = \frac{Q_{surr}^{rev}}{T_{surr}} = \frac{+1000 \text{ J}}{273.15 + 99.9999 \text{ K}} = \frac{+1000 \text{ J}}{373.1499 \text{ K}} = +2.679888 \text{ J K}^{-1}$$

and
$$\Delta S_{univ} = \Delta S_{syst} + \Delta S_{surr}$$
$$= -2.679887 + 2.679888$$
$$= +0.000001 \text{ J K}^{-1}$$

Because $\Delta S_{univ} > 0$, this process is possible, but, because the ΔT is so small, the heat transfer process would be extremely slow; so slow, in fact, that the process essentially would be reversible, as is confirmed by the fact that $\Delta S_{univ} \approx 0$.

Ans: $\boxed{\approx 0}$

To summarize: If $\Delta S_{univ} > 0$, the process is thermodynamically spontaneous and might occur.

$= 0$, the process is reversible—i.e., in equilibrium—and no net process will take place.

< 0, the process is impossible as described, and never will occur under any circumstances.

Example 10-8

Is the following process possible?

$$\begin{pmatrix} 1.00 \ g \ liquid \ toluene \\ 25°C, \ 1.00 \ atm \end{pmatrix} \rightarrow \begin{pmatrix} 1.00 \ g \ toluene \ vapor \\ 25°C, \ 1.00 \ atm \end{pmatrix}$$

The heat of vaporization of toluene at its normal boiling point (111.0°C) is 217.0 J g^{-1}. The specific heat capacities of liquid toluene and toluene vapor are 1.67 J g^{-1} K^{-1} and 1.13 J g^{-1} K^{-1}, respectively. The surroundings are maintained at 25°C and one atm pressure.

Solution

$\Delta \overline{h}_{vap} = +217.0$ J g^{-1} at the normal boiling point (111.0°C = 384.15 K)

$Q_{syst} = \Delta \overline{h}_{syst}$ (*constant pressure process*)

$\Delta \overline{h}_{25} = \Delta \overline{h}_{111} + \Delta \overline{c}_P \cdot \Delta T$ $\qquad\qquad$ $\Delta \overline{c}_P = \overline{c}_{P(V)} - \overline{c}_{P(liq)}$

$\qquad = 217.0 + (-0.54)(25 - 111)$ $\qquad\qquad = 1.13 - 1.67$

$\qquad = +263.44$ J g^{-1} $\qquad\qquad\qquad\qquad = -0.54$ J g^{-1} K^{-1}

$Q_{surr}^{rev} = Q_{surr} = -Q_{syst} = -263.44$ J   (*heat transfer in the surroundings is reversible*)

$$\Delta \overline{s}_{25} = \Delta \overline{s}_{111} + \int_{111°C}^{25°C} \frac{\Delta \overline{c}_P}{T} dT$$

$$= \Delta \overline{s}_{111} + \int_{384.15}^{298.15} \frac{-0.54}{T} dT$$

$$= \left(\frac{Q_{rev}}{T} \right)_{111} - 0.54 \ln \left(\frac{298.15}{384.15} \right)$$

$$= \frac{+217.0}{384.15} - 0.54 \ln \left(\frac{298.15}{384.15} \right)$$

$$= +0.5649 + 0.136$$

$$= +0.7018 \text{ J K}^{-1}$$

$$\Delta \bar{s}_{univ} = \Delta \bar{s}_{syst} + \Delta \bar{s}_{surr}$$

$$= +0.7018 + \frac{-263.44}{298.15}$$

$$= +0.7018 - 0.8836$$

$$= -0.1818 \text{ J K}^{-1}$$

This is an impossible process. Liquid toluene will not vaporize at 25°C if its vapor pressure is one atm. Under these conditions the phase diagram[14] shown below indicates that toluene must exist as a *liquid*.

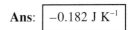

Ans: -0.182 J K^{-1}

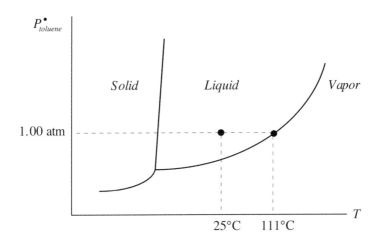

10.8 TROUTON'S RULE

Trouton's Rule is a handy formula for estimating the molar enthalpy of vaporization of a liquid at its normal boiling point:

$$\boxed{\frac{\Delta h_{vap}}{T_{nbp}} \approx +88 \text{ J K}^{-1} \text{ mol}^{-1}} \quad \textit{Trouton's Rule} \quad \ldots [14]$$

Since, for a reversible phase change, $\Delta S = \Delta H/T$, Trouton's Rule really is just stating that *the molar entropy of vaporization ΔS_{vap} is approximately the same (i.e., 88 J K^{-1} mol^{-1}) for all liquids at their normal boiling points.*

To see why this should be so, referring to Fig. 3, consider what happens on the molecular scale when a liquid vaporizes: In the liquid state, the molecules are very *compact*, and, regardless of the

---

[14] See Chapter 15 for more details on pressure–temperature phase diagrams.

liquid, all have about the same (low) degree of disorder. At the normal boiling point, by definition, the vapor pressure of *all* vapors is 101,325 Pa; furthermore,
since almost all vapors behave more or less ideally at
one atm pressure, it follows that all vapors at one atm
pressure behave the same, exhibiting approximately
the same degree of disorder (entropy). Therefore,
since

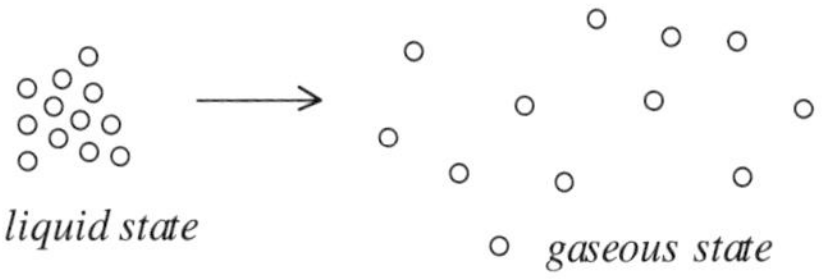

liquid state   *gaseous state*

Fig. 3

$$\Delta s_{vap} = s_{gas} - s_{liq} \qquad \qquad \ldots [15]$$

and the entropy *s* is a measure of the *disorder* of a substance, if all liquids at their boiling points
have very roughly about the same degree of disorder and all gases or vapors at the normal boiling
point also display about the same degree of disorder, it follows that at the normal boiling point the
molar entropy of vaporization will be approximately the same for all liquids, the value turning out
to be about 88 J mol^{-1} K^{-1}.

Liquids with hydrogen bonding—such as *water*—have significant structure, and therefore have
significantly more order (i.e., lower *disorder*) than liquids that don't exhibit hydrogen bonding.
Accordingly, for these kinds of liquids, s_{liq} is lower than for most "normal" liquids, and
$\Delta s_{vap} = s_{vap} - s_{liq}$ will be *greater* than for normal liquids. For example, the entropy of vaporization
Δs_{vap} for water at the normal boiling point is +109 J K^{-1} mol^{-1}, which is greater than the value of
88 J K^{-1} mol^{-1} predicted by Trouton's Rule.

Example 10-9

The normal boiling point of carbon tetrachloride is 76.75°C. Estimate the enthalpy change when
one mole of CCl$_4$ vaporizes at its normal boiling point.

Solution

Trouton's Rule predicts that $\dfrac{\Delta h_{vap}}{T_{nbp}} \approx +88$ J K^{-1} mol^{-1}

Therefore, $\Delta h_{vap} \approx (88)(T_{nbp})$
$$= (88)(76.75 + 273.15)$$
$$= 30\ 791 \text{ J mol}^{-1}$$
$$= 30.79 \text{ kJ mol}^{-1}$$

Ans: $\boxed{30.8 \text{ kJ mol}^{-1}}$

The actual value calculated earlier in Example 10-3 is 29.82 kJ mol^{-1}. In this particular example the
percentage error in the estimated value is only $\left(\frac{30.79 - 29.82}{29.82}\right) \times 100\% = +3.2\%$. Not bad!

KEY POINTS FOR CHAPTER TEN

1. **Spontaneous Process**: has a natural *tendency* to occur (doesn't necessarily mean that it *will* occur, or, if it does, at what *rate* it will occur); can happen all by itself, potentially capable of being harnessed to deliver work to the surroundings.

    Non-Spontaneous Process: *no* natural tendency to occur; requires some kind of *work input* from the surroundings in order to proceed.

    Equilibrium: no spontaneous tendency to move in either direction.

2. Spontaneity determined by (i) natural tendency for systems to move towards lower states of energy, *and* (ii) natural tendency for energy and matter to become diffuse and disordered.

    Entropy *(S)*: a measure of the disorder of matter and energy. *Entropy is a state function* (independent of path: $\Delta S = S_2 - S_1$).

3. **The Second Law of Thermodynamics**: *The entropy of the universe never decreases.*

    That is, for any real process, $\Delta S_{univ} \geq 0$.

    Defined mathematically: $dS \equiv \dfrac{\delta Q_{rev}}{T}$

    where Q_{rev} is heat that is transferred *reversibly*; i.e., heat transferred through an infinitesimally small temperature difference ($\Delta T \rightarrow 0$).

    For any given process: $\Delta S = \displaystyle\int \dfrac{\delta Q_{rev}}{T}$

    If T is kept constant: $\Delta S \equiv \dfrac{Q_{rev}}{T}$

4. Isothermal expansion of an ideal gas: $\Delta S = nR\ln\left(\dfrac{V_2}{V_1}\right)$

5. To heat or cool a substance at *constant pressure*: $\Delta S_P = \displaystyle\int_{T_1}^{T_2} \dfrac{C_P dT}{T}$

    If C_P doesn't vary with temperature: $\Delta S_P = C_P \displaystyle\int_{T_1}^{T_2} \dfrac{dT}{T} = C_P \ln\left(\dfrac{T_2}{T_1}\right)$

6. To heat or cool a substance at *constant volume*: $\Delta S_V = \displaystyle\int_{T_1}^{T_2} \dfrac{C_V dT}{T}$

If C_V doesn't vary with temperature: $\Delta S_V = C_V \int_{T_1}^{T_2} \dfrac{dT}{T} = C_V \ln\left(\dfrac{T_2}{T_1}\right)$

7. **Reversible phase change**: the change takes place when the system essentially is in equilibrium. Since at constant pressure $Q_P = \Delta H_P$, it follows that for two phases in equilibrium $Q_{rev} = \Delta H_P$

and $\Delta S = \dfrac{\Delta H}{T}$ *(Reversible phase change)*

where ΔH is the enthalpy change for the reversible phase change and T is the temperature at which the phases are in equilibrium.

8. For any process (reversible or otherwise), $\Delta S_{univ} = \Delta S_{syst} + \Delta S_{surr}$

For a spontaneous process in the system, $\Delta S_{univ} > 0$, **NOT** $\Delta S_{syst} > 0$. (ΔS_{syst} can be either positive *or* negative for a spontaneous process in the system.)

Unless otherwise specified, "surroundings" means the *local* surroundings; i.e., the surroundings immediately adjacent to the system. Furthermore, unless otherwise stated, *all processes taking place in the surroundings are* **reversible** *processes*, which also means that *the temperature of the surroundings is always constant* (i.e., heat flowing into or out of the surroundings does not change the temperature of the surroundings).

9. When calculating ΔS_{surr}, $Q_{surr}^{rev} = -Q_{syst}$ and $\Delta S_{surr} = \dfrac{Q_{surr}^{rev}}{T_{surr}}$

where Q_{syst} is the *actual*—not the reversible—heat for the *system*.

10. For any system in a *steady state*, i.e., one for which none of the parameters used to characterize the system—such as T, P, or *composition*—varies over time, the entropy of the system is *constant*; i.e., the rate of change of entropy of the system is $dS_{syst}/dt = \dot{S}_{syst} = 0$. Over a given time period, for such a steady state system $\Delta S_{syst} = 0$.

11. **Trouton's Rule**: $\dfrac{\Delta h_{vap}}{T_{nbp}} \approx +88\ J\ K^{-1}\ mol^{-1}$

where Δh_{vap} is the molar enthalpy of vaporization at the normal boiling point T_{nbp}.

PROBLEMS

1. For the process $\begin{pmatrix} 1.00\ g\ ice \\ 0.0°C,\ 1\ atm \end{pmatrix} \rightarrow \begin{pmatrix} 1.00\ g\ water\ vapor \\ 100.0°C,\ 1\ atm \end{pmatrix}$

calculate: (a) ΔH (b) ΔS.

The heat of fusion of ice at 0°C and one atm pressure is 333.5 J g^{-1}, and the heat of vaporization of water at 100°C and one atm pressure is 40.66 kJ mol^{-1}. The average specific heat capacity of liquid water between 0°C and 100°C is 4.18 J g^{-1} K^{-1}. The molar entropy of liquid water at 25°C and one atm pressure is 69.91 J K^{-1} mol^{-1}. ☺*

2. An insulated container contains 500 g of hot water at 85.0°C and atmospheric pressure. A 50-gram ice cube at -10.0°C is dropped into the container and the lid is immediately replaced. The surroundings are at 25.0°C and atmospheric pressure throughout. As a result of this process, calculate: (a) The final temperature of the system. (b) The entropy change in the system. (c) The entropy change in the universe. The specific heat capacities of liquid water and ice may be considered constant at 4.18 J g^{-1} K^{-1} and 2.05 J g^{-1} K^{-1}, respectively, over the temperature ranges involved. The heat of fusion of ice at 0°C and atmospheric pressure is 333.5 J g^{-1}. ☹*

3. 1.00 mole of liquid water at 25°C and one atm pressure is injected into an insulated rigid 1.00 m^3 tank initially containing steam at 500°C and one atm pressure. Calculate: (a) The final temperature of the system after it has reached thermal equilibrium. (b) The change of entropy of the system. Steam may be assumed to behave as an ideal gas. The molar heat capacities of liquid water and of steam may be considered constant at $c_{liq} = 75.3$ J mol^{-1} K^{-1} and $c_P = 34.4$ J mol^{-1} K^{-1}, respectively, over the temperature ranges involved. The molar enthalpy of vaporization of water at 100°C and one atm pressure is 40.66 kJ mol^{-1}. ☠*

4. (a) Using the fact that for an ideal gas $dU_T = 0$, show that the entropy change for the reversible isothermal expansion of n moles of an ideal gas from an initial pressure of P_1 to a final pressure of P_2 is given by

$$\Delta S_T^{ig} = -\frac{P_1 V_1}{T} \ln\left(\frac{P_2}{P_1}\right)$$

    (b) If the isothermal expansion were carried out irreversibly, how would you expect the value of ΔS_T to compare with the value given by the formula in part (a)?

    (c) Calculate ΔS when 10.0 moles of gaseous nitrogen expands isothermally at 300 K from 10.0 bar to 1.00 bar.

    (d) Calculate ΔS when 10.0 moles of gaseous nitrogen expands isothermally at 500 K from 10.0 bar to 1.00 bar. ☹*

5. Assuming that water has a constant heat capacity of $\bar{c}_P$, prove that the entropy of $2m$ kg of warm water at absolute temperature T_f kelvin is greater than the combined entropy of m kg of hot water at temperature T_H and m kg of cold water at temperature T_C from which it was formed. ☹*

6. Assuming air behaves as an ideal gas, calculate the minimum entropy change in the universe for the one-step isothermal compression of one mole of air at 273 K from 0.500 atm to 1.00 atm. (*One-step means constant external pressure.*) ☺

7. During the isothermal expansion of 1.00 mole of an ideal gas at 273.15 K from 1.00 atm pressure to 0.500 atm pressure, 440 J of work is done by the gas. Calculate the minimum

entropy change in the universe if the temperature of the surroundings also is at 273.15 K. ☺

8. 2000 J of work is used to compress 1.00 mole of an ideal gas isothermally at 0.00°C from 0.50 atm pressure to 1.00 atm pressure. A 10.0°C temperature drop is provided between the gas and the surroundings to facilitate the heat transfer. Calculate the minimum entropy change in the universe for the compression under these conditions. ☺

9. Calculate the minimum entropy increase in the universe for one cycle consisting of the isothermal expansion ($A \rightarrow B$) and compression ($B \rightarrow A$) of 5.00 moles of ideal gas at 50°C between one bar and 0.500 bar. During the expansion the gas delivers work equivalent to lifting 20.0 kg through 35.0 m. The compression is accomplished by a mass of 50.0 kg falling through 35.0 m. The acceleration due to gravity is 9.807 m s^{-2}. ☺

10. An electric current of 10.0 amperes flows through a thermally insulated resistor of 20.0 ohms, which initially is at a temperature of 10.0°C. The current flows for exactly one second. If the resistor has a mass of 5.00 g and a heat capacity of $\bar{c}_P = 0.85$ J g^{-1} K^{-1},

    (a) What is the entropy change of the resistor?

    (b) As a result of this process, what is the minimum entropy change of the universe ? ☺*

11. (a) Exactly 1.00 m^3 of water at 25°C (the system) is contained in an insulated[‡] vessel maintained at one atm pressure. Calculate the minimum amount of work that must be done to heat this water to 35°C if the surroundings is an infinite heat reservoir at 298.15°C. The specific heat capacity and density of water can be taken as constant at 4.184 kJ K^{-1} kg^{-1} and 1000 kg m^{-3}, respectively.

    (b) What is the minimum entropy change for the universe as a result of the process in part (a)? ☺*

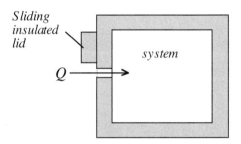

[‡] Insulation prevents heat loss, but the small opening permits heat to flow into the system.

12. A glass of ice water contains a 50.0 g ice cube at 0°C floating in 100.0 g of liquid water at the same temperature. The glass of water is placed on the table in a room at 25°C and atmospheric pressure, and eventually attains thermal equilibrium with the room. Assuming the heat capacity of the glass container is negligible and that no water is lost by evaporation, what is the change in the entropy of the universe as a result of this process? The specific heat capacities at constant pressure of ice and liquid water may be taken as 2.11 J g^{-1} K^{-1} and 4.19 J g^{-1} K^{-1}, respectively. The specific enthalpy of fusion for ice at 0°C is 333.6 J g^{-1}. ☺*

13. Two blocks of copper, each of 1.00 kg mass, one at 800 K and the other at 400 K are brought into contact inside an insulated container and allowed to reach thermal equilibrium. What is ΔS for the universe? For copper, $\bar{c}_P = 0.385$ J g^{-1} K^{-1}. ☺*

14. 100 g of ice at 0.00°C is added to 100 g of water at 100.0°C in an insulated container at a constant pressure of one atm. The surroundings is maintained at a temperature of 300K. When thermal equilibrium has been established, the final temperature of the system is T_f kelvin.

(a) What is T_f? (b) What is ΔS_{univ} for this process?

Data: Specific heat of fusion of ice: $\Delta \bar{h}_{fus} = 335$ J g^{-1}

Specific heat capacity of water: $\bar{c}_p = 4.18$ J g^{-1} K^{-1} ☺*

15. Assuming ideal gas behavior, calculate ΔS for one mole of air going from 20.0°C and 2.00 atm to -60.0 °C and 1.000 atm. c_p[air] = 29.1 J mol^{-1} K^{-1} ☺*

16. Assuming ideal gas behavior, calculate (a) ΔH, and (b) ΔS for one mole of oxygen gas undergoing the following process:

$$\left(\begin{array}{c} 220\ K \\ 5.00\ MPa \end{array} \right) \rightarrow \left(\begin{array}{c} 300\ K \\ 10.00\ MPa \end{array} \right)$$

The heat capacity for oxygen gas (with T in kelvin) is

$$c_p = 25.48 + 1.520 \times 10^{-2}T - 0.7155 \times 10^{-5}T^2 + 1.312 \times 10^{-9}T^3 \quad \text{J mol}^{-1} \text{ K}^{-1} \text{ ☺*}$$

17. Assuming methane behaves as an ideal gas, calculate ΔS for the following process:

$$\left(\begin{array}{c} 1\ mol\ CH_4 \\ 0°C,\ 25.0\ kPa \end{array} \right) \rightarrow \left(\begin{array}{c} 1\ mol\ CH_4 \\ 174.8°C,\ 500\ kPa \end{array} \right)$$

At 25 kPa the heat capacity for gaseous methane (with T in kelvin) is

$$c_p = 14.15 + 75.50 \times 10^{-3}T - 17.99 \times 10^{-6}T^2 \quad \text{J mol}^{-1} \text{ K}^{-1} \text{ ☺*}$$

18. (a) Calculate ΔS when one mole of hydrogen at 133.938 kPa and 50.0°C expands irreversibly against the external pressure of the surroundings to 101.325 kPa and 25.0°C. Under these conditions it may be assumed that hydrogen behaves as an ideal gas. The molar heat capacity for hydrogen is given by

$$\bar{c}_p = 27.280 + 3.26 \times 10^{-3}\,T + \frac{0.502 \times 10^5}{T^2}$$

where c_p is in J mol^{-1} K^{-1} and T is in kelvin.

(b) If the temperature and pressure of the surroundings are always constant at 25.0°C and 101.325 kPa, respectively, and all the processes in the surroundings are reversible, what is the change in the entropy of the universe as a result of the gas expansion?

(c) Explain why the process in the system is or is not a spontaneous process. ☹*

19. 10.0 amperes is passed continuously at steady state through a resistor having a constant electrical resistance of 5.0 Ω. Assuming reversible heat transfer in the surroundings, calculate the minimum rate of entropy production in the universe when

(a) the resistor is kept in a cryogenic container at 50 K,

(b) the resistor is kept in a flowing stream of cold water at 10°C,

(c) the resistor is kept in an oven at 1000°C. ☺

20. A beaker containing 10.0 g of water at 300 K is heated at one atm pressure to 350 K using the one-step operation of placing the beaker in a constant temperature bath at 350 K. Calculate the minimum entropy change of the universe as a result of this process. For water, $\bar{c}_P = 4.18$ J g K^{-1}. ☺*

21. An electric immersion heater is used to heat 30.0 L of water in an insulated container. The temperature of the heater is constant at 90.0°C and the water is heated from 25.0°C to 75.0°C. Calculate the minimum ΔS for the universe as a result of this process.

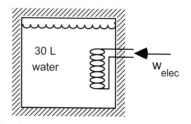

For liquid water, it may be assumed that

$\bar{c}_P = \bar{c}_V = 4.18$ J g^{-1} K^{-1} and $\rho = 1000$ g L^{-1}. ☺*

22. An electric heater requires a 2.00 kW power input to maintain a large tank of water at a constant temperature of 25.0°C in a surroundings at 15.0°C. Calculate the change in entropy in one hour for

(a) the heater, (b) the tank of water, (c) the surroundings, and (d) the universe. ☺

23. A 20.0 Ω electric resistor of 5.00 g mass and a specific heat capacity of 8.50 J g^{-1} K^{-1} is maintained at a constant temperature of 10.0°C by a flowing stream of water. Using the observation that the resistor draws a current of 10.0 amperes for 1.00 second, calculate the entropy change for (a) the resistor, and (b) the water. ☺

24. A 500 g metal bar initially at 1200 K is removed from a furnace and quenched by immersion into a well-insulated closed tank containing 10.00 kg of non-volatile oil initially at 25°C, which also is the temperature of the surroundings. The specific heat capacities of the metal and of the oil are constant at 0.32 J g^{-1} K^{-1} and 1.25 J g^{-1} K^{-1}, respectively. Neglecting any heat loss from the tank,

(a) What is the final equilibrium temperature of the bar and the water?
(b) What is the minimum change in the entropy of the universe? ☺*

25. A pan containing 10.0 g of hot water at 350 K is placed in a constant temperature bath maintained at 300 K, and allowed to come to thermal equilibrium at one atm pressure. Calculate the minimum ΔS for the universe as a result of this process.

For water, $\bar{c}_P = 4.184$ J g^{-1} K^{-1}. ☺*

26. 100.0 L of hot water at 330 K and 100.0 L of cold water at 290 K, which is also the temperature of the room, are placed into a bathtub. The result, after a certain amount of cooling has occurred, is a bath full of warm water at 305 K. Neglecting the heat capacity of the bathtub, and assuming no evaporation takes place, calculate the minimum amount by which the entropy of the universe increases as a result of this process. The specific heat capacity of water may be taken as 1.00 cal g^{-1} K^{-1}. ☺*

27. Two blocks of copper, each of 1.00 kg mass, one at 1000 K and the other at 200K, are brought into contact inside an insulated container and allowed to reach thermal equilibrium at atmospheric pressure. What is ΔS for the universe as a result of this process?

For copper, $\bar{c}_P = 0.385$ J g^{-1} K^{-1} ☺*

28. An electric current of 10.0 amperes flows through a resistor of 20.0 ohms immersed in flowing water at 20.0°C to ensure an isothermal operation. What is the rate of entropy production of the resistor? ☺

29. Use the following data and Trouton's rule to predict the entropy of vaporization of the listed liquids at their normal boiling points. Because of the approximate nature of Trouton's rule, you may neglect any small differences between the 25°C molar enthalpy values given—which are at one bar—and what the values would be at one atm. ☺

	$\left(\Delta h_f^o\right)^{liquid}_{298.15\,K}$	$\left(\Delta h_f^o\right)^{gas}_{298.15\,K}$	$\left(c_P^o\right)_{liq}$	$\left(c_P^o\right)_{gas}$	T_{nbp}
	(kJ mol^{-1})	(kJ mol^{-1})	(J mol^{-1} K^{-1})	(J mol^{-1} K^{-1})	(K)
n-pentane, C_5H_{12}	−172.75	−146.44	120.2	167.2	309.2
benzene, C_6H_6	49.21	82.93	136.0	81.67	353.2
hexane, C_6H_{14}	−198.7	−167.19	195.6	143.09	341.9
toluene, $C_6H_5CH_3$	12.22	50.0	157.3	103.64	383.8
mercury, Hg	0.00	61.317	27.983	20.786	629.9
ethylene oxide, C_2H_4O	−77.82	−52.64	87.95	47.90	283.8
bromine, Br_2	0.00	30.907	75.689	36.02	332.4
methanol, CH_3OH	−238.66	−200.66	81.6	43.89	337.2

30. One mole of an ideal gas having a constant heat capacity of $c_P = 29.0$ J mol^{-1} K^{-1} is expanded from 2.00 bar and 400 K to 1.00 bar and 300 K. During the expansion the gas delivers work to the surroundings equivalent to lifting a 10.0 kg weight through 10.0 m in the gravitational field. The gas then is expanded isothermally to 0.500 bar, during which it works against a constant external pressure of 0.40 bar. Finally, the gas is compressed to 2.00 bar and 400 K by the application of 2.50 kJ of work. If the surroundings is at a constant temperature of −12.0°C, what is the net change in the entropy of the universe as a result of the above series of processes? The acceleration due to gravity is 9.80 m s^{-2}. ☺

31. 150 g of water at 25°C and 10.0 g of ice at −15°C are mixed in a thermos bottle. Assuming that the process is adiabatic, determine (a) the final temperature and (b) ΔS_{univ}.

$\bar{c}_P[H_2O_{(liq)}] = 4.18$ J g^{-1} K^{-1}; $\bar{c}_P[ice] = 2.04$ J g^{-1} K^{-1}; $\Delta\bar{h}_{fus}[ice] = 333$ J g^{-1} at 0°C ☺

32. One mole of water undergoes the change in state

$$\begin{pmatrix} H_2O_{(liq)} \\ 1\ atm,\ T \end{pmatrix} \rightarrow \begin{pmatrix} H_2O_{(g)} \\ 1\ atm,\ T \end{pmatrix}$$

Calculate ΔH and ΔS for this change in state if (a) $T = 110°C$, and (b) $T = 90°C$.

Data: $c_P[H_2O_{(g)}] = 36.3$ J mol^{-1} K^{-1} $c_P[H_2O_{(liq)}] = 76.0$ J mol^{-1} K^{-1}

      $\Delta h_{vap} = 40.65$ kJ mol^{-1} at 100°C ☺

33. 500 g of water at 5°C is placed in a beaker on the table and allowed to heat up. The room is at a constant temperature of 25°C and constant pressure of one bar. Neglecting the heat capacity of the beaker, what would be ΔS for the universe if the water were to heat up to 35°C? Is this a spontaneous process? The heat capacity of water may be taken as constant at $c_p = 4.18$ J g^{-1} K^{-1}. ☹

34. In neutral, oxygenated water, iron spontaneously corrodes according to the following reaction:

$$2Fe_{(s)} + O_{2(g)} + 2H_2O_{(liq)} \rightarrow 2\,Fe^{2+}_{(aq)} + 4\,OH^{-}_{(aq)}$$

What is the entropy change in the universe when 1.00 mole of iron corrodes at 25°C and one bar pressure if the surroundings also are at 25°C and one bar pressure? ☺

35. A 250 mL cup of coffee is maintained at a temperature of 65°C by the use of an electrically heated cup. The cup draws a current of 125 mA at 117 volts. If the surroundings is at a constant temperature of 25°C, by how much does the entropy of the universe change when the coffee is kept hot for one hour? The density and heat capacity of coffee may be assumed to be 1.00 g mL^{-1} and 4.2 J g^{-1} K^{-1}, respectively. Assume no coffee evaporates. ☺

THERMODYNAMICS (VI)

11.1 ABSOLUTE ENTROPIES AND THE THIRD LAW OF THERMODYNAMICS

Since the temperature of a chemical system is proportional to its kinetic energy, it follows that as the temperature approaches absolute zero, the kinetic energy also approaches zero. In other words, all molecular motion stops at $T = 0$. For a perfectly crystalline solid, at $T = 0$ every atom is frozen into a well-defined location in the lattice; under these conditions, we may state that there is *zero disorder* in the crystal, that is, *its entropy is zero*. This is a statement of the *Third Law of Thermodynamics*:

| The entropy of a perfectly crystalline solid is zero at $T = 0$. | *The Third Law of Thermodynamics* | . . . [1] |

It also is observed experimentally that for all solids, $c_p \to 0$ as $T \to 0$. We showed earlier that if we heat a substance from T_1 to T_2 at constant pressure,

$$\Delta S_P = S_{T_2} - S_{T_1} = \int_{T_1}^{T_2} \frac{C_P \, dT}{T} \qquad \text{. . . [2]}$$

In Eqn [2], if we substitute zero kelvin for T_1 and any general temperature T for T_2, we get

$$\Delta S_P = S_{T_2} - S_{T_1}$$
$$= S_T - S_0$$
$$= \int_0^T \frac{C_P \, dT}{T} \qquad \text{. . . [3]}$$

But the Third Law states that $S = 0$ at $T = 0$; therefore Eqn [3] becomes

$$\Delta S_P = S_T - 0 = \int_0^T \frac{C_P \, dT}{T}$$

That is,
$$S_T = \int_0^T \frac{C_P \, dT}{T} \qquad \qquad \dots [4]$$

Since we can measure heat capacities at any temperature[1] it can be seen that, unlike other state functions such as U and H, we can calculate *absolute* values of S at the temperature T. All we have to do is plot (C_P / T) *vs.* T and take the area under the curve from $T = 0$ to $T = T$, as shown in Fig. 1. This is why tables of thermodynamic data are able to list *absolute* values s^o of standard entropies for the elements[2] and compounds, in contrast to the standard values for other state properites such as internal energy or enthalpy, which must be reported versus some reference state where the values are arbitrarily defined as zero.

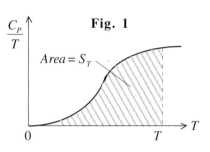

11.2 DEALING WITH PHASE TRANSITIONS

Near absolute zero all substances (except helium[3]) are solids. At constant pressure, when the temperature of a solid is raised to its melting point, the solid will melt to form liquid; and at higher temperatures still, the liquid will vaporize at its boiling point to form vapor. At each such phase transition, there is an absorption of heat without an increase in temperature, and ΔS for the transition is obtained merely by dividing the enthalpy change for the phase transition by the (equilibrium) transition temperature.

Thus, in the case of *melting*,
$$\Delta S_{fus} = \frac{\Delta H_{fus}}{T_{mp}} \qquad \qquad \dots [5]$$

while in the case of *boiling*,
$$\Delta S_{vap} = \frac{\Delta H_{vap}}{T_{bp}} \qquad \qquad \dots [6]$$

Any phase-transition entropy changes occurring to the substance before it reaches the temperature of interest T must be added to the entropy value obtained by integrating C_P / T. Thus, the molar entropy of a gas at temperature T is given by:

$$S_T = \int_0^{mp} \frac{C_{P(solid)} \, dT}{T} + \Delta S_{fus} + \int_{mp}^{bp} \frac{C_{P(liq)} \, dT}{T} + \Delta S_{vap} + \int_{bp}^{T} \frac{C_{P(gas)} \, dT}{T} \qquad \dots [7]$$

---

[1] Actually, to obtain values at temperatures very close to absolute zero we have to make an extrapolation.

[2] Note that, unlike the standard enthalpy of formation, the standard *entropy* of an *element* has a non-zero, positive value, because an element has a finite, positive level of disorder at all temperatures above absolute zero.

[3] The atoms of helium are so light that even at temperatures close to absolute zero they vibrate sufficiently vigorously to fly apart and form a gas—to solidify helium near absolute zero requires substantial pressures greater than 20 bar.

For a gas at some constant temperature T, the volume decreases as the pressure increases. Since the same amount of gas in a smaller volume has more order than in a larger volume, the entropy of the gas decreases as its pressure increases. For this reason, to meaningfully compare the entropy of different gases requires that they should be compared at the same pressure. As before, the usual standard state is taken as the pure gas at one bar pressure. Accordingly, the symbol for the standard molar entropy is s_T^o, with units of $J\ K^{-1}\ mol^{-1}$ (same units as heat capacity).[4]

Absolute molar entropy data usually are reported at 25°C. As an example of typical values of absolute molar entropies at 25°C, consider the values for H_2O: for ice, liquid water, and water vapor they are, respectively, 45, 70, and 189 $J\ K^{-1}\ mol^{-1}$. Note that because the disorder increases in passing from the solid state[5] to the liquid state to the gaseous state, $s_{(s)} < s_{(liq)} < s_{(g)}$. Note also that, because all substances always contain *some* degree of disorder (except perfect crystalline solids at absolute zero), entropies always have *positive* values.[6]

Example 11-1

Given that $c_p[\text{ice}] = 37.0\ J\ mol^{-1}\ K^{-1}$ and that $s_{298.15}^o[\text{ice}] = 45.0\ J\ K^{-1}\ mol^{-1}$, what is the molar entropy of ice at 0°C and one bar pressure?

Solution

The entropy change for the process $\left(\begin{array}{c} 1\ mole\ ice \\ 0°C,\ 1\ bar \end{array} \right) \rightarrow \left(\begin{array}{c} 1\ mole\ ice \\ 25°C,\ 1\ bar \end{array} \right)$

is
$$\Delta s^o = s_2^o - s_1^o = \int_{T_1}^{T_2} \frac{c_p\,dT}{T}$$

Therefore,
$$\Delta s^o = s_{298.15}^o - s_{273.15}^o$$

$$= \int_{273.15}^{298.15} \frac{37.0}{T}\,dT$$

$$= 37.0\ ln\left(\frac{298.15}{273.15}\right)$$

$$= 3.2\ J\ K^{-1}\ mol^{-1}$$

from which
$$s_{273.15}^o = s_{298.15}^o - 3.2$$

$$= 45.0 - 3.2$$

---

[4] $J\ K^{-1}\ mol^{-1}$ (molar entropy) is the same unit as $J\ mol^{-1}\ K^{-1}$ (molar heat capacity) for the purpose of calculation.

[5] In case you are curious, the entropy of ice at 25°C and one bar pressure is a *calculated* value—ice doesn't actually exist under these conditions, but if it did, this is what the value would be.

[6] Aqueous ions are the exception, owing to the fact that it is impossible to measure the thermodynamic properties of a single ion type, such as Cu^{2+} or Cl^-. Instead, the aqueous H^+ ion is arbitrarily assigned a value of zero. When this is done, the entropies of other ions are then measured relative to this value; since the absolute entropies of some ions are greater than, and others less than, that of the H^+ ion, this leads to both positive and negative values.

$$= 41.8 \text{ J K}^{-1} \text{ mol}^{-1}$$

Ans: $\boxed{41.8 \text{ J K}^{-1} \text{ mol}^{-1}}$

At 0°C the water molecules in ice have less vibrational motion than they do at 25°C; therefore ice at 0°C must be more ordered than ice at 25°C and must have a correspondingly lower value of entropy at 0°C.

Exercise 11-1

The standard molar entropy of solid bismuth (*Bi*) at 100 K is 29.79 J K^{-1} mol^{-1}. What is the value at 200 K? The molar heat capacity of solid bismuth in this temperature range is given by

$$c_p = 14.097 + 0.14757T - 7.0782 \times 10^{-4} T^2 + 1.2269 \times 10^{-6} T^3 \quad \text{J mol}^{-1} \text{K}^{-1}. \; \smiley$$

Example 11-2

Calculate the change in entropy when one mole of steam at 100°C and one bar pressure is cooled to produce ice at 0°C and one bar pressure. For liquid water, $\bar{c}_p^{o\,7} = 4.20$ J K^{-1} g^{-1}; for ice, $\Delta \bar{h}_{fus(273.15)}^{o} = 333.5$ J g^{-1}; for water, $\Delta \bar{h}_{vap(373.15)}^{o} = 2258.1$ J g^{-1}.

Solution

$$\left(\begin{array}{c} \text{1 mole steam} \\ \text{100°C, 1 bar} \end{array} \right) \xrightarrow[\text{condensation}]{\Delta S_1} \left(\begin{array}{c} \text{1 mole water} \\ \text{100°C, 1 bar} \end{array} \right) \xrightarrow[\text{cooling}]{\Delta S_2} \left(\begin{array}{c} \text{1 mole water} \\ \text{0°C, 1 bar} \end{array} \right) \xrightarrow[\text{freezing}]{\Delta S_3} \left(\begin{array}{c} \text{1 mole ice} \\ \text{0°C, 1 bar} \end{array} \right)$$

$$\Delta S_{total} = \Delta S_1 + \Delta S_2 + \Delta S_3$$

First calculate $\Delta \bar{s}$ for one *gram* of H$_2$O:

The condensation of steam to water is just the opposite process to the vaporization of water to steam; therefore,

$$\Delta \bar{h}_{condens(373.15)}^{o} = -\Delta \bar{h}_{vap(373.15)}^{o} = -2258.1 \text{ J g}^{-1}$$

Similarly, freezing is just the opposite process to melting, so that

$$\Delta \bar{h}_{freezing(273.15)}^{o} = -\Delta \bar{h}_{fus(273.15)}^{o} = -333.5 \text{ J g}^{-1}$$

Remembering that Q_{rev} for an equilibrium phase transition is just ΔH_p for the transition:

---

[7] $\bar{c}_p^{o}$ is the specific heat capacity at *one bar* pressure, whereas $\bar{c}_p$ is just the specific heat capacity. Since heat capacity is not significantly affected by pressure unless the pressure is *extremely* high, the two symbols usually are used interchangeably. The same goes for the molar heat capacity: c_p^{o} is usually written just as c_p.

$$\Delta \bar{s}_1 = \frac{Q^{rev}_{condens}}{T_{bp}} = \frac{\Delta \bar{h}^o_{condens\,(373.15)}}{373.15}$$

$$= \frac{-2258.1 \text{ J g}^{-1}}{373.15 \text{ K}}$$

$$= -6.051 \text{ J K}^{-1} \text{ g}^{-1}$$

and

$$\Delta \bar{s}_3 = \frac{Q^{rev}_{freezing}}{T_{fp}}$$

$$= \frac{\Delta \bar{h}^o_{freezing\,(273.15)}}{T_{fp}}$$

$$= \frac{-333.5 \text{ J g}^{-1}}{273.15 \text{ K}}$$

$$= -1.221 \text{ J K}^{-1} \text{ g}^{-1}$$

Also, from Eqn [2]:

$$\Delta \bar{s}_2 = \bar{c}^o_P \, ln\left(\frac{T_2}{T_1}\right)$$

$$= 4.20 \, ln\left(\frac{273.15}{373.15}\right)$$

$$= -1.310 \text{ J K}^{-1} \text{ g}^{-1}$$

Therefore:

$$\Delta \bar{s}_{total} = \Delta \bar{s}_1 + \Delta \bar{s}_2 + \Delta \bar{s}_3$$

$$= -6.051 - 1.310 - 1.221$$

$$= -8.582 \text{ J K}^{-1} \text{ g}^{-1}$$

For one mole, $\Delta S = \left(18.015 \, \frac{\text{g}}{\text{mol}}\right)\left(-8.582 \, \frac{\text{J / K}}{\text{g}}\right) = -154.605 \text{ J K}^{-1} \text{ mol}^{-1}$

Ans: $\boxed{-154.6 \text{ J K}^{-1} \text{ mol}^{-1}}$

Exercise 11-2

Calculate the change in entropy when one mole of metallic aluminum is heated at one bar pressure from an initial temperature of 25°C to a final temperature of 750°C. The molar heat capacities of solid and liquid aluminum at one bar pressure are 29.2 J mol^{-1} K^{-1} and 31.75 J mol^{-1} K^{-1}, respectively. The specific enthalpy of fusion of aluminum at its melting point (660.46°C) is 396.57 J g^{-1}. The molar mass of aluminum is 26.98 g mol^{-1}. ☺

11.3 ENTROPY CHANGES FOR CHEMICAL REACTIONS

The entropy change for a chemical reaction is easily calculated using tables of absolute standard molar entropies. Thus, for the reaction

$$aA + bB \rightarrow cC + dD$$

the standard reaction entropy change ΔS_R^o is given by

$$\Delta S_R^o = c\,s_C^o + d\,s_D^o - a\,s_A^o - b\,s_B^o \qquad \ldots [8]$$

Or, in a more general form:

$$\boxed{\Delta S_R^o = \sum_{Products} n_i\,s_i^o - \sum_{Reactants} n_i\,s_i^o} \qquad \ldots [9]$$

Note once again that, unlike standard *enthalpies* of formation, which have values of *zero* for the elements, the standard *entropies* of the elements have real, positive values—because at temperatures above absolute zero pure elements all have some degree of disorder, and therefore some entropy.

Example 11-3

The standard molar entropies at 25°C for nitrogen gas, hydrogen gas, and gaseous ammonia are, respectively, 191.61, 130.68, and 192.45 J K^{-1} mol^{-1}, respectively. What is the standard entropy change at 25°C for the formation of ammonia from its elements?

$$N_{2(g)} + 3H_{2(g)} \rightarrow 2NH_{3(g)}$$

Solution

$$\Delta S_R^o = \sum_{Products} n_i\,s_i^o - \sum_{Reactants} n_i\,s_i^o$$

$$= n_{NH_3}s_{NH_3}^o - n_{N_2}s_{N_2}^o - n_{H_2}s_{H_2}^o$$

$$= (2)(192.45) - (1)(191.61) - (3)(130.68)$$

$$= -198.75 \ \text{J K}^{-1} \ \text{mol}^{-1}.$$

Ans: $\boxed{-198.75 \ \text{J K}^{-1} \ \text{mol}^{-1}}$

The values for the pure elements are not zero. And note again, that, when we report the answer in J K^{-1} *mol* $^{-1}$, the term *mol* $^{-1}$ means *per mole of reaction as written*. Thus, if we had written the reaction as

$$\tfrac{1}{2}N_{2(g)} + \tfrac{3}{2}H_{2(g)} \rightarrow NH_{3(g)}$$

the answer would have been $\tfrac{1}{2}(-198.75) = -99.38$ J K^{-1} mol^{-1}.

Example 11-4

(a) Calculate the standard reaction entropy at 25°C for the complete combustion of one mole of sucrose ($C_{12}H_{22}O_{11}$) to carbon dioxide and liquid water.

(b) Is this a spontaneous reaction?

		$C_{12}H_{22}O_{11(s)}$	$O_{2(g)}$	$CO_{2(g)}$	$H_2O_{(liq)}$
$\Delta h^o_{f(298.15)}$	(kJ mol^{-1})	-2222	0.00	-393.509	-285.830
$s^o_{298.15}$	(J K^{-1} mol^{-1})	360.2	205.138	213.74	69.91

Solution

(a) The balanced reaction is: $\quad C_{12}H_{22}O_{11(s)} + 12O_{2(g)} \rightarrow 12CO_{2(g)} + 11H_2O_{(liq)}$

$$\Delta S^o_{298.15} = 12\,s^o[CO_2] + 11\,s^o[H_2O] - s^o[C_{12}H_{22}O_{11(s)}] - 12\,s^o[O_2]$$

$$= 12(213.74) + 11(69.91) - (360.2) - 12(205.138)$$

$$= +512.034 \text{ J K}^{-1} \text{ mol}^{-1}$$

Ans: $\boxed{+512.0 \text{ J K}^{-1} \text{ mol}^{-1}}$

(b) To determine if this reaction is spontaneous we shall calculate the minimum change in the entropy of the universe that results from the combustion of one mole of sucrose in the system. Since $\Delta S_{univ} = \Delta S_{syst} + \Delta S_{surr}$, in order for ΔS_{univ} to be the *minimum* change in the entropy of the universe, it is necessary that all the changes in the surroundings be *reversible* changes. If we did not stipulate reversible changes in the surroundings, then ΔS_{surr} could have almost any large positive value, such that it could completely swamp out the contribution of ΔS_{syst} to the entropy change of the universe. It follows, therefore, that the heat transfer to the surroundings must be stipulated to be *reversible*, which means that the temperature difference between the system and the surroundings must be vanishingly small. Accordingly, the surroundings must be assumed to be at the same temperature as the system; namely 25°C. First we calculate the reaction enthalpy change:

$$\Delta H^o_{298.15} = 12\,\Delta h^o_f[CO_2] + 11\,\Delta h^o_f[H_2O] - \Delta h^o_f[C_{12}H_{22}O_{11}] - 12\,\Delta h^o_f[O_2]$$

$$= 12(-393.509) + 11(-285.830) - (-2222) - 12(0)$$

$$= -5644.238 \text{ kJ mol}^{-1}$$

At constant pressure with no work done other than PV-work, $Q_P = \Delta H_P$; therefore,

$$Q_{syst} = \Delta H^o_{298.15} = -5644.238 \text{ kJ mol}^{-1}$$

The heat that *leaves* the system *enters* the surroundings, so that $\quad Q_{surr} = -Q_{syst}$

Therefore, $\qquad\qquad\qquad\qquad\qquad Q_{surr} = -Q_{syst}$

$$= -(-5644.238)$$

3. Calculate the entropy change for the formation of liquid methanol (CH_3OH) from the elements at one bar pressure and 25°C using the data provided. Is the formation process spontaneous?

$$C_{(s)} + 2H_{2(g)} + \frac{1}{2}O_{2(g)} \rightarrow CH_3OH_{(liq)} \;\ominus$$

	$\Delta h^o_{f(298.15)}$ (kJ mol^{-1})	$s^o_{298.15}$ (J K^{-1} mol^{-1})
$C_{(s)}$	0.00	5.740
$H_{2(g)}$	0.00	130.684
$O_{2(g)}$	0.00	205.138
$CH_3OH_{(liq)}$	-238.66	126.8

4. A constant pressure of one bar is applied to water in a piston/cylinder arrangement. Will the water spontaneously evaporate at 85°C? Use the data provided to do your calculation.

$$H_2O_{(liq)} \xrightarrow{\;1\ bar,\ 85°C\;} H_2O_{(g)}$$

	$\Delta h^o_{f(298.15)}$ (kJ mol^{-1})	$s^o_{298.15}$ (J K^{-1} mol^{-1})
$H_2O_{(liq)}$	-285.830	69.91
$H_2O_{(g)}$	-241.818	188.825

For water vapor, $c_P = 30.54 + 0.1029\,T$ J mol^{-1} K^{-1}, where T is in kelvin.

For liquid water, $c_P = 75.29$ J mol^{-1} K^{-1} $\;\otimes*$

5. Carbon monoxide gas deactivates the platinum-based catalysts used in fuel cells. Small amounts of this gas are found as impurities in the oxygen fed to the fuel cell cathodes. It has been proposed that one way of getting rid of these carbon monoxide impurities would be to decompose the CO to solid graphite particles and oxygen gas by heating it to 500 K. Using the data provided, determine if this idea is worth investigating.

	$\Delta h^o_{f(298.15)}$ (kJ mol^{-1})	$s^o_{298.15}$ (J K^{-1} mol^{-1})	c_P (J mol^{-1} K^{-1})
$CO_{(g)}$	-110.525	197.674	$28.41 + 0.00410\,T$
$O_{2(g)}$	0.00	205.138	$29.96 + 0.0418\,T$
$C_{(s)}$	0.00	5.740	$16.86 + 0.00477\,T - \dfrac{8.54 \times 10^5}{T^2}$ $\;\otimes*$

6. Using the following data, calculate the absolute entropy of Cl$_2$ at -200°C and one bar:

$s^o_{298.15}[Cl_{2(g)}] = 223.066$ J K^{-1} mol^{-1}; $T_{vap} = 239.0$ K; $\Delta h^o_{vap(239.0)} = 20.40$ kJ mol^{-1};

$T_{fus} = 172.12$ K; $\Delta h^o_{fus(171.12)} = 6.406$ kJ mol^{-1};

$c_P[Cl_{2(s)}] = -12.846 + 1.249\,T - 0.00980\,T^2 + 2.861 \times 10^{-5}\,T^3$

$c_P[Cl_{2(liq)}] = +63.930 + 0.04641\,T - 1.625 \times 10^{-4}\,T^2$

$c_P[Cl_{2(g)}] = +31.695 + 0.01014\,T - 4.038 \times 10^{-6}\,T^2$ $\;\odot*$

7. At very low temperatures, solid nitrogen (N_2) exists in two forms: at temperatures below 35.16 K it exists as a solid form we shall designate as S_1, and at temperatures above 35.16 K as a solid designated as S_2. The phase transition between these two solids occurs at 35.16 K, with a transition enthalpy of 228.9 J mol^{-1}. The solid form S_2 melts at 63.15 K, with an enthalpy of fusion of 710 J mol^{-1}. Liquid N_2 in turn vaporizes at 77.36 K with an enthalpy of vaporization of 5570 J mol^{-1}. The heat capacity of solid S_1 at 10 K is 6.15 J mol^{-1} K^{-1}. Because of the difficulty of accurately measuring c_p data at temperatures close to absolute zero, a formula developed by Debye usually is used to evaluate the entropy in this very low temperature region; namely:

$$s^o_{T'} \approx (c^o_P)_{T'}/3$$

where $s^o_{T'}$ is the entropy at temperature T' ($T' \leq 10\ K$) and $(c^o_P)_{T'}$ is the molar heat capacity of the solid at temperature T'. Using the above data and the appropriate heat capacities of N_2 given below as functions of temperature, calculate the absolute molar entropy of N_2 gas at 25°C and one bar pressure and compare it with the accepted value of 191.61 J K^{-1} mol^{-1}. ⊗*

Molar heat capacities in J mol^{-1} K^{-1} as functions of temperature in kelvin:

	Applicable Temperature Range
$c^o_P[S_1] = -0.26315 + 0.45397\,T + 2.927 \times 10^{-2}\,T^2 - 1.716 \times 10^{-4}T^3$	10 K ~ 35.61 K
$c^o_P[S_1] = -1.4101 + 1.9780\,T - 3.5037 \times 10^{-2}\,T^2 + 2.5243 \times 10^{-4}T^3$	35.61 K ~ 63.15 K
$c^o_P[\text{liq}] = -153.32 + 8.755\,T - 0.12305\,T^2 + 5.873 \times 10^{-4}T^3$	63.15 K ~ 77.36 K
$c^o_P[\text{gas}] = 27.4961 + 5.2298 \times 10^{-3}T$	77.36 K ~ 1000 K

TWELVE

THERMODYNAMICS (VII)

12.1 GIBBS FREE ENERGY (*G*)

When the process taking place in the system is *spontaneous*, in principle it can be harnessed to deliver *work* to the surroundings. Conversely, when the process is *non-spontaneous*, we must do work *on* the system in order to make the process occur. According to the Second Law of Thermodynamics, the process taking place in our system is spontaneous if

$$\Delta S_{univ} = \Delta S_{syst} + \Delta S_{surr} > 0 \qquad \ldots [1]$$

Calculating ΔS_{univ} is cumbersome because, although we usually are concerned primarily with what is taking place in the system—where the process of interest occurs—we also must calculate ΔS_{surr} to find out if our process is spontaneous. For this purpose a more useful function than the entropy is the **Gibbs Free Energy *G*,** defined as

$$\boxed{G \equiv H - TS} \qquad \textit{Gibbs Free Energy} \qquad \ldots [2]$$

In Eqn [2] *H* is the enthalpy of the system, *T* is its temperature, and *S* is its entropy. *G* is a state function because all its components are state functions. Furthermore, *G* is a property of the *system*, not of the surroundings. The Gibbs Free Energy is named after J. Willard Gibbs, a famous nineteenth century American thermodynamicist who is considered to be the "father" of chemical thermodynamics.

12.2 GIBBS FREE ENERGY CHANGES, "OTHER" WORK, AND SPONTANEITY

By definition, $\qquad\qquad\qquad\qquad G \equiv H - TS \qquad\qquad\qquad \ldots [3]$

But, *H* in turn is defined as $\qquad\quad H \equiv U + PV \qquad\qquad\qquad \ldots [4]$

Substituting [4] into [3] gives: $\qquad G = U + PV - TS \qquad\qquad \ldots [5]$

Differentiating: $\qquad\qquad dG = dU + PdV + VdP - TdS - SdT$

But $\qquad\qquad\qquad\qquad\qquad dU = \delta Q + \delta W$

Therefore, $\qquad dG = (\delta Q + \delta W) + PdV + VdP - TdS - SdT \qquad \ldots [6]$

Now consider any kind of work *other* than *PV*-work (e.g., electrical work, shaft work, etc.):

If we let $\qquad W_{total} = (PV\text{-work}) + (\text{"other" work}) \qquad \ldots [7]$

then $\qquad \delta W_{total} = -P_{ext}dV + \delta W^\circ \qquad \ldots [8]$

where W° is "other" work; i.e., any work *other than* PV-work.[1]

Substituting Eqn [8] into Eqn [6] gives:

$$dG = (\delta Q - P_{ext}dV + \delta W^\circ) + PdV + VdP - TdS - SdT \qquad \ldots [9]$$

Now consider a process carried out in a *reversible* manner (infinitely slowly, no friction, etc.). For this type of process, since we know that

$$P_{ext} = P_{syst}$$

we don't need to distinguish between P_{ext} and P_{syst}, so we can just replace each with P in Eqn [9] to give

$$dG = (\delta Q - PdV + \delta W^\circ) + PdV + VdP - TdS - SdT \qquad \ldots [10]$$

The two PdV's cancel out, giving

$$dG = (\delta Q + \delta W^\circ) + VdP - TdS - SdT \qquad \ldots [11]$$

But we also know that $\qquad \delta Q = \delta Q_{rev} = TdS\,^2 \qquad \ldots [12]$

and that $\qquad \delta W^\circ = \delta W^\circ_{rev}\,^3 \qquad \ldots [13]$

Therefore, substituting Eqns [12] and [13] into Eqn [11] gives

$$dG = (TdS + \delta W^o_{rev}) + VdP - TdS - SdT$$

$$= \delta W^o_{rev} + VdP - SdT \qquad \ldots [14]$$

If the process is carried out at constant temperature and constant pressure, then $dT = dP = 0$, and Eqn [14] becomes

$$\boxed{dG_{T,P} = \delta W^o_{rev}} \qquad \ldots [15]$$

Or, in integral form: $\qquad \boxed{\Delta G_{T,P} = W^o_{rev}} \qquad \ldots [16]$

---

[1] "Other" work is often called "useful" work because usually *PV*-work occurs only as a consequence of a volume change taking place in the system. In chemical reactions, for example, such volume changes usually—although not always—are not harnessed to deliver useful work.

[2] Since $dS = \delta Q_{rev}/T$, cross-multiplying gives $\delta Q_{rev} = TdS$.

[3] Since we stipulated that the process is reversible.

Along with $PV = nRT$, it can be argued that Eqn [16] is one of the two most important equations in all of physical chemistry; as we shall see, this equation to a large extent explains how the universe works!

From Eqn [16] it is seen that if $\Delta G_{T,P}$ is negative, then W^o_{rev} also will be negative. But negative work means work delivered *by* the system, and we know that work can be delivered by a system only if the process taking place in the system is a *spontaneous* process.

Conversely, if $\Delta G_{T,P}$ is positive, then W^o_{rev} also will be positive, meaning that work must be done *on* the system in order for the process in the system to take place; this means that the process taking place in the system is *not* a spontaneous process.

Furthermore, since we specified that the process taking place in the system is a *reversible* process, then any "other" work output from this process will be the *maximum* "other" work output; similarly, any "other" work input will be the *minimum* "other" work input.

To summarize:

> IF $\Delta G_{T,P} < 0$, the process is *spontaneous*, and $\left| W^o_{rev} \right|$ is the *maximum* "other" work output.
>
> IF $\Delta G_{T,P} > 0$, the process is *not* spontaneous, and W^o_{rev} is the *minimum* "other" work input.

It is very important to understand that, in order for ΔG to provide information about whether a process is spontaneous or non-spontaneous, ΔG must be evaluated for the process taking place at constant temperature and constant pressure. If the temperature or pressure vary, then ΔG doesn't give us any information about whether or not the process is spontaneous or about the maximum or minimum values of the "other" work. [4]

The advantage of $\Delta G_{T,P}$ over ΔS_{univ} as a means of determining whether or not a given process is spontaneous is that in evaluating $\Delta G_{T,P}$ we only have to consider the properties of the *system*; we don't have to bother with what is going on in the surroundings. (This is good, because we almost never know in detail what the processes taking place in the surroundings are.) Furthermore, $\Delta G_{T,P}$ immediately gives us a value for W^o_{rev}.

Eqn [16] is an extremely important equation because it permits us to determine which processes can take place spontaneously and which can't. Knowing this information gives us a powerful tool for predicting what is possible and what is not. In other words, it helps us to understand the innermost workings of the Universe itself! Furthermore, not only are we able to determine if a process is feasible or not, but also we can calculate *quantitatively* the maximum "other" work output we can ever hope to obtain from any harnessible spontaneous process, or, alternatively, the minimum

---

[4] To be more specific, when using $\Delta G_{T,P}$ to evaluate W^o_{rev}, the criterion that the temperature and pressure must be "constant" does not mean that the system temperature and pressure must be constant *throughout* the process. Rather, it means only that the initial temperature and pressure of the system must be the same as the temperature and pressure of the surroundings, and that the temperature and pressure of the *surroundings* must be constant throughout the process. The temperature and pressure of the system are permitted to vary *during* the process, provided they start and end at the values for the surroundings.

"other" work we are going to have to be prepared to do in order to carry out any non-spontaneous process in which we have an interest.

12.3 EVALUATION OF $\Delta G_{T,P}$

By definition, $G = H - TS$

Therefore $\Delta G = \Delta H - \Delta(TS)$

$$= \Delta H - (T_2 S_2 - T_1 S_1) \qquad \ldots [17]$$

At constant temperature and pressure,[5] $T_1 = T_2 = T$, and Eqn [17] becomes

$$\boxed{\Delta G_{T,P} = \Delta H - T\Delta S} \qquad \ldots [18]$$

Combining Eqns [16] and [18]:

$$\boxed{\Delta G_{T,P} = \Delta H - T\Delta S = W_{rev}^o} \qquad \ldots [19]$$

Example 12-1

For the process $\quad \left(\begin{array}{c} 1.00 \ mol \ liquid \ water \\ 0°C, \ 1.00 \ atm \end{array} \right) \rightarrow \left(\begin{array}{c} 1.00 \ mol \ ice \\ 0°C, \ 1.00 \ atm \end{array} \right)$

(a) How much work is done? The densities of water and ice at 0°C are 999.87 and 917 kg m^{-3}, respectively.

(b) Calculate the molar entropy of liquid water at 0°C and one atm pressure, given that the value at 25°C and one atm pressure is 69.91 J K^{-1} mol^{-1}, and that the average molar heat capacity of liquid water between 0°C and 25°C is $c_P = 75.64$ J mol^{-1} K^{-1}.

(c) Using the results of part (b), calculate ΔS for the process of part (a). The molar entropy of ice at 0°C and one atm pressure is $s_{ice} = 41.28$ J K^{-1} mol^{-1}.

(d) Use the results of part (c) to determine the maximum amount of "other" work (non-expansion work) that could be done by this process. The molar heat of fusion of ice at 0°C and one atm pressure is 6.01 kJ mol^{-1}.

Solution

(a) When liquid water freezes, its volume increases; therefore PV-work is done in pushing back the atmosphere.

First we calculate the volume change of the water when it freezes:

The specific volume of liquid water at 0°C is:

---

[5] See footnote [4], previous page.

$$\overline{v}_{liq} = \frac{1}{\rho_{liq}} = \frac{1}{999.87}$$

$$= 1.00013 \times 10^{-3} \text{ m}^3 \text{ kg}^{-1}$$

The specific volume of ice at 0°C is:

$$\overline{v}_{ice} = \frac{1}{\rho_{ice}} = \frac{1}{917}$$

$$= 1.0905 \times 10^{-3} \text{ m}^3 \text{ kg}^{-1}$$

Therefore,

$$\Delta \overline{v}_{syst} = \overline{v}_{ice} - \overline{v}_{liq}$$

$$= 1.0905 \times 10^{-3} - 1.00013 \times 10^{-3}$$

$$= 9.037 \times 10^{-5} \text{ m}^3 \text{ kg}^{-1}$$

and

$$\Delta V_{syst} = \left(9.037 \times 10^{-5} \tfrac{\text{m}^3}{\text{kg}}\right)\left(0.01802 \tfrac{\text{kg}}{\text{mol}}\right)$$

$$= 1.628 \times 10^{-6} \text{ m}^3 \text{ mol}^{-1}$$

The work done per mole of water when it freezes is just PV-work:

$$W_{PV} = -P_{ext}\Delta V_{syst}$$

$$= -(101\ 325 \text{ Pa})(\ 1.628 \times 10^{-6} \text{ m}^3 \text{ mol}^{-1})$$

$$= -0.1649 \text{ J mol}^{-1}$$

The negative sign indicates that work is done by the system on the surroundings.

Ans: $\boxed{W_{syst} = -0.165 \text{ J mol}^{-1}}$

(b)

$$\Delta s = s_{273.15} - s_{298.15}$$

$$= \int_{T_1}^{T_2} \frac{c_P}{T} dT = c_P \ln\left(\frac{T_2}{T_1}\right)$$

$$= (75.64)\ ln\left(\frac{273.15}{298.15}\right)$$

$$= -6.624 \text{ J K}^{-1} \text{ mol}^{-1}$$

Therefore,

$$s_{273.15} = s_{298.15} - 6.624$$

$$= 69.91 - 6.624$$

$$= 63.286 \text{ J K}^{-1} \text{ mol}^{-1}.$$

Cold water is more ordered than warm water.

Ans: $\boxed{(s_{liq})_{273.15} = 63.29 \text{ J K}^{-1} \text{ mol}^{-1}}$

(c)
$$\Delta S_{freezing} = (s_{ice} - s_{liq})_{273.15}$$
$$= 41.28 - 63.286$$
$$= -22.006 \text{ J K}^{-1} \text{ mol}^{-1}$$

Ans: $\boxed{\Delta S = -22.01 \text{ J K}^{-1} \text{ mol}^{-1}}$

(d) For the freezing of one mole of liquid water,

$$\Delta H_{freezing} = -\Delta H_{fus} = -6010 \text{ J mol}^{-1}$$

and
$$W^o_{rev} = \Delta G_{T,P} = (\Delta H - T\Delta S)_{273.15}$$
$$= (-6010) - (273.15)(-22.006)$$
$$= -6010 + 6011 \approx 0$$

No "other" work can be done by this process. Why? Because 0°C is the freezing point (*fp*) of water. That is, water at 0°C and one atm pressure is in *equilibrium* with ice at 0°C and one atm pressure. When two phases are in equilibrium, *no net process occurs*—nothing happens. That is, there is no tendency at all for the system to move spontaneously in one direction or the other—it just sits there forever and ever without changing. For any equilibrium "process," $\Delta G_{T,P} = 0$. We saw earlier that for a reversible phase transition such as freezing or melting, $Q_{rev} = \Delta H_P$, where ΔH_P is the enthalpy change for the phase transition. Thus, for water freezing at its freezing point:

$$\Delta S_{freezing} = \frac{Q_{rev}}{T_{fp}} = \frac{\Delta H_{freezing}}{T_{fp}},$$

and for this process,
$$\Delta G_{T,P} = (\Delta H - T\Delta S)_{fp}$$
$$= \Delta H_{freezing} - (T_{fp}) \left(\frac{\Delta H_{freezing}}{T_{fp}} \right)$$
$$= \Delta H_{freezing} - \Delta H_{freezing} = 0$$

as we showed above. Since $\Delta G_{T,P} = W^o_{rev}$, if $\Delta G_{T,P} = 0$, then it follows that W^o_{rev} also must be zero. In other words, no net "other" work can be delivered by a system in equilibrium.

Ans: $\boxed{W^o_{rev} = 0}$

IF $\Delta G_{T,P} = 0$, the system is in *equilibrium*, and $W^o_{rev} = 0$.

No net "other" work
can be delivered by a
system in equilibrium

Exercise 12-1

Calculate $\Delta G_{T,P} = \Delta H - T\Delta S$ when one mole of steam at one atm pressure and 100°C condenses to form liquid water at one atm pressure and 100°C. The specific entropies at one atm pressure and 100°C of liquid water and steam are 1.3069 kJ K^{-1} kg^{-1} and 7.3549 kJ K^{-1} kg^{-1}, respectively. The specific heat of vaporization of water at one atm and 100°C is 2257.0 kJ kg^{-1}. 😊

Example 12-2

Calculate the change in the Gibbs free energy when one mole of supercooled water at −3°C and one atm pressure freezes to ice at the same temperature and pressure. Is this a spontaneous process? The heat of fusion of ice at 0°C and one atm pressure is 6.01 kJ mol^{-1}. The molar heat capacities of liquid water and ice near the normal freezing point are 75.3 and 38.0 J K^{-1} mol^{-1}, respectively.

Solution

We want to find $\Delta G_{T,P}$ for the process $\begin{pmatrix} 1 \ mol \ water \\ -3°C, \ 1 \ atm \end{pmatrix} \rightarrow \begin{pmatrix} 1 \ mol \ ice \\ -3°C, \ 1 \ atm \end{pmatrix}$

To do this, we first calculate separately ΔH and ΔS for the process at −3°C, and then combine these two quantities to obtain $\Delta G = \Delta H - T\Delta S$, remembering to use the appropriate value for T.

At 0°C: $\qquad\qquad \Delta H_{freezing} = -\Delta H_{fus} = -6010 \ \text{J mol}^{-1}$

and $\qquad\qquad \Delta S_{freezing} = \dfrac{Q_{rev}^{freezing}}{T_{fp}} = \dfrac{\Delta H_{freezing}}{T_{fp}}$

$$= \frac{-6010}{273.15} = -22.003 \ \text{J K}^{-1} \ \text{mol}^{-1}$$

At −3°C: $\qquad\qquad \Delta H_{-3°} = \Delta H_{0°} + \Delta c_P \cdot \Delta T$

$$= -6010 + (c_{P,ice} - c_{P,water})\Delta T$$
$$= -6010 + (38.0 - 75.3)(-3 - 0)$$
$$= -6010 + (-37.3)(-3)$$
$$= -6010 + 111.9$$
$$= -5898.1 \ \text{J mol}^{-1}$$

We showed earlier that for a *single substance*,

$$s_{T_2} = s_{T_1} + c_P \, ln\left(\frac{T_2}{T_1}\right)$$

Similarly, for a *process* taking place at two different temperatures, we use the same equation but replace s with ΔS, and c_P with ΔC_P:

$$\boxed{\Delta S_{T_2} = \Delta S_{T_1} + \Delta C_P \, ln\left(\frac{T_2}{T_1}\right)} \quad \begin{array}{l} \textit{Constant} \\ \textit{Pressure} \\ \textit{Process} \end{array} \quad \ldots [20]$$

where, as we showed earlier,

$$\boxed{\Delta C_P \equiv \sum_{Products} n_i \, (c_P)_i - \sum_{Reactants} n_i \, (c_P)_i} \qquad \ldots [21]$$

Therefore, returning to the problem:

$$\Delta S_{T_2} = \Delta S_{T_1} + \Delta C_P \, ln\left(\frac{T_2}{T_1}\right)$$

Thus,

$$\Delta S_{-3^\circ} = \Delta S_{0^\circ} + \Delta c_P \, ln\left(\frac{-3 + 273.15}{0 + 273.15}\right)$$

$$= -22.003 + (-37.3) \, ln\left(\frac{270.15}{273.15}\right)$$

$$= -22.003 - (37.3)(-0.01104)$$

$$= -21.591 \text{ J K}^{-1} \text{ mol}^{-1}$$

Therefore:

$$(\Delta G_{T,P})_{270.15} = (\Delta H - T\Delta S)_{270.15}$$

$$= (-5898.1) - (270.15)(-21.591)$$

$$= -65.29 \text{ J mol}^{-1}$$

The negative sign indicates that this is a spontaneous process—subcooled water (liquid water cooled to below its freezing point) spontaneously freezes![6]

Ans: $\boxed{-65.3 \text{ J mol}^{-1}; \text{ spontaneous}}$

Example 12-3

Nitrogen is compressed isothermally at 100°C from an initial pressure of 1.00 bar to a final pressure of 10.0 bar. An external pressure of 25 bar is used to carry out the compression. Calculate the change in Gibbs free energy for one mole of nitrogen undergoing this process. Is this a spontaneous process?

Solution

$$\Delta G = \Delta H - T\Delta S$$

---

[6] OK, so you already knew this. But it's nice to confirm it anyway.

Since the process is isothermal, $\Delta T = 0$, and $\Delta H = C_p \Delta T = 0$ [7]

Since G, H, and S are state functions, ΔG, ΔH, and ΔS depend only on the initial and final states of the system. Thus, it doesn't matter how much external pressure was used to carry out the compression; ΔG is fixed by the initial and final states of the system.[8]

The internal energy of an ideal gas depends only on its temperature; therefore, since the process is isothermal,

$$\Delta U = Q + W = 0$$

and $$Q = -W$$

If the process is carried out *reversibly*, then

$$Q_{rev} = -W_{rev}$$

But we know that for the isothermal compression of an ideal gas,

$$W_{rev} = -nRT \ln\left(\frac{V_2}{V_1}\right) = -nRT \ln\left(\frac{P_1}{P_2}\right)$$

and we also know that for any isothermal process,

$$\Delta S = \frac{Q_{rev}}{T}$$

from which $$T\Delta S = Q_{rev}$$

Therefore, $$\Delta G = \Delta H - T\Delta S$$
$$= 0 - Q_{rev}$$
$$= -(-W_{rev})$$
$$= +W_{rev}$$
$$= -nRT \ln\left(\frac{P_1}{P_2}\right)$$

Putting in values:

$$\Delta G = -(1)(8.314)(373.15)\ln\left(\frac{1.0}{10.0}\right)$$

$$= +7143 \text{ J mol}^{-1}$$

$$= +7.14 \text{ kJ mol}^{-1}$$

_____

[7] For an *ideal gas*, $\Delta H = C_p \Delta T$ even if the pressure *changes* during the process. That is, we use the heat capacity at constant pressure even though the pressure is not constant. This is because the energy of an ideal gas depends only on its temperature, and not on its pressure. This will be proved in your thermodynamics course. For now you'll just have to take our word for it!

[8] The work and the heat, of course, *do* depend on the actual external pressure used.

This process is not spontaneous because we know from experience that a gas at a low pressure does not spontaneously compress itself. To make the process take place, we had to do work on it by the application of an external pressure.

Note that in this case the sign of ΔG does not tell us whether or not the process is spontaneous. Why? Because it is only ΔG at *constant temperature and pressure* that gives us this information; for this process the pressure was not constant.

Ans: | $+7.14$ kJ mol^{-1}; not spontaneous |

12.4 $\Delta G°$ FOR A CHEMICAL REACTION

There are two ways to calculate the standard free energy change $\Delta G_{T,P}^o$ at one bar pressure and temperature T for the reaction

$$aA + bB \rightarrow cC + dD$$

The first way involves using the standard molar enthalpies of formation Δh_f^o and the standard molar entropies $s°$ of the reactants and products. Using this method, we separately calculate $\Delta H°$ and $\Delta S°$ for the reaction and combine them to give

$$\boxed{\Delta G_{T,P}^o = \Delta H_T^o - T\Delta S°} \qquad \ldots [22]$$

In the second method, we make use of standard molar free energies of formation Δg_f^o to evaluate $\Delta G_{T,P}^o$ directly:

$$\boxed{\Delta G_{T,P}^o = \sum_{Products} n_i (\Delta g_f^o)_i - \sum_{Reactants} n_i (\Delta g_f^o)_i} \qquad \ldots [23]$$

The **standard molar free energy of formation** Δg_f^o of a compound at temperature T is defined as

> *the change in the Gibbs free energy under standard conditions (one bar pressure, most stable form of the compound) for the formation at temperature T of one mole of compound from its elements.*

Values of standard molar free energies of formation at 25°C often are found in the same tables as standard molar enthalpies of formation. As in the case of Δh_f^o, $\Delta g_f^o = 0$ for an element at all temperatures.

Example 12-4

Discuss the "cold" combustion of methane gas at 25°C and one bar pressure:

$$CH_{4(g)} + 2O_{2(g)} \rightarrow CO_{2(g)} + 2H_2O_{(liq)}$$

Solution

Putting in values of Δh_f^o and Δg_f^o obtained from the Appendices:

$$\Delta H_{298.15}^o = \Delta h_f^o[CO_2] + 2\,\Delta h_f^o[H_2O] - \Delta h_f^o[CH_4] - 2\,\Delta h_f^o[O_2]$$

$$= (-393.51) + 2(-285.83) - (-74.81) - 2(0)$$

$$= -890.36 \text{ kJ mol}^{-1}$$

$$\Delta G_{298.15}^o = \Delta g_f^o[CO_2] + 2\,\Delta g_f^o[H_2O] - \Delta g_f^o[CH_4] - 2\,\Delta g_f^o[O_2]$$

$$= (-394.36) + 2(-237.13) - (-50.72) - 2(0)$$

$$= -817.90 \text{ kJ mol}^{-1}$$

If no attempt is made to extract work from this process, we just get the release of 890.36 kJ of heat into the surroundings, since $Q_P = \Delta H_P$.

If, instead, we somehow manage to harness some of this energy to do useful work for us (for example, with an appropriate electrocatalyst we could carry out the reaction in a fuel cell), then we could obtain up to a maximum of 817.90 kJ of "other" work (electrical work in the case of a fuel cell).

Example 12-5

Which is the better reducing agent for iron ore (Fe_3O_4): zinc or aluminum?

		$Fe_{(s)}$	$Al_{(s)}$	$Zn_{(s)}$	$Fe_3O_{4(s)}$	$Al_2O_{3(s)}$	$ZnO_{(s)}$
$\Delta h_{f(298.15)}^o$	(kJ mol^{-1})	0.00	0.00	0.00	−1118.4	−1675.7	−348.28
$s_{298.15}^o$	(J K^{-1} mol^{-1})	27.28	28.33	41.63	146.4	50.92	43.64

Solution

Iron ore is reduced to pure iron by each reducing agent according to the following reactions:

$$Fe_3O_4 + 4Zn \rightarrow 3Fe + 4ZnO \qquad \ldots \text{[a]}$$

and
$$Fe_3O_4 + \tfrac{8}{3}Al \rightarrow 3Fe + \tfrac{4}{3}Al_2O_3 \qquad \ldots \text{[b]}$$

The more spontaneous a reaction, the greater is the driving force for the reaction to proceed; therefore $\Delta G_{T,P}$, which is a measure of the degree of spontaneity for a process at constant temperature and pressure, also is a measure of the tendency of the reaction to proceed. It follows then that the reducing agent that produces iron via the more negative reaction free energy change will tend to be the better reducing agent for the iron ore.

Therefore, for each reaction we first must calculate ΔS and ΔH, and then $\Delta G_{T,P} = \Delta H - T\Delta S$.

For reaction [a]:

$$(\Delta S^o_{298.15})_a = 3\,s^o\,[\text{Fe}] + 4\,s^o\,[\text{ZnO}] - s^o\,[\text{Fe}_3\text{O}_4] - 4\,s^o\,[\text{Zn}]$$
$$= 3(27.28) + 4(43.64) - (146.4) - 4(41.63)$$
$$= -56.52 \text{ J K}^{-1}\text{ mol}^{-1}$$

$$(\Delta H^o_{298.15})_a = 3\,\Delta h^o_f[\text{Fe}] + 4\,\Delta h^o_f[\text{ZnO}] - \Delta h^o_f[\text{Fe}_3\text{O}_4] - 4\,\Delta h^o_f[\text{Zn}]$$
$$= 3(0) + 4(-348.28) - (-1118.4) - 4(0)$$
$$= -274.72 \text{ kJ mol}^{-1}$$

$$(\Delta G^o_{298.15})_a = (\Delta H^o_{298.15})_a - T(\Delta S^o_{298.15})_a$$
$$= (-274\,720) - (298.15)(-56.52)$$
$$= -257\,869 \text{ J mol}^{-1}$$
$$= -257.9 \text{ kJ mol}^{-1}$$

For reaction [b]:

$$(\Delta S^o_{298.15})_b = 3\,s^o\,[\text{Fe}] + \tfrac{4}{3}\,s^o\,[\text{Al}_2\text{O}_3] - s^o\,[\text{Fe}_3\text{O}_4] - \tfrac{8}{3}\,s^o\,[\text{Al}]$$
$$= 3(27.28) + \tfrac{4}{3}(50.92) - (146.4) - \tfrac{8}{3}(28.33)$$
$$= -72.213 \text{ J K}^{-1}\text{ mol}^{-1}$$

$$(\Delta H^o_{298.15})_b = 3\,\Delta h^o_f[\text{Fe}] + \tfrac{4}{3}\,\Delta h^o_f[\text{Al}_2\text{O}_3] - \Delta h^o_f[\text{Fe}_3\text{O}_4] - \tfrac{8}{3}\,\Delta h^o_f[\text{Al}]$$
$$= 3(0) + \tfrac{4}{3}(-1675.7) - (-1118.4) - \tfrac{8}{3}(0)$$
$$= -1115.867 \text{ kJ mol}^{-1}$$

$$(\Delta G^o_{298.15})_b = (\Delta H^o_{298.15})_b - T(\Delta S^o_{298.15})_b$$
$$= (-1\,115\,867) - (298.15)(-72.213)$$
$$= -1\,094\,337 \text{ J mol}^{-1}$$
$$= -1094.3 \text{ kJ mol}^{-1}$$

It is clear that reaction [b] has a more negative Gibbs free energy change per mole of iron oxide than does reaction [a]; therefore, at 25°C and one bar pressure, aluminum is a better reducing agent than zinc.

Ans: | Aluminum is the better reducing agent.

Exercise 12-2

Given the following standard molar Gibbs free energies of formation, calculate the standard reaction Gibbs free energy change for the two reactions given in Example 12-5 and compare the values with the values obtained there using molar enthalpy of formation and molar entropy data. ☺

	$Fe_{(s)}$	$Al_{(s)}$	$Zn_{(s)}$	$Fe_3O_{4(s)}$	$Al_2O_{3(s)}$	$ZnO_{(s)}$
$\Delta g^o_{f(298.15)}$ (kJ mol^{-1})	0.00	0.00	0.00	-1015.4	-1582.3	-318.30

Exercise 12-3

A hydrogen-oxygen fuel cell operates at 25°C and one bar pressure. The cell delivers 10.0 kW of continuous electrical power to the power grid. The overall cell reaction is

$$H_{2(g)} + \tfrac{1}{2}O_{2(g)} \rightarrow H_2O_{(liq)}$$

(a) If the cell operates continuously 24 hours per day and electricity can be sold for 20.0¢ per kilowatt-hour (kWh), what is the commercial value of the electricity generated by the cell in one year (365 days)?

(b) What is the minimum mass of hydrogen fuel required to operate the cell for one year?

One kilowatt-hour of energy is the amount of energy delivered when one kilowatt flows for one hour, and is equal to exactly 3.600×10^6 J. The standard molar enthalpy of formation of liquid water at 25°C is $\Delta h^o_f = -285.830$ kJ mol^{-1}. The standard molar entropies at 25°C of hydrogen gas, oxygen gas, and liquid water are 130.684, 205.138, and 69.91 J K^{-1} mol^{-1}, respectively. ☹

12.5 THERMODYNAMIC STABILITY

For gaseous ozone (O_3), tables list a value at 25°C of $\Delta g^o_f = +163.2$ kJ mol^{-1}. This means that at one bar pressure and 298.15 K, the value of the free energy of reaction ΔG^o_R for

$$\tfrac{1}{2}O_{2(g)} + O_{2(g)} \rightarrow O_{3(g)}$$

is $\Delta G^o_{298.15} = +163.2$ kJ mol^{-1}.

The positive value of $\Delta G^o_{T,P}$ tells us that this is not a spontaneous reaction under these conditions. However, because G is a state function, the value of $\Delta G^o_{T,P}$ for the *reverse* reaction is -163.2 kJ mol^{-1}; this negative value tells us that the reverse reaction *is* spontaneous:

$$O_{3(g)} \rightarrow \tfrac{1}{2}O_{2(g)} + O_{2(g)}, \qquad \Delta G^o_{298.15} = -163.2 \text{ kJ mol}^{-1}$$

Thus, ozone at 25°C and one atm pressure tends to spontaneously decompose to its elements. In other words, O_3 is thermodynamically *unstable*. Thus, for any compound:

IF $\Delta g_f^o > 0$, the compound is thermodynamically unstable with respect to its elements;

IF $\Delta g_f^o < 0$, the compound tends to form spontaneously from its elements, and is said to be thermodynamically stable with respect to its elements (at one bar and the given temperature).

KEY POINTS FOR CHAPTER TWELVE

1. **Gibbs Free Energy G:** Defined as $G \equiv H - TS$, G is a state function because all its components are state functions. Therefore, $\Delta G = \Delta H - \Delta(TS)$

   In differential form:
   $$dG = dH - TdS - SdT = d(U + PV) - TdS - SdT$$
   $$= dU + PdV + VdP - TdS - SdT$$
   $$= (\delta Q_{rev} + \delta W_{rev}) + PdV + VdP - TdS - SdT$$
   $$= (TdS - PdV + \delta W_{rev}^o) + PdV + VdP - TdS - SdT$$
   $$= \delta W_{rev}^o + VdP - SdT$$

   where W_{rev}^o is the reversible "other" work. "Other" work is any kind of work "other" than PV-work; e.g., electrical work.

2. When no "other" work W^o is done (i.e., only PV-work is done), then, for a pure substance

   $$dG = VdP - SdT$$

3. For any process taking place at constant temperature and pressure,

   $$dG_{T,P} = \delta W_{rev}^o \quad \text{or, in integral form} \quad \Delta G_{T,P} = W_{rev}^o$$

4. $\Delta G_{T,P} < 0$, process is *spontaneous*, and $\left| W_{rev}^o \right|$ is the *maximum* "other" work *output*.

   $\Delta G_{T,P} > 0$, process is *not* spontaneous, and W_{rev}^o is the *minimum* "other" work *input*.

   If the temperature and pressure are not constant, then ΔG tells us nothing about the maximum or minimum "other" work, or whether or not the process is spontaneous.

   $\Delta G_{T,P}$ is a very useful function because it tells us if the process taking place in the system is spontaneous or not, without having to evaluate the entropy change for the universe, which entails an evaluation of ΔS for both the system *and* the surroundings. To evaluate $\Delta G_{T,P}$ we only need information for the *system*; we don't need to calculate ΔS for the surroundings.

5. A system at **equilibrium** will be at constant temperature and pressure throughout. In a system at equilibrium, nothing happens; it just sits there. Therefore a system at equilibrium at constant temperature and pressure cannot deliver any net "other" work; i.e., $W^o = 0$, and it follows that

   $$\Delta G_{T,P} = 0 \quad \text{(for a system in equilibrium)}$$

6. To evaluate $\Delta G_{T,P}$ **at different temperatures**, we must be able to evaluate ΔH at differ-

ent temperatures and also be able to evaluate ΔS at different temperatures.

We already know how to evaluate ΔH at T_2 if we know its value at T_1:

If ΔC_P does not vary with temperature, then $\quad \Delta H_{T_2} = \Delta H_{T_1} + \Delta C_P \cdot \Delta T$

where $$\Delta C_P = \sum_{Products} n_i (c_P)_i - \sum_{Reactants} n_i (c_P)_i$$

If ΔC_P does vary with temperature, then

$$\Delta H_{T_2} = \Delta H_{T_1} + \int_{T_1}^{T_2} \Delta C_P dT$$

If ΔC_P does not vary with temperature and we know ΔS at T_1, then we can evaluate ΔS at T_2 from

$$\Delta S_{T_2} = \Delta S_{T_1} + \Delta C_P \ln\left(\frac{T_2}{T_1}\right)$$

If ΔC_P does vary with temperature, then

$$\Delta S_{T_2} = \Delta S_{T_1} + \int_{T_1}^{T_2} \frac{\Delta C_P}{T} dT$$

7. **Standard Gibbs free energy change $\Delta G°$ for a chemical reaction**:

    For the reaction $\qquad aA + bB \rightarrow cC + dD$

    (i) Using standard molar enthalpies of formation Δh_f^o and standard absolute molar entropies $s°$:

    $$\Delta G_T^o = \Delta H_T^o - T\Delta S°$$

    $$or$$

    (ii) Using standard molar Gibbs free energies of formation $\Delta g°$ from tables:

    $$\Delta G_T^o = \left[\sum_{Products} n_i (\Delta g_f^o)_i - \sum_{Reactants} n_i (\Delta g_f^o)_i \right]_T$$

8. The **standard molar Gibbs free energy of formation Δg_f^o** of a compound at temperature T is defined as

    > *the change in the Gibbs free energy under standard conditions (one bar pressure, most stable form of compound) for the formation at temperature T of one mole of compound from its elements.*

    $\Delta g_f^o = 0$ for an element at all temperatures.

9. **Thermodynamic stability**: At one bar pressure and the given temperature,

    IF $\Delta g_f^o > 0$, the compound is thermodynamically unstable with respect to its elements;

    IF $\Delta g_f^o < 0$, the compound tends to form spontaneously from its elements, and is said to be thermodynamically stable with respect to its elements.

PROBLEMS

1. Calculate the entropy change at 25 °C and one bar pressure for the reaction

$$C_2H_{2(g)} + H_{2(g)} \rightarrow C_2H_{4(g)}$$

using the following data at one bar pressure:

	$C_2H_{2(g)}$	$C_2H_{4(g)}$
$\Delta h^o_{f(298.15)}$ (kJ mol^{-1})	226.73	52.26
$\Delta g^o_{f(298.15)}$ (kJ mol^{-1})	209.20	68.15	☺

2. Calculate the maximum useful work that may be obtained from the process

$$2CO_{(g)} \rightarrow CO_{2(g)} + C_{(s)}$$

at 298.15 K and one bar pressure, using the following data, at 25°C and one bar:

	$CO_{(g)}$	$CO_{2(g)}$	$C_{(s)}$
$\Delta h^o_{f(298.15)}$ (kJ mol^{-1})	−110.525	−393.509	0
$s^o_{298.15}$ (J K^{-1} mol^{-1})	197.674	213.74	5.74	☺

3. Calculate the maximum "other" work that could be obtained from the process:

1.00 kg water vapor $\rightarrow$ *1.00 kg liquid water*

at 80°C and one atm. Heat of vaporization of water at the normal boiling point = 2257 J g^{-1}; heat capacity of liquid water = 4.20 J g^{-1} K^{-1}; heat capacity of water vapor = 2.10 J g^{-1} K^{-1}. ☺

4. For the process $\left(\begin{array}{c} \text{1 mol liquid toluene} \\ \text{70.0°C, 1 atm} \end{array}\right) \rightarrow \left(\begin{array}{c} \text{1 mol toluene vapor} \\ \text{70.0°C, 1 atm} \end{array}\right)$

calculate: (a) ΔS (b) ΔG.

The normal boiling point of toluene (methyl benzene) is 111°C; its molar mass is 92.15 g mol^{-1}. For toluene at 111°C and one atm pressure: Δh_{vap} = 33.18 kJ mol^{-1}; $\bar{c}_P$ [liquid toluene] = 1.67 J g^{-1} K^{-1}; $\bar{c}_P$ [toluene vapor] = 1.13 J g^{-1} K^{-1}. Is this a spontaneous process? ☺

5. E.D. Eastman and W.C. McGavock studied the transition of rhombic sulfur to monoclinic sulfur [J. Am. Chem. Soc., **59**, 145 (1937)]. At one atmosphere pressure the following phase transition takes place at 368.5 K:

$$S_{(s)} [rhombic] \rightarrow S_{(s)} [monoclinic]$$

At the transition temperature they obtained a value of +96 cal mol^{-1} for the enthalpy of transition. They also determined that, over the temperature range of 298 K to 369 K, for this reaction

$$\Delta C_p = 0.085 + 0.66 \times 10^{-3} T$$

where ΔC_p is in cal mol^{-1} K^{-1} and T is in kelvin. At 25°C and one bar pressure, rhombic sulfur is the more stable form; therefore its standard molar free energy of formation is taken as zero at this temperature. What is the standard molar free energy of formation of monoclinic sulfur at 25°C? ☹*

6. In a certain process 100 kg s^{-1} of superheated water at 121°C and one atm pressure is flash vaporized to produce superheated steam at the same temperature and pressure. How much useful power potentially is available from this process?

    c_p [liquid water] = 75.6 J mol^{-1} K^{-1}; c_p [water vapor] = 37.8 J mol^{-1} K^{-1}; Δh_{vap} at 100°C = 40.66 kJ mol^{-1}. ☺

7. A one-gram piece of ice (the system) at a temperature of –10°C is placed on a dish in a room that is at a room temperature of 25°C and at atmospheric pressure. (a) Calculate ΔS for the system after thermal equilibrium has been reached, assuming no evaporation takes place. (b) Calculate the resulting entropy change of the universe, taking the surroundings to be at 25°C and one atm pressure. Is the process spontaneous? (c) Could we also use the value of ΔG for the process to determine if it is spontaneous?

    The specific heat of fusion of ice at 0°C and one atm pressure is 333.5 J g^{-1}; the specific heat capacities at constant pressure of liquid water and of ice are 4.2 J g^{-1} K^{-1} and 2.1 J g^{-1} K^{-1}, respectively. Note that any heat absorbed by the surroundings does not affect the temperature of the surroundings, which may be considered to be what is known as an "infinite heat reservoir." In effect, this means that no temperature gradients exist in the surroundings, so that any heat absorbed by the surroundings is considered to be absorbed *reversibly* by the surroundings. ☹

8. Calculate ΔG for the following process at –10°C and one atm pressure:

    $$\left(\begin{array}{c} 1.00 \ g \ ice \\ -10°C, \ 1 \ atm \end{array} \right) \xrightarrow{\ \Delta G\ } \left(\begin{array}{c} 1.00 \ g \ liquid \ water \\ -10°C, \ 1 \ atm \end{array} \right)$$

    Is this process spontaneous? The specific heat of fusion of ice at 0°C and one atm pressure is 333.5 J g^{-1}; the specific heat capacities at constant pressure of liquid water and of ice are 4.2 J g^{-1} K^{-1} and 2.1 J g^{-1} K^{-1}, respectively. ☺

9. Calculate ΔG for the following process:

    $$\left(\begin{array}{c} 1.00 \ g \ ice \\ 0°C, \ 1 \ atm \end{array} \right) \xrightarrow{\ \Delta G\ } \left(\begin{array}{c} 1.00 \ g \ liquid \ water \\ 0°C, \ 1 \ atm \end{array} \right)$$

    Is this process spontaneous? The specific heat of fusion of ice at 0°C and one atm pressure is 333.5 J g^{-1}; the specific heat capacities at constant pressure of liquid water and of ice are 4.2 J g^{-1} K^{-1} and 2.1 J g^{-1} K^{-1}, respectively. ☺

10. Using standard Gibbs free energy of formation data in the appendices at the back of the book, calculate the change in the Gibbs free energy at 25°C and one bar pressure when one mole of acetylene is reacted with hydrogen to produce ethylene. ☺

11. Using the data provided, calculate the Gibbs free energy change at 25°C and one bar pressure for the reaction:

$$C_2H_{2(g)} + H_{2(g)} \rightarrow C_2H_{4(g)}$$ ☺

	$\Delta h^o_{f(298.15)}$ (kJ mol^{-1})	$s^o_{298.15}$ (J K^{-1} mol^{-1})
$C_2H_{2(g)}$	226.73	200.94
$H_{2(g)}$	0.00	130.684
$C_2H_{4(g)}$	52.26	219.56

12. A beaker containing one mole of liquid water—the "system"—at 325K and one atm pressure is placed in a hot water bath that is maintained at a constant temperature of 100°C. The beaker is left in the hot water bath until the temperature of the water in the beaker just reaches 350K. For this process, calculate: (a) ΔH_{syst} (b) W_{syst} (c) ΔU_{syst} (d) ΔS_{syst} (e) ΔS_{univ} (f) ΔG_{syst} (g) if this is a spontaneous process. Why?

Data: Density of liquid water at 325K = 0.988 g cm^{-3}; at 350K = 0.972 g cm^{-3}
Molar mass of water = 18.015 g mol^{-1}
Molar entropy of liquid water at 25°C and one atm pressure = 69.939 J K^{-1} mol^{-1}
Molar heat capacity for liquid water: $c_p = 75.3$ J mol^{-1} K^{-1} ☹

13. Calculate ΔG the following process:

$$\begin{Bmatrix} 1.00 \text{ mole liquid water} \\ 25°C, \ 1.00 \text{ atm} \end{Bmatrix} \rightarrow \begin{Bmatrix} 1.00 \text{ mole water vapor} \\ 25°C, \ 1.00 \text{ atm} \end{Bmatrix}$$

The heat of vaporization for water at 100°C and one atm pressure is 40.6 kJ mol^{-1}. The molar heat capacities at constant pressure for liquid water and for water vapor may be taken as constant at 75.3 J mol^{-1} K^{-1} and 33.6 J mol^{-1} K^{-1}, respectively. ☹

14. One mole of an ideal gas is reversibly compressed from 75.00 kPa to 102.00 kPa at a constant temperature of 25.00°C. Calculate W, Q, ΔH, ΔS, and ΔG for this compression. ☺

15. Calculate the maximum "other" work available from the following process:

$$\begin{pmatrix} 1.00 \text{ mol liquid } H_2O \\ 38.9 \text{ kPa}, 75.0°C \end{pmatrix} \rightarrow \begin{pmatrix} 1.00 \text{ mol } H_2O \text{ vapor} \\ 38.9 \text{ kPa}, 75.0°C \end{pmatrix}$$

Between 0°C and 110°C the vapor pressure $P_w^\bullet$ of water (in torr) is given by

$$P_w^\bullet = -34865 + 354.2T - 1.2014T^2 + 1.3617 \times 10^{-3}T^2$$

The molar heat capacities of liquid water and water vapor as functions of temperature are:

$$c_P[H_2O_{(liq)}] = 72.43 + 0.0104T - 1.497 \times 10^{-6}T^2$$

$$c_P[H_2O_{(g)}] = 28.85 + 0.0121T + \frac{100\,603}{T^2}$$

In the above equations, c_P is in J mol^{-1} K^{-1} and T is in kelvin. The enthalpy of vaporization for liquid water at equilibrium with water vapor at 75.0°C is 41 820 J mol^{-1}. ☺

16. Calculate the maximum "other" work available from the following process:

$$\left(\begin{array}{c} 1.00\ mol\ H_2O\ vapor \\ 1.00\ atm,\ 80.0°C \end{array}\right) \rightarrow \left(\begin{array}{c} 1.00\ mol\ liquid\ H_2O \\ 1.00\ atm,\ 80.0°C \end{array}\right)$$

   Data: Enthalpy of vaporization at normal boiling point = 40.66 kJ mol^{-1}
   Heat capacity of liquid water = 75.4 J mol^{-1} K^{-1}
   Heat capacity of water vapor = 36.2 J mol^{-1} K^{-1} ☹

17. Calculate ΔG for 1.00 kg of water as it cools down at one atm pressure from 370 K to the temperature of the surroundings at 300 K. Is this process spontaneous? For liquid water, $\bar{c}_P$ = 4.184 J g^{-1} K^{-1}. The absolute molar entropy of liquid water at 25°C and one atm pressure is 69.939 J K^{-1} mol^{-1}. The molar mass of water is 18.015 g mol^{-1}. *Hint:* For a process such as this, in which there is no "other" work done and the pressure is constant, from Eqn [14] in the text,

$$dG_P = -SdT, \quad \text{from which it follows that} \quad \Delta G_P = -\int_{T_1}^{T_2} SdT.$$

Therefore, in order to solve for ΔG_P you will have to obtain an expression for S as a function of temperature and integrate. When you do this, you will find the following integration formula useful: $\int \ln x\,dx = x\ln x - x$ ☹*

18. Calculate the maximum "other" work output available from the following process:

$$\left(\begin{array}{c} 1.00\ kg\ liquid\ H_2O \\ 1.00\ atm,\ 121°C \end{array}\right) \rightarrow \left(\begin{array}{c} 1.00\ kg\ superheated\ steam \\ 1.00\ atm,\ 121°C \end{array}\right) ☺$$

   $\bar{c}_P$[liq. water] = 4.18 J g^{-1} K^{-1}; $\bar{c}_P$[steam] = 2.09 J g^{-1} K^{-1}; $\Delta\bar{h}_{vap}$[water] = 2257 J g^{-1} at 100°C

19. Using the data provided, calculate the change in the Gibbs free energy for the following process:

$$\left(\begin{array}{c} 1.00\ kg\ liquid\ toluene \\ 1.00\ atm,\ 70°C \end{array}\right) \rightarrow \left(\begin{array}{c} 1.00\ kg\ toluene\ vapor \\ 1.00\ atm,\ 70°C \end{array}\right)$$

   $c_P[CH_3C_6H_{5(liq)}]$ = 157.3 J mol^{-1} K^{-1}; $c_P[CH_3C_6H_{5(g)}]$ = 103.64 J mol^{-1} K^{-1}; at the normal boiling point (110.63°C) the enthalpy of vaporization of toluene is 33.18 kJ mol^{-1}. The molar mass of toluene is 92.14 g mol^{-1}. Compare the answer with that of Problem 4. ☺

20. The molar entropy of hydrogen at 25°C and one bar pressure is 130.684 J K^{-1} mol^{-1}. Calculate

   (a) the molar entropy of hydrogen at 100°C and one bar pressure, and

   (b) ΔG on heating 1.00 mole of hydrogen at one bar pressure from 25.0°C to 100.0°C if the molar heat capacity can be assumed to be constant at 28.87 J mol^{-1} K^{-1}. Is this a spontaneous process? *Hint:* $\int \ln T\,dT = T\ln T - T$ ☹*

21. Calculate ΔG for the process:

$$\begin{pmatrix} 1.00 \ g \ ice \\ -20°C, \ 1.00 \ atm \end{pmatrix} \rightarrow \begin{pmatrix} 1.00 \ g \ liquid \ water \\ -20°C, \ 1.00 \ atm \end{pmatrix}$$

Data: $\bar{c}_p[water] = 1.000 \ cal \ g^{-1}$; $\bar{c}_p[ice] = 0.500 \ cal \ g^{-1}$

$\Delta\bar{h}_{fus}[ice] = 80 \ cal \ g^{-1}$ at 0°C and one atm ☺

22. A certain liquid, which is in equilibrium with its vapor at 300 K, is contained in the insu-lated, frictionless, piston/cylinder arrangement at the right. An electrical resistor of 100 Ω is located in the liquid. The external pressure is constant at 1.50 bar. The system—which con-sists of the liquid, the vapor, and the resistor—has an initial volume of 1.800 m³. When an electrical current of 100 mA is passed through

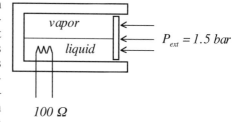

the resistor for 5.00 hours, some of the liquid is caused to vaporize isothermally, slowly pushing out the piston until the system has a final volume of 1.840 m³. If the vapor behaves as an ideal gas, calculate, for this process:

(a) Q, W, ΔU, and ΔH for the system, (b) the molar enthalpy of vaporization of the liquid at 300 K, and (c) the change in the entropy of the universe. ☹

23. (a) Calculate ΔG for 500 g of ice water as it heats up at one atm pressure from 0.00°C to the temperature of the surroundings at 25.0°C. For liquid water, $\bar{c}_p = 4.184 \ J \ g^{-1} \ K^{-1}$. The absolute molar entropy of liquid water at 25°C and one atm pressure is 69.939 J mol⁻¹

   K⁻¹. The molar mass of water is 18.015 g mol⁻¹. *Hint:* $\int \ln x \, dx = x \ln x - x$

   (b) Is this a spontaneous process? Prove it by means of a numerical calculation. ☹*

24. A pharmaceutical company wishes to develop a new diet drug C, made from A and B. The synthesis is to be carried out at 25°C and one bar pressure. Company scientists have suggested three possible procedures to synthesize this compound:

Method 1:	*Method 2:*	*Method 3:*
Step 1: $A + D \rightarrow E$	Step 1: $A + B \xrightarrow{\text{catalyst}} I$	Step 1: $A + F \rightarrow G$
Step 2: $B + E \rightarrow C + D$	Step 2: $I \rightarrow C$	Step 2: $B + H \rightarrow J$
Overall: $A + B \rightarrow C$	Overall: $A + B \rightarrow C$	Step 3: $G + J \rightarrow C + F + H$
		Overall: $A + B \rightarrow C$

Because of budget constraints, it is necessary that each of the proposed steps reacts essentially to 100% conversion. As the chief engineer in charge of research, which (if any) of the three proposed methods do you instruct your research team to start developing? The following data are available:

Compound	$\Delta g^o_{f(298.15)}$ kJ mol^{-1}	Compound	$\Delta g^o_{f(298.15)}$ kJ mol^{-1}	Compound	$\Delta g^o_{f(298.15)}$ kJ mol^{-1}
A	−200	E	−300	H	+70
B	−100	F	−130	I	−250
C	−420	G	−350	J	−110
D	+50				

☹

CHEMICAL EQUILIBRIUM (I)

13.1 EQUILIBRIUM CONSTANTS

Consider the following reaction at 25°C and one bar pressure:[1]

$$H_{2(g)} + I_{2(s)} \rightarrow 2HI_{(g)}$$

For this reaction,
$$\Delta G° = 2\, \Delta g_f^o[HI] - \Delta g_f^o[H_2] - \Delta g_f^o[I_2]$$
$$= 2(1.70) - (0) - (0)$$
$$= +3.4 \text{ kJ mol}^{-1}$$

Alternatively,
$$\Delta G° = \Delta H° - T\Delta S°$$
$$= [2(26\,480) - (0) - (0)] - (298.15)\{2(206.594) - (130.684) - (116.135)\}$$
$$= 52\,960 - (298.15)(166.369)$$
$$= 3357 \text{ J mol}^{-1} = +3.36 \text{ kJ mol}^{-1} \text{ (same as above)}$$

Since $\Delta G_{T,P}^o > 0$, this reaction is not spontaneous. This means that if H_2 and I_2 are mixed together at 25°C and one bar pressure they won't react completely to form 2HI at 25°C and one bar. BUT, they *will* form *some* HI, to produce an *equilibrium mixture* containing H_2, I_2, and HI.

$\Delta G°$ is the standard Gibbs free energy change for a reaction in which the reactants react *completely* to form products. In the case of incomplete reactions, we define another type of Gibbs free energy, the **Reaction Gibbs Free Energy** ΔG_R, which refers to the *incomplete* conversion of reactants to form products in a *mixture* (Fig. 1).

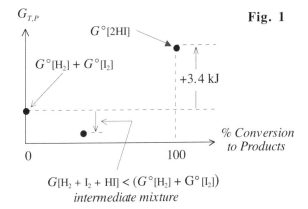

Thus, for the above reaction, although $\Delta G° > 0$, there will exist some intermediate compositions for which $\Delta G_R < 0$; which means that *some* products will form, but we won't obtain 100% conversion. Instead, we obtain *partial* conversion to form an *equilibrium mixture* that contains both reactants *and* products.

Thus, referring to the various mixtures shown in Fig. 2, for

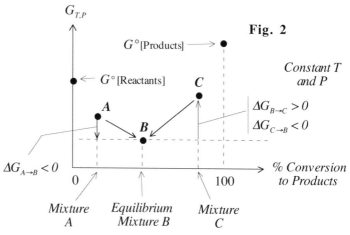

$$\text{Mixture } A \rightarrow \text{Mixture } B$$

$\Delta G_R < 0$; therefore mixture A tends spontaneously to form more products. That is, the reaction *Mixture A* → *Mixture B* occurs spontaneously. However, for *Mixture B* → *Mixture C*, $\Delta G_R > 0$; therefore *this* process will *not* take place; however, for the *reverse* process, that is, for *Mixture C* → *Mixture B*, $\Delta G_R < 0$. This means that *Mixture C* spontaneously tends to form *Mixture B*.

For some specific intermediate mixture—*Mixture B*—situated somewhere between Mixture A and Mixture C, $\Delta G_R = 0$ (exactly), and there is *no* tendency for this mixture to form either more products or more reactants. That is, *Mixture B is at equilibrium*. Thus, the criterion for a system to be at chemical equilibrium is that $(\Delta G_R)_{T,P} = 0$.

$$\boxed{(\Delta G_R)_{T,P} = 0}$$ *Criterion for Chemical Equilibrium* $\cdots$ [1]

13.2 THE REACTION QUOTIENT (Q) AND RELATIVE ACTIVITIES (a_i)

For any chemical reaction $aA + bB \rightarrow cC + dD$

we define the *Reaction Quotient Q*[2] as

$$\boxed{Q \equiv \frac{a_C^c \cdot a_D^d}{a_A^a \cdot a_B^b}}$$ *Reaction Quotient* $\cdots$ [2]

where a_i is called the **relative activity** of species i.

For an *ideal gas*, the relative activity is given by

---

[2] This Q shouldn't be confused with the Q for *heat*.

$$a_i = \left(\frac{P_i}{P^o} \right) \quad \begin{array}{l} \textit{Relative Activity} \\ \textit{of an Ideal Gas} \end{array} \qquad \dots [3]$$

Thus, the **relative activity** of an **ideal gas** i is just the *relative partial pressure* of the gas. Since this has the units of pressure/pressure, the relative activity of a gas has no units (is dimensionless); this means you have to make sure that the partial pressure P_i of the gas and the standard pressure P° both are expressed in the same units. Thus, if P_i is expressed in bar, then P° must have a value of 1.00 bar; if P_i is in Pa, then P° must have a value of 100 000 Pa; if P_i is in Torr, then P° must be 760 Torr, and so forth.

The relative activity of a **pure liquid** or of a **pure solid** is taken as unity ($a_{solid} = a_{liq} = 1$); for an **impure liquid** or **impure solid** it can be taken as the mole fraction purity ($= x_i$).

The relative activity of a **solute dissolved in a liquid solution** at very low concentration is expressed as a relative concentration: c_i /c°, where c_i is the concentration of the solute in mol L^{-1}, and c° is a standard concentration, which, for this course, we can take as 1 M (mol L^{-1}). For higher concentrations (greater than about 0.01 M) we have to make use of a correction factor known as the **activity coefficient** in order to properly evaluate the activity. (Activity coefficients will be discussed in more detail later in Chapter 19, when we consider ionic equilibrium.) Note that in all cases, a_i is *unitless* (Table 1).

To summarize:

Table 1. *Evaluation of Relative Activity for Different Types of Species*

Type of Species	Relative Activity	Type of Species	Relative Activity
Ideal gas i	$a_i = P_i/P^\circ$	Impure liquid i	$a_i = x_i < 1$
Pure liquid i	$a_i = x_i = 1$	Impure solid i	$a_i = x_i < 1$
Pure solid i	$a_i = x_i = 1$	Solute i in dilute solution	$a_i = c_i/c^\circ$

Consider again the reaction $\qquad aA + bB \rightarrow cC + dD$

As the reaction proceeds, at any point during the reaction we can calculate the value of the reaction quotient Q by taking a small sample from the reaction mixture and analyzing for the different concentrations of reactants and products present. Note from Eqn [2] that as the reaction proceeds, the value of Q constantly changes. When equilibrium is finally reached, however, the reaction stops and the concentrations (activities) of all species no longer change with time.

To show the reaction is at equilibrium we now write

$$aA + bB \rightleftharpoons cC + dD \qquad \dots [4]$$

where we have replaced the arrow $\rightarrow$ with $\rightleftharpoons$. Under these conditions, Q will have a special unique value Q_{eqm} which is *determined only by the temperature* of the equilibrium; this value is called the **thermodynamic equilibrium constant** and is given the symbol K.

Thus, for the equilibrium expressed by Eqn [4]:

$$Q_{eqm} \equiv K = \left(\frac{a_C^c \cdot a_D^d}{a_A^a \cdot a_B^b} \right)_{eqm} \quad \begin{array}{l} \textit{Equilibrium} \\ \textit{Constant} \end{array} \qquad \cdots [5]$$

The equilibrium constant K is just the value of Q for the equilibrium mixture.

Note in Eqns [2] and [5] that the relative activity of each species must be raised to the power of its stoichiometric coefficient (as given by the stoichiometry of the reaction).

13.3 FORMULATION OF EQUILIBRIUM CONSTANTS

The following examples show you how to formulate equilibrium constants in terms of activities:

For a **Gas Phase Chemical Equilibrium**, such as

$$H_{2(g)} + \tfrac{1}{2}O_{2(g)} \rightleftharpoons H_2O_{(g)}$$

$$K = \left(\frac{a_{H_2O_{(g)}}}{a_{H_{2(g)}} \cdot a_{O_{2(g)}}^{1/2}} \right)_{eqm} = \left\{ \frac{\left(P_{H_2O}/P^o \right)}{\left(P_{H_2}/P^o \right)\left(P_{O_2}/P^o \right)^{1/2}} \right\}_{eqm} = \left(\frac{P_{H_2O}}{P_{H_2} \cdot P_{O_2}^{1/2}} \right)_{eqm} \cdot (P^o)^{1/2}$$

For a **Phase Equilibrium**, such as

$$H_2O_{(liq)} \rightleftharpoons H_2O_{(V)}$$

$$K = \left(\frac{a_{H_2O_{(V)}}}{a_{H_2O_{(liq)}}} \right)_{eqm} = \left(\frac{\left(P_{H_2O_{(V)}}/P^o \right)}{1} \right)_{eqm} = \frac{P_{H_2O_{(V)}}^\bullet}{P^o}$$

Since $P_{H_2O_{(V)}}$ must be the *equilibrium* value, we use $P_{H_2O_{(V)}}^\bullet$, the *equilibrium vapor pressure* of water at the specified temperature. Note that $P_{H_2O_{(V)}}^\bullet$ is a function of temperature only.

As an example of a **Solubility Equilibrium**, consider solid AgCl (a salt of very low solubility) in equilibrium with its ions in saturated solution (Fig. 3). The equilibrium reaction is expressed as

$$AgCl_{(s)} \rightleftharpoons Ag_{(aq)}^+ + Cl_{(aq)}^-$$

The equilibrium constant for this type of *saturation equilibrium* is called the **solubility product** K_{SP}, which, for this particular equilibrium, is defined as

$$K = K_{SP} = \left(\frac{a_{Ag^+} \cdot a_{Cl^-}}{a_{AgCl_{(s)}}} \right)_{eqm} \qquad \cdots [6]$$

Since the undissolved AgCl is a pure solid, it has a relative activity of unity; since the solution is very dilute, the relative activities of the dissolved ionic species are given by their relative concentrations, as discussed above. Making these substitutions, Eqn [6] becomes

$$K_{SP} \approx \frac{(c_{Ag^+}/1)(c_{Cl^-}/1)}{(1)}$$

$$= c_{Ag^+} c_{Cl^-} = [Ag^+_{(aq)}][Cl^-_{(aq)}]$$

where the square brackets denote molar concentrations in mol L^{-1}. Note that since K_{SP} is an *equilibrium* constant, the expression only holds for *saturation* equilibrium; i.e., the concentrations of the dissolved species are at their saturated (maximum) values—i.e., their actual equilibrium *solubilities* under the given conditions.

Fig. 3

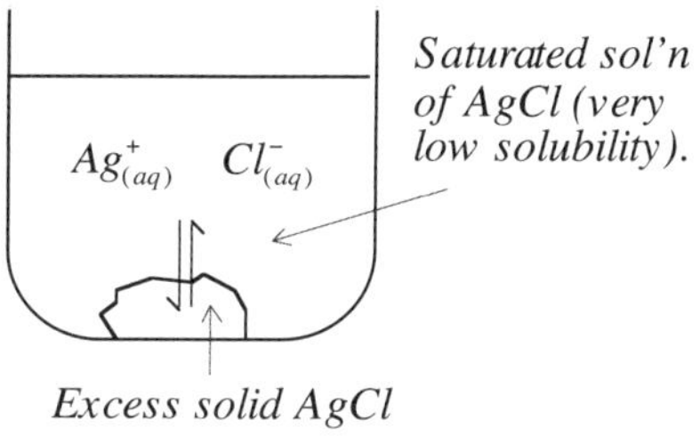

Saturated sol'n of AgCl (very low solubility).

Excess solid AgCl

Example 13-1

Formulate expressions for the equilibrium constant for each of the following reactions:

(a) $\quad 3C_{(graphite)} + 2H_2O_{(g)} \rightleftharpoons CH_{4(g)} + 2CO_{(g)}$

(b) $\quad CaCO_{3(s)} \rightleftharpoons CaO_{(s)} + CO_{2(g)}$

(c) $\quad 2Fe_2O_{3(s)} \rightleftharpoons 4Fe_{(s)} + 3O_{2(g)}$

(d) $\quad Na_{(s)} + H_2O_{(liq)} \rightleftharpoons Na^+_{(aq)} + OH^-_{(aq)} + \frac{1}{2}H_{2(g)}$

Solution

In each case, $P° = 1$ *bar*, and all partial pressures are equilibrium values, expressed in bar.

(a) $\quad K = \dfrac{a_{CH_4} \cdot a_{CO}^2}{a_C^3 \cdot a_{H_2O}^2} = \dfrac{(P_{CH_4}/P^o)(P_{CO}/P^°)^2}{(x_C)^3 \cdot (P_{H_2O}/P^°)^2} = \dfrac{(P_{CH_4}/1)(P_{CO}/1)^2}{(1)^3 \cdot (P_{H_2O}/1)^2} = \boxed{\dfrac{P_{CH_4} \cdot P_{CO}^2}{P_{H_2O}^2}}$ **Ans.**

(b) $\quad K = \dfrac{a_{CaO} \cdot a_{CO_2}}{a_{CaCO_3}} = \dfrac{(x_{CaO})(P_{CO_2}/P^o)}{x_{CaCO_3}} = \dfrac{(1)(P_{CO_2}/1)}{(1)} = \boxed{(P_{CO_2})_{eqm}}$ **Ans.**

(c) $\quad K = \dfrac{(x_{Fe})^4(P_{O_2}/P^o)^3}{(x_{Fe_2O_3})^2} = \dfrac{(1)^4(P_{O_2}/1)^3}{(1)^2} = \boxed{(P_{O_2})_{eqm}^3}$ **Ans.**

(d) $\quad K = \dfrac{a_{Na^+} \cdot a_{OH^-} \cdot \sqrt{a_{H_2}}}{a_{Na} \cdot a_{H_2O}} \approx \dfrac{(c_{Na^+}/c^o)(c_{OH^-}/c^o) \cdot \sqrt{P_{H_2}/P^o}}{(x_{Na})(x_{H_2O})} = \dfrac{[Na^+_{(aq)}][OH^-_{(aq)}]\sqrt{P_{H_2}}}{(1)(x_{H_2O})} \cdot \dfrac{1}{(c^o)^2\sqrt{P^o}}$

$$= \dfrac{[Na^+_{(aq)}][OH^-_{(aq)}]\sqrt{P_{H_2}}}{x_{H_2O}} \cdot \dfrac{1}{(c^o = 1 \; mol \; L^{-1})^2\sqrt{P^o = 1.00 \; bar}} = \boxed{\dfrac{[Na^+_{(aq)}][OH^-_{(aq)}]\sqrt{P_{H_2}}}{x_{H_2O}}} \quad \textbf{Ans.}$$

In part (d) we have assumed that the solution is sufficiently dilute that we may replace the ion activities with their concentrations.[3] Under these conditions $x_{H_2O} \approx 1$ and the expression further reduces to

$$K = [Na^+_{(aq)}][OH^-_{(aq)}]\sqrt{P_{H_2}} \; .$$

13.4 MOLAR FREE ENERGIES

For any homogeneous substance, $G = H - TS$ (by definition)

For 1 mole: $g = h - Ts$

Differentiating: $dg = dh - Tds - sdT$

At constant T, $dT = 0$: $dg_T = dh - Tds$ . . . [7]

By definition, $h = u + Pv$

therefore: $dh = du + Pdv + vdP$

                                          $= (\delta Q + \delta W) + Pdv + vdP$ . . . [8]

If we carry out the process in a reversible[4] manner, with only PV-work being done, then

$$\delta Q = \delta Q_{rev} = Tds \qquad \text{. . . [9]}$$

and $\delta W = -Pdv$ . . . [10]

Substituting Eqns [9] and [10] into Eqn [8]:

$$dh = (Tds - Pdv) + Pdv + vdP$$

$$\boxed{dh = Tds + vdP} \qquad \text{. . . [11]}$$

Substituting [11] into [7]: $dg_T = dh - Tds$

                                          $= (Tds + vdP) - Tds$

                                          $= vdP$ . . . [12]

Integrating Eqn [12]:

$$\int_{g_1}^{g_2} dg_T = \int_{P_1}^{P_2} vdP$$

that is,

$$\boxed{g_2 - g_1 = \Delta g_T = \int_{P_1}^{P_2} vdP} \quad \begin{array}{l} \textit{Constant T,} \\ \textit{PV-Work Only} \end{array} \qquad \text{. . . [13]}$$

---

[3] The solution must be very dilute to use $a_i \approx c_i$; typically less than 0.01 molar. Actually, if solid Na were present the solution would be *very* concentrated. For further discussion on the implications of this, refer to Exercise 1 later in this chapter.

[4] Remember that $ds = \delta Q_{rev}/T$, so that $\delta Q_{rev} = Tds$.

Eqn [13] tells us how the molar Gibbs free energy of a substance varies with pressure at constant temperature. Let's see how Eqn [13] is applied:

For Liquids and Solids

Since liquids and solids are essentially incompressible, then the molar volume $v \approx constant$; i.e., v is not a function of pressure and therefore can be taken outside the integral sign to give

$$\Delta g_T \approx v \int_{P_1}^{P_2} dP$$

$$\approx v(P_2 - P_1)$$

$$\approx v\Delta P$$

Therefore, $\boxed{\Delta g_T \approx v\Delta P}$ *Solids and Liquids*   . . . [14]

Eqn [14] shows how the free energy of a solid or liquid varies with pressure at constant temperature.

For Ideal Gases

For one mole of an ideal gas, $v = \dfrac{RT}{P}$

therefore: $\Delta g_T = \int_{P_1}^{P_2} v dP = \int_{P_1}^{P_2} \dfrac{RT}{P} dP$

$$= RT \int_{P_1}^{P_2} \dfrac{1}{P} dP = RT \ln\left(\dfrac{P_2}{P_1}\right)$$

Therefore for one mole of an ideal gas: $\boxed{\Delta g_T = RT \ln\left(\dfrac{P_2}{P_1}\right)}$ *Constant T,*   . . . [15]
                                                                                                Only PV-Work

For n moles of an ideal gas: $\boxed{\Delta G_T = nRT \ln\left(\dfrac{P_2}{P_1}\right)}$ . . . [16]

For one mole of an ideal gas at constant temperature, Eqn [15] says

$$g_2 - g_1 = RT \ln\left(\dfrac{P_2}{P_1}\right)   \qquad . . . [17]$$

If we let $P_1 = P°$, the standard pressure (1 bar), then $g_1 = g°$, the molar free energy of the gas in its standard state. Similarly, if we let P_2 be any general pressure P, then g_2 is just g, the corresponding value of the molar free energy at pressure P. Eqn [17] then can be written:

$$g - g^o = RT \ln\left(\frac{P}{P^o}\right)$$

that is,

$$g = g^o + RT \ln\left(\frac{P}{P^o}\right) \qquad \ldots [18]$$

But for an ideal gas, $\frac{P}{P^o}$ is just the *relative activity a* of the gas, so that Eqn [18] becomes

$$\boxed{g_T = g_T^o + RT \ln a} \quad \textit{Ideal Gas} \qquad \ldots [19]$$

Since most real gases behave as ideal gases under normal conditions, Eqn [19] also can be used for real gases, providing the pressure doesn't get too high or the temperature too low (when interactions start to become important). Eqn [19] is a very important equation in thermodynamics because it is the basis for the concept of the *chemical potential*, which is the driving force for a chemical reaction.

13.5 ΔG_R FOR CHEMICAL REACTIONS

Suppose we have the following gas phase reaction at constant temperature and pressure, for which all the gases are ideal gases: [5]

$$aA_{(g)} + bB_{(g)} \rightarrow cC_{(g)} + dD_{(g)}$$

Consider the reaction of a moles of A with b moles of B *at any intermediate stage of the reaction*. Under these conditions, each component A, B, C, and D will have some value of molar free energy g_A, g_B, g_C, and g_D, respectively. We define the **Reaction Gibbs Free Energy** ΔG_R as

$$\boxed{\Delta G_R \equiv c g_C + d g_D - a g_A - b g_B} \quad \begin{array}{l}\textit{Reaction}\\ \textit{Gibbs Free}\\ \textit{Energy}\end{array} \quad \ldots [20]$$

The molar free energy of each component can be expressed using Eqn [19]. When these expressions are substituted into Eqn [20] we obtain

$$\Delta G_R = c(g_C^o + RT \ln a_C) + d(g_D^o + RT \ln a_D) - a(g_A^o + RT \ln a_A) - b(g_B^o + RT \ln a_B)$$

$$= (c g_C^o + d g_D^o - a g_A^o - b g_B^o) + RT(c \ln a_C + d \ln a_D - a \ln a_A - b \ln a_B)$$

$$= \Delta G^o + RT \ln\left(\frac{a_C^c \cdot a_D^d}{a_A^a \cdot a_B^b}\right)$$

Therefore,

$$\boxed{\Delta G_R = \Delta G^o + RT \ln Q} \quad \begin{array}{l}\textit{Reaction Gibbs}\\ \textit{Free Energy}\end{array} \quad \ldots [21]$$

---

[5] Usually a reasonable approximation at temperatures greater than 0°C and pressures less than 5-10 bar. ("Hot vacuum," remember?)

where

$$\Delta G^o \equiv c\,g_C^o + d\,g_D^o - a\,g_A^o - b\,g_B^o$$

Standard
Reaction $\cdots$ [22]
Free Energy

ΔG_R is the *Reaction Gibbs Free Energy* at any given stage during the reaction; ΔG^o is the *Standard Reaction Gibbs Free Energy* for the reaction, and Q is the reaction quotient, $a_C^c \cdot a_D^d / a_A^a \cdot a_B^b$, at the same given stage of the reaction. As indicated on page 10 of Chapter 12, ΔG° is evaluated either from the differences in the standard molar Gibbs free energies of formation between the reaction products and reactants, or from the standard enthalpy and entropy changes using the relationship $\Delta G^\circ = \Delta H^\circ - T\Delta S^\circ$.

Exercise 13-1

(a) Is the reaction $\qquad Na_{(s)} + H_2O_{(liq)} \rightarrow Na_{(aq)}^+ + OH_{(aq)}^- + \frac{1}{2}H_{2(g)}$

spontaneous at 25°C when the relative activity of both the Na^+ ions and the OH^- ions in solution is 0.001 (on the mol L^{-1} scale) and the hydrogen gas is evolved at one bar pressure? Determine this by evaluating the reaction Gibbs free energy ΔG_R for the reaction under these conditions.

(b) As the reaction proceeds, the concentration of the Na^+ ions and the OH^- ions in solution gradually builds up. In order to evaluate the activities of highly concentrated electrolytes it becomes necessary to make use of **activity coefficients** (γ). Thus, for aqueous NaOH solution, it can be shown[6] that the product of the relative activities of the $Na_{(aq)}^+$ ions and the $OH_{(aq)}^-$ ions is given by:

$$a_{Na^+} \cdot a_{OH^-} = (\gamma_+ m_{Na^+})(\gamma_- m_{OH^-}) = \gamma_\pm^2 \cdot m_{Na^+} \cdot m_{OH^-}$$

where γ_+ and γ_- are the activity coefficients of the $Na_{(aq)}^+$ ions and the $OH_{(aq)}^-$ ions, respectively; m_{Na^+} and m_{OH^-} are the *molal* concentrations[7] of the ions, and $\gamma_\pm$ is the mean molal ionic activity coefficient for NaOH in solution at the specified concentration. When the NaOH concentration has built up to a value of 20 molal, is the reaction more spontaneous than in part (a) or less spontaneous? For 20 molal aqueous NaOH solution at 25°C, $\gamma_\pm = 19.28$. The relative activity of the water in 20 molal aqueous NaOH solution at 25°C is 0.1357.

Use the following data for your calculation:

		$Na_{(aq)}^+$	$OH_{(aq)}^-$	$H_{2(g)}$	$Na_{(s)}$	$H_2O_{(liq)}$
$\Delta h_{f(298.15)}^o$	(kJ mol^{-1})	-240.12	-229.994	0.00	0.00	-285.830
$s_{298.15}^o$	(J K^{-1} mol^{-1})	59.0	-10.75	130.684	51.21	69.91

[Note that the value listed in the above table for the absolute molar entropy of the hydroxyl ion in aqueous solution is *negative*. Although it is possible to determine the entropy of an ionic solute such as NaCl dissolved in an electrolyte solution, for theoretical reasons this entropy value cannot

---

[6] This will be treated in more detail in Chapter 19.

[7] The molal concentration of a solute is defined as the number of moles of solute per *kg of solvent* in the solution.

be separated into the contribution from the positive ion (the cation) and that from the negative ion (the anion). We can, however, measure *changes* in the entropies of single ions in solution. Therefore we arbitrarily assign a value of *zero* for the entropy of the hydrogen ion in aqueous solution and report the entropy values of all other ions relative to the zero value for the hydrogen ion. Since some ions in solution are more ordered (have lower entropy) than aqueous hydrogen ions and others are less ordered (have greater entropy) than aqueous hydrogen ions, it follows that some ions—the ones that are more ordered than the hydrogen ion—will have *negative* values for their entropies. In actuality, of course, the absolute values of the entropies of all ions will be positive.]

☹

KEY POINTS FOR CHAPTER THIRTEEN

1. $\Delta G°$ is the standard Gibbs free energy change for a chemical reaction in which the reactants react *completely* to form products. In the case of incomplete reactions, we define another type of Gibbs free energy change called the **Reaction Gibbs Free Energy ΔG_R**, which refers to the *incomplete* conversion of reactants to form products in a *mixture*.

2. There will be some intermediate mixture (situated somewhere between the starting reactant mixture and the mixture corresponding to 100% conversion of all the reactants to products) at which $\Delta G_R = 0$ (exactly). In the mixture of reactants and products having this composition there is *no* tendency for the mixture to form either more products or more reactants. That is, *this intermediate mixture is at equilibrium*. Thus, the criterion for a system to be at chemical equilibrium is that

    $$(\Delta G_R)_{T,P} = 0 \qquad \left(\begin{array}{l}\textit{Criterion for a mixture to}\\ \textit{be at chemical equilibrium}\end{array}\right)$$

3. The **relative activity a_i** of a species is a measure of its tendency to react.

For an **ideal gas i**:	$a_i = P_i/P^o$	For an **impure liquid i**: $\qquad a_i = x_i < 1$
For a **pure liquid i**:	$a_i = x_i = 1$	For an **impure solid i**: $\qquad a_i = x_i < 1$
For a **pure solid i**:	$a_i = x_i = 1$	For a **solute i in dilute solution**: $\quad a_i = c_i/c^o$

4. For the chemical reaction $\qquad aA + bB \rightarrow cC + dD$

    we define the **reaction quotient Q** as

    $$Q \equiv \frac{a_C^c \cdot a_D^d}{a_A^a \cdot a_B^b}$$

5. When the chemical reaction $\quad aA + bB \rightleftharpoons cC + dD$ has attained equilibrium, the value of the reaction quotient Q_{eqm} is called the **thermodynamic equilibrium constant K**:

    $$K = Q_{eqm} \equiv \left(\frac{a_C^c \cdot a_D^d}{a_A^a \cdot a_B^b}\right)_{eqm}$$

6. For 1 mole of homogeneous material: $h \equiv u + Pv$

    Differentiating: $dh = du + Pdv + vdP$

    $\qquad\qquad\qquad\quad = (\delta Q + \delta W) + Pdv + vdP$

    If the process is carried out in a reversible manner, with only PV-work being done, then

    $$\delta Q = \delta Q_{rev} = Tds \quad \text{and} \quad \delta W = -Pdv$$

    Therefore: $dh = (Tds - Pdv) + Pdv + vdP = Tds + vdP$

    and $\qquad\qquad dh = Tds + vdP$

7. For 1 mole of homogeneous material: $g \equiv h - Ts$

    Differentiating: $dg = dh - Tds - sdT$

    At constant T, $dT = 0$: $dg_T = dh - Tds$

    Substituting: $\qquad\quad = (Tds + vdP) - Tds = vdP$

    Therefore: $dg_T = vdP$

    Integrating: $g_2 - g_1 = \Delta g_T = \int_{P_1}^{P_2} vdP$

    For **solids** and **liquids**, $v \approx constant$, therefore $\Delta g_T = v\int_{P_1}^{P_2} dP = v\Delta P$

    For **ideal gases**, $v = RT/P$, therefore $\Delta g_T = \int_{P_1}^{P_2} \dfrac{RT}{P} dP = RT \ln\left(\dfrac{P_2}{P_1}\right)$

    For n moles of **ideal gas**, $\Delta G_T = nRT \ln\left(\dfrac{P_2}{P_1}\right)$

    If $P_2 = P$ and $P_1 = P^\circ$ (standard pressure of 1 bar), then

    $$g - g^o = RT \ln\left(\frac{P}{P^o}\right) = RT \ln\left(\frac{P}{1}\right) = RT \ln P$$

    i.e., $\qquad\qquad g = g^o + RT \ln P \qquad \begin{pmatrix} Molar\ free\ energy \\ for\ an\ ideal\ gas \end{pmatrix}$

8. For the gas phase chemical reaction $aA_{(g)} + bB_{(g)} \rightarrow cC_{(g)} + dD_{(g)}$

    we define the **Reaction Gibbs Free Energy** ΔG_R at any stage of a chemical reaction as

    $$\Delta G_R \equiv c g_C + d g_D - a g_A - b g_B$$

    Substituting $g_i = g_i^o + RT \ln a_i$ for each term in ΔG_R and rearranging gives

$$\Delta G_R = \Delta G^o + RT \ln\left(\frac{a_C^c \cdot a_D^d}{a_A^a \cdot a_B^b}\right) = \Delta G^o + RT \ln Q$$

where ΔG^o is called the **Standard Reaction Gibbs Free Energy**, and is defined as

$$\Delta G^o \equiv c\,g_C^o + d\,g_D^o - a\,g_A^o - b\,g_B^o$$

PROBLEMS

1. At 110.0°C the equilibrium vapor pressure of water is 1074.5 Torr, and the heat of vaporization is 533.5 cal g^{-1}. Assuming water vapor behaves as an ideal gas, calculate the change in Gibbs free energy for the vaporization process at 110.0°C and one atm pressure. Is this process possible? The density of liquid water at 110°C may be taken as 951 kg m^{-3} and may be considered independent of pressure.

    Data: $H_2O = 18.015$ g mol^{-1}; $\bar{c}_P$[steam] $= 0.480$ cal g^{-1} K^{-1}; $\bar{c}_P$[liquid water] $= 1.00$ cal g^{-1} K^{-1}
    ☹*

2. At 2.02 atm water boils at 121°C. Calculate the change in the Gibbs free energy when one mole of steam at 121°C and *one* atm condenses to liquid water at the same temperature and pressure. Is this process possible? Assume that steam behaves as an ideal gas and that the density of liquid water at 121°C is constant at 942 kg m^{-3}. ☹*

3. What is ΔG when one kg of liquid water at 100°C and 4.00 bar pressure vaporizes to steam at the same temperature and pressure? You may assume that water vapor behaves as an ideal gas. The molar mass of water is 18.015 g mol^{-1}. ☹*

4. The *Steam Tables* is a comprehensive compilation of accurate thermodynamic data for water and steam at various temperatures and pressures. Calculate ΔG for the following process:

    $$\left(\begin{array}{c}1.00 \ mol \ steam \\ 200 \ bar, \ 400°C\end{array}\right) \rightarrow \left(\begin{array}{c}1.00 \ mol \ steam \\ 240 \ bar, \ 400°C\end{array}\right)$$

    (a) Using entries from the Steam Tables given below, and
    (b) Assuming steam behaves as an ideal gas.

    The molar mass of water is 18.015 g mol^{-1}. ☺

	T (° C)	P (bar)	$\bar{v}_{gas}$ (cm^3 g^{-1})	$\bar{u}_{gas}$ (J g^{-1})	$\bar{h}_{gas}$ (J g^{-1})	$\bar{s}_{gas}$ (J K^{-1} g^{-1})
Superheated steam	400	200	9.94	2 619.3	2 818.1	5.5540
Superheated steam	400	240	6.73	2 477.8	2 639.4	5.2393

5. If a 1.00 L sample of ethanol (C_2H_5OH) at 0.00°C were subjected to a pressure of 100 atm, by how much would its Gibbs free energy change? The densities of ethanol at 25.00°C and 0.00°C are 0.785 g cm^{-3} and 0.806 g cm^{-3}, respectively. The molar mass of ethanol is 46.07 g mol^{-1}. Ethanol may be assumed to be an incompressible liquid. ☺

6. Derive an expression for the Gibbs free energy change when one mole of an ideal gas with constant heat capacity c_P goes from an initial temperature and pressure of T_1 and P_1 to a final temperature and pressure of T_2 and P_2. *Hint:* $\int \ln T \, dT = T \ln T - T$ ☠*

7. Calculate the Gibbs free energy change for the following process:

$$\begin{pmatrix} 500 \ mol \ N_2 \\ 300 \ K, \ 5.00 \ bar \end{pmatrix} \longrightarrow \begin{pmatrix} 500 \ mol \ N_2 \\ 400 \ K, \ 4.00 \ bar \end{pmatrix}$$

Nitrogen gas may be considered to behave as an ideal diatomic gas with a constant heat capacity of $c_P = 29.10$ J mol^{-1} K^{-1}. The standard molar entropy of gaseous nitrogen at 25°C is 191.61 J K^{-1} mol^{-1}. Comment on the significance of the value of ΔG. To solve this problem you may use the formulas developed in Problem 6. ☹

8. At 630°C and a pressure at equilibrium of 1.00 atm, SO_3 is 33.1% dissociated into SO_2 and O_2. Calculate the equilibrium constant for the following reactions at 630°C:

(a) $2SO_3 \rightleftharpoons 2SO_2 + O_2$

(b) $SO_3 \rightleftharpoons SO_2 + \frac{1}{2}O_2$

(c) $SO_2 + \frac{1}{2}O_2 \rightleftharpoons SO_3$ ☹

9. At low pressure and elevated temperature, ethane gas (C_2H_6) dehydrogenates to ethylene (C_2H_4) and hydrogen according to the reaction

$$C_2H_{6(g)} \rightleftharpoons C_2H_{4(g)} + H_{2(g)}$$

If the equilibrium constant K is 0.04053 at 900 K, calculate the partial pressure of ethylene at this temperature if a mixture of ethane (800 mol) and inert nitrogen gas (3000 mol) is passed over a catalyst that allows the reaction to attain equilibrium. The total pressure on the system is constant at 80.0 kPa. ☹

CHEMICAL EQUILIBRIUM (II)

14.1 CALCULATIONS USING THERMODYNAMIC EQUILIBRIUM CONSTANTS

In Chapter 13 we saw that when a system is in chemical equilibrium the reaction Gibbs free energy for the system is equal to zero:

$$(\Delta G_R)_{T,P} = 0$$

and the total Gibbs free energy for the system is at a *minimum* (Fig. 1). We also showed that the reaction Gibbs free energy takes the form

$$\Delta G_R = \Delta G^\circ + RT \ln Q \ \ldots [1]$$

where ΔG° is the *standard reaction Gibbs free energy* and Q is the *reaction quotient*, defined by

$$Q \equiv \frac{a_C^c \cdot a_D^d}{a_A^a \cdot a_B^b}$$

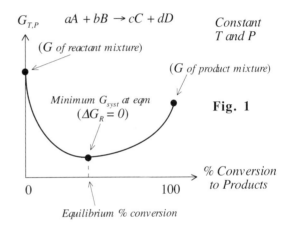

$G_{T,P}$ $aA + bB \rightarrow cC + dD$ *Constant T and P*

(G of reactant mixture)

(G of product mixture)

Minimum G_{syst} at eqn ($\Delta G_R = 0$)

Fig. 1

% Conversion to Products

0   100

Equilibrium % conversion

We also saw that at chemical equilibrium, Q becomes the *equilibrium constant, K*. Therefore, under equilibrium conditions Eqn [1] becomes:

$$0 = \Delta G^\circ + RT \ln K$$

from which

$$\boxed{\ln K = -\frac{\Delta G^o}{RT}}$$ ... [2]

That is,

$$\boxed{K = exp\left[-\frac{\Delta G^o}{RT}\right]}$$ ... [3]

Eqns [2] and [3] show that in order to calculate the equilibrium constant, all we need to know is the value of $\Delta G°$ for the reaction, which is readily calculated from the standard thermodynamic data found in the appendices. The value of K given by Eqn [3] that is based on *activities* is known as the **thermodynamic equilibrium constant**.

Example 14-1

Evaluate the thermodynamic equilibrium constant for the production at 25°C and one bar pressure of gaseous ammonia from its constituent gases:

$$N_{2(g)} + 3H_{2(g)} \rightarrow 2NH_{3(g)}$$

Solution

$$\Delta G°_{298.15} = 2\,\Delta g°_f[NH_3] - \Delta g°_f[N_2] - 3\,\Delta g°_f[H_2]$$

$$= 2(-16.45) - (0) - 3(0)$$

$$= -32.90 \text{ kJ mol}^{-1}$$

Therefore,
$$ln\,K = -\frac{\Delta G°_R}{RT} = -\frac{(-32\,900)}{(8.314)(298.15)} = 13.27$$

from which
$$K_{298.15} = exp[13.27] = 5.81 \times 10^5$$

Expanding:
$$K = \left(\frac{a^2_{NH_{3(g)}}}{a_{N_{2(g)}} \cdot a^3_{H_{2(g)}}}\right)_{eqm} = \left\{\frac{\left(P_{NH_3}/P^o\right)^2}{\left(P_{N_2}/P^o\right)\left(P_{H_2}/P^o\right)^3}\right\}_{eqm} = \left(\frac{P^2_{NH_3}}{P_{N_2} \cdot P^3_{H_2}}\right)_{eqm} \cdot (P^o)^2$$

Therefore, at equilibrium at 25°C,
$$K = \left(\frac{P^2_{NH_3}}{P_{N_2} \cdot P^3_{H_2}}\right)_{eqm} \cdot (P^o)^2 = 5.81 \times 10^5$$

Thus, for this system,
$$\Delta G_R = \Delta G° + RT\,ln\left(\frac{P^2_{NH_3} \cdot (P^o)^2}{P_{N_2} \cdot P^3_{H_2}}\right)$$

If the partial pressures are expressed in *bar*, then $P° = 1\ bar$, and

$$\Delta G_R = \Delta G° + RT\,ln\left(\frac{P^2_{NH_3}}{P_{N_2} \cdot P^3_{H_2}}\right)$$

Suppose we were to take samples of the reacting mixture and evaluate—by chemical analysis—the values of the various partial pressures at any point during the course of the reaction. As we saw in Chapter 13,

IF $\dfrac{P^2_{NH_3}}{P_{N_2} \cdot P^3_{H_2}} = 5.81 \times 10^5$,   then $\Delta G_R = 0$ and the system is at *chemical equilibrium*.

IF $\dfrac{P^2_{NH_3}}{P_{N_2} \cdot P^3_{H_2}} < 5.81 \times 10^5$,   then $\Delta G_R < 0$ and the reaction will *spontaneously proceed* to form *more products*.

IF $\dfrac{P^2_{NH_3}}{P_{N_2} \cdot P^3_{H_2}} > 5.81 \times 10^5$,   then $\Delta G_R > 0$ and the reaction to form ammonia is *not* spontaneous. However, for the *reverse* reaction $\Delta G_R < 0$, so that the system will proceed spontaneously to form *more reactants*.

Example 14-2

Ammonia is synthesized according to the reaction

$$N_{2(g)} + 3H_{2(g)} \rightarrow 2NH_{3(g)}$$

for which the equilibrium constant at 400°C is 1.60×10^{-4}.

(a) Calculate $\Delta G^o_{673.15}$ for this reaction at 400°C, (b) Calculate the reaction Gibbs free energy change for the reaction when the partial pressures of N_2 and H_2 each are maintained at 20 bar, and NH_3 is removed at a pressure of 2.0 bar. (c) Is the reaction spontaneous under the conditions of part (b)?

Solution

(a) $\Delta G^o_{673.15} = -RT \ ln \ K_{673.15} = -(8.314)(400 + 273.15) \ ln \ (1.60 \times 10^{-4}) = +48 \ 916 \ \text{J mol}^{-1}$

Ans: $\boxed{+48.92 \ \text{kJ mol}^{-1}}$

(b) $\Delta G_R = \Delta G^\circ + RT \ln Q = \Delta G^\circ + RT \ln \left[\dfrac{\left(P_{NH_3}/P^o \right)^2}{\left(P_{N_2}/P^o \right)\left(P_{H_2}/P^o \right)^3} \right]$

$= \Delta G^\circ + RT \ln \left(\dfrac{P^2_{NH_3}}{P_{N_2} \cdot P^3_{H_2}} \right) = +48 \ 916 + (8.314)(673.15) \ ln \dfrac{2^2}{(20)(20)^3}$

$= +48 \ 916 - 59 \ 304 = -10 \ 388 \ \text{J mol}^{-1} = -10.388 \ \text{kJ mol}^{-1}$

Ans: $\boxed{-10.39 \ \text{kJ mol}^{-1}}$

(c) Under these conditions the reaction is spontaneous because $\Delta G_{T,P} < 0$.

Ans: $\boxed{\text{Spontaneous}}$

Example 14-3

Calculate the equilibrium vapor pressure of water at 25°C and one bar pressure.

		$H_2O_{(V)}$	$H_2O_{(liq)}$
$\Delta h^o_{f(298.15)}$	(kJ mol^{-1})	−241.818	−285.830
$s^o_{298.15}$	(J K^{-1} mol^{-1})	188.825	69.91

Solution

We are interested in the equilibrium $H_2O_{(liq)} \rightleftharpoons H_2O_{(vap)}$

for which $K = \left(\dfrac{a_{H_2O(V)}}{a_{H_2O(liq)}} \right) = \dfrac{(P_{H_2O})_{eqm}}{x_{H_2O}} = \dfrac{(P_{H_2O})_{eqm}}{1} = (P_{H_2O})_{eqm}$

$\Delta H^o_{298.15} = \Delta h^o_f[\text{vapor}] - \Delta h^o_f[\text{liq}]$

$\qquad = -241.818 - (-285.830)$

$\qquad = +44.012 \text{ kJ mol}^{-1}$

$\Delta S^o_{298.15} = s^o[\text{vapor}] - s^o[\text{liq}]$

$\qquad = 188.825 - 69.91$

$\qquad = +118.915 \text{ J K}^{-1} \text{ mol}^{-1}$

Therefore: $\Delta G^o_{298.15} = \Delta H^o_{298.15} - T\,\Delta S^o_{298.15}$

$\qquad\qquad\qquad = (44\ 012) - (298.15)(118.915)$

$\qquad\qquad\qquad = +8557.493 \text{ J mol}^{-1}$

$\Delta G^o_{298.15} = -RT \ln K_{298.15}$

from which $\ln K_{298.15} = \dfrac{-\Delta G^o_{298.15}}{RT} = \dfrac{-(8557.493)}{(8.314)(298.15)} = -3.45225$

and $K = (P_{H_2O})_{eqm} = exp[-3.45225] = 0.03167 \text{ bar}$

Therefore, if water vapor behaves ideally (i.e., if we can use the partial pressure of the water vapor as being equal to its activity), then its equilibrium partial pressure with liquid water at 25°C must be $(P_{H_2O})_{eqm} = 0.03167$ bar $= 3167$ Pa. Tables of data list the equilibrium vapor pressure of water at 25°C as $P^{\bullet}_W = 3169$ Pa, which is essentially the same as the value of 3167 Pa calculated for $(P_{H_2O})_{eqm}$. Therefore, water at 25°C behaves ideally.

Ans: | 3167 Pa |

14.2 EFFECT OF CATALYST ON K

A catalyst increases the rate at which the system attains equilibrium by providing an alternate reaction pathway. The equilibrium constant, however, defined by $K = exp[-\Delta G^o/RT]$, depends on ΔG^o, which is a *state function*, and therefore is *independent of the reaction pathway*. Therefore, for a given chemical reaction at a specified temperature, a catalyst affects the *rate* of the reaction, but not the value of the equilibrium constant. The thermodynamic equilibrium constant K is a function only of temperature, since Δg_f^o values are functions only of temperature (the pressure already being specified—by definition—as one bar).

Example 14-4

Ammonia was formed at 450°C by passing a mixture of nitrogen gas and hydrogen gas at a 1:3 mole ratio over a catalyst. When the total pressure was held constant at 10.13 bar it was found that the product gas contained 2.04% by volume of ammonia. Calculate the value of the equilibrium constant K at 450°C for the reaction:

$$\tfrac{1}{2}N_{2(g)} + \tfrac{3}{2}H_{2(g)} \rightarrow NH_{3(g)}$$

Solution

For the reaction as written:

$$K = \left(\frac{a_{NH_3}}{a_{N_2}^{1/2} \cdot a_{H_2}^{3/2}}\right)_{eqm} = \left(\frac{(P_{NH_3}/P^o)}{(P_{N_2}/P^o)^{1/2}(P_{H_2}/P^o)^{3/2}}\right)_{eqm} = \left(\frac{P_{NH_3}}{P_{N_2}^{1/2} \cdot P_{H_2}^{3/2}}\right)_{eqm}$$

We are told that the mole ratio of nitrogen to hydrogen is 1:3; therefore, choose 1 mole of N_2 and 3 moles of H_2 as the initial amounts of gases brought together, and evaluate the reaction yield:

	$\tfrac{1}{2}N_{2(g)}$	$+$	$\tfrac{3}{2}H_{2(g)}$	$\rightarrow$	$NH_{3(g)}$	
moles, start:	1		3		0	
react:	$-x$		$-3x$		$+2x$	
equilibrium:	$1-x$		$3-3x$		$2x$	$n_{total} = (1-x) + (3-3x) + 2x = 4-2x$
mole fractions:	$\left(\dfrac{1-x}{4-2x}\right)$		$\left(\dfrac{3-3x}{4-2x}\right)$		$\left(\dfrac{2x}{4-2x}\right)$	

2.04% by volume of NH_3 in the product gas is the same as 2.04 mol %; therefore,

$$\frac{n_{NH_3}}{n_{total}} = \left(\frac{2x}{4-2x}\right) = \frac{2.04}{100} = 0.0204$$

Solving for x:
$$2x = (0.0204)(4) - (0.0204)(2x)$$
$$= 0.0816 - 0.0408x$$

$$2.0408x = 0.0816$$

$$x = \frac{0.0816}{2.0408} = 0.039984$$

The partial pressures are given by $P_i = y_i P = y_i(10.13)$ bar;

therefore:
$$P_{N_2} = \left(\frac{1-x}{4-2x}\right)P = \left(\frac{1-0.039984}{4-2(0.039984)}\right)(10.13) = 2.48084 \text{ bar}$$

$$P_{H_2} = \left(\frac{3-3x}{4-2x}\right)P = \left(\frac{3-3(0.039984)}{4-2(0.039984)}\right)(10.13) = 7.44251 \text{ bar}$$

$$P_{NH_3} = \left(\frac{2x}{4-2x}\right)P = \left(\frac{2(0.039984)}{4-2(0.039984)}\right)(10.13) = 0.20665 \text{ bar}$$

and
$$K = \left(\frac{P_{NH_3}}{P_{N_2}^{1/2} \cdot P_{H_2}^{3/2}}\right)_{eqm} = \frac{(0.20665)}{(2.48084)^{1/2}(7.44251)^{3/2}} = 6.462 \times 10^{-3}$$

Ans: $\boxed{6.46 \times 10^{-3}}$

Exercise 14-1

The principal reaction in the industrial production of iron and steel involves passing carbon monoxide gas over powdered iron oxide at 1000 K. The residence time in the reactor is sufficiently long that equilibrium is attained before the gases flow out of the reactor. What percentage of the iron oxide is converted to iron if the equilibrium constant for the reaction has a value of 0.403 at 1000 K?

$$FeO_{(s)} + CO_{(g)} \rightarrow Fe_{(s)} + CO_{2(g)} \ \ddot\frown$$

14.3 EFFECT OF TEMPERATURE ON K

We saw above that the thermodynamic equilibrium constant K is a function only of temperature. But what *is* this function? Actually, it depends on the prevailing conditions. Here, we shall deal with two common scenarios:

(i) $\Delta H°$ AND $\Delta S°$ ARE INDEPENDENT OF TEMPERATURE

Since $\Delta G° = \Delta H° - T\Delta S°$, in order to find K at some temperature other than 298.15 K it is necessary to find $\Delta H°$ and $\Delta S°$ at this new temperature. We know from our earlier work with reaction enthalpies that

$$\Delta H_T^o = \Delta H_{298.15}^o + \int_{298.15}^{T} \Delta C_p dT \qquad \dots [4]$$

From Eqn [4] it is obvious that if $\Delta C_p = 0$, then $\Delta H_T^o = \Delta H_{298.15}^o$. In other words, if $\Delta C_p = 0$ then $\Delta H°$ is not a function of temperature.

Similarly, we also saw earlier that

$$\Delta S_T^o = \Delta S_{298.15}^o + \int_{298.15}^{T} \frac{\Delta C_P}{T} dT \qquad \ldots [5]$$

Eqn [5] shows that, like ΔH°, ΔS° also is independent of temperature if $\Delta C_P = 0$. Although ΔH° and ΔS° are seldom truly independent of temperature, for many reactions the dependency is quite small; for these kinds of reactions, providing the temperature range is not too great (say less than 50°), the variation usually is not very large. Under such conditions it is usually all right to assume that ΔH° and ΔS° remain approximately constant between T_1 and T_2.

Thus, at T_1: $\ln K_{T_1} = -\dfrac{\Delta G_{T_1}^o}{RT_1}$

And at T_2: $\ln K_{T_2} = -\dfrac{\Delta G_{T_2}^o}{RT_2}$

$$= -\left(\frac{\Delta H^o - T_1 \Delta S^o}{RT_1} \right)$$

$$= -\left(\frac{\Delta H^o - T_2 \Delta S^o}{RT_2} \right)$$

$$= -\frac{\Delta H^o}{RT_1} + \frac{\Delta S^o}{R} \quad \ldots [6]$$

$$= -\frac{\Delta H^o}{RT_2} + \frac{\Delta S^o}{R} \quad \ldots [7]$$

Subtracting, [7] – [6]: $\ln K_{T_2} - \ln K_{T_1} = \left(-\dfrac{\Delta H^o}{RT_2} + \dfrac{\Delta S^o}{R} \right) - \left(-\dfrac{\Delta H^o}{RT_1} + \dfrac{\Delta S^o}{R} \right)$

$$= -\frac{\Delta H^o}{RT_2} + \frac{\Delta H^o}{RT_1}$$

Rearranging: $$\boxed{\ln\left(\frac{K_{T_2}}{K_{T_1}} \right) = -\frac{\Delta H^o}{R} \left(\frac{1}{T_2} - \frac{1}{T_1} \right)}$$ *The van't Hoff Equation* $\ldots [8]$

Eqn [8] is known as the **van't Hoff equation**, and can be used only over fairly narrow temperature ranges where it may be assumed that ΔH° and ΔS° are constant. Thus, under these conditions, if we know the value of K at one temperature T_1 and the value of ΔH° for the reaction, then we can use the van't Hoff equation to evaluate the value of K at some other temperature T_2.

Note that if $T_2 > T_1$, then $\dfrac{1}{T_2} - \dfrac{1}{T_1}$ is *negative*, and:

If the reaction is *endothermic*, then $\Delta H^\circ > 0$, $\ln(K_{T_2}/K_{T_1})$ is *positive*, and $K_{T_2} > K_{T_1}$. This means that there will be a greater yield (i.e., more products formed) at higher temperatures than at lower temperatures.

Similarly, if the reaction is *exothermic*, then $\Delta H^\circ < 0$, $\ln(K_{T_2}/K_{T_1})$ is *negative*, and $K_{T_2} < K_{T_1}$. In this case, there will be a smaller yield at higher temperatures; i.e., there will be less products formed at higher temperatures than at lower temperatures.

To summarize:

> If a reaction is **endothermic**, higher temperature favors more products.
> If a reaction is **exothermic**, higher temperature favors less products.

The van't Hoff equation is consistent with **le Chatelier's Principle**, which states:

> *"When a system at equilibrium is disturbed, the system tends to
> shift in such a way as to minimize the effect of the disturbance."*

Thus, an exothermic reaction can be written as if the heat were one of the *products* of the reaction:

$$aA + bB \rightarrow cC + dD + \textbf{\textit{heat}}$$

According to le Chatelier's Principle, "adding" heat to the right hand side of the reaction will tend to make the reaction shift to the *left*, towards more reactants and less products. But adding heat tends to raise the temperature; therefore an exothermic reaction will shift to the left at higher temperatures (K will be smaller at higher temperatures).

Similarly, an endothermic reaction can be written as if the heat were one of the *reactants*:

$$aA + bB + \textbf{\textit{heat}} \rightarrow cC + dD$$

Again, according to le Chatelier's Principle, "adding" heat to the left hand side of the reaction will tend to make the reaction shift to the *right*, towards more products and less reactants. But, as before, adding heat tends to raise the temperature; therefore an endothermic reaction will shift to the right at higher temperatures (K will be bigger at higher temperatures).

Exercise 14-2

Assuming constant $\Delta H°$, estimate the equilibrium constant for the synthesis of ammonia at 500 K:

$$N_{2(g)} + 3H_{2(g)} \rightarrow 2NH_{3(g)} \qquad \qquad ☺$$

Exercise 14-3

Silver oxide (Ag_2O) is used in high energy silver-zinc batteries. One problem encountered with this material is that when exposed to the oxygen in the air at a sufficiently high temperature, it will start to decompose according to the reaction

$$2Ag_2O_{(s)} \rightarrow 4Ag_{(s)} + O_{2(g)}$$

(a) Will this decomposition reaction take place at room temperature?

(b) If not, estimate the temperature at which the decomposition starts to occur.

The partial pressure of the oxygen in the air may be taken as $P_{O_2} = 0.20\ bar$. For this calculation you may assume that $\Delta C_P = 0$ for the decomposition reaction. ☹

(ii) $\Delta H°$ AND $\Delta S°$ VARY WITH TEMPERATURE BUT ΔC_P IS CONSTANT

If ΔC_P is constant, then
$$\Delta H_T^o = \Delta H_{298.15}^o + \int_{298.15}^{T} \Delta C_P dT \qquad \ldots [9]$$

$$= \Delta H_{298.15}^o + \Delta C_P \cdot \Delta T$$

$$= \Delta H_{298.15}^o + \Delta C_P(T - 298.15) \qquad \ldots [10]$$

Similarly,
$$\Delta S_T^o = \Delta S_{298.15}^o + \int_{298.15}^{T} \frac{\Delta C_P}{T} dT \qquad \ldots [11]$$

$$= \Delta S_{298.15}^o + \Delta C_P \cdot ln\left(\frac{T}{298.15}\right) \qquad \ldots [12]$$

The equilibrium constant K_T at any temperature T is given by

$$ln\, K_T = -\frac{\Delta G_T^o}{RT} = -\frac{\Delta H_T^o}{RT} + \frac{\Delta S_T^o}{R} \qquad \ldots [13]$$

Putting Eqns [10] and [12] into Eqn [13]:

$$ln\, K_T = -\frac{\Delta H_T^o}{RT} + \frac{\Delta S_T^o}{R}$$

$$= -\frac{1}{RT}\left[\Delta H_{298.15}^o + \Delta C_P(T - 298.15)\right] + \frac{1}{R}\left[\Delta S_{298.15}^o + \Delta C_P \cdot ln\left(\frac{T}{298.15}\right)\right]$$

$$= -\frac{\Delta H_{298.15}^o}{RT} - \frac{\Delta C_P(T - 298.15)}{RT} + \frac{\Delta S_{298.15}^o}{R} + \frac{\Delta C_P}{R} ln\left(\frac{T}{298.15}\right)$$

$$= -\frac{\Delta H_{298.15}^o}{RT} + \frac{\Delta S_{298.15}^o}{R} - \frac{\Delta C_P T}{RT} - \frac{\Delta C_P(-298.15)}{RT} + \frac{\Delta C_P}{R} ln\left(\frac{T}{298.15}\right)$$

$$\boxed{ln\, K_T = -\frac{\Delta H_{298.15}^o}{RT} + \frac{\Delta S_{298.15}^o}{R} - \frac{\Delta C_P}{R}\left[1 - \frac{298.15}{T} - ln\left(\frac{T}{298.15}\right)\right]} \qquad \ldots [14]$$

When ΔC_P is constant, Eqn [14] is a very handy equation for finding the equilibrium constant at any temperature T if its value is known at 298.15 K.

In general, the stipulation that ΔC_P be constant is not as strict as the stipulation that C_P be constant. For example, if C_P for the products and C_P for the reactants both change with temperature in the same way, then, even though they are not independent of temperature, their *difference* ΔC_P is independent of temperature. This approximation usually is fairly accurate provided the temperature range is not too large.

If ΔC_P does vary with temperature, then Eqns [9] and [11]—instead of [10] and [12]—must be put into Eqn [13] and integrations carried out between 298.15 K and T.

Example 14-5

When heated, calcium carbonate (calcite) decomposes to form calcium oxide and carbon dioxide gas:

$$CaCO_{3(s)} \rightarrow CaO_{(s)} + CO_{2(g)} \uparrow$$

Using the data given below, calculate the equilibrium partial pressure of CO_2 gas over a mixture of solid CaO and $CaCO_3$ at 500°C if ΔC_P for the reaction is independent of temperature over the temperature range 25°C ↔ 500°C.

		$CaO_{(s)}$	$CO_{2(g)}$	$CaCO_{3(s)}$
$\Delta h^o_{f(298.15)}$	(kJ mol^{-1})	−635.09	−393.51	−1206.92
$s^o_{298.15}$	(J K^{-1} mol^{-1})	39.75	213.74	92.90
c^o_P	(J mol^{-1} K^{-1})	42.80	37.11	81.88

Solution

The equilibrium constant for the equilibrium

$$CaCO_{3(s)} \rightleftharpoons CaO_{(s)} + CO_{2(g)}$$

is defined[1] as:

$$K = \frac{(x_{CaO})(P_{CO_2}/P^o)}{x_{CaCO_3}} = \frac{(1)(P_{CO_2}/P^o)}{(1)} = (P_{CO_2}/P^o)_{eqm}$$

Therefore at 500°C (= 773.15 K) the equilibrium partial pressure of carbon dioxide over a solid mixture of calcium carbonate and calcium oxide is just the equilibrium constant $K_{773.15}$ for the reaction at 500°C.

For a reaction such as this, for which ΔC_P is independent of temperature,

$$ln \, K_T = -\frac{\Delta H^o_{298.15}}{RT} + \frac{\Delta S^o_{298.15}}{R} - \frac{\Delta C_P}{R}\left[1 - \frac{298.15}{T} - ln\left(\frac{T}{298.15}\right)\right] \quad \cdots \text{[a]}$$

Thus:

$$\Delta C_P = c_P[\text{CaO}] + c_P[\text{CO}_2] - c_P[\text{CaCO}_3]$$

$$= (42.8) + (37.11) - (81.88)$$

$$= -1.97 \text{ J mol}^{-1} \text{ K}^{-1} \quad (= \text{constant})$$

---

[1] See Example 13-1 in Chapter 13.

Also, $\Delta H^o_{298.15} = \Delta h^o_f[CaO] + \Delta h^o_f[CO_2] - \Delta h^o_f[CaCO_3]$

$$= (-635.09) + (-393.51) - (-1206.92)$$

$$= +178.32 \text{ kJ mol}^{-1}$$

And, $\Delta S^o_{298.15} = s^o[CaO] + s^o[CO_2] - s^o[CaCO_3]$

$$= (39.75) + (213.74) - (92.90)$$

$$= +160.59 \text{ J K}^{-1} \text{ mol}^{-1}$$

Substituting the above values into Eqn [a] gives:

$$ln\,K_{773.15} = -\frac{178\,320}{(8.314)(773.15)} + \frac{160.59}{8.314} - \frac{(-1.97)}{8.314}\left[1 - \frac{298.15}{773.15} - ln\left(\frac{773.15}{298.15}\right)\right]$$

$$= -27.741 + 19.316 + 0.23695(1 - 0.3856 - 0.9529)$$

$$= -8.5052$$

Therefore, $(P_{CO_2}/P^o)_{eqm} = K_{773.15} = exp\,[-8.5052] = 2.024 \times 10^{-4}$

and $P_{CO_2} = 2.024 \times 10^{-4} P^o = 2.024 \times 10^{-4}\,(1 \text{ bar}) = 2.024 \times 10^{-4} \text{ bar.}$

Ans: $\boxed{2.0 \times 10^{-4} \text{ bar}}$

Exercise 14-4

When the temperature becomes sufficiently high, water vapor starts to decompose into hydrogen gas and oxygen gas according to

$$2H_2O_{(g)} \rightarrow 2H_{2(g)} + O_{2(g)}$$

Using the following data, and assuming that ΔC_P is constant for this reaction and has an effective value of $\Delta C_P = 18.88\,J\,mol^{-1}\,K^{-1}$ over the temperature range in question, calculate:

(a) the value of the decomposition constant at 1000 K

(b) the percent decomposition of water vapor at 1000 K and one bar total pressure.

		$H_{2(g)}$	$O_{2(g)}$	$H_2O_{(g)}$
$\Delta h^o_{f\,(298.15)}$	(kJ mol^{-1})	0.00	0.00	−241.82
$s^o_{298.15}$	(J K^{-1} mol^{-1})	130.684	205.138	188.825

☹*

14.4 EFFECT OF PRESSURE ON K

For the reaction $aA + bB \rightarrow cC + dD$

$$\Delta G° = c(\Delta g_f^o)_C + d(\Delta g_f^o)_D - a(\Delta g_f^o)_A - b(\Delta g_f^o)_B$$

where the Δg_f^os are the standard molar Gibbs free energies of formation of the various species.

By *definition*, $\Delta G°$ is the value at one bar pressure (the standard pressure); therefore it is independent of the pressure at which the reaction is carried out. It follows from the definition of K, namely, $K = exp[-\Delta G°/RT]$, that *K is not a function of the total pressure at which the reaction is carried out*. Although the pressure does not affect the value of the thermodynamic equilibrium constant (which is determined by the temperature only), the pressure *may* affect the equilibrium *yield*, since the yield is determined by the *partial* pressures of the various species involved in the reaction.

Le Chatelier's Principle can be used to predict the effect of increased gas pressure on a gas phase reaction (one in which all the reactants and all the products are gases). In this case, le Chatelier's Principle can be stated as follows:

> *"When the system is compressed, the composition of the gas phase*
> *adjusts to reduce the number of molecules in the gas phase."*

For example, consider the reaction

$$H_{2(gas)} + I_{2(solid)} \rightleftharpoons 2HI_{(gas)} \qquad \ldots [15]$$
$$\underset{(1\ volume\ of\ gas)}{} \qquad \underset{(2\ volumes\ of\ gas)}{}$$

There is one volume of gas on the left-hand side and two volumes of gas on the right-hand side. (Iodine is a solid, so its volume is so small compared with that of the gases that it can be ignored.) Although the equilibrium constant doesn't change with pressure, according to le Chatelier's Principle, *increasing* the pressure of the system will cause the equilibrium to shift to the *left* (towards fewer molecules in the gas phase). Similarly, *decreasing* the pressure on the system will cause the equilibrium to shift towards the *right*.

Exercise 14-5

Hydrogen gas at 25°C and 1.00 bar pressure is introduced into an evacuated tank containing a large excess of solid iodine crystals in the bottom. The temperature and pressure of the system are kept constant at 25°C and 1.00 bar pressure, respectively. Hydrogen reacts according to Eqn [15] until equilibrium is attained. (a) What is the thermodynamic equilibrium constant for the reaction? (b) What are the partial pressures of hydrogen gas and hydrogen iodide gas at equilibrium? ☹

Example 14-6

This example will illustrate the effect of lowering the pressure on the yield of a chemical reaction. Consider Reaction [15] at equilibrium at 25°C and 1.00 bar total pressure inside a reactor having a volume of 1.00 m³:

$$H_{2(g)} + I_{2(solid)} \rightleftharpoons 2HI_{(g)}$$

For this reaction, K at 25°C is easily calculated from standard molar free energies (Exercise 14-5) and found to have a value of 0.2537.

Therefore:
$$K_{298.15} = \frac{a_{HI}^2}{a_{H_2} \cdot a_{I_2}} = \frac{\left(P_{HI}/P^o\right)^2}{\left(P_{H_2}/P^o\right)(1)} = \frac{\left(P_{HI}/1\right)^2}{\left(P_{H_2}/1\right)} = \frac{P_{HI}^2}{P_{H_2}} = 0.2537$$

That is,
$$\frac{P_{HI}^2}{P_{H_2}} = 0.2537$$

where the partial pressures are expressed in *bar*.

At equilibrium, $P_{HI} = 0.3926$ bar and $P_{H_2} = 0.6074$ bar (to give a total pressure of $P = 1.00$ bar).

(Confirmed by noting that $\dfrac{P_{HI}^2}{P_{H_2}} = \dfrac{(0.3926)^2}{0.6074} = 0.2537$, satisfying the criterion for equilibrium.)

The number of moles of H_2 present is
$$n_{H_2} = \frac{P_{H_2} V}{RT} = \frac{(0.6074 \times 10^5)(1.0)}{(8.314)(298.15)} = 24.50 \text{ mol}$$

Similarly,
$$n_{HI} = \frac{P_{HI} V}{RT} = \frac{(0.3926 \times 10^5)(1.0)}{(8.314)(298.15)} = 15.84 \text{ mol}$$

Now suppose the volume is *increased* from 1.0 m³ to 10.0 m³ at the same temperature. This will have the effect of *lowering* the total pressure in the reactor. *Does the equilibrium amount of HI change?* Let's find out:

	$H_{2(g)}$	$+$	$I_{2(solid)}$	$\rightleftharpoons$	$2HI_{(g)}$
Initial equilibrium, moles:	24.50				15.84
React:	$-x$				$+2x$
New equilibrium:	$(24.50 - x)$				$(15.84 + 2x)$

We can ignore the solid iodine because, having unit activity, it is not present in the expression defining the equilibrium constant.

$$P_{H_2} = \frac{n_{H_2} RT}{V} = \frac{(24.50 - x)(8.314)(298.15)}{(10.0)} = (6073 - 248x) \text{ Pa}$$

$$P_{HI} = \frac{n_{HI} RT}{V} = \frac{(15.84 + 2x)(8.314)(298.15)}{(10.0)} = (3926 + 496x) \text{ Pa}$$

$$K_{298.15} = 0.2537 = \frac{P_{HI}^2}{P_{H_2}} = \frac{\left[(3926 + 496x)/100\,000\right]^2}{(6073 - 248x)/100\,000} = \frac{(3926 + 496x)^2}{(6073 - 248x)(100\,000)}$$

Solving[2] the resulting quadratic equation in x gives $x = 10.797$.

Thus, $\begin{array}{l} n_{HI} = 15.84 + 2x = 15.84 + 2(10.797) = 37.43 \text{ mol} \\ n_{H_2} = 24.50 - x = 24.50 - 10.797 = 13.70 \text{ mol} \end{array}$ } The yield of HI has more than *doubled*, from 15.84 mol to 37.43 mol!

$$P_{HI} = \frac{3926 + 496x}{100\ 000} = \frac{3926 + 496(10.797)}{100\ 000} = 0.09281 \text{ bar}$$

$$P_{H_2} = \frac{6073 - 248x}{100\ 000} = \frac{6073 - 248(10.797)}{100\ 000} = 0.03395 \text{ bar}$$

$$\frac{P_{HI}^2}{P_{H_2}} = \frac{(0.09281)^2}{0.03395} = 0.2537$$

(Satisfies the equilibrium constant.)

The new total pressure is $P = P_{H_2} + P_{HI} = 0.03395 + 0.09281 = 0.1268$ bar.

Thus, as predicted, decreasing the pressure shifts the equilibrium towards more products:

$$15.84 \text{ mol HI} \rightarrow 37.43 \text{ mol HI}$$

As noted above, the yield has more than *doubled*!

Boyle's Law for a constant temperature ideal gas system, which states that $P_1V_1 = P_2V_2$, would have predicted that $P_2 = 0.1P_1$; that is, $P_2 = 0.10$ *bar*. The actual value of P_2 was 0.1268 bar. This is in accordance with le Chatelier's Principle: the system shifted to *minimize* the effect of the increased volume by putting *more* molecules into the gas phase to try to compensate for the lowering of the original pressure. In this case, putting more molecules into the gas phase means increasing the yield.

Thus, although P doesn't affect the value of K, it *can* affect the product yield.

Example 14-7

When a mixture containing initial mole fractions of N_2 and H_2 of 0.25 and 0.75, respectively, was maintained at 673 K and 10.0 bar pressure in the presence of a catalyst, the mole fraction of ammonia in the resulting equilibrium mixture was 0.0385. Calculate the pressure required to produce a mole fraction of ammonia of 0.20.

$$N_{2(g)} + 3H_{2(g)} \rightarrow 2NH_{3(g)}$$

Solution

First we have to determine the value of the equilibrium constant for the reaction:

Basis: Initial 0.25 mol of N_2 and 0.75 mol of H_2:

---

[2] Make sure you can solve it! Experience has shown that you should always be skeptical of the instructor's ability to do math.

$$N_{2(g)} \quad + \quad 3H_{2(g)} \quad \rightleftharpoons \quad 2NH_{3(g)}$$

moles, start: 0.25   0.75   0

react: $-x$ $-3x$ $+2x$

equilibrium: $0.25 - x$ $0.75 - 3x$ $2x$ Total: $n = 1 - 2x$

$$P_i = y_i P: \quad \left(\frac{0.25 - x}{1 - 2x}\right)P \quad \left(\frac{0.75 - 3x}{1 - 2x}\right)P \quad \left(\frac{2x}{1 - 2x}\right)P$$

We are told that
$$y_{NH_3} = \frac{2x}{1 - 2x} = 0.0385$$

Solving for x:
$$2x = 0.0385 - 0.0770x$$
$$2.077x = 0.0385$$
$$x = 0.018536$$

Knowing x, we now can calculate the partial pressures:

$$P_{NH_3} = y_{NH_3}P = \left(\frac{2x}{1-2x}\right)P = \left(\frac{2(0.018536)}{1 - 2(0.018536)}\right)(10.0) = 0.38499 \text{ bar}$$

$$P_{N_2} = \left(\frac{0.25 - x}{1 - 2x}\right)P = \left(\frac{0.25 - 0.018536}{1 - 2(0.018536)}\right)(10.0) = 2.40375 \text{ bar}$$

$$P_{H_2} = \left(\frac{0.75 - 3x}{1 - 2x}\right)P = \left(\frac{0.75 - 3 \times 0.018536}{1 - 2(0.018536)}\right)(10.0) = 7.211256 \text{ bar}$$

Now we can evaluate the equilibrium constant:

$$K = \frac{P_{NH_3}^2}{P_{N_2} \cdot P_{H_2}^3} = \frac{(0.38499)^2}{(2.40375)(7.211256)^3} = 1.6443 \times 10^{-4}$$

Now we start over again, but this time solve for the value of P that gives $y_{NH_3} = 0.200$.

Therefore:
$$y_{NH_3} = \frac{2x}{1 - 2x} = 0.2000$$

Solving for x:
$$2x = 0.2000 - 0.4x$$
$$2.4x = 0.2000$$
$$x = 0.083333$$

Next we use this new value of x to calculate the new equilibrium mole fractions:

$$y_{N_2} = \left(\frac{0.25-x}{1-2x}\right) = \frac{0.25-0.083333}{1-2(0.083333)} = 0.2000$$

$$y_{H_2} = \left(\frac{0.75-x}{1-2x}\right) = \frac{0.75-0.083333}{1-2(0.083333)} = 0.6000$$

The partial pressures are: $\qquad P_{NH_3} = y_{NH_3}P = 0.200P$

$$P_{N_2} = y_{N_2}P = 0.200P$$

$$P_{H_2} = y_{H_2}P = 0.600P$$

Substituting these values of the partial pressures into the expression for the equilibrium constant allows us to solve for the total pressure P:

$$K = 1.6443 \times 10^{-4} = \frac{P_{NH_3}^2}{P_{N_2} \cdot P_{H_2}^3} = \frac{(0.200P)^2}{(0.200P)(0.600P)^3} = \frac{0.92593}{P^2}$$

$$P = \sqrt{\frac{0.92593}{1.6443 \times 10^{-4}}} = 75.04 \text{ bar}$$

Ans: $\boxed{75.0 \text{ bar}}$

Exercise 14-6

(a) Into an evacuated flask (flask A) of volume 1.1042 L, 1.2798 g of N_2O_4 (= 92.02 g mol^{-1}) was introduced; after equilibrium had been attained at 25°C, the pressure in the flask was 299.6 Torr. Calculate the equilibrium constant at 25°C for the reaction

$$N_2O_{4(g)} \rightleftharpoons 2NO_{2(g)}$$

(b) A second flask (flask B) having a volume of 1.1254 L contained NO (= 30.01 g mol^{-1}) at a temperature of 25°C and a pressure of 276.2 Torr. The flasks were connected by means of a tube, which was blocked by a thin glass wall. After breaking the glass wall with a magnetic hammer and waiting for equilibrium, the final gas pressure at 25°C in the combined system was measured as 294.0 Torr. Calculate the equilibrium constant for the reaction

$$N_2O_{3(g)} \rightleftharpoons NO_{2(g)} + NO_{(g)}$$

Part (a) is fairly easy; part (b) is quite difficult. Don't attempt part (b) unless you have nothing better to do on the weekend! ☠

KEY POINTS FOR CHAPTER FOURTEEN

1. The **Reaction Gibbs Free Energy** for a reaction mixture at equilibrium is

    $$\Delta G_R = 0 = \Delta G° + RT \ln K$$

    from which $\qquad \ln K = -\dfrac{\Delta G^o}{RT} \quad \text{and} \quad K = exp\left[-\dfrac{\Delta G^o}{RT}\right]$

    The value of K is based on *activities*, and is called the **thermodynamic equilibrium constant**. To calculate K, all we need to know is the value of $\Delta G°$ for the reaction, which is readily calculated from standard thermodynamic data.

2. $\Delta G_R = \Delta G° + RT \ln Q$

    If $Q = K$, then $\Delta G_R = 0$ and the system is at *chemical equilibrium*.

    If $Q < K$, then $\Delta G_R < 0$ and the reaction will *spontaneously proceed* to form *more products*.

    If $Q > K$, then $\Delta G_R > 0$ and the reaction is *not* spontaneous. However, for the *reverse reaction* $\Delta G_R < 0$; therefore the system will proceed spontaneously to form *more reactants*.

3. For a given chemical reaction at a specified temperature, a **catalyst** affects the *rate* of the reaction, but not the value of the equilibrium constant. The thermodynamic equilibrium constant K *is a function only of temperature*, since $\Delta G°$, like the values of Δg_f^o from which it is derived, is a function only of temperature (the pressure already being specified—by definition—as one bar).

4. **Effect of temperature on K:**

    If $\Delta H°$ and $\Delta S°$ do not vary with temperature, then

    $$\ln\left(\frac{K_{T_2}}{K_{T_1}}\right) = -\frac{\Delta H^o}{R}\left(\frac{1}{T_2} - \frac{1}{T_1}\right)$$

    which is known as the *van't Hoff equation*. This equation can be used only over a fairly narrow temperature range where $\Delta H°$ and $\Delta S°$ do not change much with temperature.

    If the reaction is *endothermic*, then $\Delta H° > 0$, $\ln(K_{T_2}/K_{T_1})$ is *positive*, and $K_{T_2} > K_{T_1}$. This means that there will be a greater yield (i.e., more products formed) at higher temperatures than at lower temperatures.

    Similarly, if the reaction is *exothermic*, then $\Delta H° < 0$, $\ln(K_{T_2}/K_{T_1})$ is *negative*, and $K_{T_2} < K_{T_1}$. In this case, there will be a smaller yield at higher temperatures; i.e., there will be less products formed at higher temperatures than at lower temperatures.

5. The van't Hoff equation is consistent with **le Chatelier's Principle**:

    > *"When a system at equilibrium is disturbed, the system tends to shift in such a way as to minimize the effect of the disturbance."*

Thus, an exothermic reaction can be written as if the heat were one of the *products* of the reaction: $$aA + bB \rightarrow cC + dD + \textbf{heat}$$

"Adding" heat to the right-hand side of the reaction will tend to make the reaction shift to the *left*, towards more reactants and less products. But adding heat tends to raise the temperature; therefore an exothermic reaction will shift to the left at higher temperatures.

Similarly, an endothermic reaction can be written as if the heat were one of the *reactants*:

$$aA + bB + \textbf{heat} \rightarrow cC + dD$$

"Adding" heat to the left-hand side of the reaction will tend to make the reaction shift to the *right*, towards more products and less reactants. But, as before, adding heat tends to raise the temperature; therefore an endothermic reaction will shift to the right at higher temperatures.

6. If $\Delta H°$ and $\Delta S°$ vary with temperature but ΔC_p is constant:

$$ln K_T = -\frac{\Delta H°_{298.15}}{RT} + \frac{\Delta S°_{298.15}}{R} - \frac{\Delta C_P}{R}\left[1 - \frac{298.15}{T} - ln\left(\frac{T}{298.15}\right)\right]$$

7. **Effect of pressure on K:**

By *definition*, $\Delta G°$ is the value at one bar pressure (the standard pressure); therefore it is *independent of the pressure* at which the reaction is carried out. It follows from the definition of K, namely, $K = exp[-\Delta G°/RT]$, that K is not a function of the total pressure at which the reaction is carried out. Although the total pressure does not affect the value of the thermodynamic equilibrium constant (which is determined by the temperature only), the total pressure *may* affect the equilibrium *yield*, since the yield is determined by the *partial* pressures of the various species involved in the reaction.

PROBLEMS

1. 2.0 moles of CH_4 and 1.0 mole of H_2S are held at 700°C and 1.0 bar until the equilibrium
$$CH_{4(g)} + 2H_2S_{(g)} \rightleftharpoons CS_{2(g)} + 4H_{2(g)}$$
is established. At equilibrium, the partial pressure of hydrogen is 0.16 bar. Calculate K. ☺

2. (a) Calculate the equilibrium constant at 300°C for the reaction
$$N_{2(g)} + 3H_{2(g)} \rightleftharpoons 2NH_{3(g)}$$
given that the equilibrium constant for the reaction
$$NH_{3(g)} \rightleftharpoons \frac{1}{2}N_{2(g)} + \frac{3}{2}H_{2(g)}$$
is 16.0 at 300°C.

(b) 3.0 moles of N_2 and 5.0 moles of H_2 are held at 300°C until the equilibrium

$$N_{2(g)} + 3H_{2(g)} \rightleftharpoons 2NH_{3(g)}$$

is established. If 1.0 mole of NH_3 is formed, calculate:

   (i)   the composition of the equilibrium mixture in mole %, and

   (ii) the equilibrium pressure, in atm. ☹

3.   5.0 moles of steam, 1.5 moles of CO, and 3.0 moles of CO_2 are held at 900 K until the equilibrium

$$H_2O_{(g)} + CO_{(g)} \rightleftharpoons CO_{2(g)} + H_{2(g)}$$

is established. Calculate the amount of H_2 produced. At 900 K, K for the reaction is 2.0. ☺

4.   When heated, ammonium carbamate decomposes as follows:

$$NH_4CO_2NH_{2(s)} \rightleftharpoons 2NH_{3(g)} + CO_{2(g)}$$

If the total equilibrium pressure of the system at a certain temperature is 0.318 bar, what is K for the reaction at this temperature? ☺

5.   At a certain temperature, when carbon monoxide at a partial pressure of 2.00 atm is introduced into a rigid vessel containing an excess of solid sulfur, the following equilibrium is established:

$$S_{(s)} + 2CO_{(g)} \rightleftharpoons SO_{2(g)} + 2C_{(s)}$$

At equilibrium, the final pressure in the vessel is determined to be 1.03 atm. Calculate K. ☺

6.   At 25°C the equilibrium constant for the following reaction is $K = 1.04 \times 10^{-6}$:

$$CuSO_4 \cdot 3H_2O_{(s)} \rightleftharpoons CuSO_{4(s)} + 3H_2O_{(g)}$$

What is the minimum number of moles of water vapor that must be introduced into a two-liter flask at 25°C in order to completely convert 0.0100 moles of solid $CuSO_4$ to solid $CuSO_4 \cdot 3H_2O$? ☹

7.   25.0 kg of damp cotton cloth containing 20.0% by weight total moisture content is hung out to dry in a sealed, unventilated drying chamber of 120 m³ capacity. The initial air is at a temperature of 35.0°C, has a relative humidity[3] of 20.0%, and is at a total pressure of one atm. The chamber temperature is maintained at 35.0°C. The volume occupied by the damp cloth may be ignored. The vapor pressure of pure water at 35.0°C is 42.175 Torr. At equilibrium, the relationship between the % relative humidity (% RH) and the moisture content m'_W of the cloth, expressed as kg of water per 100 kg of bone dry cloth, has been determined experimentally to be

$$m'_W = 0.15 \times (\% RH)$$

Calculate:

(a) the final % relative humidity of the air in the chamber when equilibrium has been attained,

---

[3] The relative humidity at a given temperature is the ratio of the partial pressure of the water vapor to the equilibrium vapor pressure of pure water at the same temperature; the ratio often is expressed as a percentage. Thus, a relative humidity of 40% means that $P_W/P_W^{\bullet} = 0.40$, where P_W is the actual partial pressure of the water vapor and $P_W^{\bullet}$ is the equilibrium vapor pressure of water at the same temperature.

(b) the final equilibrium moisture content m'_w of the cloth, and

(c) the final total pressure in the chamber. ☠

8. At sufficiently high temperatures water vapor will start to decompose. At 2300 K the standard Gibbs free energy change for the reaction $H_2O_{(g)} \rightleftharpoons H_{2(g)} + \frac{1}{2}O_{2(g)}$ is $\Delta G^o_{2300} = 118.08$ kJ mol^{-1}. Calculate the percentage decomposition of water vapor to hydrogen and oxygen at 2300 K and one bar pressure. ☹

9. Gaseous hydrogen and chlorine combine according to the following reaction:
$$H_{2(g)} + Cl_{2(g)} \rightleftharpoons 2HCl_{(g)}$$
How many moles of H_2 will remain when one mole of H_2 and one mole of Cl_2 are put into a reactor and allowed to react to equilibrium at 25°C and one bar pressure? The standard free energy of formation of HCl gas at 25°C is $\Delta g^o_f = -95.299$ kJ mol^{-1}. Assume that all gases behave as ideal gases. ☹

10. Calculate the equilibrium partial pressure of gaseous carbon disulfide at 25°C if the partial pressure of sulfur vapor is maintained at 0.100 bar.
$$S_{8(g)} + 4C_{graphite} \rightleftharpoons 4CS_{2(g)}$$
$\Delta g^o_{f(298.15)}[S_8] = 49.63$ kJ mol^{-1}; $\Delta g^o_{f(298.15)}[CS_2] = 67.12$ kJ mol^{-1}; $\Delta g^o_{f(298.15)}[H_2] = 0.00$ ☺

11. The equilibrium pressure of water vapor over a mixture of solid TlOH and solid Tl$_2$O at 100°C is 125 Torr. Find $\Delta G°$ for the formation of one mole of thallium hydroxide from water and oxide at 100°C. The water vapor may be assumed to behave as an ideal gas. The activities of the pure solid species each can be taken as unity.
$$Tl_2O_{(s)} + H_2O_{(v)} \rightleftharpoons 2TlOH_{(s)}$$ ☺

12. What is the standard enthalpy change of a chemical reaction for which the equilibrium constant is doubled when the temperature is raised by 10° from 25°C? ☺

13. The following gas phase reaction takes place at a constant total pressure of one bar:
$$A_{(g)} + 3B_{(g)} \rightarrow 2C_{(g)}$$
The standard enthalpies of formation and Gibbs free energies of formation at 300 K and one bar pressure are given in the accompanying table. $\Delta H°$ for the reaction (in joules per mole of A consumed) varies with temperature (in K) as follows:

	Δh^o_f (J mol^{-1})	Δg^o_f (J mol^{-1})
A	−500	−350
B	0.00	0.00
C	−300	−200

$$\Delta H° = -4000 + 10T + 0.01T^2 \quad \text{J mol}^{-1}$$

(a) Calculate K at 300 K

(b) Calculate K at 600 K ☺*

14. The following gas phase reaction takes place at a constant total pressure of one bar:

$$A_{(g)} + 3B_{(g)} \rightarrow 2C_{(g)}$$

The standard enthalpies of formation and free energies of formation at 300 K and one bar and the various molar heat capacities are given in the accompanying table.

(a) Calculate K at 300 K.

(b) Derive an expression for ΔH^o as a function of temperature.

(c) Calculate K at 500 K. ☺*

	Δh_f^o (J mol^{-1})	Δg_f^o (J mol^{-1})	c_P (J mol^{-1} K^{-1})*
A	−500	−350	20 + 0.01T
B	0.00	0.00	10 + 0.01T
C	−300	−200	30 + 0.03T

* T is in kelvin

15. It has been proposed that the following reaction be used at 25°C:

$$CH_3OH_{(liq)} \rightarrow CH_{4(g)} + \tfrac{1}{2}O_{2(g)}$$

Calculate the equilibrium constant for the reaction at 25°C. What conclusion can be drawn from this? ☺

	$s_{298.15}^o$ (J K^{-1} mol^{-1})	$(\Delta H_{combustion}^o)_{298.15}$ (kJ mol^{-1})
$CH_{4(g)}$	186.264	−890.359
$CH_3OH_{(liq)}$	126.8	−726.509
$O_{2(g)}$	205.138	

16. Evaluate the equilibrium constant for the following gas phase reaction at 1000 K:

$$SO_{2(g)} + \tfrac{1}{2}O_{2(g)} \rightleftharpoons SO_{3(g)}$$

Heat capacities (cal mol^{-1} K^{-1}):

$c_P[SO_3] = 6.077 + 23.537 \times 10^{-3}T - 9.687 \times 10^{-6}T^2$

$c_P[SO_2] = 7.116 + 9.512 \times 10^{-3}T - 3.511 \times 10^{-6}T^2$

$c_P[O_2] = 6.095 + 3.253 \times 10^{-3}T - 1.017 \times 10^{-6}T^2$ ☹*

	$\Delta h_{f(298.15)}^o$ (kcal mol^{-1})	$s_{298.15}^o$ (cal K^{-1} mol^{-1})
$SO_{2(g)}$	−70.944	59.326
$O_{2(g)}$	0	49.029
$SO_{3(g)}$	−94.579	61.367

17. One mole of solid cupric sulfate trihydrate ($CuSO_4 \cdot 3H_2O$) takes up water vapor and is converted to solid cupric sulfate pentahydrate ($CuSO_4 \cdot 5H_2O$). Given that the equilibrium vapor pressure over a mixture of the solid tri- and pentahydrates is 8.07 Torr at 26.30°C, and 21.73 Torr at 39.70°C:

(a) Calculate the equilibrium constant K at 26.30°C.

(b) Calculate ΔG°, the standard free energy of formation at 39.70°C.

(c) Calculate ΔH° for the process when one mole of cupric sulfate trihydrate reacts to produce one mole of cupric sulfate pentahydrate.

(d) Estimate ΔS° for the process at 33.0°C.

It may be assumed that water vapor behaves as an ideal gas, and that the activity of each pure solid species is unity. ☹

18. Hydrogen gas reduces solid Ag_2S to silver according to the following reaction:

$$Ag_2S_{(s)} + H_{2(g)} \rightarrow 2Ag_{(s)} + H_2S_{(g)}$$

Calculate the equilibrium constant K for this reaction at (a) 25°C, and at (b) 750°C.

	$\Delta h^o_{f(298.15)}$ (kJ mol^{-1})	$\Delta g^o_{f(298.15)}$ (kJ mol^{-1})
$H_2S_{(g)}$	−20.63	−33.56
$Ag_2S_{(s)}$	−31.80	−40.25

$\Delta H°$ for the reaction may be assumed to be constant between 25°C and 750°C. The gaseous species may be assumed to behave as ideal gases, and the activities of the pure solid species can be taken as $a_i = 1$. ☺

19. Calculate the equilibrium constant at 1000 K for the reaction

$$N_{2(g)} + 3H_{2(g)} \rightleftharpoons 2NH_{3(g)}$$

Heat capacities (in J mol^{-1} K^{-1}):

$$c_P[H_2] = 29.088 + 0.192 \times 10^{-2}T + 0.400 \times 10^{-5}T^2 - 0.870 \times 10^{-9}T^3$$

$$c_P[N_2] = 28.883 - 0.157 \times 10^{-2}T + 0.808 \times 10^{-5}T^2 - 2.871 \times 10^{-9}T^3$$

$$c_P[NH_3] = 24.619 - 3.75 \times 10^{-2}T - 0.138 \times 10^{-5}T^2$$

At 0°C, $\Delta H^o_{273.15}$ for the reaction is −91.63 kJ mol^{-1}. At 834 K, $\log K = -5.34$. Ideal gas behavior may be assumed. ☹*

20. At high temperatures water vapor starts to dissociate into hydrogen gas and oxygen gas:

$$H_2O_{(g)} \rightarrow H_{2(g)} + \tfrac{1}{2}O_{2(g)}$$

For this reaction:

(a) What is the value of K at 25°C and one bar pressure?

(b) What is the value of K at 1000°C and one bar pressure?

(c) What is the percentage dissociation (α) of water vapor at 1300 K and one bar pressure?

Hint: To avoid solving a cubic equation, note that if it is assumed that $\alpha \ll 1\%$, then
$(100 - \alpha) \approx 100, (200 + \alpha) \approx 200$, etc.

Data: For the reaction, $\Delta H^o_{298.15} = +241.818$ kJ mol^{-1}; $\Delta H^o_{1300} = +249.392$ kJ mol^{-1}
At 25°C, the standard free energy of formation for water vapor is −228.572 kJ mol^{-1} ☹

21. Calculate the equilibrium constant at 1200 K for the reaction

$$CO + \tfrac{1}{2}O_2 \rightleftharpoons CO_2$$

The standard molar free energies of formation at 25°C for CO and CO_2 are −137.3 kJ mol^{-1} and −394.4 kJ mol^{-1}, respectively, while the corresponding standard molar enthalpies of formation at 25°C are −110.5 kJ mol^{-1} and −393.5 kJ mol^{-1}. The molar heat capacities for the gases (in J mol^{-1} K^{-1}, with T in kelvin) are:

For CO and O_2: $c_P^o = 28.28 + 2.54 \times 10^{-3}T + 5.44 \times 10^{-7}T^2$

For CO_2: $c_P^o = 32.22 + 2.22 \times 10^{-2}T - 3.47 \times 10^{-6}T^2$ ☹*

22. Sulfuryl chloride is a colorless, pungent liquid that is used in the manufacture of dyes, drugs, and poison gases. Its vapor partially dissociates into sulfur dioxide and chlorine according to $SO_2Cl_{2(g)} \rightleftharpoons SO_{2(g)} + Cl_{2(g)}$.
    At 300 K the value of the equilibrium constant is 0.0144.

    (a) At what total pressure will equimolar amounts of SO_2Cl_2, SO_2, and Cl_2 exist?
    (b) At what total pressure will 95% of the SO_2Cl_2 be dissociated? ☹

23. The following gas phase chemical reaction takes place at 500 K and a constant total pressure of 2.00 bar:
$$A_{(g)} + B_{(g)} \rightarrow 2C_{(g)}$$

    When 1.00 mole of A and 1.00 mole of B are fed to the initially evacuated reactor and allowed to react to equilibrium, 1.00 mole of C is produced. What is the value of the equilibrium constant for this reaction at 500 K? ☹

24. At 250°C, $K = 1.804$ for the reaction $PCl_{5(g)} \rightleftharpoons PCl_{3(g)} + Cl_{2(g)}$.

    (a) If 0.10 mole of PCl_5 is held at 250°C in a 2-liter flask, how much Cl_2 is formed?
    (b) How much PCl_5 must be added to a 1-liter vessel, held at 250°C, in order to obtain 0.10 mole of Cl_2?
    (c) Under what pressure must an equimolar mixture of Cl_2 and PCl_3 be kept so that at equilibrium the partial pressure of PCl_5 is 1.00 atm? ☹

25. A sample of $COCl_2$ was held at 724 K until the following equilibrium was established:
$$COCl_{2(g)} \rightleftharpoons CO_{(g)} + Cl_{2(g)}$$

    At equilibrium the total pressure was 1.00 atm and the density of the gas mixture was 1.16 kg m^{-3}.

    (a) Calculate the per cent dissociation of $COCl_2$ and K for the above reaction.
    (b) Calculate the new density if the system in part (a) is compressed isothermally and slowly to a total pressure of 2.00 atm. Briefly explain. ☹

26. Can 10 kg of NH_3 be made from 196 kg of N_2 and 26 kg of H_2 if the nitrogen-hydrogen mixture is held in the presence of a suitable catalyst at 698 K and 5.0 bar? K for the reaction
$$\tfrac{1}{2}N_{2(g)} + \tfrac{3}{2}H_{2(g)} \rightleftharpoons NH_{3(g)}$$

    is 0.00908 at 698 K. Justify your answer with appropriate calculations.
    Hint: Assume that 10 kg *can* be made and calculate Q. ☹

27. A gas stream at 900 K and 2.0 atm contains 30.0 mol % CO and 70.0 mol % CO_2. At this temperature $K = 5.63$ for the reaction
$$2CO_{(g)} \rightleftharpoons C_{(gr)} + CO_{2(g)}$$

    (a) Is there a possibility that carbon deposits will form? Briefly justify your answer with an appropriate calculation.
    (b) If carbon deposits do have a tendency to form, this can be prevented by diluting the gas stream with nitrogen. What will be the composition of the gas stream that will have just

enough nitrogen to prevent the formation of carbon deposits? Assume that the temperature and total pressure do not change. ☺

28. The equilibrium constants at 1000 K are given for the two processes shown below:

$$N_{2(g)} + O_{2(g)} \rightleftharpoons 2NO_{(g)} \qquad K = 1.5 \times 10^{-4}$$
$$N_{2(g)} + 2O_{2(g)} \rightleftharpoons 2NO_{2(g)} \qquad K = 9.9 \times 10^{-6}$$

(a) Estimate the pressure at which an initial mixture of N_2 and O_2 in which the mole fraction of N_2 is 0.40 would produce equal molar amounts of NO and NO_2 at 1000 K.

(b) What are the mole fractions of all components under these conditions? ☹

29. Prof. I.M. Bogus claims that he has prepared a powerful new catalyst that makes it possible to synthesize NH_3 with a 70 percent yield in a reactor operating at 500 K and 2.0 atm pressure and having a feed consisting of N_2 and H_2 in stoichiometric ratio. Evaluate his claim by calculating the reaction free energy change for the proposed reaction under the conditions stated. For $NH_{3(g)}$ at 500 K, $\Delta g_f^o = 4.778$ kJ mol^{-1}. ☺

30. Under standard conditions, predict whether the entropy change is positive or negative for each of the following processes. Briefly justify your predictions.

(i) $N_{2(g)} + 3H_{2(g)} \rightarrow 2NH_{3(g)}$

(ii) $NH_4NO_{3(s)} \rightarrow N_{2(g)} + 2H_2O_{(g)} + \frac{1}{2}O_{2(g)}$

(iii) $H_2O_{(g)} \rightarrow H_2O_{(liq)}$

(iv) $CaCO_{3(s)} \rightarrow CaO_{(s)} + CO_{2(g)}$

(v) $H_{2(g)} + I_{2(g)} \rightarrow 2HI_{(g)}$

(vi) $C_{graphite} \rightarrow C_{diamond}$ ☺

31. At elevated temperatures $CaCO_{3(s)}$ dissociates into $CaO_{(s)}$ and $CO_{2(g)}$. Using the data below, estimate the temperature at which the equilibrium constant for this reaction is $K = 1.0$ if it is assumed that neither ΔH^o nor ΔS^o varies significantly with temperature.

	$\Delta h_{f\,(298.15)}^o$ (kJ mol^{-1})	$s_{298.15}^o$ (J K^{-1} mol^{-1})	c_P^o (J mol^{-11} K^{-1})
$CaCO_{3(s)}$	-1206.92	92.90	81.88
$CaO_{(s)}$	-635.09	39.75	42.80
$CO_{2(g)}$	-393.51	213.74	37.11 ☺

32. At elevated temperatures $CaCO_{3(s)}$ dissociates into $CaO_{(s)}$ and $CO_{2(g)}$. Using the data below, estimate the temperature at which the equilibrium constant for this reaction is $K = 1.0$ if it is assumed that ΔC_P does not vary significantly with temperature.

(Read carefully; this problem is not the same as problem 31!)

	$\Delta h^o_{f(298.15)}$ (kJ mol^{-1})	$s^o_{298.15}$ (J K^{-1} mol^{-1})	c^o_P (J mol^{-11} K^{-1})
$CaCO_{3(s)}$	−1206.92	92.90	81.88
$CaO_{(s)}$	−635.09	39.75	42.80
$CO_{2(g)}$	−393.51	213.74	37.11

FIFTEEN

PHASE EQUILIBRIUM (I)

15.1 PHASE CHANGE AND GIBBS FREE ENERGY

To understand what causes a substance to change from one phase to another requires an understanding of the Gibbs free energy change associated with the phase change. So that's what we're going to look at now.

By definition,
$$G = H - TS \qquad \qquad \ldots [1]$$

Differentiating:
$$dG = dH - TdS - SdT \qquad \qquad \ldots [2]$$

But in Chapter 13 we showed[1] that
$$dH = TdS + VdP \qquad \qquad \ldots [3]$$

Substituting Eqn [3] into Eqn [2] gives:

$$dG = (TdS + VdP) - TdS - SdT$$
$$= VdP - SdT$$

$$\boxed{dG = VdP - SdT} \quad \begin{array}{l} \textit{Homogeneous Substance} \\ \textit{of Constant Composition} \end{array} \quad \ldots [4]$$

At constant pressure, $dP = 0$, and Eqn [4] reduces to

$$dG_P = -SdT \qquad \qquad \ldots [5]$$

Dividing both sides by dT gives[2]
$$\boxed{\left(\frac{\partial G}{\partial T}\right)_P = -S} \qquad \qquad \ldots [6]$$

---

[1] Page 13-6, Eqn [11].

[2] Dividing by dT at constant P gives $\left(dG/dT\right)_P$, which is the same as dividing by ∂T if it is understood that G is a function only of T and P. Therefore $\left(dG/dT\right)_P = \left(\partial G/\partial T\right)$. To remind ourselves which variable is held constant, we often add the subscript P when writing a partial differential; i.e., $\left(\partial G/dT\right)_P$. Strictly speaking, the use of "∂" already means that every other variable that can affect the function is held constant other than the one expressed in the denominator. So, don't get worried when we divide by a dx that mysteriously becomes a ∂x.

Since the entropy of a substance is always *positive*,[3] it follows from Eqn [6] that $(\partial G/\partial T)_P$ is always *negative*; i.e., *the free energy of a substance decreases as its temperature increases.*[4] Furthermore, we know that

$$S_{solid} < S_{liquid} < S_{gas}$$

therefore, Eqn [6] tells us that

$$\left(\frac{\partial G}{\partial T}\right)_P^{solid} > \left(\frac{\partial G}{\partial T}\right)_P^{liquid} > \left(\frac{\partial G}{\partial T}\right)_P^{gas}$$

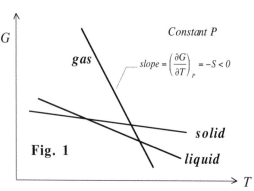

Fig. 1

But $(\partial G/\partial T)_P$ is the *slope* of the G vs. T plot at constant pressure; therefore, as indicated in Fig. 1, gases (and vapors) have more negative slopes than liquids, which in turn have more negative slopes than solids.[5]

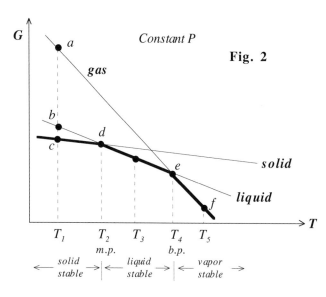

Fig. 2

We have seen that at constant temperature and pressure, systems tend to move spontaneously towards the lowest possible Gibbs free energy state; i.e., $\Delta G_{T,P} < 0$ means the process is spontaneous.

Consider the plot in Fig. 2, which shows G vs. T for a pure substance. At temperature T_1, G for the solid phase is lower than G for the liquid phase or G for the gas phase; therefore, at T_1 the solid state is the most stable, and the system will tend to exist as a solid (point c) at this temperature. As the temperature is raised the free energy of the solid moves along the line cd [6]until the point d is reached, at T_2.

At T_2 the Gibbs free energy of the solid has the *same* value as the Gibbs free energy of the liquid, which is lower than the Gibbs free energy of the gas (or vapor) at this temperature. Thus, at point d, $G_{solid} = G_{liq} < G_{gas}$, which means that at T_2 the liquid and solid are equally stable; i.e., they can

---

[3] Except at absolute zero, where it may be zero.

[4] Note that although Eqn [6] tells us that the Gibbs free energy of a substance *decreases* as the temperature increases, this does not mean that chemical reactions are less energetic at higher temperatures. In fact, we know from experience that most chemical reactions proceed *faster* at higher temperatures than at lower temperatures. For any specified temperature and pressure, it is the *change* in the Gibbs free energy—$\Delta G_{T,P}$—for the reaction that is the driving force, not G itself. For most reactions, $\Delta G_{T,P}$ becomes more negative as the temperature increases.

[5] For simplicity, the lines in Figs. 1 and 2 have been drawn as straight lines; actually, each line should be slightly curved (convex up). Do you know why? The answer is given at the bottom of page 7.

[6] Remember, we just showed that G decreases as T increases, so the lines all slope downwards to the right.

coexist in *equilibrium*—T_2 is the *melting point* of the solid (or, which is the same thing,[7] the *freezing point* of the liquid), and point *d* represents solid *in equilibrium* with liquid.

After sufficient heat input to melt all the solid to liquid, further heating will start to raise the temperature of the liquid. In order to remain in the most stable state (i.e., the lowest free energy state), when the system is further heated past T_2 it will follow the line *de*. At any intermediate temperature T_3 the system exists only as liquid. When the system temperature reaches point *e* at temperature T_4, the free energy of the liquid is the *same* as the free energy of the vapor, and $G_{liquid} = G_{gas}$, which means that both liquid and vapor are equally stable; i.e., they can coexist in *equilibrium* with each other (*liquid* $\rightleftharpoons$ *vapor*). T_4 is the *boiling point* of the liquid. After sufficient heating to convert all the liquid to vapor, further heating will cause the system to move along the line *ef* because this represents the lowest free energy of the system. All points on the line *ef* correspond to the system existing in the gaseous state. Note that as the system moves from T_1 to T_5, at all times it tends to follow the path that places it in the *lowest free energy state* for any given temperature; i.e., it proceeds along the path *cdef*.

At T_2 (the melting point), since $G_{solid} = G_{liquid}$, then $\Delta G_{T,P} = 0$ and there is no tendency for either phase to change to the other—*the system is in solid/liquid equilibrium*. Similarly, at T_4 (the boiling point), $G_{liquid} = G_{gas}$, and the system is in *liquid/vapor equilibrium*.

A gas at T_1 (point *a*) or a liquid at T_1 (point *b*) will tend spontaneously to change to a solid at T_1 (point *c*), because, at T_1, $\Delta G_{T,P} < 0$ for each of these processes. These are *tendencies*, but may not occur at significant rates. For example, any point on the line *bd* represents a *subcooled liquid*—i.e., a liquid that exists at a temperature below its freezing point—which is an unstable state. Subcooled liquids do exist, however, but they are not thermodynamically stable.

To cite another example, at 25°C and atmospheric pressure the process

$$diamond \rightarrow graphite$$

has $\Delta G_{T,P} = -2.90$ kJ mol^{-1}. This means that the diamond in your ring is unstable and has a tendency to convert spontaneously to graphite. Fortunately, this process doesn't occur at significant rates until much higher temperatures and pressures, so you don't have to rush out and sell your ring before the diamond disintegrates into a little lump of coal![8]

15.2 PHASE DIAGRAM FOR A SINGLE PURE SUBSTANCE

A plot of *P vs. T* for a single pure substance showing the regions of thermodynamic stability of the various phases is called the *P–T Phase Diagram* for the substance. A typical diagram (not to scale) is shown in Fig. 3. Such a diagram can be constructed by putting a sample of the fluid substance into a piston /cylinder arrangement and, while keeping the system at any desired temperature (by immersing it in a constant temperature bath), using the piston to apply any desired pressure. We then can observe (through a glass porthole) the *state* of the system at any value of *T* and *P*, and can plot lines delineating the various phase regions as a function of temperature and pressure.

---

[7] For a pure substance, the melting point and the freezing point are the same. For an impure substance, however, they are different. This is discussed more fully in Chapter 22.

[8] The graphite/diamond transition is explored in more detail in Chapter 16, Example 2.

Reference to the phase diagram in Fig. 3 shows that there are four phase regions: *solid*, *liquid*, *gas*, and *supercritical fluid*. The *sublimation curve* is a plot of the equilibrium vapor pressure of solids at different temperatures; the *liquid vapor pressure curve*[9] is a plot of the equilibrium vapor

pressure $P^{\bullet}$ of the liquid at different temperatures; the *melting point/freezing point curve* is a plot of the equilibrium melting point *vs*. pressure. The vertical line at T_c, the critical temperature, defines the end of the vapor pressure curve. At any location to the right of this line the system exists as a supercritical fluid. At any temperature greater than T_c, observation has shown that the system cannot be made to condense, regardless of the pressure applied.

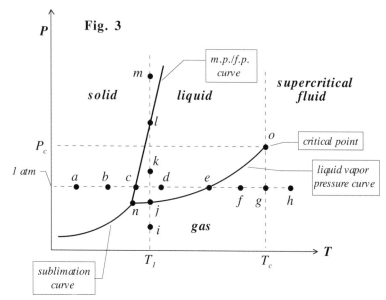

Consider point *a* on the diagram: At this low temperature and pressure the system will tend to exist as a *solid*. Keeping the pressure constant at one atm and slowly increasing the temperature, the system moves along the horizontal line *abcdefgh*. At point *b* the system is still solid, but at a higher temperature than at point *a*. Point *c* is the *normal melting point* for the solid (the equilibrium melting point at one atm pressure); further heating doesn't raise the temperature; instead, the solid starts to melt to form liquid.

At point *c* the system exists as an equilibrium mixture of solid and liquid. After sufficient heating has been applied to convert all the solid to liquid, further heating raises the temperature of the liquid. At point *d* the system is all liquid. At point *e*, which is *the normal boiling point* of the liquid (the temperature at which the equilibrium vapor pressure of the liquid is one atm), both liquid and vapor coexist in equilibrium until sufficient heating has converted all the liquid to vapor. At point *f* the system exists only as gas. Beyond point *g* the system exists only as a supercritical fluid.

Now consider a gas at T_l (point *i*): When the pressure is increased to point *j* the gas will start to condense to form liquid. At this temperature and pressure we have gas or vapor[10] in equilibrium with liquid. This pressure is the equilibrium vapor pressure of the liquid at temperature T_l.

When we try to increase the pressure at point *j*, instead of the pressure increasing, it stays at the same value until all the gas has condensed to form liquid.

---

[9] Sometimes called the boiling point curve.

[10] The terms "gas" and "vapor" often are used almost interchangeably. Strictly speaking, a vapor refers to a substance in the gaseous state that, under ordinary conditions, is usually a liquid or a solid.

After complete liquefaction, the pressure then starts to increase again. At point k, for example, the system exists as pressurized liquid. When point l is reached the liquid starts to *solidify*. T_l is the freezing point of the liquid when it is at the pressure at point l. This is a freezing point, but not the normal freezing point (which is at the temperature at point c). As we try to increase the pressure above the pressure at point l, the pressure doesn't increase. Instead, the liquid converts to *solid* at the temperature and pressure at point l. After all the liquid has been converted to solid, *then* the pressure again can rise. At point m the system exists only as a solid under high pressure.

Point n is the *triple point* for the substance—the temperature and pressure at which *all three phases can coexist in equilibrium*. The triple point is the lowest temperature at which most systems can exist as liquid (unless the melting point curve has a negative slope, as, for example, it has for water).

If the temperature of a sealed container containing liquid water in equilibrium with its vapor is slowly increased, both the temperature and the equilibrium vapor pressure increase as the system slowly proceeds along the boiling point curve. During this process, the density of the liquid slowly decreases through thermal expansion while that of the vapor slowly increases as more molecules get jammed into the vapor space.

At the *critical point*, point o, the density of the liquid becomes *equal* to the density of the vapor, and there is no longer any distinction between the liquid phase and the vapor phase. If you were to observe this process through a porthole with the naked eye,[11] you would not observe any visible change in the appearance of the liquid or of the vapor until the temperature approached the critical temperature within a small fraction of a degree. At this point the surface between the liquid and the vapor suddenly becomes indistinct and cloudy. This sudden phenomenon is called *critical opalescence*.

As the heating continues and the temperature of the system passes beyond the critical temperature, the critical opalescence disappears, no surface is visible, and only one phase—a very dense vapor called a supercritical fluid—is present. The reason the surface between a liquid phase and a vapor phase is visible at all is a consequence of the fact that a liquid has a different index of refraction from a vapor. In turn, the difference in the index of refraction is caused by the difference in density of the two phases. Therefore it follows that as the density difference disappears, so must the observation of the surface.

Another interesting observation very near the critical point is that the surface not only becomes indistinct, but actually begins to *wander*. This happens because the force of gravity is no longer sufficient to keep the slightly more dense liquid at the bottom of the container, and density fluctuations start to occur throughout the system. The critical opalescence is caused by the scattering of light as a result of these density fluctuations.

Exercise 15-1

The critical temperature, pressure, and compressibility factor for water are 374°C, 220.5 bar, and 0.230, respectively. Calculate the density of water vapor at the critical point and compare it with the value for water vapor at 25°C. The vapor pressure of water at 25°C is 23.756 Torr. ☺

---

[11] As opposed to a *clothed* eye.

Exercise 15-2

The equilibrium vapor pressures of solid and liquid uranium hexafluoride are given by: $P_i^{\bullet}$

$$ln\, P_s^{\bullet} = 29.411 - \frac{5893.5}{T} \quad \text{and} \quad ln\, P_{liq}^{\bullet} = 22.254 - \frac{3479.9}{T}$$

where the pressures are in pascal and the temperatures are in kelvin. Calculate the temperature and pressure of the triple point. ☺

15.3 GIBBS PHASE RULE

The *Phase Rule* was derived by Josiah Willard Gibbs,[12] after whom the Gibbs free energy was named. This rule states that for any chemical system in equilibrium,

$$\boxed{F = C - P + 2} \quad \textit{Gibbs Phase Rule} \qquad \dots [7]$$

• *C* is the number of *components* in the system. A **component** is one of the independent species required to set up the phases in a system.

• *P* is the number of *phases* in the system. A **phase** is any homogeneous part of the system having a definable boundary.

• *F* is the number of *degrees of freedom* for the system. This refers to the number of intensive variables or properties that can be changed independently without changing the number of phases in equilibrium.

An **intensive property** is a property that is independent of the *mass* of the system. For example, temperature, pressure, the mole fractions, and the density are all intensive properties of the system because they have the same values for a system in equilibrium whether there is one gram of system or 1000 kg of system. (Conversely, properties such as *U*, *H*, *G*, and *S* are called **extensive properties** of the system because their values are proportional to the mass of the system.)

To see how the Phase Rule is applied, consider the typical *P–T* Phase Diagram for a pure substance, shown in Fig. 4: Since the system is a pure substance, then there is only one component, and *C = 1*. (For example, if the system were water, then any equilibrium in this system could be made by starting with this one component.) Thus, for all points on the diagram, *C = 1*.

Consider point A. At this temperature and pressure the phase diagram tells us that the system will exist as *solid*. In this region, there is only one *phase* (solid phase); that is, *P = 1*. Therefore,

$$F = C - P + 2$$
$$= 1 - 1 + 2$$
$$= 2$$

---

[12] 1839–1903.

Two degrees of freedom means that we can vary both T and P independently in this region and still have the same phase. Similar arguments can be made for the points B and C, both of which also are single phase regions.

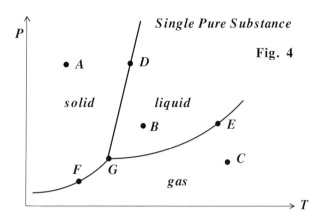

Now consider point D: This point places the system on one of the boundaries between two phases: in this case solid and liquid. *Any point on the melting point curve represents an equilibrium state* between solid and liquid. (Points A, B, and C are *not* equilibrium states because at these points the system consists of only a *single* phase, and more than one phase is required to have a phase equilibrium: e.g., *solid* $\rightleftharpoons$ *liquid*, *solid* $\rightleftharpoons$ *gas*, or *liquid* $\rightleftharpoons$ *gas*.) For point D, as before, $C = 1$; but, now we have *two* phases (solid and liquid), so $P = 2$. Putting these values into the Phase Rule gives

$$\begin{aligned} F &= C - P + 2 \\ &= 1 - 2 + 2 \\ &= 1 \end{aligned}$$

The system now has only *one* degree of freedom. In practical terms what this means is that if we want to have two phases present at the same time, we are *free* (hence, "degree of *freedom*") to independently choose *either* the pressure *or* the temperature, but *not both*. Thus, if we choose some specific value of T, then, in order to be on the melting point curve the value of P is fixed—we have no choice in the value of P. Alternatively, if we arbitrarily choose P, then we have no choice in T. *The system only has one degree of freedom.* The same arguments apply to points E and F. Thus, any point located on one of the phase boundary lines represents a condition of phase equilibrium with only one degree of freedom.

The exception is at the *triple point*, point G, where we have *three* phases—*solid*, *liquid*, and *vapor*—all existing together at the same time. Applying the Phase Rule:

$$\begin{aligned} F &= C - P + 2 \\ &= 1 - 3 + 2 \\ &= 0 \end{aligned}$$

If we want all three phases together in equilibrium at the same time we have *no choice*. There is only one T and P for which this is possible. The system has *zero* degrees of freedom. We cannot vary either the temperature or the pressure and still have all three phases present in equilibrium.

Table 1			
Point	C	P	$F = C - P + 2$
A, *B*, *C*	1	1	$1 - 1 + 2 = 2$
D, *E*, *F*	1	2	$1 - 2 + 2 = 1$
G	1	3	$1 - 3 + 2 = 0$

The above observations are summarized in Table 1.

[**Answer to Footnote 5:** As T increases, S increases, therefore the slope, which is $-S$, becomes more negative with increasing temperature; i.e., becomes increasingly steep (convex up).]

KEY POINTS FOR CHAPTER FIFTEEN

1. For a homogeneous substance of constant composition
$$dG = VdP - SdT$$
At constant pressure, $dP = 0$:
$$dG_P = -SdT$$

Dividing by dT:
$$\left(\frac{\partial G}{\partial T}\right)_P = -S$$

Since the entropy of a substance is always *positive*, it follows that $(\partial G/\partial T)_P$ is always *negative*; i.e., the free energy of a substance *decreases* as its temperature *increases*.

2. Since entropy is related to molecular disorder, we know that
$$S_{solid} < S_{liquid} < S_{gas}$$

Therefore
$$\left(\frac{\partial G}{\partial T}\right)_P^{solid} > \left(\frac{\partial G}{\partial T}\right)_P^{liquid} > \left(\frac{\partial G}{\partial T}\right)_P^{gas}$$

But $(\partial G/\partial T)_P$ is the *slope* of the G vs. T plot at constant pressure; therefore gases (or vapors) have more negative slopes than liquids, which in turn have more negative slopes than solids.

3. When two phases of a pure substance (such as water) are in equilibrium at some T and P, there is no driving force causing the system to shift from one phase to the other; therefore
$$\Delta G_{T,P} = 0.$$

For example, for the equilibrium at 0°C and one atm pressure between liquid water and solid ice
$$water \rightleftharpoons ice$$
the Gibbs free energy of one mole of liquid water will be *same* as the Gibbs free energy of one mole of ice:
$$g_{water} = g_{ice}$$

That is,
$$\Delta G_{T,P} = \left(g_{ice} - g_{water}\right)_{\substack{0°C, \\ 1\,atm}} = 0$$

The same is true for any two phases α and β of a pure substance in equilibrium:
$$\left[\, g_\alpha = g_\beta \,\right]_{T,\,P}^{phase\,eqm}$$

We also can use *specific* Gibbs free energies to express the same equilibrium:
$$\left[\, \overline{g}_\alpha = \overline{g}_\beta \,\right]_{T,\,P}^{phase\,eqm}$$

4. When a solid is heated at constant pressure, as heat is transferred into the solid the temperature of the solid rises until the melting point T_{mp} is reached. Further heating does not raise the temperature, but, instead, causes the solid to change to liquid at T_{mp}. After all the solid has been converted to liquid, further heating causes the temperature of the liquid to

rise until the boiling point T_{bp} is reached, where the temperature stays constant at T_{bp} until all the liquid has been converted to vapor; then continued heating will cause an increase in the temperature of the vapor. Once the temperature of the vapor passes the so-called *critical temperature* T_c, no amount of applied pressure can cause it to liquefy.

When a pure liquid in *equilibrium* with its vapor is heated, as the system temperature rises, the density of the liquid *decreases* and that of the vapor *increases*. Eventually the point is reached at which the two densities become *equal*. This point is called the *critical point*. At temperatures beyond that of the critical point it no longer is possible to discriminate between the liquid phase and the vapor phase. The temperature and pressure at which this takes place are called the *critical temperature* T_c and the *critical pressure* P_c, respectively. As mentioned above, this new phase is referred to as a *supercritical fluid*.

5. An **intensive property** is a property that is independent of the *mass* of the system; e.g., T, P, x, *density*. Intensive properties have the same values for a system in equilibrium whether there is one gram of system or 1000 tonnes of system. Conversely, properties such as U, H, G, and S are called **extensive properties** of the system because their values are proportional to the *mass* of the system.

6. **Gibbs Phase Rule**: $F = C - P + 2$

    - F is the number of degrees of freedom for the system; i.e., the number of intensive variables or properties that can be changed independently without changing the number of phases in equilibrium.

    - C is the number of **components** in the system. A *component* is one of the independent species required to set up the phases in a system.

    - P is the number of **phases** in the system. A *phase* is any homogeneous part of the system having a definable boundary.

    The phase rule holds for any chemical system in equilibrium (not just for a pure substance).

PROBLEMS

1. (a) The differential dB of any state function B is called an "exact" differential. If the value of B depends only on T and P, then, in general, it can be shown[13] that

    $$dB = MdT + NdP$$

    where M and N are functions of the independent variables T and P. The so-called *Reciprocity Characteristic* states that

    $$\left(\frac{dM}{dP}\right)_T = \left(\frac{dN}{dT}\right)_P$$

---

[13] You'll have to wait until your thermodynamics course for the proof.

Using the Reciprocity Characteristic and Eqn [4] , show that for an isothermal heating process in which the pressure changes from P_1 to P_2,

$$\Delta S_T = -\int_{P_1}^{P_2} \left(\frac{dV}{dT}\right)_P dP$$

(b) Calculate ΔS when 100 moles of oxygen gas expands isothermally at 25°C from 100 bar to 4.00 bar.

(c) The isobaric coefficient of thermal expansion α of a substance is defined as

$$\alpha \equiv \frac{1}{V}\left(\frac{dV}{dT}\right)_P$$

For iron at 25°C, $\alpha = 0.355 \times 10^{-4}$ K^{-1}. Calculate ΔS when 100 moles of iron expands isothermally at 25°C from 100 bar to 4.00 bar. (Same conditions as for part (b) except that now we are dealing with a solid instead of a gas.) The density of iron is 7.87 g cm^{-3} at 25°C. ☺*

2. For a pure solid component in the single phase region, sketch (on a single diagram) the qualitative behavior of G (y-axis) vs. T (x-axis) at the three different pressures P, $P + \Delta P$, and $P + 2\Delta P$, where ΔP is a fixed, positive increment. ☺

3. Sketch a plot G (y-axis) vs. T (x-axis) for a pure solid substance melting if

• C_P[solid] $\approx C_P$ [liquid]

• The melting point is at an intermediate temperature on the diagram

• The molar Gibbs free energy of the solid is taken as zero at the melting point. ☺

4. For a pure homogeneous substance, show a qualitative constant pressure plot of G (y-axis) vs. T (x-axis) from $T = 0$ (absolute) to a temperature greater than the melting point but less than the boiling point. ☹

5. When various chemical reactions can take place in a chemical system the simple Phase Rule has to be modified to the form $F = S - R - P + 2$

in which C, the minimum number of components or chemical constituents that must be specified to determine the chemical composition of every phase of the system—i.e., how many chemicals we have to throw into the beaker to form the system—has been replaced by $(S - R)$. S is the number of distinct chemical species present at equilibrium and R is the number of *restraints* on the system. R is equal to the number of chemical equilibria possible *plus* any other restraints that may be present. Any equilibrium in which the concentrations within the system are related to each other restricts the system behavior and constitutes a restraint.

Solid ammonium chloride (NH_4Cl) partially dissociates to form ammonia gas and HCl gas when heated.

(a) How many phases (P), components (C), and degrees of freedom (F) are there when a large amount of solid ammonium chloride is heated in an otherwise empty vessel?

(b) What are P, C, and F if ammonia gas is added to the vessel before the salt is heated? ☺

6. Consider a saturated aqueous solution of sodium sulfate (solid Na_2SO_4 present in excess) in an otherwise empty vessel. ☺

    (a) How many phases (P), components (C), and degrees of freedom (F) are there in this system?

    (b) Re-evaluate (a) taking the dissociation of the dissolved salt into account:

    $$Na_2SO_{4(aq)} \rightleftharpoons Na^+_{(aq)} + SO^{2-}_{4(aq)}$$

    (c) Re-evaluate (b) further taking into account the dissociation of water.

7. Solid ammonium hydrosulfide (NH_4HS) partially dissociates into ammonia gas (NH_3) and hydrogen sulfide gas (H_2S).

    (a) If a large amount of solid ammonium hydrosulfide is placed into a sealed, initially evacuated vessel, after the establishment of equilibrium, how many phases (P), components (C), and degrees of freedom (F) will there be in this system?

    (b) If large amounts of ammonia gas and hydrogen sulfide gas (with ammonia being in considerable excess of hydrogen sulfide) are placed into an initially evacuated vessel, after the attainment of equilibrium, how many phases (P), components (C), and degrees of freedom (F) will there be in this system? ☺

8. At 150°C, solid copper sulfate pentahydrate decomposes completely to solid anhydrous copper sulfate. How many phases (P), components (C), and degrees of freedom (F) are there at equilibrium when copper sulfate pentahydrate is placed in an otherwise empty vessel initially at room temperature and gradually heated to 175°C? ☺

9. A quantity of solid sodium dihydrogen orthophosphate (NaH_2PO_4) is completely dissolved in water, in a sealed, initially evacuated container. No precipitation takes place.

    (a) Neglecting any dissociation of the dissolved salt, how many degrees of freedom (F) are there in this system?

    (b) Re-evaluate F, taking into consideration that the following species can be formed by dissociation:

    $NaH_2PO_{4(aq)} \rightleftharpoons Na^+_{(aq)} + H_2PO^-_{4(aq)}$ $\qquad$ $NaH_2PO_{4(aq)} \rightleftharpoons Na^+_{(aq)} + H_2PO^-_{4(aq)}$

    $H_2PO^-_{4(aq)} \rightleftharpoons H^+_{(aq)} + HPO^{2-}_{4(aq)}$ $\qquad$ $HPO^{2-}_{4(aq)} \rightleftharpoons H^+_{(aq)} + PO^{3-}_{4(aq)}$

    $H_2O \rightleftharpoons H^+_{(aq)} + OH^-_{(aq)}$ $\qquad$ $2H_2O \rightleftharpoons (H_2O)_2$ ☺

10. In a certain high temperature, high pressure (250°C, 40 bar) metal extraction process, it is important to know the amount of $Al_2(SO_4)_3$ that can be dissolved in aqueous sulfuric acid solution that also is saturated with $MgSO_4$. At 250°C, when such a solution is saturated with aluminum sulfate the excess solid aluminum sulfate exists as the compound *alunite*, $H_3OAl_3(SO_4)_2(OH)_{6(s)}$. Neglecting the presence of air:

    (a) List the minimum set of chemical substances that must be added to the container to set up this system.

    (b) How many degrees of freedom are there in this system?

    (c) What are the variables in this system that can be adjusted independently by the experimenter? ☹

PHASE EQUILIBRIUM (II)

16.1 PHASE DIAGRAM FOR CO_2

The P-T phase diagram for CO_2, shown schematically in Fig. 1, has the shape typical for most substances, in which the melting point curve slopes steeply upward to the right. That is, the melting point increases as the pressure increases. The diagram shows that no liquid exists at pressures below 5.07 bar; therefore, at one atm pressure (1.01325 bar) the solid sublimes to form vapor at –78°C, which is called the *normal sublimation point*. This behavior makes solid CO_2 useful for keeping things (such as food) cold (at –78°C!) without any of the mess that would result from the use of ordinary ice, which melts to form puddles of water at one atm pressure. Instead, at one atm pressure solid carbon dioxide passes directly from the solid state to the

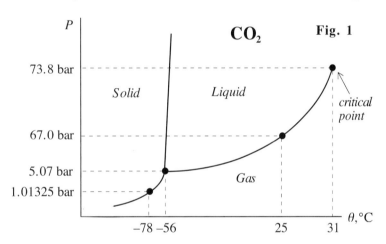

vapor state as it absorbs heat from the surroundings. Because solid CO_2 doesn't melt to form a liquid, it is often called "dry ice."

In addition, dry ice can keep things much colder than can ordinary ice. In order to form liquid CO_2 the system must be pressurized to 5.07 bar or greater. The contents of a CO_2 fire extinguisher at 25°C consist of liquid CO_2 in equilibrium with gaseous CO_2 under an equilibrium pressure of 67 bar. When the valve is opened and the contents allowed to escape, the rapid expansion causes the pressure inside the fire extinguisher to suddenly drop to about one atm with an accompanying rapid cooling effect which takes the temperature of the gas down to below –78°C. When this happens, the system is unable to absorb heat from the surroundings fast enough to restore the temperature to 25°C, and therefore the gas condenses to form the jet of solid "snow" characteristic of a CO_2 fire extinguisher.

16.2 PHASE DIAGRAM FOR WATER

The phase diagram for H_2O (Fig. 2) is atypical in that the melting point curve slopes steeply to the *left* instead of to the *right*. For most substances the solid is more dense than the liquid because the molecules are closer together in the solid state than in the liquid state. In the case of water, however, the solid is *less dense* than the liquid owing to the crystalline structure of ice, which forms loose planar sheet-like networks of puck-

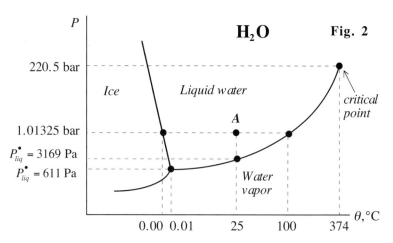

ered hexagonal rings loosely held together to other sheets by weak hydrogen bonding (Fig. 3). This density difference causes ice to float in water (for most substances, the solid sinks to the bottom of the liquid).

Under very high pressures a number of crystalline modifications of ice are observed that result from the buckling of the hydrogen bonds, causing the crys-

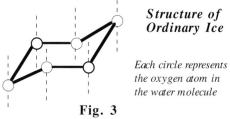

Structure of Ordinary Ice

Each circle represents the oxygen atom in the water molecule

Fig. 3

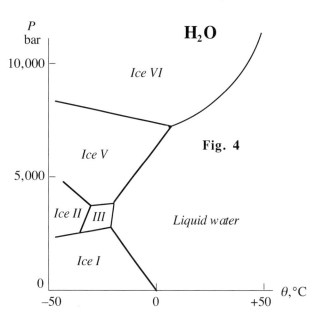

tals to adopt different structures. The high pressure equilibrium phase diagram for water is shown schematically in Fig. 4. *Ice I* is ordinary ice; ices *II*, *III*, *V*, *VI*, and *VII* are crystal modifications that are thermodynamically stable at very high pressures. In Fig. 4 the range of pressure shown on the vertical axis is so great that the ordinary sublimation and vapor pressure curves are almost compressed against the horizontal axis. One interesting observation from the diagram is that at very high pressures water freezes at quite high temperatures. For example, *Ice VII* (not shown in the diagram) melts at about 100°C under a pressure of about 25,000 bar! You can burn yourself by handling this type of ice!

16.3 THE CLAPEYRON EQUATION

Consider the following phase change taking place at temperature T_1:

$$\left(\begin{array}{c} 1\ mole\ liquid \\ water\ (liq) \end{array} \right) \rightarrow \left(\begin{array}{c} 1\ mole\ water \\ vapor\ (V) \end{array} \right)$$

If the process takes place very, very slowly, we can treat it as a reversible process, which essentially is indistinguishable from an "equilibrium" process, and write

$$\left(\begin{array}{c} 1\ mole\ liquid \\ water \end{array} \right) \rightleftharpoons \left(\begin{array}{c} 1\ mole\ water \\ vapor \end{array} \right)$$

We can locate this system on the water phase diagram (Fig. 5) at point A, which falls on the liquid/vapor equilibrium curve (the equilibrium vapor pressure curve). Point A represents liquid water in equilibrium with water vapor at temperature T_1 and equilibrium (vapor) pressure $P_1^{\bullet}$.

The condition for equilibrium is that the molar free energy difference $\Delta g_{T,P}$ between liquid and vapor be zero; that is, that the molar free energy of the liquid is the same as the molar free energy of the vapor:

$$g_1^{liq} = g_1^V \quad \ldots [1]$$

where g_1^{liq} is the molar Gibbs free energy of the liquid at T_1 and $P_1^{\bullet}$ and g_1^V is the molar Gibbs free energy of the vapor at T_1 and $P_1^{\bullet}$.

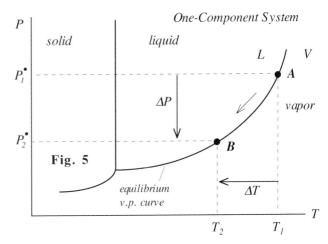

Now we consider the same equilibrium at a different temperature T_2 and pressure $P_2^{\bullet}$, located on the P-T diagram at point B. The condition for equilibrium now is

$$g_2^{liq} = g_2^V \quad \ldots [2]$$

If we "slide" down the equilibrium vapor pressure curve from point A at T_1 and $P_1^{\bullet}$ to point B at T_2 and $P_2^{\bullet}$ we go from one equilibrium state of the system to another equilibrium state of the system. Subtracting Eqn [1] from Eqn [2] gives

$$g_2^{liq} - g_1^{liq} = g_2^V - g_1^V$$

That is, $$\Delta g^{liq} = \Delta g^V \quad \ldots [3]$$

where the Δ's refer to the differences between T_1 and $P_1^{\bullet}$ and T_2 and $P_2^{\bullet}$.

If point A and point B are *very* close together, then the difference between the two equilibrium states becomes vanishingly small, such that

$$\Delta g \rightarrow dg$$

and Eqn [3] becomes
$$dg^{liq} = dg^V \qquad \qquad \cdots [4]$$

Rearranging Eqn [4]:
$$dg^V - dg^{liq} = 0$$

That is,
$$d(g^V - g^{liq}) = 0$$

which is just
$$d\Delta g_{T,P} = 0 \qquad \qquad \cdots [5]$$

Now, for a single component system, we showed earlier[1] that

$$dG = VdP - SdT$$

or, for one mole:
$$dg = vdP - sdT \qquad \qquad \cdots [6]$$

If we substitute Eqn [6] for each phase into Eqn [4], we get:

$$v^{liq}dP - s^{liq}dT = v^V dP - s^V dT$$

Rearranging:
$$(s^V - s^{liq})dT = (v^V - v^{liq})dP$$

From which
$$\left(\frac{dP}{dT}\right)_{eqm} = \frac{s^V - s^{liq}}{v^V - v^{liq}} = \frac{\Delta s}{\Delta v} \qquad \qquad \cdots [7]$$

Now, sometimes $\Delta S = \Delta H/T$, and sometimes $\Delta S \neq \Delta H/T$. When, and *only* when, can we write $\Delta S = \Delta H/T$? From $\Delta G_{T,P} = \Delta H - T\Delta S$ it is apparent that $\Delta H = T\Delta S$ only when $\Delta G_{T,P} = 0$; that is, when we have *equilibrium*. If $\Delta G_{T,P} \neq 0$, then $\Delta S \neq \Delta H/T$; instead, $\Delta H/T = Q_P/T < Q_{rev}/T$.

Consequently, under conditions of phase *equilibrium* we may substitute $\Delta s = \Delta h/T$ into Eqn [7] to obtain

$$\left(\frac{dP}{dT}\right)_{eqm} = \frac{\Delta h/T}{\Delta v} = \frac{\Delta h}{T\Delta v} \qquad \qquad \cdots [8]$$

Since $\Delta S = n\Delta s$, $\Delta H = n\Delta h$, and $\Delta V = n\Delta v$, Eqn [8] usually is written as

$$\boxed{\left(\frac{dP}{dT}\right)_{eqm} = \frac{\Delta S}{\Delta V} = \frac{\Delta H}{T\Delta V}} \quad \begin{array}{l} \textit{Clapeyron} \\ \textit{Equation} \end{array} \qquad \cdots [9]$$

Similarly, we also could write
$$\left(\frac{dP}{dT}\right)_{eqm} = \frac{\Delta \bar{s}}{\Delta \bar{v}} = \frac{\Delta \bar{h}}{T\Delta \bar{v}} \qquad \qquad \cdots [10]$$

All three forms give the same results; the form used depends on which data we have available.

---

[1] Chapter 15, Eqn [4].

Eqns [8], [9], and [10] are known as the *Clapeyron Equation* for phase equilibrium. This equation determines the change of equilibrium pressure necessary to maintain phase equilibrium when the temperature is changed; or, conversely, the change of temperature necessary when the pressure[2] is changed. It is applicable to any kind of phase equilibrium, not just liquid going to vapor.

Examination of the Clapeyron equation shows that it is just the *slope* of the equilibrium phase curve. Consider the phase equilibrium between **water** and **ice**:

$$solid \rightleftharpoons liquid$$

$$\Delta V = V_{liq} - V_{solid} = negative$$

$$\Delta H = \Delta H_{fusion} = positive$$

and
$$T = the\ melting\ point = positive$$

Therefore the Clapeyron equation tells us that

$$\left(\frac{dP}{dT}\right)_{eqm} = \frac{(+)}{(+)(-)} = negative$$

That is, *the slope of the melting point curve for water is negative.* This means that, unlike with most substances, the melting point of ice *decreases* as the pressure *increases.*

Next consider the equilibrium between **liquid** and **vapor**:

$$liquid \rightleftharpoons vapor$$

In this case,
$$\Delta V = V_V - V_{liq} = positive$$

$$\Delta H = \Delta H_{vap} = positive$$

$$T = the\ boiling\ point = positive$$

Now the Clapeyron equation tells us that

$$\left(\frac{dP}{dT}\right)_{eqm} = \frac{(+)}{(+)(+)} = positive$$

That is, *the slope of the vapor pressure curve is always positive.*

Similarly, for equilibrium between **solid** and **vapor**:

$$solid \rightleftharpoons vapor$$

$$\Delta V = V_V - V_{solid} = positive$$

---

[2] For solid/liquid or liquid/vapor equilibrium, the pressure P in the Clapeyron equation is the equilibrium vapor pressure of the solid or liquid; for solid/liquid equilibrium, since there is no vapor phase present, the pressure is the *applied* pressure (e.g., the pressure applied by a piston on the solid/liquid mixture).

$$\Delta H = \Delta H_{sub} = positive$$

$$T = the\ sublimation\ temperature = positive$$

For this case, the Clapeyron equation tells us that

$$\left(\frac{dP}{dT}\right)_{eqm} = \frac{(+)}{(+)(+)} = positive$$

That is, the *sublimation curve always has a positive slope*. BUT, since

$$\Delta H_{sub} = \Delta H_{fus} + \Delta H_{vap}$$

it follows that ΔH_{sub} must be *greater* than ΔH_{vap}; therefore, the Clapeyron equation tells us that the slope of the sublimation curve must be *more positive* than the slope of the vapor pressure curve near the triple point, where they intersect. Observation shows that this is, in fact, the case.

Example 16-1

Calculate the slopes of the (a) solid-liquid, (b) liquid-vapor, and (c) solid-vapor equilibrium curves for water in the vicinity of the triple point (0.01°C). The vapor pressure at the triple point is 611 Pa, and the latent heats of fusion and liquid vaporization are 6.01 kJ mol^{-1} and 44.9 kJ mol^{-1}, respectively. The densities of liquid water and of ice are 1.00 g cm^{-3} and 0.917 g cm^{-3}, respectively.

Solution

The molar volumes of the liquid, solid, and vapor are:

$$v_{liq} = \left(1.00\ \frac{cm^3}{g}\right)\left(\frac{1}{10^6}\ \frac{m^3}{cm^3}\right)\left(18.02\ \frac{g}{mol}\right) = 18.02 \times 10^{-6}\ m^3\ mol^{-1}$$

$$v_{s} = \left(\frac{1}{0.917}\ \frac{cm^3}{g}\right)\left(\frac{1}{10^6}\ \frac{m^3}{cm^3}\right)\left(18.02\ \frac{g}{mol}\right) = 19.65 \times 10^{-6}\ m^3\ mol^{-1}$$

$$v_{v} = \frac{RT}{P} = \frac{(8.314)(0.01 + 273.15)}{611} = 3.717\ m^3\ mol^{-1}$$

(a) *solid* $\rightleftharpoons$ *liquid*:

$$\Delta V_{fus} = v_{liq} - v_{s}$$
$$= (18.02 - 19.65) \times 10^{-6}$$
$$= -1.63 \times 10^{-6}\ m^3\ mol^{-1}$$

$$\left(\frac{dP}{dT}\right)_{fus} = \frac{\Delta H_{fus}}{T\Delta V_{fus}} = \frac{+6\,010}{(273.16)(-1.63 \times 10^{-6})} = -1.350 \times 10^7\ Pa\ K^{-1} = -135\ bar\ K^{-1}$$

Ans: $\boxed{-135 \text{ bar K}^{-1}}$

(b) *liquid* $\rightleftharpoons$ *vapor*:
$$\Delta V_{vap} = v_v - v_{liq}$$
$$= 3.717 - 18.02 \times 10^{-6}$$
$$= 3.717 \text{ m}^3 \text{ mol}^{-1}$$

$$\left(\frac{dP}{dT}\right)_{vap} = \frac{\Delta H_{vap}}{T\Delta V_{vap}} = \frac{+44\,900}{(273.16)(3.717)} = +44.22 \text{ Pa K}^{-1} = +4.422 \times 10^{-4} \text{ bar K}^{-1}$$

Ans: $\boxed{+4.42 \times 10^{-4} \text{ bar K}^{-1}}$

(c) *solid* $\rightleftharpoons$ *vapor*:
$$\Delta V_{sub} = v_v - v_s$$
$$= 3.717 - 19.65 \times 10^{-6}$$
$$= 3.717 \text{ m}^3 \text{ mol}^{-1}$$

$$\Delta H_{sub} = \Delta H_{fus} + \Delta H_{vap}$$
$$= 6010 + 44\,900$$
$$= +50\,910 \text{ J mol}^{-1}$$

$$\left(\frac{dP}{dT}\right)_{sub} = \frac{\Delta H_{sub}}{T\Delta V_{sub}} = \frac{+50\,910}{(273.16)(3.717)} = +50.14 \text{ Pa K}^{-1} = +5.014 \times 10^{-4} \text{ bar K}^{-1}$$

Ans: $\boxed{+5.01 \times 10^{-4} \text{ bar K}^{-1}}$

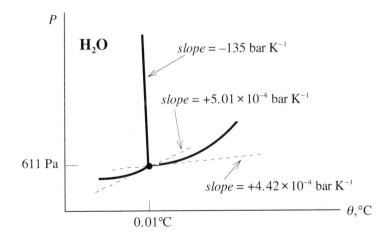

Exercise 16-1

The melting point of iron at one atm pressure is 1535°C. Calculate the melting point at 100,000 atm if Δv_{fus} and Δh_{fus} are constant at 2.30×10^{-7} m^3 mol^{-1} and 15.5 kJ mol^{-1}, respectively. ☺

Exercise 16-2

A 50-meter-high column of reinforced concrete of uniform cross-section has trapped beneath its base a layer of water at 0.00°C. The water is unable to escape or freeze without moving the column. (a) Calculate the value of dT/dP for the trapped water. (b) Calculate the temperature at which the column might be moved by the freezing of the water. ☺

Data: Heat of fusion of ice at 0°C = 6.01 kJ mol^{-1}; density of concrete, water at 0°C, and ice at 0°C = 2310 kg m^{-3}, 1000 kg m^{-3}, and 917 kg m^{-3}, respectively; acceleration due to gravity $g = 9.80$ m s^{-2}; atmospheric pressure = 1.013 bar. The densities and the heat of fusion may be assumed to be independent of temperature and pressure.

16.4 PHASE DIAGRAM FOR CARBON

The phase diagram for carbon, shown in Fig. 6, is ill-defined and incomplete because of the high temperatures required before anything happens. For example, the normal melting point of carbon is greater than 4000 K. At atmospheric pressure carbon doesn't vaporize until temperatures greater than 4000 K are reached. At a pressure of 1000 bar it vaporizes at about 4800 K. The carbon vapor formed at high temperatures is a mixture of single carbon atoms and diatomic and polyatomic molecules, C_1, C_2, C_3, etc. The diatomic and polyatomic molecules are believed to have a general structure consisting of double carbon bonds $C=C$, $C=C=C$, etc. In the temperature range 2400~2700K, C_1, C_2, and especially C_3 are the dominant species in the vapor that is in equilibrium with the solid. The contribution of C_6 and larger molecules to the vapor pressure is negligible. The molar enthalpies of vaporization of monatomic, diatomic, and triatomic carbon are 710.51, 823, and 786 kJ mol^{-1}, respectively.

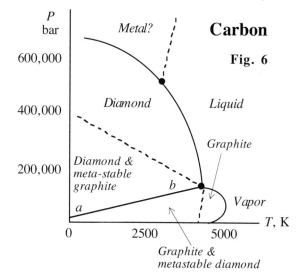

The equation of the diamond-metastable graphite/graphite-metastable diamond equilibrium line (line *ab* in Fig. 6) is given by

$$P_{eqm} = 7093 + 27.4T$$

where P_{eqm} is in bar and T is in kelvin. According to this equation, the transition from graphite to diamond should be possible at relatively low temperatures and pressures. For example, at 3300 K the equation predicts that the transition should take place at 97 500 bar. In practice, however, there are kinetic barriers that must be overcome, and temperatures greater than even

3300 K and pressures in excess of 130,000 bar are required for the direct (uncatalyzed) transformation of graphite to diamond to occur at a measurable rate. Such conditions are very difficult and costly to achieve.

Fortunately, by using iron-nickel or cobalt-iron metallic catalysts the conversion may be achieved at much lower temperatures and pressures—typically around 2000 K and 60 000 bar—which are much closer to the reversible equilibrium conditions predicted by the above equation. At 2000 K the metallic catalysts are in the molten liquid state and readily dissolve the carbon, thereby breaking the bonds between both clusters of carbon atoms as well as between individual carbon atoms, facilitating the transport of the carbon to the growing diamond surface. One way of achieving the high pressures that are required is through the use of giant hydraulic presses employing tungsten carbide pistons and electrical heating. Using this method, micron-sized diamond crystals are produced in a few minutes. To produce a much larger crystal of, say, two-carats (0.4 g) may require several weeks. Crystals as large as 1.7 cm in diameter have been produced. Another way to achieve the high pressures and temperatures required is to coat the back of a metal plate with TNT and detonate the TNT. The shock wave from the explosion causes the plate to impact a mixture of graphite and iron at a velocity in the order of 5 km s^{-1}, thereby generating temperatures and pressures in the order of 1000 K and 300,000 bar for a few microseconds. Under these conditions the transformation is almost instantaneous. With this method, however, only small synthetic polycrystalline diamonds with a maximum particle size of about 60 μm can be produced. These are limited to applications such as the formulation of polishing compounds.

Example 16-2

What is the equilibrium pressure required at 25°C to convert graphite to diamond?

		C [graphite]	C [diamond]
$\Delta h^o_{f\,(298.15)}$	(kJ mol^{-1})	0.00	1.895
$s^o_{298.15}$	(J mol^{-1} K^{-1})	5.74	2.377
ρ	(kg m^{-3})	2250	3513

Solution

C [graphite] → C [diamond]

$$\Delta H^o_{298.15} = (1.895) - (0) = 1.895 \text{ kJ mol}^{-1}$$

$$\Delta S^o_{298.15} = 2.377 - 5.74 = -3.363 \text{ J K}^{-1} \text{ mol}^{-1}$$

$$\Delta G^o_{298.15} = \Delta H^o_{298.15} - T\,\Delta S^o_{298.15} = 1895 - (298.15)(-3.363) = +2897.7 \text{ J mol}^{-1}$$

The positive value of $\Delta G_{T,P}$ at 25°C and 1 bar pressure tells us that at 25°C this reaction will not proceed spontaneously at one bar pressure.

The specific volumes of graphite and diamond are:

$$\bar{v}_{graphite} = \frac{1}{\rho_{graphite}} = \frac{1}{2250 \text{ kg}/\text{m}^3} = 4.444 \times 10^{-4} \text{ m}^3 \text{ kg}^{-1}$$

$$\bar{v}_{diamond} = \frac{1}{\rho_{diamond}} = \frac{1}{3513 \text{ kg}/\text{m}^3} = 2.847 \times 10^{-4} \text{ m}^3 \text{ kg}^{-1}$$

Since a mole of graphite has a larger volume than a mole of diamond, it follows from le Chatelier's Principle that increasing the pressure will cause the conversion reaction to shift to the right. The reaction will be on the verge of becoming spontaneous just when the pressure P becomes high enough to make $\Delta G_{T,P} = 0$. At any pressure *greater* than this pressure $\Delta G_{T,P}$ will be negative and the reaction will become spontaneous. This value of P can be viewed as the minimum pressure at which the conversion is feasible at 25°C.

The specific volume change for the reaction is:

$$\Delta \bar{v} = \bar{v}_{diamond} - \bar{v}_{graphite} = 2.847 \times 10^{-4} - 4.444 \times 10^{-4} = -1.597 \times 10^{-4} \text{ m}^3 \text{ kg}^{-1}$$

and the volume change for *one mole* of graphite is:

$$\Delta v = (-1.597 \times 10^{-4} \text{ m}^3 \text{ kg}^{-1})(0.01201 \text{ kg mol}^{-1}) = -1.9180 \times 10^{-6} \text{ m}^3 \text{ mol}^{-1}$$

We showed in Chapter 13[3] that at constant temperature the molar Gibbs free energy of a substance changes with pressure according to

$$g_{P_2} = g_{P_1} + \int_{P_1}^{P_2} v dP$$

Since solids are more or less incompressible (i.e., their volumes are essentially unaffected by pressure), the molar volume v can be taken outside the integral to give

$$g_{P_2} = g_{P_1} + v \int_{P_1}^{P_2} dP = g_{P_1} + v \Delta P$$

Thus, the molar Gibbs free energy of graphite at 25°C and P Pa is

$$(g_{graphite})_P = g^o_{graphite} + v_{graphite} \Delta P$$

$$= g^o_{graphite} + v_{graphite} (P_2 - P_1)$$

$$= g^o_{graphite} + v_{graphite} (P - 100,000)$$

where $g^o_{graphite}$ is the molar Gibbs free energy at 25°C and one bar (= 100,000 Pa) pressure. Similarly, the molar Gibbs free energy of diamond at 25°C and P is:

---

[3] Chapter 13, Eqn [13].

$$(g_{diamond})_P = g^o_{diamond} + v_{diamond}\,\Delta P$$

Therefore $\Delta G_{T,P}$ for the conversion reaction at 25°C and pressure P is:

$$\Delta G_{298.15,P} = (g_{diamond})_P - (g_{graphite})_P$$

$$= \left[g^o_{diamond} + v_{diamond}\,\Delta P\right] - \left[g^o_{graphite} + v_{graphite}\,\Delta P\right]$$

$$= \left[g^o_{diamond} - g^o_{graphite}\right] - \left[v_{diamond} - v_{graphite}\right]\Delta P$$

$$= \Delta G^o_{298.15} + \Delta v \cdot \Delta P$$

In general:

$$\boxed{\Delta G_{T_1,P_2} = \Delta G_{T_1,P_1} + \int_{P_1}^{P_2} \Delta V\,dP}$$ . . . [11]

Eqn [11] shows how ΔG for a process (including chemical reactions) varies with pressure at constant temperature. Usually T_1 is 298.15 K and P_1 is 100,000 Pa (= 1 bar).

Thus, for the conversion of graphite to diamond at 25°C:

$$\Delta G_{298.15,P} = \Delta G^o_{298.15} + \Delta v \cdot \Delta P$$

$$0 = +2897.7 + (-1.9180 \times 10^{-6})(P - 100,000)$$

$$0 = 2897.7 - 1.9180 \times 10^{-6}P + 0.1918$$

$$1.9180 \times 10^{-6}P = 2897.7 + 0.1918 = 2897.89$$

$$P = \frac{2897.89}{1.9180 \times 10^{-6}} = 1.5109 \times 10^9 \text{ Pa} = 15,109 \text{ bar (about 15,000 atm!)}$$

Ans: $\boxed{1.51 \times 10^4 \text{ bar}}$

This value is very close to the value obtained using the equilibrium pressure-temperature equation given earlier, which predicts a pressure value at 25°C of

$$P_{eqm} = 7093 + 27.4T$$

$$= 7093 + 27.4 \times 298.15$$

$$= 15\ 262 \text{ bar}$$

$$= 1.53 \times 10^4 \text{ bar.}$$

This pressure would be a little uncomfortable on your eardrums.....

16.5 PHASE DIAGRAM FOR HELIUM

Helium displays interesting behavior at low temperatures (Fig. 7). For one thing, there is no equilibrium between the solid and the gas, even at very low temperatures. The atoms of the solid are so light that even at low temperatures they vibrate sufficiently vigorously to shake apart to form liquid.

A solid phase can be obtained only by increasing the pressure to more than 10 bar. Helium is the only known substance that has two liquid phases in equilibrium: *Liquid I* is a normal type of liquid; but *Liquid II* is a "superfluid," which flows without viscosity (no friction). The critical temperature is 5.2 K; at temperatures greater than this helium exists only as a gas, regardless of how much pressure is applied. The phase diagram illustrated is that for helium-4. The phase diagram for helium-3, another isotope, differs from that for helium-4, although it too possesses a superfluid phase. Helium-3 is unusual in that the entropy of the liquid is *lower* than that of the solid, so that melting is *exothermic*. You have to *cool* the solid to make it melt! Weird....

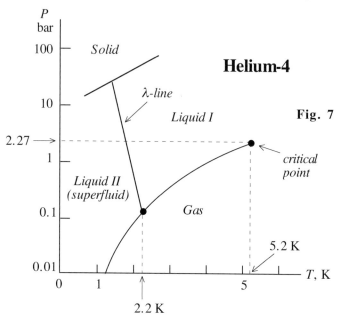

16.6 THE CLAUSIUS-CLAPEYRON EQUATION

For a change involving a *gaseous* product, i.e., *liquid* → *vapor* or *solid* → *vapor*, the molar volume of the gas is much greater than that of the condensed phase; therefore,

$$v_{gas} >> \begin{cases} v_{liquid} \\ v_{solid} \end{cases} \qquad \qquad \ldots [12]$$

and

$$\Delta v = v_{gas} - \begin{cases} v_{liquid} \\ v_{solid} \end{cases} \approx v_{gas} = \frac{RT}{P} \qquad \qquad \ldots [13]$$

This can be substituted into the Clapeyron equation.

Thus,

$$\left(\frac{dP}{dT} \right)_{eqm} = \frac{\Delta h}{T \Delta v} \qquad \qquad \ldots [14]$$

Cross-multiplying:

$$dP = \frac{\Delta h}{T \Delta v} \, dT \approx \frac{\Delta h}{T \left(\frac{RT}{P} \right)} \, dT \qquad \qquad \cdots [15]$$

Rearranging:

$$\frac{dP}{P} = \frac{\Delta h}{R} \frac{dT}{T^2} \qquad \qquad \cdots [16]$$

Integrating:

$$\int_{P_1}^{P_2} \frac{dP}{P} = \int_{T_1}^{T_2} \frac{\Delta h}{R} \frac{dT}{T^2} \qquad \qquad \cdots [17]$$

If we assume that Δh (i.e., either the enthalpy of vaporization or of sublimation) is approximately *constant* over the temperature range $T_1 \leftrightarrow T_2$, then it can be taken outside the integral sign to give

$$\int_{P_1}^{P_2} \frac{dP}{P} = \frac{\Delta h}{R} \int_{T_1}^{T_2} \frac{dT}{T^2}$$

Integrating:

$$ln \left(\frac{P_2}{P_1} \right) = \frac{\Delta h}{R} \left[-\frac{1}{T} \right]_{T_1}^{T_2}$$

That is,

$$\boxed{ln \left(\frac{P_2}{P_1} \right) = -\frac{\Delta h}{R} \left[\frac{1}{T_2} - \frac{1}{T_1} \right]} \quad \begin{array}{l} \textit{Clausius-} \\ \textit{Clapeyron} \\ \textit{Equation} \end{array} \quad \cdots [18]$$

Eqn [18] is called the *Clausius-Clapeyron Equation*, and applies to *liquid-vapor* and *solid-vapor* phase equilibrium in a single component system. Don't forget that this equation involves the assumptions that Δh—which is either Δh_{vap} or Δh_{sub}—is independent of temperature, that the molar volume of the condensed phase is negligible with respect to the molar volume of the vapor phase, and that the vapor phase behaves as an ideal gas. Remember also that the pressure terms in the equation are *equilibrium* pressures, i.e., *equilibrium vapor pressures* ($P^\bullet$) for the liquid or the solid. If the equilibrium vapor pressure and Δh are known at one temperature T_1, then the Clausius-Clapeyron equation can be used to estimate the equilibrium vapor pressure at some other temperature T_2, providing T_2 is not too far removed from T_1.

Example 16-3

Below the triple point ($-56.2°C$) the equilibrium vapor pressure of solid carbon dioxide is given by

$$log_{10} \, P_s^\bullet = 9.832 - \frac{1353}{T}$$

where $P_s^\bullet$ is in torr and T is in kelvin. The heat of fusion of solid carbon dioxide is 8.33 kJ mol^{-1}. Estimate the vapor pressure of liquid CO_2 at $0°C$. *(Note that $ln \, x = 2.303 \, log_{10} x$)*

Solution

For a liquid, if we know the heat of vaporization and the vapor pressure at one temperature, we can use the Clausius-Clapeyron equation to estimate the vapor pressure at a second temperature. Therefore we need to know the heat of vaporization of the liquid. In this problem, however, we are given the heat of *fusion* of the solid and the variation of the vapor pressure of the solid with temperature. These data can be used with the Clausius-Clapeyron equation to determine the heat of sublimation of the solid, from which we then can calculate the heat of vaporization from the fact that $\Delta h_{sub} = \Delta h_{fus} + \Delta h_{vap}$.

Thus, for the solid:
$$ln\left(\frac{P_2^\bullet}{P_1^\bullet}\right) = -\frac{\Delta h_{sub}}{R}\left[\frac{1}{T_2} - \frac{1}{T_1}\right]$$

Expressing the Clausius-Clapeyron in *base-10* logarithms:

$$2.303 log_{10}\left(\frac{P_2^\bullet}{P_1^\bullet}\right) = -\frac{\Delta h_{sub}}{R}\left[\frac{1}{T_2} - \frac{1}{T_1}\right]$$

Therefore:
$$log_{10}\left(\frac{P_2^\bullet}{P_1^\bullet}\right) = -\frac{\Delta h_{sub}}{2.303\,R}\left[\frac{1}{T_2} - \frac{1}{T_1}\right] \qquad \ldots [a]$$

Note the units of the right-hand side of Eqn [a]:

$$\frac{\left(\frac{J}{mol}\right)}{\left(\frac{J}{mol\,K}\right)}\left[\frac{1}{K}\right] = dimensionless, \text{ as required for a logarithm.}$$

Applying the vapor pressure equation for the solid at two different temperatures T_1 and T_2 gives:

$$log_{10} P_2^\bullet - log_{10} P_1^\bullet = \left[9.832 - \frac{1353}{T_2}\right] - \left[9.832 - \frac{1353}{T_1}\right]$$

That is,
$$log_{10}\left(\frac{P_2^\bullet}{P_1^\bullet}\right) = -1353\left[\frac{1}{T_2} - \frac{1}{T_1}\right] \qquad \ldots [b]$$

Comparing Eqn [b] with Eqn [a] shows that $\dfrac{\Delta h_{sub}}{2.303\,R} = 1353$

from which $\Delta h_{sub} = (2.303)(8.314)(1353) = 25\ 906$ J mol^{-1}

and $\Delta h_{vap} = \Delta h_{sub} - \Delta h_{fus} = 25\ 906 - 8330 = 17\ 576$ J mol^{-1}

We know that at the triple point ($-56.2°C = -56.2 + 273.15 = 216.95$ K) the vapor pressure of the solid is the *same* as that of the liquid; this gives us a value $P_1^\bullet$ of the vapor pressure of the liquid at one temperature T_1. The Clausius-Clapeyron equation then can be used with the liquid heat of vaporization value calculated above to estimate the vapor pressure $P_2^{\bullet}$ of the liquid at $T_2 = 0°C = 273.15$ K.

Thus, for $liquid \rightarrow vapor$:

$$ln\left(\frac{P_2^{\bullet}}{P_1^{\bullet}}\right) = -\frac{\Delta h_{sub}}{R}\left[\frac{1}{T_2} - \frac{1}{T_1}\right] \qquad \ldots [c]$$

At the triple point (216.95 K), the vapor pressure of the solid—and of the liquid—is

$$log_{10}\, P_s^{\bullet} = 9.832 - \frac{1353}{T}$$

$$= 9.832 - \frac{1353}{216.95}$$

$$= 3.5955$$

from which

$$P_{liq}^{\bullet} = P_s^{\bullet} = 10^{3.5955} = 3940.03 \text{ Torr}$$

Putting values into Eqn [c]:

$$ln\left(\frac{P_2^{\bullet}}{3940}\right) = -\frac{17\,576}{8.314}\left[\frac{1}{273.15} - \frac{1}{216.95}\right] = 2.00487$$

Therefore,

$$\frac{P_2^{\bullet}}{3940} = exp\,[2.00487] = 7.425$$

and

$$P_2^{\bullet} = (7.425)(3940) = 29\,255 \text{ Torr}$$

In SI units:

$$P_2^{\bullet} = \left(\frac{29\,255 \text{ Torr}}{760 \text{ Torr}/\text{atm}}\right)\left(101\,325\,\frac{\text{Pa}}{\text{atm}}\right) = 3.900 \times 10^6 \text{ Pa} = 3.900 \text{ MPa}$$

The actual experimental value is 3.48 MPa; the estimated value is 12% too high.

Ans: 3.9 MPa

Exercise 16-3

What is the standard boiling point (the temperature at which the equilibrium vapor pressure is 1.00 bar) of ethanol? The normal boiling point of ethanol is 78.3°C and its vapor pressure at 61.5°C is 50.0 kPa. ☺

KEY POINTS FOR CHAPTER SIXTEEN

1. **Phase Diagram for CO_2:**

   A typical phase diagram, in which the melting point curve slopes steeply upward to the *right*, indicating that the melting point of the solid *increases* as the pressure increases. The triple point is at 5.07 bar and $-78°C$; therefore, no liquid exists at pressures below 5.07 bar, which means that at atmospheric pressure, CO_2 exists as a solid that sublimates directly to a vapor without melting to a liquid. It stays "dry" when it vaporizes ("dry ice"), and is a non-messy way to keep things very cold.

2. **Phase Diagram for H_2O:**

   The phase diagram for water is atypical in that the melting point curve slopes upward to the *left* instead of to the right. This owes to the fact that the solid is *less dense* than the liquid because of the crystalline structure of ice, which forms loose planar sheet-like networks of puckered hexagonal rings loosely held together to other sheets by weak hydrogen bonding. This density difference causes ice to float in water (for most substances, the solid sinks to the bottom of the liquid). At very high pressures—thousands of bar—a number of crystalline modifications of ice are observed, because at such high pressures the hydrogen bonds buckle, causing the crystals to adopt different structures. *Ice VII*, which exists at about 25,000 bar, melts at about 100°C. If you touch it you will burn yourself!

3. **Phase Diagram for C:**

   The phase diagram for carbon is ill-defined and not completely understood owing to the very high temperatures (typically thousands of degrees) and pressures (greater than 100,000 bar) required before anything begins to happen. The normal melting point is greater than 4000 K; therefore, at atmospheric pressures carbon won't vaporize at temperatures less than about 4000 K. Carbon vapor consists of C_1, C_2, C_3, etc., with C_1, C_2, and especially C_3 being the predominant species. The molecules probably consist of double carbon bonds $C=C$, $C=C=C$, etc. At very high temperatures and pressures carbon can be made to undergo a transition to diamond.

4. **Phase Diagram for He:**

   Even at extremely low temperatures there is no solid form at atmospheric pressure—the atoms are so light and vibrate so vigorously that they fly apart to form liquid. To form solid He requires increasing the pressure to at least 10 bar. Helium is the only known substance to have two liquid phases in equilibrium; one (*Liquid II*) is a "superfluid" that flows without friction. Since the critical point is 5.2 K and 2.27 bar, at all temperatures greater than 5.2 K helium exists only as a gas—hence the designation "permanent gas." The isotope He-3 is interesting in that the entropy of the liquid is *lower* than that of the solid, so the solid must be *cooled* to make it melt!

5. The **Clapeyron Equation:**

$$\left(\frac{dP}{dT}\right)_{eqm} = \frac{\Delta S}{\Delta V} = \frac{\Delta H}{T\Delta V} = \frac{\Delta h}{T\Delta v} = \frac{\Delta \bar{h}}{T\Delta \bar{v}}$$

   For any type of phase equilibrium, this equation gives the change of equilibrium pressure necessary to maintain phase equilibrium when the temperature is changed; or, conversely, the change of temperature necessary when the pressure is changed. The Clapeyron equation is completely rigorous; there are no assumptions involved.

Gives the *slope* of the equilibrium phase curve on a *P–T* phase diagram.

6.   The **Clausius-Clapeyron Equation**:

$$ln\left(\frac{P_2}{P_1}\right)_{eqm} = -\frac{\Delta h}{R}\left[\frac{1}{T_2} - \frac{1}{T_1}\right]$$

where the pressures are *equilibrium* pressures, i.e., *equilibrium vapor pressures* $(P^{\bullet})$ for the liquid or the solid.

Applies to *liquid-vapor* and *solid-vapor* phase equilibrium in a single component system. Assumes that Δh—which is either Δh_{vap} or Δh_{sub}—is independent of temperature, that the molar volume of the condensed phase is negligible with respect to the molar volume of the vapor phase, and that the vapor phase behaves as an ideal gas.

If the equilibrium vapor pressure and Δh are known at one temperature T_1, then this equation can be used to estimate the equilibrium vapor pressure at some other temperature T_2, providing T_2 is not too far from T_1.

PROBLEMS

1.   The vapor pressure of liquid mercury is 96.5 kPa at 354°C and 107.5 kPa at 360°C. Calculate the enthalpy change when one mole of liquid Hg vaporizes, and determine its normal boiling point. ☺

2.   The vapor pressure of water at 0.01°C is 611 Pa; the vapor pressure of ice at –5.00°C is 402 Pa. Calculate the enthalpy change of sublimation of ice. ☺

3.   A railroad tank car containing liquid chlorine and its vapor is at a temperature of 15°C. What is the pressure in the tank car? A large hole is then punched in the top of the tank car, and, as a result, the liquid chlorine boils off to the surroundings, which is at a pressure of one atm. What will the temperature in the tank car be during this process? The normal boiling point of liquid Cl_2 is –34°C, and the enthalpy of vaporization is 20.4 kJ mol^{-1}. ☺

4.   Between 700 K and 739 K, the vapor pressure (in torr) of solid magnesium at temperature T (K) is given by the equation:

$$ln P^{\bullet} = \frac{1.7 \times 10^4}{T} + 19.6$$

What is the heat of sublimation of magnesium at 720 K? ☹

5.   At 100 bar pressure, the melting point of pure iron is 1535.27°C. (a) Calculate the *increase* in the melting point when the pressure is increased from 100 bar to 600 bar if Δv_{fus} and Δh_{fus} are constant at 2.30×10^{-7} m^3 mol^{-1} and 15.5 kJ mol^{-1}, respectively. (b) What is the slope (dT/dP) of the melting point curve at 1550°C? ☺

6.   Phenol (C_6H_5OH)—sometimes called carbolic acid—is a white crystalline caustic material that is used primarily as a disinfectant. The densities of solid phenol and liquid phenol at the normal melting point of 41.0°C are 1.0720 g cm^{-3} and 1.0561 g cm^{-3}, respectively. The specific enthalpy of fusion is 104.3 kJ kg^{-1}. Calculate the change in the melting point produced by an increase in pressure of 9.00 bar. ☺

7.   For the equilibrium *liquid* $\rightleftharpoons$ *vapor*, if it is assumed that (i) the vapor behaves like an ideal gas, (ii) ΔV is just the volume of the vapor, and (iii) ΔH_{vap} is independent of temperature, then the Clapeyron equation is transformed into the Clausius-Clapeyron equation. The Clausius-Clapeyron equation is not very accurate, mainly because of assumption (iii). However, if we are given ΔH_{vap} at 298.15 K and ΔC_p for the process, we can evaluate ΔH_{vap} at any other temperature T. The resulting formula can be used to express ΔH_{vap} as a function of temperature: $\Delta H_{vap} = \Delta H_{vap}(T)$. If we put this into the Clapeyron equation and assume that ΔC_p is constant, we obtain

$$\frac{dP^{\bullet}}{dT} = \frac{\Delta H_{vap}(T)}{T \Delta V},$$

which can be reworked to yield the following more accurate version of the Clausius-Clapeyron equation:

$$ln\left(\frac{P_2^{\bullet}}{P_1^{\bullet}}\right) = A\left[\frac{1}{T_2} - \frac{1}{T_1}\right] + B\, ln\left(\frac{T_2}{T_1}\right)$$

What are A and B? ☹

8.   The melting point of monoclinic sulfur (S_8) at atmospheric pressure is 119.3°C. The change in volume during fusion is +41 cm^3 kg^{-1}, and $\Delta h_{fus} = 3.376$ kcal mol^{-1}. Estimate the melting point of monoclinic sulfur when the pressure is increased to 1000 atm. S = 32.06 ☺

9.   The normal melting point of mercury is -38.87°C. At -38.87°C and one atm pressure the density of the liquid and that of the solid are 13.690 g cm^{-3} and 14.193 g cm^{-3}, respectively. The molar heat of fusion at the normal melting point is 2.42 kJ mol^{-1}. Calculate the *change* in the melting point if the pressure is increased by 100 atm. ☺

10. If the melting point of iron at one atm pressure is 1535°C, what will the melting point be at 1000 atm if Δv_{fus} and Δh_{fus} are constant at 2.30×10^{-7} m^3 mol^{-1} and 15.5 kJ mol^{-1}, respectively? ☺

11. The densities of solid and liquid bismuth at the melting point (271.0°C) are 9.673 g cm^{-3} and 10.004 g cm^{-3}, respectively. What is the heat of fusion for 1.00 g of Bi if the melting point drops by 0.00342°C for a pressure increase of 0.968 atm? ☺

12. What pressure in atm is required to make ice melt at –20°C?

   Data: $\rho_{ice} = 0.92$ g cm^{-3}; $\rho_{water} = 1.00$ g cm^{-3}; $\Delta \bar{h}_{fus} = 334$ kJ kg^{-1} ☺

13. The equilibrium vapor pressure $P^{\bullet}$ of ammonia along the sublimation curve is given by

$$ln\, P^{\bullet}_{solid} = 27.92 - \frac{3754}{T}$$

   and along the liquid vaporization curve is given by

$$ln\, P^{\bullet}_{liquid} = 24.38 - \frac{3063}{T},$$

   where $P^{\bullet}$ is in pascal and T is in kelvin.

   (a) Calculate the triple point temperature for ammonia.

   (b) What is the enthalpy change for the vaporization of one mole of liquid ammonia at the triple point?

   Hint: Make use of the mathematical relationship $\dfrac{dP}{dT} = P \dfrac{d\, ln P}{dT}$ ☹

14. Lead (*Pb*) vaporizes to form a monatomic vapor. The equilibrium vapor pressure of liquid lead is given by

$$ln\, P^{\bullet}_{liquid} = 22.903 - \frac{22\,690}{T}$$

   where $P^{\bullet}$ is in pascal and T is in kelvin. The molar enthalpy of fusion of lead at its standard melting point of 600.6 K is 4770 J mol^{-1}. For this problem, the average molar heat capacities of solid and liquid lead may be taken as constant at 28.03 and 26.36 J mol^{-1} K^{-1}, respectively. Given that the absolute standard molar entropy of solid lead at 25°C is 64.81 J K^{-1} mol^{-1}, calculate the absolute standard molar entropy of lead vapor at 25°C and 1 bar pressure. ☠

15. Estimate at what temperature water, with a normal boiling point of 100.0°C, and chloroform (*CHCl$_3$*), with a normal boiling point of 61.17°C, will have the same vapor pressures. The latent heats of vaporization for water and for chloroform at their normal boiling points are 40.656 kJ mol^{-1} and 29.79 kJ mol^{-1}, respectively. ☺

16. Estimate the boiling point of water at a place where the atmospheric pressure is 550 Torr if the latent heat of vaporization of water at the normal boiling point is 539.38 cal g^{-1}. The molar mass of water is 18.015 g mol^{-1}. ☺

17. The sublimation pressure of $CO_{2(s)} \rightleftharpoons CO_{2(g)}$ is 1008.9 Torr at -75.8°C, and 438.6 Torr at -85.0°C.

   (a) Calculate the enthalpy of sublimation.

   (b) Calculate the normal sublimation point in °C. ☺

18. In a certain temperature interval near the melting point the following formulas are valid for the equilibrium vapor pressure over solid and liquid TaBr$_5$:

$$ln\,P^{\bullet}_{solid} = 33.839 - \frac{13\,010}{T} \quad \text{and} \quad ln\,P^{\bullet}_{liquid} = 23.707 - \frac{7518}{T}$$

where $P^{\bullet}$ is in Pa, and T is in kelvin. Calculate:

(a) The melting point of $TaBr_5$.

(b) The normal boiling point of $TaBr_5$.

(c) The entropy of vaporization at the normal boiling point.

(d) The molar heat of fusion.

It may be assumed that the enthalpies of phase changes are independent of temperature in the temperature ranges considered. ☺

19. Below the triple point ($-56.2°C$) the equilibrium vapor pressure of solid carbon dioxide is given by

$$log_{10}\,P^{\bullet}_{solid} = 9.832 - \frac{1353}{T}$$

where $P^{\bullet}_{solid}$ is in torr and T is in kelvin. The heat of fusion of solid carbon dioxide is 8.33 kJ mol^{-1}. Estimate the vapor pressure of liquid CO_2 at $0°C$.
(*Note that ln x = 2.303 log$_{10}$x*) ☺

20. The following are two handy rules of thumb:

(i)   At $8°C$ the vapor pressure of water is 8 Torr and at $28°C$ it is 28 Torr.

(ii) Over the temperature range $60°C \sim 140°C$ the vapor pressure of water approximately doubles for every $20°C$ increase in temperature.

Using only these two rules of thumb, estimate the heat of vaporization of water at

(a) $18°C$ and at (b) $110°C$. ☺

21. The Saturated Steam Table lists pressures of 355.1 Torr and 1488 Torr for temperatures of $80°C$ and $120°C$, respectively. Using this information, estimate the molar heat of vaporization of water at $100°C$. ☹

22. The Saturated Steam Table lists pressures of 1.597 kPa and 1.817 kPa for temperatures of $14.0°C$ and $16.0°C$, respectively. Using this information, how much heat would be required to vaporize 1000 kg of water in equilibrium with its vapor at $14.0°C$? The molar mass of water is 18.015 g mol^{-1}. ☹

23. Given that the vapor pressures of water at $80°C$ and $120°C$ are 355.1 Torr and 1488 Torr, respectively, use the Clausius-Clapeyron equation to estimate the difference $\Delta\bar{c}_P$ between the specific heat capacities of water vapor and liquid water at $100°C$ and one atm pressure. ☹

24. Calculate the vapor pressure of water at $90.0°C$ if the enthalpy of vaporization at the normal boiling point is 40.66 kJ mol^{-1}. ☺

25. Using the data given below for saturated steam, estimate ΔS for the following process:

(1.00 kg liquid water, 250°C) → *(1.00 kg saturated steam, 250°C)*

using

(a) the Clapeyron equation

(b) the Clausius-Clapeyron equation.

(c) Which value is more accurate, (a) or (b)?

Temperature (°C)	$P^{\bullet}$ (kPa)	$\bar{v}_{liq}$ (cm^3 g^{-1})	$\bar{v}_{gas}$ (cm^3 g^{-1})
248	3844.9	1.247	51.81
252	4113.7	1.265	48.33

☺

26. A mountaineer reaches the top of a mountain and boils water to make a cup of tea. His thermometer indicates that the water boils at 75°C. How high is the mountain top above sea level? Between 0°C and 200°C the molar enthalpy of vaporization of water varies with temperature according to:

$$\Delta h_{vap} = 61\,331 - 92.068T + 0.17527T^2 - 2.0639 \times 10^{-4}T^3$$

where Δh_{vap} is the molar enthalpy of vaporization in J mol^{-1} and T is the temperature in kelvin. Up to about 20 km above sea level the atmospheric pressure varies with elevation according to

$$\log_{10} P_{atm} = 5.00584 - 5.90807 \times 10^{-5}z$$

where P is the pressure of the atmosphere in Pa and z is the elevation above sea level in meters. ☹*

27. Liquid benzene (C_6H_6) has a vapor pressure of 6.17 kPa and 15.8 kPa at 10°C and 30°C, respectively. Similarly, the solid has vapor pressures of 0.299 kPa and 3.27 kPa at −30°C and 0°C, respectively. Calculate

(a) the molar enthalpy of fusion of benzene, and (b) its triple point. ☹

28. The entropy change of water freezing to ice at −10°C and atmospheric pressure cannot be calculated from the observed heat of crystallization (− 1343 cal mol^{-1}). Why? Calculate the molar entropy change of freezing for this reaction if the normal heat of fusion of ice is 1436 cal mol^{-1}. The molar heat capacities of liquid water and ice may be considered constant at 18 and 8.7 cal mol^{-1} K^{-1}, respectively. ☺

29. Calculate $\Delta H°$, $\Delta S°$, and $\Delta G°$ for the process $\quad CH_3OH_{(liq)} \rightarrow CH_3OH_{(g)}$

at (a) 25°C and (b) 64.05°C, the standard boiling point of methanol.

	$\Delta h^o_{f(298.15)}$ (kJ mol^{-1})	$s^o_{298.15}$ (J K^{-1} mol^{-1})	c_P^o (J mol^{-1} K^{-1})
$CH_3OH_{(liq)}$	−238.7	126.8	81.6
$CH_3OH_{(g)}$	−200.7	239.7	43.9

☺

30. (a) If the vapor pressure of water is 3.168 kPa at 25°C, calculate the molar heat of vaporization using the normal boiling point as reference.

   (b) Calculate the work of expansion for one mole of water during vaporization at the normal boiling point, assuming ideal gas behavior.

   (c) What fraction of the molar heat of vaporization is attributable to the work of expansion?

   (d) On the basis of the above data, estimate the vapor pressure of water at 200°C. ☺

31. The equilibrium vapor pressures of solid and liquid SO_2 are given, respectively, as

$$ln\, P^{\bullet}_{solid} = \frac{4308}{T} + 24.39 \quad \text{and} \quad ln\, P^{\bullet}_{liquid} = \frac{-3284}{T} + 19.16$$

where $P^{\bullet}$ is in Pa, and T is in K. Estimate (a) the temperature at the triple point, and (b) the latent heat of fusion of solid SO_2. ☻

32. What total pressure must be applied to ice to reduce its melting point to −0.1°C? The densities of ice and water are, respectively, 0.917 and 1.000 g cm^{-3}. The specific heat of fusion of ice is 333 J g^{-1}. ☻

SEVENTEEN

MIXTURES

17.1 EXPRESSIONS FOR CONCENTRATION

So far we have been dealing with pure substances that consist of only one component. But the real world is a highly entropic place with chaos everywhere (just examine your own life!). Things tend spontaneously to get mixed up. In fact, everywhere we look we find mixtures: the air, a cup of coffee, blood, sweat, and tears, steel, your grandmother, your quiz marks.

A mixture is a system that consists of more than one component—a *multicomponent* system. The properties of a multicomponent system depend on its composition, as well as on its temperature and pressure. A concentrated sulfuric acid solution has properties that are quite different from those of a dilute sulfuric acid solution; if you don't believe this, notice how quickly your finger warms up when immersed in the concentrated solution, but not in the dilute solution!

First let's look at some of the various ways to express the composition of a multicomponent system.

Molarity—defined as *the number of moles of solute per liter of solution* (*not* per liter of *solvent*). The units of molarity are mol L^{-1} or mol dm^{-3}.[1] The symbol M (e.g., 4.5 M solution of NaCl) commonly is used to designate the molarity of a solution. The notation *[i]* also is used to designate the molarity of species i :

$$[i] \equiv \frac{moles\ of\ i}{L\ of\ solution} \qquad \ldots [1]$$

The molarity of a solution decreases with increasing temperature because, even though the number of moles of solute dissolved in the solution stays the same, the *volume* of the solution expands as the temperature is raised.

Mole fraction—defined as the number of moles of a given component divided by the total moles of all components (including the given component). Mole fractions are unitless. The symbol x_i is usually used to express the mole fraction of a component i in a condensed (solid or liquid) phase;

---

[1] A liter (L) is the same volume as a cubic decimeter (dm^3). In the older, now obsolete, definition, the liter—OK, the proper SI spelling is *litre*—was defined as the volume occupied by one cubic decimeter of pure water measured at the temperature of maximum density (3.98°C). In the old definition, 1 L was equal to 1.0000028001 dm^3. In 1964 it was redefined at the 12th General Conference on Weights and Measures to be *exactly* equal to 1 dm^3. Oh, a bit of trivia: Did you know that the "gram" comes from the Greek *gramma*, the weight of a pea?

the symbol y_i sometimes is used to express the mole fraction of component i in a gas phase to differentiate it from the mole fraction of the same component in the condensed phase.

$$x_i \ (or \ y_i \ for \ gas \ phases) = \frac{moles \ i}{total \ moles \ of \ all \ components \ (including \ i)} \quad \dots [2]$$

Molality—defined as *the number of moles of solute per kg of solvent* (*not* per kg of *solution*). The advantage of molality is that, because it is a mass/mass quantity, it doesn't vary with temperature; thus a solution that is 1.0 molal at 25°C also is 1.0 molal at 75°C. Another interesting feature of molality is that the addition of another component doesn't change the molality of the others (this cannot be said for mole fraction or molarity). The units of molality are mol kg^{-1}; the symbol commonly used is m, as in a 2.1 m solution of NaCl.

$$m_i = \frac{moles \ i}{kg \ of \ solvent} \quad \dots [3]$$

17.2 PARTIAL MOLAR VOLUMES

Suppose we have a two-component liquid solution consisting of component A and component B. Since both A and B will contribute to the volume of the solution, we can say that $V = f(n_A, n_B)$; i.e., the total volume is some function of the number of moles of A and of the number of moles of B. At constant temperature and pressure this is expressed mathematically as follows:

$$V = n_A (V_m)_A + n_B (V_m)_B \quad \dots [4]$$

where V is the total volume of the solution and $(V_m)_i$ is the *partial molar volume* of component i.

In general,
$$V = \sum n_i (V_m)_i \quad \dots [5]$$

Thus, the total volume of a solution consists of a partial contribution from each component in the solution (including, of course, a contribution from the solvent). Note that the partial molar volume $(V_m)_i$ is *not* the same as the molar volume v_i. If v_A is the *molar volume* of pure A (in m^3 mol^{-1}) and v_B is the *molar volume* of pure B, the total volume V is **not** just $n_A v_A + n_B v_B$. More on this later in Section 17.3.

Example 17-1

A solution of ethanol in water contains 20 mole % ethanol (C_2H_5OH). At 20°C the partial molar volume of the ethanol in this solution is 55.0 cm^3 mol^{-1}, and that of the water is 17.8 cm^3 mol^{-1}. What volumes of pure ethanol and pure water are required to make 1.000 L of this solution? The densities at 20°C of pure ethanol and pure water are 0.789 g cm^{-3} and 0.998 g cm^{-3}, respectively.

Solution

Let the number of moles of ethanol and of water required to make one liter (1000 cm^3) of the solution be n_E and n_W, respectively. Therefore, the total volume V is:

$$V = n_E(V_m)_E + n_W(V_m)_W$$
$$1000 = n_E(55.0) + n_W(17.8) \qquad \qquad \dots [a]$$

We are told that $$x_E = \frac{n_E}{n_W + n_E} = 0.20 \qquad \qquad \dots [b]$$

Cross-multiplying: $$n_E = 0.20n_W + 0.20n_E$$
$$0.80n_E = 0.20n_W$$
$$n_E = 0.25n_W \qquad \qquad \dots [c]$$

Putting [c] into [a] gives: $$1000 = (0.25n_W)(55.0) + n_W(17.8)$$
$$1000 = 13.75n_W + 17.8n_W = 31.55n_W$$
$$n_W = \frac{1000}{31.55} = 31.70 \text{ mol}$$

From Eqn [c], $$n_E = 0.25n_W = (0.25)(31.70) = 7.925 \text{ mol}$$

Volume of pure ethanol required $= V_E = \dfrac{(7.925 \text{ mol})(46.07 \text{ g}/\text{mol})}{(0.789 \text{ g}/\text{cm}^3)} = 462.7 \text{ cm}^3$

Volume of water required $= V_W = \dfrac{(31.70 \text{ mol})(18.02 \text{ g}/\text{mol})}{(0.998 \text{ g}/\text{cm}^3)} = 572.4 \text{ cm}^3$

Note that the sum of the two volumes of pure components is $462.7 + 572.4 = 1035.1$ cm^3. There is a shrinkage of 35.1 cm^3 when the two are mixed together to form 1000 cm^3 of solution.

Ans: | 462.7 cm^3 of ethanol + 572.4 cm^3 of water |

17.3 THE CHEMICAL POTENTIAL (μ)

In the previous section we introduced the partial molar volume. The concept of partial molar quantity also applies to other state properties of homogeneous mixtures besides their volumes. For example:

• the total **enthalpy** of a multicomponent system is the sum of the number of moles of each component times its *partial molar enthalpy:*

$$H = n_A(H_m)_A + n_B(H_m)_B + n_C(H_m)_C + \dots \qquad \qquad \dots [6]$$

where $(H_m)_i$ is the partial molar enthalpy of component i.

• the total **internal energy** of a multicomponent system is the sum of the number of moles of each component times its *partial molar internal energy:*

$$U = n_A (U_m)_A + n_B (U_m)_B + n_C (U_m)_C + \ldots \qquad \ldots [7]$$

where $(U_m)_i$ is the partial molar internal energy of component i.

• Of special importance is the *partial molar Gibbs free energy*, $(G_m)_i$. Thus, the total Gibbs free energy of the system is given by:

$$G = n_A (G_m)_A + n_B (G_m)_B + n_C (G_m)_C + \ldots \qquad \ldots [8]$$

The partial molar Gibbs free energy is so important[2] that it even has its own special name and symbol: the **chemical potential**, **μ**. Therefore, at any given temperature and pressure the total Gibbs free energy of a multicomponent system is given by

$$\boxed{G = n_A \mu_A + n_B \mu_B + n_B \mu_B + \ldots} \qquad \ldots [9]$$

If B is some thermodynamic property of a system (such as U, H, G, S, etc.) then the **partial molar quantity** B_m of any component i in the system is defined as the way in which the total value of B for the system changes per mole of i with an infinitesimal addition of component i at constant temperature and pressure, while keeping the amounts of all the other components j constant.

Mathematically, this is expressed by:

$$\boxed{(B_m)_i \equiv \left(\frac{\partial B}{\partial n_i} \right)_{T,P,n_j}} \qquad \ldots [10]$$

Accordingly, the **chemical potential** of species i, which is the partial molar Gibbs free energy of i, is defined as

$$\boxed{\mu_i \equiv \left(\frac{\partial G}{\partial n_i} \right)_{T,P,n_j}} \quad \begin{array}{l} \textit{Chemical} \\ \textit{Potential} \\ \textit{Defined} \end{array} \qquad \ldots [11]$$

Note that the units of μ are J mol^{-1}.

17.4 THE CHEMICAL POTENTIAL OF A COMPONENT IN AN IDEAL GAS MIXTURE

We showed earlier in the discussion of phase equilibrium that for a homogeneous pure substance,

$$dG = VdP - SdT$$

---

[2] As we shall see later, the chemical potential is important because it represents the driving force acting on a component in a mixture for both chemical and physical change.

For one mole: $$dg = vdP - sdT$$

At constant temperature: $$dg_T = vdP \qquad \ldots [12]$$

Therefore, for one mole of a *pure gas* at constant temperature, $dg = vdP$. If the gas is not pure, but instead is in a *mixture* that also contains other gases, then we must replace the *molar* Gibbs free energy of the gas in Eqn [12] with the *partial* molar Gibbs free energy of the gas (i.e., its chemical potential).

For a pure gas, Eqn [12] becomes $$d\mu_T = vdP \qquad \ldots [13]$$

Now, if the gas is a pure *ideal* gas, $$v = \frac{RT}{P} \qquad \ldots [14]$$

and, substituting Eqn [14] into Eqn [13] gives

$$d\mu_T = \frac{RT}{P} dP \qquad \ldots [15]$$

We can easily calculate the change in the chemical potential when the pressure of the gas i changes from $P_i = P^\circ$ to $P_i = P_i$ as follows:

$$\Delta\mu_i = \int_{P_i=P^\circ}^{P_i=P_i} d\mu_i = RT \int_{P^\circ}^{P_i} \frac{dP}{P}$$

Integrating: $$\mu_i - \mu_i^o = RT \, ln\left(\frac{P_i}{P^o}\right) \qquad \ldots [16]$$

Rearranging gives the chemical potential of a pure ideal gas i as

$$\boxed{\mu_i = \mu_i^o + RT \, ln\left(\frac{P_i}{P^o}\right)} \quad \begin{array}{l}\textit{Chemical Potential}\\ \textit{of an Ideal Gas}\end{array} \quad \ldots [17]$$

In this expression, μ_i^o is the chemical potential of component i in its standard state (pure i at $P^\circ = 1$ *bar*). For a gas, μ^o is the same as its standard molar Gibbs free energy of formation, Δg_f^o. In a mixture of ideal gases, since all the gases in the mixture are *ideal*, each gas behaves as if none of the other gases were present; therefore, the above expression for the chemical potential of gas i at a partial pressure P_i can be used whether other gases are present or not. This means that we can use the same expression for μ_i for both a *pure* ideal gas as well as for an ideal gas in a *mixture* of ideal gases.

17.5 MORE ABOUT ACTIVITIES

We saw earlier that the relative pressure (P_i / P°) of a gas i is an approximation for its relative activity a_i. At any specified temperature T the relative activity a_i of any species i is defined as

$$a_i \equiv exp\left[\frac{\mu_i - \mu_i^o}{RT}\right] \qquad \dots [18]$$

where μ_i is its chemical potential and μ_i^o is its chemical potential in the standard state. Taking natural logarithms of each side:

$$ln\ a_i = \frac{\mu_i - \mu_i^o}{RT}$$

Cross-multiplying: $\qquad\qquad\qquad \mu_i - \mu_i^o = RT\ ln\ a_i$

Rearranging: $\qquad\qquad\qquad \mu_i = \mu_i^o + RT\ ln\ a_i$

$$\boxed{\mu_i = \mu_i^o + RT\ ln\,a_i} \qquad \dots [19]$$

Comparison of Eqn [19] with Eqn [17] shows that for an ideal gas the relative activity is just the relative partial pressure:

$$\boxed{a_i = \frac{P_i}{P^o}} \quad \begin{array}{l}\textit{Relative Activity} \\ \textit{of an Ideal Gas}\end{array} \qquad \dots [20]$$

Eqn [19] is used when we wish to show how the chemical potential of any given species varies with composition at constant temperature T.

It should be noted that, strictly speaking, the relative activity a_i of a species i in any given state is defined as the *ratio* of its absolute activity a_i' in that state to its absolute activity $(a_i')^o$ in its *reference* state. Thus:

$$\boxed{a_i \equiv \frac{a_i'}{(a_i')^o}} \quad \begin{array}{l}\textit{Relative Activity} \\ \textit{of Any Species}\end{array} \qquad \dots [21]$$

Unlike the relative activity, the absolute activity is not dimensionless, but has the units of concentration. The concentration units chosen, however, can vary, depending on what happens to be convenient. For example, for an ideal gas, the absolute activity usually has the units of pressure; for a solute dissolved in a liquid solvent, the absolute activity may have the units of *molarity* or *molality*. Often the absolute activity is expressed as a mole fraction, in which case, like the relative activity, it will be dimensionless. Regardless of the concentration units chosen, however, in almost all cases, by definition, a value of unity is arbitrarily assigned to the concentration in the reference state. Thus,

$$(a_i')^o = 1\ bar$$
or $\qquad\qquad\qquad = 1\ mol\ L^{-1}$
or $\qquad\qquad\qquad = 1\ mol\ kg^{-1}$ etc., etc. $\qquad \dots [22]$

Thus, the relative activity is *dimensionless*, which explains why we are able to take its logarithm in Eqn [19], which, more properly should be written

$$\mu_i = \mu_i^o + RT \, ln\frac{a_i'}{(a_i')^o} \qquad \qquad \dots [23]$$

However, in view of Eqn [22], Eqn [23] becomes

$$\mu_i = \mu_i^o + RT \, ln\left(\frac{a_i'}{1}\right)$$

$$= \mu_i^o + RT \, ln \, a_i$$

which is just Eqn [19].[3]

Earlier we discovered that for a one-component system at constant temperature and pressure, the criterion for equilibrium between two phases α and β is that the molar Gibbs free energy of the substance in the α-phase must be equal to its molar Gibbs free energy in the β-phase:

$$\boxed{g^\alpha = g^\beta} \qquad \begin{array}{l}\textit{Criterion for Phase Equilibrium}\\ \textit{in a One-Component System}\end{array}$$

In the case of a multicomponent system at constant temperature and pressure, the criterion for chemical equilibrium is that *the chemical potential of each species i must be the same in each phase of the system*:

$$\boxed{\mu_i^\alpha = \mu_i^\beta} \qquad \begin{array}{l}\textit{Chemical Equilibrium in}\\ \textit{a Multicomponent System}\end{array} \qquad \dots [24]$$

17.6 IDEAL LIQUID SOLUTIONS OF VOLATILE SOLUTES: RAOULT'S LAW

Raoult[4] showed experimentally that for a liquid solution containing a *dissolved volatile component*, the partial pressure of the volatile component above and in equilibrium with the liquid phase is given by

$$\boxed{P_i = x_i P_i^\bullet} \qquad \begin{array}{l}\textit{Raoult's Law for}\\ \textit{a Volatile Solute}\end{array} \qquad \dots [25]$$

where x_i is the mole fraction of the volatile[5] component in the solution and $P_i^\bullet$ is the equilibrium vapor pressure of the *pure* volatile component. A solution for which Raoult's Law holds for all components across the complete composition range is called an ***IDEAL SOLUTION***.

---

[3] By this point you may be thoroughly confused, and have no idea what we are talking about. Don't worry! This is normal. Some day in the distant future you may look back on this and actually understand it. If not, that's OK too.

[4] François Marie Raoult, 1830–1901. French physical chemist who taught at the University of Grenoble. Remembered most for his work on the freezing points and the vapor pressures of solutions.

[5] A volatile substance is one that can readily evaporate. Thus water, ethanol, and acetone are considered volatile, whereas engine oil, sugar, and iron are considered non-volatile.

17.7 IDEAL LIQUID SOLUTIONS OF NON-VOLATILE SOLUTES

Consider a *non*volatile solute—such as sugar—dissolved in water (Fig.1). In the case of *pure* water, the pressure above the liquid is just $P_w^{\bullet}$, the equilibrium vapor pressure of the water at the temperature of the system. However, the presence of a non-volatile solute such as sugar decreases the mole fraction of the water in the liquid phase, and, as a consequence of Raoult's law— Eqn [25]—lowers its partial pressure above the solution. This in turn lowers its activity and, conse-quently, its chemical potential (see

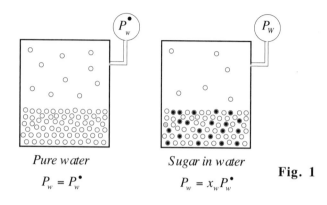

Pure water

$$P_w = P_w^{\bullet}$$

Sugar in water

$$P_w = x_w P_w^{\bullet}$$

Fig. 1

Eqn [17]). The difference in the chemical potential of the water between the liquid phase and the vapor phase is the driving force that moves water from the liquid into the vapor. If the chemical potential of the water in the liquid phase is lowered, less water will be pushed into the vapor phase, and there will be a consequent reduction in its partial pressure in the vapor phase. This is the un-derlying cause of the observations made by Raoult, who knew nothing about Gibbs's work on chemical potentials.

Example 17-2

1.5 moles of naphthalene ($C_{10}H_8$) is dissolved in one kg of liquid benzene (C_6H_6). What is the partial pressure of the benzene in the equilibrium vapor above the solution at 25°C? The vapor pressure of pure benzene at 25°C is $P_B^{\bullet} = 12.6$ kPa.

Solution:

The molar mass of C_6H_6 is $6(12.01) + 6(1.008) = 78.108$ g mol^{-1}; therefore the number of moles of benzene is $n_B = 1000/78.108 = 12.803$ mol, and the mole fraction of benzene in the solution is

$$x_B = \frac{n_B}{n_B + n_N} = \frac{12.803}{12.803 + 1.5} = 0.8951$$

Therefore, $P_B = x_B P_B^{\bullet} = (0.8951)(12.6) = 11.3$ kPa

Ans: | 11.3 kPa |

17.8 IDEAL LIQUID SOLUTIONS
OF TWO VOLATILE LIQUIDS

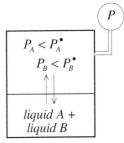

For a two-component liquid solution (a *binary* solution) of two volatile liquids A and B, through Raoult's law, the presence of liquid A lowers the chemical potential, and therefore vapor pressure of liquid B while the presence of liquid B lowers the chemical potential, and therefore vapor pressure of liquid A (Fig. 2).

The total pressure P of the system is the sum of the partial pressures of A and of B.

Thus, $$P_A = x_A\, P_A^{\bullet}$$

and $$P_B = x_B\, P_B^{\bullet}$$
$$= (1 - x_A)\, P_B^{\bullet}$$

$$\boxed{P = x_A P_A^{\bullet} + x_B P_B^{\bullet} = x_A P_A^{\bullet} + (1 - x_A) P_B^{\bullet}}$$ *Ideal Binary Mixture of Two Volatile Liquids*   . . . [26]

The above relationships are illustrated graphically in Fig. 3.

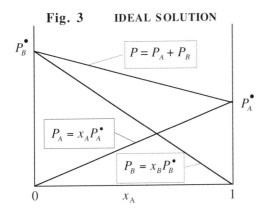

Fig. 3   IDEAL SOLUTION

$$P = P_A + P_B$$

$$P_A = x_A P_A^{\bullet}$$

$$P_B = x_B P_B^{\bullet}$$

The composition of the vapor is

$$y_i = \frac{n_i}{n_{total}} = \frac{P_i V / RT}{PV / RT} = \frac{P_i}{P} \quad \ldots [27]$$

As mentioned above, a solution whose components obey Raoult's Law across the complete concentration range (from pure A to pure B) is called an **Ideal Solution**.

Example 17-3

Methanol (CH_3OH) and ethanol (C_2H_5OH) form almost ideal solutions at 20°C; their vapor pressures at this temperature are, respectively, 11.83 kPa and 5.93 kPa.

(a) What is the mole fraction of methanol in a liquid solution at 20°C made by mixing 50.0 g of methanol with 100.0 g of ethanol?

(b) What is the mole fraction of methanol in the equilibrium vapor above this solution?

Solution

(a) The molar masses of methanol (M) and ethanol (E) are 32.04 g mol^{-1} and 46.07 g mol^{-1}, respectively.

Moles of methanol: $n_M = \dfrac{50.0 \text{ g}}{32.04 \text{ g}/\text{mol}} = 1.561$ mol

Moles of ethanol: $n_E = \dfrac{100.0 \text{ g}}{46.07 \text{ g}/\text{mol}} = 2.171$ mol

The mole fraction of *methanol* in the liquid is: $x_M = \dfrac{n_M}{n_M + n_E} = \dfrac{1.561}{1.561 + 2.171} = 0.4074$

Ans: $\boxed{x_M = 0.407}$

(b) The mole fraction of *ethanol* in the liquid is:

$$x_E = 1 - x_M = 1.0000 - 0.4074 = 0.5926$$

Using Raoult's Law, the partial pressures of methanol and ethanol above the liquid are:

$$P_M = x_M P_M^{\bullet} = (0.4074)(11.83) = 4.820 \text{ kPa}$$
$$P_E = x_E P_E^{\bullet} = (0.5926)(5.93) = 3.514 \text{ kPa}$$

The total pressure of the vapor is: $P = P_M + P_E = 4.820 + 3.514 = 8.334$ kPa

Therefore the mole fraction of methanol in the vapor is: $y_M = \dfrac{P_M}{P} = \dfrac{4.820}{8.334} = 0.5784$

Ans: $\boxed{y_M = 0.578}$

Note that, in general, when the liquid phase and the vapor phase of a binary mixture of two volatile components are in equilibrium, the composition of the vapor is *not* the same as the composition of the liquid. The vapor will be *richer* than the liquid in the more volatile component.

In the present case, the vapor is richer in methanol than the liquid by

$$\left(\frac{y_V - x_L}{x_L} \right) \times 100\% = \left(\frac{0.5784 - 0.4074}{0.4074} \right) \times 100\% = 42\%$$

Exercise 17-1

A certain benzene–toluene solution has a normal boiling point of 88.0°C. Calculate the mole fraction of benzene in the liquid and in the equilibrium vapor above the liquid. At 88.0°C the vapor pressures of benzene and toluene are 127.6 kPa and 50.7 kPa, respectively. ☺

17.9 REAL (NON-IDEAL) SOLUTIONS

Raoult's Law works best when the components have similar molecular shapes and are held together by similar molecular forces. Thus, benzene/methylbenzene, methanol/ethanol, *n*-hexane/*n*-heptane, and ethyl bromide/ethyl iodide form essentially ideal solutions whose behavior follows that shown above in Fig. 3.

Even for non-ideal solutions, Raoult's Law is obeyed when the mole fraction of volatile component *i* approaches unity:

$$P_i \rightarrow x_i P_i^{\bullet} \quad \text{as} \quad x_i \rightarrow 1 \qquad \ldots [28]$$

Consider the solution consisting of carbon disulfide (CS_2) and acetone (propanone, CH_3COCH_3) shown in Fig. 4, where the dashed lines represent ideal solution behavior. These two molecules are quite *dissimilar* in nature: carbon disulfide is *linear* and *non-polar* (no charge separation), whereas acetone is *non-linear* and *polar*. Because of these differences, it turns out that each of these substances exerts stronger attractive forces between its *own* molecules than it does between itself and the *other*.

The net result is that a mixture of carbon disulfide and acetone is *easier* to vaporize (i.e., is more volatile) than is pure carbon disulfide or pure acetone; therefore, as shown in Fig. 4, the equilibrium partial pressure of each component above such a solution is *greater* than predicted by Raoult's Law—these are *positive* deviations from Raoult's Law. Note, however, that as the mole fraction of each component approaches *unity*, its partial pressure approaches ideal Raoult's Law behavior. For dilute solutions of solute, the mole fraction of the *solvent* approaches a value of unity. Therefore, in these kinds of solutions Raoult's Law will be obeyed by the *solvent*. When $x_{CS_2} \rightarrow 1$, CS_2 becomes the solvent and obeys Raoult's Law; when $x_{CS_2} \rightarrow 0$, then $x_{acetone} \rightarrow 1$ and acetone becomes the solvent and obeys Raoult's Law. Negative deviations from Raoult's law also are observed.[6]

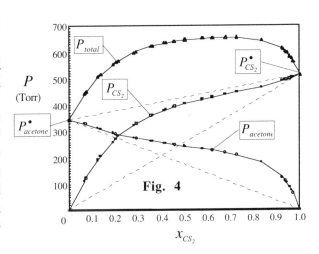

Fig. 4

17.10 CHEMICAL POTENTIAL AND SOLVENT ACTIVITY

Suppose we have an ideal solution of a nonvolatile solute—such as sugar—dissolved in water (the solvent). Because the solute is nonvolatile, none of it will be present in the vapor phase above the solution; therefore, the vapor phase will contain only water vapor, the partial pressure of which will be governed by Raoult's Law, since the solution was specified to be ideal.

---

[6] Can you guess why?

Now consider the process in this system in which water transfers from the liquid to the vapor:[7]

$$H_2O_{(solution)} \rightarrow H_2O_{(vapor)}$$

For this process to be at equilibrium, the chemical potential μ_W^{liq} of the water in the *liquid* phase must be *equal* to the chemical potential μ_W^V of the water in the *vapor* phase.

Thus, at equilibrium:
$$\mu_W^{liq} = \mu_W^V$$

$$= \mu_W^o + RT \ln\left(\frac{P_W}{P^o}\right)$$

$$= \mu_W^o + RT \ln\left(\frac{x_W P_W^{\bullet}}{P^o}\right) \quad \longleftarrow \quad \boxed{\begin{array}{c} Solvent\ ideal, \\ vapor\ obeys \\ Raoult's\ Law \end{array}}$$

$$= \mu_W^o + RT \ln\left(\frac{P_W^{\bullet}}{P^o}\right) + RT \ln x_W$$

$$= \mu_W^* + RT \ln x_W \qquad\qquad \ldots [29]$$

where $\mu_W^* = \mu_W^o + RT \ln\left(P_W^{\bullet}/P^o\right)$.

In general:
$$\boxed{\mu_{solvent} = \mu_{solvent}^* + RT \ln x_{solvent}} \quad \textit{Ideal Solution} \quad \ldots [30]$$

Eqn [30] can be used to calculate the chemical potential of the *solvent* in an ideal solution. Two assumptions were made in the above derivation: The first was that Raoult's Law can be used to express the partial pressure of the solvent; therefore, Eqn [30] is valid only under those conditions for which Raoult's Law is valid (i.e., ideal solution behavior or $x_{solvent}$ close to unity). The second assumption was that the vapor obeys the ideal gas law. This assumption was made so that we could let the (relative) activity of the gas be equal to its relative partial pressure P_i/P^o.

Note the definition of μ_W^*:
$$\boxed{\mu_W^* \equiv \mu_W^o + RT \ln\left(\frac{P_W^{\bullet}}{P^o}\right)} \qquad\qquad \ldots [31]$$

μ_W^o is the chemical potential of pure solvent *vapor* at a vapor pressure of one bar. We saw earlier that μ_W^o is a function only of temperature. $P_W^{\bullet}$ is just the vapor pressure of the pure solvent, which also is a function only of temperature. Therefore, it follows that μ_W^* also is a function only of tem-

---

[7] The vapor is assumed to behave as an ideal gas.

perature. When the mole fraction of the solvent is equal to unity, then we just have pure liquid solvent, and, from Eqn [30]:

$$\mu_w = \mu_w^* + RT \, ln \, (1) = \mu_w^*$$

In other words, μ_w^*—the chemical potential of the solvent when it is in its standard state—is just the chemical potential of the pure liquid solvent; that is,

The standard state of the solvent is just the pure liquid.

When a solute is dissolved in the solvent, the mole fraction of the solvent becomes less than unity, and Eqn [30] tells us that

The chemical potential of the solvent is
lowered when it has a solute dissolved in it.

Example 17-4

Show that the relative activity a_w of the solvent W in a solution is given by $a_w = P_w / P_w^\bullet$, where P_w is the partial pressure of the solvent in the vapor phase above the solution and $P_w^\bullet$ is the equilibrium vapor pressure of the pure solvent at the same temperature.[8]

Solution

The criterion for equilibrium is that the chemical potential of the solvent in the vapor phase must be the same as that of the solvent in the liquid phase:

$$\mu_w^V = \mu_w^{liq} \qquad \qquad \dots \text{[a]}$$

Expanding each side: $\qquad \mu_w^o + RT \, ln\left(\dfrac{P_w}{P^o}\right) = \mu_{liq}^o + RT \, ln \, a_w \qquad \dots \text{[b]}$

where μ_w^o is the chemical potential of solvent vapor in its standard state (i.e., at 1 bar pressure), and μ_{liq}^o is the chemical potential of liquid solvent in its standard state (pure liquid). For pure liquid (the standard state), the partial pressure is just the equilibrium vapor pressure $P_w^\bullet$ of the liquid at the specified temperature, and, by definition, its relative activity has a value of unity. Thus, for *pure solvent*, Eqn [b] becomes

$$\mu_w^o + RT \, ln\left(\dfrac{P_w^\bullet}{P^o}\right) = \mu_{liq}^o + RT \, ln(1) \qquad \dots \text{[c]}$$

But, $ln \, (1) = 0$, therefore Eqn [c] reduces to

---

[8] Since the most common solvent is water, we use the subscript W to denote the solvent. The equations, of course, are valid for solvents other than water.

$$\mu_W^o + RT \ln\left(\frac{P_W^\bullet}{P^o}\right) = \mu_{liq}^o \qquad \qquad \cdots \text{[d]}$$

Subtracting Eqn [d] from Eqn [b]:

$$\left\{\mu_W^o + RT \ln\left(\frac{P_W}{P^o}\right)\right\} - \left\{\mu_W^o + RT \ln\left(\frac{P_W^\bullet}{P^o}\right)\right\} = \left\{\mu_{liq}^o + RT \ln a_W\right\} - \left\{\mu_{liq}^o\right\}$$

$$RT \ln\left(\frac{P_W}{P^o}\right) - RT \ln\left(\frac{P_W^\bullet}{P^o}\right) = RT \ln a_W$$

$$RT \ln P_W - RT \ln P^o - RT \ln P_W^\bullet + RT \ln P^o = RT \ln a_W$$

The terms in P^o cancel out, leaving:

$$RT \ln P_W - RT \ln P_W^\bullet = RT \ln a_W$$

Dividing through by RT:

$$\ln P_W - \ln P_W^\bullet = \ln a_W$$

That is:

$$\ln\left(\frac{P_W}{P_W^\bullet}\right) = \ln a_W$$

or,

$$\boxed{a_W = \frac{P_W}{P_W^\bullet}} \quad \begin{array}{l}\textit{Activity of the Solvent}\\ \textit{in a Solution}\end{array} \qquad \cdots \text{[32]}$$

For an *ideal* solution—one that obeys Raoult's Law—the partial pressure of the solvent above the solution is given by

$$P_W = x_W P_W^\bullet \qquad \qquad \cdots \text{[33]}$$

Substituting Eqn [33] into Eqn [32] gives

$$a_W = \frac{x_W P_W^\bullet}{P_W^\bullet} = x_W \qquad \qquad \cdots \text{[34]}$$

Eqn [34] shows that for an ideal solution—one obeying Raoult's Law—the activity of the solvent is just its mole fraction in the solution. This is the same result expressed above in Eqn [30].

$$\boxed{a_{solvent} = x_{solvent}} \quad \textit{Ideal Solution} \qquad \cdots \text{[35]}$$

The only assumption inherent in Eqn [32] is that the vapor behaves as an ideal gas. The use of the approximation

$$a_W \approx x_W$$

involves *two* assumptions: (i) the vapor behaves as an ideal gas, and (ii) the solvent obeys Raoult's Law. Therefore, it follows that the use of Eqn [32] to calculate a_W is more accurate than the use of $a_W \approx x_W$, and Eqn [32] is to be preferred where possible.

Example 17-5

We have seen that for an ideal solution—one obeying Raoult's law—the activity of the solvent is just its mole fraction in the solution. However, most solvents do not obey Raoult's law when their mole fractions are less than about 0.90. For these solutions, if we want to do accurate calculations we must use the rigorous relative activity. The ratio between the relative activity of a solvent W and its concentration (expressed as a mole fraction) is called the **rational activity coefficient** f:

$$\boxed{f_W \equiv \frac{a_W}{x_W}} \quad \begin{array}{l}\textit{Rational Activity} \\ \textit{Coefficient}\end{array} \quad \ldots [36]$$

By way of example, at 57.2°C and a total equilibrium vapor pressure of 1.00 atm, an acetone–methanol solution containing 40.0 mole % acetone has a vapor consisting of 51.6 mole % acetone. At this temperature the equilibrium vapor pressures of acetone and methanol are 104.8 kPa and 73.5 kPa, respectively. Calculate the rational activity coefficient of each component in this solution.

Solution

The partial pressures of the acetone (A) and the methanol (M) in the solution are:

$$P_A = y_A P = (0.516)(101.325) = 52.28 \text{ kPa}$$

$$\begin{aligned} P_M = y_M P &= (1 - y_A)P \\ &= (1 - 0.516)(101.325) \\ &= (0.484)(101.325) = 49.04 \text{ kPa} \end{aligned}$$

Therefore, the activity of the acetone in the solution is:

$$a_A = \frac{P_A}{P_A^\bullet} = \frac{52.28}{104.8} = 0.4988$$

and that of the methanol is:
$$a_M = \frac{P_M}{P_M^\bullet} = \frac{49.04}{73.5} = 0.6672$$

Therefore the rational activity coefficients of the acetone and methanol are:

$$f_A = \frac{a_A}{x_A} = \frac{0.4988}{0.400} = 1.247 \qquad \text{and} \qquad f_M = \frac{a_M}{x_M} = \frac{0.6672}{0.600} = 1.112$$

If the activity coefficient of a species is known, all that has to be done to rigorously (= accurately) determine its activity is to multiply its concentration by its activity coefficient.[9]

Ans: $\boxed{f_A = 1.25;\ f_M = 1.11}$

Exercise 17-2

Calculate the activity and the rational activity coefficient of the water in a solution consisting of 122 g of a non-volatile solute (molar mass = 241 g mol^{-1}) dissolved in 0.920 kg of water at 20°C. At 20°C the vapor pressure of pure water is 2339 Pa, and the partial pressure of the water above the solution is 2269 Pa. ☹

17.11 IDEAL DILUTE SOLUTIONS OF VOLATILE SOLUTES: HENRY'S LAW

We saw above that for a volatile component i dissolved in a solvent, as $x_i \to 1$, $P_i \to x_i P_i^{\bullet}$ (this is Raoult's Law). For a *dilute* solution ($x_i \to 0$) of a volatile solute i dissolved in a solvent, it is found that the partial pressure of i again is proportional to x_i; but, this time the constant of proportionality is *not* the vapor pressure of the pure component, as it is when $x_i \to 1$; instead, it is a constant K called the *Henry's Law Constant:*

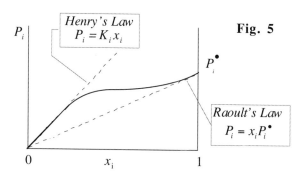

Fig. 5

$$\boxed{P_i = K_i x_i} \quad \textit{Henry's Law} \qquad \qquad \dots [37]$$

The numerical value of the Henry's Law constant K_i depends on the temperature, the solute, and the solvent. It should be noted that Henry's Law works best for *dilute* solute concentrations (i.e., as $x_i \to 0$), unlike Raoult's Law, which works best as $x_i \to 1$. Thus, as shown in Fig. 5, an ideal dilute solution obeys Henry's Law, while an ideal concentrated solution obeys Raoult's Law.

---

[9] Unfortunately, the concentration can be expressed using different concentration scales. In the present case we are using the mole fraction scale to express the concentration, and the associated activity coefficient is called the rational activity coefficient. The activity coefficients used with the molality scale and the molarity scale have slightly different values. These are discussed in detail in Chapter 19.

Exercise 17-3

The partial pressures shown in the table at the right have been experimentally determined for solutions of acetone (A) in ethyl ether (E) at 20°C.

(a) For each entry, calculate the value of the partial pressure predicted by Raoult's Law.

(b) Plot both sets of values on a P–x diagram. ☺

x_A	P_A (Torr)	P_E (Torr)
0.000	0	444
0.052	20	422
0.457	105	282
0.936	175	46
1.000	185	0

17.12 THE SOLUBILITY OF GASES

Consider a gas i dissolved in a solvent W. Since the solubility of most gases is very low, the equilibrium partial pressure of the gas above the solution can be evaluated from Henry's Law.

Thus, from Henry's Law,
$$x_i = \frac{P_i}{K_i} \qquad \dots [38]$$

But
$$x_i = \frac{n_i}{n_i + n_W} \qquad \dots [39]$$

where n_i is the number of moles of gas i in the solution and n_W is the number of moles of solvent in the solution.

Since the solubility of the gas is very low,
$$n_W \gg n_i \qquad \dots [40]$$

so that to a good approximation
$$n_i + n_W \approx n_W \qquad \dots [41]$$

Substitution of Eqn [41] into Eqn [39] gives
$$x_i \approx \frac{n_i}{n_W} \qquad \dots [42]$$

Substitution of Eqn [42] into Eqn [38] gives
$$\frac{n_i}{n_W} \approx \frac{P_i}{K_i} \qquad \dots [43]$$

The concentration C_i of a gas i dissolved in a liquid solution, in mol m^{-3}, is given by[10]

---

[10] Note that the SI unit of concentration is mol m^{-3}, not mol L^{-1}. If c_i = mol L^{-1} and C_i = mol m^{-3}, then the relation between the two is $C_i = 1000c_i$; i.e., 1 mol L^{-1} = 1000 mol m^{-3}.

$$C_i = \frac{n_i}{V_{soln}} \qquad \ldots [44]$$

But, since the solubility of the gas in the liquid is so low, the contribution of the dissolved gas to the total volume of the solution is negligibly small, so that the total volume of the solution is essentially just the volume of the solvent:

$$V_{soln} \approx V_w = n_w v_w \qquad \ldots [45]$$

where v_w is the molar volume of the pure solvent.

Putting Eqn [45] into Eqn [44]:

$$C_i \approx \frac{n_i}{n_w v_w} = \frac{1}{v_w}\left(\frac{n_i}{n_w}\right) \qquad \ldots [46]$$

Substituting the expression given in Eqn [43] for n_i/n_w into Eqn [46] gives

$$C_i \approx \frac{1}{v_w}\left(\frac{P_i}{K_i}\right)$$

Therefore:

$$\boxed{C_i \approx \frac{P_i}{K_i v_w}} \qquad \begin{array}{l}\textit{Solubility of a Gas}\\ \textit{in a Liquid Solvent}\end{array} \qquad \ldots [47]$$

Using Eqn [47] we can calculate the solubility of a gas in a liquid from a knowledge of the Henry's Law constant and of the partial pressure of the gas above the liquid.

Example 17-6

What is the molar concentration of dissolved CO_2 in water that is exposed to the air at 25°C and one atm pressure if the air contains 0.033% by volume CO_2? The mole fraction basis Henry's law constant for CO_2 in water at 25°C is 161.7 MPa. The density of water at 25°C is 997.07 kg m^{-3}.

Solution

The volume % composition of a gas is the same as its mole % composition; therefore, for air:

$$\frac{P_{CO_2}}{P_{air}} = \frac{n_{CO_2} RT/V}{n_{air} RT/V} = \frac{n_{CO_2}}{n_{air}} = \frac{0.033}{100} = 0.00033$$

from which

$$P_{CO_2} = (0.00033)(P_{air})$$

$$= (0.00033)(101\ 325)$$

$$= 33.437\ \text{Pa}$$

$$= 33.437 \times 10^{-6}\ \text{MPa}.$$

Henry's law states:
$$P_{CO_2} = K_{CO_2} \cdot x_{CO_2}$$

Therefore:
$$x_{CO_2} = \frac{P_{CO_2}}{K_{CO_2}} = \frac{33.437 \times 10^{-6} \ \text{MPa}}{161.7 \ \text{MPa}} = 2.068 \times 10^{-7}$$

That is,
$$\frac{n_{CO_2}}{n_{H_2O} + n_{CO_2}} = 2.068 \times 10^{-7}$$

Consider one liter of water: One liter of water weighs 997.07 g; therefore, the number of moles of H_2O in 1.00 L of water is:

$$n_{H_2O} = \frac{997.07 \ \text{g}}{18.02 \ \text{g/mol}} = 55.33 \ \text{mol}$$

Gases are not very soluble in water, therefore $n_{CO_2} \ll n_{H_2O}$

so that
$$x_{CO_2} = \frac{n_{CO_2}}{n_{H_2O} + n_{CO_2}} \approx \frac{n_{CO_2}}{n_{H_2O}} = 2.068 \times 10^{-7}$$

Therefore,
$$n_{CO_2} = (2.068 \times 10^{-7})(n_{H_2O})$$
$$= (2.068 \times 10^{-7})(55.33)$$
$$= 1.144 \times 10^{-5} \ \text{mol}$$

and the molar concentration of dissolved CO_2 is:

$$[CO_2] = \frac{n_{CO_2}}{V} = \frac{1.144 \times 10^{-5} \ \text{mol}}{1.00 \ \text{L}} = 1.144 \times 10^{-5} \ \text{mol L}^{-1}$$

Ans: $\boxed{1.14 \times 10^{-5} \ \text{mol L}^{-1}}$

Exercise 17-4

Verify that Eqn [47] can be used to solve Example 17–6. ☺

Exercise 17-5

At 25°C the vapor pressure of pure carbon tetrachloride (CCl_4) is 4513 Pa. Calculate the mole fraction of Br_2 in the equilibrium vapor above a solution containing 0.065 mole fraction Br_2 dissolved in carbon tetrachloride at 25°C. The Henry's law constant for bromine dissolved in carbon tetrachloride at 25°C is 16.31 kPa (mole fraction basis). ☺

17.13 DISTILLATION

Consider a mixture of two volatile liquids, A and B. Furthermore, suppose that at any given temperature the vapor pressure of pure A is greater than that of pure B (i.e., A is more *volatile* than B). Reference to Fig. 6 shows that if A is more volatile than B, then pure A has a lower normal boiling point than pure B.

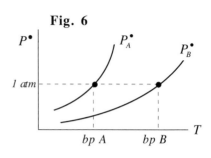

Fig. 6

If we start with one mole of pure B and slowly add A—the more volatile component—the boiling point of the *mixture* gradually *decreases*. If we eventually add such a large quantity of A that the initial one mole of B becomes negligibly small in comparison, then the solution essentially behaves like pure A, and, for all practical purposes, its boiling point will be the boiling point of pure A.

For each mixture we can calculate x_A the mole fraction of A, and plot the boiling point of the mixture *vs.* x_A. A typical plot looks like that shown in Fig. 7.

The first point corresponds to $x_A = 0$ (pure B), while the last point corresponds to $x_A = 1$ (pure A).

Consider an ideal equimolar solution of A and B:

$$x_A = x_B = 0.5$$

From Raoult's Law,

$$P_A = x_A P_A^\bullet = 0.5 P_A^\bullet$$

and

$$P_B = x_B P_B^\bullet = 0.5 P_B^\bullet$$

The total pressure will be

$$P = P_A + P_B = 0.5(P_A^\bullet + P_B^\bullet)$$

The mole fraction of A in the vapor is

$$y_A = \frac{P_A}{P} = \frac{0.5 P_A^\bullet}{0.5(P_A^\bullet + P_B^\bullet)}$$

$$= \frac{P_A^\bullet}{P_A^\bullet + P_B^\bullet}$$

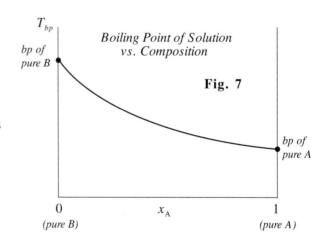

Fig. 7

Boiling Point of Solution vs. Composition

Since A is more volatile than B, $P_A^\bullet > P_B^\bullet$; therefore $y_A > 0.5$. In other words, the composition of the vapor is **not** the same as the composition of the liquid. *The vapor is enriched in the more volatile component*; i.e., at equilibrium, the mole fraction of A—the more volatile component—is greater in the *vapor* than it is in the *liquid*!

Therefore, for any boiling mixture of A and B in which A is more volatile than B,

$$y_A > x_A$$

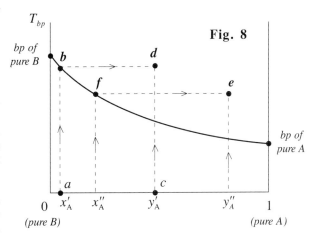

Fig. 8

Consider the liquid of composition x'_A in Fig. 8. The vapor in equilibrium with this liquid has composition y'_A, which is greater than x'_A. A vertical line upward from point a intersects the boiling point curve for the mixture at point b.

The point d is defined when a line is extended horizontally from point b until it intersects a vertical line from point c.

Point d represents the composition of the vapor that is in equilibrium with liquid of composition b.

A similar point e locates the composition of the vapor in equilibrium with liquid of composition f. The horizontal lines bd and fe link the compositions of the liquids with the compositions of the corresponding vapors and are called *tie lines*.

When tie lines are determined for many compositions, a curve is generated that gives the vapor composition as well as the liquid composition. Such a plot (Fig. 9) containing both the liquid boiling point curve and the vapor composition curve is called a *temperature–composition (T–x) diagram* for a binary mixture of two volatile components.

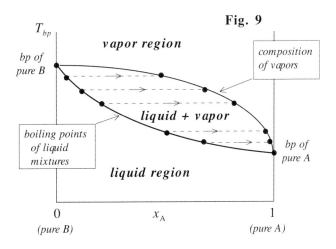

Fig. 9

Unlike the *P–T* phase diagrams we studied earlier for *single component* systems, this is a phase diagram for a *two component system* (*A* and *B*).

For any point located *below* the liquid mixture boiling point curve the system will exist as a *liquid solution*. For any point *above* the vapor curve the system will exist as a *vapor solution*. For any point located *between* the two curves, the system will exist as a *liquid in equilibrium with a vapor*, the compositions of each being given by the tie lines.

This enrichment of the vapor in the more volatile component is the basis for the process of **distillation.**

Thus, referring to Fig. 10, suppose we start with a liquid of composition x_A and boil it. The composition of the vapor in equilibrium with this liquid is found by tracing the path *abcdg*, yielding a vapor of composition y_A.

If we now collect this vapor and cool it, we can condense it into a *second* liquid of the same composition as the *first* vapor; i.e., the new liquid will have composition $x'_A = y_A$.

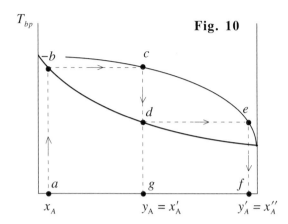

Fig. 10

We next take this second liquid (point *g*) of composition x'_A and boil *it*. The composition of the new vapor formed will be located at point *e*, and will have composition y'_A. Condensation of this second vapor yields a *third* liquid (point *f*) of composition $x''_A = y'_A$.

Thus, by carrying out two steps of vaporization-condensation we have gone from an initial liquid of composition x_A to produce a final liquid product of composition x''_A, which is almost pure *A*. This is a very important way industrially to separate two volatile liquids from a mixture. The actual number of vaporization-condensation steps that will be required to produce a final liquid of a specified composition will depend on the details of the *T–x* diagram for the components involved.

17.14 THE LEVER RULE

By making use of the so-called "lever rule," a *T–x* phase diagram can be used to obtain very useful information about the relative amounts of each phase that will exist in a given system at equilibrium.

Thus, consider two volatile liquids *A* and *B*: Suppose n_A moles of *A* and n_B moles of *B* are placed into an evacuated closed container at some initial temperature T_o. Letting *x* denote the mole fraction of *A* in the overall system,

$$x = \frac{n_A}{n_A + n_B}$$

Since the temperature T_o is below the mixture boiling curve, initially the system will exist only in the liquid state, at point *D* in Fig. 11.

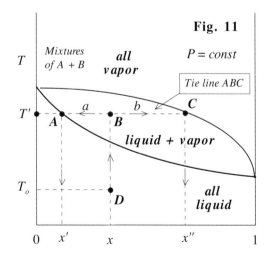

Fig. 11

Suppose the system now is heated up at constant pressure to some temperature T' to produce a boiling mixture. At the temperature T' the state of the overall system will be located at point B. The phase diagram shows that point B is located in the region in which both liquid and vapor will coexist in equilibrium. The tie line ABC indicates that at the overall composition x and temperature T' the liquid will have composition x' and the vapor in equilibrium with this liquid will have composition x''.

Note that the point B represents the mole fraction of the *total* system, including *both* the liquid *and* the vapor phases. Point B does not correspond to an actual physical phase of the system—nothing actually *physically* exists at point B: The liquid part of the system is described by point A and the vapor part of the system is described by point C.

The total moles in the *overall* system is:
$$n_{total} = n_A + n_B \qquad \ldots [48]$$

The total moles of A in the overall system is:
$$n_A^{total} = x\, n_{total} \qquad \ldots [49]$$

The moles of A in the *liquid* is
$$n_A^{liq} = x' n_{liq} \qquad \ldots [50]$$

and the moles of A in the *vapor* is
$$n_A^{vap} = x'' n_{vap} \qquad \ldots [51]$$

where n_{liq} and n_{vap} are the total moles of liquid and the total moles of vapor, respectively, in the system.

Therefore, the total moles of A in the overall system is

$$n_A^{total} = n_A^{liq} + n_A^{vap} = x' n_{liq} + x'' n_{vap} \qquad \ldots [52]$$

But
$$n_A^{total} = x n_{total} = x(n_{liq} + n_{vap}) \qquad \ldots [53]$$

Therefore, equating Eqns [52] and [53]:

$$x(n_{liq} + n_{vap}) = x' n_{liq} + x'' n_{vap}$$

Rearranging:
$$n_{liq}(x - x') = n_{vap}(x'' - x)$$

From which
$$\boxed{\frac{n_{vap}}{n_{liq}} = \frac{n_A^{vap} + n_B^{vap}}{n_A^{liq} + n_B^{liq}} = \frac{x - x'}{x'' - x} = \frac{a}{b}} \quad \begin{array}{l} \textit{The Lever} \\ \textit{Rule} \end{array} \qquad \ldots [54]$$

Eqn [54] is called the *Lever Rule*. The lever rule shows that the relative amount of liquid (of composition x') to vapor (of composition x'') in the overall system at temperature T' is given by the ratio of the *tie line* segments a:b. The lever rule can be applied to any two phases in equilibrium, not just to liquid-vapor equilibrium.

17.15 LIQUID–LIQUID PHASE DIAGRAMS

So far we have been looking at multicomponent systems in which we have phase equilibrium between the liquid phase and the vapor phase. Many interesting multicomponent systems consist of two or more partially miscible *liquid* phases in equilibrium. Now, let's examine some of these systems.

17.15.1 HEXANE–NITROBENZENE PHASE DIAGRAM

Consider the two organic liquids *hexane* (C_6H_{12}) and *nitrobenzene* $(C_6H_5NO_2)$. At 7°C (280K) the density of hexane is 0.66 g cm^{-3} and that of nitrobenzene is 1.20 g cm^{-3}.

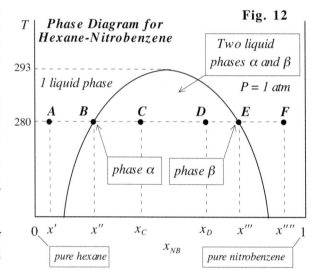

Unlike water and ethanol, which mix together in all proportions to form homogeneous solutions, depending on the temperature and the composition of the system, hexane and nitrobenzene are sometimes only *partially* miscible. The phase diagram for this system is shown in Fig. 12.

Suppose we start with n_H^o moles of hexane (H) in a beaker at 280 K and start to add nitrobenzene (NB) to it.

After we have added n_{NB}' moles of nitrobenzene we have a homogeneous liquid solution of hexane and nitrobenzene located at point A on the phase diagram. At point A the total number of moles in the beaker is

$$n_{total} = n_H^o + n_{NB}'$$

and the composition of the liquid is

$$x_{NB}' = \frac{n_{NB}'}{n_H^o + n_{NB}'}$$

At point B we have added a total of n_{NB}'' moles of nitrobenzene. Now the total number of moles in the system is

$$n_{total} = n_H^o + n_{NB}''$$

and the composition of this homogeneous liquid solution is

$$x_{NB}'' = \frac{n_{NB}''}{n_H^o + n_{NB}''}$$

At point B we have reached the *saturation limit* at 280 K. The solution at point B consists of *liquid hexane saturated with nitrobenzene*. We call this solution *phase α*.

At point C we have added a total of n_{NB}''' moles of nitrobenzene. This is more moles of nitrobenzene than will dissolve in the original n_H^o moles of hexane, and results in the formation of a second liquid phase—phase β—that consists of liquid nitrobenzene saturated with hexane. Because nitrobenzene is denser than hexane, this new liquid phase will form a layer at the bottom of the beaker, with the layer of the less dense α phase (hexane saturated with nitrobenzene) floating on top of it.

Thus, *phase α consists of liquid hexane saturated with nitrobenzene* and *phase β consists of nitrobenzene saturated with hexane*. At point C the relative amounts of α and β are given by the *lever rule*:

$$\frac{n_\alpha}{n_\beta} = \frac{\overline{CE}}{\overline{BC}}$$

where
$$n^\alpha = n_H^\alpha + n_{NB}^\alpha \quad \text{and} \quad n^\beta = n_H^\beta + n_{NB}^\beta.$$

The overall composition of the system at point C is

$$x_C = \frac{n_{NB}'''}{n_H^o + n_{NB}'''} = \frac{n_{NB}^\alpha + n_{NB}^\beta}{n_H^\alpha + n_H^\beta + n_{NB}^\alpha + n_{NB}^\beta}$$

At point D the bottom *phase β* layer (nitrobenzene saturated with hexane) has increased relative to the upper *phase α* layer of hexane saturated with nitrobenzene. Now the total number of moles in the system is

$$n_{total} = n_H^o + n_{NB}''''$$

The relative amount of the α layer with respect to the β layer is given by the lever rule as

$$\frac{n_\alpha}{n_\beta} = \frac{\overline{DE}}{\overline{BD}}$$

At point E so much nitrobenzene has been added that *all* the hexane (n_H^o) dissolves in it to form liquid nitrobenzene saturated with hexane. That is, at point E the overall system consists only of *phase β. Phase α* has disappeared, and now there is once again only one phase present (this time, *phase β*).

At point F we have one liquid phase consisting of liquid nitrobenzene containing n_H^o moles of hexane dissolved in it. This nitrobenzene solution is not yet saturated with hexane.

It can be seen from Fig. 12 that as the temperature is varied, the compositions of the two saturated phases α and β change. When the temperature reaches 293 K the two components are completely soluble in all proportions. This temperature is called the *upper critical solution temperature*, T_{UC}.[11] At temperatures equal to or greater than T_{UC} only one phase exists. As the temperature is increased, the increased thermal motion (kinetic energy) of the molecules leads to greater miscibility, such that at temperatures $\geq T_{UC}$ complete mutual miscibility is attained. From the standpoint of free energy changes, at $T \geq T_{UC}$, $\Delta G_{T,P}^{mixing} < 0$ for all compositions.

17.15.2 WATER–TRIETHYLAMINE PHASE DIAGRAM

In the case of the water-triethylamine system (Fig. 13) there is a *lower* critical solution temperature T_{LC}. At temperatures below this the two liquids mix in all proportions; at temperatures above it two immiscible phases are formed. In this particular system, at lower temperatures a weak complex is formed between the two components, which leads to greater miscibility. At higher temperatures, the complex decomposes and the two components are less miscible.

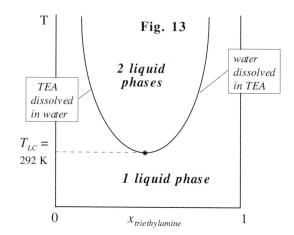

17.15.3 NICOTINE–WATER PHASE DIAGRAM

Nicotine has the structure shown Fig. 14. The phase diagram for nicotine in water (Fig. 15) is interesting because it exhibits both an *upper* and a *lower* critical solution temperature. At low temperatures, weak complexes are formed that are miscible. At mid-range temperatures the thermal energy breaks up these complexes, leading to partial miscibility. At still higher temperatures the increased thermal motion homogenizes the mixture again.

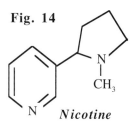

Fig. 14

Nicotine

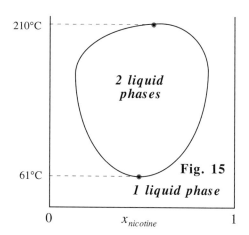

17.16 LIQUID–SOLID PHASE DIAGRAMS

Two solids, *A* and *B*, give rise to the phase diagram shown in Fig. 16. This type of phase diagram is frequently encountered in technologically important industrial materials, such as metallic alloys.

---

[11] Or, if you like big words, the *upper consolute temperature*.

If we start at a high enough temperature both solids will be in the molten state and we will have a homogeneous molten liquid solution.

So, let's start with a molten solution at point a_1 and start to cool the solution. When the temperature reaches T_2, the value at a_2, pure solid A starts to freeze out.

Further cooling to temperature T_3 takes the overall system to a_3, where the system consists of liquid of composition b_3 and pure solid A. Since solid A separates out from $T_2 \rightarrow T_3$, the

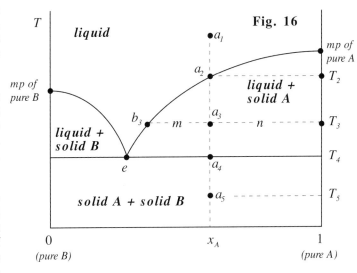

Fig. 16

remaining liquid phase gradually becomes richer in B. At the same time, as the temperature is lowered, more and more solid A freezes out and the quantity of liquid gradually decreases. Thus, at point a_2 the system is essentially all liquid of composition a_2 with an infinitesimal amount of pure solid A.

At point a_3 the system consists of liquid b_3 and pure solid A. The ratio of n^{solid}/n^{liquid} —the total number of moles of liquid in the system to the total number of moles of solid in the system—is given by the lever rule tie line ratio m/n; this ratio *increases* as T is lowered from T_2 to T_4.

At point a_4 liquid of composition e starts to freeze out of the liquid to form crystals of pure solid A and pure solid B. Thus, at T_5 (point a_5) the system consists of two separate intimately mixed solid phases, micro-crystals of pure A and of pure B. Thus, since solid A is freezing out of the liquid phase over the temperature range $T_2 \rightarrow T_4$, it can be seen that a solution does not have a constant freezing point, but rather a range of freezing points.

If the initial molten system composition is to the *left side* of composition e, as indicated in Fig. 17, then cooling results in similar behavior, except that between b_2 and b_4 the liquid phases have compositions a_3', and the solid formed is pure solid B instead of pure solid A. Thus, the solid that freezes out of the solution when the system is cooled depends on which side of point e the overall system composition lies.

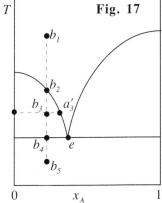

Fig. 17

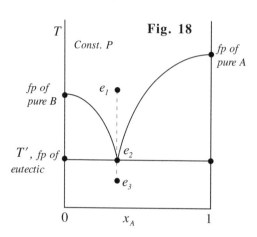

Fig. 18

Composition e (Fig. 18) is called the **eutectic** composition, from the Greek word meaning "easily melted." Cooling liquid e_1 doesn't result in the deposition of solid until e_2 is reached, when pure solid A and pure solid B form. At $T > T'$ (the eutectic temperature) there is only liquid of composition e. At $T < T'$ there is only solid (a mixture of crystals of pure solid A and pure solid B).

Therefore, the liquid of composition e freezes only at one temperature, $T,'$ as if it were a pure single component.

Frozen solid at the eutectic composition e_3 melts when heated to T' to form liquid of the same composition. T' is the lowest temperature at which liquid can exist in the system; i.e., T' is the lowest melting point of any solid mixture, which is why the name "easily melted" is applied to the eutectic composition.

Note that the melting point of a eutectic mixture is *lower* than the melting point of either pure component.

17.17 COMPOUND FORMATION

Many two-component mixtures react to form compounds; thus, in the binary system A–B shown in Fig. 19, component A can react with B to form the solid compound AB_2, which can exist in equilibrium with liquid over a range of mixture compositions.

When the formation of such a compound leads to a *maximum* in the T–x diagram at the composition of the compound ($x_B = 0.67$ in Fig. 19), the compound is called a **congruently melting compound**.

Although the type of phase diagram shown in Fig. 19 resembles two phase diagrams of the type shown in Fig. 16 squeezed side by side, it is different in that the slope of the tangent at the melting point of the congruently melting compound is *zero*, whereas this is not the case for the slopes of the two pure components in Fig. 16.

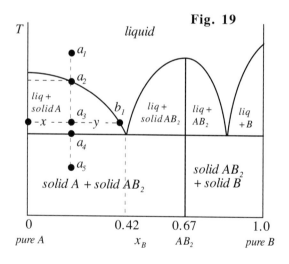

Fig. 19

This means that, in contrast to the addition of a small amount of B to pure A or of A to pure B, the additions of small amounts of A or B to the compound AB_2 will not lower its melting or freezing point. It should be noted that the compound AB_2 at $x_B = 0.67$ is a true compound, not just a 2:1 molar mixture of B:A. Such compounds are always formed at integer mole ratios of the two components, such as 2:1, 1:2, 1:3, etc.[12]

The interpretation of the information shown in Fig. 19 is made in the same way as for Fig. 16. Thus, if a molten mixture of composition a_1 is cooled, at a_2 the first solid (pure A) starts to freeze out from the liquid of composition a_2. When the system is further cooled to a_3, it will consist of pure solid A and liquid of composition b_1. The relative molar amounts of the liquid phase to the solid phase will be in the tie line ratio of x:y. Finally, at a_4, when the liquid is 42 mole % in B and 58 mole % in A, the whole solution freezes and solid A and AB_2 come out together. Further lowering of the temperature to a_5 results in no further changes in the phases: there is only solid A and solid AB_2.

Exercise 17-6

Fig. 20 shows the T–x diagram for the binary system $CaCl_2$–KCl. Identify all the one-phase areas, two-phase areas, and three-phase lines.

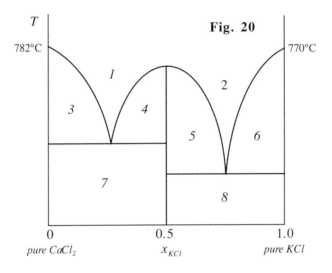

---

[12] In the case shown in Fig. 8, A:B is in the ratio of 1:2.

KEY POINTS FOR CHAPTER SEVENTEEN

1. **Mixtures**: A mixture is a system that consists of more than one component—a *multicomponent* system. The properties of a multicomponent system depend on its composition, as well as on its temperature and pressure.

2. **Molarity, M**: the number of moles of solute per liter of *solution* (*NOT* per liter of *solvent*). M or *[i]* are commonly used to designate molarity.

$$[i] = \frac{moles\ i}{L\ of\ solution} \qquad \text{Units:}\ \ mol\ L^{-1}\ \text{or}\ \ mol\ dm^{-3}$$

The molarity[13] of a solution decreases with increasing temperature because, even though the number of moles of solute dissolved in the solution stays the same, the *volume* of the solution expands as the temperature is raised.

3. **Mole fraction, x**: the number of moles of a given component divided by the total moles of all components (including the given component).

$$x_i = \frac{moles\ i}{total\ moles\ of\ all\ components\ (including\ i)} \qquad \text{Units:}\ \ \text{dimensionless}$$

For a gas phase, symbol y_i is often used instead of x_i.

4. **Molality, m**: the number of moles of solute per kg of *solvent* (*NOT* per kg of *solution*).

$$m_i = \frac{moles\ i}{kg\ of\ solvent} \qquad \text{Units:}\ \ mol\ kg^{-1}$$

Molality doesn't vary with temperature. Also, the addition of another component doesn't change the molality of the others.

5. **Partial molar volumes**: Each component A, B, C, .. in a multicomponent system contributes to the total volume V of the system according to some mathematical relationship

$$V = f(n_A, n_B, n_C, \ldots)$$
$$= n_A(V_m)_A + n_B(V_m)_B + n_B(V_m)_B + \ldots$$

where $(V_m)_i$ is the *partial molar volume* of component i. The partial molar volume $(V_m)_i$ is usually *not* the same as the molar volume v_i, although the units ($m^3\ mol^{-1}$) are the same.

6. The concept of partial molar quantity also applies to other state properties of mixtures:

Partial molar enthalpy: $\qquad H = n_A(H_m)_A + n_B(H_m)_B + n_C(H_m)_C + \ldots$

Partial molar internal energy: $\ U = n_A(U_m)_A + n_B(U_m)_B + n_C(U_m)_C + \ldots$

Partial molar entropy: $\qquad\quad S = n_A(S_m)_A + n_B(S_m)_B + n_C(S_m)_C + \ldots$

Partial molar heat capacity: $\quad C_P = n_A[(C_P)_m]_A + n_B[(C_P)_m]_B + n_C[(C_P)_m]_C + \ldots$

---

[13] Not to be confused with *morality*.

7. The **partial molar Gibbs free energy** is especially important because it relates to how much work a system can perform, equilibrium constants, rates of reaction, and other properties of mixtures.

$$G = n_A (G_m)_A + n_B (G_m)_B + n_C (G_m)_C + \ldots$$

$(G_m)_i$ has its own special name and symbol: the **chemical potential**, μ. Therefore, at any given temperature and pressure the total Gibbs free energy G of a multicomponent system is given by

$$G = n_A \mu_A + n_B \mu_B + n_C \mu_C + \ldots$$

The *partial molar quantity B_m* of any component i in the system is defined, more properly, as the way in which the total value of B for the system changes per mole of i with an infinitesimal addition of component i at *constant temperature and pressure, while keeping the amounts of all the other components j constant.*

i.e., $$\mu_i \equiv \left(\frac{\partial G}{\partial n_i} \right)_{T,P,n_j} \qquad \text{Units: J mol}^{-1}$$

8. **Chemical potential of a component i in an ideal gas mixture**:

$$\mu_i = \mu_i^o + RT \, ln \left(\frac{P_i}{P^o} \right)$$

where $P°$ is the standard pressure (= *1 bar*), T is the temperature of the mixture, P_i is the partial pressure of component i, and μ_i^o is the chemical potential of component i in its standard state (pure i at $P_i = P° = 1\ bar$). μ_i^o is the same as the standard molar Gibbs free energy of formation, $(\Delta g_f^o)_i$. The same expression for μ_i can be used for a pure ideal gas or for an ideal gas in a mixture of ideal gases, because ideal gases behave as if none of the other gases are present; i.e., there is no interaction between them.

9. The **relative activity a_i** of any species i is defined as

$$a_i \equiv exp \left[\frac{\mu_i - \mu_i^o}{RT} \right] \qquad \text{Units: dimensionless}$$

Rearranging gives $$\mu_i = \mu_i^o + RT \, ln\, a_i$$

For an **ideal gas**: $$a_i = \frac{P_i}{P^o}$$

10. **Criterion for equilibrium**: For any two phases α and β in equilibrium,

• for a *one component system* at constant T and P: $\quad g^\alpha = g^\beta$

• for a *multicomponent system* at constant T and P: $\quad \mu_i^\alpha = \mu_i^\beta$

11. **Raoult's Law**: For a liquid solution of volatile components $\qquad P_i = x_i P_i^\bullet$

where P_i is the partial pressure of component i above the solution, x_i is the mole fraction of component i in the liquid phase, and $P_i^\bullet$ is the equilibrium vapor pressure of the *pure* volatile component. A solution that obeys Raoult's law for all components at all concentrations is called an **ideal solution**.

12. For a liquid solution of a **non-volatile solute i dissolved in a volatile solvent W**,

$$P_W = x_W P_W^{\bullet}$$

For an **ideal solution of two volatile liquids** A and B, the total pressure P is

$$P = P_A + P_B = x_A P_A^{\bullet} + x_B P_B^{\bullet}$$
$$= x_A P_A^{\bullet} + (1 - x_A) P_B^{\bullet}$$
$$= (1 - x_B) P_A^{\bullet} + x_B P_B^{\bullet}$$

13. For **real (non-ideal) solutions**, Raoult's Law is obeyed when the mole fraction of component i approaches unity:

$$P_i \to x_i P_i^{\bullet} \quad \text{as} \quad x_i \to 1$$

14. The **chemical potential of the solvent W in an ideal solution** is given by

$$\mu_W = \mu_W^* + RT \ln x_W \quad \text{where} \quad \mu_W^* = \mu_W^o + RT \ln \left(\frac{P_W^{\bullet}}{P^o} \right)$$

μ_W^o, the chemical potential of the solvent when it is in its standard state, is just the chemical potential of the pure solvent, and is a function only of temperature.

The chemical potential of the *solvent* is *lowered* when it has a solute dissolved in it.

15. The **relative activity of the solvent W in a solution** is given by

$$a_W = \frac{P_W}{P_W^{\bullet}}$$

where $P_W^{\bullet}$ is the vapor pressure of the *pure* solvent.

For an ideal solution, the relative activity of the solvent is approximated by $a_W \approx x_W$.

For **non-ideal (real) solutions** that do not obey Raoult's law at all concentrations, the relative activity of the solvent more properly is given by

$$a_W = f_W x_W$$

where f_W is called the **rational activity coefficient** of the solvent. For an ideal solution, $f_W = 1$.

16. **Henry's law** for ideal dilute solutions of volatile solutes gives the partial pressure of the solute above the solution as

$$P_i = K_i x_i$$

where K_i is the Henry's law constant, which is a function of the temperature, the gas, and the solvent. Henry's law works best for *dilute* solute concentrations (i.e., as $x_i \to 0$), unlike Raoult's Law, which works best as $x_i \to 1$.

17. The **solubility of a gas** in a liquid solvent is given by

$$C_i \approx \frac{P_i}{K_i v_W}$$

where C_i is the concentration of the dissolved gas in mol m^{-3}, K_i is the Henry's law constant for the gas in the liquid in Pa, and v_W is the molar volume of the pure solvent in m^3 mol^{-1}.

18. The vapor phase in equilibrium with an ideal liquid solution containing two volatile liquid components will be enriched in the more volatile component. This is the basis for the process of **distillation**. The "more volatile component" means the component with the higher pure component vapor pressure, $P_i^\bullet$.

If the vapor pressure of pure liquid A is $P_A^\bullet$ and the vapor pressure of pure liquid B is $P_B^\bullet$ and A is more volatile than B (i.e., $P_A^\bullet > P_B^\bullet$), then the equilibrium mole fraction of component A in the vapor phase is given by

$$y_A = \frac{P_A}{P} = \frac{P_A}{P_A + P_B} = \frac{x_A P_A^\bullet}{x_A P_A^\bullet + x_B P_B^\bullet} = \frac{x_A P_A^\bullet}{x_A P_A^\bullet + (1 - x_A)P_B^\bullet} \qquad (y_A > x_A)$$

where x_A and x_B are the mole fractions of A and B, respectively, in the liquid phase.

19. The **Lever Rule**:

Referring to the T–x phase diagram at the right: For a system consisting of a liquid solution of two miscible liquids A and B in equilibrium with its vapor, if the overall system composition at temperature T' is x (point B), the composition of the liquid phase will be x' and that of the vapor phase will be x''. The ratio of the moles of vapor to the moles of liquid is given by the ratio of the segment a to the segment b of the tie line $\overline{ABC}$:

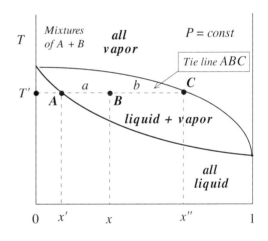

$$\frac{n_{vap}}{n_{liq}} = \frac{n_A^{vap} + n_B^{vap}}{n_A^{liq} + n_B^{liq}} = \frac{a}{b}$$

This relationship is known as the *lever rule*, and is applicable to any type of T–x phase diagram, not just liquid-vapor diagrams.

PROBLEMS

1. (a) Calculate (i) the molarity, (ii) the molality, and (iii) the mole fraction of ethanol (C_2H_5OH) in 100-proof alcohol. (100-proof alcohol is an aqueous solution containing 42.5% by mass ethanol. The solution has a density of 934 g L^{-1}.)

    (b) At 25°C the vapor pressure of pure ethanol is 59 Torr. The average person inhales 250 mL of air every 4 seconds. Assuming that all the alcohol vapor is absorbed by the lungs, how long would a person have to inhale air saturated with 100-proof alcohol vapor in order to ingest the same amount of alcohol that is contained in 30 mL of 100-proof alcohol? For this calculation you may assume that Raoult's law holds (it doesn't). ☺

2. A benzene (B)-toluene (T) solution is in equilibrium with its vapor at 80°C. If $y_B = 0.670$, calculate x_B and the total vapor pressure of the solution. The vapor pressures at 80°C of benzene and toluene are 100.4 kPa and 38.7 kPa, respectively. ☹

3. A solution of benzene (C_6H_6, molar mass = 78.11 g mol^{-1}) and toluene ($C_6H_5CH_3$, molar mass = 92.13 g mol^{-1}) is 50.0% by mass benzene. Calculate the composition of the vapor in equilibrium with the solution at 50°C. At 50°C, the vapor pressures of benzene and toluene are 36.1 kPa and 12.3 kPa, respectively. ☹

4. At 0°C and pure gas at one atm pressure, the solubilities in water of nitrogen and oxygen are 23.5 cm^3 L^{-1} and 48.9 cm^3 L^{-1}, respectively. How much nitrogen and oxygen will dissolve in ice water at 0°C that is exposed to air at 0°C and one atm pressure? Report the solubilities in mmol L^{-1}. Air may be considered to be 79.0% by volume nitrogen and 21.0% by volume oxygen. ☹

5. At 0°C and 60°C, one kg of water will dissolve 0.1589 and 2.0204 moles, respectively, of anhydrous ammonium sulfate—$(NH_4)_2SO_4$. When the ammonium sulfate crystallizes from solution it does so as the hydrate $(NH_4)_2SO_4 \cdot 24H_2O$. How many kilograms of hydrated ammonium sulfate can be recovered when 1000 kg of saturated solution at 60°C is cooled to 0°C? Molar masses (g mol^{-1}): $(NH_4)_2SO_4 = 132.15$; $(NH_4)_2SO_4 \cdot 24H_2O = 564.53$ ☠

6. At 25°C the vapor pressure of pure carbon tetrachloride (CCl_4) is 4513 Pa. Calculate the mole fraction of Br_2 in the equilibrium vapor above a solution containing 0.050 mole fraction Br_2 dissolved in carbon tetrachloride at 25°C. The Henry's law constant for bromine dissolved in carbon tetrachloride at 25°C is 16.31 kPa (mole fraction basis). ☺

7. At 35°C, pure acetone (A) has a vapor pressure of $P_A^\bullet = 46.3$ kPa; and in solution in chloroform, acetone has a Henry's law constant $K_A = 23.3$ kPa. At 35°C, pure chloroform (C) has a vapor pressure $P_C^\bullet = 39.1$ kPa; and in solution in acetone, chloroform has a Henry's law constant $K_C = 22.0$ kPa. A closed tank contains impure liquid acetone in which 8.63 moles of acetone is contaminated with 0.29 moles of chloroform. This liquid mixture is an ideal dilute solution and it is in equilibrium with vapor occupying a volume of 4.85 m^3 at 35°C. What is the amount of chloroform in the vapor? ☹

8. A mixture of chlorobenzene (C_6H_5Cl, component C) and bromobenzene (C_6H_5Br, compo-
    nent B) behaves ideally. If the vapor pressures at 137°C of the pure components are, respec-
    tively, 863 Torr and 453 Torr, calculate:

    (a) the mole fraction composition of the liquid mixture that has a normal boiling point of
         137°C;

    (b) the composition of the vapor that is in equilibrium with this liquid at its boiling point;

    (c) the composition of the vapor over a solution at 137°C containing an equal number of
         moles of chlorobenzene and bromobenzene. ☺

9. A beverage bottle having an internal volume of 1.125 L contained exactly 1.00 L of "soda
    water." The observed gauge pressure of the contents of the bottle at 25°C was 185 kPa. Cal-
    culate the weight percent of carbon dioxide dissolved in the water at this equilibrium pres-
    sure. The Henry's law constant at 25°C is 161.7 MPa/mole fraction CO_2 and the vapor pres-
    sure of the water at this temperature is 3.17 kPa. The atmospheric pressure was 98.63 kPa at
    the time of the measurement. The density of the beverage was 0.997 g cm^{-3}. ☹

10. Consider the vapor above an ideal binary solution consisting of component 1 and component
    2. For such a solution, show that

    $$V = n_1 v_1^{\bullet} + n_2 v_2^{\bullet}$$

    where V is the total volume of the vapor, n_1 and n_2 are the number of moles of component 1
    and 2, respectively, in the vapor, and $v_1^{\bullet}$ and $v_2^{\bullet}$ are the respective molar volumes of pure
    component vapors 1 and 2. ☹

11. The experimentally measured volumes (in cm^3) at 25°C and one bar pressure of aqueous
    solutions of NaCl consisting of m moles of NaCl dissolved in 1.000 kg of water have been
    fitted to the following equation:

    $$V = 1003 + 16.62m + 1.77m^{3/2} + 0.12m^2$$

    Calculate the partial molar volume of (a) the NaCl and of (b) the H_2O in a 0.5 molal aqueous
    solution of NaCl at 25°C and one bar pressure. The molar masses of water and NaCl are
    18.015 g mol^{-1} and 58.443 g mol^{-1}, respectively. ☹

12. The molar volume of solid $MgSO_4$ crystals is 45.3 cm^3 mol^{-1}. At 18°C the volume—in cm^3—
    of a solution consisting of m moles of $MgSO_4$ dissolved in 1.000 kg of water is given by:

    $$V = 1001.21 + 34.69(m - 0.07)^2$$

    Calculate the partial molar volume of (a) the $MgSO_4$, and (b) the H_2O in a 0.010 molal
    aqueous solution of magnesium sulfate at 18°C. ☹

13. The molar volume of pure methanol is 40.0 cm^3 mol^{-1}, while that of pure water is 18.02 cm^3
    mol^{-1}. It has been determined experimentally that the volume V (in cm^3) of a solution
    containing 1000 g of water and n moles of methanol is given by

    $$V = 1000 + 35n + 0.5n^2$$

    If 0.25 moles of methanol is added to 10.0 moles of water:

    (a) What is the volume of the resulting solution?

(b)   What is the partial molar volume of the methanol in this solution?

(c)   What is the partial molar volume of the water in this solution?

   Data: $H_2O = 18.015$ g mol^{-1} ☹

14.   At 20°C and atmospheric pressure the densities of solutions of *1*-propanol (component *1*) in water (component *2*) have been measured and fitted to the following equation:

$$\rho(x_1) = 0.99823 - 0.48503x_1 + 0.47518x_1^2 - 0.17163x_1^3 - 0.01387x_1^4$$

where $\rho(x_1)$ is the density of the solution in g cm^{-3} and x_1 is the mole fraction of *1*-propanol in the solution. (The mole fraction of water in the solution is x_W.)

(a)   Show that the partial molar volume of *1*-propanol in the solution is given by

$$(V_m)_1 = \frac{M_1}{\rho(x_1)} - \left[\frac{(1 - x_1)M_W + x_1 M_1}{\rho^2(x_1)}\right](1 - x_1)\frac{d\rho(x_1)}{dx_1}$$

(b)   Calculate the partial molar volumes of (i) *1*-propanol and of (ii) water in an aqueous solution containing 5 mole % *1*-propanol.

(c)   What is the volume at 20°C of a solution made by adding 50 moles of *1*-propanol to 950 moles of water? What percent shrinkage occurs when this solution is formed?

   Data: $M_{C_3H_7OH} = 60.096$ g mol^{-1} $M_{H_2O} = 18.015$ g mol^{-1} ☠

15.   For equilibrium at 25°C, what is the % by weight of bromine in the vapor phase above a liquid solution of bromine dissolved in carbon tetrachloride if the mole fraction of bromine in the liquid phase is 0.0500? Br$_2$ = 159.8 g mol^{-1}; CCl$_4$ = 153.8 g mol^{-1}. At 25°C the vapor pressure of pure CCl$_4$ is 33.85 Torr. The Henry's law constant for bromine dissolved in CCl$_4$ is 122.36 Torr. ☺

16.   If the density of a solution containing 4.45 g of pure sulfuric acid and 82.20 g of water is 1.029 g cm^{-3}, calculate: (a) the mass percent, (b) the mole fraction, (c) the mole percent, (d) the molarity, (e) the normality, and (f) the molality of the H$_2$SO$_4$ in this solution. ☺

17.   A gas mixture consisting of H$_2$ and O$_2$ in the molar ratio of H$_2$/O$_2$ = 2.00/1.00 is used at a total pressure of 5.00 atm to saturate water at 25°C. After complete saturation with the gases, the water is boiled to drive off all the dissolved gases, and the resulting gas mixture then is dried. What is the volume percent composition of this final gas mixture? ☹

18.   At 0°C and one atm partial pressure, the solubility in water of oxygen gas is 48.9 cm^3 L^{-1}. How much oxygen will dissolve in ice water at 0°C that is exposed to the air at 0°C and 1.50 bar pressure? Report the O$_2$ solubility in millimoles per liter (mmol L^{-1}). Air may be considered to be 79.0% by volume nitrogen (N_2) and 21.0% by volume oxygen (O_2). ☹

19.   One mole of liquid *A* and two moles of liquid *B* are mixed to form an ideal solution at 30°C. The total vapor pressure of this solution is 250 Torr. When an additional mole of *A* is added the vapor pressure increases to 300 Torr. Evaluate $P_A^\bullet$ and $P_B^\bullet$, the vapor pressures of the two pure components. ☺

20. An H_2SO_4 solution is 65% by weight H_2SO_4 and has a density at 25°C of 1.55 g cm^{-3}. Calculate the mole fraction, the molality, and the molarity of the solution. ☺

21. Phosphine is partly decomposed in a continuous flow reactor at 900 K and 1.00 atm into phosphorus and hydrogen according to:

$$4PH_{3(g)} \rightarrow P_{4(g)} + 6H_{2(g)}$$

(a) If 30.0% of the phosphine has been decomposed, what is the mole fraction of P_4 in the resulting gas mixture?

(b) Assuming ideal gas behavior, what is the concentration of PH_3 in the resulting gas mixture? ☺

22. An aqueous NaOH solution is 50.0% by weight NaOH. At 25°C the concentration of NaOH in this solution is 19.0 mol L^{-1}. What is the density of the solution at 25°C? ☺

23. An equimolar mixture of n-hexane vapor and helium is cooled from 62°C to 32°C at a constant pressure of one atm. Calculate: (a) the percent saturation of the n-hexane vapor at 62°C, (b) the percent of the n-hexane that condenses as a result of the cooling. The vapor pressures of n-hexane at 32°C and 62°C are 200 torr and 600 torr, respectively. ☹

24. A gaseous mixture consisting of 1.00 mole of CCl_4 and 3.00 moles of N_2 is compressed isothermally at 38.3°C from 1.00 atm to 2.00 atm. How many moles of the CCl_4 will condense? At 38.3°C the vapor pressure of CCl_4 is 200 Torr. ☹

25. N_2 saturated with benzene (C_6H_6, molar mass 78 g mol^{-1}) enters a continuous flow condenser at 50°C and 800 mm Hg pressure. It leaves the condenser at 20°C and 750 mm Hg pressure. Assuming steady state operation of the condenser, calculate the mass of benzene that condenses per 1000 m^3 of entering mixture. The vapor pressures of benzene at 20°C and 50°C are, respectively, 74.7 mm Hg and 269 mm Hg. ☹

26. At 60°C the vapor pressures of pure benzene and toluene are 385 and 39 Torr, respectively. Above a solution containing 0.60 mole fraction benzene, calculate:

(a) the partial pressures of benzene and toluene,

(b) the total vapor pressure, and

(c) the mole fraction of toluene in the vapor. ☺

27. At 90°C benzene has a vapor pressure of 1022 Torr, and toluene has a vapor pressure of 406 Torr. Calculate the composition of the benzene-toluene solution that will boil at one atm pressure and 90°C, assuming that the solution is ideal. ☺

28. (a) What is the composition (mole fraction) of a mixture of ethanol and propanol that boils at 80°C and one atm, if the vapor pressures are 52.3 kPa for propanol and 114.5 kPa for ethanol?

(b) If the vapor in equilibrium with the solution in part (a) is condensed to liquid, what will be the vapor pressure of this condensate at 80°C? ☹

29. The vapor pressures at 100°C are: benzene 1350 Torr, toluene 556 Torr, *p*-xylene 240 Torr. An ideal solution consisting of these three liquids has a normal boiling point of 100°C. The mole fraction of benzene in the vapor phase is 0.632. Calculate:

   (a) the mole fraction of benzene in the liquid phase, and

   (b) the mole fraction of *p*-xylene in the vapor phase. ☺

30. The vapor pressures of benzene and toluene at 95°C are, respectively, 155.7 kPa and 63.3 kPa. A gaseous mixture consisting of 50 moles of benzene and 50 moles of toluene was cooled to 95°C. As a result of the cooling, some of the benzene and toluene condensed. If the total pressure above the condensate was 101.3 kPa, calculate:

   (a) the mole fraction of benzene in the condensate (liquid),

   (b) the mole fraction of benzene in the gas phase after cooling, and

   (c) the number of moles of benzene in the condensate. ☹

31. At 26°C the vapor pressures of benzene and toluene are 13.33 kPa and 4.00 kPa, respectively. 100 moles of a gaseous mixture consisting of nitrogen, benzene, and toluene was cooled from 100°C to 26°C at a constant pressure of 101.3 kPa. As a result of the cooling, some of the benzene and toluene condensed. Before cooling, the gas was 60.0 mole percent N_2; after cooling it was 90.0 mole percent N_2. Calculate:

   (a) the total number of moles of condensate, and

   (b) the number of moles of benzene in the condensate. ☹

32. Vapor pressures at 80°C: benzene 100.4 kPa, toluene 30.7 kPa. An equimolar gaseous mixture of benzene and toluene is compressed isothermally at 80°C. Assuming that there is no supersaturation, calculate the total pressure

   (a) when condensation starts,

   (b) when 50.0% of the toluene has condensed, and

   (c) when condensation has just been completed. ☹

33. At 35.2°C, the vapor pressures of acetone and carbon disulfide are 45.9 kPa and 68.3 kPa, respectively. At this temperature the total pressure above a solution of acetone and carbon disulfide in which the mole fraction of the CS_2 in the liquid is 0.06 is 60.0 kPa.

   (a) Does the solution obey Raoult's law?

   (b) Making the appropriate assumptions, estimate the mole fraction of carbon disulfide in the vapor.

   (c) What is the Henry's law constant for carbon disulfide in acetone at 35.2°C? ☺

EIGHTEEN

COLLIGATIVE PROPERTIES

18.1 COLLIGATIVE PROPERTIES

We saw that nonvolatile solute molecules lower *the vapor pressure* of the solvent above the solution. Nonvolatile solutes also *raise the boiling point* of the solution, *lower the melting point* of the solution, and give rise to the phenomenon of *osmosis*. These four effects depend only on the number of solute particles present in the solution, not on their chemical identity. For this reason, such properties are called *colligative* properties, from the Latin word *colligare*, meaning "to bind together." Thus, these properties are all related to each other; once one has been measured, the others can be found by calculation—they are all "bound together." This means that a 0.01 molal solution of any (non-dissociating) solute should have the same boiling point, the same freezing point, and the same osmotic pressure as a 0.01 molal solution of any other solute in the same solvent.

18.2 VAPOR PRESSURE DEPRESSION

For an ideal solution it is easy to show that the fractional vapor pressure depression $\Delta P/P_W^\bullet$ of the solvent W brought about by the presence of a nonvolatile solute A depends only on the mole fraction of the nonvolatile solute, and not on the actual solute used; therefore $\Delta P/P_W^\bullet$ is a true colligative property.

Thus, since the solution is ideal, we can apply Raoult's law to the solvent:

$$P_W = x_W P_W^\bullet \qquad \ldots [1]$$

$$= (1 - x_A)\, P_W^\bullet \qquad \ldots [2]$$

$$= P_W^\bullet - P_W^\bullet\, x_A$$

Rearranging:
$$P_W^\bullet - P_W = P_W^\bullet\, x_A$$

i.e.,
$$\Delta P = P_W^\bullet\, x_A$$

From which
$$\boxed{\dfrac{\Delta P}{P_W^\bullet} = x_A} \qquad \textit{Vapor Pressure} \qquad \ldots [3]$$
$$\textit{Depression}$$

Example 18-1

A 0.500 molal aqueous solution of Na_2SO_4 (142.06 g mol^{-1}) has a vapor pressure of 747.4 Torr at 100°C. Estimate the degree of dissociation of the dissolved salt at this temperature.

Solution

At 100°C—the normal boiling point of water—its vapor pressure is one atm, or 760 Torr.

Thus, from Eqn [3]:
$$x_A = \frac{\Delta P}{P_W^\bullet} = \frac{760.0 - 747.4}{760.0} = \frac{12.6}{760} = 0.01658$$

A 0.50 molal solution contains 0.50 moles of solute dissolved in 1000 g of solvent. Therefore, as a basis, take 1000 g H_2O + 0.500 moles of Na_2SO_4 to give a 0.500 molal solution.

The molar mass of water is 18.015 g mol^{-1}; therefore the number of moles of solvent in this solution is
$$n_W = \frac{1000 \text{ g}}{18.015 \text{ g}/\text{mol}} = 55.509 \text{ mol}$$

Let the total moles of solute species in the solution be n_A. n_A consists of the Na^+ ions, the SO_4^{2-} ions, and the undissociated Na_2SO_4 molecules.

Therefore:
$$x_A = \frac{n_A}{n_A + n_W}$$

$$0.01658 = \frac{n_A}{n_A + 55.509}$$

$$n_A = 0.01658\, n_A + 0.92034$$

$$0.98342\, n_A = 0.92034$$

$$n_A = 0.9359$$

Therefore a total of 0.9359 moles of solute species is present in the solution at equilibrium.

Consider the dissociation of Na_2SO_4 in solution:[1]

---

[1] Neglect hydration. The water molecule is "dipolar," meaning that one end of the molecule (the H^+ end) is positively charged while the other end (the O^{2-} end) is negatively charged. The positively charged Na^+ ions attract the negative ends of water molecules, causing some water to stick (by electrostatic attraction) to the Na^+ ions; thereby in effect becoming part of the ion. This removal of some of the "free" water from the solution lowers the effective amount of solvent present. So, instead of having 1 kg of water as solvent, we end up with marginally less than 1 kg, which slightly increases the molality of the solution, since the molality is defined as the moles of solute dissolved in 1 kg of (free) solvent. The hydration effect with negative ions is less pronounced than with positive ions.

$$Na_2SO_{4(aq)} \rightarrow 2\,Na^+_{(aq)} + SO^{2-}_{4(aq)}$$

moles, start:	0.5000	0	0
dissociate:	$-x$	$+2\,x$	$+x$
equilibrium:	$0.5 - x$	$2\,x$	x	$n_{total} = (0.5000 - x) + 2\,x + x = 0.5000 + 2\,x$

Therefore, $0.9359 = 0.5000 + 2\,x$

$$x = \frac{0.9359 - 0.5000}{2} = 0.2180$$

Thus, for each 0.5000 moles of Na_2SO_4 dissolved, 0.2180 moles dissociates into ions, and the % dissociation is

$$\frac{0.2180}{0.5000} \times 100\% = 43.6\%$$

Contrary to popular belief, inorganic salts do not completely dissociate into ions in solution except at extremely low concentrations.

Ans: $\boxed{\sim 44\%}$

18.3 BOILING POINT ELEVATION AND FREEZING POINT DEPRESSION

It is observed experimentally that the increase in the boiling point of a solution over the boiling point of the pure solvent is given by an equation of the form

$$\boxed{\Delta T_b = K_b m_i} \qquad \ldots [4]$$

where ΔT_b is the boiling point elevation, K_b is the boiling point elevation constant (the *ebullioscopic constant*), and m_i is the molality of the solute.

Similarly, the lowering of the freezing point of the solution obeys a similar type of equation:

$$\boxed{\left|\Delta T_f\right| = K_f m_i} \qquad \ldots [5]$$

where $|\Delta T_f|$ is the lowering of the freezing point below the value for pure solvent, K_f is the freezing point constant (the *cryoscopic constant*), and m_i is the molality of the solute in the solution. For water as the solvent, the values of K_b and K_f are 0.513 and 1.86 K $(mol/kg)^{-1}$, respectively. Some values for other solvents are given below in Table 1.

The origin of all the colligative properties is *the lowering of the chemical potential of the solvent by the presence of a solute*.

For nonvolatile solutes it is observed that (i) the solute doesn't contribute to the vapor above the solution, and (ii) the solute doesn't dissolve in *solid* solvent; i.e., when the solution starts to freeze, the solid formed usually consists only of *pure solid solvent*.

Table 1. *Boiling Point Constants and Freezing Point Constants*

Solvent	Molecular Formula	Molar Mass (g mol^{-1})	Normal Boiling Point (°C)	K_b K (mol/kg)$^{-1}$	Normal Melting Point (°C)	K_f K (mol/kg)$^{-1}$
Acetic acid	CH_3COOH	60.05	117.9	3.07	16.6	3.57
Benzene	C_6H_6	78.12	80.1	2.53	5.45	5.07
Camphor	$C_{10}H_{16}O$	152.24	204 (subl)	--	179.8	37.7
Carbon tetrachloride	CCl_4	153.82	76.54	5.03	−22.99	--
Chloroform	$CHCl_3$	119.38	61.7	3.63	−63.5	--
Cyclohexane	C_6H_{12}	84.16	80.74	2.79	6.55	20.0
p-dibromobenzene	$C_6H_4Br_2$	235.92	218	--	87.33	12.5
p-dichlorobenzene	$C_6H_4Cl_2$	147.0	174	--	53.1	7.11
Dioxane	$C_4H_8O_2$	88.12	101	--	11.8	4.71
Ethanol	C_2H_5OH	46.07	78.5	1.22	−117.3	--
Ethyl bromide	C_2H_5Br	109.0	38.3	2.93		--
Methanol	CH_3OH	32.04	64.96	0.83	−93.9	--
Naphthalene	$C_{10}H_8$	128.19	218	--	80.55	6.98
Phenol	C_6H_5OH	94.11	181.75	3.56	43	--
Water	H_2O	18.015	100.00	0.513	0	1.86

The Gibbs free energy[2] G_W of the solvent water is plotted against temperature in Figures 1, 2, and 3. Earlier we saw that

$$(\partial G/\partial T)_P = -S \qquad \ldots [6]$$

and that S for all substances is always positive. In view of Eqn [6] it follows, therefore, that G always decreases as the temperature increases.

We also noted that, since $\qquad S_{solid} < S_{liquid} < S_{gas} \qquad \ldots [7]$

it follows that $\qquad \left(\dfrac{\partial G}{\partial T}\right)_P^{gas} < \left(\dfrac{\partial G}{\partial T}\right)_P^{liquid} < \left(\dfrac{\partial G}{\partial T}\right)_P^{solid} \qquad \ldots [8]$

In other words, the slope of the G vs. T curve is *most* negative for the gas phase and *least* negative for the solid phase, as indicated in the figures.

---

[2] There are two kinds of free energy used in thermodynamics: Gibbs free energy G and Helmholtz free energy A. The Gibbs free energy is found to be most useful under conditions of constant temperature and *pressure*, whereas the Helmholtz free energy is more applicable to conditions of constant temperature and *volume*. In this course we only deal with the Gibbs free energy.

First consider the case of **liquid-vapor equilibrium** (Fig. 1). When the liquid and vapor are in equilibrium at the normal boiling point, the equilibrium

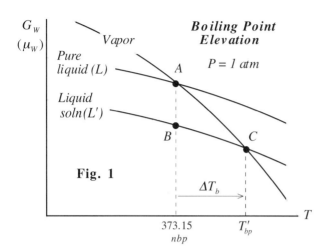

$$liquid \rightleftharpoons vapor$$

is established, and the chemical potential of the *vapor* must be the *same* as the chemical potential of the *liquid*:

$$\mu_{vapor} = \mu_{liquid} \; \cdots \; [9]$$

This condition corresponds to point *A* in Fig. 1.

The addition of *solute* to the liquid water *lowers* the mole fraction of the liquid from its original value of $x_w = 1$ to some new value $x_w < 1$.

Since $\mu_w = \mu_w^* + RT \ln x_w$

it follows that the presence of the solute also lowers the *chemical potential* of the liquid to a value μ_w *lower* than the value μ_w^* for pure liquid water. This corresponds to a lowering of μ_w to the value at point *B* so that now

$$\mu'_{liq} < \mu_{vap}$$

Under this condition the liquid phase now is more stable than the vapor phase and the vapor spontaneously will form liquid at 100°C (= 373.15 K). In order to re-establish liquid-vapor equilibrium the temperature must be raised from 373.15 K (the normal boiling point) to some new value T'_{bp} at point *C*, where vapor once again can form at one atm pressure. In other words, *the boiling point of the solution has increased by* $\Delta T_b = T'_{bp} - 373.15$. At point *C* we again have liquid-vapor equilibrium for which $\mu'_{liq} = \mu_{vap}$.

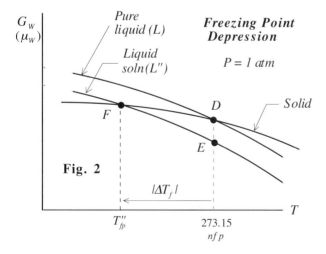

Now consider Fig. 2, for **solid-liquid equilibrium**: Point *D* corresponds to solid–liquid equilibrium between ice and water at 0°C (= 273.15 K), with

$$\mu_{solid} = \mu_{liquid}$$

The addition of solute to the liquid phase lowers the chemical potential of the liquid to μ''_{liq} at point *E*.

Now, because the chemical potential of the liquid is *less* than that of the solid, the liquid is more stable than the solid and the solid tends spontaneously to convert to the liquid:

$$\Delta G_{T,P} = \mu''_{liquid} - \mu_{solid} < 0$$

At 273.15 K the system at point E now exists only as liquid. In order to reestablish freezing equilibrium the temperature must be lowered to T''_{fp} (point F), where $\mu''_{liquid} = \mu_{solid}$. In other words, *the freezing point of the solution has been lowered.*

Fig. 3 shows both effects for water combined on the same plot: Note from Fig. 3 that, since

$$|\Delta T_f| > \Delta T_b$$

it follows that

$$K_f > K_b$$

(In fact, *1.86 > 0.51.*)

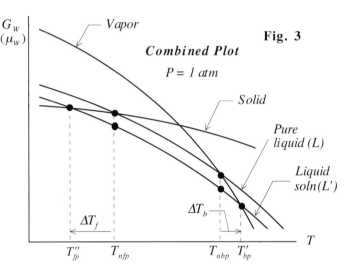

Fig. 3

Combined Plot

$P = 1$ atm

Example 18-2

The basis of all colligative properties lies in changes in the chemical potential of the solvent. Jane and Bill each purchase a 250 mL cup of hot (50°C) coffee. Jane sweetens her coffee by adding one teaspoon of sugar (sucrose, 342.30 g mol^{-1}), while Bill sweetens his by adding a one-gram package of low cal sweetener (saccharin, 2-sulfobenzoic acid imide, 183.19 g mol^{-1}). Assuming that the coffee is very weak so that the sweetener may be considered to be the only solute, what is the difference between the chemical potential of the hot water in the two cups of sweetened coffee?

The density of water at 50°C is 988.07 kg m^{-3}; the density of solid sucrose is 1580.5 kg m^{-3}; one teaspoon = 5.0 cm^3.

Solution

$$\mu_w = \mu^*_w + RT\, ln\, x_w \qquad 50°C = 50 + 273.15 = 323.15 \text{ K}$$

$$\text{Mass of water} = (250 \text{ cm}^3)\left(0.98807 \tfrac{g}{cm^3}\right) = 247.0175 \text{ g}$$

$$\text{Moles of water:} \qquad n_w = \frac{247.0175 \text{ g}}{18.015 \text{ g/mol}} = 13.7118 \text{ mol}$$

Moles of sucrose:
$$n_{suc} = \frac{(1 \text{ tsp})(5.0 \text{ cm}^3/\text{tsp})(1.5805 \text{ g}/\text{cm}^3)}{342.30 \text{ g}/\text{mol}} = 0.02309 \text{ mol}$$

Moles of saccharin:
$$n_{sac} = \frac{1.00 \text{ g}}{183.19 \text{ g}/\text{mol}} = 0.005459 \text{ mol}$$

For Jane's coffee:
$$x_w^{Jane} = \frac{n_w}{n_w + n_{suc}} = \frac{13.7118}{13.7118 + 0.02309} = 0.998319$$

For Bill's coffee:
$$x_w^{Bill} = \frac{n_w}{n_w + n_{sac}} = \frac{13.7118}{13.7118 + 0.005459} = 0.999602$$

$$\mu_w^{Jane} = \mu_w^* + RT \ln x_w^{Jane} \qquad \text{and} \qquad \mu_w^{Bill} = \mu_w^* + RT \ln x_w^{Bill}$$

Subtracting: $\mu_w^{Bill} - \mu_w^{Jane} = RT \ln\left(\dfrac{x_w^{Bill}}{x_w^{Jane}}\right) = (8.314)(323.15) \ln\left(\dfrac{0.999602}{0.998319}\right) = 3.451 \text{ J mol}^{-1}$

The chemical potential of the water in Bill's coffee is 3.45 J mol^{-1} greater than the chemical potential of the water in Jane's coffee. The hot water in Bill's coffee can deliver more work than the hot water in Jane's coffee.

Ans: 3.45 J mol^{-1} higher in Bill's coffee

Exercise 18-1

One mole of a non-dissociating nonvolatile solute was dissolved in 750 g of water at 25°C.
(a) Calculate the change in the chemical potential of the water resulting from the dissolution.
(b) What is the change in the vapor pressure of the water above the solution?
The vapor pressure of water at 25°C is 3169 Pa. ☺

Exercise 18-2

A 0.1000 molal aqueous solution of sodium chlorate ($NaClO_3$) freezes at −0.3433°C.
(a) What is the normal boiling point of this solution? (b) Estimate the percentage dissociation of the salt. ☺

Example 18-3

Camphor is a very good solvent for determining the molar mass of a substance because, owing to its very high freezing point constant, only a few milligrams of solute need to be used. A liquid solution of 7.90 mg of phenolphthalein and 129.2 mg of camphor started to crystallize at a temperature of 172.26°C. What is the molar mass of phenolphthalein from these data?

Solution

$$|\Delta T_f| = K_f m; \text{ therefore } m = \frac{|\Delta T_f|}{K_f}$$

From Table 1, the freezing point of pure camphor is 179.8°C, and its freezing point constant is $K_f = 37.7$ K (mol/kg)$^{-1}$.

Therefore the solution molality is $m = \dfrac{|\Delta T_f|}{K_f} = \dfrac{|172.26 - 179.8|}{37.7} = \dfrac{7.54}{37.7} = 0.2000$ mol kg^{-1}

The number of grams of solute in one kg of camphor is $\left(\dfrac{7.9 \times 10^{-3} \text{ g solute}}{129.2 \times 10^{-6} \text{ kg camphor}}\right) = 61.146 \frac{\text{g}}{\text{kg}}$

Since we know that the molality of the solute in the camphor is 0.2000 mol per kg of camphor, then 61.146 g of solute must correspond to 0.2000 moles of solute.

Therefore the molar mass of phenolpthalein is $\dfrac{61.146 \text{ g}}{0.2000 \text{ mol}} = 305.73$ g mol^{-1}

The actual molar mass of phenolphthalein is 318.33 g mol^{-1}.   **Ans:** $\boxed{306 \text{ g mol}^{-1}}$

Exercise 18-3

When 5.65 g of an unknown non-dissociating compound was dissolved in 110.00 g of benzene the solution froze at 4.39°C. What is the molar mass of the compound? ☺

Exercise 18-4

Naphthalene (*N*) and diphenylamine (*D*) form a eutectic[3] mixture, melting at 32.45°C. When 1.268 g of eutectic mixture was added to 18.43 g of naphthalene, the freezing point of the melt was 1.89°C lower than the freezing point of pure naphthalene. What is the percent by mass of the naphthalene in the eutectic mixture? The molar mass of diphenylamine is 165.19 g mol^{-1}. The data for naphthalene can be found in Table 1. ☹

18.4 OSMOTIC PRESSURE

The colligative properties (ΔP_i, ΔT_b, and ΔT_f) all are a consequence of the lowering of the chemical potential of the solvent by the presence of a dissolved solute. It is observed experimentally that all the colligative phenomena most closely obey the ideal equations as the solute concentration approaches zero; that is, when the solute particles become so far apart that they no longer interact with one another to give rise to "nonideal" behavior.

---

[3] A eutectic solid mixture of *A* and *B* melts without changing composition at the lowest melting point of any mixture. Only pure *A*, pure *B*, and the eutectic mixture melt and freeze at a definite temperature without changing composition. Go back to Section 17-16 for a discussion of eutectic mixtures.

Doesn't this sound familiar? It's the same argument that was made for an *ideal gas*, in which the gas more closely obeys $PV = nRT$ as the pressure approaches zero. In 1885 van't Hoff [4] discovered a fourth colligative property—the *osmotic pressure, π*—which can be approximated by an equation having *the same form as the ideal gas law*! Thus,

$$\boxed{\pi V \approx nRT} \qquad \textit{Osmotic Pressure} \qquad \ldots [10]$$

where π is the osmotic pressure of the solution, V is its volume, n is the total number of moles of solute particles in the system, and R and T have their usual meanings. If you are up to it, the derivation of Eqn [10]—and of Eqn [12]—is presented in the Addendum at the end of this chapter.

The driving force is the tendency of solvent to diffuse from a region of high solvent activity (high chemical potential) to a region of lower solvent activity (lower chemical potential). To understand this concept, consider the simple **osmometer** shown below in Fig. 4, which contains a special *semi-permeable membrane* having pores so small that hydrated solute particles are too large to be able to pass through the membrane, but smaller solvent molecules can. We know from experience that there is a spontaneous tendency for the following mixing process to occur:

$$\binom{pure}{solvent} + \binom{concentrated \ solution \ of}{nonvolatile \ solute} \rightarrow \binom{more \ dilute \ solution}{of \ nonvolatile \ solute}$$

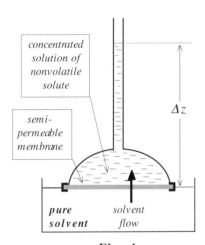

Fig. 4

Because the solute is unable to satisfy this natural mixing tendency by passing through the membrane to mix with the pure solvent to form a more dilute solution, instead, the solvent—which *can* pass through the membrane—does so and thereby *dilutes* the concentrated solution.

Thus, there is a transfer of solvent from the pure solvent compartment to the compartment containing the concentrated solution of the nonvolatile solute. This process is called **osmosis** and the net result is that a more dilute solution is formed. Osmosis can be viewed as just a spontaneous mixing process for which $\Delta G_{T,P} < 0$.

As the solvent enters the more concentrated solution, the liquid level rises up the tube until the hydrostatic downward pressure exerted by the column of liquid just balances any further tendency for pure solvent to flow through the membrane. The height Δz to which the liquid rises is proportional to the hydrostatic pressure exerted by the column of liquid:

$$\Delta P = \rho g \Delta z \qquad \ldots [11]$$

This equilibrium hydrostatic pressure is called π, the *osmotic pressure of the solution*. The osmotic pressure is a function of the solvent employed, the solute concentration, and the temperature, but is *independent* of the solute used; therefore π is a true *colligative property*.

---

[4] Jacobus Henricus van't Hoff (1852-1911), Dutch physicist.

Once we have measured the osmotic pressure of a solution containing W_A grams of dissolved non-volatile solute A, we then can use the equation[5]

$$\pi v_w = -RT\ln\,(1 - x_A) \qquad \ldots [12]$$

or its more approximate form, Eqn [10], to determine the molar mass M_A of the dissolved nonvolatile solute, since $n_A = W_A/M_A$. Fortunately, osmotic pressures can be measured very accurately down to pressures of about 0.001 bar, so it is possible to use quite dilute solutions for our measurements; this is very fortunate, because the equations hold best for dilute solutions.

A key component for osmotic pressure measurements is the special semi-permeable membrane. The kinds of membranes used include certain plastics, parchments, proteins, cellophane, etc. Because these membranes are especially suitable for retaining very *large* molecules (such as dissolved nylon or rubber), osmotic pressure measurements are commonly used in industry to determine the molar masses of dissolved polymers.

Osmosis also has special biological significance. Blood, for example, must be maintained at the same osmotic pressure as the fluid inside the cells, otherwise fluid will transfer into the cells by osmosis and cause them to swell and rupture. For this reason, when a medication is injected into your bloodstream, the drug must be dissolved in a solution of NaCl, glucose, etc. of the same osmotic pressure as your blood. Different solutions having the same osmotic pressure are called **isotonic** solutions.

Osmosis also occurs in nature. For example, ground water passes through semi-permeable membranes in the roots of plants to generate osmotic pressure inside the plant, which, along with capillary action, helps the water to rise up the stem to the leaves.

A process of great technological importance is so-called **Reverse Osmosis** (*RO*). If an *external* pressure *greater* than its osmotic pressure is applied to a solution, solvent can be forced to move in the *opposite* direction, *out of* solution and *into* a reservoir of pure solvent, thereby *concentrating* the solution. This is used to concentrate juices such as orange juice. Reverse osmosis also is used to obtain pure water from salt water (desalination), which is very important for arid countries close to an ocean but having no inland source of fresh water. *RO* membranes often are based on cellulose acetate, which lets H_2O molecules pass through but not more bulky hydrated salt ions such as $Na(H_2O)_n^+$ and $Cl(H_2O)_{n'}^-$.

Notes:

(i) Osmosis will occur with liquid rising up the tube so long as there is a *difference* $\Delta\pi$ in the osmotic pressure between the two liquids. It is not necessary that one of the liquids be *pure* solvent, just that one of the solutions has a greater osmotic pressure than the other.

(ii) As we have seen, colligative behavior is determined by the *total* concentration of all the solute particles present in the solution. For example, a 0.001 m aqueous solution of $CaCl_2$ dissociates to form 0.001 m Ca^{2+} ion and 0.002 m Cl^- ion in solution:

$$CaCl_2 \rightarrow Ca^{2+} + 2Cl^-$$

---

[5] Eqn [12] is more rigorous than Eqn [10]. Both are derived in the Addendum at the end of this chapter.

Accordingly, in this specific case the effective molality that must be used in the equations for colligative behavior will be[6]

$$m_{effective} = 0.001 + 0.002 = 0.003 \text{ m}$$

Example 18-4

The density of liquid mercury at 0°C is 13.60 g cm^{-3}. What is the osmotic pressure at 0°C of a solution made by dissolving 1.22 g of cadmium (Cd = 112.41 g mol^{-1}) in one kilogram of mercury (Hg = 200.61 g mol^{-1})?

Solution

Method I: Using $\pi = C_i RT$, where i is the solute (Cd):

Moles of Cd: $n_{Cd} = \dfrac{1.22 \text{ g}}{112.41 \text{ g}/\text{mol}} = 0.010853 \text{ mol}$

Volume of system: $V \approx V_{Hg} = \dfrac{1000 \text{ g}}{13.60 \text{ g}/\text{cm}^3} = 73.529 \text{ cm}^3 = 73.529 \times 10^{-6} \text{ m}^3$

Therefore: $\pi = C_{Cd} RT = \left(\dfrac{n_{Cd}}{V}\right) RT$

$$= \left(\dfrac{0.010853 \text{ mol}}{73.529 \times 10^{-6} \text{ m}^3}\right)(8.314)(273.15)$$

$$= 335\ 199 \text{ Pa} = 3.352 \text{ bar}$$

Ans: $\boxed{3.35 \text{ bar}}$

Method II: Using $\pi v_W = -RT \ln(1 - x_i)$, where W is the solvent (Hg) and i is the solute (Cd):

Moles of Hg: $n_{Hg} = \dfrac{1000 \text{ g}}{200.61 \text{ g}/\text{mol}} = 4.9848 \text{ mol}$

Molar volume of Hg: $v_{Hg} = \dfrac{200.61 \text{ g}/\text{mol}}{13.60 \text{ g}/\text{cm}^3} = 14.7507 \text{ cm}^3 \text{ mol}^{-1} = 14.7507 \times 10^{-6} \text{ m}^3 \text{ mol}^{-1}$

Mole fraction of Cd: $x_{Cd} = \dfrac{n_{Cd}}{n_{Cd} + n_{Hg}} = \dfrac{0.010855}{0.010855 + 4.98480} = 0.002172$

---

[6] At higher concentrations the salt will be less than 100% dissociated into ions, on account of electrostatic attractive forces causing some of the ions to stick together, forming "ion pairs" or even "ion triplets," thereby lowering $m_{effective}$. Again, for the most accurate work the effect on the solvent of ionic hydration must be taken into account.

$$\pi v_w = -RT \, ln \, (1 - x_i)$$

$$\pi = \frac{-RT \, ln(1 - x_{Cd})}{v_{Hg}} = \frac{-(8.314)(273.15) \, ln(1 - 0.002172)}{14.7507 \times 10^{-6}} = 334 \, 758 \text{ Pa} = 3.348 \text{ bar}$$

Ans: | 3.35 bar |

In this example, because the solution is very dilute, both methods give essentially the same result.

Exercise 18-5

The tube in an osmometer contained dilute solutions of the industrial polymer, Buna S-3, dissolved in the organic solvent toluene at a temperature of 30.2°C. The lower end of the tube was dipped into pure toluene. When equilibrium had been reached the liquid level in the inner tube was Δz cm higher than the outer liquid. Using the experimental data provided, estimate the molar mass of Buna S-3 by noting that the osmotic pressure equation becomes increasingly accurate as the solute concentration approaches infinite dilution.

Concentration (mg Buna S-3 in 100 mL)	Δz (cm)
730	3.57
547.5	2.26
365	1.17
182.5	0.43

The acceleration due to gravity has a value of 9.806 m s^{-2}; the density of toluene at 30.2°C is 0.8566 g cm^{-3}. The dilute solutions of Buna S-3 may be assumed to have the same density as pure toluene. ☹

KEY POINTS FOR CHAPTER EIGHTEEN

1. When a solute is dissolved in a solvent to form a solution, the physical properties of the solution—such as its boiling point, freezing point, and vapor pressure—are different from those of the pure solvent. These changes in the physical properties of solutions, and the extent of the changes, depend (approximately) only on the number of solute particles dissolved in the solvent, and not on the actual identity of the solute. Such properties are called **colligative properties** and are related to each other. Thus, once one—such as the increase in the boiling point of the solution—has been measured, all the others—such as the decrease in the freezing point or the lowering of the vapor pressure—can be calculated.

   All colligative behavior most closely obeys the ideal equations as the concentration of the solute approaches zero; i.e., when the solute particles are so far apart that there is negligible interaction among them to give rise to "nonideal" behavior.

2. The **lowering of the vapor pressure** (vapor pressure depression) by a non-volatile solute A is given by

$$\frac{\Delta P}{P_W^\bullet} = x_A$$

where $P_W^\bullet$ is the vapor pressure of the pure solvent, P_W is the vapor pressure of the solution, $\Delta P = P_W^\bullet - P_W$ is the lowering of the vapor pressure, and x_A is the mole fraction of the solute in the solution.

Note: If the solute dissociates into particles, it is the *total number of particles* that must be used to calculate the "effective" mole fraction of the solute.

3. The **boiling point elevation** ΔT_b and **freezing point depression** $|\Delta T_f|$ are given, respectively, by $\quad\quad \Delta T_b = K_b m_i \quad$ and $\quad |\Delta T_f| = K_f m_i$

    where m_i is the "effective" molality of the solute, K_b is the boiling point elevation constant (the *ebullioscopic constant*) and K_f is the freezing point depression constant (the *cryoscopic constant*).

4. The origin of all the colligative properties is *the lowering of the chemical potential of the solvent by the presence of a solute*.

    For nonvolatile solutes (i) the solute doesn't contribute to the vapor above the solution, and (ii) the solute doesn't dissolve in *solid* solvent; i.e., when the solution starts to freeze, the solid formed usually consists only of *pure solid solvent*.

5. **Osmotic pressure**: There is a natural tendency (driving force) for solute to diffuse from a region of high solute activity (high chemical potential) to a region of lower solute activity (lower chemical potential). If a membrane containing very small pores of molecular dimensions separates a solution and a pure solvent, there will be a natural tendency for solute to diffuse through the membrane into the pure solvent, thereby lowering the concentration (activity) of solute in the solution and increasing the concentration (activity) of the solute in the solvent. If, however, the pores in the membrane are too small for the *solute* molecules to pass through, but are big enough for the *solvent* molecules to pass through, then the *solvent* will spontaneously flow from the pure solvent, through the membrane, and *dilute* the solution until there is no longer any driving force. This flow of solvent through the membrane and into the solution is known as **osmosis**, and can lift the solution up a column.

    As the solvent enters the more concentrated solution, the liquid level can rise up a tube until the hydrostatic downward pressure exerted by the column of liquid just balances any further tendency for pure solvent to flow through the membrane. The height Δz to which the liquid rises is proportional to the hydrostatic pressure exerted by the column of liquid: $\quad\quad \Delta P = \rho g \Delta z$

    This equilibrium hydrostatic pressure is called π, the **osmotic pressure of the solution**. The osmotic pressure is a fourth colligative property, and is a function of the solvent employed, the solute concentration, and the temperature, but is *independent* of the solute used.

    The osmotic pressure π is given by

    $$\pi v_W = - RT \ln(1 - x_i)$$

    where v_W is the molar volume of the pure solvent and x_i is the mole fraction of the solute in the solution, taking into consideration *all* the non-solvent particles present. This equation can be manipulated into the more approximate form

$$\pi V = nRT \quad \text{or} \quad \pi \approx \frac{n}{V} RT \approx CRT$$

where C is the (effective) concentration of solute in the solution.

6. A process of great industrial importance is **reverse osmosis** (*RO*) in which an *external* pressure *greater* than the osmotic pressure of a solution is used to force solvent to move in the *opposite* direction, *out of* the solution and *into* a reservoir of pure solvent, thereby *concentrating* the solution and producing more pure solvent.

7. **Note**:

(i) Osmosis will occur with liquid rising up the tube so long as there is a *difference* $\Delta\pi$ in the osmotic pressure between the two liquids. It is not necessary that one of the liquids be *pure* solvent, just that one of the solutions have a greater osmotic pressure than the other.

(ii) Colligative behavior is determined by the *total* concentration of all the particles present in the solution. Therefore, if a 0.001m aqueous Na_2SO_4 solution dissociates 100% into 0.002 m Na^+ ions and 0.001 m SO_4^{2-} ions, the "effective" molality of the solution would be 0.003 m. Similarly, if the salt only *partly* dissociates, the "effective" molality would be calculated from the sum of the molalities of the Na^+ ions, the SO_4^{2-} ions, and the undissociated Na_2SO_4 "molecules."

ADDENDUM: OSMOTIC PRESSURE DERIVATIONS

Referring to Fig. A-1, pure water at point *1* is at pressure $P = P°$. At equilibrium, the height Δz of the solution exerts a pressure that pushes water through the membrane *out* of the solution; this pressure balances the osmotic pressure of the solution, which tends to push pure water through the membrane and *into* the solution. Therefore,

$$\pi = \rho g \Delta z$$

where ρ is the density of the solution.

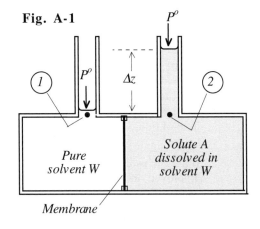

Fig. A-1

Let the solute be A, and the solvent be W. For equilibrium, the chemical potential μ_w of the pure water at point *1* at pressure $P°$ must equal the chemical potential μ_w' of the water in solution at point *2* at pressure $P° + \pi$.

Since

$$\mu_w = \mu_w^* + RT \ln x_w$$

therefore,
$$\mu_w^* + RT \, ln \, (1) = (\mu_W^*)' + RT \, ln \, x_W$$

where μ_w^* is the chemical potential of pure water at $P = P^\circ$, $(\mu_W^*)'$ is the chemical potential of pure water at $P = P^\circ + \pi$, and x_W is the mole fraction of the water in the solution.

Therefore:
$$\mu_w^* = (\mu_W^*)' + RT \, ln \, x_W$$

$$(\mu_W^*)' - \mu_w^* = -RT \, ln \, x_W \qquad \ldots \text{[A-1]}$$

We saw earlier that at constant T and P the total Gibbs free energy is just the sum of the partial molar free energies contributed by each component:

i.e.,
$$G_{T,P} = \sum n_i (G_m)_i = \sum n_i \mu_i$$

But, for a pure substance such as pure water, there is only one component, therefore

$$G_{T,P} = n_W \mu_W \quad \text{and} \quad \mu_W = \frac{G}{n} = g_W$$

where g_W is the *molar* Gibbs free energy of the solvent.

Therefore, for pure water,

$$(\mu_W^*)' - \mu_w^* = g_W' - g_W = \int\limits_{P=P^o}^{P=P^o+\pi} dg_W \qquad \ldots \text{[A-2]}$$

For 1 mole of pure water we showed that $\qquad dg = vdP - sdT$

Therefore, at constant T: $\qquad\qquad\qquad\qquad dg = vdP$

where v is the molar volume of the water.

Therefore,
$$\int\limits_{P^o}^{P^o+\pi} dg_W = \int\limits_{P^o}^{P^o+\pi} vdP$$

Since water is an incompressible fluid, it is independent of pressure, so we can write

$$\int\limits_{P^o}^{P^o+\pi} vdP = v \int\limits_{P^o}^{P^o+\pi} dP$$

$$= v\Big[(P^o + \pi) - (P^o)\Big]$$

$$= v\pi$$

Therefore, Eqn [A-2] becomes $\qquad (\mu_W^*)' - \mu_w^* = \pi v_W$

and Eqn [A-1] becomes $\qquad \pi v_W = -RT \ln x_W \qquad$ . . . [A-3]

But $\qquad\qquad\qquad x_W = 1 - x_A \qquad$ . . . [A-4]

Substituting Eqn [A-4] into Eqn [A-3] gives

$$\boxed{\pi v_W = -RT \ln (1 - x_A)} \qquad \text{. . . [A-5]}$$

Eqn [A-5] is the basic equation used to evaluate the osmotic pressure of a solution. For dilute solutions, it can be simplified as follows:

For $-1 < x < 1$, the Taylor's series expansion for $-\ln (1 - x)$ is:

$$-\ln(1 - x) = x + \frac{x^2}{2} + \frac{x^3}{3} + \frac{x^4}{4} + \ldots$$

For very dilute solutions, $x_A << 1$ and the expansion reduces to

$$-\ln (1 - x_A) \approx x_A$$

and Eqn [A-5] reduces to $\qquad \pi v_W \approx RT \, x_A \qquad$ . . . [A-6]

Since $n_W >> n_A$, $\qquad x_A = \dfrac{n_A}{n_A + n_W} \approx \dfrac{n_A}{n_W}$

And Eqn [A-6] further reduces to $\qquad \pi v_W \approx RT \left(\dfrac{n_A}{n_W} \right)$

Cross-multiplying: $\qquad \pi n_W v_W = n_A RT \qquad$ . . . [A-7]

But $\qquad\qquad n_W v_W = V_W \approx V_{total} \qquad$ . . . [A-8]

Putting Eqn [A-8] into Eqn [A-7]: $\qquad \boxed{\pi V \approx n_A RT} \qquad$ . . . [A-9]

or $\qquad\qquad \pi \approx \left(\dfrac{n_A}{V} \right) RT = C_A RT$

Therefore, $\qquad\qquad \boxed{\pi \approx C_A RT} \qquad$ . . . [A-10]

where π is the osmotic pressure in pascal and C_A is the concentration of the solute in mol m^{-3}.

Eqn [A-5] is more accurate than Eqns [A-9] and [A-10]. The latter two equations hold only for very dilute solutions.

PROBLEMS

1. A solution of 4.71% by weight sulfur in naphthalene ($C_{10}H_8$) was found to freeze at exactly 79.20°C. The freezing point of pure naphthalene is 80.55°C. Calculate the molar mass of sulfur. The cryoscopic constant for naphthalene is $K_f = 6.98$ K (mol kg^{-1})$^{-1}$. ☺

2. A 5.00% by mass aqueous solution of sucrose ($C_{12}H_{22}O_{11}$, 342.30 g mol^{-1}) is cooled to −2.00°C. Estimate the percentage of the water in the solution that is recovered as ice by this process. ☹

3. Calculate the ideal osmotic pressure (in bar) of an aqueous solution at 40°C containing 5.00 g of sodium pyrophosphate decahydrate ($Na_4P_2O_7 \cdot 10H_2O$) dissolved in one liter of solution. For this problem, you may assume that sodium pyrophosphate dissociates completely according to

$$Na_4P_2O_7 \cdot 10H_2O \xrightarrow{100\%} 4Na^+_{(aq)} + P_2O^{4-}_{7(aq)} + 10H_2O$$

Neglect hydration of the ions in solution. At 40°C the densities of pure water and of the solution are 992.2 g L^{-1} and 995.2 g L^{-1}, respectively. ☹

4. When cells of the skeletal vacuole of a frog were placed in a series of aqueous NaCl solutions of different concentrations at 25°C it was observed using a microscope that they remained unchanged in 0.70% by weight NaCl solution, shrank in more concentrated solutions, and swelled in more dilute solutions. The 0.70% salt solution freezes at −0.406°C. What is the osmotic pressure of the cell protoplasm at 25°C? The cryoscopic constant for water is $K_f = 1.86$ K (mol kg^{-1})$^{-1}$. Each Na^+ ion is hydrated with 4 water molecules, while each Cl^- ion is hydrated with 2 water molecules. At 25°C the densities of pure water and of 0.70% NaCl solution are 997.07 kg m^{-3} and 1002.0 kg m^{-3}, respectively. Molar masses: NaCl = 58.44 g mol^{-1}; H_2O = 18.015 g mol^{-1}. ☠

5. One industrial method for removing the salt from sea water is reverse osmosis, in which the natural osmotic flow of water through a semipermeable membrane separating solutions of different concentrations is reversed by applying pressure to the concentrated solution (the sea water), forcing what is essentially salt-free water to flow through the membrane. The minimum pressure required to achieve this is equal to the osmotic pressure of the sea water. Calculate the minimum applied pressure (in MPa) that is required for the desalination of sea water at 25°C using reverse osmosis.

   At 25°C, sea water may be assumed to be an aqueous solution of NaCl having a density of 1020 kg m^{-3} and a chloride content of 18 200 ppm (parts per million) Cl, based on the total mass of the solution. At 25°C the NaCl in sea water is 80% dissociated into ions. Use the appropriate data concerning hydration, etc. provided in question 4. ☹

6. One teaspoon of sugar (7.8 g) is dissolved in a glass of water (250 g H_2O) at 25°C. (a) What is the equilibrium vapor pressure above the solution? (b) What is the normal boiling point of the solution? (c) What is the normal freezing point of the solution? (d) What is the osmotic pressure of the solution? Molar masses: $C_{12}H_{22}O_{11}$ = 342.30; H_2O = 18.02. The vapor pressure of water at 25°C is 3167 Pa. For water, $K_b = 0.513$ K (mol/kg)$^{-1}$ and $K_f = 1.86$ K (mol/kg)$^{-1}$. The density of water at 25°C is 997 kg m^{-3}. ☺

7.   An aqueous solution containing 3.00% by mass sucrose ($C_{12}H_{22}O_{11}$, 342.30 g mol^{-1}) is cooled to $-1.00°C$. Estimate the percentage of the water in the solution that is recovered as ice by this process. The freezing point constant for water is $K_f = 1.86$ K (mol/kg)$^{-1}$. ☹

8.   Lead and tin form a series of low melting alloys called solders. General purpose solder is composed of 67% by weight Sn and 33% by weight Pb, and has a melting point of 183°C. What is the freezing point constant of tin if pure tin melts at 231.8°C? Relative molar masses: Pb = 207.2   Sn = 118.71 ☺

9.   For an aqueous solution containing 1.000 percent by mass common sugar (sucrose, $C_{12}H_{22}O_{11}$), calculate:

(a) the freezing point (b) the normal boiling point (c) the osmotic pressure at 20°C

(d) the vapor pressure at 100°C (e) the vapor pressure at 20°C.

The solution density at 20°C is 1.0021 g cm^{-3}. The density of water at 20°C is 0.99821 g cm^{-3}. At 20°C the vapor pressure of water is 2338.8 Pa. ☺

10. The vapor pressure of a dilute aqueous solution of a non-dissociating solid is 23.45 Torr at 25°C, whereas the vapor pressure of pure water at this temperature is 23.76 Torr.
(a) Calculate the molal concentration of solute, and (b) predict the boiling point of the solution. ☺

11. When 2.00 g of a non-volatile hydrocarbon containing 94.4% C is dissolved in 100 g of benzene at 20°C, the vapor pressure of the benzene drops from 74.66 Torr to 74.01 Torr. What is the accurate molecular mass of the hydrocarbon? Report the answer to three decimal places. $C_6H_6 = 78.113$ g mol^{-1}. ☺

12. At 100°C the vapor pressure of a solution made by the addition of 11.94 g of a substance to 100 g of water is 98.78 kPa. What is the molecular weight of the substance? ☺

13. When 6.00 g of a non-volatile non-dissociating substance is dissolved in 100 g of water, the vapor pressure of the resulting solution at 25°C is 23.332 Torr. What is the relative molar mass of the substance? For water at 25°C, $P_W^• = 23.756$ Torr. ☺

14. When 2.00 g of sulfur was dissolved in 100 g of carbon disulfide (CS_2) the resulting solution had a vapor pressure at 25°C of 848.9 Torr. What is the molecular formula for sulfur dissolved in carbon disulfide? At 25°C the vapor pressures of solid sulfur and pure liquid carbon disulfide are 3×10^{-4} Torr and 854 Torr, respectively.

Relative molar masses: $CS_2 = 76.145$ S = 32.067 ☺

15. Ehylene glycol—$C_2H_4(OH)_2$—and methanol—CH_3OH—both have been used as antifreezes in automobile cooling systems. How many grams of (a) methanol, and (b) ethylene glycol, must be added to 4.00 L of water at 25°C to form an antifreeze solution that will just prevent the formation of ice at $-10°C$? The density of water at 25°C may be taken as 0.997 g cm^{-3}. ☺

16. A 0.100 molal aqueous solution of $CuSO_4$ freezes at $-0.210°C$.

    (a) Estimate the percent dissociation of the salt at this temperature.

    (b) Estimate the dissociation constant of the salt at this temperature. State any assumptions made. ☹

17. A 0.010 molal aqueous solution of HCl freezes at $-0.036°C$. Stating any assumptions made, estimate the percent dissociation of the acid at this temperature. ☺

18. The crystalline material in a bottle having no label is known to be either NaCl, $CaCl_2$ or $AlCl_3$. When a small amount of this substance is dissolved in 150 g of benzene, the freezing point is lowered by $0.60°C$. When the same amount is dissolved in 100 g of water the freezing point is lowered by $0.998°C$. What is the substance? State any assumptions made. ☹

19. At 25°C the osmotic pressure of 100 cm^3 of an aqueous solution containing 1.346 g of the protein β-lactoglobulin is found to be 968.8 Pa. Estimate the molecular weight of this protein. At 25°C the density of water is 0.997 g cm^{-3}. The molar mass of water is 18.015 g mol^{-1}. ☺

20. Blood at 30°C has an osmotic pressure of 7.0 atm. What is the molarity of an aqueous NaCl solution that has the same osmotic pressure if the dissolved salt is 90% dissociated into ions? ☺

21. A solution containing 40.5 g L^{-1} of a non-dissociating substance has an osmotic pressure of 3.00 atm at 10°C.

    (a) What will be the osmotic pressure of a 27.3 g L^{-1} solution of the same substance at 40°C?

    (b) What is the molecular weight of the substance? ☺

22. If an m molal aqueous solution of NaCl has an osmotic pressure of 2.00 atm at 25°C, what is ΔG for the following process?

$$\begin{pmatrix} 1 \ mol \ pure \ H_2O \\ 25°C, \ 1bar \end{pmatrix} \rightarrow \begin{pmatrix} 1 \ mol \ H_2O \ in \ m \ molal \\ NaCl \ solution, \ 25°C, \ 1bar \end{pmatrix}$$

The molar mass of water is 18.015 g mol^{-1}, and its density at 25°C is 997 kg m^{-3}. ☹

23. It has been proposed that fresh water can be recovered from seawater by lowering into the ocean pipes that are capped at their lower ends with special membranes that are permeable only to water molecules. What is the minimum depth to which the pipes must be lowered in order for pure water to begin to pass through the membranes? The average temperature of the seawater may be taken as 10°C; at this temperature the density of seawater is 1027 kg m^{-3}. At 25°C the vapor pressure of pure water is 3169.0 Pa; at the same temperature, the partial pressure of water above seawater is 3105.9 Pa. The molar mass of water is 18.015 g mol^{-1}, and the density of pure water at 10°C is 999.70 kg m^{-3}. ☹

24. What is the freezing point of a solution consisting of 2.7 g of glucose (molar mass = 180 g mol^{-1}) and 250 g of water? If the solution is held at $-0.3°C$, how much ice will eventually form? For water, $K_f = 1.86$ K $(mol/kg)^{-1}$. ☹

25. The freezing point depression of a solution containing 1.00 g of a solute in 100 g of water is 0.195°C. The freezing point depression of a solution containing 3.00 g of the same solute in 100 g of water is 0.570°C. Stating any assumptions made, calculate the molar mass of the dissolved substance. For water, $K_f = 1.86$ K $(mol/kg)^{-1}$. ☹

NINETEEN

IONIC EQUILIBRIUM

19.1 IONIC EQUILIBRIUM

Many of the most important applications of chemical equilibrium are to be found in aqueous electrolyte solutions. Some of these applications include biological reactions, electrochemical processes, corrosion, and many of the instrumental methods used for chemical analysis. In these systems, dissolved salts, acids, and bases dissociate—either partly or completely—to form ions in solution.

Electrolyte solutions require special treatment because ionic species in solution—owing to their electric charge—interact much more strongly with each other and with solvent water than do non-dissociating dissolved substances such as sugar, oxygen, or ethylene glycol. As a consequence of this strong interaction, the deviations between concentration and activity are much more pronounced than for solutions of non-electrolytes, and must be taken into account for accurate calculations. Furthermore, the electrical properties of these systems give rise to many interesting phenomena that are not present in solutions of non-electrolytes.

19.2 ACTIVITIES OF DISSOLVED SPECIES IN SOLUTION

We showed earlier that, when the reaction

$$aA + bB \rightarrow cC + dD$$

is at equilibrium, the reaction free energy is zero, leading to the expression

$$\Delta G^\circ = -RT \ln K$$

where K is the thermodynamic equilibrium constant based on *relative activities*.

Thus,
$$K = \left(\frac{a_C^c \cdot a_D^d}{a_A^a \cdot a_B^b} \right)_{eqm}$$

where a_i is the *relative activity* of species i. We saw that for *gases*, the relative activity could be approximated by the relative pressure: $a_i \approx P_i/P^o$; for *pure solids and liquids*, the relative activity is assigned a value of unity: $a_{solid} = a_{liq} = 1$; the relative activity of an *impure* solid or liquid or of the *solvent* in a liquid solution is approximated by its mole fraction purity: $a_i \approx x_i$.[1]

---

[1] As we saw in Chapter 17, the relative activity of the solvent W is more accurately given by $a_w = P_W/P_W^\bullet$.

We also stated that for a *non-electrolyte solute* dissolved in a liquid solvent the relative activity of the solute in the solution was given by $a_i = \gamma_i m_i$, where m_i is the *relative molality* of the solute in the solution, and γ_i is the *molal activity coefficient* for the solute.

If the solution is dilute,[2] it usually is permissible to assume that $\gamma_i \approx 1$, so that the relative activity of the solute is just approximated by its molality.

Table 1.   *Comparison of Molarity vs. Molality for Common Aqueous Solutions*

M	m (mol/kg)		
(mol/L)	*KOH* (15°C)	*NaCl* (25°C)	*Acetic Acid* (20°C)
0.100	0.100	0.1005	0.101
0.500	0.502	0.506	0.514
1.00	1.01	1.02	1.06
5.00	5.33	5.43	6.79

Also, provided the solution is not too concentrated and the solvent is *water*, the numerical value of the *molality* of the solution is almost the same as that of its *molarity* (see Table 1). Therefore, for very dilute solutions of *nonelectrolyte* solutes, the relative activity of the solute can be approximated by its *concentration*, in mol L⁻¹:

$$a_i \approx c_i$$

Exercise 19-1

Using the data in Table 1, calculate the density of 5.00 molar acetic acid solution at 20°C. Report the answer in g L⁻¹. The molar mass of CH_3COOH is 60.05 g mol⁻¹. ☹

19.3 ACTIVITIES AND ACTIVITY COEFFICIENTS OF ELECTROLYTES IN SOLUTION

The situation is more complicated when we have to deal with solutions of *electrolytes*; now, we are *not* able to assume that the (relative) activity of an ion in solution is just the same as its molarity or molality.

Any electrolyte can be represented by the general formula $M_{v_+} A_{v_-}$. The process in which a dissolved electrolyte completely dissociates in solution to form ions can be represented by

$$M_{v_+} A_{v_-} \rightarrow v_+ M^{z+} + v_- A^{z-} \qquad \dots [1]$$

Thus, when one mole of electrolyte completely dissociates it forms v_+ moles of M^{z+} cations, each of valence $z+$, and v_- moles of A^{z-} anions, each of valence $z-$. v_+ and v_- are called the **stoichiometric coefficients** for the cation and anion, respectively.

The **molal relative activity** of an m–molal solution of an electrolyte $M_{v_+} A_{v_-}$ is *defined* as

---

[2] For *electrolyte* solutions, "dilute" means < 0.001 molal; for solutions of *non-electrolytes* "dilute" means < 0.1 molal.

$$a_{MA} \equiv \left(v_+^{v_+} \cdot v_-^{v_-} \right) \cdot \gamma_\pm^{\,v} \cdot m^v \qquad \begin{array}{l} \textit{Molal Activity of} \\ \textit{an Electrolyte} \end{array} \quad \dots [2]$$

where
$$v \equiv v_+ + v_- \qquad \qquad \dots [3]$$

and $\gamma_\pm$ is the **stoichiometric mean molal ionic activity coefficient** for the electrolyte. The relative molal activity of a *single* ionic species in solution is given by

$$a_i = \gamma_i m_i \qquad \textit{Single Ion Activity} \qquad \dots [4]$$

where γ_i is the *single ion molal ionic activity coefficient* and m_i is the relative molality of the ionic species in solution. The *stoichiometric mean molal ionic activity coefficient* $\gamma_\pm$ for the electrolyte is related to the *individual* molal activity coefficients for the ions through the formula

$$\gamma_\pm = \left(\gamma_+^{v_+} \cdot \gamma_-^{v_-} \right)^{1/v} \qquad \begin{array}{l} \textit{Mean Activity} \\ \textit{Coefficient} \end{array} \qquad \dots [5]$$

Values of $\gamma_\pm$ are obtained from tables of data, and are functions of concentration, temperature, and the nature of the individual ions involved.

Example 19-1

Tables of thermodynamic data list the value of the stoichiometric mean molal ionic activity coefficient for 2.00 molal aqueous $CaCl_2$ solution at 25°C as 0.792. What is the relative molal activity of the $CaCl_2$ dissolved in this solution?

Solution:

The salt dissociates according to $\qquad CaCl_2 \rightarrow Ca^{2+} + 2Cl^-$

Thus, when one mole of $CaCl_2$ completely dissociates into ions, it produces one mole of Ca^{2+} cations and two moles of Cl^- anions.

Therefore, $\qquad\qquad\qquad\qquad v_+ = 1 \quad v_- = 2$

and $\qquad\qquad\qquad\qquad v = v_+ + v_- = 1 + 2 = 3$

and, from Eqn [2] the relative molal activity of a 2.00 molal solution of $CaCl_2$ is

$$a_{MA}^{(m)} = \left(v_+^{v_+} \cdot v_-^{v_-} \right) \cdot \gamma_\pm^{\,v} \cdot m^v$$
$$= (1^1 \cdot 2^2)(0.792)^3 (2.00)^3 = 15.90$$

Ans: $\boxed{15.90}$

The superscript *(m)* in $a_{MA}^{(m)}$ indicates that the concentrations are expressed on the *molal* scale.

If you were to estimate the molal activity of 2.00 molal $CaCl_2$ as 2.00 you would be seriously mistaken!

Exercise 19-2

For 1.00 molal aqueous $Al_2(SO_4)_3$ solution at 25°C, $\gamma_\pm = 0.0175$. What is the molal relative activity of the $Al_2(SO_4)_3$ in this solution? ☺

Under certain conditions (when the solution is dilute, i.e., has an ionic strength of not more than about 0.001) γ_i-values for individual ions can be calculated from the theory of electrolyte solutions,[3] using the **Debye–Hückel equation**. The Debye-Hückel equation is derived by taking into account the electrostatic interactions that occur in solution between the various charged ionic species. In effect, the positive and negative ions in an electrolyte solution attract each other, which lowers their potential energy. This means that less Gibbs free energy is available for carrying out spontaneous processes than would be the case in the absence of electrostatic interactions. The net result is that the activity (the "effective" concentration) of each ionic species is *lowered*. To account for this lowering, the concentration of the ion must be multiplied by a correction factor that we call the *activity coefficient*.

The Debye-Hückel equation permits a calculation of the activity coefficients of *single* ionic species in dilute solution.[4] For dilute aqueous solutions at 25°C, the equation takes the form

$$log\gamma_i = -0.5108\,z_i^2\sqrt{I}$$

| Debye-Hückel Eqn for Aqueous Solutions at 25°C | . . . [6]

where z_i is the valence of the ion, and I is the **ionic strength** of the solution, defined[5] by

$$I \equiv \frac{1}{2}\sum_i c_i z_i^2$$

Ionic Strength Defined . . . [7]

In Eqn [7], c_i is the concentration of each ionic species i, in mol L^{-1}, and z_i is its valence.[6] Note that the ionic strength has units of mol L^{-1}.

---

[3] The derivation of this equation is beyond the level of this course. Any textbook on electrochemistry will give you more details. Note that the number "0.5108" in Eqn [6] and elsewhere is the value when water is the solvent at 25°C; it has different values at different temperatures and for different solvents. See Problem 19-8 at the end of this chapter.

[4] Single ion activity coefficients cannot be measured experimentally in the lab, only the activity coefficients for complete salts. The difficulty arises because we can't add only a single type of ion to a solution. So, if we want to measure the activity coefficient for the Na^+ ion, we have to add it to the solution as a sodium *salt*, such as NaCl. There is no jar of Na^+ ions sitting on the shelf! So we actually are adding *two* types of ion to the solution (Na^+ and Cl^-); accordingly, it's impossible to measure the properties of isolated single ions. Single ion properties *can*, however, be *calculated—if* we have a theory. That's why we develop theories such as the Debye-Hückel theory.

[5] Some authors define the ionic strength in terms of molalities instead of molarities. Strictly speaking, the form of the Debye-Hückel equation given in Eqn [6]—i.e., with the factor *0.5108*—should use the ionic strength defined in terms of molalities. However, within the concentration range in which the Debye-Hückel equation is valid there is no significant difference between using molalities or using molarities. See Problem 19-7 at the end of this chapter.

The Debye-Hückel equation also can be manipulated to yield an expression for $\gamma_\pm$, the mean molal ionic activity coefficient,[7] defined earlier by Eqn [5] as

$$\gamma_\pm = \left(\gamma_+^{\nu_+} \cdot \gamma_-^{\nu_-}\right)^{1/\nu} \qquad \ldots [5]$$

The derivation proceeds as follows:

First we take the logarithm to the *base 10* of each side of Eqn [5]:

$$log\,\gamma_\pm = \frac{1}{\nu}\left[\nu_+ log\,\gamma_+ + \nu_- log\,\gamma_-\right]$$

Substituting the Debye-Hückel expression for each of the single ions:

$$log\,\gamma_\pm = \frac{1}{\nu}\left[\nu_+\left(-0.5108\,z_+^2\,\sqrt{I}\right) + \nu_-\left(-0.5108\,z_-^2\,\sqrt{I}\right)\right]$$

$$= -0.5108\left[\frac{\nu_+ z_+^2 + \nu_- z_-^2}{\nu}\right]\sqrt{I} \qquad \ldots [8]$$

Since a dissolved salt has no net charge, when it dissociates, the total amount of positive charge in solution must equal the total amount of negative charge. This is the so-called "condition of electroneutrality," and is just a charge balance:

Thus,[8] $$\nu_+ z_+ + \nu_- z_- = 0$$

i.e., $$\nu_+ z_+ = -\nu_- z_- = \nu_-\left|z_-\right| \qquad \ldots [9]$$

Expanding the numerator of the bracketed term in Eqn [8]:

$$\nu_+ z_+^2 + \nu_- z_-^2 = \nu_+ z_+ z_+ + \nu_- z_- z_-$$

$$= \left(\nu_+ z_+\right)z_+ + \left(\nu_-\left|z_-\right|\right)\left|z_-\right| \qquad \ldots [10]$$

Now—here comes the trick!—we substitute Eqn [9] into the bracketed terms in Eqn [10]:

$$\nu_+ z_+^2 + \nu_- z_-^2 = \left(\nu_-\left|z_-\right|\right)z_+ + \left(\nu_+ z_+\right)\left|z_-\right|$$

$$= \nu_-\left|z_+ z_-\right| + \nu_+\left|z_+ z_-\right|$$

$$= \left(\nu_+ + \nu_-\right)\left|z_+ z_-\right|$$

$$= \nu\left|z_+ z_-\right| \qquad \ldots [11]$$

---

[6] In both the Debye-Hückel equation and the equation defining the ionic strength, the valences z_i of the ionic species are sign sensitive. That is, cations have a positive valence and anions have a negative valence: Thus, for Ca^{2+} ions, $z = +2$; for Cl^- ions, $z = -1$.

[7] Since $\gamma_\pm$ is the property of a complete salt, it *can* be measured. But, using the Debye-Hückel theory, $\gamma_\pm$ also can be *calculated*. If the calculated value is the same as the experimentally measured value, then we reasonably can assume that the calculated value for the single ion property also is correct. This is how we test the theory.

[8] Don't forget that z_- has a negative sign.

Substituting Eqn [11] into Eqn [8] gives

$$log\gamma_{\pm} = -0.5108\left[\frac{v|z_+z_-|}{v}\right]\sqrt{I} = -0.5108|z_+z_-|\sqrt{I}$$

$$\boxed{log\gamma_{\pm} = -0.5108\,|z_+z_-|\sqrt{I}}\quad \begin{array}{l}\textit{Debye-Hückel Eqn}\\ \textit{for Aqueous}\\ \textit{Solutions at 25°C}\end{array}\quad \cdots [12]$$

You will note from Eqn [12] that the Debye-Hückel expression for a salt is of exactly the same form as that for a single ion except that the term z_i^2 for the single ion has been replaced with $|z_+z_-|$ for the salt. Incidentally, the expression given by Eqn [12] is applicable to any electrolyte, including acids and bases, not just to salts.

As mentioned above, the Debye-Hückel equation is accurate only up to ionic strengths of about 0.001. For more concentrated solutions, a modification of the Debye-Hückel equation suggested by Guggenheim can be used:

$$\boxed{log\gamma_{\pm} = -\frac{0.5108\,|z_+z_-|\sqrt{I}}{1+\sqrt{I}}}\quad \begin{array}{l}\textit{Guggenheim Eqn}\\ \textit{for Aqueous}\\ \textit{Solutions at 25°C}\end{array}\quad \cdots [13]$$

The Guggenheim equation gives reasonably accurate values of activity coefficients up to ionic strengths of about 0.1.[9]

The Debye-Hückel equation has been derived by taking account of the electrostatic forces of attraction between positive and negative ions in solutions. From this it follows that for any species in solution that is not charged, such as an undissociated solute (e.g., sucrose), a dissolved gas (e.g., O_2) or an electrically neutral species such as an ion pair (e.g., $Ca^{2+}SO_4^{2-}$), because there will be no electrostatic interactive effects, there will be no lowering of the free energy of such species, and the activity coefficient for these species will be equal to unity; i.e., the activity will just be equal to the concentration (providing the concentration does not become extremely high, say greater than 1~2 molar).

19.4 ACIDS AND BASES: THE BRØNSTED–LOWRY CONCEPT

There are several ways to classify acids and bases; but, perhaps the most useful way for physical chemistry is the *Brønsted-Lowry* concept, in which an **acid** is a substance that can *donate protons*, whereas a **base** is substance that can *accept protons*.

A proton is just a bare hydrogen atom stripped of its electron; i.e., a *hydrogen ion is just a proton*. The concentration of hydrogen ions in solution is extremely important in many chemical, biological, geological, environmental, technological, and industrial processes.

Since this concentration can vary over an extremely wide range (15 orders of magnitude!), it is convenient to express it as a logarithmic function; namely, the negative of the logarithm (to *base 10*)[10] of the relative *molal activity* of the hydrogen ion in solution. We call this function "pH"; thus:

---

[9] For higher concentrations fancier equations are required that are beyond the scope of this text.

$$pH \equiv -log\, a_{H^+}^{(m)} = -log\,(\gamma_{H^+} m_{H^+}) \quad\middle|\quad pH\ Defined\ ^{11} \quad \ldots [14]$$

from which

$$a_{H^+}^{(m)} = 10^{-pH} \qquad \ldots [15]$$

For dilute solutions having electrolyte concentrations less than about 0.001 molal, γ_{H^+} is reasonably close to unity; furthermore, as we saw above, for dilute *aqueous* solutions the molality m is approximately the same as the molarity M; therefore, under these conditions we can say:

$$pH = -log\,(\gamma_{H^+} m_{H^+})$$
$$\approx -log\big(1 \times [H^+]\big)$$
$$\approx -log\,[H^+]$$

$$\boxed{pH \approx -log\,[H^+]} \quad\begin{array}{l}Dilute\ Aqueous\\ Solutions\end{array} \qquad \ldots [16]$$

The lower the *pH*, the more acidic the solution; a change in *pH* of *1* means a change by a factor of *10* in the hydrogen ion activity; *pH 3* means $a_{H^+}^{(m)} = 10^{-3}$.

As stated above, according to the Brønsted-Lowry concept of acids and bases, an **acid** is a substance that can *donate protons*, and a **base** is substance that can *accept protons*. Consider the following reactions:

$$\underset{acid\quad base}{HCl\ +\ NH_3} \rightarrow Cl^- + NH_4^+ \qquad\bigg|\qquad \underset{acid\qquad base}{NH_4^+\ +\ CO_3^{2-}} \rightarrow NH_3 + HCO_3^- \qquad\bigg|\qquad \underset{acid\quad base}{H_2O\ +\ CO_3^{2-}} \rightarrow OH^- + HCO_3^-$$

HCl, NH_4^+, and H_2O are all *acids* because they *donate* protons. Similarly, NH_3 and CO_3^{2-} are *bases* because they *accept* protons. Now, if a species can *donate* a proton to a second species, then the second species can donate it *back;* thus:

$\underset{acid\qquad\qquad base}{HCO_3^- \rightleftharpoons H^+ + CO_3^{2-}}$ HCO_3^- is an *acid* because it can *donate* a proton to the CO_3^{2-} ion; the CO_3^{2-} ion is a *base* because it can *accept* a proton from the HCO_3^- ion.

$\underset{acid\quad base}{NH_4^+ \rightleftharpoons NH_3 + H^+}$ Similarly, NH_4^+ is an acid because it can donate a proton to NH_3; NH_3 is a base because it can accept a proton from the NH_4^+ ion.

HCO_3^-/CO_3^{2-} and NH_4^+/NH_3 are called **conjugate acid–base pairs**. Thus, HCO_3^- is an *acid*, and CO_3^{2-} is its *conjugate base*; CO_3^{2-} is a *base*, and HCO_3^- is its *conjugate acid*.

---

[10] *Base 10 logarithms (log x) are related to natural logarithms (ln x) by: ln x = ln (10) log x ≈ 2.303 log x.*

[11] The letters *pH* stand for "10 to the negative *p*ower of the *H*ydrogen ion."

Thus, every acid-base reaction can be viewed as a proton transfer between two conjugate acid base pairs:

$$HCl + H_2O \rightleftharpoons H_3O^+ + Cl^-$$
$$\text{acid 1} \quad \text{base 2} \quad \text{acid 2} \quad \text{base 1}$$

19.5 THE SELF-DISSOCIATION OF WATER

Consider the familiar acid–base neutralization reaction:

$$H^+_{(aq)} + OH^-_{(aq)} \rightarrow H_2O \qquad \qquad \ldots [17]$$

This reaction also can be written as:

$$H_3O^+_{(aq)} + OH^-_{(aq)} \rightarrow H_2O + H_2O$$
$$\text{acid 1} \quad \text{base 2} \quad \text{acid 2} \quad \text{base 1}$$

Thus, H_3O^+/H_2O and H_2O/OH^- are each conjugate acid-base pairs; water can act either as an acid or as a base, depending on the reaction. Such a substance is said to be **amphoteric**.

The dissociation of water—called *autoprotolysis*—clearly illustrates the amphoteric nature of water:

$$2H_2O \rightleftharpoons H_3O^+_{(aq)} + OH^-_{(aq)}$$
$$\text{acid} \qquad \text{base}$$

This dissociation equilibrium is present in aqueous solutions *at all times*. Note that a proton in aqueous solution is always *hydrated*; therefore, $H^+_{(aq)}$ is more properly written $H^+(H_2O)$, i.e., H_3O^+. Actually, it would be more accurate still to write $H^+(H_2O)_n$, where $n = 3 \sim 5$. Therefore, although it's more correct to write

$$2H_2O \rightleftharpoons H_3O^+_{(aq)} + OH^-_{(aq)} \qquad \qquad \ldots [18]$$

often, for convenience, we just write

$$H_2O \rightleftharpoons H^+ + OH^-$$

Let's look at the self-dissociation of water—Eqn [18]—more closely. The equilibrium constant for this reaction can be defined as

$$K_W = \frac{a_{H_3O^+} \cdot a_{OH^-}}{a^2_{H_2O}} \qquad \qquad \ldots [19]$$

Pure water at 25°C has a density of 997.07 g L^{-1}, and contains equal concentrations of hydrogen ion and hydroxyl ion, namely:

$$[H_3O^+] = [OH^-] = 1.00 \times 10^{-7} \text{ mol L}^{-1}$$

Therefore, the number of moles of *undissociated* water in 1.00 L of water at 25°C is

$$n_{H_2O} = (\text{total moles}) - (\text{moles that are dissociated})$$

$$= \frac{997.07 \text{ g/L}}{18.015 \text{ g/mol}} - 1.00 \times 10^{-7}$$

$$= 55.3466556 - 0.0000001$$

$$= 55.3466555 \text{ mol}^{12}$$

The mole fraction of *undissociated* H_2O is

$$x_{H_2O} = \frac{n_{H_2O}}{n_{H_2O} + n_{H^+} + n_{OH^-}}$$

$$= \frac{55.3466555}{55.346655 + 0.0000001 + 0.0000001}$$

$$= 0.9999999964 \approx 1.000$$

Therefore, in "pure" water, since $x_{H_2O} = 1.000$, and, since the activity of a solvent is *defined* as its mole fraction,

$$a_{H_2O} = x_{H_2O} = 1.000$$

Substituting this value of the activity of undissociated H_2O into Eqn [19] gives

$$K_W = \frac{a_{H_3O^+} \cdot a_{OH^-}}{a_{H_2O}^2}$$

$$= \frac{a_{H_3O^+} \cdot a_{OH^-}}{(1.000)^2}$$

$$= a_{H_3O^+} \cdot a_{OH^-}$$

Therefore,
$$K_W = a_{H_3O^+} \cdot a_{OH^-}$$
$$= \gamma_{H^+} m_{H^+} \cdot \gamma_{OH^-} m_{OH^-} \qquad \dots [20]$$

We've seen that "pure" water consists of three[13] species: H^+, OH^-, and H_2O. In order to calculate the value of K_W using Eqn [20], we need to evaluate the *molality* of the hydrogen ions and of the hydroxyl ions in pure water. The number of kg of pure undissociated H_2O solvent in 1.00 L of pure water at 25°C is

$$\left(55.3466555 \text{ mol}\right)\left(0.018015 \tfrac{\text{kg}}{\text{mol}}\right) = 0.99707 \text{ kg}$$

Therefore, the molality[14] of the H^+ ions and of the OH^- ions is

---

[12] Obviously these values aren't really valid to 8 significant figures. The calculation is made for the purpose of better understanding the concepts involved.

[13] At least; there may be others, especially protons with various degrees of hydration.

[14] In case you've forgotten, molality is the moles of solute (*H*[+] or *OH*[-]) per kg of solvent (undissociated *H₂O*).

$$m_{H^+} = m_{OH^-} = \frac{1.000 \times 10^{-7} \text{ mol}}{0.99707 \text{ kg}} = 1.00294 \times 10^{-7} \text{ mol kg}^{-1}$$

For the H^+ ions and OH^- ions in pure water we calculate the ionic strength to be

$$I = \tfrac{1}{2}\left[c_{H^+} z_{H^+}^2 + c_{OH^-} z_{OH^-}^2\right]$$

$$= \tfrac{1}{2}\left[(1.00 \times 10^{-7})(+1)^2 + (1.00 \times 10^{-7})(-1)^2\right]$$

$$= 1.00 \times 10^{-7} \text{ mol L}^{-1}$$

Therefore:

$$\log \gamma_{H^+} = -0.5108 z_i^2 \sqrt{I}$$

$$= -0.5108(+1)^2 \sqrt{1.00 \times 10^{-7}}$$

$$= -1.61529 \times 10^{-4}$$

from which $\gamma_{H^+} = 10^{-1.61529 \times 10^{-4}}$

$$= 0.99963$$

Similarly,

$$\log \gamma_{OH^-} = -0.5108 z_i^2 \sqrt{I}$$

$$= -0.5108(-1)^2 \sqrt{1.00 \times 10^{-7}}$$

$$= -1.61529 \times 10^{-4}$$

and $\gamma_{OH^-} = 10^{-1.61529 \times 10^{-4}}$

$$= 0.99963$$

Note that, in this case, both activity coefficients are the same because the valence of each ion is the same.

Finally,

$$K_W = \gamma_{H^+} m_{H^+} \cdot \gamma_{OH^-} m_{OH^-}$$

$$= (0.99963)^2 (1.00294 \times 10^{-7})^2$$

$$= 1.005 \times 10^{-14} \qquad \qquad \dots [21]$$

The generally accepted value reported in the literature is 1.008×10^{-14}.

If we just *ignore* activity coefficients and molalities and proceed using the molar concentrations of the H^+ ions and the OH^- ions, we get:

$$K_W = [H^+][OH^-] \qquad \qquad \dots [22]$$

$$= (1.00 \times 10^{-7})(1.00 \times 10^{-7})$$

$$= 1.00 \times 10^{-14}$$

This latter calculation using molarities without taking into consideration activity coefficients yields the same value as the "correct" calculation (Eqn [21]) because at such low concentrations (10^{-7} M) the solution behaves ideally with $\gamma_i \approx 1$ and $m_i \approx c_i$.

Exercise 19-3

Calculate the ionic strengths of 0.50 molar solutions of KNO_3, K_2SO_4, and $La_2(SO_4)_3$, assuming complete dissociation of each salt into ions. ☺

19.6 NEUTRAL SOLUTIONS

A neutral solution is one that is neither acidic nor basic; in other words, one for which $[H^+] = [OH^-]$.[15] Therefore, for pure water, if we let $[H^+] = [OH^-] = x$,

then
$$K_W \approx [H^+][OH^-]$$
$$= (x)(x) = x^2$$

and
$$x = \sqrt{K_W}$$

At 25°C, $K_W = 1.00 \times 10^{-14}$, therefore

$$[H^+] = [OH^-] = x = \sqrt{1.00 \times 10^{-14}} = 1.00 \times 10^{-7}$$

Thus, in pure water at 25°C the hydrogen ion concentration equals the hydroxyl ion concentration equals 1.00×10^{-7} mol L^{-1}

and
$$pH = -\log a_{H^+}$$
$$= -\log (1.00 \times 10^{-7})$$
$$= 7.00$$

Thus, at 25°C, if the $pH > 7.00$, then $[OH^-] > [H^+]$ and the solution is *basic* (alkaline); if the $pH < 7.00$, then $[H^+] > [OH^-]$ and the solution is *acidic*.

Table 2 gives values of K_W as a function of temperature, and also indicates the pH of a neutral solution as a function of temperature.

Note that the pH of a neutral aqueous solution is 7.00 only at 25°C. For example, a neutral solution at body temperature (37°C) has a pH of about 6.84, not 7.00.

Table 2. *Variation of K_W with Temperature*

θ (°C)	K_W	pH of Neutral Solution
0	0.114×10^{-14}	7.471
10	0.295×10^{-14}	7.265
20	0.676×10^{-14}	7.085
25	1.00×10^{-14}	7.000
30	1.47×10^{-14}	6.966
40	2.71×10^{-14}	6.783
50	5.30×10^{-14}	6.638

Exercise 19-4

What is the concentration of the hydroxyl ions at 25°C in an aqueous solution containing 1.50×10^{-3} molar NaCl if the pH of the solution is 6.5? The NaCl may be assumed to dissociate completely. ☺

---

[15] Strictly speaking, a neutral solution is one for which the *activity* of the H^+ ions equals that of the OH^- ions.

Exercise 19-5

Assuming complete dissociation, what is the pH at 25°C of 1.00×10^{-4} molar sulfuric acid (a) In pure water? (b) In 0.001 molar aqueous calcium chloride solution? ☺

KEY POINTS FOR CHAPTER NINETEEN

1. **Activities and activity coefficients of dissolved species in solution**:

   For **non-electrolytes** dissolved in a liquid solvent, the relative activity is given by

   $$a_i = \gamma_i m_i$$

   where γ_i is the molal activity coefficient of the solute and m_i is its molality.

   For dilute solutions ($< 0.1 \ m$), we may assume that $\gamma_i \approx 1$, that $m_i = c_i$, and that

   $$a_i \approx c_i \approx m_i$$

2. **Activities and activity coefficients for electrolytes**:

   The general electrolyte dissociates into M^{z+} ions and A^{z-} ions according to

   $$M_{v_+} A_{v_-} \rightarrow v_+ M^{z+} + v_- A^{z-}$$

   where v_+ and v_- are called the *stoichiometric coefficients* of the cation and anion.

   The **molal relative activity** of an m–molal solution of an electrolyte $M_{v_+} A_{v_-}$ is *defined* as

   $$a_{MA} \equiv \left(v_+^{v_+} \cdot v_-^{v_-} \right) \cdot \gamma_\pm^v \cdot m^v$$

   where $v \equiv v_+ + v_-$ and $\gamma_\pm$ is the **stoichiometric mean molal ionic activity coefficient** of the electrolyte.

   For a single ionic species in solution, $a_i = \gamma_i m_i$

   where γ_i is the *single ion molal ionic activity coefficient* and m_i is the relative molality of the ionic species in solution.

   The *stoichiometric mean molal ionic activity coefficient* $\gamma_\pm$ for the electrolyte is related to the *individual* molal activity coefficients for the ions by the formula

   $$\gamma_\pm = \left(\gamma_+^{v_+} \cdot \gamma_-^{v_-} \right)^{1/v}$$

   Values of $\gamma_\pm$ are obtained from tables of data, and are functions of concentration, temperature, and the nature of the individual ions involved.

3. When the solution is dilute ($I \leq 0.001 \ M$) γ_i-values for individual ions in aqueous solutions at 25°C can be calculated using the **Debye–Hückel equation**:

   $$\log \gamma_i = -0.5108 \, z_i^2 \sqrt{I}$$

   where z_i is the valence of the ion, and I is the **ionic strength** of the solution, defined as

$$I \equiv \frac{1}{2} \sum_i c_i z_i^2 \qquad \text{Units: mol L}^{-1}$$

where c_i is the concentration of each ionic species i, in mol L^{-1}, and z_i is its valence.

4. For a dissolved electrolyte, the Debye-Hückel expression has exactly the same form as that for a single ion except that the term z_i^2 for the single ion is replaced with $|z_+ z_-|$ for the salt:

$$log\,\gamma_\pm = -0.5108\,|z_+ z_-|\sqrt{I}$$

This expression is applicable to any electrolyte, including acids and bases, as well as salts.

5. For solutions more concentrated than $I \approx 0.001$, the **Guggenheim equation** gives better results than the Debye-Hückel equation:

$$log\,\gamma_\pm = -\frac{0.5108\,|z_+ z_-|\sqrt{I}}{1+\sqrt{I}}$$

The Guggenheim equation can be used up to ionic strengths of about 0.1 M. At higher concentrations fancier equations are required that are beyond the scope of this text.

6. Protons (hydrogen ions) play an important role in many chemical, biological, geological, environmental, technological, and industrial processes. Because the concentration of the H$^+$ ion—more properly, written as H_3O^+, or even as $H(H_2O)_4^+$—can vary over 15 orders of magnitude, it is expressed as a logarithmic function; namely, the negative of the logarithm (to *base 10*) of the relative *molal activity* of the hydrogen ion in solution. We call this function "pH":

$$pH \equiv -log\,a_{H^+} = -log\,(\gamma_{H^+} m_{H^+})$$

from which

$$a_{H^+} = 10^{-pH}$$

In dilute aqueous solutions ($\leq 0.001\,m$ in H$^+$ ion), $\gamma_{H^+} \approx 1$ and $m_{H^+} \approx c_{H^+}$, therefore

$$pH \approx -log\,[H^+]$$

7. **Brønsted acids and bases**: An *acid* is a substance that can *donate protons*, whereas a *base* is substance that can *accept protons*. Every acid-base reaction can be viewed as a proton transfer between two acid-base pairs:

$$\begin{array}{ccccccc} HCl & + & H_2O & \rightleftharpoons & H_3O^+ & + & Cl^- \\ \textit{acid 1} & & \textit{base 2} & & \textit{acid 2} & & \textit{base 1} \end{array}$$

HCl/Cl^- and H_3O^+/H_2O are called **conjugate acid-base pairs**.

8. **Self-dissociation of H$_2$O**: $\qquad 2H_2O \rightleftharpoons H_3O^+ + OH^-$

usually just written as $\qquad H_2O \rightleftharpoons H^+ + OH^-$

This dissociation equilibrium is present in aqueous solutions *at all times*.

$$K_W = a_{H^+} \cdot a_{OH^-} = 1.00 \times 10^{-14} \text{ at } 25°C$$

9. A **neutral solution** is one for which $a_{H^+} = a_{OH^-}$

Usually we just express this as $[H^+] = [OH^-]$

At 25°C, $[H^+] = [OH^-] = \sqrt{K_W} = \sqrt{1.00 \times 10^{-14}} = 1.00 \times 10^{-7}$

Therefore, at 25°C a neutral solution has $pH = 7.00$

PROBLEMS

1. The Debye-Hückel equation is fairly accurate up to ionic strengths of about 0.001 M. For slightly higher concentrations the Guggenheim equation is more accurate.

   (a) Using the Debye-Hückel equation, estimate the single ion activity coefficients at 25°C for univalent and bivalent ions in an aqueous solution that is 7 mM in $Mg(NO_3)_2$, 3 mM in Na_2SO_4, and 10 mM in $NaNO_3$.

   (b) Repeat the calculation using the Guggenheim equation. ☹

2. Using the Guggenheim equation, estimate:

   (a) the mean molal ionic activity coefficient at 25°C of an aqueous solution of 0.01 M HCl,

   (b) the relative molal activity of the solution, and

   (c) the *pH* of the solution. ☹

3. Use the Debye-Hückel equation to estimate:

   (a) The mean molal ionic activity coefficient at 25°C of the Na_2SO_4 in an aqueous solution of 0.002 M Na_2SO_4.

   (b) The relative molal activity of the Na_2SO_4 in the solution. ☹

4. Use the Guggenheim equation to calculate:

   (a) the *pH* at 25°C of an aqueous 0.001 molar solution of HCl, and

   (b) the *pH* at 25°C of an aqueous solution that is 0.001 molar in HCl and 0.09 molar in KCl. ☹

5. An aqueous solution is 0.001 molar in HNO_3, 0.003 molar in $Cu(NO_3)_2$, and 0.0005 molar in $AgNO_3$. For this solution at 25°C, use the Debye-Hückel equation to estimate the ratio $\gamma^3_{\pm Cu(NO_3)_2} / \gamma^2_{\pm AgNO_3}$. ☹

6. At 25°C a 0.001 M aqueous solution of a certain weak acid *HA* was found to be 10% dissociated:
$$HA \rightleftharpoons H^+ + A^-$$

   (a) What is the equilibrium constant for the dissociation of this acid?

   (b) What would be the degree of dissociation if the solution also were 0.1 molar in NaCl?

Use the Guggenheim equation for both parts (a) and (b). Also, at the concentrations encountered in this problem, molalities and molarities may be assumed to have the same numerical values. ☹

7. Strictly speaking, the form of the Debye-Hückel equation given in the text, namely, $log\gamma_{\pm} = -0.5108 \left| z_{+}z_{-} \right| \sqrt{I}$, should be used with the ionic strength defined in terms of *molalities*, i.e., $I = \frac{1}{2}\Sigma m_i z_i^2$, instead of in terms of molarities. However, since concentrations are more commonly expressed in terms of molarities, it is permissible to use this equation with the ionic strength defined as $I = \frac{1}{2}\Sigma c_i z_i^2$. In the concentration range over which the Debye-Hückel theory is valid—i.e., at ionic strengths up to about 0.001 M—no significant error is incurred. Verify that this is true by using the Debye-Hückel equation to calculate the mean molal ionic activity coefficient of 0.001 M aqueous NaCl solution at 25°C using both definitions of the ionic strength. For aqueous NaCl solution at 25°C, the relationship between the molality m and the molarity c is given by the expression

$$m = -8.7078 \times 10^{-7} + 1.0029c + 1.7590 \times 10^{-2}c^2 + 1.4322 \times 10^{-3}c^3$$
$$- 5.5416 \times 10^{-5}c^4 + 3.5692 \times 10^{-6}c^5 \quad ☺$$

8. The Debye-Hückel equation often is written as $log\gamma_i = -Az_i^2\sqrt{I}$, where A is a constant having a value for aqueous solutions at 25°C of 0.5108. The value of A is a function of temperature, and is calculated from the formula

$$A = \frac{N_A^2 e_o^3 \sqrt{2\rho_o}}{8\pi (ln\,10)\left(RT\varepsilon_o\varepsilon\right)^{3/2}}$$

where N_A is Avogadro's number, e_o is the (absolute) value of the charge of the electron, ρ_o is the density of the pure solvent (water), z_i is the valence of the ion, R is the gas constant, T is the absolute temperature, ε_o is the permittivity of free space (an electrostatic term), and $\boldsymbol{\varepsilon}$ is the dielectric constant (also an electrostatic term) of the solvent. Here is a chance to see if you can use your calculator correctly. A set of internally consistent values of physical constants used to calculate A is as follows: $N_A = 6.02217 \times 10^{23}$ mol^{-1}; $e_o = 1.602192 \times 10^{-19}$ C; $R = 8.3143$ J mol^{-1} K^{-1}; $\varepsilon_o = 8.8544 \times 10^{-12}$ C V^{-1} m^{-1}.

The densities and dielectric constants for water as a function of temperature are given in the table at the right.		0°C	25°C	100°C
	ρ_o (kg m^{-3})	999.87	997.07	958.38
	$\boldsymbol{\varepsilon}$ (unitless)	87.740	78.303	55.720

(a) Work out the units of A from the formula.

(b) Calculate the value of A at 0°C, 25°C and at 100°C.

(c) What is the value of the molal activity of 0.001 molal aqueous HCl solution at 0°C, 25°C, and 100°C?

(d) What is the *pH* of the solution at each temperature? ☹

TWENTY

ACID AND BASE DISSOCIATION

20.1 ACID DISSOCIATION CONSTANTS, K_a

The dissociation of an acid *HA* can be represented as a proton transfer reaction involving a Brønsted acid:

$$HA_{(aq)} + H_2O \rightleftharpoons H_3O^+_{(aq)} + A^-_{(aq)}$$

with the equilibrium constant[1] defined by

$$K = \left(\frac{a_{H_3O^+} \cdot a_{A^-}}{a_{HA} \cdot a_{H_2O}} \right)_{eqm} \qquad \ldots [1]$$

We saw earlier that the relative activity[2] of the solvent water is given by its mole fraction, which, for most solutions, is close to unity; therefore we usually set $a_{H_2O} = 1$, so that Eqn [1] becomes

$$K_a = \frac{a_{H_3O^+} \cdot a_{A^-}}{a_{HA}} \qquad \ldots [2]$$

where K_a is called the *acid dissociation constant*. Values of the dissociation constants for several acids and bases are listed in Table 1.

We saw earlier that an acid is a substance with a tendency to transfer a proton to another molecule; therefore, it follows that a *strong* acid is a substance with a *strong* tendency to transfer a proton to another molecule. Thus, referring to Eqn [3], *HA* (*acid 1*) has a tendency to transfer a proton to H_2O (*base 2*).

$$\begin{array}{ccccccc} HA & + & H_2O & \rightleftharpoons & H_3O^+ & + & A^- \\ acid\ 1 & & base\ 2 & & acid\ 2 & & base\ 1 \end{array} \qquad \ldots [3]$$

The extent of the dissociation of *acid 1* also depends on the strength of the tendency of *base 2* to accept the proton from *acid 1*. Therefore, K_a depends not only on the *acid*, but also on the *base*. For example, HCl is a strong acid (~100% dissociated) in *water*, but weak in many non-aqueous solvents. Therefore, if we want to measure the relative strength of different acids, we should

---

[1] The relative activities for acid dissociation constants usually are based on the molal concentration scale.
[2] For a solution in which the volatile liquids obey Raoult's Law and the vapor behaves as an ideal gas.

compare them against the *same* base (*base 2*), which usually is chosen to be *water*. Reference to Table 1 shows that most K_a values are quite small; therefore, as in the case of the molal relative activity of the hydrogen ion, it's convenient to express these values in a way similar to that used for *pH*; namely, using negative logarithms to the *base 10*. Accordingly, we make use of the following ways to express dissociation constants and ionic activities:

$$pH = -log\ a_{H^+}, \quad pK_a = -log\ K_a, \quad pOH = -log\ a_{OH^-}, \quad pAg = -log\ a_{Ag^+}, \text{ etc.}$$

Table 1(a).   *Acid Dissociation Constants in Water at 25°C*

$HCl + H_2O \rightleftharpoons H_3O^+ + Cl^-$	$K_a = \dfrac{a_{H_3O^+} \cdot a_{Cl^-}}{a_{HCl}} = 1.2 \times 10^{+6}$
$H_2SO_4 + H_2O \rightleftharpoons H_3O^+ + HSO_4^-$	$K_{a1} = \dfrac{a_{H_3O^+} \cdot a_{HSO_4^-}}{a_{H_2SO_4}} = 3.89 \times 10^{+3}$
$HSO_4^- + H_2O \rightleftharpoons H_3O^+ + SO_4^{2-}$	$K_{a2} = \dfrac{a_{H_3O^+} \cdot a_{SO_4^{2-}}}{a_{HSO_4^-}} = 1.026 \times 10^{-2}$
$H_3PO_4 + H_2O \rightleftharpoons H_3O^+ + H_2PO_4^-$	$K_{a1} = 7.112 \times 10^{-3}$
$HF + H_2O \rightleftharpoons H_3O^+ + F^-$	$K_a = 5.888 \times 10^{-4}$
$CH_3COOH + H_2O \rightleftharpoons H_3O^+ + CH_3COO^-$	$K_a = 1.754 \times 10^{-5}$
$H_2CO_3 + H_2O \rightleftharpoons H_3O^+ + HCO_3^-$	$K_{a1} = 4.456 \times 10^{-7}$
$H_2PO_4^- + H_2O \rightleftharpoons H_3O^+ + HPO_4^{2-}$	$K_{a2} = 1.600 \times 10^{-7}$
$HCN + H_2O \rightleftharpoons H_3O^+ + CN^-$	$K_a = 6.081 \times 10^{-10}$
$HCO_3^- + H_2O \rightleftharpoons H_3O^+ + CO_3^{2-}$	$K_{a2} = 4.688 \times 10^{-11}$
$HPO_4^{2-} + H_2O \rightleftharpoons H_3O^+ + PO_4^{3-}$	$K_{a3} = 1.419 \times 10^{-12}$

Table 1(b).   *Base Dissociation Constants in Water at 25°C*

$KOH^3 \rightleftharpoons K^+ + OH^-$	$K_b = 1.78 \times 10^{+2}$
$NaOH \rightleftharpoons Na^+ + OH^-$	$K_b = 7.9 \times 10^{+1}$
$NH_3 + H_2O \rightleftharpoons NH_4^+ + OH^-$	$K_b = 1.770 \times 10^{-5}$

---

[3] Strictly speaking, KOH and NaOH are not Brønsted bases because they cannot accept protons. The OH⁻ that they liberate is, however, a Brønsted base. Note that "undissociated" KOH and NaOH really are ion pairs, held together by electrostatic forces of attraction: Na⁺OH⁻ and K⁺OH⁻.

Thus, for hydrofluoric acid (*HF*), $K_a = 3.5 \times 10^{-4}$ and $pK_a = 3.45$; similarly, for hydrocyanic acid (*HCN*), $K_a = 4.9 \times 10^{-10}$ and $pK_a = 9.31$.

Note that for the dissociation of water:

$$2H_2O \rightleftharpoons H_3O^+ + OH^-$$

At 25°C:
$$K_W = a_{H_3O^+} \cdot a_{OH^-} = 1.00 \times 10^{-14}$$

Taking *base-10* logarithms of both sides:

$$log\ K_W = log\ a_{H^+} + log\ a_{OH^-}$$

$$-log\ K_W = -log\ a_{H^+} - log\ a_{OH^-}$$

That is,
$$\boxed{pK_W = pH + pOH}$$ . . . [4]

Eqn [4] shows that for any aqueous solution at 25°C,

$$pH + pOH = 14.00$$

Example 20-1

Assuming 100% dissociation, given that $\gamma_{H^+} = 0.964$, what is the *pH* of 0.001 M aqueous HCl solution?

Solution:

$$pH = -log\ a_{H^+} = -log\ \gamma_{H^+} m_{H^+} \approx -log\ (0.964 \times 0.001) = 3.02$$

Ans: $\boxed{3.02}$

Exercise 20-1

What is the *pH* of 0.0005 M aqueous HCl solution at 25°C? HCl is completely dissociated at this concentration. ☺

Exercise 20-2

What is the *pH* of 1.0×10^{-8} M HCl solution at 25°C? You may assume that HCl is 100% dissociated, and that $K_W = 1.00 \times 10^{-14}$. ☺

20.2 DISSOCIATION OF A WEAK ACID[4]

A *strong acid* is one that dissociates almost 100% in solution to form ions; a *weak acid*, usually designated *HA*, is an acid that only *slightly* dissociates. As a general guideline, the acid dissociation constants for weak acids usually are $\leq 10^{-3}$. Consider the dissociation of such an acid:

$$HA + H_2O \rightleftharpoons H_3O^+ + A^- \qquad K_a = \frac{a_{H_3O^+} \cdot a_{A^-}}{a_{HA} \cdot a_{H_2O}} \approx \frac{a_{H_3O^+} \cdot a_{A^-}}{a_{HA}} \qquad \ldots [5]$$

The *degree of dissociation* α is the fraction of the acid that dissociates in solution:

$$\alpha = degree\ of\ dissociation = \frac{[A^-]}{[HA]_o} \qquad \ldots [6]$$

where $[HA]_o$ is the *stoichiometric concentration* of the acid. The stoichiometric concentration is the total acid in the solution, in all forms. For example, if a 1.00 M solution of *HA* dissociates 10%, then the concentration of the anion will be $[A^-] = 0.10$ M and the concentration of the undissociated acid will be $[HA] = 0.90$ M. The stoichiometric concentration of the acid—the concentration indicated on the label of the bottle—is $[HA]_o = [HA] + [A^-] = 0.90 + 0.10 = 1.00$ M.

Example 20-2

Estimate the *pH* of 0.001 M aqueous lactic acid ($CH_3CH(OH)COOH$) solution at 25°C, given that for this acid $K_a = 8.4 \times 10^{-4}$. All activity coefficients may be assumed to be unity.

Solution

Designate the acid by *HL*. The solution is dilute, therefore we may assume $a_i \approx c_i$.

As a basis for the calculation, take 1.00 L of the solution. By choosing 1.00 L, the number of moles of each species in the system will be the same as the molar concentration of each species, and we can use the concentration of each species as if it were the number of moles of that species.[5]
Thus:

	HL	$\rightleftharpoons$	H⁺	+	L⁻
concentration, start:	0.001		0		0
dissociation:	-0.001α		$+0.001\alpha$		$+0.001\alpha$
equilibrium:	$0.001(1-\alpha)$		$+0.001\alpha$		$+0.001\alpha$

---

[4] For the following dissociation calculations, with the exception of Example 20-3, we shall ignore activity coefficients and assume that activities are equal to molar concentrations. Life is complicated enough as it is without adding extra levels of difficulty at this stage of the game.

[5] Although the partial molar volumes of the various species in solution are not negligible, owing to their low concentrations when compared with the concentration of the solvent water (after all, pure water contains about 55 moles of H_2O per liter) the total volume of the solution essentially will be determined by the volume of the water. Therefore, shifts in the concentrations of each species as the dissociation proceeds will not have any significant effect on the total volume of the solution, which may be considered to stay constant at 1.00 L.

Substitution of the equilibrium values into the expression for the equilibrium constant gives:

$$K_a = \frac{[H^+][L^-]}{[HL]}$$

$$8.4 \times 10^{-4} = \frac{(0.001\alpha)(0.001\alpha)}{0.001(1-\alpha)} = \frac{0.001\alpha^2}{1-\alpha}$$

Cross-multiplying and rearranging:

$$\alpha^2 + 0.84\alpha - 0.84 = 0$$

This is a quadratic equation.

Solving:
$$\alpha = \frac{-b \pm \sqrt{b^2 - 4ac}}{2a} = \frac{-0.84 \pm \sqrt{(0.84)^2 - 4(1)(-0.84)}}{2(1)}$$

$$\alpha = 0.5882 \quad \text{or} \quad \alpha = -1.4282$$

The second root is impossible because α can't be negative (it also can't be greater than *1*); therefore, $\alpha = 0.5882$, and at this concentration the acid is 58.8% dissociated.

Thus,
$$[H^+] = 0.001\alpha = 0.001(0.5882) = 5.882 \times 10^{-4} \text{ mol L}^{-1}$$

and
$$pH \approx -log\ [H^+] = -log(5.882 \times 10^{-4}) = 3.23$$

Ans: $\boxed{3.23}$

In the above calculation we have assumed that all the H^+ ion came from the dissociation of the acid. This is not quite true, because the dissociation of water also contributes some H^+ ion to the system. We therefore should check to confirm our assumption that

$$[H^+]_{from\ acid} \gg [H^+]_{from\ water}$$

Since there is no other source of hydroxyl ion, all the OH^- ion present in the system comes from the dissociation of water, according to

$$H_2O \rightarrow H^+ + OH^-$$

Since the water dissociation equilibrium must *always* be obeyed in any aqueous system, then

$$K_W = a_{H_3O^+} \cdot a_{OH^-} \approx [H^+][OH^-]$$

from which
$$[OH^-] = \frac{K_W}{[H^+]} = \frac{1.00 \times 10^{-14}}{5.882 \times 10^{-4}} = 1.70 \times 10^{-11}$$

When H_2O dissociates, for each molecule of OH^- formed there will be one molecule of H^+ formed. Therefore, the concentration of H^+ ion produced *from the dissociation of water* will be the *same* as the total concentration of OH^- in the system; namely,

$$[H^+]_W = [OH^-]_W = 1.70 \times 10^{-11}$$

which is $<< 5.882 \times 10^{-4}$. Therefore the assumption that $[H^+]_{HA} >> [H^+]_W$ is valid. As a final confirmation, put all the calculated values into the expression for the acid dissociation constant:

Thus: $\qquad K_a = \dfrac{[H^+][L^-]}{[HL]} = \dfrac{(5.882 \times 10^{-4})^2}{0.001 - 5.882 \times 10^{-4}} = 8.40 \times 10^{-4}$ (Oh yeah!!)

The above calculation shows that the degree of dissociation of an acid depends on the concentration of the acid and on the value of its acid dissociation constant. An electrolyte such as NaCl is called a *1–1* electrolyte, because the valence of each ion is $z_+ = |z_-| = 1$. Similarly, $CaCl_2$ is a *2–1* salt, $CuSO_4$ is a *2–2* salt, $Al_2(SO_4)_3$ is a *3–2* salt, etc. Thus, lactic acid (*HL*) is a *1–1* weak acid. For any *1–1* weak acid such as *HA* in which only one H^+ ion dissociates, the above treatment can be generalized to give

$$\boxed{K_a \approx \dfrac{c_o \alpha^2}{1 - \alpha}} \qquad \textit{1–1 Weak Acid} \qquad \qquad \dots [7]$$

where c_o is the stoichiometric concentration of the acid. Eqn [7] also can be used for the dissociation of a *1–1* weak base; in this case, K_a is replaced with the base dissociation constant, K_b.

Example 20-3

Calculate the *pH* of 0.10 molal aqueous lactic acid (*CH₃CH(OH)COOH*) solution at 25°C

(a) if the acid is dissolved in water and activity coefficient corrections are ignored;

(b) if the acid is dissolved in water and activity coefficient corrections are taken into account using the Davies equation; and

(c) if the acid is dissolved in 1.00 molal aqueous NaCl solution, and activity coefficient corrections are taken into account using the Davies equation.[6]

Davies equation: $\qquad \qquad log\,\gamma_i = -\dfrac{0.5108\,z_i^2\,\sqrt{I}}{1 + \sqrt{I}} + 0.1\,z_i^2\,I$

For lactic acid at 25°C, $K_a = 8.4 \times 10^{-4}$. At 25°C 1.00 molal NaCl is 78.9% dissociated, and the activity of the water in this solution is 0.9669.

Solution

(a) Designating the acid by *HL*, we proceed in the usual manner:

	HL	⇌	H⁺	+	L⁻
molality, start:	0.1		0		0
dissociation:[7]	$-x$		$+x$		$+x$
equilibrium:	$0.1 - x$		x		x

---

[6] The Davies equation is useful for concentrations up to 1~2 molal.

[7] Why do we use x instead of α to show the amount of the acid that dissociates? Because we reserve α to represent the *degree* of dissociation, which is the fraction of *one mole* that dissociates. In this problem it is the fraction of *0.1 mole* that is dissociating, not 1.0 mol. So, if we wanted to, we could replace x with *0.1α*. It's easier just to use x.

Substitution of the equilibrium values into the expression for the equilibrium constant gives:

$$K_a = 8.4 \times 10^{-4} = \frac{[H^+][L^-]}{[HL]} = \frac{(x)(x)}{0.1 - x}$$

Cross-multiplying and rearranging:

$$x^2 + 8.4 \times 10^{-4} x - 8.4 \times 10^{-5} = 0$$

Solving the quadratic gives $\qquad x = m_{H^+} = 8.7548 \times 10^{-3}$

and

$$pH = -log(\gamma_{H^+} m_{H^+}) \approx -log\, m_{H^+}$$

$$= -log\,(8.7548 \times 10^{-3}) = 2.0578$$

Ans: $\boxed{2.058}$

(b) To evaluate the activity coefficients for the ions we need to know the ionic strength of the solution. Using, as a first approximation, the concentrations calculated in part (a) gives $I \approx 8.7548 \times 10^{-3}$, since for a 1–1 electrolyte the ionic strength is the same as the concentration.

Thus, $\qquad log\,\gamma_+ = log\,\gamma_- = -\dfrac{0.5108\, z_i^2 \sqrt{I}}{1 + \sqrt{I}} + 0.1 z_i^2 I$

$$= -\frac{0.5108(\pm 1)^2 \sqrt{8.7548 \times 10^{-3}}}{1 + \sqrt{8.7548 \times 10^{-3}}} + (0.1)(\pm 1)^2 (8.7548 \times 10^{-3})$$

$$= -0.043705 + 0.000876 = -0.042829$$

from which $\qquad \gamma_+ = \gamma_- = 10^{-0.042829} = 0.9061$

The undissociated acid is electrically neutral; therefore, we may assume that its activity coefficient is just unity. The acid dissociation constant is

$$K_a = \frac{a_{H^+} \cdot a_{L^-}}{a_{HL}} = \frac{(\gamma_{H^+} m_{H^+})(\gamma_{L^-} m_{L^-})}{\gamma_{HL} \cdot m_{HL}}$$

Inserting numerical values:

$$8.4 \times 10^{-4} = \frac{(0.9061x)(0.9061x)}{(1)(0.1 - x)}$$

This rearranges into the following quadratic:

$$0.82102 x^2 + 8.4 \times 10^{-4} x - 8.4 \times 10^{-5} = 0$$

Solving:

$$x = \frac{-8.4 \times 10^{-4} + \sqrt{(8.4 \times 10^{-4})^2 - 4(0.82102)(-8.4 \times 10^{-5})}}{2(0.82102)} = 9.6163 \times 10^{-3}$$

Therefore, $\qquad x = m_{H^+} = 9.6163 \times 10^{-3}$

and a more correct ionic strength of the solution is $I = 9.6163 \times 10^{-3}$, giving

$$log\gamma_i = -\frac{0.5108(\pm1)^2\sqrt{9.6163\times10^{-3}}}{1+\sqrt{9.6163\times10^{-3}}} + (0.1)(\pm1)^2(9.6163\times10^{-3}) = -0.044655$$

from which

$$\gamma_+ = \gamma_- = 10^{-0.044655} = 0.90229$$

and

$$8.4\times10^{-4} = \frac{(0.90229x)(0.90229x)}{(1)(0.1-x)}$$

which rearranges to

$$0.81412x^2 + 8.4\times10^{-4}x - 8.4\times10^{-5} = 0$$

Solving:

$$x = 9.6549\times10^{-3}$$

This value should be close enough. Check by inserting into the expression for K_a:

Thus,

$$\frac{(\gamma_{H^+}x)(\gamma_{L^-}x)}{(0.1-x)} = \frac{(0.90229\times9.6549\times10^{-3})^2}{(0.1-9.6549\times10^{-3})} = 8.400\times10^{-4} \quad \text{(Right on!)}$$

Therefore,

$$pH = -log(\gamma_{H^+}m_{H^+})$$
$$= -log(0.90229\times9.6549\times10^{-3})$$
$$= 2.05991$$

Although the value for the hydrogen ion concentration is greater by 10.3% when activity coefficient corrections are taken into account, the value for the pH is almost the same!—2.0599 $vs.$ 2.0578.

Ans: $\boxed{2.060}$

(c) In 1.00 m NaCl the activity of the water is starting to differ from unity and we must take this into account when defining the acid dissociation constant.

Thus, $HL + H_2O \rightleftharpoons H_3O^+ + L^-$ and $K_a = \dfrac{a_{H^+}\cdot a_{L^-}}{a_{HL}\cdot a_{H_2O}} = \dfrac{(\gamma_{H^+}m_{H^+})(\gamma_{L^-}m_{L^-})}{m_{HL}\cdot a_{H_2O}}$

As before, we may assume that the activity coefficient for the undissociated acid is unity. Since we are told that the 1.00 m NaCl is 78.9% dissociated, the ionic strength of the solution will be $I = 0.789\times1.00 = 0.789$ m. (The Na^+ and Cl^- ions are at much greater concentrations than the H^+ and L^- ions, so the latter two species may be neglected when determining the ionic strength.) For lack of more detailed information, we must assume that all univalent ions in the solution will have the same activity coefficient,[8] i.e.: $\gamma_{H^+} = \gamma_{Na^+} = \gamma_{L^-} = \gamma_{Cl^-}$, as predicted by the Debye-Hückel and Davies equations.

---

[8] This may not strictly be true, since the H$^+$ ion tends to have properties different from other univalent ions.

Thus, using the Davies equation:

$$log\gamma_{H^+} = log\gamma_{L^-} = -\frac{0.5108\,z_i^2\,\sqrt{I}}{1+\sqrt{I}} + 0.1z_i^2\,I$$

$$= -\frac{0.5108(\pm 1)^2\,\sqrt{0.789}}{1+\sqrt{0.789}} + (0.1)(\pm 1)^2(0.789)$$

$$= -0.16139$$

from which

$$\gamma_+ = \gamma_- = 10^{-0.16139} = 0.6896$$

and the expression for the acid dissociation constant becomes

$$8.4\times 10^{-4} = \frac{(\gamma_{H^+}m_{H^+})(\gamma_{L^-}m_{L^-})}{m_{HL}\cdot a_{H_2O}} = \frac{(0.6896x)^2}{(0.1-x)(0.9669)}$$

This rearranges to:

$$0.4775x^2 + 8.122\times 10^{-4}\,x - 8.122\times 10^{-5} = 0$$

Solving:

$$x = \frac{-8.122\times 10^{-4} + \sqrt{(8.122\times 10^{-4})^2 - 4(0.4755)(-8.122\times 10^{-5})}}{2(0.4755)} = 1.2243\times 10^{-2}$$

Therefore,

$$x = m_{H^+} = 1.2243\times 10^{-2}$$

and,

$$pH = -log(\gamma_{H^+}\cdot m_{H^+})$$

$$= -log(0.6896\times 1.2243\times 10^{-2})$$

$$= 2.07351$$

Ans: $\boxed{2.074}$

Summarizing for 0.1 m lactic acid at 25°C:

	γ_{H^+}	m_{H^+}	pH
In water, with no activity corrections:	1.0000	8.75×10^{-3}	2.0578
In water, with activity corrections:	0.9002	9.68×10^{-3}	2.0600
In 1.0 m NaCl, with activity corrections:	0.6896	12.24×10^{-3}	2.0735

20.3 POLYPROTIC ACID CALCULATION:
CHARGE AND MASS BALANCES

A *polyprotic acid* is one that can produce more than one H^+ ion per molecule of acid when the acid dissociates. Typical examples would be H_2SO_4, H_2CO_3, and H_3PO_4. Consider a dibasic[9] acid H_2A: The dissociation of this type of acid takes place in two consecutive steps, each having its own acid dissociation constant:

Step 1: $\qquad\qquad\quad H_2A \rightleftharpoons H^+ + HA^- \qquad K_1 = \dfrac{[H^+][HA^-]}{[H_2A]} \qquad\qquad \ldots [8]$

Step 2: $\qquad\qquad\quad HA^- \rightleftharpoons H^+ + A^{2-} \qquad K_2 = \dfrac{[H^+][A^{2-}]}{[HA^-]} \qquad\qquad \ldots [9]$

For these kinds of acids, usually $K_1 < \sim 10^{-3}$, and $K_2 \ll K_1$. Accordingly, the concentrations of H^+ ion may be very low; so low that the H^+ contributed by the dissociation of water may comprise a significant amount of the total H^+ ion in the system. Therefore, it is prudent to include the dissociation of water in the calculation:

$$H_2O \rightleftharpoons H^+ + OH^- \qquad\qquad K_W = [H^+][OH^-] \qquad \ldots [10]$$

In order to "solve" this equilibrium, we must determine the equilibrium concentrations of *five* species: H_2A, H^+, HA^-, A^{2-}, and OH^-. In other words, there are five unknowns; therefore, we will need to solve five equations simultaneously. (Note that since such a small fraction of the H_2O dissociates, the amount of this species in solution does not change significantly, so undissociated H_2O is *not* one of the species we have to take into consideration.)

Eqns [8], [9], and [10] provide us with three equations, so we still need two more. The final two equations are obtained from a *charge balance* and a *mass balance*.

The **charge balance** takes account of the fact that *the total solution must be electrically neutral.* You don't get a shock when you stick your finger into an electrolyte solution! This means that the solution must contain exactly the same amount of positive charge as negative charge. Thus, every mole of H^+ contributes *one* mole of positive charge, every mole of A^{2-} ion contributes *two* moles of negative charge, etc. Therefore, the charge balance states

$$\left(\begin{array}{c} \textit{The total concentration} \\ \textit{of positive charge in} \\ \textit{the solution} \end{array} \right) = \left(\begin{array}{c} \textit{The total concentration} \\ \textit{of negative charge} \\ \textit{the solution} \end{array} \right)$$

That is: $\qquad\qquad\qquad\qquad [H^+] = [OH^-] + [HA^-] + 2[A^{2-}] \qquad\qquad \ldots [11]$

Note the "2" in front of the concentration of A^{2-}; each mole of A^{2-} contributes *two* moles of negative charge per liter. With the addition of Eqn [11], we now have four equations.

---

[9] "Dibasic" means that two hydroxyl ions ("base" ions) are required for complete neutralization of one molecule of the acid. For example: H_2SO_4 is a *dibasic* acid: $H_2SO_4 + 2KOH \rightarrow K_2SO_4 + 2H_2O$. Similarly, phosphoric acid is a *tribasic* acid: $H_3PO_4 + 3KOH \rightarrow K_3PO_4 + 3H_2O$. H_2SO_4 and H_3PO_4 also may be referred to as a *diprotic* acid and a *triprotic* acid, respectively. *Polybasic* acid is a more classical term, *polyprotic* acid is a newer term.

To obtain the fifth equation, we need to do a *mass balance*: The **mass balance** is made on the component that exists in *more than one form* ("H" is not considered), which, in this case is "A." Thus, "A" is present as H_2A, HA^-, and A^{2-}. The total concentration of A that initially was put into the system is just $[H_2A]_o$ (or c_o), the stoichiometric concentration of the acid. Once the acid dissolves and partially dissociates, the "A" redistributes itself amongst the various forms, but the total amount present must be equal to the amount in the initial H_2A introduced into the system.

Thus:

Mass Balance on "A":   Total "A" = Sum of "A" in all its forms

That is, $c_o = [H_2A] + [HA^-] + [A^{2-}]$ . . . [12]

We now have five equations in five unknowns, making it possible to solve for the concentrations of the five species. This is done as follows:

From Eqn [11]: $[HA^-] = [H^+] - [OH^-] - 2[A^{2-}] = [H^+] - \dfrac{K_w}{[H^+]} - 2[A^{2-}]$ . . . [13]

From Eqn [9]: $[A^{2-}] = \dfrac{K_2[HA^-]}{[H^+]}$ . . . [14]

Putting [14] into [13]: $[HA^-] = [H^+] - \dfrac{K_w}{[H^+]} - \dfrac{2K_2[HA^-]}{[H^+]}$

Rearranging: $[HA^-]\left(1 + \dfrac{2K_2}{[H^+]}\right) = [H^+] - \dfrac{K_w}{[H^+]}$

$$[HA^-] = \frac{[H^+] - \dfrac{K_w}{[H^+]}}{1 + \dfrac{2K_2}{[H^+]}}$$ . . . [15]

From Eqn [12]: $[H_2A] = c_o - [HA^-] - [A^{2-}]$ . . . [16]

Putting Eqn [15] into Eqn [14]:

$$[A^{2-}] = \frac{K_2[HA^-]}{[H^+]} = \frac{K_2}{[H^+]}\left(\frac{[H^+] - \dfrac{K_w}{[H^+]}}{1 + \dfrac{2K_2}{[H^+]}}\right)$$ . . . [17]

Substituting Eqns [15] and [17] into Eqn [16]:

$$[H_2A] = c_o - \left(\frac{[H^+] - \dfrac{K_w}{[H^+]}}{1 + \dfrac{2K_2}{[H^+]}}\right) - \frac{K_2}{[H^+]}\left(\frac{[H^+] - \dfrac{K_w}{[H^+]}}{1 + \dfrac{2K_2}{[H^+]}}\right)$$

$$= c_o - \left(\frac{[H^+] - \frac{K_W}{[H^+]}}{1 + \frac{2K_2}{[H^+]}} \right)\left(1 + \frac{K_2}{[H^+]} \right) \qquad \ldots [18]$$

Putting [15] and [18] into [8]:

$$K_1 = \frac{[H^+]\left(\frac{[H^+] - \frac{K_W}{[H^+]}}{1 + \frac{2K_2}{[H^+]}} \right)}{c_o - \left(1 + \frac{K_2}{[H^+]} \right)\left(\frac{[H^+] - \frac{K_W}{[H^+]}}{1 + \frac{2K_2}{[H^+]}} \right)} \qquad \ldots [19]$$

Eqn [19] can be rearranged to a *quartic* equation in one unknown—$[H^+]$—of the form

$$ax^4 + bx^3 + cx^2 + dx + e = 0$$

namely:

$$[H^+]^4 + K_1[H^+]^3 + (K_1K_2 - K_W - K_1c_o)[H^+]^2 - (2K_1K_2c_o + K_1K_W)[H^+] - K_1K_2K_W = 0$$

This can be solved by iteration, by a computer, or by a programmable hand-held caclulator. However, usually it is possible to simplify it as follows:

20.4 SIMPLIFIED APPROACH

Usually $K_2 << K_1$. In other words, most of the H^+ ions are produced from the *first* dissociation

$$H_2A \rightleftharpoons H^+ + HA^- \qquad\qquad K_1 = \frac{[H^+][HA^-]}{[H_2A]} \qquad \ldots [20]$$

and only a very small amount from the second dissociation:

$$HA^- \rightleftharpoons H^+ + A^{2-} \qquad\qquad K_2 = \frac{[H^+][A^{2-}]}{[HA^-]} \qquad \ldots [21]$$

Therefore, we can treat the acid as if were a *mono*protic acid with the following concentrations at equilibrium:

$$H_2A \quad \rightleftharpoons \quad H^+ \; + \; HA^-$$

$$(1 - \alpha)c_o \qquad \alpha c_o \qquad \alpha c_o$$

Solving in the usual way for α gives $\qquad K_1 = \frac{(\alpha c_o)(\alpha c_o)}{(1 - \alpha)c_o} = \frac{c_o\alpha^2}{1 - \alpha} \qquad \ldots [22]$

which, when cross-multiplied, is just a quadratic equation that is easy to solve. Once we have solved for α, the equilibrium concentration of each species can be calculated from:

$$[H^+] \approx [HA^-] \approx \alpha c_o \qquad \ldots [23]$$

and

$$[H_2A] \approx (1 - \alpha)c_o \qquad \ldots [24]$$

Since $[H^+] \approx [HA^-] \approx \alpha c_o$, substitution into Eqn [21] gives

$$K_2 = \frac{[H^+][A^{2-}]}{[HA^-]} \approx \frac{(\alpha c_o)[A^{2-}]}{(\alpha c_o)} = [A^{2-}] \qquad \ldots [25]$$

Therefore $[A^{2-}] \approx K_2$, and Eqns [23] ~ [25] give the equilibrium concentrations of all species.

Exercise 20-3

Air is 0.033% by volume CO_2. The Henry's Law constant for CO_2 in water at 25°C is 161.7 MPa. When CO_2 from the air dissolves in water, it forms carbonic acid:

$$CO_2 + H_2O \rightarrow H_2CO_3$$

In other words, carbonic acid may be viewed as an aqueous solution of CO_2:

$$CO_{2\,(aq)} \equiv H_2CO_{3(aq)}$$

For aqueous carbonic acid at 25°C, $K_1 = 4.456 \times 10^{-7}$ and $K_2 = 4.688 \times 10^{-11}$. Calculate the equilibrium concentrations of all the ionic species in water exposed to the air at one atm pressure. What is the *pH* of this water? ☹

KEY POINTS FOR CHAPTER TWENTY

1. **Acid dissociation**: $\qquad HA_{(aq)} + H_2O \rightleftharpoons H_3O^+_{(aq)} + A^-_{(aq)}$

   Equilibrium constant for the dissociation: $\quad K = \left(\dfrac{a_{H_3O^+} \cdot a_{A^-}}{a_{HA} \cdot a_{H_2O}} \right)_{eqm}$

   Since the relative activity of the solvent water is approximated by its mole fraction, which, for most solutions is close to unity, we usually set $a_{H_2O} = 1$, and re-define K as K_a, which is called the **acid dissociation constant**:

   $$K_a = \frac{a_{H_3O^+} \cdot a_{A^-}}{a_{HA}}$$

2. The value of K_a for any given acid depends not only on the *acid*, but also on the *solvent*,

which will be a Brønsted base, since it must accept the proton from the acid. For example, HCl is a strong acid (~100% dissociated) in *water*, but weak in many non-aqueous solvents. Therefore, if we want to measure the relative strength of different acids, we should compare them against the *same* base, which usually is chosen to be *water*.

Since most K_a values are quite small, it's convenient to express these values in a way similar to that used for *pH*; namely, using negative logarithms to the base 10. For example:

$$pH = -log \ a_{H^+}, \ pK_a = -log \ K_a, \ pOH = -log \ a_{OH^-}, \ pAg = log \ a_{Ag^+}, \text{ etc.}$$

3. For the **self-dissociation of water**:

$$K_W = a_{H_3O^+} \cdot a_{OH^-}$$
$$-log \ K_W = -log \ a_{H_3O^+} - log \ a_{OH^-}$$
$$pK_W = pH + pOH \qquad \text{[= 14.00 at 25°C]}$$

4. **Weak acid *HA* dissociation**: $HA + H_2O \rightleftharpoons H_3O^+ + A^-$

A weak acid is one that only *slightly dissociates* (usually $K_a < 10^{-3}$). The *degree of dissociation* α is the fraction of the acid that dissociates in solution:

$$\alpha = \frac{[A^-]}{[HA]_o}$$

where $[HA]_o$ (also written as c_o) is the *stoichiometric concentration* of the acid (the total acid in the solution, in all forms; i.e., the amount that was weighed out from the reagent bottle when making up the solution).

5. **Contribution from water**: When H_2O dissociates, for each molecule of OH^- formed there will be one molecule of H^+ formed. Therefore, when the concentration of H^+ from the acid is very small, the concentration of H^+ ion produced *from the dissociation of water* must also be taken into account:

$$[H^+]_{total} = [H^+]_{acid} + [H^+]_{water} = [H^+]_{acid} + [OH^-]_{water} = [H^+]_{acid} + \frac{K_W}{[H^+]_{total}}$$

6. **1–1 weak acids**: α depends on the concentration of the acid and on the value of its dissociation constant. An electrolyte such as NaCl is called a *1–1* electrolyte, because the valence of each ion is $z_+ = |z_-| = 1$. Similarly, $CaCl_2$ is a *2–1* salt, $CuSO_4$ is a *2–2* salt, $Al_2(SO_4)_3$ is a *3–2* salt, etc. Thus, lactic acid (*HL*) is a *1–1* weak acid.

For any *1–1* weak acid such as *HA* in which only one H^+ ion dissociates,

$$K_a = \frac{c_o \alpha^2}{1 - \alpha} \qquad \text{(A quadratic equation in } \alpha\text{)}$$

If the acid is extremely weak, $\alpha << 1$, $(1 - \alpha) \approx 1$, and the expression further simplifies to

$$K_a \approx c_o \alpha^2 \quad \text{from which} \quad \alpha \approx \sqrt{\frac{K_a}{c_o}}$$

The same equations apply to weak bases: just substitute K_b for K_a.

7. **Polyprotic acids** are acids that can produce more than one H^+ ion per molecule of acid; e.g., H_2SO_4 ($= H_2A$), H_2CO_3 ($= H_2A$), and H_3PO_4 ($= H_3A$). Also referred to as *polybasic acids*. For these kinds of acids,

$$K_3 << K_2 << K_1$$

Typical **dibasic acid dissociation**:

$$H_2A \rightleftharpoons H^+ + HA^- \quad \textit{Step 1}, \text{ usually } \sim 100\%, \text{ with } \quad K_1 = \frac{[H^+][HA^-]}{[H_2A]}$$

$$HA^- \rightleftharpoons H^+ + A^{2-} \quad \textit{Step 2}, \text{ usually } << 100\%, \text{ with } \quad K_2 = \frac{[H^+][A^{2-}]}{[HA^-]}$$

Since there are 5 species present (H_2A, H^+, HA^-, A^{2-}, and OH^-), we need 5 equations in order to solve for the equilibrium concentration of each species. In addition to the two above equations, we also have the *water equilibrium*, which is always present in aqueous systems:

$$H_2O \rightleftharpoons H^+ + OH^- \qquad K_W = [H^+][OH^-] \qquad \text{(Now we have 3 equations)}$$

We also have a **charge balance** in the system (# positive charges = # negative charges):

$$[H^+] = [OH^-] + [HA^-] + 2[A^{2-}] \qquad \text{(Now we have 4 equations)}$$

Finally, we have a **mass balance** on the component that exists in *more than one form* ("H" is not considered), which, in this case is "A":

$$\textit{Total "A"} = \textit{Sum of "A" in all forms}$$

$$c_o = [H_2A] + [HA^-] + [A^{2-}] \qquad \text{(Now we have 5 equations)}$$

After rearranging and fooling around with the 5 equations, we end up with

$$K_1 = \frac{[H^+]\left(\dfrac{[H^+] - \dfrac{K_W}{[H^+]}}{1 + \dfrac{2K_2}{[H^+]}} \right)}{c_o - \left(1 + \dfrac{K_2}{[H^+]} \right)\left(\dfrac{[H^+] - \dfrac{K_W}{[H^+]}}{1 + \dfrac{2K_2}{[H^+]}} \right)} \qquad \begin{array}{l}\text{which is a } \textit{quartic} \text{ equation} \\ \text{that can be rearranged to}\end{array}$$

$$[H^+]^4 + K_1[H^+]^3 + \left(K_1K_2 - K_W - K_1c_o\right)[H^+]^2 - \left(2K_1K_2c_o + K_1K_W\right)[H^+] - K_1K_2K_W = 0$$

This can be solved for [H⁺] using the Newton-Raphson method, a programmable calculator, or any other method to which you have access. Once we have solved for [H⁺], we can calculate the *pH* of the system immediately, and we also can substitute back into our previous equations and solve for the equilibrium concentration of each species.

Better still, it often—but not always—is possible to simplify the above equation by assuming that, since $K_2 \ll K_1$, most of the H^+ ion comes from the first dissociation, so we can neglect the contribution from the second dissociation. Also, we can assume that, even though the acid is weak, it still contributes much more H^+ ion than does the water; therefore we also can ignore the H^+ ion contributed from the dissociation of water. We then get our usual easy-to-solve quadratic equation in α for a *1–1* weak acid:

$$K_a \approx \frac{c_o \alpha^2}{1 - \alpha}$$

Once we know α, $[H^+] \approx [HA^-] \approx \alpha c_o$

and $[H_2 A] \approx (1 - \alpha) c_o$

PROBLEMS

1. An aqueous solution at 25°C contains 2.00×10^{-7} M H_2SO_4 and 1.50×10^{-6} M HCl.

    (a) Neglecting activity coefficient corrections, calculate the concentration of the OH^- ions in this solution.

    (b) Based on the results of part (a), what is the ionic strength of the solution?

    (c) Using the results of part (b), calculate the concentration of the OH^- ions in this solution taking activity coefficient corrections into account. *Hint:* At the low concentrations encountered in this problem you may assume that molarities and molalities both have the same numerical values, and that both acids completely dissociate. ☹

2. 250 mL of 2.00×10^{-3} M aqueous NaOH is mixed with 150 mL of 3.35×10^{-3} M aqueous HCl after which water is added to bring the total volume to 1.000 L at 25°C.

    (a) Neglecting activity coefficient corrections, what is the H^+ ion concentration in this solution? What is the *pH*?

    (b) Based on the results from part (a), what is the ionic strength of the solution?

    (c) Repeat part (a) correcting for activity coefficient effects. You may assume that under these dilute conditions molarity and molality have the same numerical value.

    Data: $\log_{10} \gamma_i = -0.5108 z_i^2 \sqrt{I}$; $I = \frac{1}{2} \Sigma c_i z_i^2$; $K_w = 1.00 \times 10^{-14}$. ☺

3. Solid NaOH was added to 1.00 L of a 0.050 molar solution of H_2S at 25°C until the *pH* rose to 12.0. The first and second dissociation constants for H_2S may be taken as 1.0×10^{-9} and 1.0×10^{-14}, respectively. $K_w = 1.00 \times 10^{-14}$. The effects of activity coefficients may be neglected for this problem.

    (a) Calculate the initial *pH* of the solution before the addition of the NaOH.

(b) The neutralization can occur by either of two possible reactions:

$$[i] \quad OH^- + H_2S \rightarrow HS^- + H_2O$$

or $\quad [ii] \quad 2OH^- + H_2S \rightarrow S^{2-} + 2H_2O$

By examining the two dissociation constants, determine which of the two neutralization reactions actually takes place.

(c) Determine the concentration of each species at equilibrium.

(d) How many moles of NaOH were added? ☠

4. Calculate the *pH* of 1.00×10^{-7} molar aqueous NaOH solution at 25°C. For water at 25°C, $K_w = 1.00 \times 10^{-14}$. ☺

5. Using the Debye-Hückel equation, calculate the *pH* at 25°C of an aqueous solution that is 0.001 M in acetic acid and 0.05 M in $Ca(NO_3)_2$. The dissociation constant of acetic acid is 1.75×10^{-5}. For this calculation you may assume that the numerical values of the molalities and molarities are the same. *Hint:* The value of the activity coefficient of an undissociated species in solution may be taken as unity. ☺

6. It is usually assumed that at low concentrations the activity coefficient of an undissociated species has a value of 1.00. At high concentrations this assumption may no longer be valid. Using the Guggenheim equation and the accurate data provided that were determined by D.A. MacInnes and T. Shedlovsky for aqueous acetic acid (*HAc*) solutions at 25°C, evaluate:

(a) the acid dissociation constant for acetic acid at 25°C, and

(b) the activity coefficients of the undissociated acetic acid in each of the three solutions listed.

At these low concentrations any differences between molality and molarity may be neglected. ☹

c (mol L^{-1})	α
0.00002801	0.5393
0.100	0.013493
0.200	0.009494

7. For a 0.0100 molal aqueous solution of propionic acid at 25°C the "apparent" acid dissociation constant K'_a, i.e., the value based on using only concentrations instead of activities, is $K'_a = 1.400 \times 10^{-5}$.

$$CH_3CH_2COOH \rightleftharpoons CH_3CH_2COO^- + H^+$$

Determine:

(a) The degree of dissociation of the acid at this concentration.

(b) The ionic strength of the solution.

(c) The mean activity coefficient of the acid.

(d) The true thermodynamic acid dissociation constant. ☹

8. A 0.00100 molar aqueous solution of a certain weak acid *HA* is 5.0% dissociated at 25°C. Using the Debye-Hückel equation, estimate the degree of dissociation if the solution also contains 0.01 M $CaSO_4$. ☹

9. At $I \to 0$, the pK_a value for aqueous acetic acid at 25°C is 4.756. Using the Debye-Hückel equation, estimate the pH of 0.100 M acetic acid at 25°C. ☠

10. What is the pH at 25°C of an aqueous solution containing twice as many OH⁻ ions per liter as in pure water at the same temperature? What concentration of hydroxyl ion comes from some outside source? ☺

11. How many moles of KOH must be added to 100 L of water at 25°C so that the concentration of OH⁻ ions from the KOH is ten times the concentration from the water itself? What is the pH of the solution? ☺

12. At 25°C, 0.10109 molal aqueous acetic acid (HAc) solution is 1.3493% dissociated. Using the Davies equation—useful for concentrations up to about 0.5 molal—calculate

    (a) the pH of this solution, and

    (b) the molal activity coefficient γ_{HAc} for the undissociated acetic acid in this solution.

    For aqueous acetic acid at 25°C, $K_a = 1.7539 \times 10^{-5}$. The Davies equation for aqueous solutions at 25°C is:

    $$log\gamma_i = -\frac{0.5108\, z_i^2 \sqrt{I^{(m)}}}{1 + \sqrt{I^{(m)}}} + 0.1 z_i^2\, I^{(m)} \quad \text{and} \quad log\gamma_\pm = -\frac{0.5108 \mid z_+ z_- \mid \sqrt{I^{(m)}}}{1 + \sqrt{I^{(m)}}} + 0.1 \mid z_+ z_- \mid I^{(m)}$$

    where $I^{(m)} = \frac{1}{2}\Sigma m_i z_i^2$. Hint: $K_a = \dfrac{\gamma_\pm^2 a^2 m_0}{\gamma_{HAc}(1-a)}$ ☺

13. At $I \to 0$, the pK_a value for aqueous monobasic propanoic acid at 25°C is 4.874. Using the Debye-Hückel equation, estimate the pH of 0.0100 molal propanoic acid at 25°C. ☠

TWENTY-ONE

BASES AND THEIR SALTS

21.1 WEAK BASES: *B* or *BOH*

Consider the dissociation of an aqueous solution of the weak Brønsted base *B*:

$$B + H_2O \rightleftharpoons HB^+ + OH^- \qquad K_b = \frac{a_{HB^+} \cdot a_{OH^-}}{a_B} \qquad \textit{Example:}$$

base 1   acid 2   acid 1   base 2 $\qquad\qquad\qquad\qquad NH_3 + H_2O \rightleftharpoons NH_4^+ + OH^-$

Sometimes it is more convenient just to write *BOH* for the base instead of *B*:

$$BOH \rightleftharpoons B^+ + OH^- \qquad K_b = \frac{a_{B^+} \cdot a_{OH^-}}{a_{BOH}} \qquad \textit{Example:}$$

$(1-\alpha)c_o \quad \alpha c_o \quad \alpha c_o \qquad\qquad\qquad\qquad NH_4OH \rightleftharpoons NH_4^+ + OH^-$

In the above expression, α is the fractional degree of dissociation of the base. As in the case of the weak acid, which we discussed in Chapter 20, if the solution is dilute we can write

$$K_b \approx \frac{\alpha^2 c_o}{1-\alpha} \qquad\qquad \dots [1]$$

If the base is only very slightly dissociated, say less than about 5%, then $\alpha < 0.05$, $(1-\alpha) \approx 1$, and Eqn [1] reduces to

$$K_b = \frac{\alpha^2 c_o}{1} \qquad\qquad \dots [2]$$

from which
$$\boxed{\alpha = \sqrt{\frac{K_b}{c_o}}} \quad \textit{Weak Base } (\alpha << 1) \qquad \dots [3]$$

(A similar equation can be written for a weak acid if it dissociates less than about 5%.[1])

If the base is only slightly dissociated, the hydroxyl ion concentration is given by

---

[1] For a weak acid *HA*, the corresponding expression is $\alpha \approx \sqrt{K_a/c_o}$.

$$[OH^-] \approx [B^+] = \alpha c_o = \sqrt{\frac{K_b}{c_o}} \cdot c_o = \sqrt{c_o K_b}$$

$$\boxed{[OH^-] = \sqrt{c_o K_b}} \quad \textit{Weak Base } (\alpha << 1) \quad \cdots [4]$$

Similarly, since $[H^+][OH^-] = K_w$,

$$[H^+] = \frac{K_w}{[OH^-]} = \frac{K_w}{\sqrt{c_o K_b}}$$

$$\boxed{[H^+] = \frac{K_w}{\sqrt{c_o K_b}}} \quad \textit{Weak Base } (\alpha << 1) \quad \cdots [5]$$

Taking *base-10* logarithms of each side of Eqn [5]:

$$log[H^+] = log\, K_w - log(c_o K_b)^{1/2}$$

$$= log\, K_w - \frac{1}{2} log\, c_o - \frac{1}{2} log\, K_b$$

Multiplying both sides by -1:

$$-log[H^+] = -log\, K_w + \frac{1}{2} log\, c_o + \frac{1}{2} log\, K_b$$

That is,
$$\boxed{pH = pK_w - \frac{1}{2} pK_b + \frac{1}{2} log\, c_o} \quad \begin{array}{l} \textit{Weak Base} \\ (\alpha << 1) \end{array} \quad \cdots [6]$$

21.2 THE SALT OF A WEAK ACID: HYDROLYSIS

If *HA* is a weak acid (e.g., acetic acid), then *NaA* is the *salt* of a weak acid (e.g., sodium acetate). Providing the concentration is not too high,[2] most salts can be assumed to dissociate ~100% in solution.

Thus:
$$NaA \overset{100\%}{\to} Na^+ + A^- \qquad \cdots [7]$$

Any reaction in which water is one of the reactants is called **hydrolysis**. Some of the A^- ion will hydrolyze (react with water):

$$A^- + H_2O \rightleftharpoons OH^- + HA \qquad \textit{Hydrolysis of } A^- \qquad \cdots [8]$$
$$(1-\alpha)c_o \qquad\qquad \alpha c_o \quad\; \alpha c_o$$

Since the hydrolysis reaction produces OH^- ions, the solution of a salt of a weak acid is slightly

---

[2] Say, < 0.1 M.

alkaline. In view of Eqn [8], the **hydrolysis constant K_h** is defined as

$$K_h \equiv \frac{[OH^-][HA]}{[A^-]} \qquad \ldots [9]$$

Usually very little of the salt hydrolyzes;[3] therefore, to a good approximation $\alpha \ll 1$, so that

$$[A^-] = (1 - \alpha)c_o \approx c_o = [salt]_o \qquad \ldots [10]$$

where $[salt]_o = c_o$ is the stoichiometric concentration of the salt. Since the solution is alkaline, we also can say that[4]

$$[OH^-] \gg [H^+]$$

There are two sources of OH^- ion in this system: the hydrolysis reaction (Eqn [8]) and the self-dissociation of water:

$$H_2O \rightarrow H^+ + OH^- \qquad \ldots [11]$$

Therefore, the *total* OH^- concentration in the system will be the *sum* of the contributions from each source:

$$[OH^-]_{total} = [OH^-]_{hydrolysis} + [OH^-]_{water} \qquad \ldots [12]$$

Since the hydrolysis reaction (Eqn [8]) doesn't produce any H^+ ion, all the H^+ ion found in the system results from the dissociation of water (Eqn [11]).

From the stoichiometry of Eqn [11], for every mole of H^+ ion produced by the dissociation of water, there also will be one mole of OH^- ion produced. Thus the concentration of OH^- ion produced by the self-dissociation of water will be the *same* as the concentration of H^+ produced by the self-dissociation of water. But, since *all* the H^+ ion in the system comes from the self-dissociation of water, it follows then that the OH^- concentration from the self-dissociation of water must equal the *total* H^+ concentration in the system:[5]

$$[OH^-]_{water} = [H^+]_{total} \qquad \ldots [13]$$

However, since the solution is alkaline, we also know that

$$[OH^-]_{total} \gg [H^+]_{total}$$

and (at 25°C), in view of Eqn [13],

$$[OH^-]_{water} = [H^+]_{total} \ll [OH^-]_{total} \qquad \ldots [14]$$

Eqn [14] tells us that the concentration of OH^- ions resulting from the self-dissociation of water is insignificantly small compared with the concentration of OH^- ions produced by the hydrolysis reaction. In other words, essentially all the OH^- ion in the system comes from the hydrolysis of the salt, and almost none comes from the dissociation of water. Thus, when one mole of NaA hydro-

---

[3] If the salt hydrolyzed extensively, then the solution of a salt such as sodium acetate would be a *strong* base—which it is not: $NaAc + H_2O \rightarrow NaOH + HAc$.

[4] See Exercise 21-1.

[5] If you're getting confused, read this section over a couple more times.

lyzes, one mole of undissociated HA will be formed for each mole of OH^- formed:

$$A^- + H_2O \rightarrow OH^- + HA \qquad \qquad \text{. . . [15]}$$

That is, $$[HA] = [OH^-]_{hydrolysis}$$

But we just showed that $$[OH^-]_{hydrolysis} \approx [OH^-]_{total}$$

therefore,[6] $$[HA] \approx [OH^-]_{total} \qquad \qquad \text{. . . [16]}$$

We now have the following information from Eqns [10] and [16]:

$$\left. \begin{array}{ll} \text{From Eqn [16]:} & [HA] \approx [OH^-] \\ \text{From Eqn [10]:} & [A^-] \approx [salt]_o \end{array} \right\} \qquad \text{. . . [17]}$$

Putting these values into Eqn [9] for the hydrolysis constant gives:

$$K_h = \frac{[OH^-][HA]}{[A^-]} \approx \frac{[OH^-][OH^-]}{[salt]_o} = \frac{[OH^-]^2}{[salt]_o}$$

from which $$\boxed{[OH^-] \approx \sqrt{K_h \cdot [salt]_o}} \quad \begin{array}{l} \textit{Salt of a} \\ \textit{Weak Acid} \end{array} \qquad \text{. . . [18]}$$

Of course, once we know the value of $[OH^-]$, we can obtain $[H^+]$—and therefore the pH—from

$$[H^+] = \frac{K_W}{[OH^-]} \approx \frac{K_W}{\sqrt{K_h \cdot [salt]_o}} = \sqrt{\frac{K_W^2}{K_h \cdot c_o}} \qquad \text{. . . [19]}$$

Note that if we multiply both the top and the bottom of the expression for K_h by $[H^+]$ we get

$$K_h = \frac{[OH^-][HA]}{[A^-]} \times \frac{[H^+]}{[H^+]} = \frac{[HA]}{[H^+][A^-]} \cdot [H^+][OH^-] = \frac{1}{K_a} \cdot K_W$$

Therefore, $$\boxed{K_h = \frac{K_W}{K_a}} \quad \begin{array}{l} \textit{Hydrolysis Constant} \\ \textit{for the Salt of a Weak} \\ \textit{Acid} \end{array} \qquad \text{. . . [20]}$$

Eqn [20] shows that the hydrolysis constant is obtained merely by dividing K_W by K_a for the acid from which the salt is derived. Because they are so easy to calculate, hydrolysis constants usually are not listed in tables of thermodynamic data—all you need are the acid dissociation constants and K_W.

---

[6] The absence of a subscript accompanying a concentration means that the concentration is the *total* concentration—that from *all* sources. Thus, $[H^+]_W$ means the contribution to the H^+ concentration resulting from the dissociation of water, whereas $[H^+]$ means the *total* H^+ concentration in the solution from *all* sources.

Eqn [20] also can be substituted into Eqn [19] to give

$$[H^+] = \sqrt{\frac{K_w^2}{K_h \cdot c_o}} = \sqrt{\frac{K_w^2}{(K_w/K_a) \cdot c_o}} = \sqrt{\frac{K_w \cdot K_a}{c_o}}$$

$$\boxed{[H^+] = \sqrt{\frac{K_w \cdot K_a}{c_o}}} \quad \begin{array}{l} \textit{Salt of a Weak} \\ \textit{1--1 Acid} \end{array} \qquad \ldots [21]$$

Eqn [21] gives the concentration of the H^+ ions in a solution of the salt of a weak *1–1* acid such as acetic acid. Once the hydrogen ion concentration is known, the *pH* readily can be estimated from $pH \approx -log\ [H^+]$.

Exercise 21-1

Often in making simplifying assumptions we assume that $[H^+] \gg [OH^-]$. Taking the criterion as one part in a thousand, at what *pH* can we safely say that $[H^+] \gg [OH^-]$ in aqueous solutions at 25°C ? ☹

Exercise 21-2

Calculate the *pH* of aqueous 0.20 M NH_4Cl solution at 25°C. For NH_4OH, $K_b = 1.77 \times 10^{-5}$. ☹

21.3 THE RIGOROUS METHOD

A number of simplifying assumptions were made in the above derivation. For example, it was assumed that very little of the salt hydrolyzes, so that $[A^-] \approx [salt]_o$. Because the extent of the hydrolysis depends on the value of K_h and on the concentration of the salt, sometimes the above assumption is not valid.

Also, it was assumed that since the solution was alkaline, $[OH^-] \gg [H^+]$. "Much greater than" usually means by a factor of at least 100, and preferably 1000 (see Exercise 21-1 above). Sometimes $[OH^-]$ is just a *little* greater than $[H^+]$. For these borderline cases, the equilibrium must be solved using the rigorous method, in which five equations are needed to solve for the concentrations of the five species: H^+, OH^-, Na^+, HA, and A^-.

The approach we take is similar to that given in detail in Chapter 20 for the rigorous calculation of the polyprotic acid dissociation. Two of the five required equations are obtained from the expressions for the two equilibrium constants K_w and K_a. A third equation results from a *charge balance* on the system, and the final two equations are obtained from *mass balances* on "Na" and on "A." We will leave it with you to try to show that the resulting equation is

$$K_h = \frac{[OH^-]\left([OH^-] - \dfrac{K_W}{[OH^-]}\right)}{c_o - \left([OH^-] - \dfrac{K_W}{[OH^-]}\right)} \qquad \ldots [22]$$

Eqn [22] is a cubic equation in $[OH^-]$ and can be solved analytically[7] or by whatever tools you have available. Often some simplifying assumptions can be made. For example, we know that the solution will be alkaline; therefore, in all cases except borderline cases for which the contribution to the hydroxyl ion concentration from water is close to that from the base itself, we can state that

$$[OH^-] \gg [H^+]$$

But, from the ever-present self-dissociation of water we know that

$$[H^+] = \frac{K_W}{[OH^-]}$$

Therefore,
$$[OH^-] \gg \frac{K_W}{[OH^-]}$$

and in Eqn [22] the term $\left([OH^-] - \dfrac{K_W}{[OH^-]}\right)$ reduces to just $[OH^-]$.

This substitution in both the numerator and the denominator of Eqn [22] reduces Eqn [22] to

$$\frac{[OH^-][OH^-]}{c_o - [OH^-]} \approx K_h \qquad \ldots [23]$$

which is a quadratic equation in $[OH^-]$ and is easily solved. However, if you are too lazy to solve even a quadratic equation, it may be possible to simplify the equation even further. Since we know the solution will be only *weakly* alkaline, as long as the stoichiometric concentration c_o of the salt is not extremely low, it is usually safe to make the further assumption that

$$c_o \gg [OH^-]$$

In this case,
$$c_o - [OH^-] \approx c_o$$

and Eqn [23] further reduces to
$$\frac{[OH^-][OH^-]}{c_o} \approx K_h \qquad \ldots [24]$$

which rearranges to
$$[OH^-] = \sqrt{K_h \cdot c_o} \qquad \ldots [25]$$

Eqn [25] is just the same expression as Eqn [18], which we derived above using the simplified scenario. In general, any rigorously derived expression—such as Eqn [22]—must always reduce to the simplified expression—such as Eqn [25]—when the same simplifying assumptions are

---

[7] The analytical solution to a cubic equation is given in Appendix 6 at the end of the book.

made. This is one way to test if a rigorously derived expression has been derived correctly.

Once [OH⁻] is known, the concentrations of all the other species can be determined by substituting one at a time into the equations for the equilibrium constants, the charge balance, and the mass balances.

Exercise 21-3

Calculate the *pH* of aqueous 1.00×10^{-4} M NH₄Cl solution at 25°C. For NH₄OH, the base dissociation constant is $K_b = 1.77 \times 10^{-5}$. ☹

21.4 THE SALT OF A WEAK BASE

A solution of the salt of a weak base is treated in a manner similar to that of the salt of a weak acid. Thus, if *BOH* is a weak base (e.g., *NH₄OH*), then *BCl* is the *salt* of a weak base (e.g., *NH₄Cl*). Referring to Fig. 1, we can assume that the salt dissociates completely into ions when it dissolves.

$$BCl \xrightarrow{100\%} B^+ + Cl^-$$
$$+$$
$$H_2O \rightleftharpoons OH^- + H^+$$
$$\Updownarrow$$
$$BOH$$

Fig. 1

However, B⁺ ions produced by the dissociation of the salt combine with OH⁻ ions produced by the dissociation of water to produce BOH. Since BOH is a weak base, its equilibrium favors the undissociated BOH form, as indicated in Fig. 1 by the unequal lengths of the arrows. The net result of this process is that OH⁻ ions are removed from the solution, leaving behind a *surplus* of H⁺ ions. Therefore, *a solution of the salt of a weak base is slightly acidic.* The overall effect of these processes is just the hydrolysis of the B⁺ ion, which can be represented by:

$$B^+ + H_2O \rightleftharpoons BOH + H^+ \qquad \textit{Hydrolysis of } B^+ \quad \ldots [26]$$

The hydrolysis constant for the weak base is defined as

$$K_h \equiv \frac{[BOH][H^+]}{[B^+]} \qquad \ldots [27]$$

Multiplying both the numerator and the denominator of Eqn [27] by [OH⁻] gives

$$K_h = \frac{[BOH][H^+]}{[B^+]} \times \frac{[OH^-]}{[OH^-]}$$

$$= \frac{[BOH]}{[B^+][OH^-]} \cdot [H^+][OH^-]$$

$$= \frac{1}{K_b} \cdot K_w = \frac{K_w}{K_b}$$

Therefore,

$$K_h = \frac{K_w}{K_b}$$ *Hydrolysis Constant for the Salt of a Weak Base*   . . . [28]

Eqn [28] shows that the hydrolysis constant for the salt of a weak base can be calculated if the dissociation constant K_b of the weak base is known.[8]

As in the case of the salt of a weak acid, the equilibrium can be solved rigorously for the five species in solution (H^+, OH^-, B^+, BOH, and Cl^-) by making use of K_b for the weak base, K_w, a *charge balance*, and *mass balances* on "Cl" and on "B." When this is done, the following cubic equation in [H^+] results:[9]

$$K_h = \frac{[H^+]\left([H^+] - \dfrac{K_w}{[H^+]}\right)}{c_o - \left([H^+] - \dfrac{K_w}{[H^+]}\right)}$$ . . . [29]

In most cases this equation can be simplified as follows: We know that the solution is slightly *acidic*; therefore

$$[H^+] \gg [OH^-]$$

But we also recognize that

$$\frac{K_w}{[H^+]} = [OH^-]$$

so that in Eqn [29] the term $\left([H^+] - \dfrac{K_w}{[H^+]}\right)$ really is just $\left([H^+] - [OH^-]\right)$.

Since [H^+] $\gg$ [OH^-] this term reduces to $$[H^+] - \frac{K_w}{[H^+]} \approx [H^+]$$ . . . [30]

Using this approximation, Eqn [29] reduces to

$$K_h = \frac{[H^+][H^+]}{c_o - [H^+]}$$ . . . [31]

which is a quadratic equation in [H^+]. As a further simplification, since we know the solution is only *weakly* acidic, providing the stoichiometric concentration c_o of the salt is not too low, we usually also can assume

$$c_o \gg [H^+]$$

so that

$$c_o - [H^+] \approx c_o$$

---

[8] Compare Eqn [28] for the salt of a weak base with Eqn [20] for the salt of a weak acid.

[9] Compare Eqn [29] with Eqn [22]. Instead of going out this weekend, why not stay home and see if you can derive Eqn [29]. Just a suggestion....

Replacing the denominator of Eqn [31] with this assumption further simplifies the expression to

$$K_h = \frac{[H^+][H^+]}{c_o}$$

from which $\boxed{[H^+] \approx \sqrt{K_h \cdot [salt]_o}}$ *Salt of a Weak Base*[10]   . . . [32]

21.5 ACID–BASE TITRATIONS

Suppose we titrate a sample of acid (the *analyte* or solution to be analyzed) with a standardized solution of a base, using a burette. A plot of the *pH* of the system (*y*-axis) *vs.* the volume of the titrant added (*x*-axis) is called a **pH titration curve**. The point at which the stoichiometrically equivalent amount of base has been added to the original sample of acid is called the *equivalence point*, the *end point*, or the *stoichiometric point* of the titration (take your pick).[11]

21.5.1 TITRATION OF A STRONG ACID WITH A STRONG BASE

As an example, suppose we titrate 25.0 mL of 0.10 M HCl solution with standardized 0.20 M NaOH solution at 25°C. To simplify the calculations we shall ignore activity coefficients. The titration reaction is a familiar one:

$$HCl + NaOH \rightarrow NaCl + H_2O$$

Since HCl is a strong acid, it may be assumed to dissociate 100% into H^+ ions and Cl^- ions; similarly, the NaOH titrant is a strong base and may be assumed to dissociate 100% into Na^+ ions and OH^- ions. Thus, the real titration reaction is just

$$H^+ + OH^- \rightarrow H_2O$$

At the end point all the H^+ ions in the beaker from the original acid will have reacted with all the OH^- ions added by the base to form neutral water, so that

$$[H^+] = [OH^-] = 1.0 \times 10^{-7} \qquad \textit{At the end point}$$

and $$pH = -\log [H^+] = 7.00 \qquad \textit{At the end point}$$

The original number of moles of acid n_a^o in the sample is given by

$$n_a^o = M_a V_a^o = \left(0.100 \tfrac{mol}{L}\right)(0.025 \text{ L}) = 0.0025 \text{ mol}$$

where M_a is the molarity (mol L^{-1}) of the original sample of acid and V_a^o is its original volume (in L). The initial *pH* of the system is just the *pH* of the original acid sample in the beaker; namely,

---

[10] Compare Eqn [32] with Eqn [18].
[11] I prefers to call it the *end point*. When I was a kid I always wanted to be an engineer; now I *are* one.

$$pH \approx -log\,[H^+] = -log\,(0.10) = 1.0$$

At any point *before* the end point, the number of moles of acid n_a remaining unneutralized in the beaker is given by

$$n_a = n_a^o - M_b V_b$$

where M_b is the molarity of the NaOH titrant and V_b is the volume of NaOH added to that point. Neglecting any small errors introduced from partial molar volumes, the total volume V of the system at any point in the titration will be

$$V = V_a^o + V_b$$

Thus, at any point *before* the end point the hydrogen ion concentration will be

$$[H^+] = \frac{n_a}{V} = \frac{n_a^o - M_b V_b}{V_a^o + V_b}$$

At the end point:

$$M_b V_b = n_a^o$$

from which the volume of base added will be

$$V_b = \frac{n_a^o}{M_b} = \frac{0.0025\ \text{mol}}{0.20\ \text{mol}\,/\,\text{L}}$$

$$= 0.0125\ \text{L}$$

$$= 12.5\ \text{mL}$$

At this point $[H^+] = 1.00 \times 10^{-7}$ and $pH = 7.00$.

Table 1. *Strong Acid-Strong Base Titration Calculation*

V_b (mL)	$V = V_a^o + V_b$ (L)	$[H^+] = \dfrac{n_a^o - M_b V_b}{V}$	$[OH^-] = \dfrac{M_b V_b - n_a^o}{V}$	pH
0.00	0.025	0.100		1.00
5.00	0.030	0.050		1.30
10.00	0.035	0.0143		1.85
12.00	0.037	0.0027		2.57
12.40	0.0374	5.35×10^{-4}		3.27
12.49	0.03749	5.33×10^{-5}		4.27
12.50	0.0375	1.00×10^{-7}	1.00×10^{-7}	7.00
12.55	0.03755		2.66×10^{-4}	10.43
13.00	0.038		2.63×10^{-3}	11.42
13.60	0.0386		5.70×10^{-3}	11.76
15.00	0.040		1.25×10^{-2}	12.10
20.00	0.045		3.33×10^{-2}	12.52
25.00	0.050		5.00×10^{-2}	12.70
35.00	0.060		7.50×10^{-2}	12.88
∞	∞		0.20	13.30

Beyond the end point, the solution contains excess base n_b given by

$$n_b = M_b V_b - n_a^o$$

The hydroxyl ion concentration beyond the end point will be

$$[OH^-] = \frac{n_b}{V} = \frac{M_b V_b - n_a^o}{V_a^o + V_b}$$

and the *pH* is calculated from $pH = 14.00 - pOH$

The above equations have been used to calculate the *pH* titration data given above in Table 1.[12]

---

[12] These data are shown graphically in Fig. 2, on page 13.

21.5.2 TITRATION OF A WEAK ACID WITH A STRONG BASE

Now we consider the titration at 25°C of 25.00 mL of a *weak* acid—hypochlorous acid, *HClO*—with the same strong base used above, 0.20 M NaOH. For hypochlorous acid, at 25°C $K_a = 3.0 \times 10^{-8}$ and $pK_a = 7.52$. The titration reaction is

$$HClO + NaOH \rightarrow NaClO + H_2O \qquad \ldots [33]$$

The initial hydrogen ion concentration is that of a weak acid; namely,

$$[H^+] = \sqrt{K_a \cdot c_o} = \sqrt{(3.0 \times 10^{-8})(0.10)} = 5.48 \times 10^{-5}$$

from which the $pH = 4.26$. The titration reaction produces the *salt of a weak acid* (sodium hypochlorite) at the end point. From the stoichiometry of Eqn [33] the number of moles of NaClO at the end point is equal to n_a^o, the initial number of moles of the weak acid; namely,

$$n_a^o = M_a V_a^o = \left(0.100 \tfrac{mol}{L}\right)(0.025 \text{ L}) = 0.0025 \text{ mol}$$

At the end point, the volume of base that has been added is

$$V_b = \frac{n_a^o}{M_b} = \frac{0.0025 \text{ mol}}{0.20 \text{ mol/L}} = 0.0125 \text{ L} = 12.5 \text{ mL}$$

At the end point the concentration of the salt NaClO is

$$[salt] = \frac{n_a^o}{V_a^o + V_b} = \frac{0.0025 \text{ mol}}{(0.0125 + 0.025) \text{ L}} = 0.0667 \text{ mol L}^{-1}$$

We showed earlier[13] that the hydrogen ion concentration for a solution consisting of the salt of a weak acid is given by

$$[H^+] = \sqrt{\frac{K_w \cdot K_a}{[salt]}} = \sqrt{\frac{(1.00 \times 10^{-14})(3.0 \times 10^{-8})}{0.0667}} = 6.71 \times 10^{-11} \text{ mol L}^{-1}$$

from which the *pH* at the end point is calculated to be

$$pH = -log\,[H^+] = -log\,(6.71 \times 10^{-11}) = 10.17$$

At any point during the titration *before* the end point we have the species H⁺, HClO, and NaClO coexisting at various concentrations in equilibrium. Therefore,

$$K_a = \frac{[H^+][ClO^-]}{[HClO]} \qquad \ldots [34]$$

---

[13] Eqn [21], p. 21-5.

from which
$$[H^+] = \frac{K_a \cdot [HClO]}{[ClO^-]} = K_a \cdot \frac{[acid]}{[salt]} \qquad \ldots [35]$$

Taking negative *base-10* logarithms of each side of Eqn [35] gives

$$pH = pK_a - log\frac{[acid]}{[salt]} \quad \begin{array}{l} \textit{Henderson-} \\ \textit{Hasselbalch} \\ \textit{Equation} \end{array} \qquad \ldots [36]$$

This equation is called the *Henderson-Hasselbalch equation*, and gives the *pH* at any point before the end point for the titration of a *weak acid* with a *strong base*. Now, at any point *before* the end point the acid concentration will be

$$[acid] = \frac{n_a^o - M_b V_b}{V_a^o + V_b}$$

and the salt concentration will be
$$[salt] = \frac{M_b V_b}{V_a^o + V_b}.$$

Therefore, from Eqn [35], the hydrogen ion concentration at any point before the end point will be

$$[H^+] = K_a \cdot \frac{[acid]}{[salt]} = \frac{K_a \cdot (n_a^o - M_b V_b)/(V_a^o + V_b)}{(M_b V_b)/(V_a^o + V_b)} = K_a \cdot \frac{n_a^o - M_b V_b}{M_b V_b}$$

Beyond the end point, the solution has excess base, as in the case of the strong acid titration, and

$$[OH^-] = \frac{n_b}{V} = \frac{M_b V_b - n_a^o}{V_a^o + V_b}.$$

Note that **HALF-WAY** to the end point,

$$[acid] = [salt]$$

and

$$[H^+]_{1/2} = \frac{K_a \cdot [HClO]}{[ClO^-]}$$

$$= K_a \cdot \frac{[acid]}{[salt]}$$

$$= K_a = 3.8 \times 10^{-8}$$

so that
$$pH_{1/2} = pK_a = 7.52$$

Table 2.   *Weak Acid-Strong Base Titration*

V_b (mL)	$V = V_a^o + V_b$ (L)	$[H^+] = K_a \cdot \dfrac{(n_a^o - M_b V_b)}{M_b V_b}$	pH
0.00	0.025	5.48×10^{-5}	4.26
1.00	0.026	3.45×10^{-7}	6.46
3.00	0.028	9.5×10^{-8}	7.02
6.25	0.03125	$3.8 \times 10^{-8} = K_a$	7.52
10.00	0.035	7.50×10^{-9}	8.13
12.00	0.037	1.25×10^{-9}	8.90
12.40	0.0374	2.42×10^{-10}	9.62
12.50	0.0375	6.71×10^{-11}	10.17

This is an easy way to determine the dissociation constant for a weak acid: Just titrate it with a strong base and measure the *pH* half-way to the end point. The above equations have been used to calculate the *pH* data in Table 2.

The titration curves for both the strong acid titration and the weak acid titration are plotted in Fig. 2:

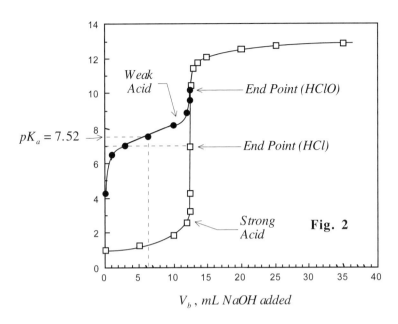

V_b, mL NaOH added

KEY POINTS FOR CHAPTER TWENTY-ONE

1. Dissociation of a **weak base BOH**: $BOH \rightleftharpoons B^+ + OH^-$

$$K_b = \frac{a_{B^+} \cdot a_{OH^-}}{a_{BOH}}$$

If the solution is dilute, then, using the same development as for a weak acid,

$$K_b \approx \frac{\alpha^2 c_o}{1-\alpha} \quad \text{[Quadratic in } \alpha\text{]} \quad \text{and, if } \alpha \ll 1, \ \alpha = \sqrt{\frac{K_b}{c_o}}$$

2. If the base is only slightly dissociated, the hydroxyl ion concentration is given by

$$[OH^-] \approx [B^+] = \alpha c_o = \sqrt{\frac{K_b}{c_o}} \cdot c_o = \sqrt{c_o K_b}$$

from which $$[H^+] = \frac{K_W}{[OH^-]} = \frac{K_W}{\sqrt{c_o K_b}}$$

3. **Salt of a weak ACID → Hydrolysis**:

If HA is a weak acid (e.g., acetic acid) then NaA is the salt of a weak acid (e.g., sodium acetate). At concentrations < 0.1 M most salts dissociate almost completely:

$$NaA \xrightarrow{100\%} Na^+ + A^-$$

Some of the A^- ion (fraction α) will hydrolyze (react with water):

$$A^- + H_2O \rightleftharpoons OH^- + HA$$
$$(1-\alpha)c \qquad \alpha c_o \quad \alpha c_o$$

Since OH^- ions are produced, the solution will be slightly alkaline. The **hydrolysis constant** K_h is defined as

$$K_h = \frac{[OH^-][HA]}{[A^-]}$$

If we assume $\alpha << 1$, then $\quad [A^-] = (1-\alpha)c_o \approx c_o = [salt]_o$

If we assume further that essentially all the OH^- ion in the system comes from the hydrolysis of the salt, and none comes from the dissociation of water, then, from the stoichiometry of the hydrolysis reaction,

$$[OH^-] \approx [HA]$$

Substituting these two assumptions into the expression for K_h gives

$$K_h = \frac{[OH^-][HA]}{[A^-]} = \frac{[OH^-][OH^-]}{[salt]_o} \quad \text{from which} \quad [OH^-] \approx \sqrt{K_h \cdot [salt]_o}$$

Knowing $[OH^-]$ we get $[H^+]$ from $\quad [H^+] = \dfrac{K_W}{[OH^-]} = \dfrac{K_W}{\sqrt{K_h \cdot [salt]_o}}$

4. Tables of hydrolysis constants are not usually found because K_h is easily calculated from a knowledge of the acid dissociation constant K_a for the weak acid:

$$K_h = \frac{[OH^-][HA]}{[A^-]} \times \frac{[H^+]}{[H^+]} = \frac{[HA]}{[H^+][A^-]} \cdot [H^+][OH^-] = \frac{K_W}{K_a}$$

5. **Rigorous method**: If neither of the two assumptions made above is valid, then the rigorous method involving charge and mass balances etc. can be used to obtain

$$K_h = \frac{[OH^-]\left([OH^-] - \dfrac{K_W}{[OH^-]}\right)}{c_o - \left([OH^-] - \dfrac{K_W}{[OH^-]}\right)}, \quad \text{which is a cubic equation in } [OH^-].$$

Recognizing that $K_W/[OH^-] = [H^+]$, and knowing that the solution should be alkaline, then we usually still can say that $[OH^-] >> [H^+]$. Therefore, the expression $[OH^-] - \left(K_W/[OH^-]\right)$ in both the numerator and the denominator of the rigorous expression becomes $\qquad [OH^-] - \dfrac{K_W}{[OH^-]} = [OH^-] - [H^+] \approx [OH^-]$

and the rigorous expression simplifies to $K_h \approx \dfrac{[OH^-][OH^-]}{c_o - [OH^-]}$, which is a quadratic equation in $[OH^-]$ that is easily solved.

6. **Salt of a weak BASE $\rightarrow$ Hydrolysis**:

   If BOH is a weak base (e.g., NH_4OH) then BCl is the salt of a weak base (e.g., NH_4Cl). In the same way that the salt of a weak acid hydrolyzes to form a slightly *basic* solution, the salt of a weak base hydrolyzes to form a slightly *acidic* solution:

   $$B^+ + H_2O \rightleftharpoons BOH + H^+$$

   Similar to the case for the salt of a weak acid,

   $$K_h = \frac{[BOH][H^+]}{[B^+]}, \qquad K_h = \frac{K_W}{K_b}$$

   and, making the usual assumptions, we get $[H^+] \approx \sqrt{K_h \cdot [salt]_o}$.

   The corresponding rigorous expression is $K_h = \dfrac{[H^+]\left([H^+] - \dfrac{K_W}{[H^+]}\right)}{c_o - \left([H^+] - \dfrac{K_W}{[H^+]}\right)}$.

7. **Titration of a STRONG acid with a strong base**:

   Suppose we titrate V_a^o liters of a strong acid of original molarity M_a with a strong base of molarity M_b. The original moles of acid n_a^o in the sample is $n_a^o = M_a V_a^o$. At any point *before* the end point, after the addition of V_b liters of base the number of moles of acid n_a remaining unneutralized in the beaker is

   $$n_a = n_a^o - M_b V_b$$

   and the total volume of the solution will be $V = V_a^o + V_b$. Therefore, at any point before the end point, the H^+ ion concentration will be $[H^+] = \dfrac{n_a}{V} = \dfrac{n_a^o - M_b V_b}{V_a^o + V_b}$ mol L^{-1}.

   At the end point, $M_b V_b = n_a^o$, and the total volume of base that will have been added will be $V_b^{EP} = \dfrac{n_a^o}{M_b}$. At this point, assuming 25°C, $[H^+] = 1.00 \times 10^{-7}$ with $pH = 7.00$.

   Beyond the end point, the solution contains excess base n_b given by $n_b = M_b V_b - n_a^o$.

   The hydroxyl ion concentration beyond the end point will be $[OH^-] = \dfrac{n_b}{V} = \dfrac{M_b V_b - n_a^o}{V_a^o + V_b}$,

   and the pH can be calculated from $pH = 14.00 - pOH$.

8. **Titration of a WEAK acid with a strong base**:

   Now we titrate V_a^o liters of a *weak* acid HA of original molarity M_a with a strong base $NaOH$ of molarity M_b. The titration reaction is

$$HA + NaOH \rightarrow NaA + H_2O$$

The initial hydrogen ion concentration is that of a weak acid; namely, $[H^+] = \sqrt{K_a \cdot c_o}$.

The titration reaction produces the *salt of a weak acid* (*NaA*) at the end point (*EP*). The number of moles of NaA at the end point is equal to n_a^o, the initial number of moles of the weak acid; namely, $n_a^o = M_a V_a^o$. At the end point, the volume of base that has been added is $V_b^{EP} = n_a^o / M_b$, and the concentration of the salt NaA is

$$[salt]_{EP} = \frac{n_a^o}{V_a^o + V_b^{EP}} = \frac{n_a^o}{V_a^o + (n_a^o / M_b)}$$

The hydrogen ion concentration for a solution consisting of the salt of a weak acid is given by

$$[H^+] = \sqrt{\frac{K_w \cdot K_a}{[salt]}} \; ; \text{ therefore at the end point } [H^+]_{EP} = \sqrt{\frac{K_w \cdot K_a}{\left(\dfrac{n_a^o}{V_a^o + (n_a^o / M_b)}\right)}}.$$

At any point during the titration *before* the end point we have the species H$^+$, HA, and NaA coexisting at various concentrations in equilibrium.

Therefore, $\quad K_a = \dfrac{[H^+][A^-]}{[HA]}, \quad$ from which $[H^+] = \dfrac{K_a \cdot [HA]}{[A^-]} = K_a \cdot \dfrac{[acid]}{[salt]}$

Taking negative logarithms, we get the **Henderson-Hasselbalch equation**

$$pH = pK_a - log \frac{[acid]}{[salt]}$$

which gives us the *pH* at any point before the end point for the titration of a weak acid with a strong base. At any point *before* the end point the acid concentration will be

$$[acid] = \frac{n_a^o - M_b V_b}{V_a^o + V_b} \text{ and the salt concentration will be } [salt] = \frac{M_b V_b}{V_a^o + V_b}.$$

Therefore the hydrogen ion concentration at any point *before* the end point will be

$$[H^+] = K_a \cdot \frac{[acid]}{[salt]} = \frac{K_a \cdot (n_a^o - M_b V_b)/(V_a^o + V_b)}{(M_b V_b)/(V_a^o + V_b)} = K_a \cdot \frac{n_a^o - M_b V_b}{M_b V_b}$$

Beyond the end point, the solution has excess base, as in the case of the strong acid titration, and

$$[OH^-] = \frac{n_b}{V} = \frac{M_b V_b - n_a^o}{V_a^o + V_b}.$$

HALF-WAY to the end point, $\qquad [acid] = [salt]$

and $$[H^+]_{1/2} = \frac{K_a \cdot [HA]}{[A^-]} = K_a \cdot \frac{[acid]}{[salt]} = K_a$$

Therefore, to determine the dissociation constant for a weak acid, just titrate it with a strong base and measure the *pH* half-way to the end point!

PROBLEMS

1. Using data from Chapter 20, calculate the *pH* at 25°C of solutions that are 0.10 molar in:
    (a) acetic acid
    (b) sodium bisulfate
    (c) sodium acetate
    (d) ammonia
    (e) ammonium chloride. ☺

2. The *pOH* of a 10^{-3} M KOH solution is 3.0. What is the *pOH* of a 1.00×10^{-8} M KOH solution? ☺

3. Calculate the hydrogen ion concentration in 1.00, 0.100, 1.00×10^{-2}, 1.00×10^{-3}, 1.00×10^{-4}, and 1.00×10^{-5} molar sulfuric acid solutions, assuming that activity equals molar concentration. Evaluate the number of moles of free hydrogen ion produced per stoichiometric mole of sulfuric acid—$[H^+]/[H_2SO_4]_o$—at each concentration. Plot $[H^+]/[H_2SO_4]_o$ (*y*-axis) vs. *log* $[H_2SO_4]$ (*x*-axis). Show that the H^+ ion contribution from water can be neglected. ☺

4. (a) Show that up to and including the end point during the titration of a weak acid with a strong base the slope of the titration curve is given by
    $$\frac{d\,pH}{d\alpha} = \frac{\left([H^+] + K_a\right)^2}{2.303\,K_a\,[H^+]}$$
    where α is the fraction of the acid neutralized at any point during the titration, K_a is the acid dissociation constant for the weak acid, and *pH* is the *pH* at any point during the titration. Neglect any activity effects.

    (b) For the example given in the text in which 25 mL of 0.10 M HClO is titrated with 0.20 M NaOH, calculate $d\,pH/d\alpha$ at (i) 0.5 mL before the end point, (ii) the end point, and (iii) 0.5 mL past the end point. ☹

5. A flask initially contains 50.0 mL of 0.100 mol L^{-1} aqueous HCl solution. This solution is titrated at 25°C with 0.100 mol L^{-1} NH_3 solution added from a burette. For aqueous ammonia at 25°C, $K_b = 1.77 \times 10^{-5}$. Neglecting activity coefficient corrections, calculate the *pH* of the contents of the flask

(a) after 20.0 mL of NH_3 solution has been added,

(b) after 50.0 mL of NH_3 solution has been added, and

(c) after 70.0 mL of NH_3 solution has been added L. ☹

6. When solving ionic equilibrium problems we often make simplifying assumptions to make solving the mathematics easier. For example, as we have seen, the most common technique is to say something such as

$$\left([H^+] + x\right) \approx [H^+] \quad \text{if} \quad [H^+] \gg x$$

Usually we report pH values to two decimal places. How much error can be tolerated in the value of the hydrogen ion concentration in order not to affect the digit in the second decimal place in the value of the pH? In other words, what is the maximum value of x that will not change the second decimal place in the value of the pH? ☹

TWENTY-TWO

BUFFER SOLUTIONS

22.1 BUFFER SOLUTIONS

In our ionic equilibrium calculations thus far, we have been solving for *pH* at equilibrium. The reason *pH* is so important is that a great number of chemical and biochemical processes only operate satisfactorily if the *pH* is held within certain narrow limits. For example, to name just a few, the *pH* of the medium affects the characteristics of electroplated deposits; the reactivity of enzymes; the rate of metallic corrosion; the permeability of cell membranes; the efficiency of fermentations to produce beer, wine, and alcohol; the precipitation of various substances; and the growth of micro-organisms and plants.

In fact, the human body itself is full of controlled *pH* processes: The *pH* of the blood should be held between 7.30 and 7.45; if your blood *pH* falls below 6.8 or rises above 7.8 your body enters a state known as—"death." The *pH* of blood plasma should be maintained between 7.38 and 7.41; the *pH* of saliva usually is about 6.8; the *pH* within the duodenum must be held between 6.0 and 6.5; for proper digestion the *pH* of the gastric juices within the stomach must be kept between 1.6 and 1.8. The body maintains these various *pH* ranges, as needed, by means of chemical constituents that resist *pH* change when small amounts of acid or base are added. Solutions with such regulatory *pH* power are called **buffer solutions**.

Buffer solutions contain relatively large amounts of either (a) a weak *acid* and its salt—this kind of buffer stabilizes *pH* < 7, or (b) a weak *base* and its salt—this kind of buffer stabilizes *pH* > 7.

Buffer action can be understood from the nature of the curve (shown in Fig. 1) for the titration of a weak acid (*HA*) with a strong base (*NaOH*):

$$HA + NaOH \rightarrow NaA + H_2O$$
weak acid   conjugate base

At any point during the titration before the end point, the *pH* is given by the Henderson-Hasselbalch equation, which we derived in Chapter 21:

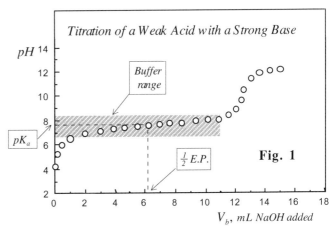

Titration of a Weak Acid with a Strong Base

Fig. 1

V_b, *mL NaOH added*

$$pH \approx pK_a - log\frac{[HA]}{[NaA]} \qquad \ldots [1]$$

It can be seen from Fig. 1 that the curve is quite flat in the region of $pH \approx pKa$; i.e., the pH is *resistant to change*, even though concentrated NaOH is being added to the system. Eqn [1] shows that when the pH is equal to pK_a, [HA] = [NaA]; i.e., [acid] = [salt]. Thus, when the concentration of the acid in the system is about the same as the concentration of the salt of the acid, the system is most resistant to change in pH. This is the basis behind buffer action.

22.2 BUFFER OF A WEAK ACID AND ITS SALT: *HA + NaA*

Consider a solution consisting of a weak acid *HA* at a concentration of $[HA]_o$ and its salt *NaA* at a concentration of $[NaA]_o$. As usual, the salt can be assumed to dissociate 100%, whereas the acid, because it is a *weak* acid, will be only *slightly* dissociated. Since HA is a weak acid, not much A⁻ will be produced from reaction [3]; rather, most of the A⁻ will come from reaction [2]. Therefore, the total concentration of A⁻ is

$$NaA \xrightarrow{100\%} Na^+ + A^- \qquad \ldots [2]$$

$$HA \rightleftharpoons H^+ + A^- \qquad \ldots [3]$$

$$[A^-] = [A^-]_{salt} + [A^-]_{acid} \approx [A^-]_{salt} = [NaA]_o \qquad \ldots [4]$$

Also, since HA is a *weak* acid, it will not undergo much dissociation at such high concentrations ($\alpha << 1$); therefore the total concentration of HA will be

$$[HA] = (1 - \alpha)[HA]_o \approx [HA]_o \qquad \ldots [5]$$

Substituting Eqns [4] and [5] into the expression for the acid dissociation constant gives

$$K_a = \frac{[H^+][A^-]}{[HA]} \approx \frac{[H^+][NaA]_o}{[HA]_o} = [H^+] \cdot \frac{[salt]_o}{[acid]_o} \qquad \ldots [6]$$

Rearranging,

$$[H^+] = \frac{[acid]_o}{[salt]_o} \cdot K_a \qquad \ldots [7]$$

Taking negative logarithms:

$$pH \approx pK_a - log\frac{[acid]_o}{[salt]_o} \qquad \ldots [8]$$

Eqn [8] is just the good old Henderson-Hasselbalch equation, and shows that by choosing appropriate values of $[acid]_o$ and $[salt]_o$ we can control the pH at any desired value by selecting a weak acid with the appropriate value of K_a.

To see how the buffering action works, consider what happens if a small amount of acid or a small amount of base is added to the system:

If a small amount of *acid* is added, it is *absorbed* by the A⁻ ion (the salt) according to

$$H^+ + \underset{in\ buffer}{A^-} \rightarrow HA \qquad \qquad \dots [9]$$

The net result is that the hydrogen ion concentration does not appreciably change; i.e., the *pH* stays constant. Similarly, if a small amount of *base* is added, it is *absorbed* by the HA (acid):

$$OH^- + \underset{in\ buffer}{HA} \rightarrow H_2O + A^- \qquad \qquad \dots [10]$$

The net result now is that the hydroxyl ion concentration does not appreciably change; since

$$[H^+] = \frac{K_w}{[OH^-]}, \qquad \qquad \dots [11]$$

if [OH⁻] doesn't change, then [H⁺] also doesn't change—that is, the *pH* again stays constant. Thus, regardless of whether acid or base is added to the system, the *pH* tends to resist change.

In order for a buffer solution to function it is necessary that the concentrations of HA and NaA be fairly high, so that their values don't change appreciably when they are stressed by acid or base additions. Thus, the greater the concentrations, the greater will be the resistance of the solution to change in *pH* upon the addition of an acid or a base. Usually the ratio $[acid]_o/[salt]_o$ ranges from 0.1 ~ 10 for useful buffer capacity.

Since

$$pH \approx pK_a - log\frac{[acid]_o}{[salt]_o}$$

if

$$\frac{[acid]_o}{[salt]_o} = \frac{10}{1}, \quad then \quad pH \approx pK_a - 1$$

Similarly, if

$$\frac{[acid]_o}{[salt]_o} = \frac{1}{10}, \quad then \quad pH \approx pK_a + 1$$

Therefore, a given weak acid or weak base can be used to prepare buffer solutions of

$$\boxed{pH \approx pK_a \pm 1} \qquad \qquad \dots [12]$$

It is only necessary to choose a weak acid with the desired pK_a-value.

Example 22-1

An aqueous buffer solution of volume 100 mL at 25°C consists of 0.10 M acetic acid (*HAc*) and 0.10 M sodium acetate (*NaAc*).

(a) What is its *pH*?

(b) Neglecting any volume changes, what is the *pH* after the addition of 3.0 mmol of NaOH to the buffer solution?

(c) Again, neglecting any volume changes, what is the *pH* after the addition of 6.0 mmol of nitric acid—*HNO₃*—to the initial buffer solution? For acetic acid, $K_a = 1.754 \times 10^{-5}$.

Solution

(a) This is a buffer solution of a weak acid and its salt: $pK_a = -log(1.754 \times 10^{-5}) = 4.756$

$$pH \approx pK_a - log\frac{[acid]_o}{[salt]_o} = 4.756 - log\left(\frac{0.10}{0.10}\right) = 4.756$$

Ans: $\boxed{4.76}$

(b) When 3 mmol of NaOH is added, it will be neutralized by the acetic acid in the buffer according to

$$NaOH + HAc \rightarrow NaAc + H_2O$$

This will *lower* the concentration of HAc and *increase* the concentration of NaAc in the solution.

The number of moles of HAc originally present was

$$n^o_{HAc} = \left(c^o_a\ \tfrac{mol}{L}\right)\left(V^o_a\ L\right) = (0.10)(0.100) = 0.0100\ mol = 10.0\ mmol$$

After the absorption of the NaOH the number of moles of HAc remaining will be

$$n_{HAc} = 10.0 - 3.0 = 7.0\ mmol$$

Similarly, the original number of moles of NaAc present in the solution was 10.0 mmol; after the addition of base there will be

$$n_{NaAc} = 10.0 + 3.0 = 13.0\ mmol$$

Therefore, the new *pH* will be

$$pH \approx 4.756 - log\frac{[acid]}{[salt]} = 4.756 - log\left(\frac{7.0}{13.0}\right) = 5.025$$

The *pH* has changed by only $\Delta pH = 5.025 - 4.756 = 0.269$ *pH* units.

Ans: $\boxed{5.025}$

(c) When 6.0 mmol of HNO_3 is added it will be neutralized by the acetate in the buffer solution according to

$$HNO_3 + NaAc \rightarrow HAc + NaNO_3$$

After the acid addition, the new values of acetic acid and sodium acetate will be

$$n_{HAc} = 10.0 + 6.0 = 16.0\ mmol$$

and $n_{NaAc} = 10.0 - 6.0 = 4.0\ mmol$

Therefore, the new *pH* will be

$$pH \approx 4.756 - log\frac{[acid]}{[salt]} = 4.756 - log\left(\frac{16.0}{4.0}\right) = 4.154$$

$$\Delta pH = 4.154 - 4.756 = -0.60$$

Ans: $\boxed{4.15}$

Exercise 22-1

Repeat Example 22-1 with a stronger buffer solution in which the initial acid and salt concentrations are 1.00 M instead of 0.10 M. Does this make the system even more resistant to change in pH? ☺

Example 22-2

A solution is made by dissolving 0.01 moles of NaOH, 0.02 moles of sodium oxalate (Na_2Ox), 0.03 moles of sodium hydrogen oxalate ($NaHOx$), and 0.04 moles of oxalic acid (H_2Ox) with enough water to form 1.00 L of solution at 25°C. (a) Solve for the equilibrium concentrations of all species in this system. (b) What is the pH of the system? K_1 and K_2 for oxalic acid at 25°C can be taken as 10^{-2} and 10^{-5}, respectively.

Solution

(a) $H_2Ox \rightleftharpoons H^+ + HOx^-$ $\qquad K_1 = \dfrac{[H^+][HOx^-]}{[H_2Ox]} = 10^{-2}$ $\qquad$ ... [a]

$\quad\;\; HOx^- \rightleftharpoons H^+ + Ox^{2-}$ $\qquad K_2 = \dfrac{[H^+][Ox^{2-}]}{[HOx^-]} = 10^{-5}$ $\qquad$ ... [b]

$\quad\;\; H_2O \rightleftharpoons H^+ + OH^-$ $\qquad K_w = [H^+][OH^-] = 1.00 \times 10^{-14}$ $\qquad$ ... [c]

Since $K_2 \ll K_1$, the extent of the second dissociation to Ox^{2-} is almost insignificant; therefore oxalic acid can be treated as if it were a monobasic acid. Some of the acid will be neutralized by the addition of NaOH according to

$$NaOH + H_2Ox \;\rightarrow\; NaHOx + H_2O$$

	NaOH	H_2Ox	NaHOx
mol L^{-1}, Start:	0.01	0.04	0.03
React:	-0.01	-0.01	$+0.01$
Eqm:	0	0.03	0.04

Therefore the initial solution essentially is a solution consisting of $[NaHOx]_0 = 0.04$ M, $[H_2Ox]_0 = 0.03$ M, and $[Na_2Ox]_0 = 0.02$ M. At equilibrium, there will be six species in solution: Na^+, H^+, OH^-, H_2Ox, HOx^-, and Ox^{2-}. The small value of K_2 tells us that HOx^- is favored over Ox^{2-}; therefore, it is unlikely that the 0.02 M Ox^{2-} from the Na_2Ox will persist. Instead, it will react with H_2Ox to form HOx^-, according to $H_2Ox + Ox^{2-} \rightarrow 2HOx^-$

For this reaction, $$K = \frac{[HOx^-]^2}{[H_2Ox][Ox^{2-}]} = \frac{[HOx^-]}{[H_2Ox]} \cdot \frac{[HOx^-]}{[Ox^{2-}]}$$

$$= \frac{K_1}{[H^+]} \cdot \frac{[H^+]}{K_2} = \frac{K_1}{K_2} = \frac{10^{-2}}{10^{-5}} = 1000$$

This tells us that HOx^- is favored over H_2Ox and Ox^{2-}. If we assume *complete* reaction, then:

	H_2Ox	+	Ox^{2-}	$\rightarrow$	$2HOx^-$
Start:	0.03		0.02[1]		0.04
React:	-0.02		-0.02		$+0.04$
Eqm:	0.01		0.00		0.08

This indicates that the solution can be regarded as just a simple buffer with composition 0.01 M H_2Ox and 0.08 M $NaHOx$, for which, according to the simple buffer formula,

$$[H^+] \approx \frac{[acid]_o}{[salt]_o} \cdot K_a = \frac{0.01}{0.08} \times 10^{-2} = 0.00125 \text{ M}$$

$$[OH^-] = \frac{K_w}{[H^+]} \approx \frac{10^{-14}}{0.00125} = 8.0 \times 10^{-12} \text{ M}$$

$[Na^+] = 0.08$ M; $[H_2Ox] \approx 0.01$ M; $[HOx^-] \approx 0.08$ M

$$[Ox^{2-}] = \frac{[HOx^-]}{[H^+]} \cdot K_2 \approx \frac{0.08}{0.00125} \cdot 10^{-5} = 6.4 \times 10^{-4} \text{ M}$$

(b) $pH \approx -log[H^+] = -log(0.00125) = 2.903$

> **Ans:** (a) $[H^+] = 0.00125$ M; $[OH^-] = 8.0 \times 10^{-12}$ M; $[Na^+] = 0.08$ M;
> $[H_2Ox] = 0.01$ M; $[Ox^{2-}] = 0.00064$ M; $[HOx^-] = 0.08$ M   (b) $pH = 2.90$

A more accurate calculation takes into account the fact that the reaction between H_2Ox and Ox^{2-} doesn't proceed 100% to completion. Thus:

	H_2Ox	+	Ox^{2-}	$\rightarrow$	$2HOx^-$
Start:	0.03		0.02		0.04
React:	$-x$		$-x$		$+2x$
Eqm:	$0.03 - x$		$0.02 - x$		$0.04 + 2x$

$$K = \frac{(0.04 + 2x)^2}{(0.03 - x)(0.02 - x)} = 1000$$

---

[1] From the 0.02 M Na_2Ox, which is 100% dissociated to $2Na^+ + Ox^{2-}$ ions.

Cross-multiplying and rearranging the expression for K yields the following quadratic equation:

$$996x^2 - 50.16x + 0.5984 = 0$$

Solving: $x = \dfrac{-(-50.16) \pm \sqrt{(-50.16)^2 - 4(996)(0.5984)}}{2(996)} = \dfrac{+50.16 \pm 11.489}{1992}$

$= 0.03095$ (extraneous: can't be greater than 0.03!)   or 0.01941 (correct root)

Therefore, $x = 0.01941$, and

$[H_2Ox] = 0.03 - x = 0.03 - 0.01941 = 0.01059$ M

$[Ox^{2-}] = 0.02 - x = 0.02 - 0.01941 = 0.00059$ M

$[HOx^-] = 0.04 + 2x = 0.04 + 2(0.01941) = 0.07882$ M

$[H^+] = \dfrac{[H_2Ox]}{[HOx^-]} \cdot K_1 = \dfrac{0.01059}{0.07882} \times 10^{-2} = 0.00134357$ M

$[OH^-] = \dfrac{K_w}{[H^+]} = \dfrac{10^{-14}}{0.00134357} = 7.44287 \times 10^{-12}$ M

$[Na^+] = [NaOH]_o + 2[Na_2Ox] + [NaHOx] = 0.01 + 2(0.02) + 0.03 = 0.08$ M

The table at the right, which compares these more accurate values with those obtained using the "simple buffer" assumption, gives an idea of how accurate such simple assumptions are. In particular, it should be noted that even though the H^+ ion concentration is in error by almost -7%, this translates to an error in the pH of only $+1\%$, owing to the "damping" nature of logarithmic functions.

Species Concentration	Simple Buffer Assumption	Accurate Calculation	Error in Simple Assumption
$[H^+]$	0.00125	0.00134	-6.7%
$[OH^-]$	8.0×10^{-12}	7.44×10^{-12}	$+7.0\%$
$[Na^+]$	0.08	0.08	0%
$[H_2Ox]$	0.01	0.01059	-5.6%
$[HOx^-]$	0.08	0.0788	$+1.5\%$
$[Ox^{2-}]$	0.00064	0.00059	$+8.6$
pH	2.903	2.873	$+1.0\%$

Exercise 22-2

Calculate the pH at 25°C of a solution produced by mixing 25.0 mL of 3.00 M NaOH with 75.0 mL of 1.00 M acetic acid (HAc). For acetic acid at 25°C, $K_a = 1.754 \times 10^{-5}$. ☺

22.3 RIGOROUS DERIVATION: *HA/NaA* BUFFER

The derivation of the buffer formula

$$\frac{[H^+][NaA]_o}{[HA]_o} = K_a \qquad \ldots [13]$$

involved two simplifying assumptions; namely, that $[H^+] \gg [OH^-]$, and that the concentrations of both the undissociated acid and the salt each were much greater than the concentrations of H^+ ion and OH^- ion. In certain borderline cases, one—or both—of these assumptions may not be valid, making it necessary to derive a rigorous expression to determine the hydrogen ion concentration.

We proceed as follows:

$$NaA \overset{100\%}{\rightarrow} Na^+ + A^- \qquad\qquad [Na^+] = [NaA]_o \qquad \ldots [14]$$

$$HA \rightleftharpoons H^+ + A^- \qquad\qquad K_a = \frac{[H^+][A^-]}{[HA]} \qquad \ldots [15]$$

$$H_2O \rightleftharpoons H^+ + OH^- \qquad\qquad K_W = [H^+][OH^-] \qquad \ldots [16]$$

Mass Balance on "A": $\quad [HA]_o + [NaA]_o = [HA] + [A^-] \qquad \ldots [17]$

Charge Balance: $\quad [H^+] + [Na^+] = [OH^-] + [A^-] \qquad \ldots [18]$

From [18]: $\quad [A^-] = [Na^+] + [H^+] - [OH^-]$

$$= [NaA]_o + [H^+] - [OH^-] \qquad \ldots [19]$$

From [17]: $\quad [HA] = [HA]_o + [NaA]_o - [A^-]$

$$= [HA]_o + [NaA]_o - \left\{ [NaA]_o + [H^+] - [OH^-] \right\}$$

$$= [HA]_o - \left\{ [H^+] - [OH^-] \right\} \qquad \ldots [20]$$

Putting [19] and [20] into [15]:

$$K_a = \frac{[H^+]\left\{ [NaA]_o + [H^+] - [OH^-] \right\}}{[HA]_o - \left\{ [H^+] - [OH^-] \right\}} \qquad \ldots [21]$$

When $[OH^-]$ is replaced with $K_W/[H^+]$, Eqn [21] becomes a cubic equation in $[H^+]$, which can be solved (maybe with some difficulty) for $[H^+]$. Once $[H^+]$ is known, the concentrations of all the other species can be calculated using Eqns [19], [20], and K_W.

SIMPLIFYING ASSUMPTIONS:

Usually the concentrations of the salt and of the acid are much greater than the concentrations of the hydrogen ions and the hydroxyl ions, so that

$$\left\{ [NaA]_o + [H^+] - [OH^-] \right\} \approx [NaA]_o \qquad \ldots [22]$$

and
$$[HA]_o - \left\{ [H^+] - [OH^-] \right\} \approx [HA]_o \qquad \ldots [23]$$

Substituting Eqns [22] and [23] into Eqn [21] simplifies Eqn [21]:

$$K_a = \frac{[H^+]\left\{ [NaA]_o + [H^+] - [OH^-] \right\}}{[HA]_o - \left\{ [H^+] - [OH^-] \right\}} \approx \frac{[H^+][NaA]_o}{[HA]_o} \qquad \ldots [24]$$

from which
$$[H^+] \approx \frac{[HA]_o}{[NaA]_o} \cdot K_a \qquad \ldots [25]$$

and
$$\boxed{pH \approx pK_a - log \frac{[acid]_o}{[salt]_o}}$$

which is just the simplified equation we derived earlier as Eqn [8].

22.4 BUFFER OF A WEAK BASE AND ITS SALT: *BOH + BCl*

The simplified expression for a buffer consisting of a weak base *BOH* and its salt *BCl* is derived in the same way as the derivation for a weak acid and its salt. The final equation also is similar, with the only differences being that $[OH^-]$ replaces $[H^+]$, $[base]_o$ replaces $[acid]_o$, and K_b replaces K_a, giving

$$[OH^-] = \frac{[base]_o}{[salt]_o} \cdot K_b \qquad \ldots [26]$$

from which
$$[H^+] = \frac{K_w}{[OH^-]} \approx \frac{[salt]_o}{[base]_o} \cdot \frac{K_w}{K_b} \qquad \ldots [27]$$

Taking negative logarithms:
$$\boxed{pH \approx pK_w - pK_b + log \frac{[base]_o}{[salt]_o}} \qquad \ldots [28]$$

Eqn [28] is the simplified expression for a buffer solution consisting of a weak base and its salt. If a small amount of *acid* is added to this system, it is absorbed by the *weak base*, with a negligible change in $[H^+]$:

$$H^+ + \underset{in\,buffer}{BOH} \rightarrow B^+ + H_2O \qquad \ldots [29]$$

Similarly, if a small amount of *base* is added, it is absorbed by the *salt*, this time with a negligible change in $[OH^-]$:

$$OH^- + \underset{in\,buffer}{B^+} \rightarrow BOH \qquad \ldots [30]$$

Exercise 22-3

Calculate the *pH* at 25°C of an aqueous solution that is 0.15 M in NH_3 and 0.10 M in NH_4Cl. If 20 mL of a 0.12 M HCl solution is added to 80 mL of the solution, by how much will the *pH* change? For aqueous NH_3, $K_b = 1.77 \times 10^{-5}$ at 25°C. ☺

22.5 RIGOROUS EXPRESSION FOR *BOH/BCl* BUFFER

Using the usual mass balances, charge balances, and equilibrium constants, the rigorous expression for a buffer solution consisting of a weak base *BOH* and its salt *BCl* can be shown to be

$$K_b = \frac{\left\{ [salt]_o + [OH^-] - [H^+] \right\} \cdot [OH^-]}{[base]_o + [H^+] - [OH^-]} \qquad \ldots [31]$$

This expression, which is a cubic equation in [OH⁻], reduces to the simplified version when the appropriate assumptions are made.

Example 22-3

Calculate:

(a) The *pOH* of a solution made by dissolving 0.100 moles of NH_4Cl in one liter of 0.100 M aqueous NH_3 solution at 25°C. Assume no increase in volume.

(b) The *pOH* of this solution when diluted with water to 1000 times its original volume. Justify any assumptions made.

For aqueous ammonia, $K_b = 1.77 \times 10^{-5}$ at 25°C. Assume complete dissociation of the salt.

Solution

(a)   An aqueous ammonia solution is just a solution of NH_4OH:

$$NH_{3(aq)} + H_2O \equiv NH_4OH_{(aq)}$$

$NH_4OH_{(aq)}$ is a weak base:

$$NH_4OH \rightleftharpoons NH_4^+ + OH^- \qquad K_b = 1.77 \times 10^{-5}$$

When 0.100 moles of NH_4Cl is dissolved in the solution, it dissociates completely:[2]

$$NH_4Cl \overset{100\%}{\rightarrow} NH_4^+ + Cl^-$$

---

[2] Provided they are not too concentrated (say < 0.1 M), salts may be assumed to dissociate completely in aqueous solution.

Therefore we have a solution containing a weak base (NH_4OH) and its salt (NH_4Cl); i.e., we have a *buffer* solution. For this kind of buffer, the appropriate expression is Eqn [26]:

$$[OH^-] \approx \frac{[base]_o}{[salt]_o} \cdot K_b = \frac{(0.100)}{(0.100)} \cdot (1.77 \times 10^{-5}) = 1.77 \times 10^{-5}$$

$$pOH \approx -log\,[OH^-] = -log\,(1.77 \times 10^{-5}) = 4.752$$

We should always check the assumptions inherent in the approximate solution, namely:

$$\left.\begin{array}{l} [base]_o = 0.100 \\ [salt]_o = 0.100 \end{array}\right\} \gg \left.\begin{array}{l} [OH^-] = 1.77 \times 10^{-5} \\ [H^+] = \frac{K_w}{[OH^-]} = \frac{10^{-14}}{1.77 \times 10^{-5}} = 5.6 \times 10^{-10} \end{array}\right\} \quad \text{OK!!}$$

Ans: $\boxed{4.75}$

(b) When the solution is diluted by a factor of 1000, the base and salt concentrations become

$$[base]_o = [salt]_o \approx \frac{0.100}{1000} = 10^{-4}\,M$$

The approximate formula suggests

$$[OH^-] \approx \frac{[base]_o}{[salt]_o} \cdot K_b = \frac{10^{-4}}{10^{-4}} \cdot (1.77 \times 10^{-5}) = 1.77 \times 10^{-5} \quad \text{(the same as before)}$$

with
$$[H^+] = \frac{K_w}{[OH^-]} = \frac{10^{-14}}{1.77 \times 10^{-5}} = 5.6 \times 10^{-10}$$

Checking the assumptions inherent in the formula shows that we still safely can say that

$$[OH^-] \approx 1.77 \times 10^{-5} \gg [H^+] \approx 5.6 \times 10^{-10}$$

BUT, we *no longer can assume that* $[OH^-] \approx 1.77 \times 10^{-5} \ll \left\{\begin{array}{l} [base]_o = 10^{-4} \\ [salt]_o = 10^{-4} \end{array}\right\}$

However, it *is* still true that $\left\{\begin{array}{l} [base]_o = 10^{-4} \\ [salt]_o = 10^{-4} \end{array}\right\} \gg [H^+] \approx 5.6 \times 10^{-10}.$

This means that we must turn to the rigorous solution, Eqn [31]:

$$K_b = \frac{[OH^-]\left\{[salt]_o + [OH^-] - [H^+]\right\}}{[base]_o - \left\{[OH^-] - [H^+]\right\}} \quad \ldots [a]$$

Fortunately, the [H^+] still is insignificant compared with the concentrations of the other species, so that, to a good approximation

$$[salt]_o + [OH^-] - [H^+] \approx [salt]_o + [OH^-] \quad \ldots [b]$$

Substituting Eqn [b] into Eqn [a] simplifies Eqn [a]:

$$K_b = \frac{[OH^-]\{[salt]_o + [OH^-] - [H^+]\}}{[base]_o - \{[OH^-] - [H^+]\}} \quad \rightarrow \quad K_b \approx \frac{[OH^-]\{[salt]_o + [OH^-]\}}{[base]_o - [OH^-]} \quad \cdots [c]$$

Putting values into Eqn [c]: $1.77 \times 10^{-5} = \dfrac{[OH^-]\{10^{-4} + [OH^-]\}}{10^{-4} - [OH^-]} \quad \cdots [d]$

When cross-multiplied and rearranged, Eqn [d] gives a quadratic equation in $[OH^-]$:

$$1.77 \times 10^{-5} \times 10^{-4} - 1.77 \times 10^{-5}[OH^-] = 10^{-4}[OH^-] + [OH^-]^2$$

Rearranging: $[OH^-]^2 + 1.177 \times 10^{-4}[OH^-] - 1.77 \times 10^{-9} = 0 \qquad \cdots [e]$

Solving:

$$[OH^-] = \frac{-1.177 \times 10^{-4} \pm \sqrt{(1.177 \times 10^{-4})^2 - 4(1)(-1.77 \times 10^{-9})}}{2(1)}$$

$$= \frac{-1.177 \times 10^{-4} \pm 1.4468 \times 10^{-4}}{2}$$

$$= 1.3492 \times 10^{-5} \quad or \quad -1.3119 \times 10^{-4}$$

The second root is extraneous because $[OH^-]$ must be greater than zero.

Therefore, $[OH^-] = 1.3492 \times 10^{-5}$

Checking the assumptions: $[H^+] = \dfrac{K_W}{[OH^-]} = \dfrac{10^{-14}}{1.3492 \times 10^{-5}} = 7.41 \times 10^{-10}$

$$[H^+] = 7.41 \times 10^{-10} \ll \begin{cases} [base]_o = 10^{-4} \\ [salt]_o = 10^{-4} \\ [OH^-] = 1.35 \times 10^{-5} \end{cases} \quad \text{Good, good...the assumptions are valid!}$$

Therefore, $pOH = -log\,[OH^-] = -log\,(1.3492 \times 10^{-5}) = 4.8699$

This answer is *not* the same as that for part (a), which was 4.75.   **Ans**: | 4.87 |

22.6 ACID–BASE INDICATORS: *HIn*

Acid-base indicators are large water-soluble organic molecules with an acid form *HIn* and a conjugate base form *In⁻* that has a different color. Small amounts of these substances are added to the analyte during an acid-base titration; the characteristic color change that occurs at the end point of the titration indicates when the titration is complete. Thus, the dissociation of the indicator and its equilibrium constant are given by

$$HIn_{(aq)} \rightleftharpoons H^+_{(aq)} + In^-_{(aq)} \qquad K_{In} = \frac{a_{H^+} \cdot a_{In^-}}{a_{HIn}} \qquad \ldots [32]$$

from which

$$\frac{K_{In}}{a_{H^+}} = \frac{a_{In^-}}{a_{HIn}} \approx \frac{[In^-]}{[HIn]}$$

Taking *base-10* logarithms of each side:

$$\boxed{log\frac{[In^-]}{[HIn]} \approx pH - pK_{In}} \qquad \ldots [33]$$

When *[In⁻]/[HIn] > 10*, the color of the *basic* form predominates; when *[In⁻]/[HIn] < 0.1*, the color of the *acid* form predominates.

To be useful as an indicator of the end point of the titration, the indicator must be chosen such that pK_{In} has approximately the same value as the *pH* at the stoichiometric end point. For strong acid-strong base titrations, choose an indicator with $pK_{In} \approx 7$; for weak acid-strong base titrations, choose an indicator with $pK_{In} > 7$; for strong acid-weak base titrations, choose an indictor with $pK_{In} < 7$.

For example, suppose we are titrating 25.0 mL of 0.100 M acetic acid (a weak acid) with 0.200 M NaOH (a strong base).

The titration reaction is $\qquad HAc + NaOH \rightarrow NaAc + H_2O$

At the end point (with 12.5 mL of base added to give a total volume of 37.5 mL), we will have a solution of 2.5/37.5 = 0.0667 M sodium acetate—the salt of a weak acid—for which the hydrogen ion concentration is given by

$$[H^+] \approx \sqrt{\frac{K_w \cdot K_a}{[salt]}} = \sqrt{\frac{(1.00 \times 10^{-14})(1.754 \times 10^{-5})}{0.0667}} = 1.6216 \times 10^{-9}$$

The *pH* at the end point will be

$$pH \approx -log\,[H^+]$$
$$= -log\,(1.6216 \times 10^{-9})$$
$$= 8.79$$

A good choice of indicator would be *thymol blue*, with $pK_{In} = 8.9$. The acid form of thymol blue is yellow, while the basic form is blue; therefore, at the end point the color of the titrated solution will change from yellow to blue (see Fig. 2 below).

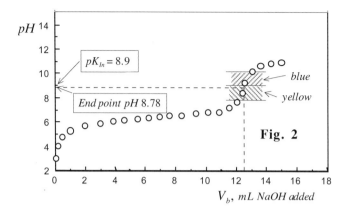

Fig. 2

KEY POINTS FOR CHAPTER TWENTY-TWO

1. Solutions that are able to resist *pH* changes when small amounts of acid or base are added are called **buffer solutions**. Buffer solutions contain relatively large amounts of either (a) a weak *acid* and its salt—this kind of buffer stabilizes *pH* < 7, or (b) a weak *base* and its salt—this kind of buffer stabilizes *pH* > 7.

2. Buffer action can be understood by considering the titration of a weak acid *HA* with a strong base *NaOH*:

$$HA + NaOH \rightarrow NaA + H_2O$$

At any point during the titration before the end point, the *pH* is given by the Henderson-Hasselbalch equation,

$$pH \approx pK_a - log \frac{[HA]}{[NaA]}$$

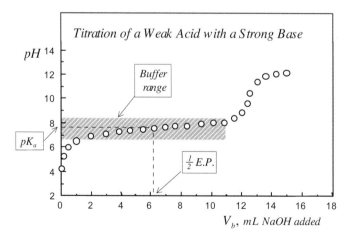

The titration curve is quite flat in the region of $pH \approx pK_a$; i.e., the pH is *resistant to change*, even though concentrated NaOH is being added to the system. The Henderson-Hasselbalch equation shows that when the pH is equal to pK_a, [HA] = [NaA]; i.e., [acid] = [salt].

Thus, when the concentration of the acid is about the same as the concentration of the salt of the acid, the system is most resistant to change in pH. This is the basis behind buffer action.

3. **Buffer of a weak ACID and its salt:** $HA + NaA$

    Consider a fairly concentrated solution consisting of a weak acid HA at concentration $[HA]_o$ and its salt NaA at concentration $[NaA]_o$. The salt dissociates 100% but the weak acid only slightly dissociates:

    $$NaA \xrightarrow{\ 100\%\ } Na^+ + A^-$$

    $$HA \rightleftharpoons H^+ + A^-$$

    Since HA is a weak acid, not much A^- comes from the acid; most will come from the NaA. Also, very little of the acid will be dissociated at such high concentrations ($\alpha << 1$):

    $$[A^-] = [A^-]_{salt} + [A^-]_{acid} \approx [A^-]_{salt} = [NaA]_o$$

    $$[HA] = (1 - \alpha)[HA]_o \approx [HA]_o$$

    Therefore:

    from $\quad K_a = \dfrac{[H^+][A^-]}{[HA]} \quad$ we get $\quad [H^+] = K_a \cdot \dfrac{[HA]}{[A^-]} \approx K_a \cdot \dfrac{[HA]_o}{[NaA]_o} = K_a \cdot \dfrac{[acid]_o}{[salt]_o}$

    Taking negative logarithms: $\qquad pH \approx pK_a - log\dfrac{[acid]_o}{[salt]_o}$

    Thus, to buffer at a desired pH, you must choose a weak acid having a pK_a that is close to the desired pH, and then fine-tune by adjusting the ratio of the acid to the salt.

4. **Mechanism of buffering action**:

    If a small amount of *acid* is added, it is *absorbed* by the A^- ion (the salt), so the hydrogen ion concentration does not appreciably change; i.e., the pH stays approximately constant:

    $$H^+ + \underset{in\,buffer}{A^-} \rightarrow HA$$

    If a small amount of *base* is added, it is *absorbed* by the HA (acid), so the hydroxyl ion concentration does not appreciably change. If [OH$^-$] doesn't change, then [H$^+$] also doesn't change—that is, the pH again stays constant.

    $$OH^- + \underset{in\,buffer}{HA} \rightarrow H_2O + A^-$$

    So, regardless of whether acid or base is added, the pH tends to resist change.

    For effective buffering action the concentrations of HA and NaA should be high, so their values don't change appreciably with small acid or base additions. Usually the $[acid]_o/[salt]_o$ ratio ranges from 0.1 ~ 10 for useful buffer capacity, giving a useful buffering range for any given weak acid of

$$pH \approx pK_a \pm 1$$

5. **Rigorous *HA/NaA* buffer derivation**:

    Making no assumptions and using mass and charge balances, etc., the rigorous equation is

    $$K_a = \frac{[H^+]\{[NaA]_o + [H^+] - [OH^-]\}}{[HA]_o - \{[H^+] - [OH^-]\}}$$
    which, when $[OH^-]$ is replaced with $K_w/[H^+]$, becomes a cubic equation in $[H^+]$ that can be solved using various methods available.

6. **Buffer of a weak BASE and its salt: *BOH + BCl***

    The derivation is the same as for a weak acid and its salt except that $[H^+]$ is replaced with $[OH^-]$, $[acid]_o$ is replaced with $[base]_o$, and K_a is replaced with K_b:

    $$[OH^-] \approx \frac{[base]_o}{[salt]_o} \cdot K_b$$

    If we know $[OH^-]$, then $[H^+]$ is readily calculated from $[H^+] = K_w/[OH^-]$.

    If a small amount of *acid* is added, it is absorbed by the *weak base*, with a negligible change in $[H^+]$:

    $$H^+ + \underset{in\ buffer}{BOH} \rightarrow B^+ + H_2O$$

    Similarly, if a small amount of *base* is added, it is absorbed by the *salt*, this time with a negligible change in $[OH^-]$:

    $$OH^- + \underset{in\ buffer}{B^+} \rightarrow BOH$$

    The rigorous expression is

    $$K_b = \frac{\{[salt]_o + [OH^-] - [H^+]\} \cdot [OH^-]}{[base]_o + [H^+] - [OH^-]}$$

7. **Acid-Base indicators: *HIn***

    These are large water-soluble organic molecules with an acid form *HIn* and a conjugate base form *In⁻* that has a different color. When added in small amounts to an acid-base titration, the characteristic color change indicates the end point of the titration.

    $$HIn_{(aq)} \rightleftharpoons H^+_{(aq)} + In^-_{(aq)} \qquad K_{In} = \frac{a_{H^+} \cdot a_{In^-}}{a_{HIn}} \rightarrow \frac{K_{In}}{a_{H^+}} \approx \frac{a_{In^-}}{a_{HIn}} \approx \frac{[In^-]}{[HIn]}$$

    Taking *base-10* logarithms of each side: $\qquad log\frac{[In^-]}{[HIn]} \approx pH - pK_{In}$

    When $[In^-]/[HIn] > 10$, the color of the *basic* form predominates; when $[In^-]/[HIn] < 0.1$, the color of the *acid* form predominates. The indicator must be chosen so that its color change takes place at a *pH* that is close to the *pH* of the end point of the titration.

PROBLEMS

1. 40.0 mL of 0.50 M aqueous NH_3 solution and x mL of aqueous 0.50 M NH_4Cl, both at 25°C, are added to a beaker. Then 10.0 mL of 0.10 M HCl, also at 25°C, is added. If the final pH at 25°C is to be 9.20, how many mL of NH_4Cl must be added? That is, what is x? You may assume all volumes are additive, and may neglect activity coefficient corrections. For aqueous ammonia at 25°C, $K_b = 1.77 \times 10^{-5}$; for water at 25°C, $K_w = 1.00 \times 10^{-14}$. ☹

2. Aqueous nitrous acid (HNO_2) is a weak acid with $K_a = 4.27 \times 10^{-4}$ at 25°C.

   (a) 50.0 moles of HNO_2 is dissolved in enough water to make 418 L of solution at 25°C. What will be the pH of this solution?

   (b) How many moles of solid $NaNO_2$ must be added to this system to bring its pH to 3.00?

   (c) If 15.0 L of aqueous 1.00 M HCl is added to this system, what will be the final pH at 25°C?

   For this problem you may neglect any activity coefficient corrections. ☺

3. 500 mL of aqueous 0.300 M acetic acid (CH_3COOH) is added to 500 mL of aqueous 0.100 M NaOH at 25°C. What is the pH of this solution? For acetic acid at 25°C, $K_a = 1.754 \times 10^{-5}$. For this problem you may neglect activity coefficient corrections, may assume that molarity = molality, and may assume that solution volumes are additive. ☺

4. An aqueous 25°C solution is 0.150 molar in ammonia (NH_3) and 0.100 molar in ammonium chloride (NH_4Cl).

   (a) Calculate the pH of the solution.

   (b) If 20 mL of a 0.15 mol L^{-1} solution of hydrochloric acid (HCl) is added to 80 mL of the solution of part (a), by how much will the pH change?

   For this problem you may neglect activity coefficient corrections. At 25°C, for water $K_W = 1.00 \times 10^{-14}$ and for ammonia $K_b = 1.77 \times 10^{-5}$. ☺

5. How many mL of 0.150 M HCl must be added to 100.0 mL of 0.200 M NH_3 solution to prepare an aqueous buffer solution having a pH of 9.20 at 25°C? For ammonia at 25°C, $K_b = 1.77 \times 10^{-5}$. You may disregard any activity coefficient corrections. ☹

6. How many mL of 0.150 M HCl must be added to 100.0 mL of 0.200 M sodium acetate (CH_3COONa) solution to prepare an aqueous buffer solution having a pH of 4.80 at 25°C? For acetic acid at 25°C, $K_a = 1.774 \times 10^{-5}$. Neglect activity coefficient corrections. ☺

7. How many mL of 0.200 M NaOH must be added to 50.0 mL of 0.150 M acetic acid (CH_3COOH) solution to prepare an aqueous buffer solution having a pH of 5.20 at 25°C? For acetic acid at 25°C, $K_a = 1.774 \times 10^{-5}$. Neglect activity coefficient corrections. ☺

8. At 25°C, formic acid ($HCOOH$) has a dissociation constant of $K_a = 1.77 \times 10^{-4}$. Neglecting activity coefficient corrections:

(a) Calculate the *pH* of a solution made by dissolving 0.100 moles of potassium formate (*HCOOK*) in 1.00 L of 0.100 mol L^{-1} formic acid solution, assuming no change in volume.

(b) What is the *pH* of the solution resulting from diluting the solution in part (a) by a factor of 1000?

Justify any assumptions you make in calculating your answers. ☹

9.   Propanoic acid (C_2H_5COOH) is a weak acid, with $pK_a = 4.874$ at 25°C. 56.0 moles of this acid is put into a tank, which then is filled with pure water to a total volume of 7.00 m^3. Neglecting activity coefficient effects, at 25°C what is

(a) the *pH* of this solution?

(b) the *pH* of this solution after 35.0 moles of sodium hydroxide is added to the tank? You may assume that this addition does not change the volume of the solution. ☺

10. What are the concentrations of the undissociated acetic acid and of the free acetate ion in a "0.3 M acetate" buffer of *pH* 5.20? (A "0.3 M acetate" buffer is a buffer solution that contains a total of 0.3 moles of acetate per liter; i.e., [HAc] + [Ac⁻] = 0.3 M.) For acetic acid, $K_a = 1.754 \times 10^{-5}$. ☺

11. An aqueous solution at 25°C contains 1.00×10^{-3} mol L^{-1} hydrofluoric acid (*HF*) and 1.00×10^{-3} mol L^{-1} free fluoride ion (*F⁻*). For hydrofluoric acid at 25°C, $K_a = 5.888 \times 10^{-4}$.

(a) If 5.0×10^{-4} moles of HCl is added to one liter of the above solution, calculate the final concentrations of H⁺, HF, and F⁻ and the *pH*. Carry out the calculation making use of the most simplifying assumption.

(b) Repeat the calculation in part (a) using a more rigorous method to calculate the exact values.

For the above calculations you may ignore activity coefficient corrections. Also, you may assume that the addition of the HCl does not change the volume of the solution. ☹

12. Blood plasma contains dissolved CO_2 and HCO_3^-, which exert a buffering action. At 37°C (human body temperature) the overall equilibrium process controlling the dissolved CO_2, HCO_3^-, and H⁺ is

$$CO_{2(aq)} + H_2O \rightleftharpoons H^+ + HCO_3^- \qquad K = \frac{[H^+][HCO^-]}{[CO_2]} = 7.9 \times 10^{-7} \text{ at 37°C}$$

Blood plasma differs from conventional buffer solutions in that the concentration of dissolved CO_2 is maintained at a constant value by the release of any excess CO_2 to the lungs.

(a) Blood plasma contains a total carbonate pool (i.e., $HCO_3^- + CO_2$) of 0.0252 mol L^{-1} and has a *pH* of 7.40. What are the concentrations of dissolved CO_2 and HCO_3^-?

(b) What will be the new *pH* if 0.001 mol L^{-1} of H⁺ ion is introduced into the blood?

When solving this problem you may neglect any activity coefficient corrections. ☹

TWENTY-THREE

SOLUBILITY EQUILIBRIA

23.1 SOLUBILITY EQUILIBRIA

Salts such as silver iodide (AgI), which have extremely low solubilities, are called *slightly soluble salts*. Consider the equilibrium (shown in Fig. 1) between excess crystals of solid AgI in the bottom of a beaker and a saturated solution of dissolved silver iodide:

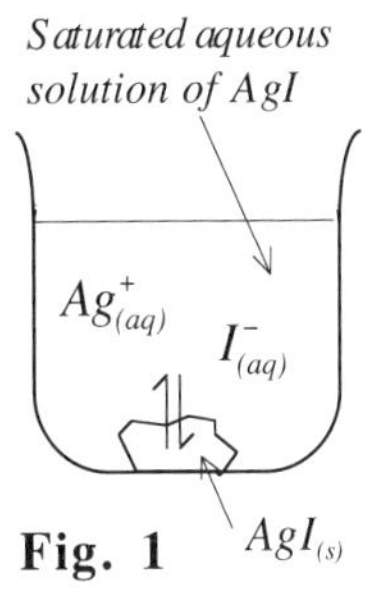

$$AgI_{(s)} \rightleftharpoons Ag^+_{(aq)} + I^-_{(aq)}$$

An equilibrium constant K_{SP} can be defined for this equilibrium:

$$K_{SP} = \frac{a_{Ag^+} \cdot a_{I^-}}{a_{AgI_{(s)}}} = \frac{a_{Ag^+} \cdot a_{I^-}}{1}$$

Saturated aqueous solution of AgI

Fig. 1

$$\boxed{K_{SP} = a_{Ag^+} \cdot a_{I^-}} \quad Solubility \quad Product \qquad \ldots [1]$$

where the a's are the relative activities of the various ionic species. This type of equilibrium constant is called a *solubility product* (hence the "SP" in K_{SP}), because, as we shall see, its value is related to the solubility s of the salt.

At 25°C, K_{SP} for AgI has a value of 1.5×10^{-16}. For salts with such low solubilities, the concentrations in solution are so dilute that we can safely set $\gamma_i \approx 1$, and $a_i \approx m_i \approx c_i$. Therefore,

$$K_{SP}(\text{AgI}) = [Ag^+][I^-] = 1.5 \times 10^{-16}$$

For every mole of AgI that dissolves, one mole of $Ag^+_{(aq)}$ and one mole of $I^-_{(aq)}$ is formed:

$$AgI_{(aq)} \rightarrow Ag^+_{(aq)} + I^-_{(aq)}$$
$$s \quad \rightarrow \quad s \qquad s$$

If s is the solubility of AgI, in mol L^{-1}, then $\quad [Ag^+_{(aq)}] = [I^-_{(aq)}] = s$.

and
$$K_{SP} = [Ag^+_{(aq)}] \, [I^-_{(aq)}] = (s)(s) = s^2 \qquad \ldots [2]$$

from which
$$s = \sqrt{K_{SP}} = \sqrt{1.5 \times 10^{-16}} = 1.22 \times 10^{-8} \qquad \ldots [3]$$

Therefore the solubility of AgI in water at 25°C is 1.22×10^{-8} mol L^{-1}.

You may recall that the Debye-Hückel equation gives the following expression for the single ion activity coefficient in water at 25°C:

$$log\gamma_i = -0.5108 \, z_i^2 \sqrt{I}$$

where I is the ionic strength of the solution, defined as $I \equiv \frac{1}{2}\Sigma c_i z_i^2$. Using this equation, we can calculate γ_{Ag^+} and γ_{I^-} in saturated AgI solution as follows:

$$I = \frac{1}{2}\sum c_i z_i^2 = \frac{1}{2}\left[c_+ z_+^2 + c_- z_-^2\right]$$

$$= \frac{1}{2}\left[(s)(+1)^2 + (s)(-1)^2\right]$$

$$= s = 1.22 \times 10^{-8}$$

Thus, for a *1–1* electrolyte, the ionic strength is just the same as its molarity; however, this is *not* true for other types of electrolytes.[1]

Therefore,
$$log\gamma_{Ag^+} = log\gamma_{I^-}$$

$$= -0.5108(\pm 1)^2\sqrt{1.22 \times 10^{-8}}$$

$$= -5.64 \times 10^{-5}$$

from which
$$\gamma_{Ag^+} = \gamma_{I^-} = 10^{-5.64 \times 10^{-5}} = 0.99987 \approx 1$$

Therefore our earlier assumption that the activity coefficients are equal to unity is justified.

Exercise 23-1

We might pause at this point to consider at what ionic strength we have to start worrying about using activity coefficient corrections. In general, for most purposes, if $\gamma_i > 0.95$ we can ignore activity coefficients. At what ionic strengths do we have to start making activity coefficient corrections for the following ions: (a) Na^+ (b) SO_4^{2-} (c) Al^{3+} ? ☺

---

[1] NaCl is a *1–1* salt; the numbers refer to the valences of the ions in the salt. Thus, $CuSO_4$ is a *2–2* salt, and $AlPO_4$ is a *3–3* salt. If the cation and the anion both have the same valence, as in these three examples, the salt is said to be a *symmetrical* salt. $CaCl_2$ is a *2–1* salt; Na_2SO_4 is a *1–2* salt; $AlCl_3$ is a *3–1* salt; H_3PO_4 is a *1–3* acid, etc.

23.2 RELATIONSHIP BETWEEN K_{SP} AND SOLUBILITY FOR DIFFERENT SALT TYPES

For the following discussion we first shall assume that the solutions are sufficiently dilute that all the ionic activity coefficients are equal to unity. Then, following that we will look at activity effects.

Consider a slightly soluble *2-1* salt such as CaF_2: Let the solubility of this salt in water be s mol L^{-1}. Each mole of this salt that dissolves dissociates to produce one mole of Ca^{2+} ions and two moles of F^- ions; therefore, if s moles of the salt dissolves,

$$CaF_2 \overset{100\%}{\rightarrow} Ca^{2+} + 2F^-$$
$$s \rightarrow s \quad 2s$$

$$[Ca^{2+}] = s \quad \text{and} \quad [F^-] = 2s$$

The solubility product then can be calculated as

$$K_{SP} = [Ca^{2+}][F^-]^2 = (s)(2s)^2 = 4s^3$$

from which the solubility of the salt is
$$s = \sqrt[3]{\frac{K_{SP}}{4}} \qquad \ldots [4]$$

Similar expressions are readily derived for other salt types.[2]

Exercise 23-2

How many grams of lead iodate—$Pb(IO_3)_2$— can be dissolved in one liter of water at 25°C if the solubility product of lead iodate in water has a value of 3.69×10^{-13} at 25°C? ☺

Example 23-1

Will a precipitate form if 60 mL of 2.0×10^{-3} M NaCl solution is added to 40 mL of 3.0×10^{-5} M $AgNO_3$ solution? If so, how much, and what will be the final concentrations of the Ag^+ ions and Cl^- ions in the solution? For AgCl, $K_{SP} = 1.6 \times 10^{-10}$. For this problem, activity coefficient corrections may be ignored, and additive volumes may be assumed.

Solution

AgCl may precipitate from solution according to: $Ag^+ + Cl^- \rightarrow AgCl\downarrow$ $\ldots [a]$

If the reaction quotient $Q = [Ag^+][Cl^-]$

is less than K_{SP}, the solution is not saturated, and no precipitation of AgCl will take place. If Q equals K_{SP}, then the solution is saturated, and precipitation is on the verge of taking place, but doesn't. However, if Q is *greater* than K_{SP}, then the solution is supersaturated, and the precipitation of AgCl will tend to take place until the solution attains a saturated state, with

---

[2] See Section 2 in Key Points for more details.

$$Q = \left\{ [Ag^+][Cl^-] \right\}_{\substack{saturation \\ equilibrium}} = K_{SP}.$$

The total number of moles n_{Ag} of Ag^+ ion in the system will be equal to the total number of moles of $AgNO_3$ that dissolved:

$$n_{Ag} = (3.0 \times 10^{-5}\ mol\ L^{-1})(0.040\ L) = 1.20 \times 10^{-6}\ mol\ Ag^+$$

The total number of moles n_{Cl} of Cl^- ion in the system will be equal to the total number of moles of NaCl added to the system:

$$n_{Cl} = (2.0 \times 10^{-3}\ mol\ L^{-1})(0.060\ L) = 1.20 \times 10^{-4}\ mol\ Cl^-$$

After the addition of the NaCl solution, the total volume will be

$$V = 40 + 60 = 100\ mL = 0.10\ L$$

Therefore, the new concentration $[Ag^+]'$ of Ag^+ ion and $[Cl^-]'$ of Cl^- ion will be[3]

$$[Ag^+]' = \frac{1.20 \times 10^{-6}\ mol}{0.10\ L} = 1.20 \times 10^{-5}\ M$$

$$[Cl^-]' = \frac{1.20 \times 10^{-4}}{0.10} = 1.20 \times 10^{-3}\ M$$

and the reaction quotient upon the addition will be

$$Q = [Ag^+]'[Cl^-]' = (1.20 \times 10^{-5})(1.20 \times 10^{-3}) = 1.44 \times 10^{-8}$$

This value exceeds $K_{SP} = 1.6 \times 10^{-10}$; therefore, precipitation of AgCl is favored to take place.[4]

The stoichiometry of reaction [a] shows that for every mole of AgCl formed, one mole of Ag^+ and one mole of Cl^- will be removed from solution. If we let the number of moles of AgCl that precipitates from solution be x, then:

No. of moles of Ag^+ ion remaining in solution $= (1.2 \times 10^{-6} - x)$ mol

No. of moles of Cl^- ion remaining in solution $= (1.2 \times 10^{-4} - x)$ mol

Therefore, the final concentrations of the two ions will be

$$[Ag^+]_{final} = \frac{1.20 \times 10^{-6} - x}{0.10} \quad \text{and} \quad [Cl^-]_{final} = \frac{1.20 \times 10^{-4} - x}{0.10}$$

These will be saturation concentrations, and therefore must satisfy K_{SP}:

---

[3] You also can get the new concentration M_2 by noting that $M_1 V_1 = M_2 V_2$; therefore $M_2 = M_1 V_1 / V_2$, where M is the solution molarity and V is its volume.

[4] We say "favored" because sometimes the solution becomes supersaturated; however, such solutions are thermodynamically unstable and usually spontaneously decompose to saturated concentration levels when the system is slightly stressed mechanically by tapping, vibration, etc.

$$[Ag^+]_{final}[Cl^-]_{final} = K_{SP}$$

i.e., $$\left(\frac{1.20\times10^{-6} - x}{0.10}\right)\left(\frac{1.20\times10^{-4} - x}{0.10}\right) = 1.6\times10^{-10}$$

This rearranges to $x^2 - 1.212\times10^{-4}x + 1.424\times10^{-10} = 0$ (a quadratic equation)

Solving: $$x = \frac{-(-1.212\times10^{-4}) \pm \sqrt{(-1.212\times10^{-4})^2 - 4(1)(1.424\times10^{-10})}}{2}$$

$$= \frac{1.212\times10^{-4} \pm 1.188269\times10^{-4}}{2} = 1.200134\times10^{-4} \ or \ 1.186533\times10^{-6}$$

The first root is impossible because you can't remove more ions from solution than were originally present; therefore the second root is the correct one:

$$x = 1.1865\times10^{-6}$$

and $$[Ag^+]_{final} = \frac{1.20\times10^{-6} - x}{0.10} = \frac{1.20\times10^{-6} - 1.186533\times10^{-6}}{0.10} = 1.34665\times10^{-7} \text{ M}$$

$$[Cl^-]_{final} = \frac{1.20\times10^{-4} - x}{0.10} = \frac{1.20\times10^{-4} - 1.186533\times10^{-6}}{0.10} = 1.188134\times10^{-3} \text{ M}$$

In order to check our result, we recognize that, because the solution is saturated with AgCl, the new value of the ion product must be equal to the saturated value, namely, K_{SP}.

Thus: $$\left\{[Ag^+][Cl^-]\right\}_{final} = (1.34665\times10^{-7})(1.188134\times10^{-3})$$

$$= 1.600000\times10^{-10} = K_{SP}   \text{(Right on!)}$$

Ans: | Precipitation occurs, with 1.19×10^{-6} mol of AgCl coming out of solution; $[Ag^+] = 1.35\times10^{-7}$ M; $[Cl^-] = 1.19\times10^{-3}$ M.

23.3 ACTIVITY EFFECTS AND SOLUBILITY

Now let's consider how to take activity coefficients into account when carrying out solubility calculations. We have seen that the complete dissociation of any electrolyte $M_{v_+}A_{v_-}$ can be represented by

$$M_{v_+}A_{v_-} \rightarrow v_+M^{z+} + v_-A^{z-}   \text{. . . [5]}$$

If the solubility of the salt is s mol L^{-1}, then for every s moles of salt that dissolves, there will be released into the solution v_+s moles of M^{z+} cations and v_-s moles of A^{z-} anions.

The solubility product of the salt is defined as:

$$K_{SP} = (a_+^{v_+})(a_-^{v_-})$$

$$= (\gamma_+ m_+)^{\nu_+} (\gamma_- m_-)^{\nu_-}$$

$$\approx (\gamma_+ c_+)^{\nu_+} (\gamma_- c_-)^{\nu_-}$$

$$= [\gamma_+ (\nu_+ s)]^{\nu_+} [\gamma_- (\nu_- s)]^{\nu_-}$$

$$= \gamma_+^{\nu_+} \cdot \gamma_-^{\nu_-} \cdot \nu_+^{\nu_+} \cdot \nu_-^{\nu_-} \cdot s^\nu \qquad \qquad \ldots [6]$$

where $\nu = \nu_+ + \nu_-$. The **mean activity coefficient** $\gamma_\pm$ for the salt is defined as

$$\gamma_\pm \equiv \left(\gamma_+^{\nu_+} \cdot \gamma_-^{\nu_-} \right)^{1/\nu} \qquad \qquad \ldots [7]$$

Therefore, it can be seen that $\qquad \gamma_+^{\nu_+} \cdot \gamma_-^{\nu_-} = \gamma_\pm^\nu \qquad \qquad \ldots [8]$

Substituting Eqn [8] into Eqn [6] gives

$$K_{SP} = \gamma_\pm^\nu \cdot \nu_+^{\nu_+} \cdot \nu_-^{\nu_-} \cdot s^\nu$$

Rearranging: $\qquad \qquad s^\nu = \dfrac{K_{SP}}{\gamma_\pm^\nu \cdot \nu_+^{\nu_+} \cdot \nu_-^{\nu_-}} = \left(\dfrac{1}{\gamma_\pm} \right)^\nu \dfrac{K_{SP}}{\nu_+^{\nu_+} \cdot \nu_-^{\nu_-}}$

$$\boxed{ s = \frac{1}{\gamma_\pm} \left(\frac{K_{SP}}{\nu_+^{\nu_+} \cdot \nu_-^{\nu_-}} \right)^{1/\nu} } \qquad \qquad \ldots [9]$$

Eqn [9] gives the general relationship between the solubility s of a salt and its solubility product K_{SP}, corrected for nonideal behavior in the solution. The mean activity coefficient can be obtained using the Debye-Hückel equation for a complete salt:

$$\boxed{ log\,\gamma_\pm = -0.5108 \, | z_+ z_- | \sqrt{I} } \qquad \qquad \ldots [10]$$

Note that Eqn [10], the Debye-Hückel equation for a complete salt, essentially is the same as the Debye-Hückel equation for a single ion, except that the factor z_i^2 has been replaced with the absolute value $| z_+ z_- |$, which incorporates the valences of both ions. As in the case of the equation for a single ion, Eqn [10] is valid only in dilute aqueous solutions at 25°C.[5]

Example 23-2

The solubility of Ag_2SO_4 in water at 25°C is 0.030 mol L^{-1}. Calculate its solubility product, taking account of activity effects.

---

[5] Strictly speaking, when the Debye-Hückel equation contains the factor *0.5108*, the ionic strength should be calculated using *molalities* rather than molarities. In dilute solutions, however, the two are essentially the same.

Solution

(a)
$$Ag_2SO_4 \xrightarrow{100\%} 2Ag^+ + SO_4^{2-}$$
$$0.03 \text{ M} \rightarrow 0.06 \text{ M} \quad 0.03 \text{ M}$$

$$K_{SP} = a_{Ag^+}^2 \cdot a_{SO_4^{2-}} = (\gamma_+ m_+)^2(\gamma_- m_-)$$

$$= (0.06)^2(0.03)\gamma_+^2\gamma_- = 1.08 \times 10^{-4}\gamma_\pm^3$$

$$I = \frac{1}{2}\left[c_+ z_+^2 + c_- z_-^2\right]$$

$$= \frac{1}{2}[(0.06)(+1)^2 + (0.03)(-2)^2]$$

$$= 0.09$$

$$\log \gamma_\pm = -0.5108\left|z_+ z_-\right|\sqrt{I}$$

$$= -0.5108(1 \times 2)\sqrt{0.09}$$

$$= -0.3065$$

$$\gamma_\pm = 10^{-0.3065} = 0.4937$$

Therefore,
$$K_{SP} = 1.08 \times 10^{-4}\gamma_\pm^3$$
$$= 1.08 \times 10^{-4}(0.4937)^3$$
$$= 1.30 \times 10^{-5}$$

Ans: $\boxed{1.3 \times 10^{-5}}$

Example 23-3

Calculate the solubility of PbI_2 in water at 25°C, given the following molar free energies of formation (in kJ mol^{-1}) at 25°C: $PbI_{2(s)} = -173.76$; $Pb_{(aq)}^{2+} = -24.43$; $I_{(aq)}^- = -51.57$.

Solution

First we formulate the solubility product for saturation equilibrium:

$$PbI_{2(s)} \rightleftharpoons Pb_{(aq)}^{2+} + 2 I_{(aq)}^-$$
$$s \quad \rightarrow \quad s \quad\quad 2s$$

$$K_{SP} = \frac{a_{Pb^{2+}} \cdot a_{I^-}^2}{a_{PbI_2}} = \frac{a_{Pb^{2+}} \cdot a_{I^-}^2}{1} = a_{Pb^{2+}} \cdot a_{I^-}^2$$

Next we calculate the standard free energy change for the reaction; we need this to evaluate the equilibrium constant:

$$\Delta G^o_{298.15} = \Delta g^0_f [Pb^{2+}_{(aq)}] + 2\, \Delta g^0_f [I^-_{(aq)}] - \Delta g^0_f [PbI_{2(s)}]$$

$$= (-24.43) + 2(-51.57) - (-173.76)$$

$$= +46.19 \text{ kJ mol}^{-1}$$

Since $\qquad\qquad\qquad\qquad \Delta G^\circ = -RT \ln K_{SP}$

therefore $\qquad\qquad\qquad \ln K_{SP} = \dfrac{-\Delta G^o}{RT} = \dfrac{-46\,190}{(8.314)(298.15)} = -18.6339$

and $\qquad\qquad\qquad\qquad K_{SP} = exp\,[-18.6339] = 8.080 \times 10^{-9}$

Let the solubility of PbI_2 be s mol L^{-1}. We may assume molarity and molarity are the same in this solution, since it will be dilute.

From the definition of K_{SP}: $\qquad\qquad K_{SP} = a_{Pb^{2+}} \cdot a^2_{I^-}$

$$= (\gamma_+ m_+)(\gamma_- m_-)^2$$

$$\approx (\gamma_+ c_+)(\gamma_- c_-)^2$$

$$= [\gamma_+ s][\gamma_-(2s)]^2$$

$$= \gamma_+ \gamma^2_-(4s^3) \qquad\qquad\qquad \dots \text{[a]}$$

For PbI_2 in solution, by definition, $\qquad \gamma_\pm = \left(\gamma^1_+ \cdot \gamma^2_-\right)^{1/3}$

therefore, $\qquad\qquad\qquad\qquad \gamma^1_+ \cdot \gamma^2_- = \gamma^3_\pm \qquad\qquad\qquad \dots \text{[b]}$

Substituting Eqn [b] into Eqn [a] gives $\qquad K_{SP} = 4\gamma^3_\pm s^3 \qquad\qquad \dots \text{[c]}$

Now we evaluate the mean activity coefficient for the salt:

$$\log \gamma_\pm = -0.5108\,|z_+ z_-|\sqrt{I}$$

$$= -0.5108(2 \times 1)\sqrt{I}$$

$$= -1.0216\sqrt{I} \qquad\qquad\qquad \dots \text{[d]}$$

The ionic strength of the solution is $\qquad I = \frac{1}{2}\sum c_i z^2_i$

$$= \tfrac{1}{2}\left[[Pb^{2+}](+2)^2 + [I^-](-1)^2\right]$$

$$= \tfrac{1}{2}\left[(s)(4) + (2s)(1)\right]$$

$$= 3s \qquad\qquad\qquad\qquad \dots \text{[e]}$$

Substituting Eqn [e] into Eqn [d]:

$$\log \gamma_\pm = -1.0216\sqrt{I}$$

$$= -1.0216\sqrt{3s}$$

$$= -1.0216\sqrt{3}\sqrt{s}$$

$$= -1.7695\sqrt{s} \qquad \qquad \cdots \text{[f]}$$

Thus the ionic strength depends on the solubility, so it follows that the solubility determines the mean activity coefficient; but the mean activity coefficient depends on the ionic strength, and the ionic strength depends on the solubility! Yuk!! We have to use an iterative approach:

From Eqn [c]:
$$\frac{K_{SP}}{4} = \gamma_{\pm}^{3}s^{3}$$

Taking *base-10* logarithms:
$$log\left(\frac{K_{SP}}{4}\right) = 3(log\gamma_{\pm} + log\,s)$$

Rearranging:
$$\frac{1}{3}log\left(\frac{K_{SP}}{4}\right) = log\gamma_{\pm} + log\,s \qquad \qquad \cdots \text{[g]}$$

Putting Eqn [f] into Eqn [g]:
$$\frac{1}{3}log\left(\frac{K_{SP}}{4}\right) = -1.7695\sqrt{s} + log\,s$$

i.e.,
$$\frac{1}{3}log\left(\frac{8.080 \times 10^{-9}}{4}\right) = -2.8982 = -1.7695\sqrt{s} + log\,s$$

Rearranging:
$$1.7695\sqrt{s} - log\,s - 2.8982 = 0 \qquad \qquad \cdots \text{[h]}$$

Recognizing that
$$log_{10}s = \frac{ln\,s}{ln\,10} = \frac{ln\,s}{2.302585}$$

and dividing both sides through by *1.7695*, Eqn [h] becomes

$$\sqrt{s} - 0.24543\,ln\,s - 1.6379 = 0 \qquad \qquad \cdots \text{[i]}$$

Eqn [i] is solved using the Newton-Raphson method outlined in Appendix 7.

Thus,
$$f(s) = \sqrt{s} - 0.24543\,ln\,s - 1.6379$$

and
$$f'(s) = \frac{0.5}{\sqrt{s}} - \frac{0.24543}{s} \qquad {}^{6}$$

As a first guess, use the value of s that we would calculate if we ignored activity coefficients; namely,

$$s_{o} = \sqrt[3]{\frac{K_{SP}}{4}} = \sqrt[3]{\frac{8.080 \times 10^{-9}}{4}} = 0.00126\ \text{mol L}^{-1}$$

---

[6] In case you forgot (or never knew in the first place), $d\,ln\,x/dx = 1/x$. See Appendix 9 for details galore on natural logarithms.

n	s_n	$f(s_n)$	$f'(s_n)$	$s_{n+1} = s_n - \dfrac{f(s_n)}{f'(s_n)}$	
0	1.26×10^{-3}	3.6245×10^{-2}	-1.8070×10^2	1.4606×10^{-3}	
1	1.4606×10^{-3}	2.7105×10^{-3}	-1.5495×10^2	1.4781×10^{-3}	
2	1.4781×10^{-3}	1.6781×10^{-5}	-1.5304×10^2	1.4782×10^{-3}	
3	1.4782×10^{-3}	0.00000	-1.5303×10^2	1.4782×10^{-3}	$\leftarrow$ *converged!*

Therefore, the solubility is $s = 1.4782 \times 10^{-3}$ mol L^{-1}. The value of 1.26×10^{-3} calculated without considering activity coefficients is in error by $\dfrac{1.26 - 1.48}{1.48} \times 100\% = -15\%$.

Ans: $\boxed{s = 1.48 \text{ mM}}$

Check: The ionic strength is $I = 3s = 3(0.0014782) = 0.0044346$

$$log \; \gamma_\pm = -1.7695\sqrt{s} = -1.7695\sqrt{0.0014782} = -0.0680326$$

$$\gamma_\pm = 10^{-0.0680326} = 0.8550025$$

$$K_{SP} = 4\gamma_\pm^3 s^3 = 4(0.8550025)^3(0.0014782)^3 = 8.08 \times 10^{-9} \quad \text{(OK!)}$$

Exercise 23-3

Consider the slightly soluble salt lead bromide: The solubility product of $PbBr_2$ at 25°C is 7.9×10^{-5}. Calculate the solubility of $PbBr_2$ in water at 25°C, (a) neglecting activity coefficients, and (b) taking activity coefficients into account. In a saturated aqueous $PbBr_2$ solution at 25°C, the activity coefficient of the Pb^{2+} ions is 0.0823, and that of the Br^- ions is 0.5356. ☺

23.4 THE COMMON ION EFFECT

The solubility of a slightly soluble salt is *lowered* by the presence of a *common ion* in the solution. Consider the saturation equilibrium of AgI, a slightly soluble salt:

$$AgI_{(s)} \rightleftharpoons Ag^+_{(aq)} + I^-_{(aq)} \qquad \qquad \ldots [11]$$

For this equilibrium, $\qquad \qquad K_{SP} \approx [\, Ag^+_{(aq)}][\, I^-_{(aq)}]$

where the ion concentrations are *saturated* values. Suppose we now add additional iodide to the system in the form of *sodium* iodide, which is a *highly* soluble salt. Now, in effect, we are trying to *exceed* the product of the saturated ion concentrations because, by adding extra I^- ion to a system that already is saturated, we are attempting to increase [I⁻] *beyond* the saturation value; in other words, now $[Ag^+][I^-] > K_{SP}$. That is, the solution now is *supersaturated*.

In accordance with le Chatelier's Principle, if we add extra I⁻ ion to the right hand side of equilibrium [11], the equilibrium will shift to the *left*. However, when this reaction shifts to the left, not only I⁻ is removed, but also, in order to satisfy electric charge neutrality, an equal number of moles of Ag⁺ ion also leaves the solution along with the iodide. The two ions precipitate out of solution together as solid AgI until the concentrations of the ions remaining in solution have re-adjusted so that their product once again is equal to K_{SP}.

The new concentration of Ag⁺ ion in the solution is considered to be the new solubility of AgI in this solution; since this is lower than before the addition of the NaI, the effect of adding additional iodide ion is to *lower* the solubility of the AgI. Thus, AgI *is less soluble in a solution containing I⁻ ions than it is in pure water*. This lowering of the solubility is called the **common ion effect**.

Exercise 23-4

The solubility product for CaF_2 in water is 4.0×10^{-11} at 25°C. Calculate the solubility of CaF_2 (a) in pure water, (b) in 0.10 M $Ca(NO_3)_2$ solution, and (c) in 0.10 M NaF solution. You may neglect activity coefficient corrections for this problem. ☺

Exercise 23-5

Calculate the solubility of $PbCl_2$ in aqueous 0.1 M NaCl solution, if the solubility in water is 0.01 M at 25°C, (a) neglecting activity coefficient effects, and (b) including them, using the Debye–Hückel equation. ☹

23.5 SALTING IN

Is the solubility of AgI affected by dissolving it in KNO_3 solution, which does *not* contain a common ion? You may think not, but you are wrong! To understand this, consider the following:

$$K_{SP} = a_{Ag^+} \cdot a_{I^-}$$

$$= \gamma_+ m_+ \cdot \gamma_- m_-$$

$$\approx \gamma_{Ag^+} \cdot \gamma_{I^-} [Ag^+][I^-]$$

$$= \gamma_\pm^2 [Ag^+][I^-] \qquad \qquad \dots [12]$$

From the Debye-Hückel equation,

$$log\ \gamma_\pm (AgI) = -0.5108\ |z_+ z_-|\sqrt{I} = -0.5108(1 \times 1)\sqrt{I}$$

The presence of KNO_3 *increases* the ionic strength of the solution, and therefore causes $log\gamma_\pm$ for AgI to become *more negative*; i.e., $\gamma_\pm$ becomes *smaller*.

Since
$$K_{SP} = constant = \gamma_\pm^2 [Ag^+][I^-]$$
$$= \gamma_\pm^2 (s)(s)$$
$$= \gamma_\pm^2 s^2,$$

if $\gamma_\pm^2$ gets *smaller*, then s^2 —and therefore s—must get *bigger*, since the product $\gamma_\pm^2 s^2$ must stay constant at the value of K_{SP}. It follows therefore that AgI is *more* soluble in KNO_3 solution than it is in pure water. This phenomenon is called **salting in**: the presence of another (non-common ion) *salt* causes more AgI to dissolve *in* to the solution than would dissolve in pure water.

23.6 SELECTIVE PRECIPITATION

Solubility differences can be used to selectively remove only one of several species from solution. For example, consider a solution that is 0.1 M in Cl^- ion and 0.1 M in CrO_4^{2-} ion. Furthermore, let's suppose that we want to selectively remove the *chloride* but leave the *chromate* in solution. If we add silver nitrate ($AgNO_3$) solution we can precipitate AgCl and Ag_2CrO_4, both of which are slightly soluble salts:

$$K_{SP}(AgCl) = [Ag^+][Cl^-] = 1.6 \times 10^{-10} \qquad \ldots [13]$$

$$K_{SP}(Ag_2CrO_4) = [Ag^+]^2 [CrO_4^{2-}] = 1.9 \times 10^{-12} \qquad \ldots [14]$$

Precipitation of each salt will begin when that salt reaches its saturation conditions (i.e., when the solubility product is satisfied).

Therefore, for precipitation of AgCl:

$$[Ag^+] = \frac{K_{SP(AgCl)}}{[Cl^-]} = \frac{1.6 \times 10^{-10}}{0.1} = 1.6 \times 10^{-9} \text{ M}$$

Similarly, for precipitation of Ag_2CrO_4:

$$[Ag^+] = \sqrt{\frac{K_{SP(Ag_2CrO_4)}}{[CrO_4^{2-}]}} = \sqrt{\frac{1.9 \times 10^{-12}}{0.1}} = 4.36 \times 10^{-6} \text{ M}$$

Therefore, AgCl starts to precipitate out of solution *first*, since $[Ag^+]$ reaches 1.6×10^{-9} M long before it reaches 4.36×10^{-6} M. Ag_2CrO_4 won't start to precipitate out of solution until $[Ag^+] = 4.36 \times 10^{-6}$ M. When $[Ag^+]$ reaches 4.36×10^{-6} M, the remaining Cl^- ion concentration in solution will be

$$[Cl^-] = \frac{K_{SP(AgCl)}}{[Ag^+]} = \frac{1.6 \times 10^{-10}}{4.36 \times 10^{-6}} = 3.7 \times 10^{-5} \text{ M}$$

Therefore CrO_4^{2-} doesn't start to precipitate from solution until

$$\left(\frac{0.1 - 0.000037}{0.1}\right) \times 100\% = 99.96\%$$

of the Cl^- has been removed. Thus we selectively have removed the chloride from solution while leaving the chromate still dissolved in the solution.

The above reactions are the basis for the **Mohr Titration**, which is used to analyze for Cl^- in solution (e.g., analysis of dissolved NaCl). The method involves titrating an NaCl solution of unknown concentration using standardized $AgNO_3$ as the titrant. A small amount of K_2CrO_4 is added to the system as the *indicator*. Dissolved CrO_4^{2-} ion in solution is *yellow*, whereas precipitated Ag_2CrO_4 is *reddish*. The titration stoichiometry is:

$$AgNO_3 + NaCl \rightarrow AgCl \downarrow + NaNO_3$$

This reaction really is just $\qquad Ag^+ + Cl^- \rightarrow AgCl \downarrow$

Thus, the $AgNO_3$ titrant precipitates the Cl^- from solution as insoluble AgCl. As shown above, Ag_2CrO_4 doesn't start to come out of solution until the NaCl has essentially all been removed. Then, at the *end point*, the color changes suddenly from yellow to a reddish pink, indicating the formation of Ag_2CrO_4 precipitate.

Exercise 23-6

0.01 moles of solid $AgNO_3$ is added to 1.00 L of a solution containing 0.0050 M $NaIO_3$ and 0.020 M Na_2CrO_4. (a) Which solid precipitates out first? (b) Calculate the number of moles of each solid formed and the final concentrations of $Ag_{(aq)}^+$, $IO_{3(aq)}^-$, and $CrO_{4(aq)}^{2-}$.

$K_{SP}(Ag_2CrO_4) = 1.9 \times 10^{-12}$; $K_{SP}(AgIO_3) = 1.0 \times 10^{-8}$. ☠

KEY POINTS FOR CHAPTER TWENTY-THREE

1. Salts such as AgI that have extremely low solubilities are called **slightly soluble salts**. Consider the saturation equilibrium between excess crystals of solid AgI in the bottom of a beaker and a saturated solution of dissolved silver iodide:

    $$AgI_{(s)} \rightleftharpoons Ag_{(aq)}^+ + I_{(aq)}^-$$

    An equilibrium constant called the **solubility product** K_{SP} can be defined for this equilibrium:

    $$K_{SP} = \frac{a_{Ag^+} \cdot a_{I^-}}{a_{AgI_{(s)}}} = \frac{a_{Ag^+} \cdot a_{I^-}}{1} = a_{Ag^+} \cdot a_{I^-}$$

    For salts such as these, which have very low solubilities, the concentrations in solution are so dilute that we can safely set $\gamma_i \approx 1$, and $a_i \approx m_i \approx c_i$; therefore,

$$K_{SP}(AgI) = [Ag^+][I^-]$$

For every mole of AgI that dissolves, one mole of $Ag^+_{(aq)}$ and one mole of $I^-_{(aq)}$ is formed:

$$AgI_{(aq)} \rightarrow Ag^+_{(aq)} + I^-_{(aq)}$$
$$s \quad\rightarrow\quad s \quad\quad s$$

If s is the solubility in mol L^{-1} of the AgI, then, $[Ag^+_{(aq)}] = [I^-_{(aq)}] = s$

and $K_{SP} = [Ag^+_{(aq)}][I^-_{(aq)}] = s \times s = s^2$ from which $s = \sqrt{K_{SP}}$

2. **Relationship between K_{SP} and s for different salt types**:

   1–1 salts: e.g., AgI $s = \sqrt{K_{SP}}$

   2–1 salts or 1–2 salts: e.g., CaF_2, Ag_2CrO_4 $CaF_{2(aq)} \xrightarrow{100\%} Ca^{2+}_{(aq)} + 2F^-_{(aq)}$
   $$s \quad\quad\quad \rightarrow \quad\quad s \quad\quad 2s$$

   $$K_{SP} = [Ca^{2+}_{(aq)}][F^-_{(aq)}]^2 = (s)(2s)^2 = 4s^3 \quad\rightarrow\quad s = \sqrt[3]{\frac{K_{SP}}{4}}$$

   3–1 salts: e.g., $Al(OH)_3$ $Al(OH)_{3(aq)} \xrightarrow{100\%} Al^{3+}_{(aq)} + 3OH^-_{(aq)}$
   $$s \quad\quad \rightarrow \quad\quad s \quad\quad 3s$$

   $$K_{SP} = [Al^{3+}_{(aq)}][OH^-_{(aq)}]^3 = (s)(3s)^3 = 27s^4 \quad\rightarrow\quad s = \sqrt[4]{\frac{K_{SP}}{27}}$$

3. **Activity effects**:

   The complete dissociation of any electrolyte $M_{v+}A_{v-}$ can be represented by

   $$M_{v+}A_{v-} \xrightarrow{100\%} v_+ M^{z+} + v_- A^{z-}$$

   If the solubility of the salt is s mol L^{-1}, then for every s moles of salt that dissolve, there will be released into the solution $v_+ s$ moles of M^{z+} cations and $v_- s$ moles of A^{z-} anions.

   $$K_{SP} = (a_+^{v_+})(a_-^{v_-}) = (\gamma_+ m_+)^{v_+}(\gamma_- m_-)^{v_-} \approx (\gamma_+ c_+)^{v_+}(\gamma_- c_-)^{v_-}$$
   $$= [\gamma_+(v_+ s)]^{v_+}[\gamma_-(v_- s)]^{v_-} = \gamma_+^{v_+} \cdot \gamma_-^{v_-} \cdot v_+^{v_+} \cdot v_-^{v_-} \cdot s^v$$

   where $v = v_+ + v_-$. The **mean activity coefficient** $\gamma_\pm$ for the salt is defined as

   $$\gamma_\pm \equiv \left(\gamma_+^{v_+} \cdot \gamma_-^{v_-}\right)^{1/v}$$

   Therefore $\gamma_+^{v_+} \cdot \gamma_-^{v_-} = \gamma_\pm^v$ and the above equation for K_{SP} becomes

$$K_{SP} = \gamma_{\pm}^{\nu} \cdot v_{+}^{\nu_+} \cdot v_{-}^{\nu_-} \cdot s^{\nu} \quad \rightarrow \quad s^{\nu} = \frac{K_{SP}}{\gamma_{\pm}^{\nu} \cdot v_{+}^{\nu_+} \cdot v_{-}^{\nu_-}} = \left(\frac{1}{\gamma_{\pm}}\right)^{\nu} \frac{K_{SP}}{v_{+}^{\nu_+} \cdot v_{-}^{\nu_-}}$$

Taking roots:
$$s = \frac{1}{\gamma_{\pm}} \left(\frac{K_{SP}}{v_{+}^{\nu_+} \cdot v_{-}^{\nu_-}}\right)^{1/\nu}$$

This gives the general relationship between the solubility s of a salt and its solubility product K_{SP}, corrected for nonideal behavior in the solution. The mean activity coefficient (at 25°C) can be obtained using the Debye-Hückel equation for a complete salt:

$$log\,\gamma_{\pm} = -0.5108\,|z_+z_-|\,\sqrt{I} \quad \text{where} \quad I = \tfrac{1}{2}\sum_i c_i z_i^2$$

4.   The **common ion effect**: The solubility of a slightly soluble salt is *lowered* by the presence of a *common ion* in the solution. Thus, the solubility of AgCl will be lower in an aqueous solution of NaCl than it will be in pure water.

$$K_{SP}(AgCl) = [Ag^+][Cl^-]$$

Since K_{SP} is an equilibrium constant, it must be obeyed for any solution in equilibrium. Thus, if the Cl^- ion concentration increases (by the addition of, say, NaCl, which is very soluble), then the Ag^+ ion concentration must *decrease*, in order that the product of the two ion concentrations stays equal to the value of K_{SP}. This means that the solution is able to hold less AgCl—i.e., the solubility of AgCl is lowered.

5.   **Salting in**: The solubility of a slightly soluble salt is *increased* when it is dissolved in a solution containing other ions, none of which is in common with either ion of the slightly soluble salt. This effect is called *salting in*. Thus, the slightly soluble salt AgCl has a *higher* solubility in an aqueous solution of, say, KNO_3 (a highly soluble salt) than it does in pure water.

This can be understood as follows: For AgCl,

$$K_{SP}(AgCl) = a_{Ag^+} \cdot a_{Cl^-} = \gamma_+ m_+ \cdot \gamma_- m_- \approx \gamma_{Ag^+} \cdot \gamma_{Cl^-} [Ag^+][Cl^-] = \gamma_{\pm}^2 [Ag^+][Cl^-]$$

The more ions that are dissolved in a solution, the greater will be the ionic strength I of the solution. But, from the Debye-Hückel equation we know that as the ionic strength *increases*, $\gamma_{\pm}$ *decreases*. Therefore, reference to the expression for K_{SP} shows that if $\gamma_{\pm}^2$ decreases, then $[Ag^+][Cl^-] = s \times s = s^2$ must *increase*. That is, the solubility of AgCl becomes *greater* when the ionic strength of the solution is *increased* by the presence of any ions other than Ag^+ or Cl^-.

6.   **Selective precipitation**:

Solubility differences can be used to selectively remove only one of several species from solution. For example, suppose that we want to selectively remove the *chloride* but leave the *chromate* in a solution 0.1 M in Cl^- and 0.1 M in CrO_4^{2-}. If we add silver nitrate ($AgNO_3$) we can precipitate AgCl and Ag_2CrO_4, both of which are slightly soluble salts:

$$K_{SP}(AgCl) = [Ag^+][Cl^-] = 1.6 \times 10^{-10}$$

$$K_{SP}(Ag_2CrO_4) = [Ag^+]^2[CrO_4^{2-}] = 1.9 \times 10^{-12}$$

Precipitation of each salt will begin when that salt reaches its saturation conditions (i.e., when the solubility product is satisfied).

Therefore, for precipitation of AgCl:

$$[Ag^+] = \frac{K_{SP}}{[Cl^-]} = \frac{1.6 \times 10^{-10}}{0.1} = 1.6 \times 10^{-9} \text{ M}$$

Similarly, for precipitation of Ag_2CrO_4:

$$[Ag^+] = \sqrt{\frac{K_{SP}}{[CrO_4^{2-}]}} = \sqrt{\frac{1.9 \times 10^{-12}}{0.1}} = 4.36 \times 10^{-6} \text{ M}$$

Therefore, AgCl starts to precipitate out of solution *first*, since $[Ag^+]$ reaches 1.6×10^{-9} M long before it reaches 4.36×10^{-6} M. Ag_2CrO_4 won't start to precipitate out of solution until $[Ag^+] = 4.36 \times 10^{-6}$ M. When $[Ag^+]$ reaches 4.36×10^{-6} M, the remaining Cl^- ion concentration in solution will be

$$[Cl^-] = \frac{K_{SP}}{[Ag^+]} = \frac{1.6 \times 10^{-10}}{4.36 \times 10^{-6}} = 3.7 \times 10^{-5} \text{ M}$$

Therefore CrO_4^{2-} doesn't start to precipitate from solution until

$$\left(\frac{0.1 - 0.000037}{0.1}\right) \times 100\% = 99.96\%$$

of the Cl^- has been removed. Thus we selectively have removed the chloride from solution while leaving the chromate still dissolved in the solution.

PROBLEMS

1. At 25°C, the solubility of calcium hydroxide in water is 1.59 g L^{-1}, and the *pH* of a saturated aqueous solution of calcium hydroxide is 12.454. The molar mass of $Ca(OH)_2$ is 74.095 g mol^{-1}.

(a) Assuming that the dissolved $Ca(OH)_2$ dissociates 100%, what is the concentration of the OH^- ion in this solution?

(b) What is the activity of the OH^- ion?

(c) Based on the experimentally measured *pH* of the saturated solution, what is the value of γ_-, the activity coefficient of the OH^- ion?

(d) What is the value of γ_- calculated from the expression $\log \gamma_i = -0.5108 z_i^2 \sqrt{I}$?

(e) Taking account of activity coefficients, calculate the "true" value of K_{SP}.

(The value reported in the literature is $K_{SP} = 5.5 \times 10^{-6}$.)

(f) What is the value of K_{SP} if activity corrections are *ignored*?

(g) Given that $K_{SP} = 5.5 \times 10^{-6}$, calculate the % error that would result in the predicted solubility of $Ca(OH)_2$ if activity corrections were not taken into account. ☺

2. Sea water contains 1.27×10^3 ppm of magnesium (grams of Mg per million grams of sea water) and has a density of 1025 g L^{-1} at 25°C. The magnesium can be recovered by adjustment of the *pH* by adding calcium hydroxide to precipitate solid $Mg(OH)_2$. K_{SP} for $Mg(OH)_2$ is 1.80×10^{-11} at 25°C. Activity coefficient corrections may be neglected in this problem. Mg = 24.31 g mol^{-1}.

(a) At what *pH* does $Mg(OH)_2$ start to precipitate?

In the commercial recovery of magnesium from sea water only 90% of the Mg^{2+} is precipitated on account of economic considerations and to control product purity (if it is attempted to remove all the Mg^{2+}, some $Ca(OH)_2$ starts to co-precipitate).

(b) Calculate the *pH* by which 90% of the Mg^{2+} ion has been recovered.

(c) Calculate the *pH* by which 99.9% of the Mg^{2+} has been recovered. ☺

3. Calculate the molarity of the hydrogen ion in aqueous 0.01 M Na_2SO_4 solution at 25°C. What is the *pH* of the solution? Take activity effects into account. ☺

4. At 25°C the solubilities of TlCl (thallium chloride) in water and in 0.100 M aqueous NaCl solution are 0.01607 mol L^{-1} and 0.00395 mol L^{-1}, respectively. The solubility product of TlCl at the same temperature is 2.02×10^{-4}. What is the mean activity coefficient $\gamma_\pm$ of TlCl at 25°C (a) in pure water, and (b) in 0.100 M NaCl solution? At these dilutions it may be assumed that molality and molarity have the same numerical value. ☺

5. The solubility product for Ag_2CrO_4 at 25°C is 4.05×10^{-12}. What is the solubility (in mg L^{-1}) of Ag_2CrO_4 in 10 mM K_2CrO_4 at 25°C? ☺

6. In aqueous solutions, gypsum—$CaSO_4 \cdot 2H_2O$—dissolves according to

$$CaSO_4 \cdot 2H_2O_{(s)} \rightleftharpoons Ca^{2+}_{(aq)} + SO^{2-}_{4(aq)} + 2H_2O_{(liq)}$$

The solubility product for gypsum is defined as

$$K_{SP} = \frac{a_{Ca^{2+}} \cdot a_{SO_4^{2-}} \cdot a_{H_2O}^2}{a_{CaSO_4 \cdot 2H_2O}} = \frac{a_{Ca^{2+}} \cdot a_{SO_4^{2-}} (1)^2}{(1)} = a_{Ca^{2+}} \cdot a_{SO_4^{2-}}$$

(a) Calculate K_{SP} for gypsum at 25°C. The standard molar Gibbs free energy of formation at 25°C for solid gypsum is $\Delta g_f^\circ = -1797.45$ kJ mol^{-1}.

(b) Taking account of activity corrections, calculate the molal solubility of gypsum in aqueous 0.10 M Na_2SO_4 solution at 25°C. Be sure to verify any assumptions you make. For solutions of this ionic strength, use the Guggenheim equation. ☹

7. Calomel[7] (mercurous chloride, Hg_2Cl_2) is used in the manufacture of calomel reference electrodes, which are widely used in electrochemical research. Calomel is a sparingly soluble salt which, when dissolved, dissociates into mercurous ions, Hg_2^{2+}, and Cl^- ions.

    (a) Using data from the appendix, calculate the solubility product for calomel at 25°C.

    (b) What is the molal solubility of calomel in aqueous 0.10 M NaCl solution at 25°C? For the concentrations involved in this problem, you may assume that molal concentrations are essentially the same as molar concentrations. Use the Davies equation to take account of activity coefficient corrections.

$$log_{10} \gamma_{\pm} = \frac{-0.51159 \, | \, z_+ z_- \, | \sqrt{I}}{1 + \sqrt{I}} - 0.10 \, | \, z_+ z_- \, | \, I \quad \text{(Davies equation)} \; \smile$$

8. How much AgCl (in mol L^{-1}) can be dissolved in a saturated aqueous solution of AgI at 25°C? For this problem you may neglect activity corrections.

    Data at 25°C: $K_{SP}(AgCl) = 1.6 \times 10^{-10}$; $K_{SP}(AgI) = 1.5 \times 10^{-16}$ $\smile$

9. What is the solubility of $AgBrO_3$ in 5 mM KNO_3 at 25°C if the solubility product of silver bromate has a value of 5.20×10^{-5}? $\otimes$

10. How many grams of $Al(OH)_3$ can be dissolved in a liter of water at 25°C if the solubility product of $Al(OH)_3$ in water has a value of 9.918×10^{-36} and the only species formed during the dissolution of $Al(OH)_3$ were $Al_{(aq)}^{3+}$ and $OH_{(aq)}^-$? $\otimes$

11. The solubility product of $Al(OH)_3$ in water at 25°C is $K_{SP} = 9.918 \times 10^{-36}$. However, the aluminum ion has a very strong affinity for hydroxyl ions; consequently, in addition to $Al_{(aq)}^{3+}$, a number of other aluminum-containing species (complexes) are formed when solid aluminum hydroxide dissolves in water. Each complex has its own equilibrium constant (called a *formation constant*); at equilibrium the equilibrium constants for all species must be satisfied simultaneously. When solid $Al(OH)_3$ is dissolved in water at 25°C, the following equilibria must be considered:

$Al(OH)_{3(s)} \rightleftharpoons Al_{(aq)}^{3+} + 3 \, OH_{(aq)}^-$	$K_{SP} = 9.918 \times 10^{-36}$
$Al_{(aq)}^{3+} + OH_{(aq)}^- \rightleftharpoons AlOH_{(aq)}^{2+}$	$K = 9.908 \times 10^8$
$AlOH_{(aq)}^{2+} + OH_{(aq)}^- \rightleftharpoons Al(OH)_{2(aq)}^+$	$K' = 4.535 \times 10^8$
$Al(OH)_{2(aq)}^+ + OH_{(aq)}^- \rightleftharpoons Al(OH)_{3(aq)}$	$K'' = 5.395 \times 10^8$
$Al(OH)_{3(aq)} + OH_{(aq)}^- \rightleftharpoons Al(OH)_{4(aq)}^-$	$K''' = 2.468 \times 10^7$
$H_2O_{(liq)} \rightleftharpoons H_{(aq)}^+ + OH_{(aq)}^-$	$K_W = 1.00 \times 10^{-14}$

In view of the above equilibria, calculate (a) how much solid $Al(OH)_3$ will dissolve in water at 25°C, and (b) the *pH* of the resulting solution. Because of the low concentrations involved, you may neglect activity coefficient corrections for this problem.

Hint: As a first guess, assume that the solution is neutral. $\skull$

---

[7] Don't confuse poisonous *calomel* (Hg_2Cl_2) with *caramel* (the candy). You will make this mistake only once!

12. At 25°C the solubility product in water of MgF_2 is 7.0×10^{-9}. Calculate the solubility of this salt in 0.001 M aqueous $Mg(NO_3)_2$ solution at 25°C. For this calculation you may neglect any activity effects and you may assume complete dissociation of all salts. ☹

TWENTY-FOUR

OXIDATION-REDUCTION REACTIONS

24.1 OXIDATION-REDUCTION (REDOX) REACTIONS

When a substance reacts with oxygen to form an oxide, we say the substance is *oxidized*. Thus, iron is oxidized to form ferrous oxide. In this process (Fig. 1), one could say that the iron *loses* two electrons to become Fe^{2+} while the oxygen *gains* two electrons to become O^{2-}.[1]

$$\overbrace{Fe + \tfrac{1}{2}O_2}^{2e^-} \to FeO$$

Fig. 1

The same change can occur in Fe without reaction with oxygen; e.g., the reaction

$$Fe_{(s)} + Cu^{2+}_{(aq)} \to Fe^{2+}_{(aq)} + Cu_{(s)}$$

This led to the definition that *oxidation* is a process in which a substance *loses* electrons

$$Fe \to Fe^{2+} + 2e^- \qquad \text{(Oxidation of } Fe\text{)}$$

while *reduction* is a process in which a substance *gains* electrons

$$Fe^{2+} + 2e^- \to Fe \qquad \text{(Reduction of } Fe^{2+}\text{)}$$

Thus a **redox reaction** came to be defined as a reaction in which electrons are *transferred*.

However, in many cases—such as the example shown in Fig. 2—the electrons are *shared*, not transferred, so the above definition doesn't hold up. Furthermore, how do we explain a case such as

$$NO^-_{3(aq)} + 4H_3O^+_{(aq)} + 3e^- \to NO_{2(g)} + 6H_2O_{(liq)}?$$

$$2H_2 + \tfrac{1}{2}O_2 \to 2H_2O$$

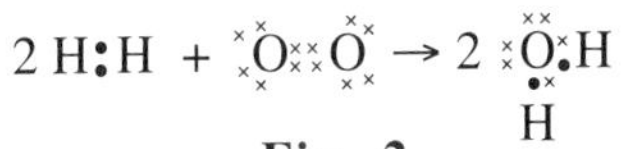

$$2\ H\!:\!H\ +\ \overset{\times\times}{\underset{\times\times}{O}}\!\!\times\!\!\overset{\times}{\underset{\times}{O}}\ \to 2\ \overset{\times\times}{\underset{\bullet}{O}}\!\!:\!\!H$$
H

Fig. 2

Here, three electrons are consumed, but we don't know where they end up: on the *H*, *N*, or *O*? Electron transfer usually occurs between *groups* of atoms or ions, not just by the transfer of electrons from one atom to another.

In order to avoid the above difficulties, it has been found more convenient to introduce the concept of *oxidation numbers (ON)* or *oxidation states*. Thus, the **oxidation number** (or state) is just a

[1] We now know that this is an oversimplification of the true charge distribution, but it is still a useful way to look at it.

fictitious charge assigned to an atom in a molecule or in an ion according to an **arbitrary** *set of rules.* It should be viewed as just a "bookkeeping device" that may or may not be related to the actual transfer of electrical charge.

24.2 RULES FOR ASSIGNMENT OF OXIDATION NUMBERS

Oxidation numbers are assigned according to the following rules:

Rule	Examples	Exceptions
1. The oxidation number of any free element is *zero*.	H_2, O_2, Ne	None.
2. Oxygen in compounds has an oxidation number of –2.	H_2O, MgO	Oxygen in peroxides ($O–O$ link) has an oxidation number of –1: e.g., H_2O_2.
3. Hydrogen in compounds has an oxidation number of +1.	H_2O, CH_4	Hydrogen in metallic hydrides has an oxidation number of –1: e.g., NaH, LiH.
4. The sum of the oxidation numbers must be *zero* for a neutral molecule and equal to the *net charge* for an ion.	H_2O: $2(+1) + (-2) = 0$ NO_3^-: $(+5) + 3(-2) = -1$	None.

Exercise 24-1

Assign oxidation numbers to the elements in each of the following: ☺

Br_2	$Br =$		
S_8	$S =$		
Fe^{3+}	$Fe =$		
NO	$O =$	$N =$	
NO_2	$O =$	$N =$	
MnO_4^-	$O =$	$Mn =$	
$HONO_2$	$O =$	$H =$	$N =$

$HCrO_4^-$	$O =$	$H =$	$Cr =$
$Na_2Cr_2O_7$	$O =$	$Na =$	$Cr =$
C_6H_5CHO	$H =$	$C =$	$O =$
C_6H_6	$H =$	$C =$	
CH_4	$H =$	$C =$	
CO	$O =$	$C =$	
CO_2	$O =$	$C =$	

Perhaps the best definition is just the following:

> A **redox reaction** is a process in which oxidation numbers change.

(1)   In a balanced redox equation, the oxidation state of one element *increases*, while the oxidation state of another element *decreases*.

(2)   Oxidation ≡ an *increase* in oxidation state (electrons on the right-hand side *RHS*).

(3)   Reduction ≡ a *decrease* in oxidation state (electrons on the left-hand side *LHS*).

An overall balanced redox reaction can be broken down into the sum of two half-reactions, one of which will be an oxidation half-reaction, and the other a reduction half-reaction. No electrons will show in the overall balanced reaction; the electrons released by the oxidation half-reaction will be consumed by the reduction half-reaction. The following examples illustrate these points:

$2Fe + \frac{3}{2}O_2 \rightarrow Fe_2O_3$ An overall balanced redox reaction—no electrons showing. This reaction consists of the following changes in oxidation states:

$2Fe(0) \rightarrow 2Fe(+3)$ Δ(oxidation state) = 2(+3) − 2(0) = +6   Oxidation—increase in oxidation state by six units.

$3O(0) \rightarrow 3O(-2)$ Δ(oxidation state) = 3(−2) − 3(0) = −6   Reduction—decrease in oxidation state by six units.

Net change = +6 − 6 = 0   No net change in oxidation state for overall reaction.

$NO_3^- + 4H_3O^+ + 3e^- \rightarrow NO + 6H_2O$ A balanced *half*-reaction.

$N(+5) \rightarrow N(+2)$ Δ(oxidation state) = (+2) − (+5) = −3   (Reduction)

$7O(-2) \rightarrow 7O(-2)$ Δ(oxidation state) = 7(−2) − 7(−2) = 0   (No change)

$12H(+1) \rightarrow 12H(+1)$ Δ(oxidation state) = 12(+1) − 12(+1) = 0   (No change)

Net change = −3 + 0 + 0 = −3   A reduction half-reaction—electrons on the *left-hand side*.

$Mn^{2+} + 6H_2O \rightarrow MnO_2 + 4H_3O^+ + 2e^-$ A balanced *half*-reaction.

$Mn(+2) \rightarrow Mn(+4)$ Δ(oxidation state) = (+4) − (+2) = +2   (Oxidation)

$6O(-2) \rightarrow 6O(-2)$ Δ(oxidation state) = 6(−2) − 6(−2) = 0   (No change)

$12H(+1) \rightarrow 12H(+1)$ Δ(oxidation state) = 12(+1) − 12(+1) = 0   (No change)

Net change = +2 + 0 + 0 = +2   An oxidation half-reaction—electrons on the *right-hand side*.

24.3 STEPS FOR BALANCING REDOX REACTIONS

Some redox reactions are quite complicated and can't readily be balanced just by visual inspection. The following steps—applied in the order given—will enable you to balance such equations.

1. By assigning oxidation numbers to the elements present, determine which species is being oxidized, and which is being reduced.
2. Write out each skeleton half-reaction; then, for each skeleton half-reaction:
3. Do a mass balance on the reducing or oxidizing species.
4. Add electrons as required to account for the amount of oxidation or reduction taking place.
5. Carry out a charge balance by adding H^+ as required in acid media or OH^- as required in alkaline media.
6. Add H_2O as required to complete the mass balance.
7. Multiply each half-reaction as required to balance the overall exchange of electrons, and combine the two half-reactions.
8. Tidy up the H_2Os and H^+s or OH^-s appearing on both sides.

Example 24-1

Thiosulfate ion ($S_2O_3^{2-}$) reacts in aqueous acid solution with dichromate ion ($Cr_2O_7^{2-}$) to produce chromic ion (Cr^{3+}) and sulfate ion (SO_4^{2-}). Write out the balanced chemical reaction for this process.

Solution

The step numbers below correspond to the above steps for balancing redox reactions.

(1) S goes from +2 to +6 (oxidation).
    Cr goes from +6 to +3 (reduction).

$$\overset{(+2)}{\underset{+4\ -6}{S_2O_3^{2-}}} + \overset{(+6)}{\underset{+12\ -14}{Cr_2O_7^{2-}}} \xrightarrow{acid} \overset{+6\ -8}{SO_4^{2-}} + \overset{+3}{Cr^3}$$

(2) First consider the skeleton half-reaction for the oxidation process:
    Skeleton half-reaction—not balanced: $S_2O_3^{2-} \rightarrow SO_4^{2-}$

(3) Mass balance on S, the oxidizing species: $S_2O_3^{2-} \rightarrow 2\,SO_4^{2-}$

(4) Add 8 electrons to the right-hand side for the $S_2O_3^{2-} \rightarrow 2\,SO_4^{2-} + 8\,e^-$
    oxidation of the sulfur:
    $(+4) \qquad 2(+6) = +12 \quad \Delta ON = (+12) - (+4) = +8$

(5) total charge on LHS = −2
    total charge on RHS   therefore add $10H^+$ to right-hand side:
    $= 2(-2) - (8) = -12$

$$S_2O_3^{2-} \rightarrow 2\,SO_4^{2-} + 8\,e^- + 10H^+$$

(6) There are 8 oxygens on the right, but only 3 on the left; for a mass balance on the oxygen, we need to add 5 oxygens to the left-hand side

by adding $5H_2O$ to the left-hand side: $S_2O_3^{2-} + 5H_2O \rightarrow 2SO_4^{2-} + 8e^- + 10H^+$

Check for mass balance by counting the hydrogens: LHS = 10, RHS = 10   (OK)
The oxidation half-reaction is now balanced both for mass and charge.

Next we do the same for the *reduction* half-reaction, starting with step 2:

(2) Skeleton half-reaction: $Cr_2O_7^{2-} \rightarrow Cr^{3+}$

(3) Mass balance on the reducing species: $Cr_2O_7^{2-} \rightarrow 2Cr^{3+}$

(4) Determine change in oxidation number:
$Cr_2(2\times[+6] = +12) \rightarrow 2Cr(2\times[+3] = +6)$; therefore, $\Delta ON = (+6) - (+12) = -6$
There are 6 units of reduction; therefore add 6 electrons to the left-hand side:
$$Cr_2O_7^{2-} + 6e^- \rightarrow 2Cr^{3+}$$

(5) total charge on LHS = $-2 - 6 = -8$ Therefore we have to add an additional +14 positive
     total charge on RHS = $2(+3) = +6$ charges to the left-hand side; do this by adding
                                             $14H^+$ ions. (In alkaline solution we would add OH^-
                                             ions.) $Cr_2O_7^{2-} + 14H^+ + 6e^- \rightarrow 2Cr^{3+}$

(6) For a mass balance on the oxygen, we need
     to add 7 oxygens to the right-hand side;
     therefore add $7H_2O$ to the right-hand side: $Cr_2O_7^{2-} + 14H^+ + 6e^- \rightarrow 2Cr^{3+} + 7H_2O$

Check for mass balance by counting the hydrogen atoms: LHS = 14, RHS = 14,   (OK)
The reduction half-reaction is now balanced for both mass and charge.

(7) No electrons should show in the overall balanced chemical reaction. The electrons released by the oxidation half-reaction (8 electrons) must be taken up by the reduction half-reaction (6 electrons). Therefore, to get an electron balance, we have to multiply the oxidation half-reaction by 3 (to get 24 electrons released), and the reduction half-reaction by 4 (to get 24 electrons consumed):

$S_2O_3^{2-} + 5H_2O \rightarrow 2SO_4^{2-} + 8e^- + 10H^+$ $\times 3$
$Cr_2O_7^{2-} + 14H^+ + 6e^- \rightarrow 2Cr^{3+} + 7H_2O$ $\times 4$

Oxidation: $3S_2O_3^{2-} + 15H_2O \rightarrow 6SO_4^{2-} + 24e^- + 30H^+$

Reduction: $4Cr_2O_7^{2-} + 56H^+ + 24e^- \rightarrow 8Cr^{3+} + 28H_2O$

Combining: $3S_2O_3^{2-} + 4Cr_2O_7^{2-} + 56H^+ + 15H_2O \rightarrow 6SO_4^{2-} + 8Cr^{3+} + 30H^+ + 28H_2O$

(8) Tidying up: **Ans:** $\boxed{3S_2O_3^{2-} + 4Cr_2O_7^{2-} + 26H^+ \rightarrow 6SO_4^{2-} + 8Cr^{3+} + 13H_2O}$

Exercise 24-2

Using the method of half-reactions, balance the following unbalanced redox reactions, adding H^+, OH^-, and H_2O, as required:

(a) $CrO_4^{2-} + HSnO_2^- \xrightarrow{base} Cr(OH)_3 + Sn(OH)_6^{2-}$ ☺

(b) $ClO_2 \xrightarrow{base} ClO_3^- + ClO_2^-$ ☺

(c) $AsO_3^{3-} + MnO_4^- \xrightarrow{acid} AsO_4^{3-} + Mn^{2+}$ ☺

(d) $SO_2 + MnO_4^- \xrightarrow{acid} SO_4^{2-} + Mn^{2+}$ ☺

(e) $Cl_2 \xrightarrow{base} Cl^- + ClO_3^-$ ☺

(f) $CrO_4^{2-} + H_2S \xrightarrow{acid} Cr^{3+} + S$ ☺

(g) $Pb + PbO_2 + SO_4^{2-} \xrightarrow{acid} PbSO_4$ ☺

(h) $Cr(OH)_4^- + HO_2^- \xrightarrow{base} CrO_4^{2-}$ ☹

(i) $Ag + NO_3^- \xrightarrow{acid} Ag^+ + NO$ ☺

(j) $I^- + MnO_4^- \xrightarrow{base} I_2 + MnO_2$ ☺

(k) $Zn + NO_3^- \xrightarrow{acid} Zn^{2+} + NH_4^+$ ☺

(l) $Fe(OH)_2 + O_2 \xrightarrow{base} Fe(OH)_3$ ☹

(m) $PbO_2 + Cl^- \xrightarrow{acid} PbCl_2 + Cl_2$ ☹

(n) $MnO_4^- + CH_3OH \xrightarrow{base} MnO_4^{2-} + CO_3^{2-}$ ☺

(o) $H_2C_2O_4 + MoO_4^{2-} \xrightarrow{acid} CO_2 + Mo^{3+}$ ☺

(p) $MnO_4^- + H_2O_2 \xrightarrow{acid} Mn^{2+} + O_2$ ☺

(q) $Zn + OH^- \xrightarrow{base} H_2 + Zn(OH)_4^{2-}$ ☺

(r) $Cu(NH_3)_4^{2+} + S_2O_4^{2-} \xrightarrow{base} Cu + SO_4^{2-} + NH_3$ ☺

Example 24-2

There is much interest in developing fuel cells as an environmentally clean and highly efficient source of electrical energy. In most fuel cells, hydrogen gas is fed to the anode,[2] where it undergoes an oxidation reaction under acidic conditions, while oxygen (from the air) is fed to the cathode, where it is reduced in an acidic environment. The overall process taking place in the fuel cell is

---

[2] The anode is the electrode at which oxidation occurs; the cathode is the electrode at which reduction occurs.

$$H_2 + \tfrac{1}{2}O_2 \xrightarrow{\ acidic\ } H_2O$$

What are the half-reactions taking place at each electrode?

Solution

Assigning oxidation numbers shows that hydrogen in the zero oxidation state in H_2 is oxidized to the +1 oxidation state in H_2O. Similarly, oxygen in the zero oxidation state in O_2 is reduced to the −2 oxidation state in H_2O.

For the oxidation half-reaction:

$H_2 \rightarrow H_2O$

$2H(0) \rightarrow 2H(+1)$

Total Δ(oxidation state) = +2 − 0 = +2; therefore add two electron's worth of oxidation:

$H_2 \rightarrow H_2O + \mathbf{2e^-}$

Next add $2H^+$ to the right-hand side to get a charge balance:

$H_2 \rightarrow H_2O + 2e^- + \mathbf{2H^+}$

Now add H_2O to the left-hand side to balance the oxygen atom on the right-hand side:

$H_2 + \mathbf{\mathit{H_2O}} \rightarrow H_2O + 2e^- + 2H^+$

This is now a balanced half-reaction; but we note that the water on each side is redundant, so drop it, leaving:

$H_2 \rightarrow 2H^+ + 2e^-$

This is the *anode reaction.*

For the reduction half-reaction:

$\tfrac{1}{2}O_2 \rightarrow H_2O$

$\tfrac{1}{2}O_2(0) \rightarrow O(-2)$

Total Δ(oxidation state) = $(-2) - \tfrac{1}{2}(0) = -2$; therefore add two electron's worth of reduction:

$\tfrac{1}{2}O_2 + 2e^- \rightarrow H_2O$

Next add $2H^+$ to the left-hand side to get a charge balance:

$\tfrac{1}{2}O_2 + \mathbf{2H^+} + 2e^- \rightarrow H_2O$

This is now a balanced half-reaction, and is the *cathode reaction.*

Check by adding together the two half-reactions:

anode: $H_2 \rightarrow 2H^+ + 2e^-$

cathode: $\tfrac{1}{2}O_2 + 2H^+ + 2e^- \rightarrow H_2O$

---

overall: $H_2 + \tfrac{1}{2}O_2 \rightarrow H_2O$ (OK!)

Ans:
anode: $H_2 \rightarrow 2H^+ + 2e^-$
cathode: $\tfrac{1}{2}O_2 + 2H^+ + 2e^- \rightarrow H_2O$

KEY POINTS FOR CHAPTER TWENTY-FOUR

1. **Oxidation**: • a process in which a substance *loses* electrons (electrons on RHS).

    $$M \rightarrow N + ze^- \quad (M \text{ is oxidized to } N)$$

    • an *increase* in the oxidation state of an element

    Reduction: • a process in which a substance *gains* electrons (electrons on the LHS).

    $$N + ze^- \rightarrow M \quad (N \text{ is reduced to } M)$$

    • a *decrease* in the oxidation state of an element

2. **Oxidation Number—*ON*** (oxidation state): A *fictitious charge assigned to an atom in a molecule or in an ion according to an* **arbitrary** *set of rules*. (A "bookkeeping device" that may or may not be related to the actual transfer of electrical charge.)

    Rules:
    • The oxidation number of any free element is *zero* (e.g., N_2, C, He, Cu).
    • O in compounds has an oxidation number of -2 (but in peroxides $= -1$; e.g., H_2O_2).
    • H in compounds has an oxidation number of $+1$ (but in metal hydrides $= -1$, e.g., NaH).
    • The sum of the oxidation numbers must be *zero* for a neutral molecule and equal to the *net charge* for an ion (e.g., in H_2SO_4, $H = +1$, $S = +6$, $O = -2$; in NO_3^-, $N = +5$, $O = -2$).

4. A **redox reaction**: a chemical reaction in which oxidation numbers change. In a balanced redox equation, the oxidation state (oxidation number) of one element *increases*, while the oxidation state (oxidation number) of another element *decreases*. An overall balanced redox reaction can be broken down into the sum of two half-reactions, one of which will be an oxidation half-reaction, and the other a reduction half-reaction. No electrons will show in the overall balanced reaction; the electrons released by the oxidation half-reaction will be consumed by the reduction half-reaction.

5. **Steps for balancing redox reactions**:

    1. By assigning oxidation numbers to the elements present, determine which species is being oxidized, and which is being reduced.
    2. Write out each skeleton half-reaction; then, for each skeleton half-reaction :
    3. • Do a mass balance on the reducing or oxidizing species.
    4. • Add electrons as required to account for the amount of oxidation or reduction taking place.
    5. • Do a charge balance by adding H^+ as required in acid media or OH^- as required in basic media.
    6. • Add H_2O as required to complete the mass balance.
    7. Multiply each half-reaction as required to balance the overall exchange of electrons, and combine the two half-reactions.
    8. Tidy up the H_2Os and H^+s or OH^-s appearing on both sides.

PROBLEMS

1. Balance the following unbalanced equation using the method of oxidation numbers and half-reactions:

$$(CN)_2 \xrightarrow{\text{basic}} CN^- + OCN^- \ominus$$

2. Redo example 24-2, but this time for a fuel cell that operates in an alkaline environment:

$$H_2 + \tfrac{1}{2}O_2 \xrightarrow{\text{basic}} H_2O \ominus$$

3. "The Energizer" battery that features that stupid pink rabbit beating the drum is an alkaline zinc–manganese dioxide cell. The overall cell reaction can be represented as:

$$Zn_{(s)} + 2MnO_{2(s)} \xrightarrow{\text{basic}} ZnO_{(s)} + Mn_2O_{3(s)}$$

What is the oxidation half-reaction at the anode and the reduction half-reaction at the cathode?

TWENTY-FIVE

ELECTROCHEMISTRY

25.1 HALF-REACTIONS, ELECTRODES, AND ELECTROCHEMICAL CELLS

When a piece of metallic zinc is dropped into an aqueous $CuSO_4$ solution (Fig. 1), the following chemical reaction occurs:

Fig. 1

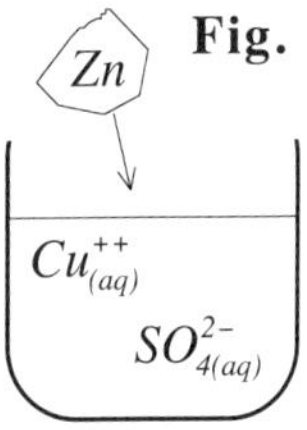

$$Zn_{(s)} + CuSO_{4(aq)} \rightarrow ZnSO_{4(aq)} + Cu_{(s)}\downarrow$$

i.e., $$Zn_{(s)} + Cu^{2+}_{(aq)} + SO^{2-}_{4(aq)} \rightarrow Zn^{2+}_{(aq)} + SO^{2-}_{4(aq)} + Cu_{(s)}\downarrow$$

or, $$Zn_{(s)} + Cu^{2+}_{(aq)} \rightarrow Zn^{2+}_{(aq)} + Cu_{(s)}\downarrow$$

For this reaction, $$\Delta H^o_{298.15} = -218.66 \text{ kJ mol}^{-1}$$

and $$\Delta G^o_{298.15} = -212.55 \text{ kJ mol}^{-1}$$

Therefore, this process will proceed spontaneously under standard conditions at 25°C, with copper depositing from the solution, and zinc metal dissolving into the solution until an equilibrium is reached.

Since no "other" work (non-PV-work) is performed,

$$Q_P = \Delta H = -218.66 \text{ kJ mol}^{-1}$$

That is, for every mole of zinc that dissolves, 218.66 kJ of heat will be given off.

Inspection shows that this reaction is a *redox* reaction, which, as we showed in Chapter 24, can be broken down into two half-reactions:

$$Zn_{(s)} \rightarrow Zn^{2+}_{(aq)} + 2e^- \qquad oxidation\text{—Zn(0) is oxidized to Zn(+2)}$$

and $$Cu^{2+}_{(aq)} + 2e^- \rightarrow Cu_{(s)}\downarrow \qquad reduction\text{—Cu(+2) is reduced to Cu(0)}$$

We say that the Zn is *oxidized* by the Cu^{2+}, which is an *oxidizing agent*, while the Cu^{2+} is *reduced* by the Zn, which is a *reducing* agent. Thus the oxidizing agent (Cu^{2+}) is *reduced*, and the reducing agent (Zn) is *oxidized*.

Since $\Delta G^o_{298.15}$ for the reaction at constant temperature and pressure is negative, in principle this reaction is capable of delivering "other" work. However, when the reaction is carried out in a beaker, as indicated above in Fig. 1, the electrons transfer *directly* from the metallic Zn to the Cu^{2+} ions, and all the energy is released as *heat* ($Q_P = \Delta H^o_{298.15} = -218.66$ kJ mol^{-1}), with no useful work extracted from the system.

Fig. 2

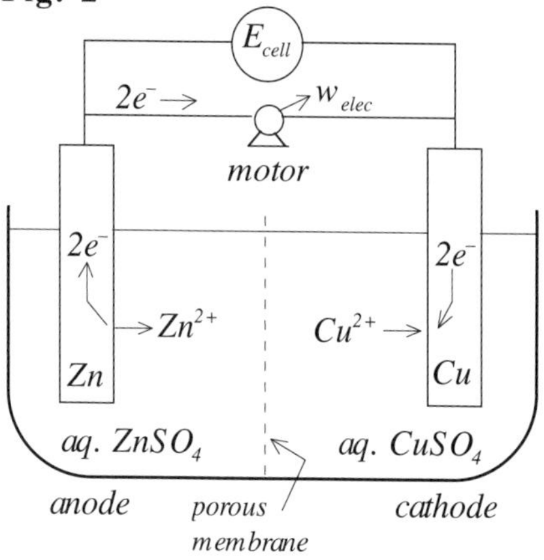

aq. $ZnSO_4$ aq. $CuSO_4$

anode   porous   cathode
          membrane

BUT, the same overall reaction also can be carried out by having each half-reaction take place in a separate compartment in an **electrochemical cell**, as shown in Fig. 2.

By *definition*:
The **anode** is the electrode at which *oxidation* occurs (the *Zn* electrode).

The **cathode** is the electrode at which *reduction* occurs (the *Cu* electrode).[1]

Therefore, the anode and cathode reactions are, respectively:

anode: $Zn_{(s)} \rightarrow Zn^{2+}_{(aq)} + 2e^-$

cathode: $Cu^{2+}_{(aq)} + 2e^- \rightarrow Cu_{(s)}$

Thus, instead of reacting directly in solution as in Fig. 1, by separating the beaker into two compartments and providing an external electrical pathway between the two metals, the driving force for the reaction causes the electrons to transfer from the Zn to the Cu^{2+} ions by flowing through the external circuit from the anode to the cathode.

The porous membrane is needed to prevent the two solutions from mixing; if they mixed, the reaction would occur directly between the Zn metal and the Cu^{2+} ions in solution. The membrane also is required to make electrical contact between the two electrode compartments in order for the electric circuit to be completed.[2]

The electrons leaving the anode are at a higher energy than those entering the cathode; in flowing through a resistive electrical load in the external circuit—such as a motor—they give up this energy to the load, making the energy available as useful electrical work.

25.2 ELECTRICAL WORK

Consider the two charged plates shown in Fig. 3, one of which is negatively charged and the other positively charged, separated by a vacuum: There will exist an electrical voltage gradient E (an *electric field*) between the two plates, of E V m^{-1} (volts per meter). If we introduce a small particle carrying a negative charge of Q coulombs between the two plates, it will spontaneously move *away*

---

[1] Don't make the mistake of thinking that the anode is the positive electrode and the cathode the negative electrode. Although this is true for an electrolytic cell, the opposite is true for a galvanic cell.
[2] If there is no complete electrical pathway or loop in the cell, no steady current can flow.

from the negative plate and *toward* the positive plate. If we wish to *reverse* this process and make the particle move in the opposite direction (right to left), then we will have to overcome the electrostatic forces that tend to make it move in the spontaneous direction (left to right). To do this, we will have to apply an opposing *force* *through a displacement.* That is, *we must do work on the charged* *particle.*

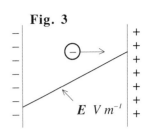

Fig. 3

In order to move the charge a distance of z meters against the electric field, the total work we must do on the particle is given by

$$w_{elec} = \left[\left(E\ \tfrac{volt}{m}\right)(z\,m)\right]\!\left(Q\ \text{coulombs}\right)$$
$$= \left[\Delta\phi\ \text{volts}\right]\!\left(Q\ \text{coulombs}\right)$$
$$= \Delta\phi \cdot Q\ \text{joules}^3$$

It follows that if we must do $\Delta\phi \cdot Q$ joules of work *on* the particle to move it through a voltage difference of $\Delta\phi$ volts when moving *against* the field, then, when the particle moves spontaneously *with* the field through a voltage difference of $\Delta\phi$ volts, an amount of electrical work equal to $\Delta\phi \cdot Q$ joules will be delivered *by* the system.

Therefore the electrical work output delivered by an electrochemical cell when a charge of Q coulombs passes through the cell voltage E_{cell} is given by

$$w_{elec} = E_{cell} \cdot Q\ \text{joules} \qquad\qquad \ldots [1]$$

25.3 TYPES OF CELLS

An electrochemical cell such as that shown in Fig. 2 that produces electrical energy via a spontaneous chemical reaction is called a **galvanic cell**—typical examples are a flashlight cell or a car battery. Note that in a galvanic cell the anode is the *negative* electrode. A galvanic cell in which reactants such as hydrogen (a fuel) and oxygen (an oxidant) are supplied in a steady stream to the cell from an outside supply is called a **fuel cell**.

The other type of cell, in which electrical energy must be put *into* the cell to cause a (nonspontaneous) chemical reaction to occur is called an **electrolysis cell** or an **electrolytic cell**—an example would be a cell in which water is electrolyzed to produce gaseous hydrogen and oxygen through the use of an external d.c. power supply.

25.4 LIQUID JUNCTIONS AND SALT BRIDGES

Whenever two solutions of different composition come into electrical contact—as occurs within the porous membrane in our galvanic cell—there will exist a small voltage difference across the interface known as a **liquid junction potential** ($\Delta\phi_{lj}$). There is considerable uncertainty in accurately

---

[3] Power = watts = (volts)(amperes) = $V \cdot C/s$; but watts also are J/s; therefore $V \cdot C/s = J/s$, or $V \cdot C = J$.

calculating the exact magnitude of such liquid junction potentials, which may range from a few millivolts to several hundred millivolts. This adds a complicating feature to the measurement of cell voltages.

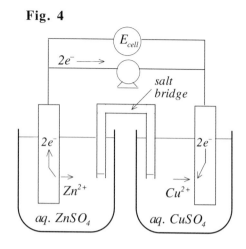

Fig. 4

To alleviate this problem, the liquid junction potential between the two solutions is eliminated[4] by the use of what is known as a **salt bridge** (Fig. 4). A salt bridge typically consists of a tube containing a jelly-like material such as *agar* (made from seaweed) that has been saturated with concentrated KCl solution or NH_4NO_3 (ammonium nitrate) solution. The salt bridge actually replaces the *single* liquid junction potential across the porous membrane with *two* liquid junction potentials[5] of equal magnitude, but opposite sign, so that they in effect cancel each other.

25.5 CELL NOTATION

We use the following short-hand notation to describe the cell of Fig. 4:

$$Zn \left| Zn^{2+}_{(aq)} \right\| Cu^{2+}_{(aq)} \left| Cu \right.$$

$$\quad\ \ anode \qquad\qquad cathode$$

The single vertical lines denote boundaries between two different phases; the double vertical line represents a salt bridge (which contains a *double* phase boundary). If we had used a porous membrane rather than a salt bridge—such as the cell in Fig. 2—we would write

$$Zn \left| Zn^{2+}_{(aq)} \mid Cu^{2+}_{(aq)} \right| Cu$$

$$\quad\ \ anode \qquad\qquad cathode$$

Note that when this notation is used, the *anode* is *always* the electrode on the *left-hand side*.

25.6 EQUILIBRIUM VOLTAGES

The overall process that takes place in an electrochemical cell is a redox reaction; the cell has been constructed in such a way that the oxidation half-reaction takes place at an anode, often located in a separate anode compartment, while the reduction half-reaction takes place at a cathode, often located in a separate cathode compartment.

---

[4] Or, at least rendered sufficiently small to be insignificant.
[5] One at the interface between the gel and the $ZnSO_4$ solution, and the other at the interface between the gel and the $CuSO_4$ solution.

The cell reaction taking place in a galvanic cell is a spontaneous process, for which, at constant temperature and pressure,

$$\Delta G_{T,P} < 0$$

It should be noted that no electrons appear in the overall cell reaction, which is just an ordinary redox chemical reaction, but one that has been carried out in a clever way. We saw earlier in our discussion of thermodynamics that such a chemical reaction is capable of delivering non-expansion work ("other" work) $W°$, and that the maximum "other" work is the work that is delivered when the cell operates *reversibly*. This is given by

$$W_{rev}^{o} = \Delta G_{T,P}$$

or,
$$-\Delta G_{T,P} = -W_{rev}^{o}$$

In the case of a galvanic cell, $\Delta G_{T,P}$ is *negative*, so that W_{rev}^{o} also will be negative, indicating that work is being delivered *by* the cell. Thus, $-\Delta G_{T,P}$ will be *positive*, as will $-W_{rev}^{o}$. For a galvanic cell, $-W_{rev}^{o}$ represents the maximum *electrical* work that can be delivered by the cell:[6]

$$-\Delta G_{T,P} = w_{elec}^{max} \qquad \ldots [2]$$

We showed earlier that the work delivered when Q coulombs spontaneously flows through a voltage difference of $\Delta\phi$ volts is given by

$$w_{elec} = Q \cdot \Delta\phi \qquad \ldots [3]$$

The charge on a single electron is $e_o = 1.602192 \times 10^{-19}$ C. The charge on *one mole* of electrons is called the **faraday** (F), and has a value of

$$\begin{aligned} F &= N_A e_o \\ &= (6.02217 \times 10^{23} \text{ mol}^{-1})(1.602192 \times 10^{-19} \text{ C}) \\ &= 96\,487 \text{ C mol}^{-1} \end{aligned}$$

For the redox reaction taking place in the cell, if n moles of electrons are transferred during the overall cell reaction, then the corresponding number of coulombs of charge transferred is

$$Q = \left(n \text{ mol} \right)\left(F \, \tfrac{C}{\text{mol}} \right) = nF \text{ coulombs} \qquad \ldots [4]$$

Since this charge flows through the cell voltage E_{cell}, then the electrical work that is delivered by a galvanic cell is

$$\begin{aligned} w_{elec} &= Q \cdot \Delta\phi \\ &= nFE_{cell} \qquad \ldots [5] \end{aligned}$$

If the cell is operated *reversibly*, then the cell voltage will be the *equilibrium* (reversible) voltage, E_{cell}^{eqm}, and the electrical work delivered by the cell will be the *maximum* work:

---

[6] Don't forget that w is a scalar quantity (always positive); W can be positive or negative.

$$w_{elec}^{max} = nFE_{cell}^{eqm} \qquad \qquad \ldots [6]$$

but, from Eqn [2]:

$$w_{elec}^{max} = -\Delta G_{T,P}$$

Therefore, with this substitution, Eqn [6] becomes

$$-\Delta G_{T,P} = nFE_{cell}^{eqm}$$

$$\boxed{\begin{array}{l} w_{elec} = nFE_{cell} \\[2mm] w_{elec}^{max} = -\Delta G_{T,P} = nFE_{cell}^{eqm} \end{array}} \qquad Galvanic\ Cell \qquad \ldots [7]$$

Redox couples—when discussing the half-cell reactions in an electrochemical cell, it is customary to write each half-cell reaction as if it were a *reduction* reaction. The cathode reaction, of course, already *is* a reduction reaction; but, for the purpose of calculation, the direction of the anode reaction is changed and it too is written as if it were a reduction reaction.

The oxidized and reduced substances in each half-reaction comprise what is called a *redox couple*. In general, for the half-reaction

$$Ox + ne^- \rightarrow Red$$

the redox couple is represented by *Ox/Red*, where *Ox* is the oxidized form of the species and *Red* is the reduced form. Thus, the two redox couples in the cell shown in Fig. 4 are the Zn^{2+}/Zn couple and the Cu^{2+}/Cu couple.

Example 25-1

Given the following standard reduction potentials at 25°C, calculate the molar Gibbs free energy of formation Δg_f^o for the $OH_{(aq)}^-$ ion at 25°C.

Data:

$$2H^+ + \tfrac{1}{2}O_2 + 2e^- \rightleftharpoons H_2O \qquad E_1^o = +1.2288\ V$$

$$2H^+ + 2e^- \rightleftharpoons H_2 \qquad E_2^o = 0.00\ V$$

$$H_2O + e^- \rightleftharpoons OH^- + \tfrac{1}{2}H_2 \qquad E_3^o = -0.8279\ V$$

Solution

The formation reaction is
$$\tfrac{1}{2}H_{2(g)} + \tfrac{1}{2}O_{2(g)} + e^- \xrightarrow{H_2O} OH_{(aq)}^-$$

We are going to manipulate the reduction reactions given as data in order to obtain the correct sequence such that everything cancels out except the desired reaction.[7] First we evaluate the standard

---

[7] This is just like using Hess's law, and makes use of the fact that the Gibbs free energy is a state function.

Gibbs free energies of the above reactions through the relation $\Delta G° = -nFE°$; thus:[8]

(1) $2H^+ + \frac{1}{2}O_2 + 2e^- \rightleftharpoons H_2O$ $\quad \Delta G_1^o = -n_1FE_1^o = -(2)(96\,487)(+1.2288) = -237\,126$ J mol^{-1}

(2) $2H^+ + 2e^- \rightleftharpoons H_2$ $\qquad\qquad \Delta G_2^o = -n_2FE_2^o = -(2)(96\,487)(0.00) = 0$ J mol^{-1} (by definition)[9]

(3) $H_2O + e^- \rightleftharpoons OH^- + \frac{1}{2}H_2$ $\quad \Delta G_3^o = -n_3FE_3^o = -(1)(96\,487)(-0.8279) = +79\,882$ J mol^{-1}

Now we start to assemble the reactions so they add up to the desired reaction:

First we want a single hydroxyl ion on the right-hand side; therefore we write reaction (3) as is:

(3) $H_2O + e^- \rightleftharpoons OH^- + \frac{1}{2}H_2$ $\qquad\qquad \Delta G_3^o$

Next, eliminate the single H_2O on the left-hand side of (3). We do this by adding reaction (1):

$$
\begin{array}{lll}
(3) & H_2O + e^- \rightleftharpoons OH^- + \frac{1}{2}H_2 & \Delta G_3^o \\
(1) & 2H^+ + \frac{1}{2}O_2 + 2e^- \rightleftharpoons H_2O & \Delta G_1^o \\
\hline
\text{Adding:} & 2H^+ \; \frac{1}{2}O_2 + 3e^- \rightleftharpoons OH^- + \frac{1}{2}H_2 & \Delta G_3^o + \Delta G_1^o \quad (4)
\end{array}
$$

Next we want to eliminate the two unwanted H^+ ions on the left-hand side of reaction (4). We do this by adding the *reverse* of reaction (2):

$$
\begin{array}{lll}
(4) & 2H^+ + \frac{1}{2}O_2 + 3e^- \rightleftharpoons OH^- + \frac{1}{2}H_2 & \Delta G_3^o + \Delta G_1^o \\
-(2) & H_2 \rightleftharpoons 2H^+ + 2e^- & -\Delta G_2^o \\
\hline
\text{Adding:} & \frac{1}{2}H_2 + \frac{1}{2}O_2 + e^- \rightleftharpoons OH^- & \Delta G_3^o + \Delta G_1^o - \Delta G_2^o \quad (5)
\end{array}
$$

Reaction (5) is the desired reaction; therefore,

$$
\begin{aligned}
\Delta g_f^o [OH_{(aq)}^-] &= \Delta G_3^o + \Delta G_1^o - \Delta G_2^o \\
&= (+79882) + (-237\,126) - (0) \\
&= -157\,244 \text{ J mol}^{-1}
\end{aligned}
$$

Ans: $\boxed{-157.24 \text{ kJ mol}^{-1}}$

$\boxed{\textbf{Exercise } \textbf{25-1}}$

Given that the standard reduction potential at 25°C for the $Na_{(aq)}^+/Na$ redox couple is -2.714 V, calculate the standard Gibbs free energy of formation of the $Na_{(aq)}^+$ ion at 25°C. ☺

---

[8] $E°$ is the potential under *standard* conditions; i.e., 1 bar pressure, unit activities. See page 25-9 for more details.
[9] The standard hydrogen electrode is discussed later in this chapter, in Section 25.8.

25.7 THE NERNST EQUATION

We saw earlier that, for the reaction

$$aA + bB \overset{nF}{\rightarrow} cC + dD$$

the reaction free energy ΔG_R varies with the composition of the reaction mixture as

$$\Delta G_R = \Delta G_R^o + RT \, ln \, Q$$

$$= \Delta G_R^o + RT \, ln \left(\frac{a_C^c \cdot a_D^d}{a_A^a \cdot a_B^b} \right) \qquad \ldots [8]$$

Substituting $\Delta G_{T,P} = -nFE_{cell}$ into Eqn [8] when the cell is at equilibrium (no current flow) gives

$$- nFE_{cell}^{eqm} = -nFE_{cell}^o + RT \, ln \left(\frac{a_C^c \cdot a_D^d}{a_A^a \cdot a_B^b} \right)$$

That is,

$$\boxed{E_{cell}^{eqm} = E_{cell}^o - \frac{RT}{nF} ln \left(\frac{a_C^c \cdot a_D^d}{a_A^a \cdot a_B^b} \right)} \quad \begin{array}{l} \textit{The Nernst} \\ \textit{Equation} \end{array} \qquad \ldots [9]$$

Eqn [9] is called the *Nernst equation*, and shows how the equilibrium cell voltage varies under non-standard conditions of composition. This equation holds for both galvanic and electrolytic cells.[10]

In Eqn [9], the term E_{cell}^o is called the **standard cell potential**, and is defined by

$$\boxed{E_{cell}^o = \frac{-\Delta G_{T,P}^o}{nF}} \quad \begin{array}{l} \textit{Standard Cell} \\ \textit{Potential} \end{array} \qquad \ldots [10]$$

25.8 SINGLE ELECTRODE POTENTIALS

Although it is not possible to measure the absolute value of a *single* electrode potential difference,[11] if one electrode is *arbitrarily* assigned a value of *zero* volts, then we can measure the value of all other single electrodes *relative* to that electrode.

25.8.1 THE STANDARD HYDROGEN ELECTRODE

The electrode chosen as the standard against which all others are measured is the **standard hydrogen electrode** (*SHE*):

---

[10] E_{cell}^{eqm} will be positive for a galvanic cell ($\Delta G_{T,P} < 0$) and negative for an electrolytic cell ($\Delta G_{T,P} > 0$).

[11] This is related to the fact that a voltmeter can only measure the *difference* in potential between two electrodes.

$$Pt \; |H_{2(g)}, \; H^+_{(aq)}$$

To be a *standard* hydrogen electrode, the partial pressure of the hydrogen gas must be *one* bar and the hydrogen ions must be at *unit* relative activity.

The reaction at the hydrogen electrode is

$$2\,H^+_{(aq)} + 2\,e^-_{(Pt)} \rightleftharpoons H_{2(g)} \qquad\qquad \dots [11]$$

For this reaction, $\Delta G^o_{298.15} = \Delta g^o_f[H_{2(g)}] - 2\,\Delta g^o_f[H^+_{(aq)}] - 2\,\Delta g^o_f[e^-_{(Pt)}]$

Because we are unable to measure experimentally the molar free energy of an electron or of any single ionic species, by *definition* Δg^o_f for an electron is arbitrarily assigned a value of *zero*; similarly, Δg^o_f for aqueous hydrogen ion also is arbitrarily assigned a value of *zero*, as is Δg^o_f for a pure element. Accordingly, for Reaction [11],

$$\Delta G^o_{298.15} \equiv 0 - 2(0) - 2(0) = 0$$

That is, $\Delta G^o_{298.15}$ for Reaction [11] is *zero* by definition—*at all temperatures*. From Eqn [10] the standard electrode potential for the hydrogen electrode at all temperatures is thus

$$E^o_H = \frac{-\Delta G^o}{nF} = \frac{(0)}{(2)(96\,487)} = 0 \text{ volts} \qquad\qquad \dots [12]$$

As stated earlier, it is customary to write half-cell reactions as reduction reactions. **Standard conditions** for an electrode reaction means *one bar pressure* and *unit activity for all species*. The half-cell potential under these conditions is called the **standard reduction potential**, and is given the symbol $E°$. Tables of standard reduction potentials usually are tabulated for half-reactions at 25°C, but the temperature is not part of the definition of standard conditions. Tabulated values of standard reduction potentials are reported *vs.* the standard hydrogen electrode (*SHE*), which has a value, as discussed above, of zero, by definition.

The Nernst equation can be applied to half-cell reactions as well as to complete overall cell reactions. Thus, for the hydrogen electrode (Eqn [11]):[12]

$$\Delta\phi_H = E^o_H - \frac{RT}{2F} ln\left(\frac{P_{H_2}}{a^2_{H^+}} \right) \qquad\qquad \dots [13]$$

Since, *by definition*, $E^o_H = 0$, Eqn [13] becomes

$$\Delta\phi_H = -\frac{RT}{2F} ln\left(\frac{P_{H_2}}{a^2_{H^+}} \right) \qquad\qquad \dots [14]$$

If the partial pressure of the hydrogen gas is one bar, then Eqn [14] further reduces to

---

[12] The relative activity of an electron always is arbitrarily assigned a value of unity. And of course, at non-extreme pressures we know that H_2 behaves ideally, so that $a_{H_2} = P_{H_2}/1 = P_{H_2}$.

$$\Delta\phi_H = -\frac{RT}{2F}ln\left(\frac{1}{a_{H^+}^2}\right) = +\frac{RT}{F}ln\,a_{H^+} = \frac{2.303\,RT}{F}log\,a_{H^+}$$

That is,

$$\boxed{\Delta\phi_H = -\frac{2.303\,RT}{F}pH}$$ ... [15][13]

At 25°C:

$$\Delta\phi_H = -\frac{2.303(8.314)(298.15)}{96\,487}pH$$

$$= -0.0592\,pH$$ ... [16]

Eqns [15] and [16] show that the electric potential established at the hydrogen electrode is (negatively) directly proportional to the *pH* of the solution. This, in fact, is the principle on which *pH* meters operate.

25.8.2 TABLES OF STANDARD REDUCTION POTENTIALS

Consider the half-cell reduction reaction

$$Zn_{(aq)}^{2+} + 2e^- \rightleftharpoons Zn$$ ... [17]

We can calculate the standard Gibbs free energy change for this reaction using data from tables in the appendices listing standard molar free energies of formation:[14]

Thus, for reaction [17], $\Delta G_{298.15}^o = (0) - (-147.06) - 2(0)$

$$= +147.06\ \text{kJ mol}^{-1}$$

from which $E_{298.15}^o = \frac{-\Delta G_{298.15}^o}{nF} = \frac{-(147\,060)}{(2)(96\,487)} = -0.76\ \text{V}$ ... [18]

Similarly, for $Cu_{(aq)}^{2+} + 2e^- \rightleftharpoons Cu$ ... [19]

$$\Delta G_{298.15}^o = (0) - (+65.49) - 2(0)$$

$$= -65.49\ \text{kJ mol}^{-1}$$

and $E_{298.15}^o = \frac{-\Delta G_{298.15}^o}{nF} = \frac{-(-65\,490)}{(2)(96\,487)} = +0.34\ \text{V}$ ... [20]

---

[13] Unfortunately, most of the standard reduction potentials listed in the literature (including the value of *zero* for the standard hydrogen electrode) are based on the older standard pressure of one *atm*, rather than one *bar*. Fortunately, the values at one atm are so close to those at one bar that, for most purposes, we can ignore the difference.
[14] Remember that the Gibbs free energies of formation for pure elements such as *Zn* arbitrarily are assigned values of zero, as is the value for electrons.

By doing similar calculations for different half-cell reduction reactions we can construct a table of *Standard Reduction Potentials* such as that shown in Table 1. (A more extensive table can be found in Appendix 10.)

Table 1. *Standard Reduction Potentials at 25°C*

Reaction	$E°$ (V vs. SHE)	Reaction	$E°$ (V vs. SHE)
$Cl_{2(g)} + 2e^- \rightleftharpoons 2Cl^-_{(aq)}$	+1.358	$2H^+_{(aq)} + 2e^- \rightleftharpoons H_{2(g)}$	0.000
$Ag^+_{(aq)} + e^- \rightleftharpoons Ag_{(s)}$	+0.799	$Pb^{2+}_{(aq)} + 2e^- \rightleftharpoons Pb_{(s)}$	−0.126
$Cu^{2+}_{(aq)} + 2e^- \rightleftharpoons Cu_{(s)}$	+0.337	$Zn^{2+}_{(aq)} + 2e^- \rightleftharpoons Zn_{(s)}$	−0.763

The more positive the value of $E°$, the stronger the tendency to *reduce* ($\Delta G^o_{T,P}$ becomes increasingly *negative* as $E°$ becomes increasingly *positive*). Thus, in a galvanic cell, **the electrode with the more positive reduction potential will be the cathode** (reduction), while the other electrode, by default, will be the anode (oxidation).

Example 25-2

At 25°C, the standard reduction potentials (vs. *SHE*) for the Fe^{2+}/Fe and Fe^{3+}/Fe^{2+} redox couples are −0.440 V and +0.771 V, respectively. Calculate $E°$ for the Fe^{3+}/Fe couple.

Solution

We want $E°$ for $\qquad Fe^{3+} + 3e^- \rightleftharpoons Fe \qquad E^o_x = ?$

Given: $\qquad\qquad Fe^{2+} + 2e^- \rightleftharpoons Fe \qquad E^o_1 = -0.440$ V $\quad$ ... (1)

and $\qquad\qquad\quad Fe^{3+} + e^- \rightleftharpoons Fe^{2+} \qquad E^o_2 = +0.771$ V $\quad$ ... (2)

The desired half-cell reaction is the *sum* of the other two reactions in series; thus,

(2) $\quad Fe^{3+} + e^- \rightarrow Fe^{2+} \qquad \Delta G^o_2 = -n_2 F E^o_2 = -(1)F(+0.771) = -0.771F$ J mol^{-1}

(1) $\quad Fe^{2+} + 2e^- \rightarrow Fe \qquad \Delta G^o_1 = -n_1 F E^o_1 = -(2)F(-0.440) = +0.886F$ J mol^{-1}

Adding: $\quad Fe^{3+} + 3e^- \rightarrow Fe \qquad \Delta G^o_x = -0.771F + 0.886F = +0.115F$ J mol^{-1}

Therefore, $\qquad\qquad E^o_x = -\dfrac{\Delta G^o_x}{n_x F} = -\dfrac{+0.115F}{(3)F} = \dfrac{-0.115}{3} = -0.0383$ V

Ans: $\boxed{-0.0383 \text{ V}}$

If you try "averaging" the other $E°$ values to obtain the desired $E°$ you are asking for trouble! (Try it.) $E°$s always should be calculated from $\Delta G°$s using the relation $\Delta G° = -nFE°$, being careful to note the correct value of n for each half-reaction.

Exercise 25-2

At 25°C, the standard reduction potentials for the Cu^{2+}/Cu and Cu^+/Cu redox couples are +0.3394 V and +0.521 V, respectively. Calculate $E°$ for the Cu^{2+}/Cu^+ couple. ☺

25.9 CALCULATION OF EQUILIBRIUM CELL VOLTAGES FROM HALF-CELL POTENTIALS

When a voltmeter is used to measure the voltage of an electrochemical cell, what is measured is the *difference* in electrical potential between the two electrodes. As we saw earlier, in a galvanic cell the cathode is positive and the anode is negative; thus the equilibrium[15] cell voltage is

$$E_{cell}^{eqm} = \Delta\phi_{cat}^{eqm} - \Delta\phi_{an}^{eqm}$$

...[21]

where $\Delta\phi_{cat}^{eqm}$ is the equilibrium reduction potential of the cathode, and $\Delta\phi_{an}^{eqm}$ is the equilibrium reduction potential of the anode, as determined from the Nernst equation.

Thus, to calculate the equilibrium voltage of an electrochemical cell:

1. Write both the cathode and the anode reaction as *reduction* reactions;
2. Apply the Nernst equation to each; and
3. Obtain the cell voltage by subtracting the reduction potential of the anode from the reduction potential of the cathode.

Example 25-3

Calculate the equilibrium cell voltage at 25°C for the cell shown at the right. Ignore activity coefficient corrections.

$$Pt \left| \begin{array}{l} Sn(OH)_6^{2-}, \ (0.1\,m) \\ HSnO_2^-, \ \ \ (0.05\,m) \\ KOH_{(aq)}, \ pH \ 12 \end{array} \right\| \left. \begin{array}{l} Cr_2O_7^{2-}, \ (0.01\,m) \\ Cr^{3+}, \ \ \ (0.2\,m) \\ H_2SO_{4(aq)}, \ pH \ 3 \end{array} \right| Pt$$

Data: $Cr_2O_7^{2-} + 14H^+ + 6e^- \rightleftharpoons 2Cr^{3+} + 7H_2O$ $E_{298.15}^o = +1.33$ V

$Sn(OH)_6^{2-} + 2e^- \rightleftharpoons HSnO_2^- + 3OH^- + H_2O$ $E_{298.15}^o = -0.93$ V

---

[15] The equilibrium cell voltage is the cell voltage when no net process takes place in the cell; that is, the voltage when no current is flowing. Thus, the equilibrium cell voltage is the *open circuit voltage (OCV)*; i.e., the voltage when the switch is open and no current flows.

Solution

The cell notation (anode on the left-hand side) tells us that the $Sn(OH)_6^{2-}/HSnO_2^-$ electrode is the anode and the $Cr_2O_7^{2-}/Cr^{3+}$ is the cathode. Applying the Nernst equation to each electrode:

For the cathode:
$$\Delta\phi_{Cr} \approx E_{Cr}^o - \frac{RT}{6F} \ln \left(\frac{a_{Cr^{3+}}^2 \cdot a_{H_2O}^7}{a_{Cr_2O_7^{2-}} \cdot a_{H^+}^{14}} \right)$$

$$\approx +1.33 - \frac{RT}{6F} \ln \frac{[Cr^{3+}]^2 (1)^7}{[Cr_2O_7^{2-}] \cdot a_{H^+}^{14}}$$

$$= 1.33 - \frac{(8.314)(298.15)}{(6)(96\ 487)} \ln \frac{(0.2)^2}{(0.01)(10^{-3})^{14}}$$

$$= +0.9100 \text{ V}$$

For the anode:[16]
$$\Delta\phi_{Sn} \approx -0.93 - \frac{RT}{2F} \ln \frac{[HSnO_2^-] \cdot a_{OH^-}^3 \cdot (1)}{[Sn(OH)_6^{2-}]}$$

$$= -0.93 - \frac{(8.314)(298.15)}{(2)(96\ 487)} \ln \frac{(0.05)(10^{-14}/10^{-12})^3}{(0.10)}$$

$$= -0.7436 \text{ V}$$

Therefore,
$$E_{cell}^{eqm} = \Delta\phi_{cathode} - \Delta\phi_{anode}$$

$$= (+0.9100) - (-0.7436)$$

$$= 1.6536 \text{ V}$$

Ans: $\boxed{1.654 \text{ V}}$

The reactions are:

cathode: $\quad Cr_2O_7^{2-} + 14H^+ + 6e^- \rightarrow 2Cr^{3+} + 7H_2O$

anode: $\quad 3HSnO_2^- + 9OH^- + 3H_2O \rightarrow 3Sn(OH)_6^{2-} + 6e^-$

cell: $\quad Cr_2O_7^{2-} + 5H_2O + 5H^+ + 3HSnO_2^- \rightarrow 2Cr^{3+} + 3Sn(OH)_6^{2-}$

or, tidying up terms: $\quad H_2Cr_2O_7 + 3H_2SnO_2 + 5H_2O \rightarrow Cr_2[Sn(OH)_6]_3$

You should be careful to note that, even though we "tidied up" terms by combining the two Cr^{3+} ions with the three $Sn(OH)_6^{2-}$ ions to give $Cr_2(Sn(OH)_6)_3$, and three H^+ ions with the three

---

[16] Don't forget that we apply the Nernst equation to the anode half-reaction written as if it were a reduction reaction.

$HSnO_2^-$ ions to give $3H_2SnO_2$, the $Cr_2(Sn(OH)_6)_3$ and the H_2SnO_2 actually don't physically exist in the cell, because the cations are in different electrode compartments from the anions, and, in addition, are at different concentrations.

Example 25-4

The "lead-acid" cell used in 12-volt automotive batteries consists of a lead electrode and a lead dioxide electrode immersed in concentrated sulfuric acid.

(a)   Using the data given below, write the cell reaction and calculate E_{cell}^o at 25°C.

(b)   A typical lead-acid battery contains 4.37 molal sulfuric acid electrolyte. At 25°C the stoichiometric mean activity coefficient for this acid is 0.186, and the relative molal activity of the water in the acid is 0.752.[17] Calculate the reversible (equilibrium) voltage for this battery.

$$PbO_{2(s)} + 4H_{(aq)}^+ + SO_{4(aq)}^{2-} + 2e^- \rightleftharpoons PbSO_{4(s)} + 2H_2O_{(liq)} \qquad E_1^o = +1.685 \text{ V}$$

$$PbSO_{4(s)} + 2e^- \rightleftharpoons Pb_{(s)} + SO_{4(aq)}^{2-} \qquad\qquad\qquad\qquad E_2^o = -0.356 \text{ V}$$

Solution

(a)   The electrode with the more positive reduction potential will be the cathode; by default, the other electrode will be the anode.[18] Therefore, the PbO_2 electrode is cathode, and the Pb electrode is anode:

$$\textit{cathode:}\ PbO_{2(s)} + 4H_{(aq)}^+ + SO_{4(aq)}^{2-} + 2e^- \rightarrow PbSO_{4(s)} + 2H_2O_{(liq)} \qquad E_{cat}^o = +1.685$$

$$\textit{anode:}\ Pb_{(s)} + SO_{4(aq)}^{2-} \rightarrow PbSO_{4(s)} + 2e^- \qquad\qquad\qquad\qquad E_{an}^o = -0.356$$

$$\begin{aligned}\textit{overall} \\ \textit{cell:}\end{aligned}\ Pb_{(s)} + PbO_{2(s)} + 4H_{(aq)}^+ + 2SO_{4(aq)}^{2-} \xrightarrow{2F} 2PbSO_{4(s)} + 2H_2O_{(liq)} \qquad \begin{aligned} E_{cell}^o &= E_{cat}^o - E_{an}^o \\ &= 1.685 - (-0.356) \\ &= +2.041 \text{ V}\end{aligned}$$

$$\textit{i.e.}\ Pb_{(s)} + PbO_{2(s)} + 2H_2SO_{4(aq)} \xrightarrow{2F} 2PbSO_{4(s)} + 2H_2O_{(liq)}$$

Ans: $\boxed{E_{cell}^o = +2.041 \text{ V}}$

Note in the above calculation that, although we write the reaction taking place at the lead electrode as an *anodic* reaction, we do *not* change the sign of $E°$ ($E_{an}^o = -0.356$ V is the *reduction* potential

---

[17] Note that when "stoichiometric" activity coefficients are provided, you don't have to worry about the incomplete dissociation of the electrolyte. The stoichiometric activity coefficient already takes this into account; therefore, when using stoichiometric activity coefficients the electrolyte can be treated as if it dissociates completely. The Debye-Hückel equation, on the other hand, calculates the "true" activity coefficient, and the ion concentrations used to calculate the ionic strength of the solution for use with the Debye-Hückel equation must be the "true" concentrations of free ions in the system; i.e., incomplete dissociation must be taken into account. Sorry to confuse you, but life isn't always as simple as we would like it to be.

[18] Usually inspection of the $E°$s determines which is cathode and which is anode. Occasionally, when the two $E°$s have similar values, the logarithmic term can decide which is more positive and which is less positive.

for the anode).

(b) Applying the Nernst equation to the overall cell reaction:

$$E_{cell}^{eqm} = E_{cell}^o - \frac{RT}{2F} ln\left(\frac{a_{H_2O}^2}{a_{H_2SO_4}^2}\right) \quad \textit{(All the solids are assigned unit activities.)}$$

As mentioned in the footnote on stoichiometric activity coefficients, even though the sulfuric acid is very concentrated, we are permitted to "pretend" that it dissociates completely:

$$H_2SO_4 \rightarrow 2H^+ + SO_4^{2-}$$

If m is the molality of the sulfuric acid, then

$$a_{H_2SO_4} = a_{H^+}^2 \cdot a_{SO_4^{2-}}$$

$$= (\gamma_+ m_+)^2 (\gamma_- m_-)$$

$$= \gamma_+^2 \gamma_- (2m)^2 (m)$$

$$= 4\gamma_\pm^3 m^3$$

where $\gamma_\pm$ is the stoichiometric mean molal activity coefficient, which has a value of 0.186 for 4.37 molal sulfuric acid solution at 25°C.

Therefore the relative activity of the acid is:

$$a_{H_2SO_4} = 4(0.186)^3(4.37)^3 = 2.148$$

and
$$E_{cell}^{eqm} = E_{cell}^o - \frac{RT}{2F} ln\left(\frac{a_{H_2O}^2}{a_{H_2SO_4}^2}\right)$$

$$= E_{cell}^o - \frac{RT}{F} ln\left(\frac{a_{H_2O}}{a_{H_2SO_4}}\right)$$

$$= 2.041 - \frac{(8.314)(298.15)}{96\,487} ln\left(\frac{0.752}{2.148}\right)$$

$$= 2.041 + 0.02696$$

$$= 2.068 \text{ V}$$

Thus, each cell in the battery generates an equilibrium voltage of 2.068 V. An automotive battery consists of six lead-acid cells in series; therefore the terminal voltage of the battery will be $E_{battery} = 6 \times 2.068 = 12.41$ V. This explains why it is called a 12-volt battery.

Ans: 12.41 V

Exercise 25-3

Taking activity effects into account using the Debye-Hückel equation, calculate the equilibrium voltage at 25°C for the cell

$$Zn \mid ZnCl_{2(aq)} \ (0.02 \ molal) \mid AgCl_{(s)} \mid Ag$$

Zinc chloride may be assumed to dissociate completely at this concentration. ☺

Data at 25°C: $Zn^{2+}_{(aq)} + 2e^- \rightleftharpoons Zn_{(s)}$ $E^o_1 = -0.763$ V

$AgCl_{(s)} + e^- \rightleftharpoons Ag_{(s)} + Cl^-_{(aq)}$ $E^o_2 = +0.2224$ V

25.10 EQUILIBRIUM CONSTANTS FROM CELL VOLTAGES

Consider the cell shown in Fig. 5: One compartment of the cell contains a silver electrode immersed in a solution containing Ag^+ ions, while the other compartment contains a platinum electrode immersed in a solution containing a mixture of ferrous and ferric ions. Everything is at 25°C and one bar pressure. For the sake of convenience, we shall assume (incorrectly) that molar concentrations are equal to relative activities. Before the switch is closed, all the ions in the cell are at unit concentration:

$$[Ag^+] = [Fe^{2+}] = [Fe^{3+}] = 1 \text{ mol L}^{-1}$$

Fig. 5

The two half-cell reactions and the corresponding standard reduction potentials (*vs. SHE*) are:

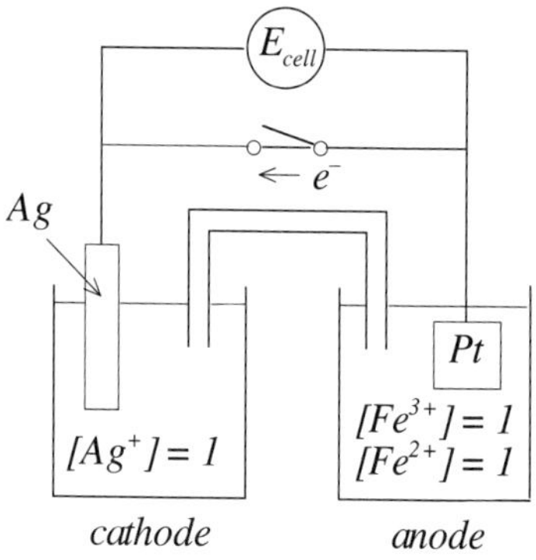

$$Ag^+ + e^- \rightleftharpoons Ag \qquad E^o_{Ag} = +0.799 \text{ V}$$

$$Fe^{3+} + e^- \rightleftharpoons Fe^{2+} \qquad E^o_{Fe} = +0.771 \text{ V}$$

The reduction potentials for the electrodes are:

$$\Delta\phi_{Ag} = E^o_{Ag} - \frac{RT}{nF} ln \frac{1}{[Ag^+]}$$

$$= +0.799 + \frac{(8.314)(298.15)}{(1)(96\ 487)} ln[Ag^+]$$

$$= 0.799 + 0.0257 \, ln[Ag^+]$$

and $$\Delta\phi_{Fe} = E^o_{Fe} - \frac{RT}{nF} ln \frac{[Fe^{2+}]}{[Fe^{3+}]}$$

$$= +0.771 - \frac{(8.314)(298.15)}{(1)(96\ 487)} \ln \frac{[Fe^{2+}]}{[Fe^{3+}]}$$

$$= 0.771 - 0.0257 \ln \frac{[Fe^{2+}]}{[Fe^{3+}]}$$

Since all the species are at unit concentration (unit relative activity), then each electrode is at standard conditions and the initial values of the two electrode potentials are just the *standard* electrode potentials:

$$\Delta\phi_{Ag} = 0.799 + 0.0257 \ln [Ag^+] = 0.799 + 0.0257 \ln (1)$$

$$= 0.799 + 0 = 0.799 \text{ V}$$

$$= E^o_{Ag}$$

$$\Delta\phi_{Fe} = 0.771 - 0.0257 \ln \frac{[Fe^{2+}]}{[Fe^{3+}]} = 0.771 - 0.0257 \ln \frac{(1)}{(1)}$$

$$= 0.771 - 0 = 0.771 \text{ V}$$

$$= E^o_{Fe}$$

The potential difference resulting from these two voltages acts as the driving force for a spontaneous galvanic cell reaction to take place when the switch is closed and electrons are allowed to flow. Since $\Delta\phi_{Ag} > \Delta\phi_{Fe}$, the silver electrode will be the *cathode* and the platinum electrode will be the *anode*. Therefore, the cell reactions and the expression for the cell voltage will be:

cathode: $Ag^+ + e^- \rightarrow Ag$	$E^o_{cell} = \Delta\phi_{cat} - \Delta\phi_{an}$
anode: $Fe^{2+} \rightarrow Fe^{3+} + e^-$	$= (E^o_{Ag} - E^o_{Fe}) - \dfrac{RT}{F} \ln \dfrac{[Fe^{3+}]}{[Ag^+][Fe^{2+}]}$
cell: $Ag^+ + Fe^{2+} \rightarrow Ag + Fe^{3+}$	$= (0.799 - 0.771) - 0.0257 \ln \dfrac{[Fe^{3+}]}{[Ag^+][Fe^{2+}]}$

Therefore:

$$\boxed{E_{cell} = 0.028 - 0.0257 \ln \frac{[Fe^{3+}]}{[Ag^+][Fe^{2+}]}} \qquad \ldots [22]$$

Since all the species initially are at unit concentration, the initial cell voltage will be 0.028 V. Once we close the switch, the cell reactions start to take place, with the removal from solution of Ag^+ ions and Fe^{2+} ions, and the production of Fe^{3+} ions and deposited metallic Ag. So long as the cell voltage is greater than zero, the cell processes will continue. However, as can be seen from the data in Table 2, as the concentrations of the various ions change, the cell voltage slowly decreases until finally, it becomes *zero*. At this point all reaction stops—the cell has run down and is now in a state of overall *cell equilibrium*.

Table 2   *Calculated Data for Progression of Ag-Fe Chemical Reaction*

	Ag^+	$+$	Fe^{2+}	$\rightarrow$	$Ag \downarrow$	$+$	Fe^{3+}	$\Delta\phi_{cat}$	$\Delta\phi_{an}$	E_{cell}	$\dfrac{[Fe^{3+}]}{[Ag^+][Fe^{2+}]}$
concn, start	1		1				1	0.799	0.771	0.028	1.000
	0.9		0.9				1.1	0.796	0.776	0.020	1.358
time	0.8		0.8				1.2	0.793	0.781	0.012	1.875
	0.7		0.7				1.3	0.790	0.787	0.003	2.653
	0.68		0.68				1.32	0.789	0.788	0.001	2.855
	0.6682		0.6682				1.3318	0.7887	0.7887	0	2.983 = K

Fig. 6 shows a plot of the data in Table 2 *vs.* time.

When the cell is at equilibrium, the concentrations of all the species are the same as they would be if we just mixed all the reactants together in the same beaker and allowed them to react until they reached chemical equilibrium. At this point the reaction quotient Q becomes the equilibrium constant K for the reaction.

Thus, the value of $Q = \dfrac{[Fe^{3+}]}{[Ag^+][Fe^{2+}]}$

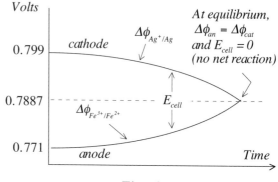

Fig. 6

becomes equal to the equilibrium constant K for the cell reaction *only* when $E_{cell} = 0$.

When this happens, the reaction

$$Ag^+ + Fe^{2+} \rightleftharpoons Ag + Fe^{3+}$$

will have attained a state of overall chemical equilibrium, at which point nothing further happens. Under these conditions, the expression

$$E_{cell} = E_{cell}^o - \frac{RT}{nF} \ln \frac{[Fe^{3+}]}{[Ag^+][Fe^{2+}]} \qquad \dots [23]$$

becomes

$$0 = E_{cell}^o - \frac{RT}{nF} \ln K \qquad \dots [24]$$

from which

$$\boxed{\ln K = \frac{nFE_{cell}^o}{RT}} \qquad \dots [25]$$

Thus, by setting up a cell with 1 molar (unit relative activity) solutions—i.e., a cell under standard conditions—and measuring the resulting cell voltage, we are able to determine K for the cell reaction. Thus, for the reaction

$$Ag^+ + Fe^{2+} \rightleftharpoons Ag + Fe^{3+}$$

we can see that

$$K_{298.15} = exp\left[\frac{nFE^o_{cell}}{RT}\right] = exp\left[\frac{(1)(96\ 487)(0.028)}{(8.314)(298.15)}\right] = 2.983$$

as confirmed by the last entry in the tabulated data in Table 2.

Example 25-5

Find the equilibrium constant at 25°C for the reaction

$$2Ag^+_{(aq)} + Zn_{(s)} \rightarrow Zn^{2+}_{(aq)} + 2Ag_{(s)}$$

given the following data at 25°C: $E^o_{Ag^+/Ag} = +0.799$ V; $E^o_{Zn^{2+}/Zn} = -0.763$ V

Solution

The standard reduction potential for the Ag electrode is more positive than that for the Zn electrode; therefore, the silver electrode will be the *cathode* and the zinc electrode will be the *anode*:

$$
\begin{array}{lll}
cathode: & 2Ag^+ + 2e^- \rightarrow 2Ag & E^o_{cat} = +0.799 \text{ V} \\
anode: & Zn \rightarrow Zn^{2+} + 2e^- & E^o_{an} = -0.763 \text{ V} \\
\hline
cell: & 2Ag^+ + Zn \rightarrow 2Ag + Zn^{2+} &
\end{array}
$$

$$E^o_{cell} = E^o_{cat} - E^o_{an} = (+0.799) - (-0.763) = 1.562 \text{ V}$$

$$ln\ K_{298.15} = \frac{nFE^o_{cell}}{RT} = \frac{(2)(96\ 487)(1.562)}{(8.314)(298.15)} = 121.6$$

Therefore, $K_{298.15} = exp[121.6] = 6.4 \times 10^{52}$.   **Ans:** $\boxed{6.4 \times 10^{52}}$

This extremely high value tells us that this reaction goes essentially to completion. The Ag^+ ion concentration is negligible—and impossible to measure. Yet a simple voltage measurement in a standard cell—which we *can* measure—tells all!

Exercise 25-4

Using the standard reduction potentials at 25°C given below, calculate the equilibrium constant at 25°C for the reaction

$$2 Fe^{2+}_{(aq)} + Au^{3+}_{(aq)} \rightleftharpoons 2 Fe^{3+}_{(aq)} + Au^{+}_{(aq)}$$

Given: $Au^+ + e^- \rightleftharpoons Au$ $E^o_{298.15} = +1.68$ V

               $Au^{3+} + 3e^- \rightleftharpoons Au$ $E^o_{298.15} = +1.50$ V

               $Fe^{3+} + e^- \rightleftharpoons Fe^{2+}$ $E^o_{298.15} = +0.77$ V ☺

Example 25-6

We saw earlier that when an ion such as a Na^+ ion enters aqueous solution, water molecules "stick" to it through electrostatic attractive forces. When the solvent is water, we say that the ion is *hydrated*. We also saw that a hydrated H^+ ion really is just a hydrated *proton*: $H(H_2O)^+_n$. The term "hydrated" is reserved for use with water as the solvent; if the solvent is not water, we use the more general term and say that the ion is *solvated*. Unlike solvated protons, solvated electrons do not exist to any appreciable extent in aqueous solutions; however, in nonaqueous solutions such species sometimes are observed. For example, if alkali metals such as sodium or potassium are dissolved in liquid ammonia, bluish-colored solutions are obtained. This blue color is attributed to the presence of solvated electrons: $(NH_3)^-_n$, or more simply, $e^-_{(soln)}$. Such electrons also can be injected into the solution from an inert electrode if its potential is made sufficiently negative.

When sodium metal is dissolved in liquid ammonia at $-34°C$, the following solubility equilibrium develops:

$$Na_{(s)} \rightleftharpoons Na^+_{(soln)} + e^-_{(soln)} \qquad \ldots [a]$$

Careful measurements involving the vapor pressure depression over various solutions of Na metal in liquid ammonia at $-34°C$ have shown that the solubility product for reaction [a] is

$$K_{SP} = a_{Na^+} \cdot a_{e^-} = 0.0696 \qquad \ldots [b]$$

Given that the standard reduction potential for the Na^+/Na couple in liquid ammonia at $-34°C$ is -2.018 V, calculate the standard potential for the $Pt/e^-_{(soln)}$ couple at this temperature.

Solution

The pertinent half-reactions and their standard reduction potentials in liquid ammonia at $-34°C$ are:

$$Na^+_{(soln)} + e^-_{(Na)} \rightleftharpoons Na_{(s)} \qquad E^o_{Na} = -2.018 \text{ V}$$

$$e^-_{(Pt)} \rightleftharpoons e^-_{(soln)} \qquad E^o_{e^-} = ?$$

We are seeking the value of $E^o_{e^-}$. This can be determined by considering the following cell (in liquid ammonia at $-34°C = 239.15$ K):

$$Na_{(s)} \mid Na^+_{(soln)}, \; e^-_{(soln)} \mid Pt_{(s)}$$

The reactions taking place in this cell are:

$$\textit{anode:} \qquad Na_{(s)} \rightarrow Na^+_{(soln)} + e^-_{(Na)} \qquad\qquad E^o_{Na} = -2.018 \text{ V}$$

$$\textit{cathode:} \qquad\qquad e^-_{(Pt)} \rightarrow e^-_{(soln)} \qquad\qquad\qquad E^o_{e^-} = ?$$

$$\textit{cell:} \quad Na_{(s)} + e^-_{(Pt)} \rightarrow Na^+_{(soln)} + e^-_{(soln)} + e^-_{(Na)} \qquad E^o_{cell} = E^o_{cat} - E^o_{an}$$

$$= E^o_{e^-} - (-2.018)$$

$$= E^o_{e^-} + 2.018$$

Remembering that , by convention, $\qquad a_{e^-(Pt)} = a_{e^-(Na)} = a_{Na} = 1$

we apply the Nernst equation to the cell reaction to obtain:

$$E_{cell} = E^o_{cell} - \frac{RT}{F} ln \left(\frac{a_{Na^+(soln)} \cdot a_{e^-(soln)} \cdot a_{e^-(Na)}}{a_{Na} \cdot a_{e^-(Pt)}} \right)$$

$$= E^o_{cell} - \frac{RT}{F} ln \left(\frac{a_{Na^+(soln)} \cdot a_{e^-(soln)} \cdot (1)}{(1) \cdot (1)} \right)$$

$$= (E^o_{e^-} + 2.018) - \frac{RT}{F} ln(a_{Na^+(soln)} \cdot a_{e^-(soln)}) \qquad \dots [c]$$

At overall cell equilibrium, $\qquad\qquad\qquad E_{cell} = 0$

and $\qquad\qquad\qquad\qquad a_{Na^+(soln)} \cdot a_{e^-(soln)} = K_{SP} = 0.0696$

Putting these values into Eqn [c] gives:

$$0 = E^o_{e^-} + 2.018 - \frac{RT}{F} ln K_{SP}$$

Rearranging: $\qquad\qquad E^o_{e^-} = -2.018 + \frac{RT}{F} ln K_{SP}$

$$= -2.018 + \frac{(8.314)(239.15)}{96\ 487} ln(0.0696)$$

$$= -2.018 - 0.0549$$

$$= -2.0729 \text{ V}$$

Thus under standard conditions in liquid NH_3 at $-34°C$, solvated electrons in the solution are 2.073 V more negative than the Pt metal electrode.

$$\textbf{Ans:} \quad \boxed{E^o_{e^-} = \phi_{Pt} - \phi_{soln} = -2.073 \text{ V}}$$

Example 25-7

Using the fact that the water dissociation constant for water at 25°C is $K_w = 1.00 \times 10^{-14}$, calculate the standard reduction potential E_x^o at 25°C for the half-reaction

$$H_2O_{(liq)} + e^- \rightleftharpoons OH^-_{(aq)} + \tfrac{1}{2}H_{2(g)}$$

Solution A

The standard reduction potential of an electrode is defined as the value of the equilibrium cell voltage for a cell in which the electrode of interest (at standard conditions) is the cathode and a standard hydrogen electrode is the anode. Therefore the equilibrium cell voltage of the following cell will be E_x^o:

$$Pt \left| \begin{matrix} H_{2(g)} \\ P_{H_2} = 1.00\ bar \end{matrix} \right| \begin{matrix} H^+_{(aq)} \\ a_{H^+} = 1.00\,m \end{matrix} \left\| \begin{matrix} OH^-_{(aq)} \\ a_{OH^-} = 1.00\,m \end{matrix} \right| \begin{matrix} H_{2(g)} \\ P_{H_2} = 1.00\ bar \end{matrix} \right| Pt$$

$$\text{Anode (SHE, } E_H^o = 0) \qquad\qquad\qquad\qquad\qquad \text{Cathode } (E_x^o)$$

For this cell:

$$\text{anode:} \quad \tfrac{1}{2}H_2 \xrightarrow{\ acidic\ } H^+ + e^- \qquad\qquad E_{an}^o = 0.00 \text{ volt}$$

$$\text{cathode:} \quad H_2O + e^- \xrightarrow{\ alkaline\ } OH^- + \tfrac{1}{2}H_2 \quad E_x^o \text{ volt}$$

$$\overline{\qquad\qquad\qquad\qquad\qquad\qquad\qquad\qquad\qquad\qquad\qquad\qquad\qquad\qquad}$$

$$\text{overall cell:} \quad H_2O \xrightarrow{\ 1F\ } H^+ + OH^- \qquad\qquad \begin{aligned} E_{cell}^o &= E_{cat}^o - E_{an}^o \\ &= E_x^o - 0.00 \\ &= E_x^o \end{aligned}$$

Applying the Nernst equation to the overall cell reaction:

$$E_{cell} = E_{cell}^o - \frac{RT}{F} \ln\left(\frac{a_{H^+} \cdot a_{OH^-}}{a_{H_2O}} \right) = E_x^o - \frac{RT}{F} \ln\left(\frac{a_{H^+} \cdot a_{OH^-}}{a_{H_2O}} \right)$$

When the cell has run down and attained final equilibrium, the reaction quotient will be the equilibrium constant for the cell reaction, which in this case is K_w.

Thus:

$$E_{cell} = E_x^o - \frac{RT}{F} \ln\left(\frac{a_{H^+} \cdot a_{OH^-}}{a_{H_2O}} \right)$$

At equilibrium:

$$0 = E_x^o - \frac{RT}{F} \ln\left(\frac{a_{H^+} \cdot a_{OH^-}}{a_{H_2O}} \right)_{eqm}$$

$$= E_x^o - \frac{RT}{F} \ln K_w$$

Rearranging:
$$E_x^o = + \frac{RT}{F} \ln K_w$$

$$= \frac{(8.314)(298.15)}{96487} \ln(1.00 \times 10^{-14})$$

$$= -0.8282 \text{ V}$$

Ans: $\boxed{-0.828 \text{ V}}$

Now we present an alternate solution:

Solution B

Consider the hydrogen electrode shown in the figure below. The half-cell potential of this electrode can be calculated using either of the following equilibria:

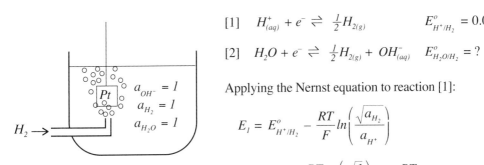

[1] $H_{(aq)}^+ + e^- \rightleftharpoons \frac{1}{2} H_{2(g)}$ $E_{H^+/H_2}^o = 0.00$

[2] $H_2O + e^- \rightleftharpoons \frac{1}{2} H_{2(g)} + OH_{(aq)}^-$ $E_{H_2O/H_2}^o = ?$

Applying the Nernst equation to reaction [1]:

$$E_1 = E_{H^+/H_2}^o - \frac{RT}{F} \ln \left(\frac{\sqrt{a_{H_2}}}{a_{H^+}} \right)$$

$$= 0.00 - \frac{RT}{F} \ln \left(\frac{\sqrt{1}}{a_{H^+}} \right) = + \frac{RT}{F} \ln a_{H^+}$$

But
$$a_{H^+} = \frac{K_w}{a_{OH^-}}$$

Therefore,
$$E_1 = \frac{RT}{F} \ln \left(\frac{K_w}{a_{OH^-}} \right) = \frac{RT}{F} \ln \left(\frac{K_w}{1} \right) \qquad \ldots \text{[a]}$$

Similarly, applying the Nernst equation to reaction [2]:

$$E_2 = E_{H_2O/H_2}^o - \frac{RT}{F} \ln \left(\frac{\sqrt{a_{H_2}} \cdot a_{OH^-}}{a_{H_2O}} \right) = E_{H_2O/H_2}^o - \frac{RT}{F} \ln \left(\frac{\sqrt{1} \cdot 1}{1} \right) = E_{H_2O/H_2}^o \qquad \ldots \text{[b]}$$

The activities of the various species in the beaker put the electrode under standard conditions for reaction [2] but not for reaction [1] (because $a_{H_2} \neq 1$).

However, a voltmeter gives the same voltage reading for this electrode regardless of how we choose to write the reaction.

Therefore

$$E_2 = E_1$$

and we can equate equation [a] with equation [b]:

i.e.,

$$E^o_{H_2O/H_2} = \frac{RT}{F} \ln K_W = \frac{(8.314)(298.15)}{(96\ 487)} \ln\left(1.00 \times 10^{-14}\right)$$

$$= -0.828 \text{ V}$$

Ans: $\boxed{E^o_{H_2O/H_2} = -0.828 \text{ V } (vs.\ SHE)}$

25.11 THERMODYNAMIC FUNCTIONS FROM CELL VOLTAGES

We already have shown that the Gibbs free energy change for a redox reaction is easily determined from the relationship

$$\boxed{\Delta G_{T,P} = -nFE^{eqm}_{cell}} \qquad \dots [26]$$

Earlier, we also showed that
$$\Delta S = -\left(\frac{\partial \Delta G}{\partial T}\right)_P \qquad \dots [27]$$

from which it immediately follows that

$$\Delta S = -\frac{\partial(-nFE^{eqm}_{cell})}{\partial T} = +nF\frac{\partial E^{eqm}_{cell}}{\partial T}$$

$$\boxed{\Delta S = nF\frac{\partial E^{eqm}_{cell}}{\partial T}} \qquad \dots [28]$$

Eqn [28] shows that we can obtain ΔS for the cell reaction just by measuring the way the cell voltage varies with temperature at constant total pressure!

We also know that $\Delta G = \Delta H - T\Delta S$

rearranging: $\Delta H = \Delta G + T\Delta S$ $\qquad \dots [29]$

Substituting Eqns [26] and [28] into Eqn [29]:

$$\boxed{\Delta H = -nFE^{eqm}_{cell} + nFT\frac{\partial E^{eqm}_{cell}}{\partial T}} \qquad \dots [30]$$

Eqn [30] shows how easily we also can evaluate ΔH for the cell reaction from voltage data.

Again, in our discussion of chemical equilibrium we showed that

$$\Delta G° = -RT \ln K \qquad \qquad \dots [31]$$

Substituting $\Delta G° = -nFE°_{cell}$ into Eqn [31] gives

$$\boxed{\ln K = \frac{nFE°_{cell}}{RT}} \qquad \qquad \dots [32]$$

which we have already shown earlier (Eqn [25]).

Exercise 25-5

The equilibrium voltage for the cell

$$Ag \mid AgCl_{(s)} \mid ZnCl_{2(aq)} \ (0.0010 \ m) \mid Zn_{(s)}$$

is -1.240 V at 25°C and -1.260 V at 35°C. Write the cell reaction and calculate
(a) ΔG, (b) ΔH, and (c) ΔS per mole of $ZnCl_2$ consumed by this reaction at 30°C. ☺

25.12 CONCENTRATION CELLS

Suppose we set up a cell with two Cu^{2+}/Cu half-cells with different concentrations of Cu^{2+}, as shown in Fig. 7. What is the cell voltage and what is the cell reaction at 25°C?

Data: $Cu^{2+}_{(aq)} + 2e^- \rightleftharpoons Cu$ $E°_{Cu^{2+}/Cu} = +0.337$ V

For the left-hand side (*LHS*) compartment:

$$\Delta\phi^{LHS}_{Cu} = 0.337 - \frac{RT}{2F}\ln\left(\frac{1}{a_1}\right)$$

$$= 0.337 + \frac{RT}{2F}\ln a_1$$

Similarly, for the right-hand side (*RHS*) compartment:

$$\Delta\phi^{RHS}_{Cu} = 0.337 + \frac{RT}{2F}\ln a_2$$

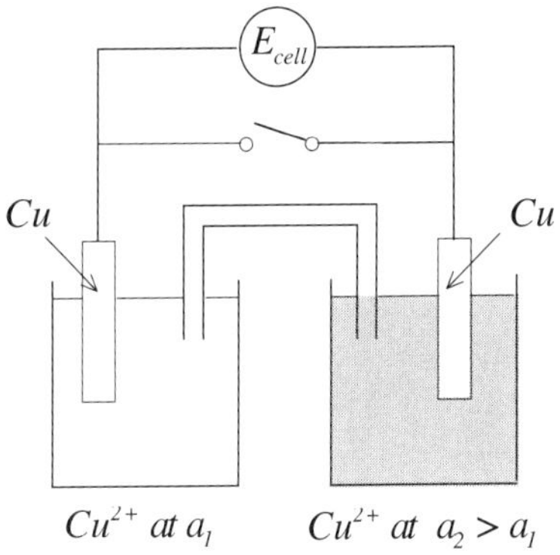

Cu^{2+} at a_1 Cu^{2+} at $a_2 > a_1$

Fig. 7

Note that both half-cell reactions are treated as if they were *reduction* reactions (electrons on the left-hand side). Since $a_2 > a_1$, it follows that $\Delta\phi^{RHS}_{Cu} > \Delta\phi^{LHS}_{Cu}$; therefore, the *RHS* is the *cathode*, and, by default, the *LHS* will be the *anode*. The electrode reactions are therefore:

$$
\begin{aligned}
&cathode: &&Cu^{2+}(a_2) + 2e^- \rightarrow Cu &&E^o_{Cu^{2+}/Cu} = +0.337 \text{ V}\\
&anode: &&Cu \rightarrow Cu^{2+}(a_1) + 2e^- &&E^o_{Cu^{2+}/Cu} = +0.337 \text{ V}\\
\hline
&cell: &&Cu^{2+}(a_2) \rightarrow Cu^{2+}(a_1)
\end{aligned}
$$

$$
\begin{aligned}
E_{cell} &= \Delta\phi_{cat} - \Delta\phi_{an}\\
&= \left\{0.337 + \frac{RT}{2F}\ln a_2\right\} - \left\{0.337 + \frac{RT}{2F}\ln a_1\right\}\\
&= \frac{RT}{2F}\ln\left(\frac{a_2}{a_1}\right)
\end{aligned}
$$

The E^os cancel out! This kind of cell in which the same electrode couple is present in each half-cell, is called a **concentration cell**; its potential is determined only by[19] the relative concentrations (relative activities) of the reacting species:

$$
\boxed{E_{cell} = \frac{RT}{2F}\ln\left(\frac{a_2}{a_1}\right)} \quad \begin{array}{l}Concentration\\ Cell\ (a_2 > a_1)\end{array} \qquad \cdots [33]
$$

Note that when the switch is closed and current is allowed to flow, the concentration of $Cu^{2+}_{(aq)}$ ion in side 2 will gradually *decrease*, by Cu depositing on the cathode, while the concentration in side 1 will gradually *increase* by the dissolution of some of the Cu anode, until eventually the activities in each side will be the same, so that $\Delta\phi_{Cu}^{RHS} = \Delta\phi_{Cu}^{LHS}$, at which point $E_{cell} = 0$. Thus, as the concentrations slowly change, the cell voltage also slowly decreases until final chemical equilibrium is reached, when $E_{cell} = 0$.

The overall cell reaction is just the *spontaneous dilution process* that would occur if we set up the beakers in the manner shown in Fig. 8. This simple spontaneous dilution process can be used to generate electricity and deliver useful work!

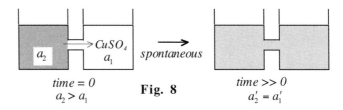

time = 0
$a_2 > a_1$

Fig. 8

time >> 0
$a_2' = a_1'$

Thus, initially, at time $t = 0$, before the concentration in either side has started to change, the cell voltage will be

$$
\left(E_{cell}\right)_{t=0} = \frac{RT}{2F}\ln\left(\frac{a_2}{a_1}\right)_{initial}
$$

After the mixing process has finished, and the cell has run down, reaching overall equilibrium, the concentration—and hence relative activity—in each half-cell compartment will be the same, and, as indicated in Fig. 9, the cell voltage will be

---

[19] Assuming any liquid junction potentials that may be present are negligible.

$$\left(E_{cell}\right)^{t\gg0}_{(eqm)} = \frac{RT}{2F} ln(1) = 0$$

As long as $a_2 > a_1$ the cell process

$$Cu^{2+}(a_2) \rightarrow Cu^{2+}(a_1) \qquad \ldots [34]$$

will continue.[20] At any time before equilibrium has been attained the instantaneous cell voltage will be:

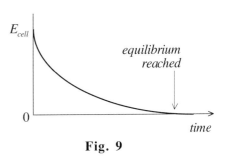

Fig. 9

$$E_{cell} = \frac{RT}{2F} ln\left(\frac{a_2}{a_1}\right)$$

At equilibrium the reaction quotient will just be the equilibrium constant K for the spontaneous dilution process of Eqn [34]:

$$\left(E_{cell}\right)_{eqm} = \frac{RT}{2F} ln\left(\frac{a_2}{a_1}\right)_{eqm} = \frac{RT}{2F} ln K$$

What is the value of this "equilibrium constant"? We know that at equilibrium $E_{cell} = 0$; therefore, at equilibrium

$$0 = \frac{RT}{2F} ln K$$

that is, $ln\ K = 0$

and therefore $K = exp[0] = 1$

In other words, $K = \left(\frac{a_2}{a_1}\right)_{eqm} = 1$ which just means that $a_1 = a_2$!

Exercise 25-6

Show how the cell in Fig. 10 can be used to determine K_{SP} for AgCl.

At 25°C the cell voltage is 0.564 V.

☹

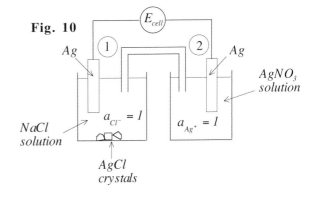

Fig. 10

---

[20] With a_1 and a_2 constantly changing (a_2 keeps getting smaller, and a_1 keeps getting bigger, until they are equal).

KEY POINTS FOR CHAPTER TWENTY-FIVE

1. **Electrochemical reactions**: When a piece of metallic zinc is dropped into a beaker containing aqueous $CuSO_4$ solution the following spontaneous chemical redox reaction occurs:

    $$Zn_{(s)} + CuSO_{4(aq)} \rightarrow ZnSO_{4(aq)} + Cu_{(s)}\downarrow$$

    which really is just $\quad Zn_{(s)} + Cu^{2+}_{(aq)} \rightarrow Zn^{2+}_{(aq)} + Cu_{(s)}\downarrow$

    We know that this redox reaction consists of two half-reactions:

    $$Zn_{(s)} \rightarrow Zn^{2+}_{(aq)} + 2e^- \quad Oxidation\text{---}Zn(0) \text{ is oxidized to } Zn(+2)$$
    $$Cu^{2+}_{(aq)} + 2e^- \rightarrow Cu_{(s)} \quad Reduction\text{---}Cu(+2) \text{ is reduced to } Cu(0)$$

    Because the reaction is spontaneous, it is capable of delivering "other" work $W°$. However, when the reaction is carried out in a beaker the electrons transfer *directly* from the metallic Zn to the Cu^{2+} ions and all the energy is released as *heat* $(Q_P = \Delta H^o_{298.15})$, with no useful work extracted from the system.

2. But by separating the two half-reactions in separate compartments of an **electrochemical cell** we can cause the more energetic electrons from the zinc to flow through an external electrical circuit before they recombine with Cu^{2+} ions in the other compartment.

    The porous membrane is needed to prevent the two solutions from mixing; if they mixed, the reaction would occur directly between the Zn metal and the Cu^{2+} ions in solution. The membrane also is required to make electrical contact between the two electrode compartments in order for the electric circuit to be completed.

    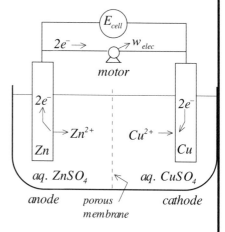

3. By definition, the **anode** is the electrode at which *oxidation* occurs (the Zn electrode), and the **cathode** is the electrode at which *reduction* occurs (the Cu electrode).

    $\quad$ **Anode** $\equiv$ *oxidation* (electrons on RHS): $\qquad Zn \rightarrow Zn^{2+} + 2e^-$

    $\quad$ **Cathode** $\equiv$ *reduction* (electrons on LHS): $\qquad Cu^{2+} + 2e^- \rightarrow Cu$

4. If the external circuit contains an electric motor, then the energy of the electrons will be given up to the motor as **electrical work** W_{elec}, which will then be converted by the motor to mechanical work.

    Electrical work w_{elec}: The electrical work output delivered by an electrochemical cell when a charge of Q coulombs passes through the cell voltage E_{cell} is given by

    $$w_{elec} = Q \times E_{cell} \quad joules$$

5. A small voltage difference known as a **liquid junction potential** $(\Delta\phi_{lj})$, which is diffi-

cult to evaluate, is generated within the membrane where the two different solutions meet. This is effectively eliminated by the use of a **salt bridge**, which typically consists of a tube containing a jelly-like material that has been saturated with concentrated KCl solution or NH_4NO_3 solution. The salt bridge actually replaces the *single* liquid junction potential across the porous membrane with *two* liquid junction potentials of equal magnitude, but opposite sign, so that in effect they cancel.

6. **Types of cells:** GALVANIC: redox chemical reaction produces electricity.

    ELECTROLYTIC (**ELECTROLYSIS**): electrical work input causes redox chemical reaction to occur.

7. **Cell notation**:
    - A single vertical solid line denotes a boundary between two different phases.
    - A double solid vertical line denotes a salt bridge.
    - A single vertical dotted line denotes a porous boundary with a liquid junction.
    - By convention, *the anode is always on the left-hand side*. Accordingly, the *Zn-Cu* cell

    with a salt bridge is written as $Zn\,|\,Zn^{2+}_{(aq)}\,\|\,Cu^{2+}_{(aq)}\,|\,Cu$

    If the separator were a porous membrane, the double line representing the salt bridge would be replaced by a single dotted or dashed line.

8. **Galvanic cell at equilibrium**: $w^{max}_{elec} = nFE^{eqm}_{cell} = -\Delta G_{T,P}$

    Operating galvanic cell (non-equilibrium): $w_{elec} = nFE_{cell}$

    where w_{elec} is the electrical work obtained per mole of cell reaction, n is the number of moles of electrons transferred per mole of cell reaction, F is the faraday, and $\Delta G_{T,P}$ is the Gibbs free energy change for the cell reaction at the specified constant temperature and pressure.

    The **faraday** is the charge contained in one mole of electrons:
    $$F = N_A e_o = (6.02217 \times 10^{23}\ mol^{-1})(1.602192 \times 10^{-19}\ C) = 96\,487\ C\ mol^{-1}$$

9. **Redox couples**:   The oxidized and reduced substances in each half-reaction form what is called a *redox couple*. In general, for the half-reaction

    $$Ox + ne^- \rightarrow Red$$

    the redox couple is represented by *Ox/Red*, where *Ox* is the oxidized form of the species and *Red* is the reduced form. Thus, the two redox couples in the Zn-Cu cell are the Zn^{2+}/Zn couple and the Cu^{2+}/Cu couple.

10. The **Nernst**[21] **equation** shows the relationship between the equilibrium cell voltage E^{eqm}_{cell} and the relative activities of the various chemical species involved in the cell reaction.

    For the overall cell reaction $aA + bB \xrightarrow{nF} cC + dD$

---

[21] Caution! A common spelling mistake is to eliminate the middle "n" in *Nernst* by writing *Nerst*.

$$\Delta G_R = \Delta G_R^o + RT\,ln\left(\frac{a_C^c \cdot a_D^d}{a_A^a \cdot a_B^b}\right). \text{ Putting } \Delta G_{T,P} = -nFE_{cell}^{eqm} \text{ gives the } \textbf{Nernst equation}:$$

$$E_{cell}^{eqm} = E_{cell}^o - \frac{RT}{nF}ln\left(\frac{a_C^c \cdot a_D^d}{a_A^a \cdot a_B^b}\right)$$

where E_{cell}^o is the **standard cell potential**, given by $E_{cell}^o = \dfrac{-\Delta G_T^o}{nF}$.

11. **Single electrode potentials** and the **hydrogen electrode**: The equilibrium cell volt-age consists of the difference between two equilibrium half-cell voltages, one for the cath-ode reaction and the other for the anode reaction. The Nernst equation can be applied to half-cell reactions in the same manner as to complete cell reactions.

Because it is not possible to measure the absolute value of a single electrode potential dif-ference, one electrode is *arbitrarily* assigned a value of *zero* volts, and the potentials of all other single electrodes are measured *relative* to that electrode. The *hydrogen electrode* is the standard against which all other half-cell potentials are measured.

The **hydrogen electrode** reaction is $2H_{(aq)}^+ + 2e^- \rightleftharpoons H_{2(g)}$

Applying the Nernst equation: $\Delta\phi_H = E_H^o - \dfrac{RT}{2F}ln\left(\dfrac{P_{H_2}}{a_{H^+}^2}\right)$

Note that we use $\Delta\phi$ for a half-cell potential to enable us to distinguish it from an overall cell potential E_{cell}. In the case of electrochemical cells, **standard conditions** means that all the reactants and products are at unit relative activity. For gases, this essentially means one bar pressure. Therefore, when the partial pressure of the hydrogen gas is one bar, and the activity of the H^+ ions is unity, then the hydrogen electrode becomes a **standard hy-drogen electrode** (*SHE*), and

$$\Delta\phi_H^o = E_H^o - \frac{RT}{2F}ln\left(\frac{1}{1^2}\right) = E_H^o$$

For the hydrogen reaction, $(\Delta G_{298.15}^o)_H = \Delta g_f^o[H_{2(g)}] - 2\Delta g_f^o[H_{(aq)}^+] - 2\Delta g_f^o[e^-]$

Because we are unable to measure experimentally the molar free energy of an electron or of a single ionic species, by *definition* Δg_f^o for an electron is arbitrarily assigned a value of *zero*; similarly, Δg_f^o for aqueous hydrogen ion also is arbitrarily assigned a value of *zero*.

Therefore $(\Delta G_{298.15}^o)_H \equiv 0$ and $E_H^o = \dfrac{-(\Delta G_{298.15}^o)_H}{nF} = \dfrac{-(0)}{2F} = 0$ (at *all* temperatures)

The half-cell potential of any reduction reaction under standard conditions—one bar, unit relative activity of all species—is called the **standard reduction potential** for that half-cell reaction, and is given the symbol $E°$. Tables of standard reduction potentials usually are tabulated for half-reactions at 25°C, but the temperature is not part of the definition of stan-dard conditions. Tabulated values of standard reduction potentials are reported *vs.* the standard hydrogen electrode (*SHE*), which has a value of zero, by definition.

If the partial pressure of the hydrogen gas is 1 bar, then

$$\Delta\phi_H = -\frac{RT}{2F}\ln\left(\frac{1}{a_{H^+}^2}\right) = +\frac{2.303\,RT}{F}\log a_{H^+} = -\frac{2.303\,RT}{F}\,pH$$

At 25°C, $\Delta\phi_H = -\dfrac{2.303(8.314)(298.15)}{96\,487}\,pH = -0.0592\,pH$

Therefore the electric potential established at the hydrogen electrode is directly (negatively) proportional to the *pH* of the solution. This is the principle on which *pH* meters operate. A *pH* meter really is just a fancy voltmeter; in fact, the scale can be read as millivolts as well as *pH*.

12. **Tables of standard reduction potentials**:

When comparing different redox couples, it is convenient to write each reaction as a *reduction* reaction. Using $\Delta E_T^o = -\Delta G_T^o/nF$, we can generate a table of *standard reduction potentials* for each reduction half-reaction, and list these from the most positive values to the most negative values. The more positive the value of $E°$, the stronger the tendency to *reduce* (ΔG_T^o becomes increasingly *negative* as $E°$ becomes increasingly *positive*). Thus, in a galvanic cell, *the electrode with the more positive reduction potential will be the cathode* (reduction), while the other electrode, by default, will be the *anode* (oxidation).

13. **Calculation of equilibrium cell voltages**:

In a galvanic cell the cathode is positive and the anode is negative (or less positive); thus the equilibrium cell voltage is

$$E_{cell}^{eqm} = \Delta\phi_{cat}^{eqm} - \Delta\phi_{an}^{eqm}$$

where $\Delta\phi_{cat}^{eqm}$ is the equilibrium *reduction* potential of the cathode, and $\Delta\phi_{an}^{eqm}$ is the equilibrium *reduction* potential of the anode, as determined from the Nernst equation.

Thus, to calculate the equilibrium voltage of an electrochemical cell:

1. Write both the cathode and the anode reaction as *reduction* reactions;
2. Apply the Nernst equation to each; and
3. Obtain the cell voltage by subtracting the reduction potential of the anode from the reduction potential of the cathode.

14. **Stoichiometric activity coefficients**:

When using the Nernst equation you need to know the activities of various ions in the cell electrolyte solution. In the case of an incompletely dissociated electrolyte you don't have to worry about the value of the "true" concentration of free ions if you are provided with a value for the "stoichiometric" activity coefficient, because the stochiometric activity coefficient already *includes* a correction for incomplete dissociation. For example, if you need to evaluate the activity of sulfuric acid in solution, you don't have to worry about the incomplete second dissociation of the HSO_4^- ion. Instead, just "pretend" that the acid dissociates

100% as $H_2SO_4 \rightarrow 2H^+ + SO_4^{2-}$ and do the calculation accordingly.

For example, at 25°C the stoichiometric activity coefficient $\gamma_\pm$ for 4.37 molal H_2SO_4 is 0.186. The *activity* of the H_2SO_4 is calculated as follows:

"Assume" that $\qquad\qquad H_2SO_4 \xrightarrow{\ 100\% \ } 2H^+ + SO_4^{2-}$

Therefore: $\qquad\qquad a_{H_2SO_4} = a_{H^+}^2 \cdot a_{SO_4^{2-}} = (\gamma_+ m_+)^2 (\gamma_- m_-)$

$$= \gamma_+^2 \gamma_- (2m)^2 (m) = 4\gamma_\pm^3 m^3$$

and $\qquad\qquad a_{H_2SO_4} = 4(0.186)^3 (4.37)^3 = 2.148$

15. **Equilibrium constants from cell voltages**:

Consider an electrochemical cell having the overall cell reaction

$$aA + bB \xrightarrow{\ nF\ } cC + dD$$

with the two half-cell reactions: $\quad$ *cathode*: $\ aA + ne^- \rightarrow cC$

$$\textit{anode}: \ bB \rightarrow dD + ne^-$$

Applying the Nernst equation to evaluate the **reduction** potential of each half-reaction:

For the cathode: $\qquad\qquad \Delta\phi_{cat} = E_{cat}^o - \dfrac{RT}{nF} \ln\left(\dfrac{a_C^c}{a_A^a} \right)$

For the anode, we first write the reaction *as if* it were a cathode: $\quad dD + ne^- \rightarrow bB$

and *then* apply the Nernst equation: $\quad \Delta\phi_{an} = E_{an}^o - \dfrac{RT}{nF} \ln\left(\dfrac{a_B^b}{a_D^d} \right)$

Note: **don't** change the sign of E^o when doing this.

The cell voltage is the *difference* between these two potentials:

$$E_{cell} = \Delta\phi_{cat} - \Delta\phi_{an}$$

As long as there is a difference of potential between the two electrodes, the cell will continue to run, with a flow of electric current producing power. As the cell gradually runs down, the concentrations of the reactants A and B slowly *decrease*, while the concentrations of the products C and D slowly *increase*. These changes in concentrations (and, of course, activities) cause the cathodic potential $\Delta\phi_{cat}$ to become less positive (i.e., less cathodic) and the anodic potential $\Delta\phi_{an}$ to become more positive (i.e., less anodic), until eventually, the two potentials become *equal*: $\quad \Delta\phi_{cat} = \Delta\phi_{an}$

At this point, $\qquad\qquad E_{cell} = \Delta\phi_{cat} - \Delta\phi_{an} = 0$

and the cell has completely run down, with no further driving force; in other words, the complete system has reached overall *EQUILIBRIUM*.

$$E_{cell}^{eqm} = \left(\Delta\phi_{cat} - \Delta\phi_{an} \right)_{eqm} = 0$$

i.e., $\left\{ E_{cat}^{o} - \dfrac{RT}{nF} ln\left(\dfrac{a_C^c}{a_A^a} \right) \right\}_{cell\,eqm} - \left\{ E_{an}^{o} - \dfrac{RT}{nF} ln\left(\dfrac{a_B^b}{a_D^d} \right) \right\}_{cell\,eqm} = 0$

Rearranging: $E_{cat}^{o} - E_{an}^{o} = \dfrac{RT}{nF} ln\left(\dfrac{a_C^c}{a_A^a} \right) - \dfrac{RT}{nF} ln\left(\dfrac{a_B^b}{a_D^d} \right) = \dfrac{RT}{nF} ln\left(\dfrac{a_C^c \cdot a_D^d}{a_A^a \cdot a_B^b} \right)_{cell\,eqm}$

but $\left(\dfrac{a_C^c \cdot a_D^d}{a_A^a \cdot a_B^b} \right)_{cell\,eqm}$ is just the equilibrium constant, K, for the overall reaction.

Therefore, $E_{cat}^{o} - E_{an}^{o} = \dfrac{RT}{nF} ln\,K$

or, $ln\,K = \dfrac{nFE_{cell}^{o}}{RT}$ where $E_{cell}^{o} = E_{cat}^{o} - E_{an}^{o}$.

Thus, by setting up a cell in which all the gaseous species are at one bar pressure and all the dissolved species are at one molal concentration—more correctly, unit relative activity– we will produce a cell *under standard conditions*. The resulting *experimentally measurable* voltage of this cell will be the standard cell potential E_{cell}^{o}, which we then can use with the above equation to determine K for the overall cell reaction. Note that this cell is a *standard* cell, not an *equilibrium* cell, even though its voltage is used to determine the equilibrium constant for the cell reaction. On the other hand, if we already have the E^o data for each of the cell half-reactions, we don't even have to construct the cell to obtain K for the cell reaction. Pretty nifty, eh!

16.   **Thermodynamic functions from cell voltages**:

Many thermodynamic functions for chemical processes are determined from measurements using electrochemical cells. As we already have seen,

$$\Delta G_{T,P} = -nFE_{cell}^{eqm}$$

Starting with this, other thermodynamic functions based on electrochemical measurements also can be derived:

$$\Delta S = -\frac{\partial \Delta G}{\partial T} = -\frac{\partial(-nFE_{cell}^{eqm})}{\partial T} = nF\frac{\partial E_{cell}^{eqm}}{\partial T}$$

$\Delta G = \Delta H - T\Delta S \quad \rightarrow \quad \Delta H = \Delta G + T\Delta S$

$$= -nFE_{cell}^{eqm} + T \times nF\frac{\partial E_{cell}^{eqm}}{\partial T}$$

Therefore, $\Delta H = -nFE_{cell}^{eqm} + nFT\frac{\partial E_{cell}^{eqm}}{\partial T}$

Finally, there is
$$K = exp\left[\frac{nFE^o_{cell}}{RT}\right]$$

17. **Metal/Metal Ion Concentration Cells**:

A cell containing the same redox couple in each half-cell but at different concentrations is called a *concentration cell*. For such a cell, the $E°$s for each half-cell are the same; therefore, in the expression for the equilibrium cell voltage, $E^o_{cell} = 0$ and the cell voltage is determined only by the logarithmic term in the Nernst equation.

For example, consider a cell in which each half-cell consists of a metal M in contact with a solution containing the metal ion M^{z+} but at two different concentrations, m_1 and m_2, with corresponding relative molal activities, a_1 and a_2, respectively. Let $a_2 > a_1$.

Each half-cell system consists of the equilibrium $M^{z+} + ze^- \rightleftharpoons M$

For the half-cell at concentration m_1:
$$\Delta\phi_1 = E^o_{M^{z+}/M} - \frac{RT}{zF}ln\frac{1}{a_1}$$

For the half-cell at concentration m_2:
$$\Delta\phi_2 = E^o_{M^{z+}/M} - \frac{RT}{zF}ln\frac{1}{a_2}$$

Since $a_2 > a_1$, it follows that $\Delta\phi_2 > \Delta\phi_1$. That is, half-cell 2 will be the cathode, with the following cathodic process taking place: $M^{z+}(at\,a_2) + ze^- \rightarrow M$

and half-cell 1 will be the anode with $M \rightarrow M^{z+}(at\,a_1) + ze^-$

The overall cell reaction, obtained by adding the two half-cell reactions and noting that M cancels out from each side, is
$$M^{z+}(a_2) \rightarrow M^{z+}(a_1)$$

In other words, metal deposits from the solution at the higher concentration, thereby lowering the concentration in half-cell 2, while in half-cell 1, at the lower concentration, the metal electrode starts to dissolve, producing M^{z+} ions, thereby increasing the concentration in half-cell 1. The overall process is just the transfer of metal ion from a higher concentration to a lower concentration; in other words, a natural spontaneous process.

The initial cell voltage, before the concentrations have changed significantly, is
$$E_{cell} = \Delta\phi_{cat} - \Delta\phi_{an} = \Delta\phi_2 - \Delta\phi_1$$

$$= \left\{E^o_{M^{z+}/M} - \frac{RT}{zF}ln\frac{1}{a_2}\right\} - \left\{E^o_{M^{z+}/M} - \frac{RT}{zF}ln\frac{1}{a_1}\right\} = +\frac{RT}{zF}ln\frac{a_2}{a_1}$$

As the activities in each side slowly change, E_{cell} slowly gets smaller and smaller, until, when the concentrations are the same in each half-cell, the overall system will have reached overall cell equilibrium with $E_{cell} = 0$.

The equilibrium constant for the cell reaction $M^{z+}(a_2) \rightarrow M^{z+}(a_1)$

is just

$$K = exp\left[\frac{zFE^o_{cell}}{RT}\right] = exp\left[\frac{zF(0)}{RT}\right] = 1$$

That is,

$$K = \frac{a_{M^{z+}}(a_2)}{a_{M^{z+}}(a_1)} = 1$$

so that, at equilibrium,

$$(a_{M^{z+}})_2 = (a_{M^{z+}})_1$$

PROBLEMS

1. Given the following two half-cell reduction potentials at 25°C, calculate the solubility product of $PbSO_4$ at this temperature.

    $$PbSO_{4(s)}/Pb_{(s)}/SO^{2-}_{4(aq)} \qquad E^o_{298.15} = -0.356 \text{ V } vs. \text{ SHE}$$

    $$Pb^{2+}_{(aq)}/Pb_{(s)} \qquad E^o_{298.15} = -0.126 \text{ V } vs. \text{ SHE} \; \ominus$$

2. The cell

    $$Zn(Hg) \,(a_{Zn} = x) \mid aq. \; Zn(NO_3)_2 \,(a_{Zn^{++}} = 0.15 \; m) \mid Zn(Hg) \,(a_{Zn} = y)$$

    consists of two zinc amalgam[22] electrodes in contact with aqueous $Zn(NO_3)_2$ solution in which the relative molal activity of the Zn^{2+} ions is 0.15. If the equilibrium cell voltage at 35°C is 35.95 mV, what is the ratio of x/y? $\ominus$

3. A hydrogen electrode is placed into an aqueous solution of 0.0050 molar HBr at 25°C. The partial pressure of the hydrogen gas is 1.45 bar. By how much will the potential of the hydrogen electrode change if the HBr concentration is increased to 0.025 mol L^{-1}? For this problem, when using the Debye-Hückel equation you may assume that HBr dissociates 100% and that molarity is the same as molality. $\ominus$

4. For the following cell at 25°C:

    $$Hg_{(liq)} \mid aq. \; HgCl_2 \parallel aq. \; TlNO_3 \mid Tl_{(s)}$$

    (a) What is the standard cell potential?

    (b) What is the cell potential if the concentrations of the $HgCl_2$ and the $TlNO_3$ solutions are 0.0100 molal and 0.0025 molal, respectively?

    (You may assume that at these concentrations each salt dissociates completely, and that the difference between molarity and molality is negligible.)

---

[22] A metal dissolved in liquid mercury is called an "amalgam." Tooth fillings usually are made from silver amalgams. Yes! you have mercury in your mouth. Silver amalgams are considered to be safe because of their great chemical stability, and have been used for many years with no apparent associated health problems.

$$Tl^+ + e^- \rightleftharpoons Tl \qquad E^o_{298.15} = -0.34 \text{ V}$$
$$Hg^{2+} + 2e^- \rightleftharpoons Hg \qquad E^o_{298.15} = +0.86 \text{ V} \; \text{☹}$$

5. At 25°C, the *emf* for the cell[23]

$$Hg_{(liq)} \mid Hg_2Br_{2\,(s)} \parallel Br^-_{(aq)} \mid AgBr_{(s)} \mid Ag_{(s)}$$

is independent of the concentration of the bromide solution, and is given by the expression

$$E_{cell} = 68.04 + 0.312 \, (t - 25) \text{ mV}$$

where *t* is the temperature, in °C. For the cell operating at 25°C, write the cell reaction (for the passage of 2 *F*) and evaluate: (a) ΔG (b) ΔS (c) ΔH. ☺

6. The potential of the following cell is 0.43783 V at 25°C:

$$Pt_{(s)} \mid H_{2(g)} \, (P_{H_2} = 1.00 \text{ bar}) \mid 0.0171 \, m \, HCl_{(aq)} \mid AgCl_{(s)} \mid Ag_{(s)}$$

(a) What is the overall cell reaction?

(b) What is the mean activity coefficient $\gamma_\pm$ for the 0.0171 molal aqueous HCl in this cell?

(c) What is the value of the equilibrium constant for the cell reaction at 25°C?

$$AgCl_{(s)} + e^- \rightleftharpoons Ag_{(s)} + Cl^-_{(aq)} \qquad E^o_{298.15} = +0.2225 \text{ V} \; \text{☺}$$

7. In the industrial electrochemical production of caustic soda (*NaOH*) and chlorine, an aqueous solution of concentrated NaCl is electrolyzed. The desired reaction at the anode is the generation of chlorine gas according to

$$2Cl^- \rightarrow Cl_2\uparrow + 2e^- \qquad E^o_{298.15} = +1.36 \text{ V}$$

but an unwanted side reaction that also takes place at the anode evolves oxygen:

$$4OH^- \rightarrow 2H_2O + O_2\uparrow + 4e^- \qquad E^o_{298.15} = -0.827 \text{ V}$$

At the cathode, only hydrogen gas is evolved:

$$4H_2O + 4e^- \rightarrow 2H_2\uparrow + 4OH^- \qquad E^o_{298.15} = -0.827 \text{ V}$$

The cells operate at a total current of 10,000 amperes and at a cell voltage of 2.50 V.

After a certain cell was operated for exactly one hour, it was found that 89.6 grams of oxygen gas had been generated at the anode.

(a) What is the percentage current efficiency for chlorine production in this cell? (That is, what percentage of the total current contributes towards the production of chlorine?)

(b) How many tonnes of NaOH are produced by this cell in one year if it is operated 24 hours per day for 320 days of the year?

(c) What is the cost of electricity to operate this cell for one year, if electricity sells for 8.00¢ per kilowatt-hour?

---

[23] *emf* = *electromotive force*; i.e., the driving force for the reaction. The greater E_{cell}, the greater the driving force.

(d)   What is the cost of electricity to produce one kilogram of NaOH? ☹

8.   Consider the following cell:

$$Ag_{(s)} \mid AgI_{(s)} \mid aq. \; KI \; (0.01 \; m) \mid\mid aq. \; KCl \; (0.001 \; m) \mid Cl_{2(g)} \; (P_{Cl_2} = 1 \; bar) \mid Pt_{(s)}$$

At 15°C and 25°C the emf of this cell is, respectively, 1.5773 V and 1.5678 V.
Determine the following, using the Guggenheim equation

$$log \; \gamma_{\pm} = - \frac{0.5108 \mid z_+ z_- \mid \sqrt{I^{(m)}}}{1 + \sqrt{I^{(m)}}}$$

for any activity coefficient corrections that may be required:[24]

(a)   The overall cell reaction for the passage of 2 faradays of electricity.

(b) ΔG for the overall cell reaction at 25°C.

(c)   The standard reduction potential $E^o_{298.15}$ for the electrode on the left-hand side.

(d) $\Delta G°$ for the overall cell reaction at 25°C.

(e)   The solubility product at 25°C of AgI: $K_{SP} = a_{Ag^+} \cdot a_{I^-}$.

(f) ΔS for the overall cell reaction at 25°C.

(g) ΔH for the overall cell reaction at 25°C.

      Data: $Ag^+_{(aq)} + e^- \rightleftharpoons Ag_{(s)}$ $E^o_{298.15} = +0.7996$ V

                          $Cl_{2(g)} + 2e^- \rightleftharpoons 2Cl^-_{(aq)}$ $E^o_{298.15} = +1.3583$ V ☹

9.   Calculate $E°$ for the reaction $2Na + 3S \rightleftharpoons Na_2S_3$

      Given: $2Na + 5S \rightleftharpoons Na_2S_5$ $E^o_1 = 2.08$ V

                     $2Na + 4Na_2S_5 \rightleftharpoons 5 \; Na_2S_4$ $E^o_2 = 1.97$ V

                     $2Na + 3Na_2S_4 \rightleftharpoons 4Na_2S_3$ $E^o_3 = 1.81$ V   ☺

10.   Calculate the value of K_W at 25°C using the following data:

$$2H_2O_{(liq)} + 2e^- \rightleftharpoons 2OH^-_{(aq)} + H_{2(g)}$$ $E^o_{298.15} = -0.828$ V

$$2H^+_{(aq)} + 2e^- \rightleftharpoons H_2$$ $E^o_{298.15} = 0.000$ V ☺

11.   At 200°C the emf of the solid state cell

$$Ag \mid AgCl_{(s)} \mid Au\text{-}Ag \; alloy$$

was measured to be 86.4 mV. The mole fraction of Ag in the Au-Ag alloy was 0.400.

(a)   Write the anode, cathode, and overall cell reaction.

(b)   What is the activity of the silver in the alloy?

---

[24] $I^{(m)}$ is the ionic strength defined using molalities instead of molarities.

(c)   What is the activity coefficient of the silver in the alloy? ☺

12.   A scientist often had to compare the *pH* of different solutions with great accuracy. She used
      the cell
$$Pt \mid H_2 \mid unknown\ solution \parallel aq.\ KCl\ (1.0\ M) \mid Hg_2Cl_{2(s)} \mid Hg_{(liq)}$$
      —always at 25°C—and had computed a table that gave for each *E*–value in this cell the corre-
      sponding *pH* of the solution, provided the hydrogen partial pressure was 1.00 atm. One day
      for a certain solution she measured an *E* which, according to the table, corresponded to *pH*
      5.873. However, unknown to the scientist, after correcting for the vapor pressure of water,
      the partial pressure that day of the hydrogen at the electrode was only 720 Torr. What was
      the actual *pH* of the solution (using the *pH* scale defined by her tables)?

      The calomel half-cell reaction is $Hg_2Cl_{2(s)} + 2e^- \rightleftharpoons 2Hg_{(liq)} + 2Cl^-_{(aq)}$. ☺

13.   For the cell $Pt_{(s)} \mid H_{2(g)}\ (1\ bar) \mid dilute\ NaOH_{(aq)} \mid Ag_2O_{(s)} \mid Ag_{(s)}$

      $E_{cell} = 1.172$ V at 25°C [Luther and Pokorny, Zanorg Ch. **57**, 290 (1908)]. Moreover, $\Delta G°$ for the
      reaction $H_{2(g)} + \frac{1}{2}O_{2(g)} \rightarrow H_2O_{(liq)}$ has been calculated to be –237.316 kJ at the same tem-
      perature. At what partial pressure will oxygen be in equilibrium with silver oxide and silver at
      25°C? ☹

14.   An aqueous solution, containing 0.010 molal $CdSO_4$, 0.010 molal $ZnSO_4$, and 0.50 molal
      H_2SO_4 is electrolyzed at 25°C between Pt electrodes with good stirring and low current
      density (to maintain near-equilibrium conditions, so that the Nernst equation can be assumed
      to apply at all times). At 25°C the standard reduction potentials for Zn^{2+}/Zn and Cd^{2+}/Cd are
      -0.763 V and -0.403 V, respectively, *vs. SHE*.

      (a)   As the cathode is made more negative, which of the two metals will start to deposit first?
      (b)   What will be the concentration of this metal remaining in the solution when the second
            metal begins to deposit? ☹

15.   Calculate *E°*, the standard reduction potential at 25°C, for the half-cell reaction
$$2\ Hg^{2+}_{(aq)} + 2e^- \rightarrow Hg_2^{2+}_{(aq)}$$
      given the following *E°* data at 25°C:

$$Hg^{2+}_{(aq)} + 2e^- \rightleftharpoons Hg_{(liq)} \qquad E°_{298.15} = +0.851\ V$$
$$Hg_2^{2+}_{(aq)} + 2e^- \rightleftharpoons 2Hg_{(liq)} \qquad E°_{298.15} = +0.7961\ V \ ☺$$

16.   Calculate the standard reduction potential for the couple Cu^{2+}/Cu^+, given the following two
      standard reduction potentials:

$$E°_{Cu^{2+}/Cu} = +0.337\ V \qquad E°_{Cu^+/Cu} = +0.522\ V \ ☺$$

17.   Consider the following cell at 25°C:
$$Au \mid AuI_{(s)} \mid aq.\ HI\ (m) \mid H_{2(g)}\ (1\ atm) \mid Pt$$

The equilibrium voltage of this cell is -0.970 V and -0.410 V when the molality of the hydroiodic acid is $m = 1.00 \times 10^{-4}$ mol kg^{-1} and 3.00 mol kg^{-1}, respectively.

(a) Write the anode, cathode, and overall cell reaction.

(b) Determine the value of the standard reduction potential for the Au–AuI electrode at 25°C.

(c) Determine the mean molal stoichiometric activity coefficient for 3.00 molal aqueous HI at 25°C.

(d) Using the additional information—not available for parts (a), (b), and (c)—that the standard reduction potential at 25°C for the $Au^+_{(aq)}/Au$ couple is $+1.692$ V, calculate the solubility product K_{SP} for AuI at 25°C. ☹

18. Consider the following two cells:

$$Cell\ 1: \quad Pt_{(s)}\ |\ H_{2(g)}\ (a_{H_2} = 1)\ |\ 7.000\ molal\ H_2SO_{4(aq)}\ |\ Hg_2SO_{4(s)}\ |\ Hg_{(liq)}$$

$$Cell\ 2: \quad Pb_{(s)}\ |\ PbSO_{4(s)}\ |\ 7.000\ molal\ H_2SO_{4(aq)}\ |\ PbSO_{4(s)}\ |\ PbO_{2(s)}$$

At 25°C the emf's of cell *1* and cell *2* are 0.5666 V and 2.1525 V, respectively. Determine the activity of the water in 7 molal aqueous sulfuric acid solution at 25°C.

Standard Reduction Potentials at 25°C:

$$PbSO_{4(s)} + 2e^- \rightleftharpoons Pb_{(s)} + SO_{4(aq)}^{2-} \qquad\qquad -0.3553\ V$$

$$PbO_{2(s)} + 4H^+_{(aq)} + SO_{4(aq)}^{2-} + 2e^- \rightleftharpoons PbSO_{4(s)} + 2H_2O_{(liq)} \qquad +1.6849\ V$$

$$Hg_2SO_{4(s)} + 2e^- \rightleftharpoons 2Hg_{(liq)} + SO_{4(aq)}^{2-} \qquad\qquad +0.6151\ V$$

$$2H^+_{(aq)} + 2e^- \rightleftharpoons H_{2(g)} \qquad\qquad 0.000\ V\ ☠$$

19. Determine the standard cell potential E^o_{cell}, the standard Gibbs free energy change ΔG^o, and the equilibrium constant K for the following reaction at 25°C:

$$Pb_{(s)} + 2H^+_{(aq)} \rightleftharpoons Pb^{2+}_{(aq)} + H_{2(g)}$$

The standard reduction potential for $Pb^{2+}_{(aq)}$ is -0.1262 V. ☺

20. Calculate the standard reduction potential for the reduction of $Zn(OH)_{2(s)}$ to $Zn_{(s)}$ in alkaline solution according to

$$Zn(OH)_{2(s)} + 2e^- \rightarrow Zn_{(s)} + 2OH^-_{(aq)}$$

given: $\qquad Zn(OH)_{2(s)} \rightleftharpoons Zn^{2+}_{(aq)} + 2OH^-_{(aq)} \qquad K_{SP} = 3.0 \times 10^{-17}$

At 25°C the standard reduction potential for $Zn^{2+}_{(aq)}$ is -0.7618 V. ☺

21. The aluminum-air cell is being actively developed for use as a battery system to power electric automobiles. The half-cell reactions are:

$$H_2AlO^-_{3(aq)} + H_2O_{(liq)} + 3e^- \rightarrow Al_{(s)} + 4OH^-_{(aq)} \qquad E^o_{298.15} = -2.33\ V$$

$$O_{2(g)} + 2H_2O_{(liq)} + 4e^- \rightarrow 4OH^-_{(aq)} \qquad E^o_{298.15} = +0.401\ V$$

(a) What is the overall cell reaction?

(b) What is the standard cell potential of this cell at 25°C?

(c) Under standard conditions, how much energy is potentially available from one kg of aluminum?

(d) What is the equilibrium constant for the cell reaction at 25°C?

(e) How much aluminum would remain at equilibrium? ☺

22. A commonly used reference electrode is the silver-silver chloride electrode, which consists of a silver wire coated with a porous layer of solid AgCl and immersed in aqueous KCl solution. By comparing under standard conditions the potential of this electrode with that of the silver-silver ion electrode, determine the solubility product at 25°C for $AgCl_{(s)}$, assuming molar concentrations equal activities.

$$AgCl_{(s)} + e^- \rightleftharpoons Ag_{(s)} + Cl^-_{(aq)} \qquad E^o_{298.15} = +0.2223 \text{ V}$$
$$Ag^+_{(aq)} + e^- \rightleftharpoons Ag_{(s)} \qquad E^o_{298.15} = +0.7996 \text{ V} \ ☺$$

23. A dilute solution of dissolved ore contains free zinc ions $Zn^{2+}_{(aq)}$ and free cobalt ions $Co^{2+}_{(aq)}$ at concentrations of 150 mg L^{-1} and 250 mg L^{-1}, respectively. If this solution is electrolyzed at 25°C between inert electrodes, what is the molar concentration of the cobalt ions when the zinc ions just begin to be deposited? Assume molar concentrations are equivalent to activities and that the ions begin to deposit at the reversible potentials.

Molar masses: $Zn = 65.39 \quad Co = 58.93$

$$Zn^{2+}_{(aq)} + 2e^- \rightleftharpoons Zn_{(s)} \qquad E^o_{298.15} = -0.7618 \text{ V}$$
$$Co^{2+}_{(aq)} + 2e^- \rightleftharpoons Co_{(s)} \qquad E^o_{298.15} = -0.277 \text{ V} \ ☺$$

TWENTY-SIX

CHEMICAL REACTION KINETICS

26.1 RATES OF REACTIONS: CHEMICAL KINETICS

The two most important characteristics of a chemical reaction are: (1) the position of the equilibrium (i.e., the *yield*), and (2) the reaction *rate*. In other words, we are not interested only in the fact that the equilibrium favors products over reactants; but we also want to know how fast the reaction proceeds: if a reaction takes 75,000 years to reach equilibrium it's not of any practical use![1]

In general, when reactants are put together, more than one reaction is possible. The reaction that is actually observed is that which occurs the *fastest*; this is because the reactants are all consumed by this fast reaction before the slow reactions have enough time to produce any significant amount of *their* products.

Factors affecting reaction rate:

1. *nature of reactants*
2. *temperature*—usually faster at higher temperatures
3. *concentration*—usually faster at higher reactant concentration
4. *surface area*—important for heterogeneous reactions such as combustions, dissolutions
5. *agitation*—especially important for heterogeneous reactions: stirring speeds up the supply of reactants to the interface, where the reaction takes place

26.2 CONCENTRATION PROFILES

For the reaction

$$A + B \rightarrow C + D$$

a typical concentration profile with time[2] looks like that shown in Fig. 1.

In general, the concentrations initially change quite rapidly with time, and then level off as they

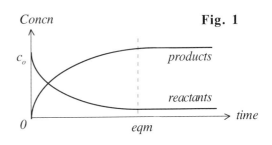

Fig. 1

---

[1] Unless the reaction started 75,000 years ago!
[2] Sometimes, for gas phase reactions, we plot pressure *vs.* time instead of concentration *vs.* time; but plotting concentration is more common.

approach equilibrium.[3] Usually, as products begin to form, there is the possibility that some of them will start reacting back to reactants again:

$$C + D \rightarrow A + B$$

In general the process is a *dynamic* one in which the *net* rate is given by:

$$\left\langle \begin{array}{c} Net \\ Reaction \\ Rate \end{array} \right\rangle = \langle Forward\ Rate \rangle - \langle Reverse\ Rate \rangle \qquad \ldots [1]$$

At equilibrium, the net rate equals zero; i.e., we have a *dynamic equilibrium* for which

$$\langle Forward\ Rate \rangle = \langle Reverse\ Rate \rangle \qquad \ldots [2]$$

The reverse rate generally starts to become significant as equilibrium is approached.

26.3 EXPRESSION OF REACTION RATES

When the reactants are first brought together (i.e., before any significant amount of product has formed), the net rate is approximately just the forward rate.

Consider the reaction (in a constant volume reactor) of hydrogen with oxygen to form water:

$$2H_{2(g)} + O_{2(g)} \rightarrow 2H_2O_{(g)}$$

The rates of change with time of the concentrations of the various reactants and products are related according to

$$-\frac{1}{2}\frac{d[H_2]}{dt} = -\frac{d[O_2]}{dt} = +\frac{1}{2}\frac{d[H_2O]}{dt} \qquad \ldots [3]$$

That is, H_2 is used up twice as fast as O_2, and H_2O is produced at the same rate at which H_2 is consumed. In general, for the reaction

$$aA + bB \rightarrow cC + dD$$

$$\begin{array}{c} Rate \\ (mol\ L^{-1}\ s^{-1}) \end{array} = r = -\frac{1}{a}\frac{d[A]}{dt} = -\frac{1}{b}\frac{d[B]}{dt} = +\frac{1}{c}\frac{d[C]}{dt} = +\frac{1}{d}\frac{d[D]}{dt} \qquad \ldots [4]$$

Example 26-1

The rate of the reaction $\qquad\qquad A + 2B \rightarrow 3C + D$

was reported as 2.0 mol L^{-1} s^{-1}. State the rates of consumption or formation of each participant.

---

[3] The introductory treatment of chemical kinetics given in this chapter deals mainly with constant volume batch processes in closed systems.

Solution

The rate of the reaction is $\quad r = -\dfrac{1}{1}\dfrac{d[A]}{dt} = -\dfrac{1}{2}\dfrac{d[B]}{dt} = +\dfrac{1}{3}\dfrac{d[C]}{dt} = +\dfrac{1}{1}\dfrac{d[D]}{dt} = 2.0 \text{ mol L}^{-1}\text{s}^{-1}$

For species A: $\quad r = -\dfrac{1}{1}\dfrac{d[A]}{dt}$; therefore $\dfrac{d[A]}{dt} = -r = -2.0 \text{ mol L}^{-1}\text{s}^{-1}$ [**Ans.**]

For species B: $\quad r = -\dfrac{1}{2}\dfrac{d[B]}{dt}$; therefore $\dfrac{d[B]}{dt} = -2r = -4.0 \text{ mol L}^{-1}\text{s}^{-1}$ [**Ans.**]

For species C: $\quad r = +\dfrac{1}{3}\dfrac{d[C]}{dt}$; therefore $\dfrac{d[C]}{dt} = +3r = +6.0 \text{ mol L}^{-1}\text{s}^{-1}$ [**Ans.**]

For species D: $\quad r = +\dfrac{1}{1}\dfrac{d[D]}{dt}$; therefore $\dfrac{d[D]}{dt} = +r = +2.0 \text{ mol L}^{-1}\text{s}^{-1}$ [**Ans.**]

26.4 DIFFERENTIAL RATE LAWS

It is usually possible to express the rate of a chemical reaction as a product of the reactant concentrations, each raised to some power: For example, for

$$3A + 2B \rightarrow C + D$$

we can put $\quad r = -\dfrac{1}{3}\dfrac{d[A]}{dt} = -\dfrac{1}{2}\dfrac{d[B]}{dt} = \dfrac{d[C]}{dt} = \dfrac{d[D]}{dt} = k[A]^m[B]^n$

where m and n are generally integers or half-integers. We say that m is the *order* of the reaction with respect to A, and n is the *order* of the reaction with respect to B. The *overall order* of the reaction is $(m + n)$. m and n must be determined *experimentally*; they are sometimes—but not always—equal to the stoichiometric coefficients.

Thus, for $\qquad H_2 + I_2 \rightarrow 2HI \qquad\qquad r = -\dfrac{d[H_2]}{dt} = k[H_2][I_2]$

but for $\qquad H_2 + Br_2 \rightarrow 2HBr \qquad\qquad r = -\dfrac{d[H_2]}{dt} = k'[H_2][Br_2]^{1/2}$

k and k' are called the **rate constants** for the reactions, and indicate whether the reactions are fast or slow. The greater rate constant, the faster the reaction. Like the equilibrium constant, the rate constant is a function of temperature, but not of concentration.

26.5 FIRST ORDER REACTIONS

Consider the gas phase reaction: $\qquad N_2O_5 \rightarrow 2NO_2 + \frac{1}{2}O_2$

This reaction is a *first order* reaction:

$$-\frac{d[N_2O_5]}{dt} = k[N_2O_5]$$

That is,
$$\frac{dc}{dt} = -kc$$

Rearranging:
$$\frac{dc}{c} = -kdt$$

Integrating:
$$\int_{c=c_o}^{c=c} \frac{dc}{c} = -k\int_{t=0}^{t=t} dt$$

$$\boxed{ln\left(\frac{c}{c_o}\right) = -kt}$$ *First Order Reaction*   . . . [5]

or $$\boxed{ln\, c = ln\, c_o - kt}$$ *First Order Reaction* . . . [6]

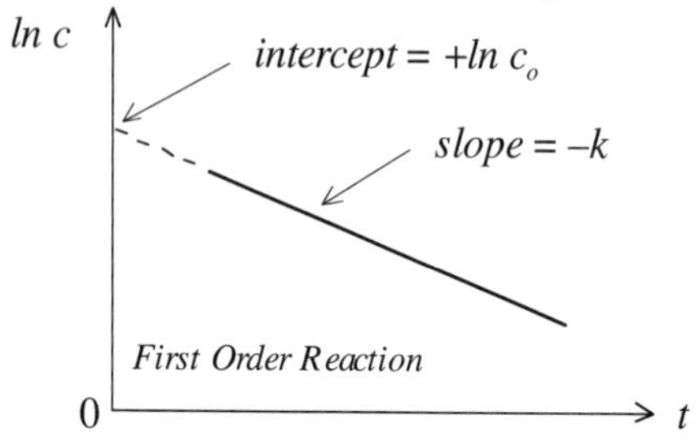

As shown in Fig. 2, a plot of $ln\, c$ *vs.* t has a slope of $-k$ and an intercept of $ln\, c_o$.

Alternatively, from Eqn [5]

$$ln\left(\frac{c}{c_o}\right) = -kt$$

from which

$$\left(\frac{c}{c_o}\right) = exp\left[-kt\right]$$ . . . [7]

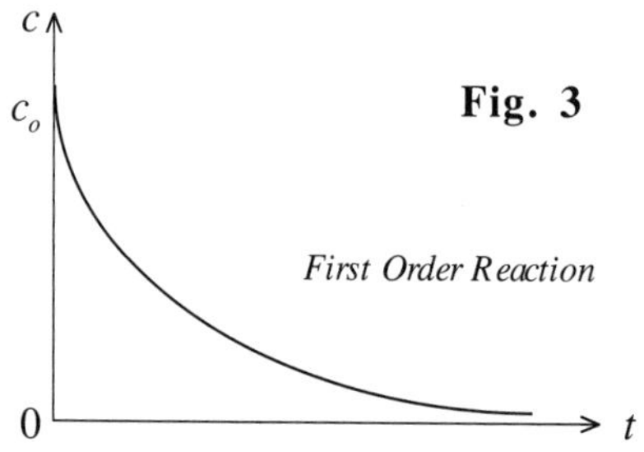

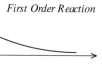

Fig. 2

intercept $= +ln\, c_o$

slope $= -k$

First Order Reaction

Fig. 3

First Order Reaction

that is,    $$\boxed{c = c_o exp\left[-kt\right]}$$ *First Order Reaction*   . . . [8]

This behavior is shown in Fig. 3.

Thus *the concentration of the reactant decays exponentially with time for a first order reaction.*

Example 26-2

At 67°C the decomposition of N_2O_5 to NO_2 and O_2 is known to be first order in N_2O_5. In 3.00 minutes the N_2O_5 concentration fell from an initial value of 1.00 mol L^{-1} to 0.399 mol L^{-1}.

Calculate   (a) the rate constant for this reaction at 67°C, and

(b) the time required for 99.5% of the N_2O_5 to decompose.

Solution

(a) The expression for the rate of a first order reaction is

$$\frac{dc}{dt} = -kc$$

This can be rearranged to
$$k = -\frac{dc}{dt}\bigg/c$$

from which it can be seen that k has the units of $\left(\text{mol L}^{-1}\,\text{s}^{-1}\right)\big/\left(\text{mol L}^{-1}\right) = \text{s}^{-1}$.

For a first order reaction, we showed that

$$ln\,c = ln\,c_o - kt$$

Rearranging and putting in values gives

$$k = \frac{ln\,c_o - ln\,c}{t} = \frac{\ln(1.00) - \ln(0.399)}{3.00 \times 60} = 5.10 \times 10^{-3}\,\text{s}^{-1}$$

Ans: $\boxed{5.10 \times 10^{-3}\,\text{s}^{-1}}$

(b) When 99.5% of the N_2O_5 has decomposed, the concentration of remaining N_2O_5 will be

$$c = 0.005c_o = (0.005)(1.00) = 0.005\,\text{mol L}^{-1}$$

For a first order reaction, $\qquad ln\,c = ln\,c_o - kt$

which rearranges to
$$t = \frac{ln\,c_o - ln\,c}{k}$$

Inserting values gives $\qquad t = \dfrac{ln(1.00) - ln(0.005)}{5.10 \times 10^{-3}} = 1039\,\text{s} = 17.3\,\text{min}$

Ans: $\boxed{17.3\,\text{min}}$

Exercise 26-1

In aqueous solution, ordinary table sugar (sucrose, $C_{12}H_{22}O_{11}$) hydrolyzes to produce two simpler sugars, glucose and fructose (both of which have the same formula, but different structures):

$$\underset{sucrose}{C_{12}H_{22}O_{11}} + H_2O \rightarrow \underset{glucose}{C_6H_{12}O_6} + \underset{fructose}{C_6H_{12}O_6}$$

In a series of experiments at 35°C the following data were obtained:

Initial Sucrose Concentration (mol L^{-1})	Initial Rate of Formation of Glucose (mol m^{-3} s^{-1})
0.10	6.17×10^{-3}
0.20	1.23×10^{-2}
0.50	3.09×10^{-2}

(a) What is the reaction order with respect to sucrose?

(b) After reacting for one hour, the concentration of a sucrose solution having an initial concentration of 0.40 mol L^{-1} had dropped to 0.32 M. What is the rate constant for the reaction?

(c) How many minutes will it take for the concentration to fall to 0.10 M? ☺

Example 26-3

At 553 K the following gas phase decomposition reaction is a first order reaction:

$$SO_2Cl_2 \rightarrow SO_2 + Cl_2$$

If, at 553 K, 50.0% of the initial SO_2Cl_2 has decomposed after 8.75 h,

(a) How much *additional* time will it take for 99.9% of the initial SO_2Cl_2 to have decomposed?

(b) How much time is required for the extent of the reaction to go from 99.9% to 99.99%?

Solution

(a) For a first order reaction, $\qquad ln\left(\dfrac{c}{c_o}\right) = -kt$

We are told that at time $t = 8.75$ h, $c = \frac{1}{2}c_o$

Therefore, putting values into the above equation:

$$ln\left(\frac{c_o/2}{c_o}\right) = ln\,\tfrac{1}{2} = -k\,(8.75)$$

from which $\qquad k = \dfrac{ln\left(\frac{1}{2}\right)}{-8.75} = +0.07922 \text{ h}^{-1}$

For 99.9% to react, $\qquad c = 0.001 c_o$

therefore $\qquad -kt = ln\left(\dfrac{c}{c_o}\right)$

$$-(0.07922)\,t = ln\,(0.001)$$

from which $\qquad t = \dfrac{ln(0.001)}{-0.07922} = \dfrac{-6.9078}{-0.07922} = 87.20 \text{ h}$

The additional time required is $\Delta t = t_{99.9} - t_{0.5}$

$$= 87.20 - 8.75$$

$$= 78.45 \text{ h}$$

Ans: | 78.5 h |

(b) The total time required for 99.99% reaction is

$$t_{99.99} = \frac{ln(0.0001)}{-0.07922} = 116.26 \text{ h};$$

therefore, $\Delta t = t_{99.99} - t_{99.9}$

$$= 116.26 - 87.20$$

$$= 29.06 \text{ h}$$

Ans: | 29.1 h |

It takes a long time to get rid of that last little bit!

26.6 HALF-LIVES

The half-life of a chemical reaction is the time it takes for the concentration of the reactant to fall to *half* its original value. Thus, for a *first order* reaction, we substitute $c = \frac{1}{2}c_o$ into Eqn [5] as follows:

$$ln\left(\frac{c}{c_o}\right) = -kt \qquad \qquad \text{. . . [5]}$$

giving $$ln\left(\frac{0.5c_o}{c_o}\right) = ln\frac{1}{2} = -ln\,2 = -kt_{1/2}$$

which rearranges to | $t_{1/2} = \dfrac{ln\,2}{k}$ | *Half-life for a First Order Reaction* | . . . [9]

Eqn [9] shows that for a first order reaction the half-life is *independent* of the initial concentration of reactant. During each successive duration of $t_{1/2}$, the concentration of the reactant in a first order reaction decays to half its value at the start of that period. After n such periods, the reactant concentration will be $\left(\frac{1}{2}\right)^n$ of its initial concentration.

Example 26-4

There is a constant flux of particles falling on the earth as a result of nuclear processes occurring in the sun and other parts of the universe; these particles, which are mostly protons, are called *cosmic rays*. When cosmic rays collide with N_2 molecules in the earth's atmosphere they produce carbon-14, an unstable radioactive isotope of carbon that has a half-life of 5720 years. Accordingly, when a living tree takes in CO_2 from the air, a certain amount of C^{14} becomes incorporated into its wood. When the tree dies, it no longer ingests CO_2, and the radioactivity present in the wood gradually disappears through the first-order process of radioactive disintegration. Thus, by measuring the residual amount of C^{14} present in an ancient sample of wood and comparing it with the amount found in living wood, we are able to estimate how long the tree has been dead, and therefore the age of the sample.[4] This is the principle behind the technique of *carbon-14 dating*.

An archeological sample contained wood that had only 72% of the C^{14} found in living trees. Estimate the age of the sample.

Solution

The process of radioactive decay is a first order process, so we can write

$$\frac{dm}{dt} = -km$$

where k is the first order rate constant and m is the mass of C^{14} present in the sample.

Furthermore,
$$\frac{m}{m_0} = exp[-kt]$$

where m is the amount at time t and m_0 is the initial amount (at $t = 0$).

Also, for a first order process we showed that the half-life is given by $t_{1/2} = \frac{ln\ 2}{k}$.

Thus, since the half-life of C^{14} is 5720 years,

$$k = \frac{ln\ 2}{t_{1/2}} = \frac{0.6931}{5720\ y} = 1.2118 \times 10^{-4}\ y^{-1}$$

Therefore: $\dfrac{m}{m_0} = 0.72 = exp[-kt] = exp\ [-1.2118 \times 10^{-4}t]$

$$ln\ (0.72) = -1.2118 \times 10^{-4}t$$

$$t = \frac{ln(0.72)}{-1.2118 \times 10^{-4}\ y^{-1}} = \frac{-0.3285}{-1.2118 \times 10^{-4}} = 2.7108 \times 10^3\ y$$

---

[4] We have to assume that the level of cosmic ray bombardment is constant throughout history, so that living trees in antiquity contained the same concentration of C^{14} as they do now.

Ans: ~2710 years

| Exercise 26-2 |

In the 1950s the United States and Russia carried out many above-ground nuclear tests. One of the hazards of such testing is the atmospheric generation of the unstable radioactive isotope strontium-90, which emits β-rays when it decays and has a half-life of 28.1 years. Strontium-90 has the ability to enter the body and replace the calcium in your bones with radioactive strontium. For a few years, every day the radio and newspapers would report the level of Sr^{90} in the atmosphere. During a testing period when the levels were especially high, Sr^{90} would find its way into cow's milk and warnings were given not to give milk to children.

If 1.0 μg of Sr^{90} is absorbed by a newly born baby, how much will remain in its body after (a) 19 years? (b) 80 years? ☺

Example 26-5

Derive an expression for the half-life of a reaction which follows the rate expression

$$\frac{dc}{dt} = - kc^{5/2}$$

Solution

Rearranging the rate expression: $\dfrac{dc}{c^{5/2}} = - kt$

Integrating:

$$\int_{c_o}^{c} \frac{dc}{c^{5/2}} = \int_{c_o}^{c} c^{-5/2} dc$$

$$= \left[\frac{c^{-3/2}}{-3/2}\right]_{c_o}^{c} = -\frac{2}{3}\left[\frac{1}{c^{3/2}}\right]_{c_o}^{c}$$

$$= -\frac{2}{3}\left[\frac{1}{c^{3/2}} - \frac{1}{c_o^{3/2}}\right] = -kt \qquad \ldots [a]$$

When $t = t_{1/2}, \quad c = \frac{1}{2}c_o$

Substituting these values into Eqn [a] gives:

$$-k\,t_{1/2} = -\frac{2}{3}\left[\frac{1}{\left(c_o/2\right)^{3/2}} - \frac{1}{c_o^{3/2}}\right]$$

$$= -\frac{2}{3}\left[\frac{1}{c_o^{3/2}2^{-3/2}} - \frac{1}{c_o^{3/2}}\right]$$

$$= -\frac{2}{3c_o^{3/2}}\left[2^{3/2} - 1\right]$$

$$= -\frac{2}{3c_o^{3/2}}\left[2.828 - 1\right]$$

$$= -\frac{1.219}{c_o^{3/2}}$$

Therefore: $t_{1/2} = \dfrac{-1.219}{-kc_o^{3/2}} = \dfrac{1.219}{kc_o^{3/2}}$

Ans: $\boxed{t_{1/2} = \dfrac{1.219}{kc_o^{3/2}}}$

Exercise 26-3

Derive an expression for the half-life of a reaction that follows the rate expression

$$\frac{dc}{dt} = -kc^{3/2} \ \ \text{☺}$$

26.7 SECOND ORDER REACTIONS

26.7.1 REACTIONS THAT ARE SECOND ORDER IN ONE REACTANT

Cantharene is produced from butadiene according to the following reaction:

$$2C_4H_{6(g)} \rightarrow C_8H_{12(g)}$$

This is a *second order* reaction, which we can represent as $2A \rightarrow B$.

Therefore, $-\dfrac{1}{2}\dfrac{d[A]}{dt} = k[A]^2$

that is, $-\dfrac{1}{2}\dfrac{dc}{dt} = kc^2$ . . . [10]

Rearranging: $-\dfrac{dc}{c^2} = 2k\,dt$

Integrating:

$$-\int_{c_o}^{c}\frac{dc}{c^2} = 2k\int_0^t dt = 2kt$$

That is,

$$-\left[-\frac{1}{c}\right]_{c_o}^{c} = \left[\frac{1}{c}\right]_{c_o}^{c} = 2kt$$

$$\boxed{\frac{1}{c} - \frac{1}{c_o} = 2kt} \quad Second\ Order\ Reaction\ \dots [11]$$

Eqn [11] rearranges to

$$\frac{1}{c} = \frac{1}{c_o} + 2kt \qquad \dots [12]$$

Fig. 4 shows that a plot of $1/c$ *vs.* t has a slope of $+2k$ and an intercept of $1/c_o$.

The integrated rate law can be rearranged to show how the concentration of reactant varies with time for a second order reaction:

$$\frac{1}{c} = \frac{1}{c_o} + 2kt$$

$$c = \frac{1}{\left(\dfrac{1}{c_o} + 2kt\right)} = \frac{1}{\left(\dfrac{1+2c_okt}{c_o}\right)} = \frac{c_o}{1+2c_okt}$$

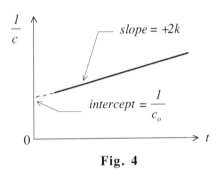

Fig. 4

$$\boxed{c = \frac{c_o}{1+2c_okt}} \quad Second\ Order\ Reaction\ \dots [13]$$

Example 26-6

The following gas phase reaction is second order with respect to atomic iodine (I):

$$2I + Ar \rightarrow I_2 + Ar$$

At 293 K the rate constant for the reaction is $k = 0.59 \times 10^{16}$ cm^6 mol^{-2} s^{-1}. If the initial concentration of atomic iodine is $[I]_o = 2.0 \times 10^{-5}$ mol L^{-1} and the argon concentration is $[Ar] = 5.0 \times 10^{-3}$ mol L^{-1}, what is the half-life of the atomic iodine?

Solution

First, since the argon concentration is given in moles per *liter*, let's change the concentration units of the rate constant from mol cm^{-3} to mol L^{-1}.

Thus,
$$k = 0.59 \times 10^{16} \left(\frac{cm^3}{mol}\right)^2 s^{-1} \times \left(\frac{1 \ L}{1000 \ cm^3}\right)^2$$

$$= 0.59 \times 10^{10} \left(\frac{L}{mol}\right)^2 s^{-1}$$

The rate law is given by
$$\frac{1}{2} \frac{d[I]}{dt} = -k[I]^2[Ar] \qquad \ldots \text{[a]}$$

Note that although no argon is consumed in the overall reaction, its concentration affects the kinetics of the reaction. It can be confirmed that an argon concentration term must be included in Eqn [a] by examining the units of the rate constant: Thus, from Eqn [a]—which includes a term for the argon concentration—we get the required units for k:

$$k = \frac{1}{2} \frac{d[I]}{dt} \bigg/ ([I]^2[Ar]) = \frac{(mol/L)/s}{(mol/L)^2(mol/L)} = \left(\frac{L}{mol}\right)^2 s^{-1} \qquad \text{[The required units]}$$

Letting c represent the molar concentration of atomic iodine, Eqn [a] can be rearranged to give

$$\frac{dc}{c^2} = -2k[Ar]dt$$

$$= -(2)(0.59 \times 10^{10})(5.0 \times 10^{-3})dt$$

$$= -5.90 \times 10^7 \, dt$$

Integrating:
$$\int_{c_o}^{c} \frac{dc}{c^2} = \left[-\frac{1}{c}\right]_{c_o}^{c} = -\left[\frac{1}{c}\right]_{c_o}^{c}$$

$$= -\left[\frac{1}{c} - \frac{1}{c_o}\right] = \frac{1}{c_o} - \frac{1}{c}$$

$$= -5.90 \times 10^7 \, t$$

When $\quad t = t_{1/2}, \quad c = \frac{1}{2}c_o$

Putting these values into the above equation gives

$$\left[\frac{1}{c_o} - \frac{1}{c_o/2}\right] = \frac{1}{c_o} - \frac{2}{c_o} = -\frac{1}{c_o} = -5.90 \times 10^7 \, t_{1/2}$$

from which
$$t_{1/2} = \frac{-1}{c_o(-5.90 \times 10^7)} = \frac{1}{(2 \times 10^{-5})(5.90 \times 10^7)}$$

$$= 0.848 \times 10^{-3} \text{ s} = 0.848 \text{ ms}$$

Ans: $\boxed{0.85 \text{ ms}}$

Example 26-7

At 600 K the decomposition of gaseous hydrogen iodide (HI) obeys the following rate law:

$$\frac{dc}{dt} = -kc^2$$

with a rate constant of $k = 4.00 \times 10^{-6}$ L mol^{-1} s^{-1}.

How many molecules of HI decompose per second per liter at 600 K and one atm pressure when t is close to $t = 0$?

Solution

From $PV = nRT$, the initial concentration of gaseous HI is given by

$$\frac{n}{V} = \frac{P}{RT} = \frac{101\,325}{(8.314)(600)} = 20.312 \text{ mol m}^{-3} = 20.312 \times 10^{-3} \text{ mol L}^{-1}$$

Therefore the initial rate of the reaction is

$$\left(\frac{dc}{dt}\right)_{t=0} = -kc_o^2$$

$$= -\left(4.00 \times 10^{-6} \; \tfrac{\text{L}/\text{mol}}{\text{s}}\right)\left(20.312 \times 10^{-3} \; \tfrac{\text{mol}}{\text{L}}\right)^2$$

$$= -1.65 \times 10^{-9} \; \tfrac{\text{mol}/\text{L}}{\text{s}}$$

One mole of gas contains $N_A = 6.02217 \times 10^{23}$ molecules of gas; therefore the number of molecules of HI decomposing per second initially is

$$\left(1.65 \times 10^{-9} \; \tfrac{\text{mol}/\text{L}}{\text{s}}\right)\left(6.02217 \times 10^{23} \text{ mol}^{-1}\right) = 9.94 \times 10^{14} \text{ L}^{-1} \text{ s}^{-1}$$

Ans: $\boxed{9.94 \times 10^{14} \text{ L}^{-1} \text{ s}^{-1}}$

Exercise 26-4

Gaseous hydrogen iodide decomposes according to the following reaction:

$$HI_{(g)} \rightarrow \tfrac{1}{2}H_{2(g)} + \tfrac{1}{2}I_{2(g)}$$

At 700°C the reaction is second order, with a rate constant of 1.6×10^{-3} L mol^{-1} s^{-1}. In a certain experiment at 700°C, after reacting for 2500 min the HI concentration had decreased to 4.5×10^{-4} mol L^{-1}. What was the initial concentration? 😐

Example 26-8

Butadiene dimerizes[5] according to $2C_4H_{6(g)} \rightarrow C_8H_{12(g)}$

The data at the right were obtained for this reaction at 500 K.

(a) Is this reaction first order or second order? What is the value of the rate constant?

(b) If the initial concentration of butadiene were 0.025 mol L^{-1}, how long would it take for the concentration to fall to half the initial value?

time (s)	$[C_4H_6]$ (mol L^{-1})
195	1.6×10^{-2}
604	1.5×10^{-2}
1246	1.3×10^{-2}
2180	1.1×10^{-2}
6210	0.68×10^{-2}

Solution

(a) To determine if the reaction is first or second order we note that if it is a *first* order reaction, then, according to Eqn [6], a plot of *ln c vs. time* should yield a straight line with a slope of $-2k$, where k is the rate constant. On the other hand, if the reaction is *second* order, then, as shown in Fig. 4, a plot of *1/c vs. time* should yield a straight line with a slope of $+2k$.

In order to make these plots we construct the table at the right from the original data:

The plot—shown below—of *ln c vs. time* clearly is not linear, so the reaction is not first order. The plot of *1/c vs. time*, however, *is* linear; therefore the reaction is second order. The straight line fit to the data has a slope of $+1.417 \times 10^{-2}$ L mol^{-1} s^{-1}, which is $2k$; therefore $k = 7.085 \times 10^{-3}$.

time (s)	c	ln c	1/c
195	0.016	−4.1352	62.50
604	0.015	−4.1997	66.67
1246	0.013	−4.3428	76.92
2180	0.011	−4.5099	90.91
6210	0.0068	−4.9908	147.05

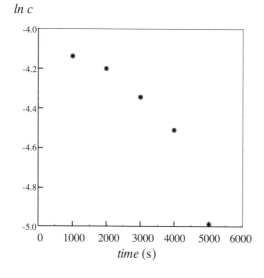

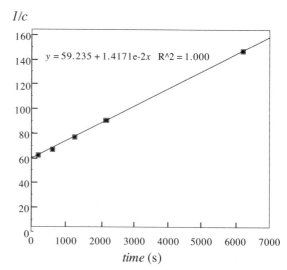

---

[5] A dimer is a molecule formed by the joining together of two identical molecules.

Ans: | Second order; $k = 7.085 \times 10^{-3}$ L mol^{-1} s^{-1} |

(b) If the reaction is second order, $\dfrac{1}{c} - \dfrac{1}{c_o} = 2kt$

therefore, putting in $c_{1/2} = 0.5c_o$:

$$\left(\frac{1}{0.5c_o} - \frac{1}{c_o} \right) = \left(\frac{2}{c_o} - \frac{1}{c_o} \right) = \frac{1}{c_o} = 2\,k\,t_{1/2}$$

and $t_{1/2} = \dfrac{1}{2\,kc_o} = \dfrac{1}{2(7.085 \times 10^{-3})(0.025)} = 2823$ s $= 47.05$ min

Ans: | 47 min |

| Exercise 26-5 |

The following gas phase reaction is second order in NO_2:

$$2NO_2 \rightarrow 2NO + O_2$$

At 600 K the rate constant for this reaction is 630 mL mol^{-1} s^{-1}. How long will it take for the decomposition of (a) 10% and (b) 90% of a sample of NO_2 initially at 0.526 atm pressure? ☺

26.7.2 REACTIONS THAT ARE FIRST ORDER IN TWO REACTANTS

Some reactions such as $aA + bB \rightarrow cC + dD$

have reaction rates that are first order in both A and B:

Thus, $r = -\dfrac{1}{a}\dfrac{d[A]}{dt} = -\dfrac{1}{b}\dfrac{d[B]}{dt} = k[A][B]$ . . . [14]

Consider the consumption of A:

For every a moles of A that react, b moles of B also react; therefore, for every *one* mole of A that reacts, b/a moles of B react. If, at time $= t$, x moles of A have reacted, then $\frac{b}{a}x$ moles of B also will have reacted; therefore, at time $= t$

$c_A = c_A^o - x$ . . . [15-a]   and $c_B = c_B^o - \dfrac{b}{a}x$ . . . [15-b]

where c_A^o and c_B^o are the initial concentrations of A and B, respectively, at time $t = 0$.

From Eqn [15-a], $x = c_A^o - c_A$

Substituting this into Eqn [15-b] gives

$$c_B = c_B^o - \frac{b}{a}[c_A^o - c_A] = c_B^o - \frac{b}{a}c_A^o + \frac{b}{a}c_A \qquad \qquad \dots [16]$$

and the rate equation [14] becomes

$$-\frac{1}{a}\frac{dc_A}{dt} = k c_A c_B = k c_A \left\{ c_B^o - \frac{b}{a}[c_A^o - c_A] \right\}$$

Rearranging:

$$-\frac{1}{a}\frac{dc_A}{c_A \left\{ c_B^o - \frac{b}{a}[c_A^o - c_A] \right\}} = k\,dt$$

Integrating:

$$-\frac{1}{a}\int_{c_A^o}^{c_A}\frac{dc_A}{c_A \left\{ c_B^o - \frac{b}{a}[c_A^o - c_A] \right\}} = k\int_0^t dt = kt \qquad \dots [17]$$

From tables of integrals,

$$\int \frac{dx}{x(A_o + B_o x)} = -\frac{1}{A_o} ln\left(\frac{A_o + B_o x}{x} \right)$$

where A_o and B_o are constants.

With $A_o = c_B^o - \frac{b}{a}c_A^o$, $B_o = \frac{b}{a}$, and $x = c_A$, the integral in Eqn [17] becomes

$$\int \frac{dc_A}{c_A \left\{ c_B^o - \frac{b}{a}[c_A^o - c_A] \right\}} = -\frac{1}{c_B^o - \frac{b}{a}c_A^o} ln\left(\frac{c_B^o - \frac{b}{a}c_A^o + \frac{b}{a}c_A}{c_A} \right)$$

and Eqn [17] becomes

$$-\frac{1}{a}\left(-\frac{1}{c_B^o - \frac{b}{a}c_A^o} \right)\left[ln\frac{c_B^o - \frac{b}{a}c_A^o + \frac{b}{a}c_A}{c_A} \right]_{c_A^o}^{c_A} = kt$$

i.e.,

$$\frac{1}{ac_B^o - bc_A^o}\left[ln\frac{c_B^o - \frac{b}{a}c_A^o + \frac{b}{a}c_A}{c_A} - ln\frac{c_B^o - \frac{b}{a}c_A^o + \frac{b}{a}c_A^o}{c_A^o} \right] = kt$$

$$\frac{1}{ac_B^o - bc_A^o}\left[ln\frac{c_B^o - \frac{b}{a}c_A^o + \frac{b}{a}c_A}{c_A} - ln\frac{c_B^o}{c_A^o} \right] = kt$$

$$\frac{1}{ac_B^o - bc_A^o}\left[ln\frac{c_A^o(c_B^o - \frac{b}{a}c_A^o + \frac{b}{a}c_A)}{c_B^o c_A} \right] = kt$$

$$ln\frac{c_A^o(c_B^o - \frac{b}{a}c_A^o + \frac{b}{a}c_A)}{c_B^o c_A} = (ac_B^o - bc_A^o)kt$$

$$\frac{c_A^o(c_B^o - \frac{b}{a}c_A^o + \frac{b}{a}c_A)}{c_B^o c_A} = exp\left[(ac_B^o - bc_A^o)kt \right]$$

$$\boxed{\frac{c_B^o - \frac{b}{a}c_A^o + \frac{b}{a}c_A}{c_A} = \frac{c_B^o}{c_A^o}\cdot exp\left[(ac_B^o - bc_A^o)kt \right]} \qquad \dots [18]$$

If we know the rate constant k, Eqn [18] can be used to solve for c_A at any time t.

Noting from Eqn [16] that $\qquad c_B^o - \frac{b}{a}c_A^o + \frac{b}{a}c_A = c_B$

and substituting this into Eqn [18] gives an alternative form of Eqn [18]:

$$\frac{c_B}{c_A} = \frac{c_B^o}{c_A^o} \cdot exp\left[(ac_B^o - bc_A^o)kt\right]$$

i.e.,
$$\boxed{\frac{c_B/c_B^o}{c_A/c_A^o} = exp\left[(ac_B^o - bc_A^o)kt\right]} \qquad \qquad \cdots [19]$$

Example 26-9

The gas phase reaction between hydrogen and iodine produces hydrogen iodide according to

$$H_{2(g)} + I_{2(g)} \rightarrow 2HI_{(g)}$$

This reaction has been found experimentally to be a second order reaction:

$$\frac{d[H_2]}{dt} = \frac{d[I_2]}{dt} = -k[H_2][I_2]$$

At 350°C the rate constant for this reaction is $k = 1.85 \times 10^{-3}$ L mol^{-1} s^{-1}. At this temperature:

(a) What will be the iodine concentration after one hour if the initial concentrations of H_2 and I_2 are 2.00×10^{-2} mol L^{-1} and 1.00×10^{-2} mol L^{-1}, respectively?

(b) What will be the iodine concentration after one hour if the initial concentrations of H_2 and I_2 are each 1.00×10^{-2} mol L^{-1}?

Solution

(a) Eqn [18] gives the following expression for the reaction

$$aA + bB \rightarrow cC + dD$$

that is first order in both A and B:

$$\frac{c_B^o - \frac{b}{a}c_A^o + \frac{b}{a}c_A}{c_A} = \frac{c_B^o}{c_A^o} \cdot exp\left[(ac_B^o - bc_A^o)kt\right] \qquad \qquad \cdots [a]$$

For the case under consideration, if we let I_2 be reactant "A" and H_2 be reactant "B," then $a = b = 1$, $c_A^o = 1.00 \times 10^{-2}$ mol L^{-1}, $c_B^o = 2.00 \times 10^{-2}$ mol L^{-1}, and Eqn [a] becomes

$$\frac{c_{H_2}^o - c_{I_2}^o + c_{I_2}}{c_{I_2}} = \frac{c_{H_2}^o}{c_{I_2}^o} \cdot exp\left[(c_{H_2}^o - c_{I_2}^o)(1.85 \times 10^{-3})(1 \times 3600)\right]$$

$$\frac{0.020 - 0.010 + c_{I_2}}{c_{I_2}} = \frac{0.020}{0.010} \cdot exp\left[(0.020 - 0.010)(1.85 \times 10^{-3})(1 \times 3600)\right]$$

i.e.,
$$\frac{0.010 + c_{I_2}}{c_{I_2}} = 2 \cdot exp[0.0666] = 2.1377$$

$$2.1377 \, c_{I_2} = c_{I_2} + 0.010$$

and
$$c_{I_2} = \frac{0.010}{1.1377} = 8.790 \times 10^{-3} \text{ mol L}^{-1}$$

Ans: $\boxed{8.79 \times 10^{-3} \text{ mol L}^{-1}}$

(b) Now $c_A^o = c_B^o$ and $a c_B^o = b c_A^o$. When these values are put into Eqn [a] we get

$$\frac{c_{I_2}}{c_{I_2}} = 1 \cdot exp[0]$$

i.e.,
$$1 = exp[0] = 1$$

which is of no help to us at all. In order to solve for c_{I_2} we have to recognize that since $c_{I_2}^o = c_{H_2}^o$,
then at any time t, $c_{I_2} = c_{H_2}$ and the reaction rate becomes just

$$\frac{dc_{I_2}}{dt} = k c_{I_2} c_{I_2} = k c_{I_2}^2$$

which is a simple second order reaction, for which

$$\frac{1}{c_{I_2}} - \frac{1}{c_{I_2}^o} = k t$$

and therefore
$$\frac{1}{c_{I_2}} = \frac{1}{c_{I_2}^o} + kt = \frac{1}{0.010} + (1.85 \times 10^{-3})(3600) = 106.66$$

giving
$$c_{I_2} = \frac{1}{106.66} = 9.3759 \times 10^{-3}$$

Ans: $\boxed{9.36 \times 10^{-3} \text{ mol L}^{-1}}$

26.8 ZEROTH ORDER REACTIONS

Occasionally the rate of a reaction appears to be independent of the reactant concentration.
Thus, if the reaction $A \rightarrow B$
is independent of the concentration of A, then

$$-\frac{d[A]}{dt} = k [A]^o = k \qquad\qquad \dots [20]$$

which is a *constant rate*. Such a reaction is a *zeroth order* reaction. The catalytic decomposition of phosphine (PH_3) on hot tungsten at high pressures is an example of a zeroth order reaction—the phosphine decomposes at a constant rate until it has almost entirely disappeared. Only hetero-geneous[6] reactions can have rate laws that are zero-order overall.

26.9 EXPERIMENTAL DETERMINATION OF *k*, *m*, and *n*

26.9.1 THE ISOLATION METHOD

In this method the experiment is carried out by ensuring that the initial concentrations of all the reactants except one are present in large excess. For example, for the reaction

$$aA + bB \rightarrow cC + dD$$

the experiment is arranged so that the initial concentration of B is much greater than the initial concentration of A:

$$[B]_o >> [A]_o$$

Under these conditions, as the reaction proceeds and the concentration of A steadily decreases, the concentration of B remains essentially unchanged, so that we may say

$$[B] \approx [B]_o = constant$$

at any time during the reaction. The rate equation then becomes

$$r = -\frac{d[A]}{dt} = k\,[A]^m\,[B]^n$$

$$\approx k\,[A]^m\,[B]_o^n = \left\{k\,[B]_o^n\right\}[A]^m$$

$$= k'\,[A]^m$$

Taking logarithms of each side:

$$ln\,r = ln\,k' + m\,ln\,[A]$$

Plotting *ln r vs. ln [A]* gives a straight line with a slope of *m*, permitting a determination of *m* (see Fig. 5). The experiment is then repeated, only this time with *A* in great excess, permitting determination of *n*. Knowing *n*, we then can determine *k* from the intercept of the first plot, which is

$$ln\,k' = ln\left(k\,[B]_o^n\right)$$

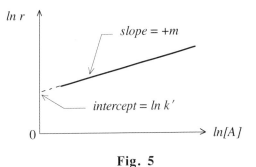

Fig. 5

---

[6] A heterogeneous reaction is a reaction involving more than one phase that does not involve the equilibrium of a species present in more than one phase. The reaction $CaCO_{3(s)} \rightarrow CaO_{(s)} + CO_{2(g)}$ is an example.

26.9.2 FROM INITIAL RATES

In this method, which often is used in conjunction with the isolation method, the "instantaneous" initial rate is measured at the beginning of the reaction for several different initial concentrations of reactant A (keeping the initial concentration of B constant) by measuring the change $\Delta[A]$ in the concentration of A that takes place over a small time interval Δt.

Thus, for the reaction

$$aA + bB \rightarrow cC + dD$$

$$\langle Initial\ Rate \rangle = r_o = -\frac{1}{a}\left(\frac{d[A]}{dt}\right)_o \approx -\frac{1}{a}\frac{\Delta[A]}{\Delta t}$$

$$= k[A]_o^m[B]_o^n \qquad \ldots [21]$$

$$= k'[A]_o^m \qquad \ldots [22]$$

Thus,

$$r_o = k'[A]_o^m$$

and

$$\ln r_o = \ln k' + m\ln[A]_o$$

Plotting $\ln r_o$ vs. $\ln[A]_o$ gives a straight line with a slope of m and an intercept of $\ln k'$, permitting an evaluation of m and k'. Then the experiment is repeated keeping $[A]_o$ fixed and changing $[B]_o$, which permits an evaluation of n. A comparison of Eqns [21] and [22] shows that

$$k[B]_o^n = k'$$

That is,

$$k = \frac{k'}{[B]_o^n} \qquad \ldots [23]$$

Thus, the rate constant k can be obtained from Eqn [23].

26.9.3 FROM INTEGRATED RATE LAWS

Sometimes it's difficult to measure initial rates because the reaction proceeds very quickly; for example, suppose the following reaction is very fast:

$$aA \rightarrow cC + dD$$

The rate law is

$$r = -\frac{1}{a}\frac{d[A]}{dt} = k[A]^n$$

that is,

$$r = -\frac{1}{a}\frac{dc}{dt} = kc^n \qquad \ldots [24]$$

Assuming $n \neq 1$,[7] Eqn [24] can be rearranged and integrated as follows:

---

[7] Examination of Eqn [25] shows why n cannot be equal to one.

$$-\frac{1}{a}\frac{dc}{dt} = k\,c^n$$

Rearranging:

$$-\frac{1}{a}\frac{dc}{c^n} = k\,dt$$

Integrating:

$$-\frac{1}{a}\int_{c_o}^{c}\frac{dc}{c^n} = k\int_{0}^{t}dt$$

$$-\frac{1}{a}\left[\frac{c^{-n+1}}{-n+1}\right]_{c_o}^{c} = k\,t$$

That is,

$$\frac{1}{a(n-1)}\left[c^{1-n} - c_o^{1-n}\right] = k\,t$$

$$\frac{c^{1-n}}{a(n-1)} = \frac{c_o^{1-n}}{a(n-1)} + k\,t \qquad \ldots [25]$$

A plot of $c^{1-n}/a(n-1)$ vs. t gives a straight line with a slope of k and an intercept of $c_o^{1-n}/a(n-1)$. Thus, the idea is to *guess* a value of n and plot $c^{1-n}/a(n-1)$ vs. t. Keep guessing different values of n until the plot yields a straight line. A straight line tells us that we have the correct value of n, and, from the slope of the straight line, we can get the rate constant, k.

26.10 REACTION MECHANISMS

Most reactions pass through several steps in progressing from reactants to products. For example, consider the decomposition of N_2O_5:

$$2N_2O_5 \to 4NO_2 + O_2 \qquad \ldots [26]$$

This is observed to be a first order reaction, with $\quad r = -\frac{1}{2}\frac{d[N_2O_5]}{dt} = k\,[N_2O_5]$

This seemingly simple reaction actually consists of the sequence of steps shown in Table 1. Some of the N_2O_5 molecules borrow energy from the others to form the "activated" species $N_2O_5^*$, which has more energy than the average.

This unstable species then decomposes to form NO_2 and NO_3. Each step in the sequence is called an **elementary step**. (Note that the first elementary step occurs twice.) The overall sequence of elementary processes is called the **reaction mechanism**.

Table 1   *Reaction Mechanism for Reaction [26]*

$$2\,(\,N_2O_5^* \to NO_2 + NO_3\,)$$
$$NO_2 + NO_3 \to NO + NO_2 + O_2$$
$$NO + NO_3 \to 2NO_2$$

$$2\,N_2O_5^* \to 4NO_2 + O_2$$

There are three types of elementary process:

Unimolecular—Only one molecule participates:

Examples:

$$CH_2\!\!\underset{\displaystyle CH_2}{\overset{\displaystyle CH_2^*}{\diagdown}}\!\!\!CH_2 \rightarrow CH_3CH = CH_2, \qquad O_3^* \rightarrow O_2 + O$$

Bimolecular—Two molelcules participate:

Examples: $NO + O_3 \rightarrow NO_2 + O_2,$ $Cl + CH_4 \rightarrow CH_3 + HCl$

Termolecular—Three molecules participate:

Examples: $O + O_2 + N_2 \rightarrow O_3 + N_2,$ $O + N_2 + NO \rightarrow NO_2 + N_2$

(N_2 absorbs energy to prevent NO_2 from dissociating)

Elementary processes with molecularities greater than three are not known, since collisions in which more than three molecules happen to come together at the same time are very rare. Note that for an elementary process, the *order* and the *molecularity* are both the same, because the rates at which collisions occur are directly proportional to the concentrations of the reacting molecules.

Thus, if $A + B + C \rightarrow D + E$

is an elementary process, this reaction will be a third order reaction, with

$$r = -\frac{d[A]}{dt} = k[A][B][C]$$

For example, the elementary reaction $NO + O_3 \rightarrow NO_2 + O_2$

is second order with $-\frac{d[NO]}{dt} = k[NO][O_3]$

Although for an elementary process the molecularity is the same as the order, the converse is not generally true. Thus, a second order overall reaction doesn't necessarily have a bimolecular mechanism. For example, the reaction

$$2N_2O_5 \rightarrow 4NO_2 + O_2$$

is a *first order* reaction; but, as we saw above, the mechanism consists of both unimolecular and bimolecular elementary steps.

26.11 THE RATE-DETERMINING STEP

Consider the gas phase reaction $\qquad 2NO_2 + F_2 \rightarrow 2NO_2F$

It is observed experimentally that this reaction follows a second order rate law:

$$-\frac{1}{2}\frac{d[NO_2]}{dt} = k[NO_2][F_2]$$

The mechanism is believed to be

$$NO_2 + F_2 \xrightarrow{k_1} NO_2F + F \qquad \text{(slow)} \qquad \ldots \text{Step 1}$$

$$F + NO_2 \xrightarrow{k_2} NO_2F \qquad \text{(fast)} \qquad \ldots \text{Step 2}$$

Overall reaction: $\qquad 2NO_2 + F_2 \rightarrow 2NO_2F$

The rate of the overall reaction is controlled by the *slow* step. The second step, which is fast, must wait for the production of the F from the first step before it can proceed. The overall reaction cannot proceed faster than the slowest step, which is called the **rate-determining step** (*RDS*).

Thus, the overall process occurs at the same rate as that of the rate-determining step:

$$r_{overall} = -\frac{1}{2}\frac{d[NO_2]}{dt} = k_1[NO_2][F_2]$$

and ***not*** $\qquad\qquad r_{overall} = k[NO_2]^2[F_2]$

as might be thought from the stoichiometry of the overall reaction.

26.12 REACTION RATES AND EQUILIBRIUM

The overall reaction discussed in the previous section proceeds and eventually slows down until an equilibrium is established. At equilibrium, the rates of the forward and reverse processes are *equal*. If we designate the rate constant for a *reverse* process as k', then we can write out rate expressions for both the forward and reverse processes at equilibrium as follows, for each step equating the forward and reverse rates:

For Step 1: $\qquad k_1[NO_2][F_2] = k_1'[NO_2F][F] \qquad\qquad \ldots [27]$

For Step 2: $\qquad k_2[F][NO_2] = k_2'[NO_2F] \qquad\qquad \ldots [28]$

All the concentrations now are the *equilibrium* values. Multiplying the two left hand sides and the two right hand sides of Eqns [27] and [28] and equating the two products of the multiplication gives

$$k_1 k_2 [NO_2]^2 [F_2][F] = k_1' k_2' [NO_2F]^2 [F]$$

Rearranging:
$$\frac{k_1 k_2}{k_1' k_2'} = \left(\frac{[NO_2F]^2}{[NO_2]^2 [F_2]} \right)_{eqm}$$

But $\left(\dfrac{[NO_2F]^2}{[NO_2]^2 [F_2]} \right)_{eqm}$ is just K, the equilibrium constant for the overall reaction at equilibrium:

$$2NO_2 + F_2 \rightleftharpoons 2NO_2F \qquad K = \left(\frac{[NO_2F]^2}{[NO_2]^2 [F_2]} \right)_{eqm} = \frac{k_1 k_2}{k_1' k_2'} \qquad \ldots [29]$$

Thus, the equilibrium constant for an overall chemical reaction is a function of the various rate constants for the individual elementary steps involved in the mechanism of the reaction. If the product of the forward rate constants ($k_1 k_2$) is much greater than that of the reverse rate constants ($k_1' k_2'$), then K is big, and products are favored over reactants (i.e., we will have a high yield of products at equilibrium).

26.13 REACTION INTERMEDIATES

If a reaction mechanism consists of several elementary steps, in general it is found that one of the steps—the rate-determining step—is *much* slower than any of the other steps, and therefore controls the rate of the overall reaction. However, sometimes the rates of all the steps are of *comparable* values. How do we find the reaction rate for this kind of reaction?

Consider the earlier-discussed decomposition of N_2O_5, which is a first order reaction in N_2O_5:

$$2N_2O_5 \rightarrow 4NO_2 + O_2 \qquad r = -\frac{1}{2}\frac{d[N_2O_5]}{dt} = k[N_2O_5] \qquad \ldots [30]$$

We saw (see Table 1) that the mechanism for this reaction is:

$$2(N_2O_5 \underset{k_1'}{\overset{k_1}{\rightleftharpoons}} NO_2 + \underline{\underline{NO_3}}) \qquad \ldots \text{Step 1}$$

$$\underline{\underline{NO_3}} + NO_2 \overset{k_2}{\rightarrow} \underline{NO} + NO_2 + O_2 \qquad \ldots \text{Step 2}$$

$$\underline{\underline{NO_3}} + \underline{NO} \overset{k_3}{\rightarrow} 2NO_2 \qquad \ldots \text{Step 3}$$

The species $\underline{NO}$ and $\underline{\underline{NO_3}}$ are called **reaction intermediates**. Intermediates are involved in the mechanism of a reaction but don't show up in the overall reaction. These intermediates are used up as fast as they are produced, so they have a *small*, but *constant*, steady state concentration.

Thus:

$$\langle Rate\ of\ \textbf{production}\ of\ NO\ in\ Step\ 2 \rangle = \langle Rate\ of\ \textbf{consumption}\ of\ NO\ in\ Step\ 3 \rangle$$

$$k_2 [NO_3][NO_2] = k_3 [NO_3][NO]$$

Rearranging:
$$[NO] = \left(\frac{k_2}{k_3}\right)[NO_2] \qquad (= \text{steady state concentration of NO}) \quad \ldots [31]$$

Similarly, since the species NO_3 doesn't appear in the overall reaction, it follows that

$$\langle \textit{Net rate of } \textbf{production } \textit{of } NO_3 \rangle = \langle \textit{Net rate of } \textbf{consumption } \textit{of } NO_3 \rangle$$

i.e.,
$$k_1[N_2O_5] = k_1'[NO_2][NO_3] + k_2[NO_3][NO_2] + k_3[NO_3][NO]$$
$$= [NO_3]\{k_1'[NO_2] + k_2[NO_2] + k_3[NO]\} \qquad \ldots [32]$$

Substituting $[NO] = \left(\frac{k_2}{k_3}\right)[NO_2]$ from Eqn [31] into Eqn [32] gives:

$$k_1[N_2O_5] = [NO_3]\left\{k_1'[NO_2] + k_2[NO_2] + k_3\left(\frac{k_2}{k_3}\right)[NO_2]\right\}$$
$$= [NO_3][NO_2](k_1' + 2k_2) \qquad \ldots [33]$$

From which
$$[NO_3] = \frac{k_1[N_2O_5]}{(k_1' + 2k_2)[NO_2]} \qquad \ldots [34]$$

From its stoichiometry, the rate of the overall reaction—reaction [30]—is

$$r = -\frac{1}{2}\frac{d[N_2O_5]}{dt} = +\frac{d[O_2]}{dt}$$

But, from Step 2,
$$\frac{d[O_2]}{dt} = k_2[NO_3][NO_2]$$

Therefore,
$$r = k_2[NO_2][NO_3] \qquad \ldots [35]$$

Putting Eqn [34] into Eqn [35]:
$$r = k_2[NO_2]\left\{\frac{k_1[N_2O_5]}{(k_1' + 2k_2)[NO_2]}\right\}$$
$$= \frac{k_1 k_2[N_2O_5]}{(k_1' + 2k_2)} \qquad \ldots [36]$$

Therefore, the rate of the overall reaction is

$$\boxed{-\frac{1}{2}\frac{d[N_2O_5]}{dt} = k[N_2O_5]} \quad \text{where} \quad \boxed{k = \frac{k_1 k_2}{k_1' + 2k_2}} \qquad \begin{array}{l}\text{which is a first order reaction, exactly}\\ \text{as observed experimentally!}\end{array}$$

26.14 EFFECT OF TEMPERATURE

Many chemical reactions occur with the reactants forming what is called an **activated complex**. This *activated complex* then decomposes into the products. For example, the overall reaction

$$A + B \rightarrow C + D$$

may actually proceed by $\qquad A + B \rightarrow A\text{–}B^* \rightarrow C + D$

where $A\text{–}B^*$ is an intermediate *activated complex*. In this kind of reaction, the rate of formation of the activated complex is often the rate-determining step. Thus, the reactants must attain a certain amount of extra energy over and above the average energy in order to form the activated complex.

This extra amount of energy that is required to form the activated complex is called the **activation energy**, ΔG_a^*. ΔG_a^* sometimes is called the *Arrhenius activation energy*. There are many different intermediate states that the system might, in principle, pass through as it proceeds from its initial state $(A + B)$ to its final state $(C + D)$. The path that actually is followed will be the *energetically least demanding path*, which is called the **reaction coordinate**.

Fig. 6 shows a plot of the Gibbs free energy of the system *vs.* the *reaction coordinate* for the above reaction. Even though the free energy of the products is lower than that of the reactants, before they are able to react to form products the reactants must first attain excess free energy ΔG_a^* over and above the average free energy of the reactants.

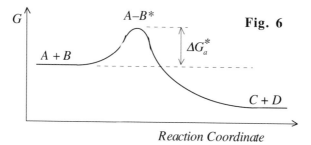

Reaction Coordinate

The rate constant can be shown to take the form

$$k = A \cdot exp\left[-\frac{\Delta G_a^*}{RT} \right] \qquad\qquad \cdots [37]$$

where A is a constant[8]—called the **pre-exponential factor**—and is essentially *independent of temperature*. It can be seen that k increases as the temperature increases because the term $exp\left(-\Delta G_a^*/RT\right)$ increases[9] with increasing temperature. Thus, the reaction goes *faster* at higher temperatures. Basically this is because at higher temperatures, owing to the *Maxwell-Boltzmann*

---

[8] Not to be confused with the reactant A.
[9] Because $-\Delta G/RT$ becomes *less* negative as T increases.

Distribution of kinetic energy among the molecules, a larger fraction of the molecules will possess enough kinetic energy to overcome the activation energy "barrier", and therefore react to form the activated complex.

Taking logarithms of both sides of Eqn [37]:

$$ln\ k = ln\ A - \frac{\Delta G_a^*}{RT} \qquad \ldots [38]$$

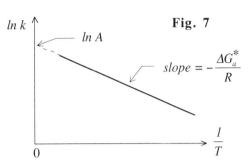

Eqn [38] indicates that a plot of $ln\ k$ vs. $1/T$ should give a straight line with a slope of $-\Delta G_a^*/R$ and an intercept of $ln\ A$ (Fig. 7). This type of plot is used experimentally to determine ΔG_a^* and A.

For two different temperatures T_1 and T_2:

$$ln\,k_{T_2} = ln\ A - \frac{\Delta G_a^*}{RT_2} \qquad \ldots [39]$$

and

$$ln\,k_{T_1} = ln\ A - \frac{\Delta G_a^*}{RT_1} \qquad \ldots [40]$$

Subtracting Eqn [40] from Eqn [39]:

$$ln\,k_{T_2} - ln\,k_{T_1} = -\frac{\Delta G_a^*}{RT_2} + \frac{\Delta G_a^*}{RT_1} = -\frac{\Delta G_a^*}{R}\left[\frac{1}{T_2} - \frac{1}{T_1}\right]$$

That is,

$$ln\left(\frac{k_{T_2}}{k_{T_1}}\right) = -\frac{\Delta G_a^*}{R}\left[\frac{1}{T_2} - \frac{1}{T_1}\right] \qquad \ldots [41]$$

Eqn [41] is very similar to the *van't Hoff equation* for the variation of the equilibrium constant with temperature. This is not surprising, since we showed above that the equilibrium constant can be expressed as a ratio of rate constants.

Example 26-10

When we were children, our parents always nagged us to put any uneaten food back into the refrigerator during the summer because it would quickly go bad if left out on the kitchen counter. Of course we ignored their advice and a lot of good food had to be thrown into the garbage. If your lunch spoils in five days when left out during the winter, when the average indoor temperature is 20°C, in how many days would it spoil if left out during the summer if the average indoor temperature were 30°C? Experiments have shown that food spoils 40 times faster at 25°C than it does when refrigerated at 4°C.

Solution

First we have to determine the activation energy for the "rotting reaction"; then we can calculate how much faster the reaction proceeds at 30°C than it does at 20°C.

We are told that the reaction is 40 times faster at $T_2 = 25 + 273.15 = 298.15$ K than it is at $T_1 = 4 + 273.05 = 277.15$ K. Since the rate of a reaction is proportional to its rate constant, this means that the ratio of the rate constants for the reaction at these two temperatures also equals 40.

That is,
$$k_2 = 40k_1$$

Thus:
$$\ln\frac{k_2}{k_1} = -\frac{\Delta G_a^*}{R}\left[\frac{1}{T_2} - \frac{1}{T_1}\right]$$

Putting in values:
$$\ln\frac{40k_1}{k_1} = \ln(40) = -\frac{\Delta G_a^*}{8.314}\left[\frac{1}{298.15} - \frac{1}{277.15}\right]$$

$$3.6888 = -\frac{\Delta G_a^*}{8.314}[-2.54138 \times 10^{-4}]$$

$$\Delta G_a^* = \frac{-(8.314)(3.6888)}{(-2.54138 \times 10^{-4})}$$

$$= 120\,677 \text{ J mol}^{-1}$$

$$= 120.68 \text{ kJ mol}^{-1}$$

Now that we know ΔG_a^* we can calculate the ratio of the rate constant—and therefore the rate of the reaction—at 30°C to that at 20°C.

Thus, $T_2 = 30 + 273.15 = 303.15$ K   and $T_1 = 20 + 273.15 = 293.15$ K

$$\ln\frac{k_{30}}{k_{20}} = -\frac{120\,677}{8.314}\left[\frac{1}{303.15} - \frac{1}{293.15}\right] = 1.6333$$

$$\frac{k_{30}}{k_{20}} = \exp[1.6333] = 5.12$$

Therefore the reaction is 5.12 times faster at 30°C than it is at 20°C. This means that food that goes bad in 5 days at 20°C will spoil in about only one day at 30°C.

Ans: ~one day

Exercise 26-6

A rule of thumb often used is that the rate of a chemical reaction doubles for every ten degree increase in temperature. What is the activation energy of such a reaction? ☺

Example 26-11

When your body temperature increases, many normal physiological processes become faster. A relationship between body temperature and the subjective perception of time was first demonstrated during an influenza epidemic in the 1930s, when a sick woman having a high fever (39°C) was asked to estimate the duration of a minute by counting to 60 at a rate of one number per second.[10] It turned out that the woman's estimates of a minute were only 38 seconds long; her high body temperature had caused her biological clock to "tick" more rapidly than usual. Noting that the normal body temperature of a human is 37°C, determine the activation energy of the processes that are involved in the biological clock.

Solution

We note that the biological (mental) processes involved in counting—which, in this case normally should require 60 seconds at a body temperature of 37°C (= 310.15 K)—only required 38 seconds at a body temperature of 39°C (= 312.15 K). If the rate of reaction of the process at 37°C is r_1, then the rate r_2 at 39°C is 60/38 = 1.579 times faster than at 37°C.

For an activation-controlled process,

$$ln\left(\frac{r_2}{r_1}\right) = -\frac{\Delta G_a^*}{R}\left[\frac{1}{T_2} - \frac{1}{T_1}\right]$$

Putting in values:

$$ln(1.579) = -\frac{\Delta G_a^*}{(8.314)}\left[\frac{1}{312.15} - \frac{1}{310.15}\right]$$

$$0.4568 = -\frac{\Delta G_a^*}{8.314}[-2.0658 \times 10^{-5}]$$

$$\Delta G_a^* = \frac{-(8.314)(0.4568)}{(-2.0658 \times 10^{-5})}$$

$$= 183\ 843\ \text{J mol}^{-1}$$

$$= 183.8\ \text{kJ mol}^{-1}$$

Activation energies for *chemical* processes in which chemical bonds are broken typically have values of one to several hundred kilojoules per mole, whereas activation energies for *physical* processes such as adsorption or diffusion typically are in the order of tens of kilojoules per mole. It therefore would appear that the processes involved in our biological clocks and our mental functioning are chemical (or electrochemical) processes. Chemistry and electrochemistry rule!!

Ans: $\boxed{184\ \text{kJ mol}^{-1}}$

---

[10] This incident is related by Stanley Coren, in his fascinating book, *Sleep Thieves: An Eye-opening Exploration into the Science and Mysteries of Sleep*, published by The Free Press, New York, 1996.

Exercise 26-7

When you are cold, time seems to pass faster. Estimate how many seconds your subjective estimate of the duration of one minute might be if your body temperature were lowered to 34.5°C.

☺

Exercise 26-8

What is the activation energy for a reaction that triples its rate when the temperature is increased from 600 K to 610 K? ☺

Example 26-12

A certain first order reaction has an activation energy of 104.6 kJ mol^{-1} and a pre-exponential factor of 5×10^{13} s^{-1}. At what temperature will the half-life be (a) 1.0 min (b) 30 days?

Solution

We are given that
$$k = A \cdot exp\left[-\frac{\Delta G_a^*}{RT}\right]$$

$$= 5 \times 10^{13} exp\left[-\frac{104\ 600}{8.314\ T}\right]$$

$$= 5 \times 10^{13} exp\left[-\frac{12\ 581}{T}\right]$$

Since the reaction is first order:
$$\frac{dc}{dt} = -kc = -5 \times 10^{13} exp\left[-\frac{12\ 581}{T}\right] \cdot c$$

Rearranging and integrating:
$$-ln\left(\frac{c}{c_o}\right) = -kt = -5 \times 10^{13} exp\left[-\frac{12\ 581}{T}\right] \cdot t$$

(a) If $t_{1/2} = 1.0$ min $= 60$ s:
$$ln\,(0.5) = -5 \times 10^{13} exp\left[-\frac{12\ 581}{T}\right] \cdot 60$$

$$exp\left[-\frac{12\ 581}{T}\right] = \frac{ln(0.5)}{(-5 \times 10^{13})(60)} = 2.3105 \times 10^{-16}$$

Taking logarithms:
$$-\frac{12\ 581}{T} = -36.00$$

From which $$T = \frac{-12\,581}{-36.00} = 349.47 \text{ K} = 76.3°\text{C}$$

Ans: $\boxed{76.3°\text{C}}$

(b) If $t_{1/2} = 30$ days: $$ln\,(0.5) = -5 \times 10^{13}\,exp\left[-\frac{12\,581}{T}\right] \cdot (30 \times 24 \times 60 \times 60)$$

$$exp\left[-\frac{12\,581}{T}\right] = \frac{ln(0.5)}{(-5 \times 10^{13})(2592)} = 5.348 \times 10^{-21}$$

$$-\frac{12\,581}{T} = -46.6775$$

$$T = 269.53 \text{ K} = -3.62°\text{C}$$

Ans: $\boxed{-3.62°\text{C}}$

Exercise 26-9

A rate constant is 1.78×10^{-4} L mol^{-1} s^{-1} at 19°C and 1.38×10^{-3} L mol^{-1} s^{-1} at 37°C. Evaluate the Arrhenius parameters A and ΔG_a^* for the reaction. ☺

Exercise 26-10

At 455°C the decomposition of nitrous oxide (N_2O)—"laughing gas"—into N_2 and $\frac{1}{2}O_2$ follows first-order kinetics with an activation energy of 251 kJ mol^{-1} and a reactant half-life of 6.5 Ms. What is the half-life of the decomposition at 550°C? ☺

26.15 RATES OF IONIC REACTIONS

It is commonly observed that the rates of reaction between ions in solution are dependent on the ionic strength of the solution. This dependency is known as the **primary kinetic salt effect**, and first was observed by Brønsted and Bjerrum. Consider the following elementary reaction between two ions in solution:

$$A^{z_A} + B^{z_B} \rightarrow Products \qquad \ldots [42]$$

where z_A and z_B are the valences (with signs) of the two ions. Since this is an elementary reaction it will be first order in each reactant, and the reaction rate will be

$$r = -\frac{d[A^{z_A}]}{dt} = -\frac{d[B^{z_B}]}{dt} = k\,[A^{z_A}][B^{z_B}] \qquad \ldots [43]$$

Earlier in this chapter we introduced the idea that the elementary steps in the mechanisms of many chemical reactions often involve the formation of an unstable activated species. It usually is assumed that the formation from the reactants of this activated complex is very fast; so fast, in fact, that it may be considered to be in *equilibrium* with the reactants. Thus, in the case of ionic reaction [42] we can write

$$A^{z_A} + B^{z_B} \rightleftharpoons (A \cdot B)^{z_A + z_B} \qquad \ldots [44]$$

where $(A \cdot B)^{z_A + z_B}$ is the activated complex. The equilibrium constant K^* for the formation of the activated complex is given by

$$K^* = \frac{a_{AB}}{a_A \cdot a_B} = \frac{\gamma^* [(A \cdot B)^{z_A + z_B}]}{(\gamma_A [A^{z_A}])(\gamma_B [B^{z_B}])} \qquad \ldots [45]$$

where γ^* is the activity coefficient of the activated complex. Although we don't know very much about the activated complex, we do know that it has a net charge of $(z_A + z_B)$, which could be positive, negative, or zero, depending on the ions A^{z_A} and B^{z_B}.

Although the activated complex is very quickly formed, its decomposition to form product is not so fast; in fact, it is so slow that usually it is the rate determining step that controls the rate of the overall reaction. Thus, we can express the overall reaction sequence as

$$A^{z_A} + B^{z_B} \rightleftharpoons (A \cdot B)^{z_A + z_B} \xrightarrow{\ k^*\ } Products \qquad \ldots [46]$$

where k^* is the rate constant for the decomposition of the activated complex to form products.

Thus, the rate of the overall reaction is

$$-\frac{d[A^{z_A}]}{dt} = -\frac{d[B^{z_B}]}{dt} = k^* [(A \cdot B)^{z_A + z_B}] \qquad \ldots [47]$$

From Eqn [45],

$$[(A \cdot B)^{z_A + z_B}] = \frac{K^* \gamma_A \gamma_B [A^{z_A}][B^{z_B}]}{\gamma^*} \qquad \ldots [48]$$

Substituting Eqn [48] into Eqn [47] gives

$$-\frac{d[A^{z_A}]}{dt} = -\frac{d[B^{z_B}]}{dt} = \frac{k^* K^* \gamma_A \gamma_B [A^{z_A}][B^{z_B}]}{\gamma^*} = k_{obs}[A^{z_A}][B^{z_B}] \quad \ldots [49]$$

where k_{obs} is the experimentally observed rate constant for the reaction, defined from Eqn [49] as

$$k_{obs} = \frac{k^* K^* \gamma_A \gamma_B}{\gamma^*} \qquad \ldots [50]$$

Eqn [50] indicates that the experimentally observed rate "constant" k_{obs} will not, in fact, be constant; but, instead, will depend on the concentrations of the species in the solution, since we know that the activity coefficients of ionic species are functions of concentration.[11]

---

[11] We know this from the Debye-Hückel equation and its various modifications, such as the Guggenheim equation and the Davies equation.

We also know that as the ionic strength of the solution becomes lower and lower, all the activity coefficients approach a value of unity. Thus, at infinite dilution k_{obs} becomes k^o, where

$$k^o = \frac{k * K * (1)(1)}{(1)} = k * K * \qquad \ldots [51]$$

Therefore, k^o, which is the value of the observed rate constant when the ionic strength of the solution approaches $I = 0$, may be considered to be the inherently "true" rate constant for the reaction in the absence of any activity effects.

Putting Eqn [51] into Eqn [50] gives $\qquad k_{obs} = \frac{k^o \gamma_A \gamma_B}{\gamma *} \qquad \ldots [52]$

Providing the ionic strength of the solution is not greater than about 0.001, the Debye-Hückel equation (Eqn [53]) can be used to evaluate each of the activity coefficients in Eqn [52]:

$$log\,\gamma_i = - Az_i^2 \sqrt{I} \qquad \ldots [53]^{12}$$

At higher ionic strengths—up to say about 0.05 molar—the Guggenheim equation (Eqn [54]) may be used instead of the Debye-Hückel equation:

$$log\,\gamma_i = - \frac{Az_i^2 \sqrt{I}}{1 + \sqrt{I}} \qquad \ldots [54]$$

Taking *base-10* logarithms of both sides of Eqn [52]:

$$log\,k_{obs} = log\,k^o + log\,\gamma_A + log\,\gamma_B - log\,\gamma * \qquad \ldots [55]$$

Noting that the valences of A^{z_A}, B^{z_B}, and $(A \cdot B)^{z_A + z_B}$ are z_A, z_B, and $(z_A + z_B)$, respectively, substitution of the γ-values given by Eqn [53] into Eqn [55] gives:

$$log\,k_{obs} = log\,k^o - Az_A^2 \sqrt{I} - Az_B^2 \sqrt{I} + A(z_A + z_B)^2 \sqrt{I}$$

$$= log\,k^o - A\left[z_A^2 + z_B^2 - (z_A + z_B)^2\right]\sqrt{I}$$

$$= log\,k^o - A\left[z_A^2 + z_B^2 - z_A^2 - 2z_A z_B - z_B^2\right]\sqrt{I}$$

$$= log\,k^o + 2z_A z_B A\sqrt{I}$$

Therefore, $\qquad \boxed{log\,k_{obs} = log\,k^o + 2z_A z_B A\sqrt{I}} \qquad \ldots [56]$

Eqn [56] is known as the *Brønsted–Bjerrum equation*.[13] This equation also can be expressed in exponential form:

---

[12] Careful!! This "A" is the constant in the Debye-Hückel equation, not the reactant A. The value of A at 25°C for aqueous solutions is 0.5108.

[13] J.N. Brønsted, Z. *Phys. Chem.*, **102**, 169 (1922); **115**, 337 (1925); N. Bjerrum, Z. *Phys. Chem.*, **108**, 82 (1924).

Thus, $$\log k_{obs} - \log k^o = 2z_A z_B A\sqrt{I}$$

i.e., $$\log\left(k_{obs}/k^o\right) = 2z_A z_B A\sqrt{I}$$

Therefore, $$\frac{k_{obs}}{k^o} = 10^{2z_A z_B A\sqrt{I}}$$

and $$\boxed{k_{obs} = k^o \cdot 10^{2z_A z_B A\sqrt{I}}}$$ ... [57]

Eqn [57] yields some interesting information:

(1) If the two reacting ions are oppositely charged, then $z_A z_B < 0$, and increasing the ionic strength of the solution lowers the observed rate constant, and therefore *lowers* the rate of the reaction.

(2) Conversely, if the reacting ions have the same sign (e.g., two cations or two anions), then $z_A z_B > 0$, and increasing the ionic strength of the solution will *increase* the rate of the reaction.

(3) Finally, if one of the reactants is an uncharged species ($z = 0$), then $z_A z_B = 0$, and the rate constant should be independent of the ionic strength of the solution.

Example 26-13

The oxidation in aqueous Na_2SO_4 solutions at 25°C of tris(1,10-phenanthroline)Fe(II) by the periodate ion was studied by Cyfert, Latko, and Wawrzeczyk.[14] The measured rate constants for the reaction were found to be 0.224 and 0.252 (L/mol)$^{1/2}$ s^{-1} in 0.0050 M and 0.0125 M Na_2SO_4 solution, respectively. (These salt concentrations were both much greater than the reactant concentrations.) What is the charge of the activated complex in the rate determining step and what is the rate constant when the activity coefficients are all unity?

Solution

Since the sodium sulfate concentrations are much greater than the reactant concentrations, the ionic strengths of the solutions will be controlled by the Na_2SO_4 concentrations. Assuming complete dissociation of the Na_2SO_4 at these low sodium sulfate concentrations, the ionic strength of the solution is calculated from

$$I = \frac{1}{2}\sum c_i z_i^2 = \frac{1}{2}\left\{[Na^+](+1)^2 + [SO_4^{2-}](-2)^2\right\} = \frac{1}{2}\left\{(2c^o)\times 1 + (c^o)\times 4\right\} = 3c^o$$

where c^o is the stoichiometric concentration of the Na_2SO_4 in the solution.

These ionic strengths, which are shown in the accompanying table, are somewhat too concentrated to use the Debye-Hückel equation; therefore we will use the Guggenheim equation.

---

[14] *International Journal of Chemical Kinetics*, **28**, 103 (1996).

$[Na_2SO_4]$ mol L^{-1}	k_{obs} (L/mol)$^{1/2}$ s^{-1}	$I = 3[Na_2SO_4]$	$\sqrt{I}$	$\dfrac{\sqrt{I}}{1+\sqrt{I}}$
0.0050	0.224	0.0150	0.12247	0.10911
0.0125	0.252	0.0375	0.19365	0.16223

When the Guggenheim equation at 25°C is used instead of the Debye-Hückel equation, Eqn [55] becomes

$$log\,k_{obs} = log\,k^o + \frac{2z_A z_B A\sqrt{I}}{1+\sqrt{I}}$$

$$= log\,k^o + 2z_A z_B(0.5108)\frac{\sqrt{I}}{1+\sqrt{I}}$$

$$= log\,k^o + 1.0216\,z_A z_B\frac{\sqrt{I}}{1+\sqrt{I}} \qquad \ldots [a]$$

Inserting numerical values into Eqn [a] for each of the two sets of data gives

$$log\,(0.252) = log\,k^o + 1.0216\,z_A z_B(0.16223) \qquad \ldots [b]$$

and $\qquad log\,(0.224) = log\,k^o + 1.0216\,z_A z_B(0.10911) \qquad \ldots [c]$

Subtracting Eqn [c] from Eqn [b]:

$$log\frac{0.252}{0.224} = 1.0216 \times (0.16223 - 0.10911)\,z_A z_B$$

$$z_A z_B = \frac{0.051153}{1.0216 \times 0.05312} = 0.9426 \approx 1$$

Therefore $z_A z_B = 1$, which means that each reactant that forms the activated complex has the same valence of either $+1$ or -1.

Ans: $\boxed{z_A z_B = 1}$

The value of the rate constant when the activity coefficients are all unity is just k^o, which can be calculated using either Eqn [b] or Eqn [c] with $z_A z_B = 1$.

Thus, using Eqn [b]:

$$log\,k^o = log\,(0.252) - 1.0216 \times 1 \times 0.16223 = -0.76433$$

from which $\qquad k^o = 10^{-0.76433} = 0.1721$ (L/mol)$^{1/2}$ s^{-1}

Alternatively, using Eqn [c]:

$$log\,k^o = log\,(0.224) - 1.0216 \times 1 \times 0.10911 = -0.76122$$

giving $\qquad\qquad\qquad k^o = 10^{-0.76122} = 0.1733 \;(\text{L/mol})^{1/2}\;\text{s}^{-1}$

Taking the average value gives $\qquad k^o = 0.1727 \;(\text{L/mol})^{1/2}\;\text{s}^{-1}$

Ans: $\quad\boxed{k^o = 0.173 \;(\text{L/mol})^{1/2}\;\text{s}^{-1}}$

KEY POINTS FOR CHAPTER TWENTY-SIX

1. **Factors affecting reaction rate**:
    1. *nature of reactants*
    2. *temperature*—usually faster at higher temperatures
    3. *concentration*—usually faster at higher reactant concentration
    4. *surface area*—important for heterogeneous reactions such as combustions, dissolutions
    5. *agitation*—especially important for heterogeneous reactions: stirring speeds up the supply of reactants to the interface, where the reaction takes place

2. **Chemical reactions** are **dynamic processes**, in which the *net* rate is given by

$$\left(\begin{array}{c} Net \\ Reaction \\ Rate \end{array}\right) = \langle Forward\;\;Rate\rangle - \langle Reverse\;Rate\rangle$$

At equilibrium, the net rate equals zero; i.e., we have a *dynamic equilibrium* for which

$$\langle Forward\;\;Rate\rangle = \langle Reverse\;Rate\rangle$$

The reverse rate generally starts to become significant as equilibrium is approached.

For the reaction $\qquad\qquad aA + bB \rightarrow cC + dD$

the rate r in mol L^{-1} s^{-1} is given by

$$r = -\frac{1}{a}\frac{d[A]}{dt} = -\frac{1}{b}\frac{d[B]}{dt} = +\frac{1}{c}\frac{d[C]}{dt} = +\frac{1}{d}\frac{d[D]}{dt}$$

3. **Differential rate laws**:

It is usually possible to express the rate as a product of the reactant concentrations, each raised to some power.

Thus, for the reaction $3A + 2B \rightarrow C + D$

$$r = -\frac{1}{3}\frac{d[A]}{dt} = -\frac{1}{3}\frac{d[B]}{dt} = +\frac{d[C]}{dt} = +\frac{d[D]}{dt} = k[A]^m[B]^n$$

where m and n are generally integers or half-integers.

- m is the *order* of the reaction *with respect to A*
- n is the *order* of the reaction *with respect to B*
- $m + n$ = the *overall order* of the reaction
- m and n must be determined *experimentally*; they are sometimes—but not always—equal to the stoichiometric coefficients

4. **First order reactions**:

If the reaction $\qquad\qquad A \rightarrow 2B + C$

is *first order* in A, then $m = 1$, and $\quad -\dfrac{d[A]}{dt} = k[A]$

If we denote $[A]$ by c: $\qquad\qquad -\dfrac{dc}{dt} = kc$

Separating variables: $\qquad\qquad \dfrac{dc}{c} = -kt$

Integrating: $\quad \displaystyle\int_{c_o}^{c}\frac{dc}{c} = -\int_0^t dt \quad \rightarrow \quad ln\left(\frac{c}{c_o}\right) = kt$ i.e. $ln\,c = ln\,c_o - kt$

Plotting the results:

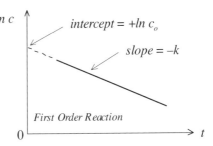

First Order Reaction

The integrated rate expression also can be written as

$$c = c_o\,exp\,[-kt]$$

which, when plotted, looks like

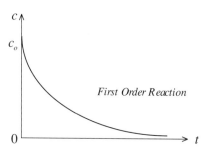

First Order Reaction

The concentration of the reactant decays exponentially with time for a first order reaction.

5. **Half-lives**:

The half-life of a chemical reaction is the time it takes for the concentration of the reactant to fall to *half* its original value.

For a *first order* reaction, putting $c = c_o/2$ into the integrated rate expression gives

$$ln\left(\frac{c_o/2}{c_o}\right) = ln\left(\frac{1}{2}\right) = -ln\,2 = -k\,t_{1/2}$$

From which
$$t_{1/2} = \frac{ln\,2}{k}$$

where $t_{1/2}$ is the **half-life** for the reaction, which is independent of the initial concentration of reactant. During each successive duration of $t_{1/2}$, the concentration of the reactant in a first order reaction decays to half its value at the start of that period.

After n such periods, the reactant concentration will be $\left(\frac{1}{2}\right)^n$ of its initial concentration.

6. **Second order reactions**:

If the reaction $2A \rightarrow B$

is *second order* in A, then $m = 2$, and $r = -\frac{1}{2}\frac{d[A]}{dt} = k[A]^2$ or $-\frac{1}{2}\frac{dc}{dt} = kc^2$

Rearranging: $-\frac{dc}{c^2} = 2kdt$

Integrating: $-\int_{c_o}^{c}\frac{dc}{c^2} = 2k\int_0^t dt = 2kt$

i.e., $\frac{1}{c} - \frac{1}{c_o} = 2kt,\quad \frac{1}{c} = \frac{1}{c_o} + 2kt,\quad$ or $\quad c = \frac{c_o}{1 + 2c_o kt}$

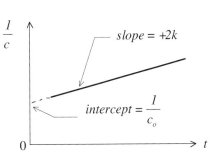

A plot of the results looks like

7. **Reactions that are first order in two reactants**:

Some reactions such as $\qquad aA + bB \rightarrow cC + dD$

have reaction rates that are first order in both A and B:

$$r = -\frac{1}{a}\frac{dc_A}{dt} = -\frac{1}{b}\frac{dc_B}{dt} = k\,c_A c_B$$

Solving the resulting mess results in two alternative equations for the integrated rate constant for this type of situation:

$$\frac{c_B^o - \frac{b}{a}c_A^o + \frac{b}{a}c_A}{c_A} = \frac{c_B^o}{c_A^o} \cdot exp\left[(ac_B^o - bc_A^o)kt\right]$$

Alternatively: $\qquad \dfrac{c_B/c_B^o}{c_A/c_A^o} = exp\left[(ac_B^o - bc_A^o)kt\right]$

8. **Zeroth order reactions**:

Some reaction rates appear to be independent of the reactant concentration.

Thus, if the reaction $\qquad\qquad A \rightarrow B$

is independent of the concentration of A, then

$$-\frac{dc_A}{dt} = kc_A^o = k$$

which is a constant rate until all the reactant is used up. Such a reaction is a *zeroth order* reaction. Only heterogeneous reactions can have rate laws that are zero-order overall.

A heterogeneous reaction is a reaction involving more than one phase that does not involve the equilibrium of a species present in more than one phase. The reaction

$$CaCO_{3(s)} \rightarrow CaO_{(s)} + CO_{2(g)}$$

is an example.

9. **Experimental determination of k, m, and n:**

    (i) The **ISOLATION METHOD**:

    The experiment is carried out by ensuring that the initial concentrations of all the reactants except one are present in large excess. For example, for the reaction

    $$aA + bB \rightarrow cC + dD$$

    the experiment is arranged so that the initial concentration of B is much greater than the initial concentration of A:

    $$[B]_o >> [A]_o$$

    Under these conditions, as the reaction proceeds and the concentration of A steadily decreases, the concentration of B remains essentially unchanged, so that we may say

    $$[B] \approx [B]_o = constant$$

    at any time during the reaction. The rate equation then becomes

    $$-\frac{d[A]}{dt} = k[A]^m [B]^n \approx k[A]^m [B]_o^n = \left\{ k[B]_o^n \right\} [A]^m = k'[A]^m$$

    Taking logarithms of each side:

    $$ln\, r = ln\, k' + m\, ln\, [A]$$

    Plotting $ln\, r$ vs. $ln\, [A]$ gives a straight line with slope m, permitting a determination of m:

    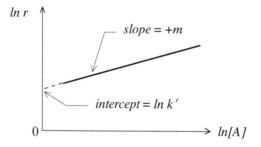

    The experiment is then repeated, this time with A in great excess, permitting determination of n. Knowing n, we then can determine k from the intercept of the first plot, which is

    $$ln\, k' = ln\left(k[B]_o^n \right)$$

    (ii) **FROM INITIAL RATES**:

    The "instantaneous" initial rate is measured at the beginning of the reaction for several different initial concentrations of reactant A (keeping the initial concentration of B constant) by measuring the change Δc_A in the concentration of A that takes place over a small time interval Δt.

    Thus, for the reaction $aA + bB \rightarrow cC + dD$

    $$\langle Initial\ Rate \rangle = r_o = -\frac{1}{a}\left(\frac{dc_A}{dt} \right)_o \approx -\frac{1}{a}\frac{\Delta c_A}{\Delta t}$$

    $$= k(c_A)_o^m (c_B)_o^n = k'(c_A)_o^m$$

Thus, $r_o = k'(c_A)_o^m$

where $k' = k(c_B)_o^n$, from which $k = \dfrac{k'}{(c_B)_o^n}$.

Taking logarithms of both sides: $\ln r_o = \ln k' + m \ln(c_A)_o$

Plotting $\ln r_o$ vs. $\ln(c_A)_o$ gives a straight line with a slope of m and an intercept of $\ln k'$, permitting an evaluation of m and k'. Then the experiment is repeated, only this time keeping $(c_A)_o$ fixed and changing $(c_B)_o$, which permits an evaluation of n. The rate constant k then can be obtained from $k = \dfrac{k'}{(c_B)_o^n}$.

(iii) FROM INTEGRATED RATE LAWS:

If the reaction proceeds too fast to readily measure the initial rates, then we can use an integrated rate law to evaluate the rate parameters.

For example, consider the very fast reaction $aA \rightarrow cC + dD$

for which the rate is given by $r = -\dfrac{1}{a}\dfrac{dc_A}{dt} = kc_A^n$

Rearranging and integrating: $-\dfrac{1}{a}\dfrac{dc_A}{c_A^n} = k\,dt$

$$-\frac{1}{a}\int_{c_o}^{c}\frac{dc}{c^n} = k\int_0^t dt \quad \rightarrow \quad -\frac{1}{a}\left[\frac{c^{1-n}}{1-n} - \frac{c_o^{1-n}}{1-n}\right] = kt \qquad [n \neq 1]$$

Therefore: $$\frac{c^{1-n}}{a(n-1)} = \frac{c_o^{1-n}}{a(n-1)} + kt$$

A plot of $c^{1-n}/a(n-1)$ vs. t gives a straight line with slope of k and an intercept of $c_o^{1-n}/a(n-1)$. Thus, the idea is to *guess* a value of n and plot $c^{1-n}/a(n-1)$ vs. t. Keep guessing different values of n until the plot yields a straight line. A straight line tells us that we have the correct value of n. The rate constan k then can be obtained from the slope of this straight line.

10. **Reaction mechanisms**:

Most reactions pass through several steps in passing from reactants to products. Each step in the sequence is called an **elementary step**; the overall sequence of elementary steps is called the **reaction mechanism**.

There are three types of elementary process: **unimolecular**, **bimolecular**, or **termolecular**, in which one, two, or three molecules participate, respectively. Elementary processes with molecularities greater than three are not known, since collisions in which more than three molecules come together at the same time are very rare. Note that for an elementary process, the *order* and the *molecularity* are both the same, because the rates at which collisions occur are directly proportional to the concentrations of the reacting molecules.

Thus, if $\qquad A + B + C \rightarrow Products$

is an elementary process, then the overall reaction order will be *3*:

$$r = -\frac{dc_A}{dt} = -\frac{dc_B}{dt} = -\frac{dc_C}{dt} = kc_A c_B c_C$$

The converse is not generally true. Thus, a second order overall reaction doesn't necessarily have a bimolecular mechanism.

11. The **rate-determining step**:

In most reaction mechanisms consisting of more than one step, the rate of the overall reaction is controlled by the *slowest* step. This step represents the "bottleneck," and is called the "rate-determining step" (*RDS*). The rate of the overall reaction will be that of the *RDS*.

12. **Reaction rates and equilibrium**:

Consider the reaction $\qquad aA + bB \overset{k}{\rightarrow} cC + dD$

consisting of elementary steps, and having a rate constant k. As the reaction proceeds, c_A and c_B decrease, while c_C and c_D increase. As the products C and D accumulate, eventually the rate of the *reverse* reaction

$$cC + dD \overset{k'}{\rightarrow} aA + bB$$

with rate constant k' starts to take place at a significant rate.

The rate of the forward reaction is $\qquad r = kc_A^a c_B^b$

and the rate of the backward reaction is $\qquad r' = k'c_C^c c_D^d$

Eventually the two rates become equal, at which point there is a net **dynamic equilibrium** for which all the concentrations are the equilibrium values, and there is no net process:

$$kc_A^a c_B^b = k'c_C^c c_D^d$$

i.e, $\dfrac{k}{k'} = \left(\dfrac{c_C^c c_D^d}{c_A^a c_B^b} \right)_{eqm}$ is recognized as the equilibrium constant K for the overall process:

$$K = \frac{k}{k'}$$

Thus, the equilibrium constant K for an overall chemical reaction is a function of the rate constants for the individual elementary steps involved in the mechanism of the reaction. If the forward rate constant k is much greater than the reverse rate constant k', then K is big, and products are favored over reactants (i.e., there will be a high yield of products at equilibrium).

13. **Reaction intermediates**:

If a reaction mechanism consists of several elementary steps, usually one of the steps—the rate-determining step—is *much* slower than any of the other steps, and therefore controls the rate of the overall reaction. However, sometimes the rates of all the steps have *comparable* values. How do we find the reaction rate for this kind of reaction?

Consider the reaction $$2A \overset{k}{\rightarrow} 4B + C$$

which is observed experimentally be a first order reaction in A with rate constant k:

$$r = -\frac{1}{2}\frac{dc_A}{dt} = kc_A$$

Since there are *two* As in the reaction, why isn't it *second* order in A? To understand this, we need to understand the concept of **reaction intermediates**. A *reaction intermediate* is a species that is involved in the mechanism of a reaction but doesn't show up in the overall reaction. Intermediates are used up as fast as they are produced, so they have a *small*, but *constant* steady state concentration during the course of the reaction.

Let us suppose that the reaction mechanism of the above reaction actually consists of the following series of elementary steps:

$$2\left[A \underset{k_1'}{\overset{k_1}{\rightleftharpoons}} B + D\right] \qquad \ldots \text{Step 1}$$

$$D + B \overset{k_2}{\rightarrow} E + B + C \quad \ldots \text{Step 2}$$

$$D + E \overset{k_3}{\rightarrow} 2B \qquad \ldots \text{Step 3}$$

In the above scheme, species D and E are **reaction intermediates**, and don't appear in the overall reaction. The process in *Step 1* is so fast that in effect it is in equilibrium. The processes in *Steps 2* and *3*, however, are somewhat slow and of comparable rates. Because they are of comparable rates, there is no single rate-determining step for this overall process. If you add up steps *1* to *3* you will see that they add up to the overall reaction.

Since there is no net E produced by the overall reaction, then

⟨*Rate of **production** of E in Step 2*⟩ = ⟨*Rate of **consumption** of E in Step 3*⟩

i.e. $$k_2 c_D c_B = k_3 c_D c_E$$

from which $$c_E = \frac{k_2}{k_3}\frac{c_B c_D}{c_D} = \frac{k_2}{k_3}c_B$$

Similarly, the net production of D in step *1* must be equal to the net consumption of D in the reverse direction of step *1* and in steps *2* and *3*:

$$k_1 c_A = k_1' c_B c_D + k_2 c_D c_B + k_3 c_D c_E = k_1' c_B c_D + k_2 c_D c_B + k_3 c_D\left(\frac{k_2}{k_3}c_B\right)$$

Rearranging: $$k_1 c_A = c_D c_B\left[k_1' + 2k_2\right] \quad \rightarrow \quad c_D = \frac{k_1 c_A}{c_B\left[k_1' + 2k_2\right]}$$

From its stoichiometry, the rate of the overall reaction is

$$r = -\frac{1}{2}\frac{dc_A}{dt} = +\frac{dc_C}{dt}$$

But from step 2,
$$\frac{dc_C}{dt} = k_2 c_D c_B$$

Therefore
$$r = k_2 c_D c_B = k_2 \left(\frac{k_1 c_A}{c_B \left[k'_1 + 2k_2 \right]} \right) c_B = \left(\frac{k_1 k_2}{k'_1 + 2k_2} \right) c_A$$

And we see that the overall reaction is indeed first order in A, with a rate constant of

$$k = \frac{k_1 k_2}{k'_1 + 2k_2}$$

14. **Effect of temperature**:

Many chemical reactions occur with the reactants forming what is called an **activated complex**. This activated complex then decomposes into the products. For example, the overall reaction

$$A + B \rightarrow C + D$$

may actually proceed by $A + B \rightarrow A\text{–}B^* \rightarrow C + D$

where $A\text{–}B^*$ is an intermediate activated complex. In this kind of reaction, the rate of formation of the activated complex is often the rate-determining step. Thus, the reactants must attain a certain amount of extra energy over and above the average energy in order to form the activated complex.

This extra amount of energy that is required to form the activated complex is called the **activation energy**, ΔG_a^*. ΔG_a^* also is called the *Arrhenius activation energy*. The reactants must first attain excess free energy ΔG_a^* over and above the average free energy of the reactants before they can react to form products. There are many different intermediate states that the system might, in principle, pass through as it proceeds from its initial state $(A + B)$ to its final state $(C + D)$. The path that is actually followed will be the *energetically least demanding* path, which is called the **reaction coordinate**.

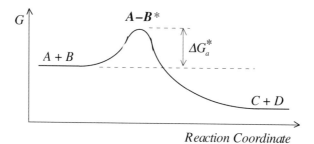

The rate constant can be shown to take the form

$$k = A \cdot exp\left[-\frac{\Delta G_a^*}{RT}\right]$$

where A is a constant—called the **pre-exponential factor**—that is essentially *independent of temperature*. It can be seen that k increases as the temperature increases because the term $exp\left(-\Delta G_a^*/RT\right)$ increases with increasing temperature. Thus, the reaction goes *faster* at higher temperatures. Basically this is because at higher temperatures, owing to the *Maxwell-Boltzmann Distribution* of kinetic energy among the molecules, a larger fraction of the molecules will possess enough kinetic energy to overcome the activation energy "barrier," and therefore react to form the activated complex.

Taking logarithms of both sides: $\quad ln\,k = ln\,A - \frac{\Delta G_a^*}{RT}$

A plot of $ln\,k$ vs. $1/T$ should give a straight line with slope $-\Delta G_a^*/R$ and intercept $ln\,A$. This type of plot is used experimentally to determine ΔG_a^* and A.

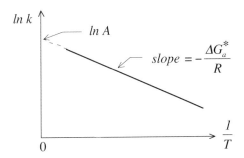

We have seen this type of behavior before when we studied the Clausius-Clapeyron equation for phase equilibrium and the van't Hoff equation for equilibrium constants. If we know the rate constant at T_1, then, knowing ΔG_a^* for the reaction, we can evaluate the rate constant at T_2 from

$$ln\left(\frac{k_{T_2}}{k_{T_1}}\right) = -\frac{\Delta G_a^*}{R}\left[\frac{1}{T_2} - \frac{1}{T_1}\right]$$

15. **Rates of ionic reactions**:

Often the rates of reaction between ions in solution are dependent on the ionic strength of the solution. This dependency is known as the **primary kinetic salt effect**.

Consider the following elementary reaction between two ions in solution:

$$A^{z_A} + B^{z_B} \rightarrow Products$$

where z_A and z_B are the valences (with signs) of the two ions. Since this is an elementary reaction it will be first order in each reactant, and the reaction rate will be

$$r = -\frac{d[A^{z_A}]}{dt} = -\frac{d[B^{z_B}]}{dt} = k[A^{z_A}][B^{z_B}] \qquad \ldots [a]$$

This kind of reaction often involves the formation of an unstable activated species known as an **activated complex** $(A \cdot B)^{z_A + z_B}$, which is a type of reaction intermediate. The formation from the reactants of this "activated complex" is very fast; so fast, in fact, that it may be considered to be in *equilibrium* with the reactants:

Thus, $$A^{z_A} + B^{z_B} \rightleftharpoons (A \cdot B)^{z_A + z_B}$$

The equilibrium constant $K*$ for the formation of the activated complex is given by

$$K* = \frac{a_{AB}}{a_A \cdot a_B} = \frac{\gamma * [(A \cdot B)^{z_A + z_B}]}{(\gamma_A [A^{z_A}])(\gamma_B [B^{z_B}])} \qquad \ldots [b]$$

where $\gamma *$ is the activity coefficient of the activated complex. Although we don't know very much about the activated complex, we do know that it has a net charge of $(z_A + z_B)$, which could be positive, negative, or zero, depending on the ions A^{z_A} and B^{z_B}.

Although the activated complex is very quickly formed, its decomposition to form product is not so fast; in fact, it is so slow that usually it is the rate determining step that controls the rate of the overall reaction. Thus, we can express the overall reaction sequence as

$$A^{z_A} + B^{z_B} \rightleftharpoons (A \cdot B)^{z_A + z_B} \xrightarrow{k*} Products$$

where $k*$ is the rate constant for the decomposition of the activated complex to form products. Thus, the rate of the overall reaction is

$$-\frac{d[A^{z_A}]}{dt} = -\frac{d[B^{z_B}]}{dt} = k*[(A \cdot B)^{z_A + z_B}]$$

From Eqn [b], $$[(A \cdot B)^{z_A + z_B}] = \frac{K* \gamma_A \gamma_B [A^{z_A}][B^{z_B}]}{\gamma *} \qquad \ldots [c]$$

Substituting this into Eqn [c] gives

$$-\frac{d[A^{z_A}]}{dt} = -\frac{d[B^{z_B}]}{dt} = \frac{k* K* \gamma_A \gamma_B [A^{z_A}][B^{z_B}]}{\gamma *} = k_{obs}[A^{z_A}][B^{z_B}] \qquad \ldots [d]$$

where k_{obs} is the experimentally observed rate constant for the reaction, defined from Eqn [d] as

$$k_{obs} = \frac{k* K* \gamma_A \gamma_B}{\gamma *} \qquad \ldots [e]$$

Eqn [e] indicates that the experimentally observed rate "constant" k_{obs} will not, in fact, be constant; but, instead, will depend on the concentrations of the species in the solution, since we know that the activity coefficients of ionic species are functions of concentration.

We also know that as the ionic strength of the solution becomes lower and lower, all the activity coefficients approach a value of unity. Thus, at infinite dilution k_{obs} becomes k^o,

where
$$k^o = \frac{k * K * (1)(1)}{(1)} = k * K * \qquad \cdots [f]$$

Therefore, k^o, which is the value of the observed rate constant when the ionic strength of the solution is $I = 0$, may be considered to be the inherently "true" rate constant for the reaction in the absence of any activity effects.

Putting Eqn [f] into Eqn [e] gives
$$k_{obs} = \frac{k^o \gamma_A \gamma_B}{\gamma *} \qquad \cdots [g]$$

Taking *base-10* logarithms:
$$log\, k_{obs} = log\, k^o + log\, \gamma_A + log\, \gamma_B - log\, \gamma * \quad \cdots [h]$$

From the Debye-Hückel equation:
$$log\, \gamma_i = -Az_i^2 \sqrt{I}$$

where $A = 0.5108$ for aqueous solution at 25°C.

Therefore, noting that the valences of A^{z_A}, B^{z_B}, and $(A \cdot B)^{z_A + z_B}$ are z_A, z_B, and $(z_A + z_B)$, respectively, substitution of the Debye-Hückel equation into Eqn [h] gives

$$log\, k_{obs} = log\, k^o - Az_A^2 \sqrt{I} - Az_B^2 \sqrt{I} + A(z_A + z_B)^2 \sqrt{I}$$

which rearranges to what is called the *Brønsted–Bjerrum equation*

$$log\, k_{obs} = log\, k^o + 2z_A z_B A\sqrt{I}$$

Expressed in exponential form:

$$k_{obs} = k^o \cdot 10^{2z_A z_B A\sqrt{I}}$$

This equation tells us the following:

(1) If the two reacting ions are oppositely charged, then $z_A z_B < 0$, and increasing the ionic strength of the solution lowers the observed rate constant, and therefore *lowers* the rate of the reaction.

(2) Conversely, if the reacting ions have the same sign (e.g., two cations or two anions), then $z_A z_B > 0$, and increasing the ionic strength of the solution will *increase* the rate of the reaction.

(3) Finally, if one of the reactants is an uncharged species ($z = 0$), then $z_A z_B = 0$, and the rate constant should be *independent* of the ionic strength of the solution.

PROBLEMS

1. The radioactive decay of naturally occurring potassium-40 (K^{40}) is a first order process in which a stable mixture of argon-40 (Ar^{40}) and calcium-40 (Ca^{40}) is produced. (These products are not produced in equal molar quantities.) The half-life of K^{40} is 1.25×10^9 years.
    (a) What is the decay rate constant for this process?
    (b) When a sample of Moon rock was analyzed by mass spectrometry, it was found that for every atom of K^{40} in the sample there were 10.5 atoms of stable products ($Ar^{40} + Ca^{40}$). Assuming that all the Ar^{40} atoms and Ca^{40} atoms resulted from the decay of the K^{40} that was present at the time the Moon solidified from a molten state, and that all the decay products so produced were retained in the rock, use the above analytical results to estimate the age of the Moon, and therefore of the solar system. *Hint:* If the number of K^{40} atoms originally in the rock sample was N_K^o, and the combined number of atoms of decay products at any time t is N_{DP}, then N_K, the number of K^{40} atoms remaining at time t, will be $N_K = N_K^o - N_{DP}$. ☹

2. As stated in the previous problem, radioactively unstable potassium-40 (K^{40}) decays to produce a stable mixture of argon-40 (Ar^{40}) and calcium-40 (Ca^{40}). The potassium-40 decays *via* two simultaneous parallel processes to form these two decay products. The individual decay rate constants for the disintegration of K^{40} to Ar^{40} and to Ca^{40} are $k_1 = 0.581 \times 10^{-10}$ y^{-1} and $k_2 = 4.962 \times 10^{-10}$ y^{-1}, respectively.
    (a) Calculate the half-life of K^{40}.
    (b) A sample of Moon rock was analyzed and found to contain Ar^{40} and K^{40} in an atomic ratio of 1:16 to 1. Assuming that all the Ar^{40} atoms present in the sample resulted from the decay of K^{40}, and that all the Ar^{40} so produced was retained in the rock, develop an expression giving the age t' of the rock as a function of k_1, k_2, and N_{Ar}/N_K, the ratio of the number of atoms of Ar^{40} to K^{40} in the sample.
    (c) Based on the above sample analysis, use the expression to estimate the age of the Moon.
    Hint: $dN_K/dt = -(k_1 + k_2)N_K$, where N_K is the number of atoms of K^{40} in the sample at any time t. ☻

3. Ethylene oxide (C_2H_4O) reacts with water to produce ethylene glycol (CH_2OHCH_2OH):

    $$C_2H_4O_{(aq)} + H_2O_{(liq)} \rightarrow CH_2OHCH_2OH_{(aq)}$$

    In an experiment at 20°C, the concentration of C_2H_4O dissolved in 0.0076 mol L^{-1} aqueous perchloric acid was observed to decrease over time as follows:

Time, (ks)	0	3	10	30
$[C_2H_4O]$, mol L^{-1}	0.150	0.132	0.099	0.043

    Determine:
    (a) the order of the reaction
    (b) the rate constant k for the consumption of ethylene oxide
    (c) the half-life of ethylene oxide, and

(d) the percentage of the initial ethylene oxide that remains after one day. ☺

4. A catalyst speeds up a reaction without itself being consumed. Tungsten metal acts as a heterogeneous catalyst for the decomposition of ammonia:

$$NH_{3(g)} \xrightarrow{\text{tungsten}} \tfrac{1}{2}N_{2(g)} + \tfrac{3}{2}H_{2(g)}$$

The tungsten is a solid phase while the ammonia is a gas. If NH_3 molecules must attach to sites on the tungsten surface before they react, then the rate of reaction depends only on the number of these sites that are available, and is independent of the concentration of NH_3 molecules in the gas phase. Pure NH_3 is pumped into an empty reactor with a packing having a tungsten surface of fixed area. As the NH_3 decomposes to N_2 and H_2 at 856°C, the following decrease in the partial pressure of NH_3 with time is observed:

Time, (s)	0	105	312	504
P_{NH_3} (kPa)	12.9	11.6	9.0	6.4

Determine:

(a) the order of the reaction
(b) the rate constant k for the consumption of ammonia
(c) the time at which the concentration of hydrogen (H_2) reaches 0.0015 mol L^{-1}
(d) the value of the rate constant if the reaction were written

$$2NH_{3(g)} \xrightarrow{\text{tungsten}} N_{2(g)} + 3H_{2(g)} \ \ ☹$$

5. In acid solution, bromate ion (BrO_3^-) oxidizes bromide ion (Br^-) to produce elemental bromine according to the following reaction:

$$5\,Br^-_{(aq)} + BrO^-_{3(aq)} + 6\,H^+_{(aq)} \rightarrow 3Br_{2(aq)} + 3H_2O_{(liq)}$$

The kinetics of this reaction were studied using the method of initial rates. When the concentrations of the three reactants were held constant at the values tabulated in each column, the indicated initial rates of consumption of Br$^-$ ion were observed:

Expt. No.	$[Br^-]_0$ (mol L^{-1})	$[BrO_3^-]_0$ (mol L^{-1})	$[H^+]_0$ (mol L^{-1})	$-(d[Br^-]/dt)_0$ (mol L^{-1} s^{-1})
1	0.120	0.150	0.080	0.79×10^{-3}
2	0.250	0.150	0.080	1.64×10^{-3}
3	0.120	0.280	0.080	1.47×10^{-3}
4	0.120	0.150	0.130	2.08×10^{-3}

(a) What are the orders of the reaction in each reactant?
(b) Which orders are the same as or different from the corresponding stoichiometric

coefficients?

(c) What is the value of the rate constant k for the reaction?

(d) What is the rate constant k_{H^+} for the consumption of H^+ ion? ⊛

6. Hydrogen ions and acetate ions combine rapidly in aqueous solution to produce acetic acid, which has $pK_a = 4.76$ at 25°C:

$$H^+_{(aq)} + CH_3COO^-_{(aq)} \xrightarrow{k} CH_3COOH_{(aq)}$$

This reaction is second order overall, first order in each reactant. For the consumption of either H^+ or CH_3COO^- the rate constant is $k = 4.5 \times 10^{10}$ $(mol/L)^{-1}$ s^{-1} at 25°C. However, the reverse reaction also occurs.

(a) Suppose 2.0×10^{-8} moles of undissociated CH_3COOH were dissolved instantaneously in 1.00 mm^3 of pure water to produce a solution of undissociated CH_3COOH, which then begins to dissociate. What would be the initial rate of change of pH?

(b) What is the pH when equilibrium is reached?

(c) In practice, would it be reasonable to assume that equilibrium is reached quickly enough for the pH always effectively to be at its equilibrium value? ⊛

7. Ozone (O_3), a reactive form of oxygen, decomposes to ordinary oxygen (O_2) according to the reaction

$$2O_{3(g)} \xrightarrow{k} 3O_{2(g)}$$

In the presence of hydrogen sulfide (H_2S), which catalyzes the decomposition in the gas phase, the reaction is second order in O_3. At 15°C the rate constant for this reaction is $k = 9.0 \times 10^4$ $(mol/L)^{-1}$ s^{-1}. A tank of fixed volume initially contains gaseous O_3 at the low partial pressure of 2.00 Pa, and also some H_2S, but no O_2. After 10.0 s at 15°C, what are

(a) the concentrations of O_3 and O_2, and

(b) the rates of change of the concentrations of O_3 and O_2? ☺

8. Tritium-3 ($^1H^3$) is an isotope of hydrogen that is formed at a steady state in nature by the irradiation of water vapor in the upper atmosphere by cosmic rays. It undergoes first order radioactive decay with a half-life of 12.32 years. A glass bottle at the surface of the earth provides sufficient shielding to prevent the formation of $^1H^3$ inside the bottle. This fact can be used to determine the age of vintage wines. Estimate the age of a vintage wine that is 17% as radioactive in tritium as a freshly bottled sample of the same type of wine. ☺

9. Iodine-131, with a half-life of 8.04 days, is used in the treatment of thyroid disorders. What percentage of this radioactive isotope still will be present in the body (a) one week, and (b) 30 days after ingesting a small amount of this material? You may assume that none of the material is excreted from the body. ☺

10. The gas phase reaction between hydrogen and iodine produces hydrogen iodide according to

$$H_{2(g)} + I_{2(g)} \rightarrow 2HI_{(g)}$$

and obeys the following rate expression:

$$\frac{d[H_2]}{dt} = \frac{d[I_2]}{dt} = -k[H_2][I_2]$$

The activation energy has been found experimentally to be 162.2 kJ mol^{-1}.

When hydrogen and iodine are introduced into an evacuated reactor each at an initial concentration of 3.00×10^{-2} mol L^{-1} at 375°C, it is found that at this temperature 10% of the hydrogen is consumed after 10.0 minutes. If the reactor were charged at 450°C with hydrogen and iodine, each at an initial concentration of 3.00×10^{-2} mol L^{-1}, how long would it take at this new temperature for the same fraction of hydrogen to be consumed? ☺

11. The gas phase reaction $\qquad CO_{(g)} + Cl_{2(g)} \rightarrow COCl_{2(g)}$

is first order in each reactant. If the initial partial pressure of each reactant is 0.100 bar, at 30°C the time required for the reaction to proceed to 50% conversion is 43 minutes, while at 40°C only 22 minutes are required.

(a) Determine the activation energy and the Arrhenius pre-exponential factor for this reaction.

(b) Calculate the rate constant in bar^{-1} s^{-1} at 20°C. ☺

12. Two ions, A^+ and B^+, react in aqueous solution as follows:

$$A^+ + B^+ \xrightarrow{\ k\ } cC \qquad\qquad \ldots [a]$$

with $\qquad\qquad\qquad r = k[A^+][B^+]$

At infinite dilution the value of the rate constant is $k°$. Two other ions, M^{2+} and N^{2-}, react as follows:

$$M^{2+} + N^{2-} \xrightarrow{\ k'\ } dD + eE \qquad\qquad \ldots [b]$$

with $\qquad\qquad\qquad r' = k'[M^{2+}][N^{2-}]$

The rate constant for reaction [b] at infinite dilution is three times as great as that for reaction [a] at infinite dilution. Estimate the relative rates of these two reactions in an aqueous solution of ionic strength 0.001 at 25°C. ☺

13. In aqueous solution bromacetate ion reacts with thiosulfate ion according to

$$BrCH_2COO^- + S_2O_3^{2-} \rightarrow S_2O_3CH_2COO^{2-} + Br^-$$

Two experiments were carried out at 25°C using the sodium salts of each reactant: *Run A* in which the initial normality of each reactant was 5.0×10^{-4} eq L^{-1}, and *Run B* in which the initial normality of each reactant was 1.50×10^{-3} eq L^{-1}. In *Run A*, after 3356 min the bromacetate concentration had fallen to 3.33×10^{-4} N, while in *Run B* the bromacetate concentration had fallen to 5.34×10^{-4} N after 3517 min. The reaction rate is known to be second order in bromacetate:

$$r = -\frac{d[BrCH_2COO^-]}{dt} = k[BrCH_2COO^-]$$

(a) Using the experimental data, determine the rate constant for this reaction at zero ionic strength.

(b) Comment on the charge of the activated complex. ☹

14. The table at the right lists values of the population of the world for various years. It has been suggested that the rate of growth of the world population follows a second order rate process:

$$dN/dt = kN^2$$

where N is the population in billions, t is the date in years AD, and k is the "rate constant."

Year	N (billions)
1650	0.510
1700	0.625
1750	0.710
1800	0.910
1850	1.130
1900	1.600
1950	2.565
1960	3.050
1970	3.721
1980	4.476
1990	5.320

(a) Use the data in the table to verify this assumption.

(b) Using the results of (a), estimate the world population for the following years:

   (i)   one million years BC, when Uncle Frank was born,

   (ii) 2600 BC, when the pyramids of Egypt were built,

   (iii) 563 BC, the year attributed to the birth of Buddha,

   (iv) 0 AD, the assumed year of the birth of Jesus,

   (v)   570 AD, when the Prophet Mohammed was born,

   (vi) 1980, (vi) 1990, (vii) 2000, (viii) 2004, (ix) 2018.

(c) Comment on the validity of the model by comparing the population values predicted by the model to the following generally accepted values: *One million years BC = 125×10^3; 0 AD = 100×10^6; 1995 = 5.73×10^9, 2000 = 6.22×10^9, 2004 = 6.67×10^9.* ☹

APPENDICES

APPENDIX 1. PROPERTIES OF THE ELEMENTS

Name	Symbol	Atomic No.	Relative Atomic Mass (C-12 = 12)	ρ (20°C) g cm^{-3}	T_{mp} °C	T_{bp} °C	$c_P^{(1)}$(25°C) J g^{-1} K^{-1}
Actinium	Ac	89	227.0	10.06	1323	3473	0.092
Aluminum	Al	13	26.981 539(5)	2.699	660	2450	0.900
Americium	Am	95	241.1	13.67	1541		
Antimony	Sb	51	121.75(3)	6.691	630.5	1380	0.205
Argon	Ar	18	39.948(1)	1.663×10^{-3}	-189.4	-185.8	0.523
Arsenic	As	33	74.921 59(2)	5.78	817$^{(2)}$	613	0.331
Astatine	At	85	210.0		302		
Barium	Ba	56	137.327(7)	3.594	729	1640	0.205
Berkelium	Bk	97	249.1	14.79			
Beryllium	Be	4	9.012 182(3)	1.848	1287	2770	1.83
Bismuth	Bi	83	208.980 37(3)	9.747	271.37	1560	0.122
Boron	B	5	10.811(5)	2.34	2030		1.11
Bromine	Br	35	79.904(1)	3.12 (liq)	-7.2	58	0.293
Cadmium	Cd	48	112.411(8)	8.65	321.03	765	0.226
Calcium	Ca	20	40.078(4)	1.55	838	1440	0.624
Californium	Cf	98	252.1				
Carbon	C	6	12.011(1)	2.26	3727	4830	0.691
Cerium	Ce	58	140.115(4)	6.768	804	3470	0.188
Cesium	Cs	55	132.905 43(5)	1.873	28.40	690	0.243
Chlorine	Cl	17	35.452 7(9)	2.995×10^{-3}	-101	-34.7	0.486
Chromium	Cr	24	51.996 1(6)	7.19	1857	2665	0.448
Cobalt	Co	27	58.933 20(1)	8.85	1495	2900	0.423
Copper	Cu	29	63.546(3)	8.96	1083.4	2595	0.385
Curium	Cm	96	244.1	13.3			
Dysprosium	Dy	66	162.50(3)	8.55	1409	2330	0.172
Einsteinium	Es	99	252.1				
Erbium	Er	68	167.26(3)	9.15	1522	2630	0.167
Europium	Eu	63	151.965(9)	5.243	817	1490	0.163
Fermium	Fm	100	257.1				
Fluorine	F	9	18.998 403 2(9)	1.580×10^{-3}	-219.6	-188.2	0.753
Francium	Fr	87	223.0		27		
Gadolinium	Gd	64	157.25(3)	7.90	1312	2730	0.234
Gallium	Ga	31	69.723(1)	5.907	29.75	2237	0.377
Germanium	Ge	32	72.61(2)	5.323	937.25	2830	0.322
Gold	Au	79	196.966 54(3)	19.32	1064.43	2970	0.131
Hafnium	Hf	72	178.49(2)	13.31	2227	5400	0.144
Hahnium	Ha	105					

Continues...

Properties of the Elements *(Continued)*

Name	Symbol	Atomic No.	Relative Atomic Mass (C-12 = 12)	ρ (20°C) g cm^{-3}	T_{mp} °C	T_{bp} °C	$c_P^{(1)}$(25°C) J g^{-1} K^{-1}
Hassium	Hs	108					
Helium	He	2	4.002 602(2)	1.664×10^{-4}	−269.7	−268.9	5.23
Holmium	Ho	67	164.930 32(3)	8.79	1470	2330	0.165
Hydrogen	H	1	1.007 94(7)	8.375×10^{-5}	−259.19	−252.7	14.4
Indium	In	49	114.88(3)	7.31	156.634	2000	0.233
Iodine	I	53	126.904 47(3)	4.93	113.7	183	0.218
Iridium	Ir	77	192.22(3)	22.5	2447	5300	0.130
Iron	Fe	26	55.847(3)	7.874	1536.5	3000	0.447
Krypton	Kr	36	83.80(1)	3.488×10^{-3}	−157.37	−152	0.247
Lanthanum	Ln	57	138.905 5(2)	6.189	920	3470	0.195
Lawrencium	Lr	103	260.1				
Lead	Pb	82	207.2(1)	11.35	327.45	1725	0.129
Lithium	Li	3	6.941(2)	0.534	180.55	1300	3.58
Lutetium	Lu	71	174.967(1)	9.849	1663	1930	0.155
Magnesium	Mg	12	24.305 0(6)	1.738	650	1107	1.03
Manganese	Mn	25	54.938 05(1)	7.44	1244	2150	0.481
Meitnerium	Mt	109					
Mendelevium	Md	101	256.1				
Mercury	Hg	80	200.59(3)	13.55	−38.87	357	0.138
Molybdenum	Mo	42	95.94(1)	10.22	2617	5560	0.251
Neodymium	Nd	60	144.24(3)	7.007	1016	3180	0.188
Neon	Ne	10	20.179 7(6)	8.387×10^{-4}	−248.597	−246.0	1.03
Neptunium	Np	93	237.0	20.25	637		
Nickel	Ni	28	58.693 4(2)	8.902	1453	2730	0.444
Nielsbohrium	Ns	107					
Niobium	Nb	41	92.906 38(2)	8.57	2468	4927	0.264
Nitrogen	N	7	14.006 74(7)	1.165×10^{-3}	−210	−195.8	1.03
Nobelium	No	102	259.1				
Osmium	Os	76	190.23(3)	22.59	3027	5500	0.130
Oxygen	O	8	15.999 4(3)	1.332×10^{-3}	−218.80	−183.0	0.913
Palladium	Pd	46	106.42(1)	12.02	1552	3980	0.243
Phosphorus	P	15	30.973 762(4)	1.83	44.25	280	0.741
Platinum	Pt	78	195.08(3)	21.45	1769	4530	0.134
Plutonium	Pu	94	239.1	19.8	640	3235	0.130
Polonium	Po	84	210.0	9.32	254		
Potassium	K	19	39.098 3(1)	0.862	63.20	760	0.758
Praseodymium	Pr	59	140.907 65(3)	6.773	931	3020	0.197

Continues...

Properties of the Elements *(Continued)*

Name	*Symbol*	*Atomic No.*	*Relative Atomic Mass (C-12 = 12)*	ρ *(20°C)* g cm^{-3}	T_{mp} °C	T_{bp} °C	$c_P^{(1)}$ *(25°C)* J g^{-1} K^{-1}
Promethium	Pm	61	146.9	7.22	1027		
Protactinium	Pa	91	231.035 88(2)	15.37 (est)	1230		
Radium	Ra	88	226.0	5.0	700		
Radon	Rn	86	222.0	9.28×10^{-3}	−71	−61.8	0.092
Rhenium	Re	75	186.207(1)	21.02	3180	5900	0.134
Rhodium	Rh	45	102.905 50(3)	12.41	1963	4500	0.243
Rubidium	Rb	37	85.467 8(3)	1.532	39.49	688	0.364
Ruthenium	Ru	44	101.07(2)	12.37	2250	4900	0.239
Rutherfordium	Rf	104					
Samarium	Sm	62	150.36(3)	7.52	1072	1630	0.197
Scandium	Sc	21	44.955 910(9)	2.99	1539	2730	0.569
Seaborgium	Sg	106					
Selenium	Se	34	78.96(3)	4.79	221	685	0.318
Silicon	Si	14	28.085 5(3)	2.33	1412	2680	0.712
Silver	Ag	47	107.868 2(2)	10.49	960.8	2210	0.234
Sodium	Na	11	22.989 768(6)	0.9712	97.85	892	1.23
Strontium	Sr	38	87.62(1)	2.54	768	1380	0.737
Sulfur	S	16	32.066(6)	2.07	119.0	444.6	0.707
Tantalum	Ta	73	180.947 9(1)	16.6	3014	5425	0.138
Technetium	Tc	43	98.91	11.46	2200		0.209
Tellurium	Te	52	127.60(3)	6.24	449.5	990	0.201
Terbium	Tb	65	158.925 34(3)	8.229	1357	2530	0.180
Thallium	Tl	81	204.383 3(2)	11.85	304	1457	0.130
Thorium	Th	90	232.038 1(1)	11.72	1755	3850	0.117
Thulium	Tm	69	168.934 21(3)	9.32	1545	1720	0.159
Tin			118.710(7)	7.2984	231.868	2270	0.226
Titanium	Ti	22	47.88(3)	4.54	1670	3260	0.523
Tungsten	W	74	183.84(1)	19.3	3380	5930	0.134
Uranium	U	92	238.028 9(1)	18.95	1132	3818	0.117
Vanadium	V	23	50.941 5(1)	6.11	1902	3400	0.4900
Xenon	Xe	54	131.29(2)	5.495×10^{-3}	−111.79	−108	0.159
Ytterbium	Yb	70	173.04(3)	6.965	824	1530	0.155
Yttrium	Y	39	88.905 85(2)	4.469	1526	3030	0.297
Zinc	Zn	30	65.39(2)	7.133	419.58	906	0.389
Zirconium	Zr	40	91.224(2)	6.506	1852	3580	0.276

[1] Heat capacities are at one atm pressure. [2] Arsenic melting point at 28 atm. Sources: Molar Masses: IUPAC Commission on Atomic Weights, *Pure Appl. Chem.*, **64**, 1520 (1992); Physical data: D. Halliday, R. Resnick, and J. Walker, *Fundamentals of Physics: Extended*, John Wiley and Sons, Inc., New York (1997).

APPENDIX 2. THERMODYNAMIC DATA FOR SELECTED GASES AND VAPORS at $P° = 1$ bar and 298.15 K

Substance	Δh_f^o $\left(\dfrac{kJ}{mol}\right)$	Δg_f^o $\left(\dfrac{kJ}{mol}\right)$	s^o $\left(\dfrac{J}{K \cdot mol}\right)$	c_P^o $\left(\dfrac{J}{K \cdot mol}\right)$	a $\left(\dfrac{Pa\ m^6}{mol^2}\right)$	$10^6 b$ $\left(\dfrac{m^3}{mol}\right)$	Henry's Law Constant (H_2O)
Acetaldehyde (Ethanal), CH_3CHO	−166.19	−128.86	250.3	57.3	1.11	86.0	427 kPa
Acetic acid, CH_3COOH	−434.84	−376.59	282.61	66.53	1.78	107	1.01 kPa
Acetylene (Ethyne), C_2H_2	226.73	209.20	200.94	43.93	0.452	52.2	135 MPa
Ammonia, NH_3	−46.11	−16.45	192.45	35.06	0.4225	37.07	96 kPa
Antimony hydride (Stibine), SbH_3	145.11	147.75	232.78	41.05			
Argon, Ar	0	0	154.843	20.786	0.1363	32.19	4.01 GPa
Arsenic, As	302.5	261.0	174.21	20.79			
Arsenic, As_4	143.9	92.4	314				
Arsenic hydride (Arsine), AsH_3	66.44	68.93	222.78	38.07			630 MPa
Barium, Ba	180	146	170.24	20.79			
Benzene, C_6H_6	82.93	129.72	269.31	81.67	1.824	115.4	31.3 MPa
Bismuth, Bi	207.1	168.2	187.00	20.79			
Bromine, Br_2	30.907	3.110	245.46	36.02			8.01 MPa
Bromine (atomic), Br	111.88	82.396	175.02	20.786			165 MPa
Butane (n), C_4H_{10}	−126.15	−17.03	310.23	97.45	1.466	122.6	5.10 GPa
Butane (iso), C_4H_{10}	−134.52	−20.76	294.75	96.82	1.304	114.2	6.68 GPa
Butene (1-butene), C_4H_8	−0.13	71.39	305.71	85.65			1.40 GPa
Carbon, C	716.68	671.26	158.10	20.838			
Carbon dioxide, CO_2	−393.509	−394.359	213.74	37.11	0.3640	42.67	161.7 MPa
Carbon monoxide, CO	−110.525	−137.168	197.674	29.142	0.1505	39.85	6.12 GPa
Chlorine, Cl_2	0	0	223.066	33.907	0.6579	56.22	68.7 MPa
Chlorine (atomic), Cl	121.679	105.680	165.198	21.840			52 MPa
Copper, Cu	338.32	298.58	166.38	20.79			
Cyclohexane, C_6H_{12}	−123.14	−31.91	298.35	106.27			1.01 GPa
Cyclopropane, C_3H_6	53.30	104.45	237.55	55.94			467 MPa
Deuterium, D_2	0	0	144.96	29.20			
Deuterium oxide, D_2O	−249.20	−234.54	198.34	34.27			
Dimethyl ether, $(CH_3)_2O$	−184.05	−112.59	266.38	64.39	0.876	78.0	5.61 MPa
Dinitrogen tetraoxide, N_2O_4	9.16	97.89	304.29	77.28			3.74 MPa
Dinitrogen pentoxide, N_2O_5	11.3	115.1	355.7	84.5			
Ethane, C_2H_6	−84.68	−32.82	229.60	52.63	0.5562	63.8	2.91 GPa
Ethanol, C_2H_5OH	−235.10	−168.49	282.70	65.44	1.22	84.3	29.2 kPa
Ethylbenzene, C_8H_{10}	29.79	130.70	360.56	128.41			42.9 MPa
Ethylene (Ethene), C_2H_4	52.26	68.15	219.56	43.56	0.4530	57.14	1.18 GPa
Ethylene glycol, HOC_2H_4OH	−389.32	−304.35	323.66	97.07	2.74	123.0	1.39 Pa

Continues...

THERMODYNAMIC DATA FOR SELECTED GASES AND VAPORS at $P° = 1$ bar and 298.15 K (Continued)

Substance	Δh_f^o $\left(\dfrac{kJ}{mol}\right)$	Δg_f^o $\left(\dfrac{kJ}{mol}\right)$	s^o $\left(\dfrac{J}{K \cdot mol}\right)$	c_P^o $\left(\dfrac{J}{K \cdot mol}\right)$	a $\left(\dfrac{Pa \; m^6}{mol^2}\right)$	$10^6 b$ $\left(\dfrac{m^3}{mol}\right)$	Henry's Law Constant (H_2O)
Ethylene oxide, C_2H_4O	−52.64	−13.16	243.01	47.90	0.892	67.8	reacts
Fluorine, F_2	0	0	202.78	31.30			
Fluorine (atomic), F	78.99	61.91	158.75	22.74			2.67 MPa
Formaldehyde (Methanal), HCHO	−117	−113	218.77	35.40	0.718	63.5	reacts
Formic acid, HCOOH	−378.61	−350.97	248.85	45.23	1.78	110.	1.17 kPa
Helium, He	0	0	126.150	20.786	0.003457	23.70	14.95 GPa
Hexane, C_6H_{14}	−167.19	−0.07	388.51	143.09			8.76 GPa
Hydrogen, H_2	0	0	130.684	28.824	0.02476	26.61	7.19 GPa
Hydrogen (atomic), H	217.97	203.25	114.71	20.784			
Hydrogen (ion), H^+	1536.20						
Hydrogen bromide, HBr	−36.40	−53.45	198.70	29.142			431 kPa
Hydrogen chloride, HCl	−92.307	−95.299	186.908	29.12			295 kPa
Hydrogen cyanide, HCN	135.1	124.7	201.78	35.86			584 kPa
Hydrogen fluoride, HF	−271.1	−273.2	173.779	29.133			204 Pa
Hydrogen iodide, HI	26.48	1.70	206.594	29.158			239 MPa
Hydrogen sulfide, H_2S	−20.63	−33.56	205.79	34.23	0.4490	42.87	56.1 MPa
Iodine, I_2	62.438	19.317	260.69	36.90			1.81 MPa
Iodine (atomic), I	106.838	70.250	180.791	20.786			130 MPa
Iron, Fe	416.3	370.7	180.49	25.68			
Ketene, $CH_2C=O$	−61.09	−60.24	241.90	51.76	0.648	60.8	reacts
Krypton, Kr	0	0	164.082	20.786	0.2349	39.78	2.29 GPa
Lithium, Li	159.37	126.66	138.77	20.79			
Mercury, Hg	61.317	31.820	174.96	20.786			60.3 MPa
Methane, CH_4	−74.81	−50.72	186.264	35.309	0.2283	42.78	4.18 GPa
Methanol, CH_3OH	−200.66	−161.96	239.81	43.89	0.958	66.6	25.2 kPa
Methylamine, CH_3NH_2	−22.91	32.16	243.41	53.1			63 kPa
Neon, Ne	0	0	146.328	20.786	0.02135	17.09	12.46 GPa
Nitrogen, N_2	0	0	191.61	29.125	0.1408	39.13	8.90 GPa
Nitrogen (atomic), N	472.70	455.56	153.30	20.786			
Nitrogen monoxide, NO	90.25	86.57	210.761	29.844	0.1358	27.89	2.95 GPa
Nitrogen dioxide, NO_2	33.18	51.31	240.06	37.20	0.5354	44.24	231 MPa
Nitrous oxide, N_2O	82.05	104.20	219.85	38.45			227 MPa
Octane, C_8H_{18}	−208.45	16.64	466.84	188.87			19.8 GPa
Oxygen, O_2	0	0	205.138	29.355	0.1378	31.83	4.31 GPa

Continues...

THERMODYNAMIC DATA FOR SELECTED GASES
AND VAPORS at $P° = 1$ bar and 298.15 K (Continued)

Substance	Δh_f^o $\left(\dfrac{kJ}{mol}\right)$	Δg_f^o $\left(\dfrac{kJ}{mol}\right)$	s^o $\left(\dfrac{J}{K \cdot mol}\right)$	c_P^o $\left(\dfrac{J}{K \cdot mol}\right)$	a $\left(\dfrac{Pa\ m^6}{mol^2}\right)$	$10^6 b$ $\left(\dfrac{m^3}{mol}\right)$	Henry's Law Constant (H_2O)
Oxygen (atomic), O	249.17	231.73	161.06	21.912			
Ozone, O_3	142.67	163.18	238.93	39.24	0.357	48.7	502 MPa
Pentane (n), C_5H_{12}	−146.44	−8.20	348.40	120.2	1.926	146.0	7.01 GPa
Phosphine, PH_3	5.4	13.4	210.23	37.11			692 MPa
Phosphorus (atomic), P	314.64	278.25	163.193	20.786			
Phosphorus, P_2	144.3	103.7	218.129	32.05			
Phosphorus, P_4	58.91	24.44	279.98	67.15			
Phosphorus pentachloride, PCl_5	−374.9	−305.0	364.58	112.8			
Phosphorus trichloride, PCl_3	−287.0	−267.8	311.78	71.84			
Propane, C_3H_8	−103.89	−23.38	270.02	73.51	0.8779	84.45	3.89 GPa
Propene, C_3H_6	20.42	62.78	267.05	63.89			1.17 GPa
Styrene (Vinylbenzene), C_8H_8	147.22	213.89	345.21	122.09			16.2 MPa
Sulfur dioxide, SO_2	−296.830	−300.194	248.22	39.87	0.6803	56.36	4.67 MPa
Sulfur trioxide, SO_3	−395.72	−371.06	256.76	50.67			
Toluene (Methyl benzene), $CH_3C_6H_5$	50.00	122.10	320.77	103.64			35.5 MPa
Water, H_2O	−241.818	−228.572	188.825	35.577	0.5536	30.49	
Xenon, Xe	0	0	169.683	20.786	0.4250	51.05	1.30 GPa

Source: Most of the thermochemical data are from the National Bureau of Standards *Tables of Chemical Thermodynamic Properties*, published as a supplement to Vol. II of the *Journal of Physical and Chemical Reference Data* (1982). The Henry's Law Constants are from *Compilation of Henry's Law Constants for Inorganic and Organic Species of Potential Importance in environmental Chemistry*, by R. Sander, Max-Planck Institure of Chemistry (1999) [http://www.mpch-mainz.mpg.de/~sander/res/henry.html]

APPENDIX 3. THERMODYNAMIC DATA FOR SELECTED LIQUIDS
at $P° = 1$ bar and 298.15 K

Substance	Δh_f^o $\left(\dfrac{kJ}{mol}\right)$	Δg_f^o $\left(\dfrac{kJ}{mol}\right)$	s^o $\left(\dfrac{J}{K \cdot mol}\right)$	c_P^o $\left(\dfrac{J}{K \cdot mol}\right)$	ρ [1] $\left(\dfrac{kg}{m^3}\right)$	T_{bp} (K)	Δh_{vap}^{o} [2] $\left(\dfrac{kJ}{mol}\right)$
Al (Aluminum, molten)	10.56	7.20	39.55	24.21	2375[3]	2792.2	294
Br_2 (Bromine)	0	0	152.231	75.689	3102.8	332.4	29.45
CCl_4 (Carbon tetrachloride)	−135.44	−65.21	216.40	131.75	1594.0	349.9	30.00
CH_3CHO (Acetaldehyde)	−192.30	−128.12	160.2	89.0	783.4	293.3	25.76
CH_3COOH (Acetic acid)	−484.5	−389.9	159.8	124.3	1047.7	391.1	23.70
CH_3OH (Methanol)	−238.66	−166.27	126.8	81.6	786.7	337.2	35.27
$(CH_3)_2CO$ (Acetone)	−248.1	−155.4	200.4	124.7	784.5[4]	329.2	29.10
C_2H_4O (Ethylene oxide)	−77.82	−11.76	153.85	87.95	882.1[5]	283.8	25.54
C_2H_5OH (Ethanol)	−277.69	−174.78	160.7	111.46	784.9	351.4	43.5
$C_4H_8O_2$ (Ethyl acetate)	−479.0	−332.7	259.4	170.1	900.3	350.3	31.94
C_5H_{12} (n-pentane)	−172.75	−9.11	263.94	167.2	621.2	309.2	25.79
$C_6H_5NH_2$ (Aniline)	31.1			191.9	1021.7	457.3	42.44
C_6H_6 (Benzene)	49.21	124.75	173.13	136.0	871.1	353.2	30.72
C_6H_{12} (Cyclohexane)	−155.82	27.01	205.32	154.9	773.9	353.9	29.97
C_6H_{14} (Hexane)	−198.7		204.3	195.6	654.8[4]	341.9	28.85
C_7H_8 (Toluene)	12.22	114.23	220.77	157.3	861.6	383.8	33.18
C_7H_{16} (n-heptane)	−223.86	1.55	329.58	224.7	679.3	371.7	31.77
C_8H_{18} (Isooctane)	−258.66	7.47	329.47	239.1	687.7	372.4	30.79
CS_2 (Carbon disulfide)	89.70	65.27	151.34	75.7	1263.2	319.4	25.23
HCN (Hydrogen cyanide)	108.87	124.97	112.84	70.63	684	298.9	25.22
HCOOH (Formic acid)	−424.72	−361.35	128.95	99.04	1220	374.2	22.69
HNO_3 (Nitric acid)	−174.1	−80.71	155.60	109.87	1550	356.2	39.1[4]
H_2O (Water)	−285.830	−237.129	69.91	75.291	997.0	373.15	40.656
H_2O_2 (Hydrogen peroxide)	−187.78	−120.35	109.6	89.1	1440	423.4	51.6[4]
H_2SO_4 (Sulfuric acid)	−813.989	−690.003	156.904	138.91	1830.5	610.2	
Hg (Mercury)	0	0	76.02	27.983	13 550	629.9	59.11
N_2H_4 (Hydrazine)	50.63	149.43	121.21	139.3	1003.6	386.7	41.8
N_2O_4 (Nitrogen tetroxide)	−19.50	97.54	209.2	142.7	1450	294.3	38.12
PCl_3 (Phosphorus trichloride)	−319.7	−272.3	217.1	120.1	1574	349.1	30.5

Source: Most of the thermochemical data are from the National Bureau of Standards *Tables of Chemical Thermodynamic Properties*, published as a supplement to Vol. II of the *Journal of Physical and Chemical Reference Data* (1982).
[1] At 20°C. [2] At the normal boiling point. [3] At the melting point. [4] At 25°C. [5] At 10°C.

APPENDIX 4. THERMODYNAMIC DATA FOR SELECTED SOLIDS
at P° = 1 bar and 298.15 K

Substance	Δh_f^o $\left(\dfrac{kJ}{mol}\right)$	Δg_f^o $\left(\dfrac{kJ}{mol}\right)$	s^o $\left(\dfrac{J}{K \cdot mol}\right)$	c_P^o $\left(\dfrac{J}{K \cdot mol}\right)$	$\rho^{(1)}$ $\left(\dfrac{kg}{m^3}\right)$	T_{mp} (K)	$\Delta h_{fus}^{o\ (2)}$ $\left(\dfrac{kJ}{mol}\right)$
Ag	0	0	42.55	25.351	10 490	1234	11.30
AgBr	−100.37	−96.90	107.1	52.38	6470	705	9.12
AgCl	−127.068	−109.789	96.2	50.79	5560	728	13.2
$AgNO_3$	−129.39	−33.41	140.92	93.05	4350	485	11.5
Ag_2O	−31.05	−11.20	121.3	65.86	7200	~200[3]	
Al	0	0	28.33	24.35	2699	933	10.71
Al_2O_3 (α)	−1675.7	−1582.3	50.92	79.04	3970	2326	111.4
$AlCl_3$	−704.2	−628.8	110.67	91.84	2480	466	35.4
As (α)	0	0	35.1	24.64	5780	1090	24.44
Au	0	0	47.40	25.42	19 320	1338	12.55
Ba	0	0	62.8	28.07	3594	1002	7.12
$BaCl_2$	−858.6	−810.4	123.68	75.14	3900	1235	15.85
BaO	−553.5	−525.1	70.43	47.78	5720	1972	46
Be	0	0	9.50	16.44	1848	1560	7.895
Benzoic acid, C_6H_5COOH	−385.1	−245.3	167.6	146.8	1265	396	18.02
Bi	0	0	56.74	25.52	9747	545	11.30
C (graphite)	0	0	5.740	8.527	2250	4765[4]	117
C (diamond)	1.895	2.900	2.377	6.113	3513	4713	
Ca	0	0	41.42	25.31	1550	1115	8.54
$CaBr_2$	−682.8	−663.6	430		3380	1015	29.1
CaC_2	−59.80	−64.89	69.96	62.72	2220	2570	
$CaCl_2$	−795.8	−748.1	104.6	72.59	2150	1048	28.05
$CaCO_3$ (calcite)	−1206.92	−1128.79	92.9	81.88	2710	1603	
$CaCO_3$ (aragonite)	−1207.13	−1127.75	88.7	81.25	2830	1098[3]	36
CaF_2	−1219.6	−1167.3	68.87	67.03	3180	1691	30
CaH_2	−176.98	−138.01	41.40	41.00	1700	1273	6.7
CaO	−635.09	−604.03	39.75	42.80	3340	3172	80
$Ca(OH)_2$	−986.09	−898.47	83.39	87.49	2200	1023	29
Cd (γ)	0	0	51.76	25.98	8650	594	6.19
$CdCO_3$	−750.6	−669.4	92.5		4258	~500[3]	
CdO	−258.2	−−228.4	54.8	43.43	8150	1700	
$CdSO_4 \cdot \frac{8}{3}H_2O$	−1729.4	−1465.141	229.630	213.26			
Cr	0	0	23.77	23.35	7190	2180	21.0
Cs	0	0	85.23	32.17	1873	302	2.09
Cu	0	0	33.150	24.435	8960	1358	13.26
Cu_2O	−168.6	−146.0	93.14	63.64	6000	1508	

Continues...

THERMODYNAMIC DATA FOR SELECTED SOLIDS
at $P° = 1$ bar and 298.15 K (Continued)

Substance	Δh_f^o $\left(\dfrac{kJ}{mol}\right)$	Δg_f^o $\left(\dfrac{kJ}{mol}\right)$	s^o $\left(\dfrac{J}{K \cdot mol}\right)$	c_P^o $\left(\dfrac{J}{K \cdot mol}\right)$	$\rho^{(1)}$ $\left(\dfrac{kg}{m^3}\right)$	T_{mp} (K)	$\Delta h_{fus}^{o\,(2)}$ $\left(\dfrac{kJ}{mol}\right)$
CuO	−157.3	−129.7	42.63	42.30	6310	1719	11.8
$CuSO_4$	−771.36	−661.8	109	100.0	3600	833[3]	
$CuSO_4 \cdot H_2O$	−1085.8	−918.11	146.0	134			
$CuSO_4 \cdot 5H_2O$	−2279.7	−1879.7	300.4	280	2280	383[3]	
Fe	0	0	27.28	25.10	7874	1810	13.81
Fe_2O_3 (hematite)	−824.2	−742.2	87.40	103.85	5250	1838	
Fe_3O_4 (magnetite)	−1118.4	−1015.4	146.4	143.43		1870	138
FeS (α)	−100.0	−100.4	60.29	50.54	4700	1461	31.5
FeS_2	−178.2	−166.9	52.93	62.17	5020	875[3]	
α-D-glucose, $C_6H_{12}O_6$	−1274						
β-D-glucose, $C_6H_{12}O_6$	−1268	−910	212		1544	415	
β-D-fructose, $C_6H_{12}O_6$	−1266				1600	376[3]	
H_3PO_3	−964.4				1650	348	12.8
H_3PO_4	−1279.0	−1119.1	110.50	106.06		316	13.4
HgO (red)	−90.83	−58.539	70.29	44.06	11 140	773[3]	
Hg_2Cl_2	−265.22	−210.745	192.5	102	7160	798[4]	
$HgCl_2$	−224.3	−178.6	146.0	73.5	5600	549	19.41
HgS (black)	−53.6	−47.7	88.3	49.9	7700	856	
I_2	0	0	116.135	54.438	4930	387	15.52
K	0	0	64.18	29.58	862	337	2.33
KBr	−393.80	−380.66	95.90	52.30	2750	1003	25.3
KCl	−436.747	−409.14	82.59	51.30	1984	1049	26.53
KF	−576.27	−537.75	66.57	49.04	2480	1131	27.2
KI	−327.90	−324.89	106.32	52.93	3120	954	24
KOH	−424.764	−379.08	78.9	64.9	2044	679	7.9
Lactic acid, $C_3H_6O_3$	−694.0						
Li	0	0	29.12	24.77	534	454	3.00
Mg	0	0	32.68	24.89	1738	923	8.48
$MgCl_2$	−641.32	−591.79	89.62	71.38	2325	549	19.41
$MgCO_3$	−1095.8	−1012.1	65.7	75.52	3050	1263	59
MgO	−601.70	−569.43	26.94	37.15	3600	3098	77
Na	0	0	51.21	28.24	971	371	2.601
NaBr	−361.06	−348.98	86.82	51.38	3200	1020	26.11
NaCl	−411.153	−384.138	72.13	50.50	2165	1074	28.16

Continues...

THERMODYNAMIC DATA FOR SELECTED SOLIDS

at $P° = 1$ bar and 298.15 K (Continued)

Substance	Δh_f^o $\left(\dfrac{kJ}{mol}\right)$	Δg_f^o $\left(\dfrac{kJ}{mol}\right)$	s^o $\left(\dfrac{J}{K\cdot mol}\right)$	c_P^o $\left(\dfrac{J}{K\cdot mol}\right)$	ρ [1] $\left(\dfrac{kg}{m^3}\right)$	T_{mp} (K)	Δh_{fus}^o [2] $\left(\dfrac{kJ}{mol}\right)$
NaOH	−425.609	−379.494	64.455	59.54	2130	596	6.60
Naphthalene, $C_{10}H_8$	78.5	201.6	224.1	165.7	1025	353	19.01
NH_4Cl	−314.43	−202.87	94.6	84.1	1519	793[3,4]	
NH_4NO_3	−365.56	−183.87	151.08	84.1	1720	483[3]	6.40
N_2O_5	−43.1	113.9	178.2	143.1	2000	303	
Oxalic acid, $(COOH)_2$	−821.7	−442.8	109.8	91.0			
P (white)	0	0	41.09	23.84	1830	317	0.66
PCl_5	−443.5				2100	440[4]	
P_4O_6	−1640.1				2130	297	
P_4O_{10}	−2984.0	−2697.0	228.86	211.71	2300	835	
Pb	0	0	64.81	26.44	11 350	601	4.77
PbO (yellow)	−217.32	−187.89	68.70	45.77	9640		
PbO (red)	−218.99	−188.93	66.5	45.81	9350	1161	
PbO_2	−277.4	−217.33	68.6	64.64	9640	563[3]	
Phenol, C_6H_5OH	−165	−50.9	146.0		1073	314	11.51
S (rhombic)	0	0	31.80	22.64	2070	392	
S (monoclinic)	0.33	0.1	32.6	23.6	1819	388	1.72
Sb	0	0	45.69	25.23	6680	904	19.87
Si	0	0	18.83	20.0	2330	1687	50.21
$SiO_2(\alpha)$	−910.94	−856.64	41.84	44.43	1995	9.6	
Sn (β, white)	0	0	51.55	26.99	7298	505	7.03
SnO	−285.8	−256.9	56.5	44.31	6450	1353[3]	
SnO_2	−580.7	−519.6	52.3	52.59	6850	1903	
Sucrose, $C_{12}H_{22}O_{11}$	−2222	−1543	360.2		1588	457[3]	
Ti	0	0	30.63	25.02	4540	1941	14.15
TiO_2	−939.7	−884.5	49.92	55.48	4230	2116	
Urea, $CO(NH_2)_2$	−333.51	−197.33	104.60	93.14	1323	406	13.9
Zn	0	0	41.63	25.40	7133	693	7.32
ZnO	−348.28	−318.30	43.64	40.25	5600	2247	52.3

Source: Most of the thermochemical data are from the National Bureau of Standards *Tables of Chemical Thermodynamic Properties*, published as a supplement to Vol. II of the *Journal of Physical and Chemical Reference Data* (1982). Densities and melting points: *CRC Handbook of Chemistry and Physics, 80th Edition*, D.R. Lide (ed.-in-chief), CRC Press, Boca Raton (1999)
[1] At room temperature (20~25°C). [2] At the normal melting point. [3] Decomposes. [4] Triple point.

APPENDIX 5. THERMODYNAMIC DATA FOR SELECTED AQUEOUS IONS[1] at P^o = 1 bar and 298.15 K

	Δh_f^o $\left(\dfrac{kJ}{mol}\right)$	Δg_f^o $\left(\dfrac{kJ}{mol}\right)$	s^o $\left(\dfrac{J}{K \cdot mol}\right)$	Substance	Δh_f^o $\left(\dfrac{kJ}{mol}\right)$	Δg_f^o $\left(\dfrac{kJ}{mol}\right)$	s^o $\left(\dfrac{J}{K \cdot mol}\right)$
$Ag^+_{(aq)}$	105.579	77.107	72.68	$ClO^-_{2(aq)}$	−66.5	17.2	101.3
$Al^{3+}_{(aq)}$	−531	−485	−321.7	$ClO^-_{3(aq)}$	−104.0	−8.0	162.3
$AlCl_{3(aq)}$	−1033.0	−879.0	−152.3	$ClO^-_{4(aq)}$	−129.3	−8.5	182.0
$AlO^-_{2(aq)}$	−930.9	−830.9	−36.8	$Cr^{2+}_{(aq)}$	−143.5	−180.7	−23.8
$Al(OH)^-_{4(aq)}$	−1502.5	−1305.3	102.9	$CrO^{2-}_{4(aq)}$	−881.2	−727.8	50.2
$AsO^-_{2(aq)}$	−429.0	−350.0	40.6	$Cr_2O^{2-}_{7(aq)}$	−1490.3	−1301.1	261.9
$AsO^-_{4(aq)}$	−888.1	−648.4	−162.8	$Cs^+_{(aq)}$	−258.28	−292.02	133.05
$BF^-_{4(aq)}$	−1574.9	−1486.9	180.0	$Cu^+_{(aq)}$	71.67	49.98	40.60
$BH^-_{4(aq)}$	48.2	114.4	110.5	$Cu^{2+}_{(aq)}$	64.77	65.49	−99.6
$BO^-_{2(aq)}$	−772.4	−678.9	−37.2	$F^-_{(aq)}$	−332.63	−278.79	−13.8
$Ba^{2+}_{(aq)}$	−537.64	−560.77	9.6	$Fe^{2+}_{(aq)}$	−89.1	−78.90	−137.7
$Be^{2+}_{(aq)}$	−382.8	−379.7	−129.7	$Fe^{3+}_{(aq)}$	−48.5	−4.7	−315.9
$BeO^-_{2(aq)}$	−790.8	−640.1	−159.0	$Fe(CN)^{3-}_{6(aq)}$	561.9	729.4	270.3
$Br^-_{(aq)}$	−121.6	−104.0	82.4	$Fe(CN)^{4-}_{6(aq)}$	455.6	695.1	95.0
$BrO^-_{(aq)}$	−94.1	−33.4	42.0	$FeOH^+_{(aq)}$	−324.7	−277.4	−29.0
$BrO^-_{3(aq)}$	−67.1	18.6	161.7	$FeOH^{2+}_{(aq)}$	−290.8	−229.4	−142.0
$BrO^-_{4(aq)}$	13.0	118.1	199.6	$H^+_{(aq)}$	0	0	0
$Ca^{2+}_{(aq)}$	−542.83	−553.58	−53.1	$HCO^-_{3(aq)}$	−691.99	−586.77	91.2
$Cd^{2+}_{(aq)}$	−75.90	−77.612	−73.2	$HC_2O^-_{4(aq)}$	−818.4	−698.3	149.4
$Ce^{3+}_{(aq)}$	−696.2	−672.0	−205.0	$H_2CO_{3(aq)}$	−699.65	−623.08	187.4
$Ce^{4+}_{(aq)}$	−537.2	−503.8	−301.0	$HF^-_{2(aq)}$	−649.9	−578.1	92.5
$Co^{2+}_{(aq)}$	−58.2	−54.4	−113.0	$HPO^{2-}_{4(aq)}$	−1292.1	−1089.2	−33.5
$Co^{3+}_{(aq)}$	92.0	134.0	−305.0	$HP_2O^{2-}_{7(aq)}$	−2274.8	−1972.2	46.0
$CH_3COO^-_{(aq)}$	−486.01	−369.31	86.6	$HS^-_{(aq)}$	−17.6	12.08	62.8
$CN^-_{(aq)}$	150.6	172.4	94.1	$HSO^-_{3(aq)}$	−626.2	−527.7	139.7
$CNO^-_{(aq)}$	−146.0	−97.4	106.7	$HS_2O^-_{4(aq)}$		−614.5	
$CO_{2(aq)}$	−413.80	−385.98	117.6	$HSO^-_{4(aq)}$	−887.34	−755.91	131.8
$CHOO^-_{(aq)}$	−425.6	−351.0	92.0	$HSe^-_{(aq)}$	15.9	44.0	79.0
$CO^{2-}_{3(aq)}$	−677.14	−527.81	−56.9	$HSeO^-_{3(aq)}$	−514.6	−411.5	135.1
$C_2O^{2-}_{4(aq)}$	−825.1	−673.9	45.6	$HSeO^-_{4(aq)}$	−581.6	−452.2	149.4
$Cl^-_{(aq)}$	−167.159	−131.228	56.5	$H_2AsO^-_{3(aq)}$	−714.8	−587.1	110.5
$ClO^-_{(aq)}$	−107.1	−36.8	42.0	$H_2AsO^-_{4(aq)}$	−909.6	−753.2	117.0

Continues...

THERMODYNAMIC DATA FOR SELECTED AQUEOUS IONS
at $P^o = 1$ bar and 298.15 K (Continued)

Substance	Δh_f^o $\left(\dfrac{kJ}{mol}\right)$	Δg_f^o $\left(\dfrac{kJ}{mol}\right)$	s^o $\left(\dfrac{J}{K \cdot mol}\right)$	Substance	Δh_f^o $\left(\dfrac{kJ}{mol}\right)$	Δg_f^o $\left(\dfrac{kJ}{mol}\right)$	s^o $\left(\dfrac{J}{K \cdot mol}\right)$
$H_2PO_{4(aq)}^-$	−1296.3	−1130.2	90.4	$NiOH_{(aq)}^+$	−287.9	−227.6	−71.0
$Hg_{(aq)}^{2+}$	171.1	164.40	−32.2	$OH_{(aq)}^-$	−229.99	−157.24	−10.75
$HgOH_{(aq)}^+$	−84.5	−52.3	71.0	$PO_{4(aq)}^{3-}$	−1277.4	−1018.7	−221.8
$Hg_{2(aq)}^{2+}$	172.4	153.52	84.5	$P_2O_{7(aq)}^{4-}$	−2271.1	−1919.0	−117.0
$I_{(aq)}^-$	−55.19	−51.57	111.3	$Pb_{(aq)}^{2+}$	−1.7	−24.43	10.5
$IO_{(aq)}^-$	−107.5	−38.5	−5.4	$Rb_{(aq)}^+$	−251.17	−283.98	121.50
$IO_{3(aq)}^-$	−221.3	−128.0	118.4	$Re_{(aq)}^-$	46.0	10.1	230.0
$IO_{4(aq)}^-$	−151.5	−58.5	222.0	$S_{(aq)}^{2-}$	33.1	85.8	−14.6
$K_{(aq)}^+$	−252.38	−283.27	102.5	$SCN_{(aq)}^-$	76.4	92.7	144.3
$La_{(aq)}^{3+}$	−707.	−683.7	−217.6	$SO_{3(aq)}^{2-}$	−635.5	−486.5	−29.0
$Li_{(aq)}^+$	−278.49	−293.31	13.4	$SO_{4(aq)}^{2-}$	−909.27	−744.53	20.1
$Mg_{2(aq)}^{2+}$	−466.85	−454.8	−138.1	$S_2O_{3(aq)}^{2-}$	−652.3	−522.5	67.0
$Mn_{(aq)}^{2+}$	−220.8	−228.1	−73.6	$S_2O_{4(aq)}^{2-}$	−753.5	−600.3	92.0
$MnO_{4(aq)}^-$	−541.4	−447.2	191.2	$S_2O_{8(aq)}^{2-}$	−1344.7	−1114.9	244.3
$MnO_{4(aq)}^{2-}$	−653.0	−500.7	59.0	$SeO_{3(aq)}^{2-}$	−509.2	−369.8	13.0
$MnOH_{(aq)}^+$	−450.6	−405.0	−17.0	$SeO_{4(aq)}^{2-}$	−599.1	−441.3	54.0
$MoO_{4(aq)}^{2-}$	−997.9	−836.3	27.2	$Sn_{(aq)}^{2+}$	−8.8	−27.2	−17.0
$N_{3(aq)}^-$	275.1	348.2	107.9	$SnOH_{(aq)}^+$	−286.2	−254.8	50.0
$NH_{3(aq)}$	−80.29	−26.50	111.3	$Sr_{(aq)}^{2+}$	−545.8	−559.5	−32.6
$NH_{4(aq)}^+$	−132.51	−79.31	113.4	$Tl_{(aq)}^+$	5.4	−32.4	125.5
$NO_{2(aq)}^-$	−104.6	−32.2	123.0	$Tl_{(aq)}^{3+}$	196.6	214.6	−192.0
$NO_{3(aq)}^-$	−205.0	−108.74	146.4	$UO_{2(aq)}^{2+}$	−1019.6	−953.5	−97.5
$Na_{(aq)}^+$	−240.12	−261.905	59.0	$VO_{3(aq)}^-$	−888.3	−783.6	50.0
$Ni_{(aq)}^{2+}$	−54.0	−45.6	−128.9	$Zn_{(aq)}^{2+}$	−153.89	−147.06	−112.1

Sources: National Bureau of Standards *Tables of Chemical Thermodynamic Properties*, published as a supplement to Vol. II of the *Journal of Physical and Chemical Reference Data* (1982); *CRC Handbook of Chemistry and Physics, 80th Edition*, D.R. Lide (ed.-in-chief), CRC Press, Boca Raton (1999); J.F. Zemaitis, D.M. Clark, and N.C. Scrivner, *Handbook of Aqueous Electrolyte Thermodynamics*, American Institute of Chemical Engineers, New York (1986).
[1] At infinite dilution

APPENDIX 6. ANALYTICAL SOLUTION OF A CUBIC EQUATION

Given the equation: $\qquad x^3 + px^2 + qx + r = 0,\qquad$ solve for x.

Let
$$a = \tfrac{1}{3}\left(3q - p^2\right)$$

$$b = \tfrac{1}{27}\left(2p^3 - 9pq + 27r\right)$$

$$A = \sqrt[3]{-\frac{b}{2} + \sqrt{\frac{b^2}{4} + \frac{a^3}{27}}}$$

$$B = \sqrt[3]{-\frac{b}{2} - \sqrt{\frac{b^2}{4} + \frac{a^3}{27}}}$$

The three roots are:
$$x = A + B - \frac{p}{3}$$

$$x = -\left(\frac{A+B}{2}\right) + \left(\frac{A-B}{2}\right)\sqrt{-3} - \frac{p}{3}$$

$$x = -\left(\frac{A+B}{2}\right) - \left(\frac{A-B}{2}\right)\sqrt{-3} - \frac{p}{3}$$

If $\left(\dfrac{b^2}{4} + \dfrac{a^3}{27}\right) > 0,$ there is one *real* root, and two *imaginary* roots;

$\qquad\qquad\qquad = 0,$ there are 3 *real* roots, of which at least 2 are *equal*;

$\qquad\qquad\qquad < 0,$ there are 3 *real* and *unequal* roots.

In the last case, i.e., $\left(\dfrac{b^2}{4} + \dfrac{a^3}{27}\right) < 0,$ a trigonometric solution often is useful:

Thus, compute the value (in *degrees*) of the angle ϕ in the expression $\qquad \cos\phi = \dfrac{-b/2}{\sqrt{-a^3/27}}.$

The three roots of x are: $\qquad x = \left[2\sqrt{-\frac{a}{3}} \cdot \cos\left(\frac{\phi}{3}\right)\right] - \frac{p}{3}$

$$x = \left[2\sqrt{-\frac{a}{3}} \cdot \cos\left(\frac{\phi}{3} + 120°\right)\right] - \frac{p}{3}$$

$$x = \left[2\sqrt{-\frac{a}{3}} \cdot \cos\left(\frac{\phi}{3} + 240°\right)\right] - \frac{p}{3}$$

Example

Solve for the roots of the cubic equation:

$$x^3 - 6x^2 + 11x - 6 = 0$$

Solution

Comparing the given equation $x^3 - 6x^2 + 11x - 6 = 0$ with $x^3 + px^2 + qx + r = 0$ shows that $\qquad$ $p = -6, \; q = +11, \; r = -6$

Therefore: $\quad a = \frac{1}{3}\left(3q - p^2\right) = \frac{1}{3}\left[3(11) - (-6)^2\right] = \frac{1}{3}[33 - 36] = \mathbf{-1}$

$$b = \frac{1}{27}\left(2p^3 - 9pq + 27r\right) = \frac{1}{27}\left[2(-6)^3 - 9(-6)(11) + 27(-6)\right]$$

$$= \frac{1}{27}[-432 + 594 - 162] = \frac{1}{27}[0] = \mathbf{0}$$

$$A = \sqrt[3]{-\frac{b}{2} + \sqrt{\frac{b^2}{4} + \frac{a^3}{27}}} = \sqrt[3]{-\frac{0}{2} + \sqrt{\frac{0^2}{4} + \frac{(-1)^3}{27}}} = \left[\sqrt{\left(\frac{-1}{3}\right)^3}\right]^{\frac{1}{3}} = \left[\left(-\frac{1}{3}\right)^{\frac{3}{2}}\right]^{\frac{1}{3}}$$

$$= \left(-\frac{1}{3}\right)^{\frac{1}{2}} = (-1)^{\frac{1}{2}}\left(\frac{1}{3}\right)^{\frac{1}{2}} = \frac{\sqrt{-1}}{\sqrt{3}}$$

$$B = \sqrt[3]{-\frac{b}{2} - \sqrt{\frac{b^2}{4} + \frac{a^3}{27}}} = \sqrt[3]{-\frac{0}{2} - \sqrt{\frac{0^2}{4} + \frac{(-1)^3}{27}}} = \left[-\left(-\frac{1}{3}\right)^{\frac{3}{2}}\right]^{\frac{1}{3}} = (-1)^{\frac{1}{3}}\left[\left(-\frac{1}{3}\right)^{\frac{3}{2}}\right]^{\frac{1}{3}}$$

$$= (-1)\left[(-1)^{\frac{3}{2}}\left(\frac{1}{3}\right)^{\frac{3}{2}}\right]^{\frac{1}{3}} = -(-1)^{\frac{1}{2}}\left(\frac{1}{3}\right)^{\frac{1}{2}} = -\frac{\sqrt{-1}}{\sqrt{3}} = \mathbf{-A}$$

The value of $\left(\dfrac{b^2}{4} + \dfrac{a^3}{27}\right) = \left(-\dfrac{1}{3}\right)^3 = -\dfrac{1}{27} < 0$; therefore there are 3 *real* and *unequal* roots.

The three roots are:

$$x_1 = A + B - \frac{p}{3} = \frac{\sqrt{-1}}{\sqrt{3}} - \frac{\sqrt{-1}}{\sqrt{3}} - \frac{(-6)}{3} = \mathbf{2}$$

$$x_2 = -\left(\frac{A + B}{2}\right) + \left(\frac{A - B}{2}\right)\sqrt{-3} - \frac{p}{3} = -\left(\frac{A - A}{2}\right) + \left(\frac{A - (-A)}{2}\right)\sqrt{-3} - \frac{p}{3}$$

$$= 0 + \left(\frac{2A}{2}\right)\sqrt{-3} - \frac{(-6)}{3} = A\sqrt{-3} + 2 = \frac{\sqrt{-1}}{\sqrt{3}} \cdot \sqrt{-3} + 2 = \frac{\sqrt{(-1)(-3)}}{\sqrt{3}} + 2$$

$$= \frac{\sqrt{3}}{\sqrt{3}} + 2 = 1 + 2 = \mathbf{3}$$

$$x_3 = -\left(\frac{A+B}{2}\right) - \left(\frac{A-B}{2}\right)\sqrt{-3} - \frac{p}{3} = 0 - \left(\frac{2A}{2}\right)\sqrt{-3} - \frac{(-6)}{3}$$

$$= -A\sqrt{-3} + 2 = -\frac{\sqrt{-1}}{\sqrt{3}} \cdot \sqrt{-3} + 2 = -1 + 2 = \mathbf{1}$$

Ans: $\boxed{x = 1,\ 2,\ 3}$

APPENDIX 7.   THE NEWTON-RAPHSON METHOD

The Newton-Raphson Method (or *the method of tangents* or, just *Newton's method*) is a method of finding the roots of equations of the general form $f(x) = 0$, where $f(x)$ and its derivative $f'(x)$ are continuous functions. Of the various methods put forward for the numerical solution of equations, this method seems generally to be the most satisfactory. As will be shown below, the method is based on the premise that a line drawn tangent to the curve $f(x)$ at some point x_1 will intersect the *x*-axis at a point x_2 where the value of x_2 is closer to the root than is the value x_1. A useful feature of the method is that it is self-correcting for minor errors.

The method can be illustrated by considering the van der Waals equation of state for one mole of gas:

$$\left(P + \frac{a}{v^2} \right)(v - b) = RT \qquad \dots \text{[A7-1]}$$

Given the molar volume v and the temperature T, the pressure P is easily determined as follows:

Rearranging:
$$P + \frac{a}{v^2} = \frac{RT}{v - b}$$

$$P = \frac{RT}{v - b} - \frac{a}{v^2} \qquad \dots \text{[A7-2]}$$

Consider a rigid one liter tank containing four moles of gaseous methane (CH_4) at 0°C. Determine the pressure of this gas according to the van der Waals equation. The van der Waals constants for methane gas are $a = 0.23026$ Pa m^6 mol^{-2} and $b = 4.3067 \times 10^{-5}$ m^3 mol^{-1}.

The molar volume of the gas is

$$v = \frac{1.00 \times 10^{-3} \text{ m}^3}{4 \text{ mol}} = 2.5 \times 10^{-4} \text{ m}^3 \text{ mol}^{-1}$$

Therefore, from Eqn [A7-2]:

$$P = \frac{RT}{v - b} - \frac{a}{v^2}$$

$$= \frac{(8.3145)(273.15)}{2.5 \times 10^{-4} - 4.3067 \times 10^{-5}} - \frac{0.23026}{(2.5 \times 10^{-4})^2}$$

$$= 10\ 975\ 077 - 3\ 684\ 160$$

$$= 7\ 290\ 917 \text{ Pa}$$

$$= 72.909 \text{ bar}^1$$

---

[1] The actual (experimental) value is 78.6 bar; the value predicted by the ideal gas equation is $P = RT/v = 90.84$ bar. The error using the van der Waals equation is −7.25%, whereas the error using the ideal gas equation is +15.6%. The van der Waals equation is definitely better than the ideal gas equation at such moderately high pressures.

Now suppose we were given that the pressure was 72.909 bar at 0°C and asked to use the van der Waals equation to calculate the molar volume v. When you try to rearrange Eqn [A7-1] to solve for v, you get the following:

$$\left(P + \frac{a}{v^2}\right)(v - b) = RT$$

Multiplying:
$$Pv + \frac{a}{v} - Pb - \frac{ab}{v^2} = RT$$

Multiplying through by v^2:
$$Pv^3 + av - Pbv^2 - ab = RTv^2$$

Rearranging:
$$Pv^3 - Pbv^2 - RTv^2 + av - ab = 0$$

Dividing through by P:
$$v^3 - \left(b + \frac{RT}{P}\right)v^2 + \left(\frac{a}{P}\right)v - \left(\frac{ab}{P}\right) = 0 \qquad \ldots \text{[A7-3]}$$

Eqn [A7-3] is a cubic equation in v. Although there are analytical formulas to solve cubic[2] equations, these formulas are very cumbersome to use.

Polynomial equations such as Eqn [A7-3] are often encountered in science and engineer–ing, so we may as well learn how to solve them, and, as men-tioned above, we are going to do this by making use of the *Newton-Raphson* method. To understand how the method works, consider some general-ized function $y = f(x)$, shown in Fig. 1.

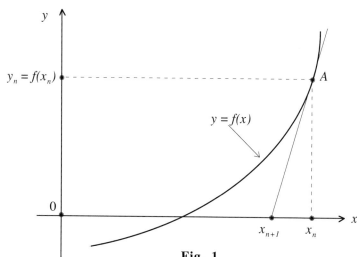

Fig. 1

Consider some point A on the curve at $x = x_n$:

The tangent to the curve at this point intersects the x–axis at $x = x_{n+1}$, and the slope of this tangent is given by

$$f'(x) = \frac{dy}{dx} = \frac{y_n - 0}{x_n - x_{n+1}} = \frac{f(x_n) - 0}{x_n - x_{n+1}}$$

---

[2] Formulas also are available to solve quartic (4th power) equations; but there are no formulas that can be used to solve fifth and higher power equations.

Therefore:
$$f'(x) = \frac{f(x_n)}{x_n - x_{n+1}}$$

Rearranging:
$$x_n - x_{n+1} = \frac{f(x_n)}{f'(x)}$$

From which:
$$x_{n+1} = x_n - \frac{f(x_n)}{f'(x)}$$
... [A7-4]

Now suppose we are given the same equation

$$y = f(x) = 0$$

and we wish to solve for x. Solving for x means we must find the value (or values) of x for which $y = f(x) = 0$. Graphically, as shown in Fig. 2, this means we must find x_*, which is the value[3] of x at which the curve intersects the x-axis, where $y = f(x_*) = 0$.

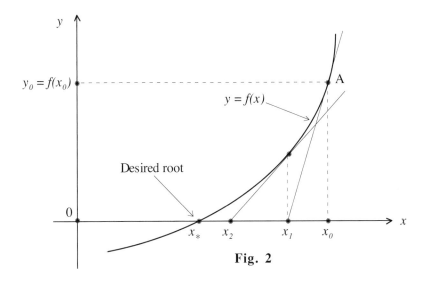

Fig. 2

APPLICATION OF THE METHOD

1. First, we guess some initial value x_0 for x that we have reason to believe is somewhat close to the actual root x_*.[4] In the case of solving for P, v, or T for a non-ideal gas, we usually use the value predicted by the ideal gas equation as the initial guess.

---

[3] There may be more than one intersection point, depending on $f(x)$.

[4] If the initial guess is too far from the root we want, we may end up solving for some other, extraneous, root of $f(x)$.

2. Next we work out the expression for the derivative $f'(x) = \dfrac{dy}{dx}$ of our function $y = f(x)$, and calculate the value $f'(x_0)$ of the derivative when $x = x_0$.

3. Then we use Eqn [A7-4] to evaluate a value of the second guess x_1 for x_* as follows:

$$x_1 = x_0 - \frac{f(x_0)}{f'(x_0)}$$

As can be seen from Fig. 2, the second value x_1 is closer to x_* than the initial guess of x_0.

4. Eqn [A7-3] is then used again to obtain a third guess x_2 using the second value x_1:

$$x_2 = x_1 - \frac{f(x_1)}{f'(x_1)}$$

Reference to Fig. 2 indicates that the value of each successive guess quickly approaches the desired value x_*, and that each successively computed value of $f(x)$ becomes closer to zero. The root is found when $f(x) = 0$ to the degree of accuracy desired such that two successive guesses yield the same value for x. A convenient rule of thumb is that, if the correction term $f(x_i)/f'(x_i)$ begins with n zeros after the decimal point, then the result is correct to about $2n$ decimals; i.e., the number of correct decimals approximately doubles at each stage.

Example 1

Above we used the van der Waals equation to calculate a pressure of 7 290 917 Pa for methane gas at 0°C and a molar volume of 2.500×10^{-4} m³ mol⁻¹. Suppose instead that we were given the *pressure* of the gas at 0°C and asked to calculate the *molar volume*. We can do this using the Newton-Raphson method as follows:

Eqn [A7-3] shows that the molar volume of a van der Waals gas is a cubic equation in v;

$$f(v) = v^3 - \left(b + \frac{RT}{P}\right)v^2 + \left(\frac{a}{P}\right)v - \left(\frac{ab}{P}\right) \qquad \ldots \text{[A7-3]}$$

Inserting values:

$$f(v) = v^3 - \left(4.3067 \times 10^{-5} + \frac{(8.3145)(273.15)}{7\,290\,917}\right)v^2 + \left(\frac{0.23026}{7\,290\,917}\right)v - \left(\frac{(0.23026)(4.3067 \times 10^{-5})}{7\,290\,917}\right)$$

$$= v^3 - 3.54565 \times 10^{-4}\,v^2 + 3.15818 \times 10^{-8}\,v - 1.36013 \times 10^{-12} \qquad \ldots \text{[a]}$$

Next we evaluate the first derivative of Eqn [a]:

$$f'(v) = 3v^2 - 7.09130 \times 10^{-4}\,v + 3.15818 \times 10^{-8} \qquad \ldots \text{[b]}$$

As a first guess v_0 we use the value predicted by the ideal gas law:

$$v_0 = \frac{RT}{P} = \frac{(8.3145)(273.15)}{7\,290\,917} = 3.1150 \times 10^{-4} \text{ m}^3 \text{ mol}^{-1}$$

Then we generate the following table:

n	v_n	$f(v_n)$	$f'(v_n)$	$v_{n+1} = v_n - \dfrac{f(v_n)}{f'(v_n)}$
0	3.1150×10^{-4}	4.29890×10^{-12}	1.01785×10^{-7}	2.69265×10^{-4}
1	2.69265×10^{-4}	9.59159×10^{-13}	5.81486×10^{-8}	2.52770×10^{-4}
2	2.52770×10^{-4}	1.18828×10^{-13}	4.40128×10^{-8}	2.50070×10^{-4}
3	2.50070×10^{-4}	2.92330×10^{-15}	4.18546×10^{-8}	2.50000×10^{-4}
4	2.50000×10^{-4}	1.92900×10^{-18}	4.17993×10^{-8}	2.50000×10^{-4}
5	2.50000×10^{-4}	3.00000×10^{-21}	4.17993×10^{-8}	2.50000×10^{-4}
6	2.50000×10^{-4}	0.00000	4.17993×10^{-8}	2.50000×10^{-4}

It is seen that the function converges to within 5 decimal points after only 4 iterations, and that with each iteration $f(v_n)$ gets closer to 0 as v_n gets closer to v_*, the true value. After 6 iterations the function has completely converged. Note also that $f'(v_n)$, the slope of the curve where it crosses the v-axis, also must approach a constant value.

Ans: $\boxed{v = 2.50000 \times 10^{-4} \text{ m}^3 \text{ mol}^{-1}}$

The Newton-Raphson method works very well and usually converges within 10 iterations for the kinds of problems we are interested in. However, it doesn't always work, and when it doesn't it will be pretty obvious that the function is not converging. Another problem to be aware of is that unless the initial guess is close to the root sought, the process may actually yield some other root. In such cases other methods[5] must be used.

Occasionally, an equation will take more than 10 iterations to solve. For example, using the Newton-Raphson method to solve the equation

$$exp\,[-x] + sin\,x + \frac{x}{2} = 1$$

required more than 50 iterations to get a solution of $x = 1.50235$.

Fortunately, the method is readily carried out using a spreadsheet.

---

[5] Such as Simpson's rule, or even brute force trial and error. Software capable of solving such equations also is readily available.

If you want to practice using the Newton-Raphson method, try solving the following exercises without looking at the solutions, which are given following the problems:

1. $x^3 + 3x^2 + 3x - 1 = 0$

    Solve for x to five decimal places, if the desired root lies somewhere between 0 and 1.

    [**Ans**: 0.25992]

2. $x^5 + 2x^4 + 4x = 5$

    Solve for x to four decimal places for the root that lies between 0 and 1.

    [**Ans**: 0.8596]

3. $e^{-x} + \dfrac{x}{5} = 1$

    Solve for x to four significant figures.

    [**Ans**: 4.965]

SOLUTIONS TO THE EXERCISES

| Exercise 1 |

$$f(x) = x^3 + 3x^2 + 3x - 1$$

and $\qquad f'(x) = 3x^2 + 6x + 3$

We are told that the root lies between 0 and 1; therefore, as a first guess, try $x_0 = 0.5$:

n	x_n	$f(x_n)$	$f'(x_n)$	$x_{n+1} = x_n - \dfrac{f(x_n)}{f'(x_n)}$
0	0.50000	1.37500	6.7500	0.296300
1	0.296300	0.178294	5.04118	0.260930
2	0.260930	0.004809	4.76983	0.259920
3	0.259920	−0.000005	4.76220	0.259920
4	0.259920			

Therefore the answer to five decimal places is 0.25992.

Ans: | 0.25992 |

Exercise 2

$$f(x) = x^5 + 2x^4 + 4x - 5$$

and $$f'(x) = 5x^4 + 8x^3 + 4$$

We are told that the root lies between 0 and 1; therefore, as a first guess, try $x_0 = 0.5$:

n	x_n	$f(x_n)$	$f'(x_n)$	$x_{n+1} = x_n - \dfrac{f(x_n)}{f'(x_n)}$
0	0.50000	-2.84375	5.3125	1.03529
1	1.03529	2.62821	18.62145	0.894155
2	0.894155	0.426631	12.91523	0.861122
3	0.861122	0.0177315	11.85774	0.859627
4	0.859627	3.4128×10^{-5}	11.81213	0.859624
5	0.859624			

Therefore the answer to four decimal places is 0.8596.

Ans: 0.8596

Exercise 3

$$f(x) = exp[-x] + 0.2x - 1$$

and $$f'(x) = -exp[-x] + 0.2$$

If $x > 1$, then $exp[-x]$ will be smaller than $0.2x$, and the equation would be approximated by

$$0.2x \approx 1$$

That is, $$x \approx 5$$

Therefore, as a first guess, try $x = 5$:

n	x_n	$f(x_n)$	$f'(x_n)$	$x_{n+1} = x_n - \dfrac{f(x_n)}{f'(x_n)}$
0	5.0000	6.73795×10^{-2}	0.19326	4.96514
1	4.96514	4.143×10^{-6}	0.19302	4.96511
2	4.96511			

Therefore the answer to four significant figures is 4.965.   **Ans**: 4.965

APPENDIX 8. UNDERSTANDING BASIC INTEGRATION

A8.1 AREA UNDER THE LINE $y = constant$

In Chapter 7 we saw that the heat required to heat a substance from T_1 to T_2 is given by

$$Q = C\Delta T \qquad \qquad \ldots [\text{A8-1}]$$

where C is the heat capacity,[1] in J K^{-1}, and $\Delta T = T_2 - T_1$. Eqn [A8-1] assumes that the heat capacity does not change with temperature. In the case shown it is of the form

$$C = a \qquad \ldots [\text{A8-2}]$$

where a is a constant. If we plot the heat capacity $vs.$ temperature, as shown in Fig. 1, we can see that the plot is just a horizontal straight line, and Q is just the cross-hatched area under the line between T_1 and T_2.

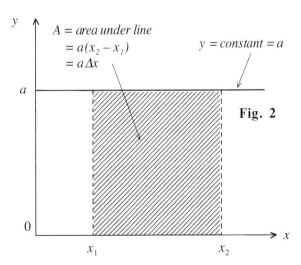

Fig. 1

Thus, $$Q = C\Delta T = a\left(T_2 - T_1\right) \qquad \qquad \ldots [\text{A8-3}]$$

We can extend this to the more familiar math terminology in which we use the letters x and y to denote the variables. Thus, for this case, the equation becomes

$$y = a \qquad \ldots [\text{A8-4}]$$

where a is a positive constant. A plot on the x and y axes now looks like Fig. 2, and the area under the line between $x = x_1$ and $x = x_2$ is given by

$$A = a(x_2 - x_1) \qquad \ldots [\text{A8-5}]$$

---

[1] Depending on the heating method, C may be the heat capacity at constant pressure C_P or at constant volume C_V.

A8.2 AREA UNDER THE LINE $y = bx$

Sometimes the heat capacity changes with the temperature. As an example, suppose the heat capacity varies linearly with T as

$$C = bT \qquad\qquad \dots \text{[A8-6]}$$

where b is a constant.

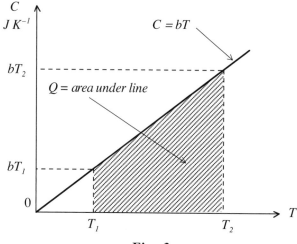

Now a plot of the heat capacity $vs.$ temperature exhibits the variation shown in Fig. 3, and, as before, the heat is just the *area under the line* between T_1 and T_2.

Now, the heat capacity at T_1 is $C_1 = bT_1$ and at T_2 is $C_2 = bT_2$. So what value of C do we use in Eqn [A8-1]?

Because the variation is *linear*, we just use the *average* value of C; namely

Fig. 3

$$C_{av} = \frac{C_2 + C_1}{2} = \frac{bT_2 + bT_1}{2} \qquad\qquad \dots \text{[A8-7]}$$

So, as shown in Fig. 4, the heat now will be

$$Q = C_{av}\Delta T$$

$$= \left(\frac{bT_2 + bT_1}{2}\right)(T_2 - T_1)$$

$$= b\left(\frac{T_2^2}{2} - \frac{T_1^2}{2}\right) \quad \dots \text{[A8-8]}$$

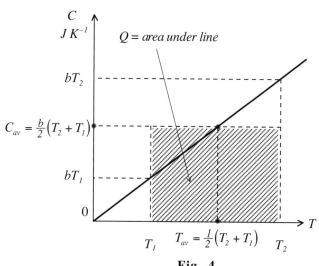

The cross-hatched area in Fig. 4 is the same as that in Fig. 3. So, once again, regardless of the equation used to describe C, the heat is just the area under the C-line between T_1 and T_2.

Fig. 4

As before, to make the result mathematically more general— so we can use these ideas for courses other than just physical chemistry—let's redo the above using x and y as the independ–ent and dependent variables, respectively.[2] When we do this, the equation becomes

$$y = bx \quad \ldots [A8\text{-}9]$$

where b is a constant. Now we want to find the area under the line between $x = x_1$ and $x = x_2$, as indicated in Fig. 5.

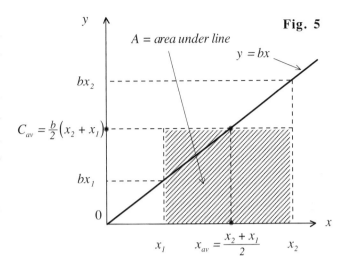

Fig. 5

$A = area\ under\ line$

$y = bx$

bx_2

$C_{av} = \dfrac{b}{2}\left(x_2 + x_1\right)$

bx_1

$x_1 \qquad x_{av} = \dfrac{x_2 + x_1}{2} \qquad x_2$

Using the results from above, this area will be

$$A = b\left(\frac{x_2^2}{2} - \frac{x_1^2}{2}\right) \qquad \ldots [A8\text{-}10]$$

A8.3 AREA UNDER THE LINE $y = a + bx$

Now let's suppose the heat capacity equation is given as

$$C = a + bT \qquad \ldots [A8\text{-}11]$$

where both a and b are positive constants. This is a very common form that describes the variation with temperature of the heat capacity of many substances. The plot is shown in Fig. 6, where, as before, the heat is just the cross-hatched area. Again, because the C vs. T line is linear, as in the previous case, to calculate the area under the line we can use the mid-point value of C as the correct average value:

$$C_{av} = \frac{C_2 + C_1}{2} = \frac{\left(a + bT_2\right) + \left(a + bT_1\right)}{2} = a + \frac{b}{2}\left(T_2 + T_1\right) \quad \ldots [A8\text{-}12]$$

[2] x is called the "independent" variable because you are free to give it any value you like; its value doesn't "depend" on anything other than the value you choose to give it. However, once you independently choose a value for x, the corresponding value of y is automatically determined by the relationship $y = bx$; that is, the value of y "depends" on the value chosen for x, so y is called the *dependent* variable.

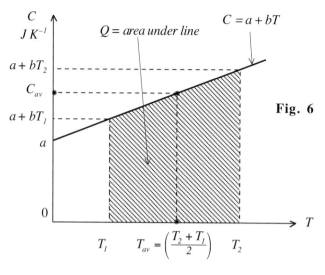

Fig. 6

Therefore

$$Q = C_{av}\Delta T = \left\{a + \frac{b}{2}(T_2 + T_1)\right\}(T_2 - T_1)$$

$$= a(T_2 - T_1) + b\left(\frac{T_2^2}{2} - \frac{T_1^2}{2}\right) \qquad \ldots [A8\text{-}13]$$

Thus, when the heat capacity is given by an equation of the type $C = a + bT$, Eqn [A8-13] shows that the heat is calculated by breaking down the expression for the heat capacity into its two components, the term a and the term bT, computing the heat for each term separately, and then adding the two contributions to get the total heat. The first term is evaluated using Eqn [A8-3] and the second term is evaluated using Eqn [A8-8].

In x and y notation, the area under the curve $y = a + bx$ between $x = x_1$ and $x = x_2$ is given by

$$A = a(x_2 - x_1) + b\left(\frac{x_2^2}{2} - \frac{x_1^2}{2}\right) \qquad \ldots [A8\text{-}14]$$

A8.4 AREA UNDER THE LINE $y = a + bx + cx^2$

The next case we shall consider is the case when the heat capacity of the substance is given by an expression of the type

$$C = a + bT + cT^2 \qquad \ldots [A8\text{-}15]$$

where a, b, and c all are constants. As you have probably guessed, the heat required to raise the temperature of this substance from T_1 to T_2 is just going to be the sum of three components, one evaluated for each of the three terms in Eqn [A8-15]. The first two are exactly the same as we evaluated above, in Eqn [A8-13]. The next term—the one involving the T^2—is trickier, and we will use a more sophisticated, mathematical method to evaluate the contribution to the heat from this term. Also, to save time and effort, let's proceed directly to the x and y notation.

Thus, we start by considering the expression

$$y = x^2 \qquad \qquad \ldots \text{[A8-16]}$$

As before, our task is to evaluate the area under this curve between two different values of x. But before we do that, first we will evaluate the area under the curve from $x = 0$ to some value $x = a$. Since the curve is not linear, we can't just use some arithmetic average value of x to evaluate the area under the curve. As can clearly be seen from Fig. 7, this will give an *inaccurate* value for the area.

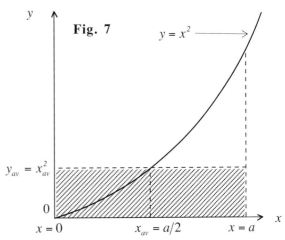

Fig. 7

Instead, as shown in Fig. 8, it makes more sense to break up the x-interval into a *series* of little rectangles, each of width Δx and of some representative height $\bar{y} = (x^*)^2$ taken at some representative x-value x^* within each rectangle. The total area under the curve then can be estimated as the *sum* of all the little component rectangles.

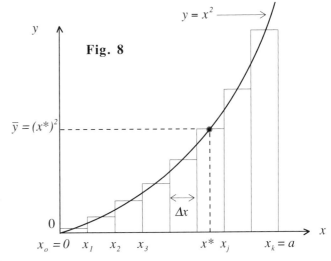

Fig. 8

But even here we seemingly encounter a difficulty: In order to evaluate the representative height $\bar{y}$ of each rectangle we need to know what value to choose for x^*. As in the case of y_{av} it would appear that we can't just choose as a representative value of x the point in the exact middle of each interval. Fortunately, in this case it doesn't matter what value of x we choose as x^*, provided that it lies somewhere *within* the interval of interest. This is true because as the value chosen for the interval Δx gets smaller and smaller, there is less and less variation in the resulting value of $\bar{y}$. As we will see below, to evaluate accurately the area under the curve we will chop up the curve into an *infinite* number of little rectangles.

Since we are free to choose *any* x-value within the base of each little rectangle as our "representative" value x^*, it has been found to be most convenient[3] to choose the x-value at the *right hand side* of each rectangle to be x^*, as indicated in Fig. 9.

First we divide the distance a into k equal-sized intervals, each of size Δx, as follows,

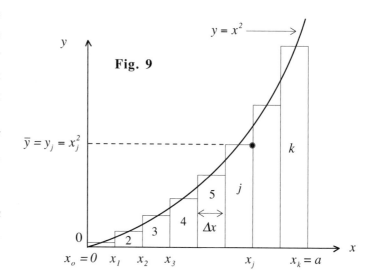

Fig. 9

where

$$\Delta x = \frac{a-0}{k} = \frac{a}{k} \qquad \text{. . . [A8-17]}$$

This defines $k + 1$ points: $x_o, \ x_1, \ x_2, \ . \ . \ . \ . \ x_j, \ . \ . \ . \ x_k$

You may notice that there is a little mismatch between the number of intervals and the number of x-points, in that there are k intervals but $k + 1$ points. For this reason we usually call the first point x_o instead of x_1. By so doing, we have nicely arranged things so that the labeling of the points becomes the same as the labeling of the intervals. That is, the representative x-values for the 3rd, 4th, and j-th intervals are x_3, x_4, and x_j, respectively.

We then *approximate* the total area under the curve as the sum of the k small rectangles, each of width Δx and the corresponding appropriate height. Thus, referring to Fig. 9, in the case of the j-th rectangle, the height and width are $\bar{y}_j = y_j = x_j^2$ and Δx, respectively, and the area is

$$A_j = y_j \cdot \Delta x = x_j^2 \Delta x \qquad \text{. . . [A8-18]}$$

The total area under the curve from $x = 0$ to $x = a$ is approximated by summing all the little rectangles:

---

[3] Most convenient because it gives the easiest formulas in terms of j.

$$A \approx \sum_{j=1}^{k} x_j^2 \cdot \Delta x \approx \sum_{j=1}^{k} x_j^2 \cdot \frac{a}{k} \qquad \dots [A8-19]\ ^4$$

Reference to Fig. 9 shows that to get to the right-hand endpoint of the j-th interval, we start at $x = 0$ and move to the right by the amount $\Delta x = a/k$ for a total of j times to obtain

$$x_j = j \cdot \Delta x = j \cdot \frac{a}{k} \qquad \dots [A8-20]$$

Therefore

$$y_j = x_j^2 = \left(j \cdot \frac{a}{k} \right)^2 = \frac{a^2 \cdot j^2}{k^2} \qquad \dots [A8-21]$$

Substituting [A8-21] into Eqn [A8-19] gives

$$A \approx \sum_{j=1}^{k} \left(\frac{a^2 \cdot j^2}{k^2} \right) \cdot \left(\frac{a}{k} \right) \approx \sum_{j=1}^{k} \frac{a^3 \cdot j^2}{k^3} = \frac{a^3}{k^3} \sum_{j=1}^{k} j^2 \qquad \dots [A8-22]\ ^5$$

Now, from mathematics it can be shown[6] that

$$\sum_{j=1}^{k} j^2 = 1^2 + 2^2 + 3^2 + \dots + k^2 = \frac{k(k+1)(2k+1)}{6} \qquad \dots [A8-23]$$

For example, 8 intervals are shown in Fig. 9, so that

$$\sum_{j=1}^{8} j^2 = 1^2 + 2^2 + 3^2 + 4^2 + 5^2 + 6^2 + 7^2 + 8^2$$
$$= 1 + 4 + 9 + 16 + 25 + 36 + 49 + 64$$
$$= 204$$

Applying the formula given by Eqn [A8-23] with $k = 8$ gives:

$$\sum_{j=1}^{8} j^2 = \frac{k(k+1)(2k+1)}{6} = \frac{8(8+1)(2 \times 8 + 1)}{6} = \frac{8 \times 9 \times 17}{6} = 204$$

which is the same! Right on!

Substitution of Eqn [A8-23] into Eqn [A8-22] gives

---

[4] The capital Greek letter $\sum$ (sigma) is called the *summation symbol*, and means that a whole bunch of terms are to be added together. The letter j is called the *counter* or the *index*.

[5] Equations such as [A8-22] are called "approximating sums" or "Riemann sums" (the latter after the famous German mathematician who developed these ideas in the nineteenth century).

[6] *I* certainly can't show it, but we'll have to take the mathematicians' word for it. You should certainly check it out for yourself to satisfy yourself that this formula is correct. Try it for a few values of k in addition to the one in the example.

$$A \approx \frac{a^3}{k^3}\left(\frac{k(k+1)(2k+1)}{6}\right) = \frac{a^3(k^2+k)(2k+1)}{6k^3} = \frac{a^3(2k^3+3k^2+k)}{6k^3} = \frac{2a^3k^3+3a^3k^2+a^3k}{6k^3}$$

Therefore,

$$A \approx \frac{a^3}{3} + \frac{a^3}{2k} + \frac{a^3}{6k^2} \qquad \ldots [A8\text{-}24]$$

As k gets bigger and bigger, the intervals Δx get smaller and smaller, as do the areas of the small rectangles, until, in the limit, when $k \to \infty$, the width of each rectangle becomes infinitesimally small such that the rectangles *exactly* fit the curve, with no jagged protrusions, as indicated in Fig. 10.

When this happens, the sum of all the little areas *exactly* equals the true area under the curve.

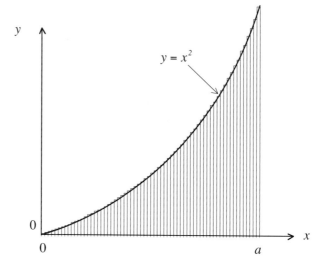

Fig. 10

Expressed mathematically,

$$A_{exact} = \lim_{k \to \infty}\left[\frac{a^3}{3} + \frac{a^3}{2k} + \frac{a^3}{6k^2}\right] = \left[\frac{a^3}{3} + \frac{a^3}{2 \times \infty} + \frac{a^3}{6(\infty)^2}\right] = \left[\frac{a^3}{3} + 0 + 0\right]$$

$$= \frac{a^3}{3} \qquad \ldots [A8\text{-}25]$$

Therefore the area under the curve of $y = x^2$ from $x = 0$ to $x = a$ is just $a^3/3$.

The process of summing up all the little infinitesimally small rectangles under the curve of some function $y = f(x)$ between two values of x (x_1 and x_2) to yield the overall area under the curve between these two points is called **integration**. We sum up all the parts to form the *integrated* whole. Thus, the true area under the curve is

$$A_{exact} = \lim_{\Delta x \to 0} \sum_{x=x_1}^{x=x_2} f(x) \cdot \Delta x \qquad \ldots [A8\text{-}26]$$

When Δx becomes vanishingly small, we designate it as dx. The term dx, which is called an **in-finitesimal**, can be thought of as what Δx becomes when it becomes closer and closer to zero. Under this condition the summation of the series represented by Eqn [A8-26] becomes so finely divided that the upper ends of the little rectangles in effect completely smooth out to form the continuous curve of the function $y = f(x)$. To denote this we replace the summation symbol "Σ" with the integral symbol "$\int$"—an elongated letter "S" [7]—which is called the **integration sign**. In mathematical symbols our area is expressed as

$$A_{exact} = \int_{x=0}^{x=a} f(x)dx \qquad \ldots [A8\text{-}27]$$

The number $x = 0$ and $x = a$ are called **the limits of integration**.[8] The function $f(x)$ to the right of the integration sign is called the **integrand**. $\int_{x=a}^{x=b} f(x)dx$ is called the **definite integral of $f(x)$ from a to b**. Usually it is just written as $\int_{a}^{b} f(x)dx$.

Using this new notation, the information given by Eqn [A8-26] is expressed as

$$A_{exact} = \int_{0}^{a} x^2 dx = \frac{x^3}{3} \qquad \ldots [A8\text{-}28]$$

It now becomes a simple matter to evaluate the area under the curve $y = x^2$ between the two values $x_1 = a$ and $x_2 = b$. Using [A8-28] we merely evaluate the area A_1 between $x = 0$ and $x = a$, and subtract this from the area A_2 between $x = 0$ and $x = b$, as indicated in Fig. 11.

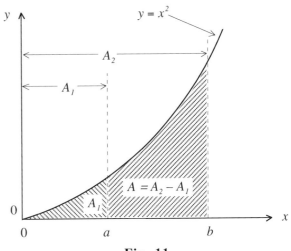

Fig. 11

Thus,
$$A_1 = \int_{x=0}^{x=a} x^2 dx = \frac{a^3}{3} \quad \text{and} \quad A_2 = \int_{x=0}^{x=b} x^2 dx = \frac{b^3}{3}$$

And the desired area A is
$$A = A_2 - A_1 = \frac{b^3}{3} - \frac{a^3}{3} \qquad \ldots \text{[A8-29]}$$

Using mathematical notation:

$$A = \int_a^b x^2 dx = \left[\frac{x^3}{3}\right]_a^b = \frac{b^3}{3} - \frac{a^3}{3} \qquad \ldots \text{[A8-30]}$$

If we wish to integrate the function $y = 5x^2$ between, say, $x = 2$ and $x = 6$, we merely recognize that the integral will just be 5 times the value for the integral of $y = x^2$. Thus,

$$\int_2^6 5x^2 dx = 5\int_2^6 x^2 dx = 5 \times \left[\frac{x^3}{3}\right]_2^6 = 5\left[\frac{6^3}{3} - \frac{2^3}{3}\right] = \frac{5}{3}(216 - 8) = 346.67$$

We are now in a position where we can summarize our results. In each case, a, b, and k are constants. From the pattern, it is apparent that for the general exponent n:

$$\int x^n dx = \frac{x^{n+1}}{n+1} \qquad \ldots \text{[A8-31]}$$

Summary of Simple Integration Formulas

$y = f(x)$	Indefinite Integral (general form)	Definite Integral (with integration limits)
$y = k = kx^0$	$\int k\,dx = k\int dx = kx$	$\int_a^b k\,dx = k(b - a)$
$y = x$	$\int x\,dx = \frac{x^2}{2}$	$\int_a^b x\,dx = \frac{b^2}{2} - \frac{a^2}{2}$
$y = x^2$	$\int x^2 dx = \frac{x^3}{3}$	$\int_a^b x^2 dx = \frac{b^3}{3} - \frac{a^3}{3}$
$y = x^n$ $(n \neq -1)$	$\int x^n dx = \frac{x^{n+1}}{n+1}$	$\int_a^b x^n dx = \frac{b^{n+1}}{n+1} - \frac{a^{n+1}}{n+1}$

Note that n doesn't have to be a positive whole number in order to use Eqn [A8-31]; n it can be any fraction, decimal, or even a negative number. The only exception is $n = -1$ because if this value is substituted into Eqn [A8-31] we get a value of infinity, which is not very useful. We deal with this special case of $n = -1$ in Appendix 9.

Each of the above formulas is applicable to our original calculation of heat.

Example 1

The heat capacity of a certain substance has the form

$$C = 12 - 0.006T + 2.0 \times 10^{-5} T^2 \quad \text{J K}^{-1}$$

Calculate the heat required to heat this substance from 300 K to 350 K.

Solution

Using the definite integrals for $\int k\,dx$, $\int x\,dx$, and $\int x^2\,dx$ for each respective term of the heat capacity equation, we get

$$Q = \int_{300}^{350} C\,dT = \int_{300}^{350} 12\,dT - \int_{300}^{350} 0.006T\,dT + \int_{300}^{350} 2.0 \times 10^{-5}\,T^2\,dT$$

$$= 12 \int_{300}^{350} dT - 0.006 \int_{300}^{350} T\,dT + 2.0 \times 10^{-5} \int_{300}^{350} T^2\,dT$$

$$= 12(350 - 300) - 0.006\left(\frac{350^2}{2} - \frac{300^2}{2}\right) + 2.0 \times 10^{-5}\left(\frac{350^3}{3} - \frac{300^3}{3}\right)$$

$$= 600 - 97.5 + 105.83$$

$$= 608.33 \text{ J}$$

Ans: 608 J

Exercise 1

If you feel that you understood the above derivation for $\int x^2\,dx = x^3/3$ using Riemann series, you might try to derive one on your own and check your result with the expressions given by [A8-31].

The following table lists the formulas for a number of power series up to j^{10}:

Power Series Summation Formulas

Summation	Equivalent Value
$\displaystyle\sum_{j=1}^{k} j^0 = \sum_{j=1}^{k} 1 = 1+1+1+1+\ldots+1 \ [k \text{ times}]$	$\displaystyle\sum_{j=1}^{k} 1 = k$
$\displaystyle\sum_{j=1}^{k} j = 1+2+3+4+\ldots+k$	$\displaystyle\sum_{j=1}^{k} j = \frac{n(n+1)}{2}$
$\displaystyle\sum_{j=1}^{k} j^2 = 1+4+9+16+\ldots+k^2$	$\displaystyle\sum_{j=1}^{k} j^2 = \frac{k(k+1)(2k+1)}{6}$
$\displaystyle\sum_{j=1}^{k} j^3 = 1+8+27+64+\ldots+k^3$	$\displaystyle\sum_{j=1}^{k} j^3 = \frac{k(k+1)^2}{4}$
$\displaystyle\sum_{j=1}^{k} j^4 = 1+16+81+256+\ldots+k^4$	$\displaystyle\sum_{j=1}^{k} j^4 = \frac{k^5}{5}+\frac{k^4}{2}+\frac{k^3}{3}-\frac{k}{30}$
$\displaystyle\sum_{j=1}^{k} j^5 = 1+32+243+1024+\ldots+k^5$	$\displaystyle\sum_{j=1}^{k} j^5 = \frac{k^6}{6}+\frac{k^5}{2}+\frac{5k^4}{12}-\frac{k^2}{12}$
$\displaystyle\sum_{j=1}^{k} j^6 = 1+64+729+4096+\ldots+k^6$	$\displaystyle\sum_{j=1}^{k} j^6 = \frac{k^7}{7}+\frac{k^6}{2}+\frac{k^5}{2}-\frac{k^3}{6}+\frac{k}{42}$
$\displaystyle\sum_{j=1}^{k} j^7 = 1+127+2187+16384+\ldots+k^7$	$\displaystyle\sum_{j=1}^{k} j^7 = \frac{k^8}{8}+\frac{k^7}{2}+\frac{7k^6}{12}-\frac{7k^4}{24}+\frac{k^2}{12}$
$\displaystyle\sum_{j=1}^{k} j^8 = 1+256+6561+65536+\ldots+k^8$	$\displaystyle\sum_{j=1}^{k} j^8 = \frac{k^9}{9}+\frac{k^8}{2}+\frac{2k^7}{3}-\frac{7k^5}{15}+\frac{2k^3}{9}-\frac{k}{30}$
$\displaystyle\sum_{j=1}^{k} j^9 = 1+512+19683+262144+\ldots+k^9$	$\displaystyle\sum_{j=1}^{k} j^9 = \frac{k^{10}}{10}+\frac{k^9}{2}+\frac{3k^8}{4}-\frac{7k^6}{10}+\frac{k^4}{2}-\frac{3k^2}{20}$
$\displaystyle\sum_{j=1}^{k} j^{10} = 1+1024+59049+1048576+\ldots+k^{10}$	$\displaystyle\sum_{j=1}^{k} j^{10} = \frac{k^{11}}{11}+\frac{k^{10}}{2}+\frac{5k^9}{6}-k^7+k^5-\frac{k^3}{2}+\frac{5k}{66}$

Source: http://www.math2.org/math/expansion/power.htm

APPENDIX 9.   UNDERSTANDING LOGARITHMS, EXPONENTIALS, AND $\int dx/x$ [1]

A9.1   AREA UNDER THE LINE $y = 1/x$

Sometimes the heat capacity of a substance takes the form

$$C = a + bT + cT^2 + \frac{d}{T} \qquad \dots [\text{A9-1}]$$

where a, b, c, and d are constants. In this case, the heat required to raise the substance from T_1 to T_2 is given by

$$Q = \int_{T_1}^{T_2} \left(a + bT + cT^2 + \frac{d}{T} \right) dT$$

$$= a\int_{T_1}^{T_2} dT + b\int_{T_1}^{T_2} TdT + c\int_{T_1}^{T_2} T^2 dT + d\int_{T_1}^{T_2} \left(\frac{1}{T} \right) dT \qquad \dots [\text{A9-2}]$$

In Appendix 8 we saw how to evaluate the first three terms of Eqn [A9-2]; but, to evaluate the last term requires a knowledge of the area A under the curve $y = 1/x$ between $x = a$ and $x = b$. Fig. 1 shows a plot of this function. In calculus notation, this area A is represented by

$$A = \int_{a}^{b} \frac{dx}{x} \qquad \dots [\text{A9-3}]$$

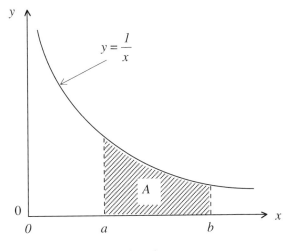

Fig. 1

The function $y = 1/x$ is just $y = x^{-1}$, with the value of the x-exponent being $n = -1$. In Appendix 8 we saw that the general rule (the "power rule") for simple integration is given by

$$\int x^n dx = \frac{x^{n+1}}{n+1} \qquad \dots [\text{A9-4}]$$

Unfortunately we can't use this expression when $n = -1$ because the resulting value would be in-

---

[1] Please believe me when I tell you that it is worth your while to spend the time required to digest this material. If you do, then you will understand one of the most important concepts in mathematics and science, and will be much less confused in your future studies. I kid you not!

finity, which we definitely know is *not* the area A under the curve in Fig. 1.

In Appendix 8, when we were evaluating the area under the curve $y = x^2$, you may recall that we split up the distance along the x-axis into k equal-sized intervals, each of width Δx, then drew little rectangles up to the curve, developed a formula expressing the area of a representative rectangle, and then approximated the total area under the curve as the sum of the k little rectangles. Next, we saw that as $\Delta x \rightarrow 0$, the number of subintervals and rectangles approached infinity, and the sum of the area of all the infinitesimally small rectangles became the *exact* area under the curve. The key was in finding an expression we could use to evaluate the sum of an infinite series. In the case of $\int x^2 dx$, the expression we required was

$$\sum_{j=1}^{k} j^2 = 1^2 + 2^2 + 3^2 + \ldots + k^2 = \frac{k(k+1)(2k+1)}{6} \qquad \ldots [A9\text{-}5]$$

In a table at the end of Appendix 8 we presented a table of power series summation formulas that would permit you (should you be ambitious) to use the same method to evaluate $\int x^n dx$ for all values of n from $n = 0$ to $n = 10$. Unfortunately, for the special case of $n = -1$, there is no such summation formula that can be used to evaluate $\int (1/x) dx$. So we have to use a different approach.

A9.2 COMPOUND INTEREST AND $\left(1 + \frac{1}{n}\right)^n$

The solution to evaluating $\int (1/x) dx$ probably had its origin in the financial calculation of the interest earned by investing some principal amount of money P at a given compounded interest rate. For example, suppose we invest $1 at an annual interest rate of $R = 10\%$ per year, or, expressed as a fraction, $r = \frac{10}{100} = 0.10$. After one year, at 10% interest compounded once annually at the end of the year, our initial investment of $1.00 (the principal) will be worth $1.10:

Thus, after 1 year, the value is $1.00 \times (1 + 0.10) \quad = \$1.10$

After 2 years the value would be $1.00 \times (1 + 0.10)^2 \quad = \$1.21$

After 10 years, the value would be $1.00 \times (1 + 0.10)^{10} = \$2.59374$

And after t years the value would be $1.00 \times (1 + 0.10)^t \quad = P(1 + r)^t$

where P is the principal (the initial investment) and r is the yearly (fractional) interest rate.

Now, instead of compounding the interest only *once* per year, suppose it were compounded *every month*, i.e., *12 times* per year. Now the number of compounding intervals is increased by a factor of 12; but, since the yearly interest rate is still 10%, the interest rate per interval will be correspondingly smaller by a factor of 12; namely, $r = \frac{0.10}{12} = 0.008333$. As shown in Table 1 as calculation 2, now the amount after 1 year and 10 years will be \$1.10471 and \$2.70704, respectively. Compounding monthly instead of annually doesn't make much difference in one year, but in 10 years the difference starts to become significant (7.1% greater earnings!).

Since we seem to do better with increased compounding intervals, now (calculation 3) let's compound the interest on a *daily* basis; i.e., 365 times per year, with a daily interest rate of $R = \frac{10}{365} = 0.0273973\%$. Surprisingly, compounding the interest on a daily basis doesn't earn much more than compounding on a monthly basis!

OK, let's try compounding on an *hourly* basis (calculation 4). Hey, what a disappointment! We're hardly doing better at all.

Let's try compounding the interest on a *per minute* basis (calculation 5). Nope.... still doesn't help much.

Table 1.   *Money Earned by an Initial Investment of \$1.00 for One Year and for Ten Years at an Interest Rate of 10% Per Year.*

No.	Value After One Year at 10% Interest per Year	Value After Ten Years at 10% Interest per Year
1	1 Compounding Interval: $$1.00(1+0.10) = 1.10000000$$	$$1.00(1+0.10)^{10} = 2.59374246$$
2	12 Compounding Intervals: $$1.00\left(1+\frac{0.10}{12}\right)^{12} = 1.10471307$$	$$1.00\left(1+\frac{0.10}{12}\right)^{12\times10} = 2.70704149$$
3	365 Compounding Intervals: $$1.00\left(1+\frac{0.10}{365}\right)^{365} = 1.10515578$$	$$1.00\left(1+\frac{0.10}{365}\right)^{365\times10} = 2.71790955$$
4	365×24 Compounding Intervals: $$1.00\left(1+\frac{0.10}{365\times24}\right)^{365\times24} = 1.10517029$$	$$1.00\left(1+\frac{0.10}{365\times24}\right)^{365\times24\times10} = 2.71826631$$
5	$365 \times 24 \times 60$ Compounding Intervals: $$1.00\left(1+\frac{0.10}{365\times24\times60}\right)^{365\times24\times60} = 1.10517091$$	$$1.00\left(1+\frac{0.10}{365\times24\times60}\right)^{365\times24\times60\times10} = 2.71828157$$

6	$365 \times 24 \times 3600$ Compounding Intervals:
	$1.00\left(1 + \dfrac{0.10}{365 \times 24 \times 3600}\right)^{365 \times 24 \times 3600} = 1.10517092$ $\quad$ $1.00\left(1 + \dfrac{0.10}{365 \times 24 \times 3600}\right)^{365 \times 24 \times 3600 \times 10} = 2.71828187$

OK, in desperation (our greed for money knows no bounds!) let's try one final shot and com—pound on a *per second* basis (calculation 6).

One would think that by compounding the overall yearly interest rate over more intervals, even though we decrease the rate of interest earned for each interval, in general, we still should earn considerably more on our initial investment. Intuitively you might be inclined to think that by in-creasing the number of intervals *indefinitely* you should be able to squeeze out an ever increasing amount of interest per year from your initial one dollar investment. Unfortunately, a glance at Table 1 indicates otherwise; the amount earned seems *naturally* to level off to some *finite, fixed, limiting value.* For an initial investment of $1 for one year at 10% it would appear that this value is about $1.11, and in ten years about $2.72.

As we shall see, this natural limit on compounding is of the utmost importance to mathematics, sci-ence, and engineering—to say nothing of your bank balance. In fact, it is a key factor in under-standing how many of the processes in the universe as a whole can be expected to behave.

So you'd better pay attention.

We now are going to evaluate this *exact* limit.

A9.3   CONTINUOUS GROWTH

From Table 1 it is apparent that if we invest P dollars at a yearly interest rate of r (fractional basis) compounded over k intervals, then, after t years the value S of the initial P will be given by the general formula

$$S = P\left(1 + \frac{r}{k}\right)^{kt} \qquad \ldots \text{[A9-6]}$$

Rearranging: $\qquad$ $$S = P\left(1 + \frac{1}{k/r}\right)^{\frac{k}{r} \cdot rt} \qquad \ldots \text{[A9-7]}$$

We saw above that as the number of intervals k increases, the value of S seems to head toward some finite, natural limit. For any given value of the interest rate r, when k approaches *infinity*, we will have the limit of *continuous growth*. Continuous growth (or decay) often is encountered in the real world. Examples would include population growth, production of products during a chemical reaction, radioactive decay, the spread of disease, and the proliferation of bacterial cultures. Since r is constant, when k approaches infinity, k/r also will approach infinity. Thus, the maximum our money can grow in t years with continuous growth at an annual interest rate of r will be given by

$$S_{max} = \lim_{k/r \to \infty}\left[P\left(1+\frac{1}{k/r}\right)^{\frac{k}{r}\cdot rt}\right] = P\left[\lim_{k/r \to \infty}\left(1+\frac{1}{k/r}\right)^{\frac{k}{r}\cdot rt}\right]$$

$$= P\left[\lim_{k/r \to \infty}\left(1+\frac{1}{k/r}\right)^{\frac{k}{r}}\right]^{rt} \qquad \dots [A9\text{-}8]$$

To simplify the terminology, we will call $k/r = n$, so Eqn [A9-8] then becomes

$$S_{max} = P\left[\lim_{n \to \infty}\left(1+\frac{1}{n}\right)^{n}\right]^{rt} \qquad \dots [A9\text{-}10]$$

A9.4 EVALUATION OF $\displaystyle\lim_{n \to \infty}\left(1+\frac{1}{n}\right)^{n}$

The first step is to expand $\left(1+\frac{1}{n}\right)^{n}$. Do you recall from high school algebra that you studied something called the *Binomial Theorem*? You've probably forgotten it, but what it does is show us how to expand the binomial[2] expression $(a+b)^{n}$, where $n = 0, 1, 2, 3, \dots$

For $a^{2} > b^{2}$ the Binomial Theorem gives the expansion as

$$(a+b)^{n} = a^{n} + na^{n-1}b + \frac{n(n-1)}{2!}a^{n-2}b^{2} + \frac{n(n-1)(n-2)}{3!}a^{n-3}b^{3} + \dots \qquad \dots [A9\text{-}11]$$

Or, expressed as a series summation:

$$(a+b)^{n} = \sum_{k=0}^{n}\frac{n!}{k!(n-k)!}a^{n-k}b^{k}n \qquad \dots [A9\text{-}12]$$

where $n!$ is the n factorial, defined as the product

$$n! \equiv 1\cdot2\cdot3\cdot4\cdot\dots\cdot n$$
$$= n\cdot(n-1)\cdots3\cdot2\cdot1 \qquad \dots [A9\text{-}13]$$

---

[2] A *binomial* is any expression consisting of the sum of two terms.

Note that the factorial of 1 is 1, and, by definition, the factorial of zero also is 1:

$$1! = 1 \quad\quad 0! = 1 \quad\quad\quad\quad \ldots [A9\text{-}14]$$

Since, in view of Eqns [13] and [14]

$$\frac{n!}{0!(n-0)!} = \frac{n!}{1(n!)} = 1 \quad\quad\quad \ldots [A9\text{-}15]$$

and

$$\frac{n!}{1!(n-1)!} = \frac{n(n-1)!}{(1)(n-1)!} = n, \quad\quad\quad \ldots [A9\text{-}16]$$

we may use these two expressions to replace the first two coefficients 1 and n in Eqn [A9-11] and re-write the expansion using the notation of the series given by Eqn [A9-12] to get

$$(a+b)^n = \frac{n!}{0!(n-0)!}a^{n-0}b^0 + \frac{n!}{1!(n-1)!}a^{n-1}b^1 + \frac{n!}{2!(n-2)!}a^{n-2}b^2 + \ldots$$

$$\ldots + \frac{n!}{3!(n-3)!}a^{n-3}b^3 + \ldots + \frac{n!}{(n-1)!\,1!}ab^{n-1} + \frac{n!}{n!(n-n)!}a^{n-n}b^n$$

$$= \frac{n!}{0!\cdot n!}a^n b^0 + \frac{n(n-1)!}{1!(n-1)!}a^{n-1}b + \frac{n(n-1)(n-2)!}{2!(n-2)!}a^{n-2}b^2 + \frac{n(n-1)(n-2)(n-3)!}{3!(n-3)!}a^{n-3}b^3 + \ldots$$

$$\ldots + \frac{n(n-1)!}{(n-1)!\,1!}ab^{n-1} + \frac{n!}{n!\,0!}a^0 b^n$$

$$= \frac{1}{0!}a^n b^0 + \frac{n}{1!}a^{n-1}b + \frac{n(n-1)}{2!}a^{n-2}b^2 + \frac{n(n-1)(n-2)}{3!}a^{n-3}b^3 + \ldots + \frac{n}{1!}ab^{n-1} + \frac{1}{0!}a^0 b^n$$

$$\ldots [A9\text{-}17]^3$$

Example 1

Use the formula given in Eqn [A9-17] to expand $(a+b)^5$.

Solution

$$(a+b)^5 = \sum_{k=0}^{5} \frac{5!}{(5-k)!\,k!}a^{5-k}b^k$$

---

[3] I know most of you probably didn't check out these equations for mistakes. Thank you for your trust.

$$= \frac{5!}{(5-0)!\,0!}a^{5-0}b^0 + \frac{5!}{(5-1)!\,1!}a^{5-1}b^1 + \frac{5!}{(5-2)!\,2!}a^{5-2}b^2 + \frac{5!}{(5-3)!\,3!}a^{5-3}b^3$$

$$+ \frac{5!}{(5-4)!\,4!}a^{5-4}b^4 + \frac{5!}{(5-5)!\,5!}a^{5-5}b^5$$

$$= \frac{5!}{5!\,0!}a^5 + \frac{5!}{4!\,1!}a^4b^1 + \frac{5!}{3!\,2!}a^3b^2 + \frac{5!}{2!\,3!}a^5b^3 + \frac{5!}{1!\,4!}ab^4 + \frac{5!}{0!\,5!}b^5$$

$$= \frac{1}{0!}a^5 + \frac{5\cdot 4!}{4!\,1!}a^4b + \frac{5\cdot 4\cdot 3!}{3!\,2!}a^3b^2 + \frac{5\cdot 4\cdot 3!}{2!\,3!}a^5b^3 + \frac{5\cdot 4!}{1!\,4!}ab^4 + \frac{1}{0!}b^5$$

$$= a^5 + \frac{5}{1!}a^4b + \frac{5\cdot 4}{2!}a^3b^2 + \frac{5\cdot 4}{2!}a^2b^3 + \frac{5}{1!}ab^4 + b^5$$

Ans: $\boxed{(a+b)^5 = a^5 + 5a^4b + 10a^3b^2 + 10a^2b^3 + 5ab^4 + b^5}$

There are several points to note about the binomial expansion:

- The symmetry of the coefficients. For the above example of $(a + b)^5$: *1, 5, 10, 10, 5, 1.*
- The coefficient of the first and last terms is always *1.*
- The coefficient of the second and the second-to-the-last coefficient is always the same as the exponent *n.*
- For each term the sum of the exponents of *a* and *b* always adds to *n.*

We now are in a position to evaluate $\displaystyle\lim_{n\to\infty}\left(1+\frac{1}{n}\right)^n$.

We do this by using Eqn [A9-17] to expand $\left(1+\frac{1}{n}\right)^n$ with $a = 1$ and $b = 1/n$, noting that all the powers of *a* are always *1*, since $a = 1$ and *1* raised to any power always equals *1*. Thus:

$$\left(1+\frac{1}{n}\right)^n = \frac{1}{0!}a^nb^0 + \frac{n}{1!}a^{n-1}b + \frac{n(n-1)}{2!}a^{n-2}b^2 + \frac{n(n-1)(n-2)}{3!}a^{n-3}b^3 + \ldots + \frac{1}{0!}a^0b^n$$

$$= \frac{1}{0!}\left(\frac{1}{n}\right)^0 + \frac{n}{1!}\left(\frac{1}{n}\right) + \frac{n(n-1)}{2!}\left(\frac{1}{n}\right)^2 + \frac{n(n-1)(n-2)}{3!}\left(\frac{1}{n}\right)^3 + \ldots + \frac{1}{0!}\left(\frac{1}{n}\right)^n$$

$$= \frac{1}{0!} + \frac{1}{1!} + \frac{n(n-1)}{2!\,n^2} + \frac{n(n-1)(n-2)}{n^3\,3!} + \ldots + \frac{1}{0!\,n^n}$$

$$= \frac{1}{0!} + \frac{1}{1!} + \frac{\frac{n}{n}\left(\frac{n-1}{n}\right)}{2!} + \frac{\frac{n}{n}\left(\frac{n-1}{n}\right)\left(\frac{n-2}{n}\right)}{3!} + \ldots + \frac{1}{0!\,n^n}$$

$$= \frac{1}{0!} + \frac{1}{1!} + \frac{\frac{n}{n}\left(\frac{n}{n} - \frac{1}{n}\right)}{2!} + \frac{\frac{n}{n}\left(\frac{n}{n} - \frac{1}{n}\right)\left(\frac{n}{n} - \frac{2}{n}\right)}{3!} + \ldots + \frac{1}{n^n}$$

$$= \frac{1}{0!} + \frac{1}{1!} + \frac{\left(1 - \frac{1}{n}\right)}{2!} + \frac{\left(1 - \frac{1}{n}\right)\left(1 - \frac{2}{n}\right)}{3!} + \ldots + \frac{1}{n^n} \qquad \ldots \text{[A9-18]}$$

The last term in Eqn [A9-18] can be rearranged as follows:

$$\frac{1}{n^n} = \frac{n!}{n!} \cdot \frac{1}{n^n} = \frac{n(n-1)(n-2)\cdots(1)}{n!} \cdot \frac{1}{n^n} = \frac{n(n-1)(n-2)\cdots(n-[n-1])}{n!} \cdot \frac{1}{n^n}$$

$$= \frac{n(n-1)(n-2)\cdots(n-[n-1])}{n^n} \cdot \frac{1}{n!} = \frac{n}{n}\left(\frac{n-1}{n}\right)\left(\frac{n-2}{n}\right)\cdots\left(\frac{n}{n} - \frac{n-1}{n}\right) \cdot \frac{1}{n!}$$

$$= \left(1 - \frac{1}{n}\right)\left(1 - \frac{2}{n}\right)\cdots\left(1 - \frac{n-1}{n}\right) \cdot \frac{1}{n!} \qquad \ldots \text{[A9-19]}$$

Replacing the last term of Eqn [A9-18] with [A9-19] gives

$$\left(1 + \frac{1}{n}\right)^n = \frac{1}{0!} + \frac{1}{1!} + \frac{\left(1 - \frac{1}{n}\right)}{2!} + \frac{\left(1 - \frac{1}{n}\right)\left(1 - \frac{2}{n}\right)}{3!} + \ldots + \left(1 - \frac{1}{n}\right)\left(1 - \frac{2}{n}\right)\cdots\left(1 - \frac{n-1}{n}\right) \cdot \frac{1}{n!}$$

$$\ldots \text{[A9-20]}$$

By now you probably have forgotten what the whole purpose of this exercise was! To refresh your memory, it started when we noticed from Table 1 that the values of $\left(1 + \frac{1}{n}\right)^n$ appeared to be heading towards some finite limit slightly greater than 2.7 as n got larger and larger. Therefore, we have been trying to determine what this limit is, and we now are ready—finally!— to do this.

Thus, we want to evaluate $$\lim_{n \to \infty} \left(1 + \frac{1}{n}\right)^n \qquad \ldots \text{[A9-21]}$$

We tediously worked through the binomial expansion of $\left(1 + \frac{1}{n}\right)^n$ to give us Eqn [A9-20], which we now substitute into Eqn [A9-21] to give

$$\lim_{n \to \infty} \left(1 + \frac{1}{n}\right)^n$$

$$= \lim_{n \to \infty} \left[\frac{1}{0!} + \frac{1}{1!} + \frac{\left(1 - \frac{1}{n}\right)}{2!} + \frac{\left(1 - \frac{1}{n}\right)\left(1 - \frac{2}{n}\right)}{3!} + \ldots + \left(1 - \frac{1}{n}\right)\left(1 - \frac{2}{n}\right) \cdots \left(1 - \frac{n-1}{n}\right) \cdot \frac{1}{n!} \right]$$

$$\ldots [A9\text{-}22]$$

As $n \to \infty$ each of the terms $\frac{1}{n}$, $\frac{2}{n}$, etc. goes to zero.

When the expression $\frac{n-1}{n}$ found in the final term is rearranged to $1 - \frac{1}{n}$, it is obvious that the final term also goes to zero as $n \to \infty$.

Therefore Eqn [A9-22] becomes the sum of the infinite series

$$\lim_{n \to \infty} \left(1 + \frac{1}{n}\right)^n = \sum_{n=0}^{\infty} \frac{1}{n!} = \frac{1}{0!} + \frac{1}{1!} + \frac{1}{2!} + \frac{1}{3!} + \frac{1}{4!} + \frac{1}{5!} + \ldots$$

$$= 1 + 1 + \frac{1}{2!} + \frac{1}{3!} + \frac{1}{4!} + \frac{1}{5!} + \ldots \qquad \ldots [A9\text{-}23]$$

This infinite series[4] leads to a finite value that we denote by the symbol e, sometimes known as *Euler's number*,[5] in honor of the Swiss mathematician Leonhard Euler (1707-1783), who was the first to denote this number by the letter e, and also was the first to prove that e, like π, is a finite irrational number[6] that never terminates and never repeats itself.

$$e \text{ defined} \qquad \boxed{e \equiv \sum_{n=0}^{\infty} \frac{1}{n!} = 2 + \frac{1}{2!} + \frac{1}{3!} + \frac{1}{4!} + \frac{1}{5!} + \ldots} \qquad \ldots [A9\text{-}24]$$

Like π, e can be calculated to almost as much accuracy as we wish merely by evaluating increasing numbers of terms in the series Eqn [A9-24]. With the possible exception of π, e is the most important number in mathematics since it appears in countless mathematical expressions involving growth rates, limits and derivatives. In addition to being an irrational number, like π, e also is a *transcendental* number, which means that it cannot be the solution of any polynomial algebraic equation of the form

$$a_n x^n + a_{n-1} x^{n-1} + \ldots + a_1 x + a_0 = 0$$

---

[4] This series for e was first discovered by Sir Isaac Newton in 1665.
[5] Not to be confused with *Euler's constant*, which is a different number. Maybe they honored Euler too much.
[6] An irrational number is one that cannot be expressed by the ratio of two integers.

whose coefficients a_i are integers. The idea is that equations containing transcendental numbers transcend (go beyond) the equations studied in elementary algebra.[7]

Euler's number e also plays a very important role in statistics in the function

$$f(x) = \frac{e^{-x^2/2}}{\sqrt{2\pi}}$$

which is the equation of the familiar *bell curve* used to fudge your marks. One of the special characteristics of this equation is that its integral from $-\infty$ to $+\infty$ (i.e., the area under the curve) has a value of exactly 1:

$$\int_{-\infty}^{+\infty} \frac{e^{-x^2/2}}{\sqrt{2\pi}} dx = 1$$

The number e commonly is interpreted geometrically in either of two ways (Fig. 6, p. A9-25):

    (1) The area under the curve $y = 1/x$ between $x = 1$ and $x = e$ is equal exactly to 1, or

    (2) The area under the curve $y = e^x$ between $x = -\infty$ and $x = +1$ is equal exactly to e.

As an exercise, let's calculate e using the first 11 terms of the series using 12 decimal point accuracy for our numbers (Table 2):

Table 2. *Estimate of the Value of e by Summation of Terms*

n	$1/n!$		Progressive Sums
0	$1/0! = 1/1$	$= 1.000000000000$	$\sum_{n=0}^{0} \frac{1}{n!} = 1.000000000000$
1	$1/1! = 1/1$	$= 1.000000000000$	$\sum_{n=0}^{1} \frac{1}{n!} = 2.000000000000$
2	$1/2! = 1/2$	$= 0.500000000000$	$\sum_{n=0}^{2} \frac{1}{n!} = 2.500000000000$
3	$1/3! = 1/6$	$= 0.166666666666667$	$\sum_{n=0}^{3} \frac{1}{n!} = 2.666666666667$
4	$1/4! = 1/24$	$= 0.041666666666667$	$\sum_{n=0}^{4} \frac{1}{n!} = 2.708333333333$

---

[7] In 1873 the French mathematician Charles Hermite was the first to prove that e is a transcendental number. It has since been proved that most real numbers are, in fact, irrational, and that most irrational numbers are transcendental!

5	$1/5! = 1/120$	$= 0.008333333333333$	$\displaystyle\sum_{n=0}^{5} \frac{1}{n!} = 2.716666666667$
6	$1/6! = 1/720$	$= 0.001388888888889$	$\displaystyle\sum_{n=0}^{6} \frac{1}{n!} = 2.718055555556$
7	$1/7! = 1/5040$	$= 0.000198412698413$	$\displaystyle\sum_{n=0}^{7} \frac{1}{n!} = 2.718253968254$
8	$1/8! = 1/40\,320$	$= 0.000024801587302$	$\displaystyle\sum_{n=0}^{8} \frac{1}{n!} = 2.718278769841$
9	$1/9! = 1/362\,880$	$= 0.000002755731922$	$\displaystyle\sum_{n=0}^{9} \frac{1}{n!} = 2.718281525573$
10	$1/10! = 1/3628800$	$= 0.000000275573192$	$\displaystyle\sum_{n=0}^{10} \frac{1}{n!} = 2.718281801146$

Do you remember way back in Table 1 when we were calculating compound interest that we said as n approached infinity the value of $\left(1 + \frac{1}{n}\right)^{n}$ seemed naturally to be approaching some fixed value around 2.72? Well, we now have determined that this value—called e—is defined *exactly* by the infinite non-repeating series given by Eqn [A9-24]. We have calculated this value to ten decimal places[8] and have obtained the number 2.7182818011. The true value to ten decimal places is 2.7182818284, so our calculation was in error by only

$$\frac{2.7182818011 - 2.7182818284}{2.7182818284} \times 100\% = -0.00000099\%$$

This tells us that the series converges quite rapidly. The value after the addition of only the first six terms is 2.7180555556, which itself is in error by only about −0.0083%. If you require more accuracy, the value of e to 2000 decimal places is

```
2.71828 18284 59045 23536 02874 71352 66249 77572 47093 69995
95749 66967 62772 40766 30353 54759 45713 82178 52516 64274
27466 39193 20030 59921 81741 35966 29043 57290 03342 95260
59563 07381 32328 62794 34907 63233 82988 07531 95251 01901
15738 34187 93070 21540 89149 93488 41675 09244 76146 06680
82264 80016 84774 11853 74234 54424 37107 53907 77449 92069
55170 27618 38606 26133 13845 83000 75204 49338 26560 29760
67371 13200 70932 87091 27443 74704 72306 96977 20931 01416
92836 81902 55151 08657 46377 21112 52389 78442 50569 53696
77078 54499 69967 94686 44549 05987 93163 68892 30098 79312
77361 78215 42499 92295 76351 48220 82698 95193 66803 31825
```

[8] Because of rounding errors, you need to use 12 decimal places in your calculation of each term to obtain 10 decimal point accuracy in the final result.

```
28869 39849 64651 05820 93923 98294 88793 32036 25094 43117
30123 81970 68416 14039 70198 37679 32068 32823 76464 80429
53118 02328 78250 98194 55815 30175 67173 61332 06981 12509
96181 88159 30416 90351 59888 85193 45807 27386 67385 89422
87922 84998 92086 80582 57492 79610 48419 84443 63463 24496
84875 60233 62482 70419 78623 20900 21609 90235 30436 99418
49146 31409 34317 38143 64054 62531 52096 18369 08887 07016
76839 64243 78140 59271 45635 49061 30310 72085 10383 75051
01157 47704 17189 86106 87396 96552 12671 54688 95703 50354
02123 40784 98193 34321 06817 01210 05627 88023 51930 33224
74501 58539 04730 41995 77770 93503 66041 69973 29725 08868
76966 40355 57071 62268 44716 25607 98826 51787 13419 51246
65201 03059 21236 67719 43252 78675 39855 89448 96970 96409
75459 18569 56380 23637 01621 12047 74272 28364 89613 42251
64450 78182 44235 29486 36372 14174 02388 93441 24796 35743
70263 75529 44483 37998 01612 54922 78509 25778 25620 92622
64832 62779 33386 56648 16277 25164 01910 59004 91644 99828
93150 56604 72580 27786 31864 15519 56532 44258 69829 46959
30801 91529 87211 72556 34754 63964 47910 14590 40905 86298
49679 12874 06870 50489 58586 71747 98546 67757 57320 56812
88459 20541 33405 39220 00113 78630 09455 60688 16674 00169
84205 58040 33637 95376 45203 04024 32256 61352 78369 51177
88386 38744 39662 53224 98506 54995 88623 42818 99707 73327
61717 83928 03494 65014 34558 89707 19425 86398 77275 47109
62953 74152 11151 36835 06275 26023 26484 72870 39207 64310
05958 41166 12054 52970 30236 47254 92966 69381 15137 32275
36450 98889 03136 02057 24817 65851 18063 03644 28123 14965
50704 75102 54465 01172 72115 55194 86685 08003 68532 28183
15219 60037 35625 27944 95158 28418 82947 87610 85263 98139
```

If you know any American history, a handy way to remember the value of e to ten decimal places is to remember it as $2.7 + (Andrew\ Jackson)^2$, i.e., $2.7 + 1828 + 1828 = 2.718281828$. Andrew Jackson was elected president of the United States in 1828.

A9.5   THE NATURE OF THE FUNCTION $y = e^x$

We showed above that e was connected with the phenomenon of continuous growth. In the case of an initial investment of P dollars invested at an annual interest rate of $100r\%$ compounded *continuously*, we formulated the following expression—Eqn [A9-8]—for the value S after t years:

$$S_{max} = P\left[\lim_{k/r \to \infty}\left(1 + \frac{1}{k/r}\right)^{\frac{k}{r}}\right]^{rt}$$

We now know that the limit in the square brackets is just e; therefore the expression becomes

$$S_{max} = P \cdot e^{rt} \qquad \dots [A9\text{-}25]$$

When the exponent is simple, such as rt in the above expression, we can use the notation shown in Eqn [A9-25]. However, when, as often is the case, the exponent is more complicated, such as

$\left(\dfrac{1-\alpha}{RT}\right)\eta$, it becomes awkward to write something like $y = A \cdot e^{\left(\frac{1-\alpha}{RT}\right)\eta}$. Instead we adopt the no-

tation "exp," standing for "exponential." Thus, $y = A \cdot e^{\left(\frac{1-\alpha}{RT}\right)\eta}$ is more commonly written as

$y = A \cdot exp\left[\left(\dfrac{1-\alpha}{RT}\right)\eta\right]$.

The sum S in equation [A9-25] is said to exhibit *exponential growth*[9] because the equation describing S *vs.* time involves e to some *exponent*; in this case the exponent is rt. To observe exponential growth, suppose we initially invest $P = \$1$ at 10% interest per year with continuous exponential growth. Table 3 shows the values after various times calculated using Eqn [A9-25]. The data are plotted in Fig. 2, which shows that exponential growth usually starts slowly and then increases very rapidly.

Table 3. *Continuous Exponential Growth of $1 at 10% Interest Per Year.*

Number of Years t	Value S_t at time t $S_t = P \cdot e^{rt} = 1.00 \times e^{0.10t}$
0	1.00
1	1.11
5	1.65
10	2.72
15	4.48
20	7.39
25	12.18
30	20.09
35	33.12

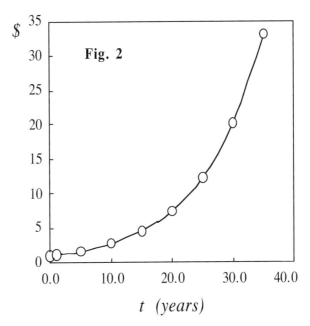

Fig. 2

t (years)

---

[9] Or, *exponential decay*, if the value of rt is negative.

Example 2

Superman is flying along at 50 km h^{-1} and decides to test his powers of acceleration. He decides to smoothly and continuously increase his speed by 50% every minute. After 34 minutes how fast will he be going?

Solution

The Man of Steel's initial speed is $v_o = 50 \frac{km}{h} \times \frac{1000\ m}{km} \times \frac{1\ h}{60\ min} = 833.333$ m min^{-1}

This problem is similar to the continuous growth in interest calculation above; but, this time, speed instead of principal is increasing at a continuous rate of 50% ($r = 0.50$) per minute. The initial investment "principal" is the initial speed v_o. Thus, after 34 minutes, Superman's speed will have increased to

$$v = v_o e^{0.5t} = 833.333 \times (2.71828)^{0.5 \times 34}$$
$$= 2.0129 \times 10^{10} \text{ m min}^{-1} = 3.355 \times 10^8 \text{ m s}^{-1}$$

Ans: | 3.355×10^8 m s^{-1} |

Hey! Wait a minute! This is faster than the speed of light (which is 3.00×10^8 m s^{-1}). Even Superman would have trouble with this, and so we can conclude that his powers have limitations.

Note that we usually don't write $(2.71828)^{0.5 \times 34}$. In the first place, what numerical value do we use for e? Do we take it to 5 decimal places as here? or do we choose 10 decimal places? Instead, we just write "e" or "exp": $\qquad e^{0.5 \times 34} \quad$ or $\quad exp[0.5 \times 34]$

and push the *exp*-button on our calculator, which probably uses a value to 10 or 12 significant digits.

What does a plot of $y = e^x$ look like? Table 4 and Fig. 3 show some values of e^x and the resulting plot, respectively. Note from Fig. 3 that the value of e^x never becomes negative, and that at $x = 1$ the value is just

Table 4. *Values of e^x*

x	e^x	x	e^x
-10.0	0.0000454	$+0.1$	1.1051
-5.0	0.00674	$+0.2$	1.221
-2.0	0.1353	$+0.5$	1.648
-1.0	0.3679	$+1.0$	2.718
-0.5	0.6065	$+2.0$	7.389
0	1.0000	$+3.0$	20.09

e = 2.71828..., while at *x = 0* the value is *1*, since any number raised to the zeroth power equals *1*.

You also should note that as *x* becomes greater than about *1* the value of e^x starts to increase rapidly. This type of rapid increase in behavior is what is meant by "exponential growth."

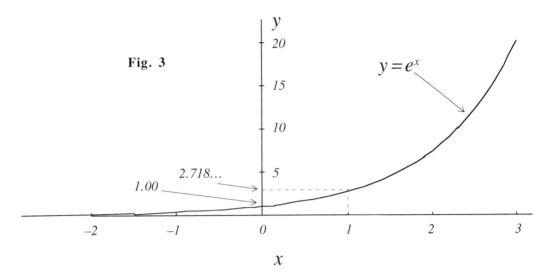

Fig. 3

A9.6   EXPONENTS AND LOGARITHMS

You certainly will be familiar with numbers raised to various powers, in particular the base *10*. Thus $10^1 = 10$, $10^2 = 100$, $10^3 = 1000$, etc. We define any number raised to the *zeroth* power as having a value of *1*; therefore $10^0 = 1$, $a^0 = 1$, and of course, $e^0 = 1$. The function defined by $y = e^x$ is just the base number *e* raised to the power *x*.

By *definition*, the *logarithm of x to the base a* is the exponent to which *a* must be raised to give the number *x*. Consider the base *10*:

$10^2 = 100$ and we say $log_{10} 100 = 2$; but, if $100 = 10^2$ and $log_{10} 100$ equals 2, then we can replace the "2" in 10^2 with "$log_{10} 100$" to give $10^2 = 10^{log_{10} 100}$.

Therefore, $100 = 10^{log_{10} 100}$; similarly $1000 = 10^{log_{10} 1000}$ and $x = 10^{log_{10} x}$.

We can do the same for any other base as well. For example, $8 = 2^3$; therefore $log_2 8 = 3$ and $8 = 2^{log_2 8}$. Similarly, $100 = 2^{6.64386}$, therefore $log_2 100 = 6.64386$ and $100 = 2^{log_2 100}$. Therefore, the general formula for the logarithm of any number *x* to the base *a* is

$$x = a^{log_a x}$$

Now let's consider the base e:

$e^2 = (2.71828..)^2 = 7.389056$, therefore $log_e 7.389056 = 2$, and $7.389056 = e^{log_2 7.389056} = e^2$

$e^3 = (2.71828..)^3 = 20.0855..$, therefore $log_e 20.0855 = 3$, and $20.0855 = e^{log_2 20.0855} = e^3$

In general, if $x = e^a$, then $log_e x = a$ and

$$x = e^{log_e x}$$ ... [A9-26]

Logarithms to the base e are so often encountered that we usually designate them by the more convenient designation "ln":

$$ln\, x \equiv log_e x$$

$ln\, x$ is called the *natural logarithm* of x, or the *Napierian logarithm* of x, in honor of the Scottish mathematician John Napier (1550-1617), who did much of the early work on logarithms.[10] The first explicit recognition of the significance of the number e in mathematics is found in an appendix of a 1618 translation of Napier's *Descriptio*, where there appears the equivalent of the statement that $log_e 10 = 2.302585$.[11]

The "**rules of exponents and logarithms**," which apply to any type of logarithm and exponent, are summarized as follows:

$$log\,(xy) = log\, x + log\, y \qquad\qquad \frac{a^x}{a^y} = a^{x-y}$$

$$log\, \frac{x}{y} = log\, x - log\, y \qquad\qquad \left(a^x\right)^y = a^{xy}$$

$$log\, x^y = y \cdot log\, x \qquad\qquad log\, 1 = log\, a^0 = 0$$

$$a^x \cdot a^y = a^{x+y} \qquad\qquad a^{log_a x} = x = log_a a^x$$

The relationship between logarithms to the *base-10* ($log\, x$) and natural logarithms ($ln\, x$) is easily derived as follows. Consider some number x:

---

[10] The word *logarithm* was coined by Napier and derives from the Greek "logos" and "arithmes." *Arithmes* means *number*, whereas *logos* has many meanings, one of which means *ratio*. Accordingly, a *logarithm* is a "ratio number," possibly so-called because logarithms are useful when calculating ratios. Napier originally called the exponent of each power its "artificial number," but later changed it to *logarithm*. Napier's definition of logarithms was slightly different from the definition used today; the modern definition was introduced in 1728 by Leonhard Euler. One of the oddities in the history of mathematics is that the logarithm actually predates the exponential function.

[11] Quoted from *e: The Story of a Number*, by E. Maor, Princeton University Press, Princeton, 1994.

Expressed using *base-10*:

$$10^{log\,x} = x$$

Expressed using *base-e*:

$$e^{ln\,x} = x$$

Equating the two expressions for *x*:

$$10^{log\,x} = e^{ln\,x} \qquad \text{. . . [A9-27]}$$

But

$$10 = e^{ln\,10}$$

Substituting this into Eqn [27]:

$$\left(e^{ln\,10}\right)^{log\,x} = e^{ln\,x}$$

But

$$\left(e^{ln\,10}\right)^{log\,x} = e^{ln\,10 \cdot log\,x}$$

Therefore

$$e^{ln\,10 \cdot log\,x} = e^{ln\,x}$$

from which it is seen that

$$(ln\,10) \cdot (log\,x) = ln\,x$$

Therefore,

$$ln\,x = (ln\,10) \cdot (log\,x) = 2.3026\,log\,x$$

$$\boxed{\begin{aligned} ln\,x &= (ln\,10) \cdot \left(log_{10}x\right) \\ &= (2.3026..) \cdot log_{10}x \end{aligned}} \qquad \text{. . . [A9-28]}$$

Just as the mathematicians—and we too—worked out an infinite series to define the value of *e*, they also have found infinite series to evaluate *ln x*. Values of *ln x*—including the values generated by your calculator—are computed by using these infinite series formulas. Since most values of *ln x* are irrational numbers, the series formulas can be evaluated to as many terms as desired to achieve any degree of accuracy that is required.

The following formulas are most commonly used to define *ln x*:

$$ln\,x = 2\left[\frac{x-1}{x+1} + \frac{1}{3}\left(\frac{x-1}{x+1}\right)^2 + \frac{1}{5}\left(\frac{x-1}{x+1}\right)^3 + \frac{1}{7}\left(\frac{x-1}{x+1}\right)^4 + ...\right] \quad (x > 0)$$

$$ln\,x = \frac{x-1}{x} + \frac{1}{2}\left(\frac{x-1}{x}\right)^2 + \frac{1}{3}\left(\frac{x-1}{x}\right)^3 + \frac{1}{4}\left(\frac{x-1}{x}\right)^4 + ... \quad (x > 0.5) \qquad \text{. . [A9-29]}$$

$$ln\,x = \lim_{n \to \infty} n\left(x^{1/n} - 1\right)$$

A9.7 THE AREA UNDER THE CURVE $y = 1/x$ EXHIBITS LOGARITHMIC PROPERTIES!

You may recall that the purpose of this appendix is to evaluate the integral $\int_a^b \left(\frac{1}{x}\right) dx$, which

is the area under the curve $y = 1/x$ between $x = a$ and $x = b$ (refer back to Fig. 1). This whole is-
sue has become increasingly messy because we saw that the simple power rule of integration

$$\int x^n dx = \frac{x^{n+1}}{n+1}$$

doesn't work when $x = -1$. You needn't feel too confused about this because for many years
mathematicians struggled with trying to solve this integral. Then, in about 1630, a Belgian Jesuit
by the name of Grégoire de Saint-Vincent[12] noticed that the area under the curve $y = 1/x$ from any
reference point $x_o > 0$ to any value $x = b$ where $b > a$ is *directly proportional to ln b*, the natural
logarithm of b. You'll recall that there's nothing particularly mysterious about logarithms. The
natural logarithm of b is just *defined* as the number to which the base e must be raised to give b;

i.e., $e^{ln b} \equiv b$

It's easiest to understand de Saint-
Vincent's discovery if we take the
above-mentioned reference point x_o as x_o
$= 1$, because, as we know from the rules
of exponents and logarithms, $ln\ 1 = 0$,
and this happens to be an especially
convenient value for our reference
point.[13] Fig. 4 shows the areas we are
talking about. These areas could be
measured by de Saint-Vincent fairly easily to two decimal places by accurately
plotting the curve $y = 1/x$ on graph
paper having a linear grid pattern and
actually counting the number of squares
contained in the cross-hatched area under the curve.

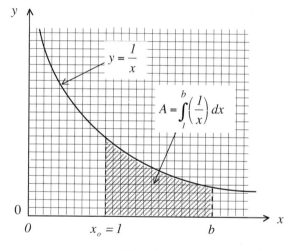

Fig. 4

Table 4 shows the kind of results de Saint-Vincent probably obtained. When plotting the measured
areas on the y-axis against the natural logarithm of b on the x-axis he observed to his surprise that

---

[12] 1584-1667.

[13] Note that x can be less than 1, but must be greater than zero because at $x = 0$ the value of $ln\ x$ becomes negatively
infinite.

Table 4. *Values of* $A = \int_{1}^{b}(1/x)\,dx$		
b	$A = \int_{1}^{b}(1/x)\,dx$	*ln b*
1	0	0
2	0.69	0.6931
3	1.10	1.0986
4	1.39	1.3863
5	1.61	1.6094

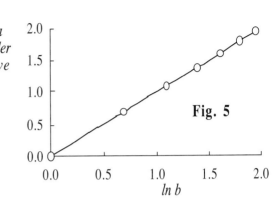

Area Under Curve

Fig. 5

there was a direct proportionality between the two, as indicated in Fig. 5. As the value of b gets bigger, the area A under the curve from $x_o = 1$ to $x = b$ also gets bigger, not in direct proportion to b, but rather in proportion to the *logarithm* of b. Thus $\int_{1}^{b}(1/x)\,dx$ exhibits *logarithmic properties*.

A9.8 THE DERIVATIVE OF THE FUNCTION $y = e^x$

The function $y = e^x$ has a very interesting derivative. We showed earlier that

$$e = \lim_{n \to \infty}\left(1 + \frac{1}{n}\right)^n$$

Therefore
$$e^x = \left[\lim_{n \to \infty}\left(1 + \frac{1}{n}\right)^n\right]^x = \lim_{n \to \infty}\left(1 + \frac{1}{n}\right)^{nx} \qquad \ldots \text{[A9-30]}$$

We can expand this expression for e^x using the—here we go again!—Binomial Theorem:

$$\left(1 + \frac{1}{n}\right)^{nx} = \frac{1}{0!}\left(\frac{1}{n}\right)^0 + \frac{nx}{1!}\left(\frac{1}{n}\right) + \frac{nx(nx-1)}{2!}\left(\frac{1}{n}\right)^2 + \frac{nx(nx-1)(nx-2)}{3!}\left(\frac{1}{n}\right)^3 + \ldots$$

$$= 1 + nx\left(\frac{1}{n}\right) + \frac{nx(nx-1)}{2!}\left(\frac{1}{n}\right)^2 + \frac{nx(nx-1)(nx-2)}{3!}\left(\frac{1}{n}\right)^3 + \ldots$$

$$= 1 + x + \frac{nx(nx-1)}{2!\,n\cdot n} + \frac{nx(nx-1)(nx-2)}{3!\,n\cdot n\cdot n} + \ldots$$

$$= 1 + x + \frac{\frac{nx}{n}\left(\frac{nx-1}{n}\right)}{2!} + \frac{\frac{nx}{n}\left(\frac{nx-1}{n}\right)\left(\frac{nx-2}{n}\right)}{3!} + \ldots$$

$$= 1 + x + \frac{\frac{nx}{n}\left(\frac{nx}{n} - \frac{1}{n}\right)}{2!} + \frac{\frac{nx}{n}\left(\frac{nx}{n} - \frac{1}{n}\right)\left(\frac{nx}{n} - \frac{2}{n}\right)}{3!} + \dots$$

$$= 1 + x + \frac{x\left(x - \frac{1}{n}\right)}{2!} + \frac{x\left(x - \frac{1}{n}\right)\left(x - \frac{2}{n}\right)}{3!} + \dots \qquad \dots [A9\text{-}31]$$

Therefore

$$e^x = \left[\lim_{n \to \infty}\left(1 + \frac{1}{n}\right)^n\right]^x = \lim_{n \to \infty}\left[1 + x + \frac{x\left(x - \frac{1}{n}\right)}{2!} + \frac{x\left(x - \frac{1}{n}\right)\left(x - \frac{2}{n}\right)}{3!} + \dots\right] \dots [A9\text{-}32]$$

As $n \to \infty$ each of the terms $\frac{1}{n}$, $\frac{2}{n}$, etc. goes to zero and we get

$$\boxed{e^x = 1 + x + \frac{x^2}{2!} + \frac{x^3}{3!} + \dots} \qquad \dots [A9\text{-}33]$$

Now let's take the derivative of this expression:

$$\frac{d\left(e^x\right)}{dx} = \frac{d}{dx}\left(1 + x + \frac{x^2}{2!} + \frac{x^3}{3!} + \dots\right)$$

$$= 0 + 1 + \frac{2x}{2!} + \frac{3x^2}{3!} + \dots$$

$$= 1 + x + \frac{x^2}{2!} + \dots$$

Hey... reference to Eqn [33] shows that this is just e^x!

So *the derivative of the function $y = e^x$ is just the function itself.*

$$\boxed{\frac{d\left(e^x\right)}{dx} = e^x} \qquad \dots [A9\text{-}34]$$

Since we know that the derivative dy/dx of any function $y = f(x)$ is the *slope* of the curve at any given value of x, then the slope of $y = e^x$ at any given value of x is just the value of the function itself for that particular value of x.

Thus, referring back to Table 4 and Fig. 3, at $x = 0$, $e^x = 1$. Therefore the slope of the curve where it crosses the y-axis also is $dy/dx = 1$. Similarly, at $x = 3$, the slope of the curve is 20.09, and so on.

e^x *is unique because it is the only function whose derivative is the same as the function itself.* This is not true for exponents taken from any other base. For example, consider the function $y = 10^x$. The general formula for evaluating the differential of the function $y = a^u$ is

$$\frac{d(a^u)}{dx} = a^u \ln a \frac{du}{dx}.$$

Applying this formula to $y = 10^x$ with $a = 10$ and $u = x$ gives

$$\frac{d(10^x)}{dx} = 10^x (\ln 10) \frac{dx}{dx} = 10^x (2.3026)(1) = 2.3026 \times 10^x$$

which is different from 10^x. However, if we apply the formula to $y = e^x$, with $a = e$ and $u = x$, we get

$$\frac{d(e^x)}{dx} = e^x (\ln e) \frac{dx}{dx} = e^x (1)(1) = e^x$$

which is just the function itself.

A9.9   THE DERIVATIVE OF THE FUNCTION $y = \ln x$

We saw in section A9-7 that the area under the curve $y = 1/x$ exhibits logarithmic behavior, and in section A9-8 we found that the derivative dy/dx of the function $y = e^x$ is unique in that it is the same as the function itself, namely

$$\frac{d(e^x)}{dx} = e^x \qquad \ldots \text{[A9-35]}$$

Guess what? The derivative dy/dx of the function $y = \ln x$ also is interesting!

Do you remember the *Chain Rule* of differential calculus? That's the one that says if some function u is a function of x, and some other function v also is a function of x, then

(The Chain Rule) $$\frac{d(uv)}{dx} = u \cdot \frac{dv}{dx} + v \cdot \frac{du}{dx} \qquad \ldots \text{[A9-36]}$$

We can make use of the chain rule to evaluate $d(\ln y)/dx$ as follows:

Let $y = e^{\ln x}$ and let $\ln x = u$.

Therefore
$$y = e^{\ln x} = e^u$$

From the chain rule:
$$\frac{dy}{dx} = \frac{dy}{du} \cdot \frac{du}{dx} \qquad \ldots [A9\text{-}37]$$

But, since $y = e^u$ and $u = \ln x$, Eqn [A9-37] is just

$$\frac{d(e^u)}{dx} = \frac{d(e^u)}{du} \cdot \frac{d(\ln x)}{dx}$$

But we know that the differential of e^u with respect to u is just e^u; therefore Eqn [A9-37] is

$$\frac{d(e^u)}{dx} = e^u \cdot \frac{d(\ln x)}{dx}$$

but, since $u = \ln x$, this also is
$$\frac{d(e^{\ln x})}{dx} = e^{\ln x} \cdot \frac{d(\ln x)}{dx} \qquad \ldots [A9\text{-}38]$$

But, from the *definition* of a logarithm
$$e^{\ln x} = x$$

Therefore,
$$\frac{d(e^{\ln x})}{dx} = \frac{d(x)}{dx} = 1 \qquad \ldots [A9\text{-}39]$$

We now have two expressions for $d(e^{\ln x})/dx$—Equations [A9-38] and [A9-39]. Equating the two:

$$e^{\ln x} \cdot \frac{d(\ln x)}{dx} = 1 \qquad \ldots [A9\text{-}40]$$

But again, by definition, $x = e^{\ln x}$. Therefore, substituting this into Eqn [A9-40]:

$$x \cdot \frac{d(\ln x)}{dx} = 1$$

from which
$$\frac{d(\ln x)}{dx} = \frac{1}{x}$$

So the slope dy/dx of the curve $y = \ln x$ is just our "good old" familiar function $1/x$!

$$\boxed{\frac{d(\ln x)}{dx} = \frac{1}{x}} \qquad \ldots [A9\text{-}41]$$

A9.10 INVERSE FUNCTIONS

Consider the function of x
$$y = f(x) = x^3 \qquad \ldots \text{[A9-42]}$$

The same information could be conveyed by expressing x as a function of y:

Thus, we could write
$$x = g(y) = \sqrt[3]{x} \qquad \ldots \text{[A9-43]}$$

If the function $y = f(x)$ in Eqn [A9-42] takes a value of $x = a$ and assigns it a value of b, then the function $g = g(y)$ in Eqn [A9-43] takes a value of $y = b$ and assigns it a value of a. In other words, $f(a) = b$ expresses exactly the same information as $g(b) = a$. The function $g(x)$ is called the **inverse function** of $f(x)$.

Not every function has an inverse function. For example, the function $y = f(x) = x^2$ would appear to have the inverse function $x = g(y) = \sqrt{y}$. However, for the value $y = f(x) = 4$, there exist *two* values of $x = g(y) = \sqrt{y}$; namely $x = +2$ and $x = -2$. This kind of uncertainty is not permitted for a function, because part of the definition of a function is that it must have a unique value at each point. In the case of $f(x) = x^2$ the problem can be resolved by defining the function $f(x)$ only for values of $x \geq 0$. By doing this there is only one value for $g(y)$ when $y = 4$, and that is $x = +2$.

In view of the above it can be seen that, providing $a > 1$, the function $f(x) = a^x$ has an inverse function, namely, $g(x) = \log_a x$. Thus *the inverses of exponential functions are logarithmic functions*. Therefore the function $ln\,x$ is the inverse of the function e^x.

Earlier—Eqn [A9-30]—we showed that

$$e^x = \lim_{n \to \infty} \left(1 + \frac{1}{n} \right)^{nx} \qquad \ldots \text{[A9-44]}$$

We can manipulate this equation as follows:[14]

Let $nx = m$, therefore
$$n = \frac{m}{x} \quad \text{and} \quad \frac{1}{n} = \frac{x}{m}$$

Now, providing $x > 0$, as $n \to \infty$, $m = nx$ also goes to infinity, and Eqn [A9-44] can be written

$$e^x = \lim_{m \to \infty} \left(1 + \frac{x}{m} \right)^{m} \qquad \ldots \text{[A9-45]}$$

---

[14] Bear with me; there actually is a reason I am doing this!

But, since it doesn't matter which letter we use in the limit expression, let's go back to n, the letter we are more familiar with, and replace the m's in Eqn [A9-45] with n's again, to get

$$e^x = \lim_{n \to \infty} \left(1 + \frac{x}{n}\right)^n \qquad \qquad \ldots \text{[A9-46]}$$

Eqn [A9-46] is just an alternative way of expressing Eqn [A9-44].

Now, if we just forget about the "limit as n goes to infinity" in Eqn [A9-46] and just go ahead and solve Eqn [A9-46] for x, what do we get?

We start with
$$y = \left(1 + \frac{x}{n}\right)^n$$

Rearranging: $\quad 1 + \frac{x}{n} = y^{1/n} \quad \to \quad \frac{x}{n} = y^{1/n} - 1 \quad \to \quad x = n\left(y^{1/n} - 1\right)$

If you compare this last expression with the expression given by Eqn [A9-29] for the definition of $ln\,x$, namely

$$ln\,x = \lim_{n \to \infty} n\left(x^{1/n} - 1\right) \qquad \qquad \ldots \text{[A9-29]}$$

you will see that they are the same, indicating[15] that the two functions e^x and $ln\,x$ are indeed inverse functions!

Inverse functions are mirror images of each other in the 45° line $y = x$. Thus, as shown below in Fig. 6, the function $f(x) = e^x$ is the mirror image of the function $g(x) = ln\,x$ about the dashed 45° line. Fig. 6 provides a neat summary of the information we have learned about the functions $y = 1/x$, $y = e^x$, and $y = ln\,x$:

$y = 1/x$:

• For all positive real values of x the curve is located in only one quadrant.
• Intersects both axes only at $+\infty$.
• All slopes dy/dx are negative.

$y = e^x$:

• Considering its importance, very striking for its simplicity.
• Never crosses the x-axis (no x-intercepts).
• Providing x is a real number, $y = e^x$ is always a positive number.
• Exhibits no maxima or minima.

---

[15] I use the word *indicate* because we didn't take into account the part expressing the limit as n goes to infinity. To deal with that little wrinkle involves the use of some slightly more subtle mathematical arguments. However, the two functions are, in fact, inverse functions.

• All slopes dy/dx are positive
• No inflection points.
• No vertical asymptote.
• Is the only function that equals its own derivative.

The important feature of an exponential function is that its derivative is proportional to the function itself. Thus, if $y = b^x$, then $dy/dx = kb^x = ky$, or, as we have shown in the case of base e, if $y = e^x$, then $dy/dx = e^x = y$. Thus, the bigger the value of x, the steeper the slope of the curve. The rate of increase "feeds on itself" and rapidly can attain extremely high values, hence the phrase "exponential growth."

$y = \ln x$:

• For all real x logarithms of negative numbers do not exist.
• As x approaches zero from the right-hand side, the function heads towards $-\infty$.
• At $x = 1, \ln x = 0$.
• The function has no upper limit: as x goes to infinity, $\ln x$ also goes to infinity, it just goes at an
  increasingly slower rate.
• Among all the logarithmic functions, the curve $y = \ln x$ is the only one that has a tangent at $(1,0)$
  with a slope of 1.

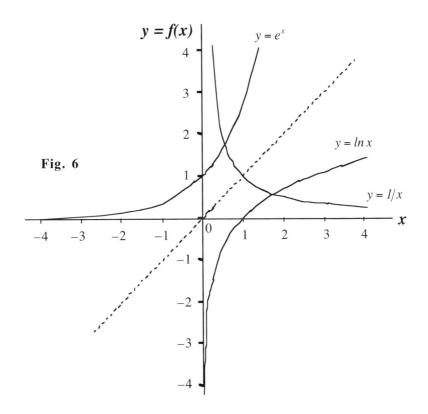

Fig. 6

A9.11   ANTIDERIVATIVES AND THE FUNDAMENTAL THEOREM OF THE CALCULUS

We saw from the power rule in Appendix 8 that the integral (or area A under the curve) of the function $y = x$ between x_1 and x_2 is

$$\int_{x_1}^{x_2} x\,dx = \frac{x_2^2}{2} - \frac{x_1^2}{2}$$

This type of integral is called a **definite integral** because we have specified a *definite* lower limit x_1 and upper limit x_2 between which we want to know the value of the area. On the other hand, if we don't specify the actual values of x_1 and x_2 and instead just write

$$\int x\,dx = \frac{x^2}{2}$$

then the integral is called an **indefinite integral**. Now here's the key point: If we take the derivative of the indefinite integral, we just get back the original function $y = f(x)$.

Thus,
$$\frac{d\left(\int x\,dx\right)}{dx} = \frac{d}{dx}\left(\frac{x^2}{2}\right) = \frac{2x}{2} = x = y$$

Differentiation just *undoes* integration.

Similarly, if our original function is

$$y = 3x^4 + 9x^2$$

then the indefinite integral is $\int y\,dx = 3\dfrac{x^5}{5} + 9\dfrac{x^3}{3} = \tfrac{3}{5}x^5 + 3x^3.$

Again, if we differentiate the integral, we retrieve the original function $y(x)$:

Thus:
$$\frac{d\left(\int x\,dx\right)}{dx} = \frac{d}{dx}\left(\tfrac{3}{5}x^5 + 3x^3\right) = \frac{3 \times 5x^4}{5} + 3 \times 3x^2 = 3x^4 + 9x^2 = y$$

Note that, strictly speaking, we should add some constant C to the expression for our indefinite integral. This C is called a *constant of integration* or an *integration constant*.

Thus,
$$\int x\,dx = \frac{x^2}{2} + C$$

and
$$\int \left(3x^4 + 9x^2\right)dx = \tfrac{3}{5}x^5 + 3x^3 + C$$

The reason we have to add the integration constant C is that just from looking at the expressions $y = x$ and $y = 3x^4 + 9x^2$ we can't tell whether the correct indefinite integrals from which these functions were derived were $\dfrac{x^2}{2}$ or $\dfrac{x^2}{2} + C$ in the case of the first function, or $\left(\frac{3}{5}x^5 + 3x^3\right)$ or $\left(\frac{3}{5}x^5 + 3x^3 + C\right)$ in the case of the second function. This difficulty arises because d/dx for a constant is zero; therefore upon differentiating the indefinite integral any constant that may be part of the indefinite integral will disappear. So, just to be on the safe side, we add the constant C to our indefinite integral to cover all our options just in case the correct expression for the indefinite integral has a constant in it. If, in fact, the correct indefinite integral does *not* have a constant, then the value of C is just $C = 0$.

Newton, when developing the calculus,[16] recognized that there is some kind of an inverse mutual relationship between the area under a curve (integration) and the slope of its tangent (differentiation). This inverse mutual relationship is known as the *Fundamental Theorem of the Calculus*. As we have seen, given the function $y = f(x)$, we can define a new function $A(x)$, which defines the *area* under the graph of $f(x)$ between any two x-values $x_1 = a$ and $x_2 = b$. **The Fundamental Theorem of Calculus** states that

At any point the rate of change $\dfrac{d\{A(x)\}}{dx}$ *of the area function A(x) is equal to the value*

of the original function y = f(x) at the same point. In symbols: $\dfrac{d\{A(x)\}}{dx} = f(x)$

This means that the area function $A(x)$ is an **antiderivative** of the function $y = f(x)$. Therefore, to find the area under the curve $y = f(x)$ we must find the antiderivative of $f(x)$. In other words, finding the area and finding the derivative are inverses of each other. In symbols, the antiderivative of the function $y = f(x)$ is denoted as $\int y\, dx$ or $\int f(x)\, dx$. As we have seen, the antiderivative also is called the indefinite integral of $y = f(x)$. Accordingly, the Fundamental Theorem also can be written

$$\frac{d\left\{\int f(x)\, dx\right\}}{dx} = f(x) \quad \text{or} \quad \frac{d\left\{\int y\, dx\right\}}{dx} = y \qquad \ldots \text{[A9-47]}$$

---

[16] The word *calculus* comes from the Latin word for pebble. Pebbles were used for performing calculations in a primitive form of the abacus. The use of the word *calculus* to refer specifically to differentiation and integration in mathematics was first adopted by the German philosopher and mathematician Gottfried Wilhelm Leibniz (1646-1716). Newton and Leibniz sustained a rather long-running feud because each claimed to have invented the calculus. Nowadays we give equal credit to both men. It was Leibniz who proposed both the notations dy/dx as well as $\int y\, dx$. The term *integral* was coined in 1690 by the Swiss mathematician Jakob Bernouilli (1654-1705). The term *derivative* and the notation fx—now written $f(x)$—for a function of x were first proposed by the French mathematician Joseph Louis Lagrange (1736-1813) in 1797.

Eqn [A9-47] confirms what we already intuitively knew: that differentiating the integral of a function just returns the original function to us. Taking the differential d of an integral sign $\int$ just cancels out both operations.

For example, by now you probably have figured out that

$$\int \frac{1}{x} dx = \ln x$$

If we differentiate both sides of this equation, we get

$$d\left[\int \frac{1}{x} dx\right] = d \ln x$$

The $d\int$ just cancel each other out, leaving $\quad \frac{1}{x} dx = d \ln x$

and dividing both sides by dx gives $\quad \dfrac{d \ln x}{dx} = \dfrac{1}{x}.$

But the *real* utility of the Fundamental Theorem is that it relates integrals to derivatives, and tells us that in order to evaluate a definite integral $\int_a^b f(x)\,dx$ we don't have to go through all the effort of working out a Riemann series and summing it up over an infinite number of infinitesimally small intervals Δx. Instead, providing that the function $y = f(x)$ is continuous over the interval from a to b, all we have to do is find some function $A(x)$—an antiderivative—which, when differentiated, gives us our original function $y = f(x)$. Then the value of the area under the curve between $x = a$ and $x = b$ will be given by

$$\int_a^b f(x)\,dx = A(b) - A(a)^{17}$$

So, the Fundamental Theorem of Calculus tells us that if we want to find $\int_a^b \left(\frac{1}{x}\right) dx$, we only need to find a function whose derivative gives us our original function $y = 1/x$. But we have already done this in Section A9-9 (Eqn [A9-41]) where we showed that $\dfrac{d(\ln x)}{dx} = \dfrac{1}{x}$! Therefore our sought after antiderivative $A(x)$ is $A(x) = \ln x$, and it follows from the Fundamental Theorem that

$$\int_a^b \left(\frac{1}{x}\right) dx = \ln b - \ln a = \ln\left(\frac{b}{a}\right).$$

---

[17] Doesn't this look familiar? Yes, the mathematical basis for the formulation of the state function!!

To return to our initial task, which was to evaluate the heat Q required to heat a substance from temperature T_1 to T_2 when the substance has a heat capacity of the form

$$C = a + bT + cT^2 + \frac{d}{T}$$

where a, b, c, and d are constants, we found that Q was given by

$$Q = a\int_{T_1}^{T_2} dT + b\int_{T_1}^{T_2} T dT + c\int_{T_1}^{T_2} T^2 dT + d\int_{T_1}^{T_2}\left(\frac{1}{T}\right) dT \qquad \ldots [A9\text{-}48]$$

Using the integration techniques explained in Appendix 8, we had no problem evaluating the first three integrals in Eqn [A9-48] using the power rule for integrals. But the power rule didn't work for the integral in the last term,

$$\int_{T_1}^{T_2}\left(\frac{1}{T}\right) dT \qquad \ldots [A9\text{-}49]$$

so we embarked on this long, tedious, but (I hope) enlightening journey. Now we know that this last integral is just

$$\int_{T_1}^{T_2}\left(\frac{1}{T}\right) dT = \ln\left(\frac{T_2}{T_1}\right) \qquad \ldots [A9\text{-}50]$$

A9.12   FURTHER DELVING INTO THE MYSTERY OF NATURAL LOGARITHMS

It is our hope that by now you are no longer mystified with the concept of natural logarithms and how to evaluate the integral $\int_{T_1}^{T_2}\left(\frac{1}{T}\right) dT$, which also is written $\int_{T_1}^{T_2}\frac{dT}{T}$. In this section we will look a little closer at the meaning of this integral. Suppose we want to heat something from $T_1 = 300$ K to $T_2 = 350$ K, and one of the terms in the heat capacity equation is of the form of Eqn [A9-49].

You may recall from Appendix 8 when we were using Riemann series to evaluate integrals that the procedure was to divide up the interval between 300 K and 350 K into a bunch of subintervals of size Δx, evaluate the area of each little rectangle in each interval, and add up all the little rectangles to give us an approximation to the area under the curve between T_1 and T_2.

Although, as we have seen, we can't use this method to obtain an exact antiderivative of the integral $\int_{T_1}^{T_2}\frac{dT}{T}$, we *can* apply the *concept* to this integral to get a clearer understanding of what this integral means. We saw that when a subinterval ΔT was made smaller and smaller, eventually it approached the differential dT. We also saw that integration just means a summing up of an infinite number of infinitely small rectangles. With this in mind it becomes apparent that the term dT/T can be approximated by $\Delta T/T$, and that the integral $\int dT/T$ can be approximated by summing up all the little $\Delta T/T$'s in the interval between T_1 and T_2. Thus,

$$\int \frac{dT}{T} \approx \sum \frac{\Delta T}{T} \qquad \ldots \text{[A9-51]}$$

Or, in the limit when we make ΔT infinitesimally small and increase the number of n subintervals to infinity, the expression becomes *exact*:

$$\int \frac{dT}{T} = \sum_{\substack{n \to \infty \\ \Delta T \to 0}} \frac{\Delta T}{T} \qquad \ldots \text{[A9-52]}$$

The next question that arises is: "What temperature do we use for T in the denominator of the term $\Delta T/T$?" Common sense dictates that we naturally use the *average* temperature T_{av} in the middle of the interval; i.e.,

$$\frac{\Delta T}{T} \to \frac{\Delta T}{T_{av}} \qquad \ldots \text{[A9-53]}$$

Therefore the integral should be approximated by

$$\int \frac{dT}{T} \approx \sum \frac{\Delta T}{T_{av}} \qquad \ldots \text{[A9-54]}$$

and the smaller that we make ΔT, the more closely $\sum \Delta T/T_{av}$ should approximate the true value of $\int dT/T$. If we have only *one* (albeit small) interval ΔT, then $\sum \dfrac{\Delta T}{T_{av}} = \dfrac{\Delta T}{T_{av}}$, and Eqn [A9-54] reduces to $\int \dfrac{dT}{T} \approx \dfrac{\Delta T}{T_{av}}$. Let's test out this hypothesis by using this method to approximate $\int_{300}^{350} \dfrac{dT}{T}$.

First of all, the *exact* value (to six decimal places) is $\int_{300}^{350} \dfrac{dT}{T} = ln\left(\dfrac{350}{300}\right) = 0.154151$.

For our *first* calculation, we let ΔT equal the *whole* interval; i.e., $\Delta T = 350 - 300 = 50$ K.

In this case, the average temperature in the interval is $T_{av} = \dfrac{300 + 350}{2} = 325$ K, and

$$\int_{300}^{350} \frac{dT}{T} \approx \frac{\Delta T}{T_{av}} = \frac{350 - 300}{325} = \frac{50}{325} = 0.153846$$

which is pretty good! Compared with the actual value of 0.154151, this estimate is only in error by

$$\frac{0.153846 - 0.154151}{0.154151} \times 100\% = -0.20\%!$$

Let's try to refine the calculation by decreasing the size of the subintervals: For our *second* calculation, let's divide the temperature interval into *two* subintervals, each of $\Delta T = 25$ K.

$\Delta T = 25\,K$	Interval	T_{av}	$(\Delta T/T_{av})_i$
	1	$\frac{300 + 325}{2} = 312.5$	$(\Delta T/T_{av})_1 = \frac{325 - 300}{312.5} = \frac{25}{312.5} = 0.080000$
	2	$\frac{325 + 350}{2} = 337.5$	$(\Delta T/T_{av})_2 = \frac{350 - 325}{337.5} = \frac{25}{337.5} = 0.074074$

$$\sum \Delta T/T_{av} = 0.080000 + 0.074074 = 0.154074$$

$$\% \text{ error} = \frac{0.154074 - 0.154151}{0.154151} \times 100\% = -0.050\%$$

This is even more accurate!

Finally, let's divide up the temperature interval into 5 equal subintervals of $\Delta T = 10$ K each.

$\Delta T = 10\,K$	Interval	T_{av}	$(\Delta T/T_{av})_i$
	1	$\frac{300 + 310}{2} = 305$	$(\Delta T/T_{av})_1 = \frac{310 - 300}{305} = \frac{10}{305} = 0.032787$
	2	$\frac{310 + 320}{2} = 315$	$(\Delta T/T_{av})_2 = \frac{320 - 310}{315} = \frac{10}{315} = 0.031746$
	3	$\frac{320 + 330}{2} = 325$	$(\Delta T/T_{av})_3 = \frac{330 - 320}{325} = \frac{10}{325} = 0.030769$
	4	$\frac{330 + 340}{2} = 335$	$(\Delta T/T_{av})_4 = \frac{340 - 330}{335} = \frac{10}{325} = 0.029851$
	5	$\frac{340 + 350}{2} = 345$	$(\Delta T/T_{av})_5 = \frac{350 - 340}{345} = \frac{10}{345} = 0.028986$

$$\sum \Delta T/T_{av} = 0.032787 + 0.031746 + 0.030769 + 0.029851 + 0.028986 = 0.154139$$

$$\% \text{ error} = \frac{0.154139 - 0.154151}{0.154151} \times 100\% = -0.0078\%$$

Wow! Less than one hundredth of one percent error!

The exact calculation using accurate logarithms also can be divided into 5 subintervals of 10 K each. Thus, from the rules of exponents and logarithms,

$$ln\left(\frac{350}{300}\right) = ln\left(\frac{310}{300}\right) + ln\left(\frac{320}{310}\right) + ln\left(\frac{330}{320}\right) + ln\left(\frac{340}{330}\right) + ln\left(\frac{350}{340}\right)$$

$$= 0.032790 + 0.031749 + 0.030772 + 0.029853 + 0.028988$$

$$= 0.154151$$

Since de Saint-Vincent showed that the areas under the curve of $y = 1/x$ display logarithmic properties, we can see that the same rules of exponents and logarithms also should apply to our approximate calculation.

Thus, since
$$ln\left(\frac{T_2}{T_1}\right) = \sum_i ln\left(\frac{T_{i+1}}{T_i}\right)_i$$

we would expect that
$$\frac{T_2 - T_1}{T_{av}} \approx \sum_i \left(\frac{\Delta T_i}{T_{av}}\right)_i$$

i.e., $$\frac{350 - 300}{325} \approx \left(\frac{310 - 300}{305}\right) + \left(\frac{320 - 310}{315}\right) + \left(\frac{330 - 320}{325}\right) + \left(\frac{340 - 330}{335}\right) + \left(\frac{350 - 340}{345}\right)$$

$$0.153846 \approx 0.032787 + 0.031746 + 0.030769 + 0.029851 + 0.028986$$
$$\approx 0.154139$$

Yes, as expected, it checks out!

Therefore we can see that really there is nothing particularly mysterious about the integral $\int_{x_1}^{x_2}\frac{dx}{x}$.

It is merely the sum of the quantities $\frac{\Delta x}{x_{av}}$, which, for each subinterval Δx, are just the fractional changes of the average x in the subinterval.

A9.13   EULER'S AMAZING FORMULA

As one final little tidbit, for those of you who have not heard of it, I offer Euler's famous formula:

$$e^{\pi i} + 1 = 0 \qquad\qquad \text{. . . [A9-55]}$$

Euler's formula is considered to be the most famous formula in all of mathematics. It connects not only the five fundamental constants of mathematics: 0, 1, e, π, and $i = \sqrt{-1}$; but also the three most fundamental operations of mathematics: addition, multiplication, and exponentiation, *and*, the most fundamental concept of all: *equality*.

And check this out: Rearranging Eqn [A9-55], $e^{\pi i} = -1$

Taking logarithms: $ln\, e^{\pi i} = \pi i = ln(-1)$

Thus the principal value of $ln(-1)$ is πi! This is not something that could have been foreseen.

Here's something to think about:

> Of the infinity of real numbers, the five most important numbers in mathematics, namely, 0, 1, $\sqrt{-1}$, e, and π, are all located within less than four units on the number line! Everything stems from the basics!

Under Ben Bulben
by W.B. Yeats

Under bare Ben Bulben's head
In Drumcliffe churchyard
Yeats is laid.
An ancestor was rector there
Long years ago, a church stands near,
by the road an ancient cross.
No marble, no conventional phrase;
On limestone quarried near the spot
by his command these words are cut:

Cast a cold eye on life, on death.
Horseman, pass by!

Reaction	Potential
$XeF + e^- \rightarrow Xe + F^-$	+3.4
$Tb^{4+} + e^- \rightarrow Tb^{3+}$	+3.1
$F_2 + 2H^+ + 2e^- \rightarrow 2HF$	+3.053
$F_2 + 2e^- \rightarrow 2F^-$	+2.866
$O_{(g)} + 2H^+ + 2e^- \rightarrow H_2O$	+2.421
$Cu^{3+} + e^- \rightarrow Cu^{2+}$	+2.4
$HFeO_4^- + 8H^+ + 3e^- \rightarrow Fe^{3+} + 4H_2O$	+2.20
$S_2O_8^{2-} + 2H^+ + 2e^- \rightarrow 2HSO_4^-$	+2.123
$2HFeO_4^- + 8H^+ + 6e^- \rightarrow Fe_2O_3 + 5H_2O$	+2.09
$HFeO_4^- + 4H^+ + 3e^- \rightarrow FeOOH + 2H_2O$	+2.08
$O_3 + 2H^+ + 2e^- \rightarrow O_2 + H_2O$	+2.076
$HFeO_4^- + 7H^+ + 3e^- \rightarrow Fe^{3+} + 4H_2O$	+2.07
$OH + e^- \rightarrow OH^-$	+2.02
$S_2O_8^{2-} + 2e^- \rightarrow 2SO_4^{2-}$	+2.010
$Ag^{2+} + e^- \rightarrow Ag^+$	+1.980
$Co^{3+} + e^- \rightarrow Co^{2+}$	+1.92
$Au^{2+} + e^- \rightarrow Au^+$	+1.8
$H_2O_2 + 2H^+ + 2e^- \rightarrow 2H_2O$	+1.776
$N_2O + 2H^+ + 2e^- \rightarrow N_2 + H_2O$	+1.766
$Ce^{4+} + e^- \rightarrow Ce^{3+}$	+1.72
$Au^+ + e^- \rightarrow Au$	+1.692
$PbO_2 + SO_4^{2-} + 4H^+ + 2e^- \rightarrow PbSO_4 + 2H_2O$	+1.6913
$MnO_4^- + 4H^+ + 3e^- \rightarrow MnO_2 + 2H_2O$	+1.679
$NiO_2 + 4H^+ + 2e^- \rightarrow Ni^{2+} + 2H_2O$	+1.678
$HClO_2 + 2H^+ + 2e^- \rightarrow HClO + H_2O$	+1.645
$HClO_2 + 3H^+ + 3e^- \rightarrow \frac{1}{2}Cl_2 + 2H_2O$	+1.628
$HClO + H^+ + e^- \rightarrow \frac{1}{2}Cl_2 + H_2O$	+1.611
$2NO + 2H^+ + 2e^- \rightarrow N_2O + H_2O$	+1.591
$HClO_2 + 3H^+ + 4e^- \rightarrow Cl^- + 2H_2O$	+1.570
$Mn^{3+} + e^- \rightarrow Mn^{2+}$	+1.5415
$MnO_4^- + 8H^+ + 5e^- \rightarrow Mn^{2+} + 4H_2O$	+1.507
$PtO_3 + 4H^+ + 2e^- \rightarrow Pt(OH)_2^{2+} + H_2O$	+1.5
$Au^{3+} + 3e^- \rightarrow Au$	+1.498
$HO_2 + H^+ + e^- \rightarrow H_2O_2$	+1.495
$Mn_2O_3 + 6H^+ + e^- \rightarrow 2Mn^{2+} + 3H_2O$	+1.485
$BrO_3^- + 6H^+ + 5e^- \rightarrow \frac{1}{2}Br_2 + 3H_2O$	+1.482
$HClO + H^+ + 2e^- \rightarrow Cl^- + H_2O$	+1.482
$ClO_3^- + 6H^+ + 5e^- \rightarrow \frac{1}{2}Cl_2 + 3H_2O$	+1.47
$PbO_2 + 4H^+ + 2e^- \rightarrow Pb^{2+} + 2H_2O$	+1.455
$3IO_3^- + 6H^+ + 6e^- \rightarrow Cl^- + 3H_2O$	+1.451
$Au(OH)_3 + 3H^+ + 3e^- \rightarrow Au^- + 3H_2O$	+1.45
$2HIO + 2H^+ + 2e^- \rightarrow I_2 + 2H_2O$	+1.439
$BrO_3^- + 6H^+ + 6e^- \rightarrow Br^- + 3H_2O$	+1.423
$Au^{3+} + 2e^- \rightarrow Au^+$	+1.401
$ClO_4^- + 8H^+ + 7e^- \rightarrow \frac{1}{2}Cl_2 + 4H_2O$	+1.39
$ClO_4^- + 8H^+ + 8e^- \rightarrow Cl^- + 4H_2O$	+1.389
$Cl_{2(g)} + 2e^- \rightarrow 2Cl^-$	+1.35827
$HCrO_4^- + 7H^+ + 3e^- \rightarrow Cr^{3+} + 4H_2O$	+1.350
$AuOH^{2+} + H^+ + 2e^- \rightarrow Au^+ + H_2O$	+1.32
$2HNO_2 + 4H^+ + 4e^- \rightarrow N_2O + 3H_2O$	+1.297
$PdCl_6^{2-} + 2e^- \rightarrow PdCl_4^{2-} + 2Cl^-$	+1.288
$ClO_2 + H^+ + 2e^- \rightarrow HClO_2$	+1.277
$Tl^{3+} + 2e^- \rightarrow Tl$	+1.252
$O_3 + H_2O + 2e^- \rightarrow O_2 + 2OH^-$	+1.24
$Cr_2O_7^{2-} + 14H^+ + 6e^- \rightarrow 2Cr^{3+} + 7H_2O$	+1.232
$O_2 + 4H^+ + 4e^- \rightarrow 2H_2O$	+1.229
$MnO_2 + 4H^+ + 2e^- \rightarrow Mn^{2+} + 2H_2O$	+1.224
$ClO_3^- + 3H^+ + 2e^- \rightarrow HClO_2 + H_2O$	+1.214
$PtOH^+ + H^+ + 2e^- \rightarrow Pt + 2H_2O$	+1.2
$2IO_3^- + 12H^+ + 10e^- \rightarrow I_2 + 6H_2O$	+1.195
$ClO_4^- + 2H^+ + 2e^- \rightarrow ClO_3^- + H_2O$	+1.189
$Pt^{2+} + 2e^- \rightarrow Pt$	+1.18
$Ir^{3+} + 3e^- \rightarrow Ir$	+1.156
$ClO_3^- + 2H^+ + e^- \rightarrow ClO_2 + H_2O$	+1.152
$Cu^{2+} + 2CN^- + e^- \rightarrow Cu(CN)_2^-$	+1.103
$Pu^{5+} + e^- \rightarrow Pu^{4+}$	+1.099
$Br_{2(aq)} + 2e^- \rightarrow 2Br^-$	+1.0873
$IO_3^- + 6H^+ + 6e^- \rightarrow I^- + 3H_2O$	+1.085
$Hg(OH)_2 + 2H^+ + 2e^- \rightarrow Hg + 2H_2O$	+1.034
$PtO_2 + 2H^+ + 2e^- \rightarrow PtO + 2H_2O$	+1.01
$Pu^{4+} + e^- \rightarrow Pu^{3+}$	+1.006
$PtO_2 + 4H^+ + 4e^- \rightarrow Pt + 2H_2O$	+1.00
$HNO_2 + H^+ + e^- \rightarrow NO + H_2O$	+0.983
$NO_3^- + 4H^+ + 3e^- \rightarrow NO + 2H_2O$	+0.957
$ClO_{2(aq)} + e^- \rightarrow ClO_2^-$	+0.954
$Pd^{2+} + 2e^- \rightarrow Pd$	+0.951
$NO_3^- + 3H^+ + 2e^- \rightarrow HNO_2 + H_2O$	+0.934
$2Hg^+ + 2e^- \rightarrow Hg_2^{2+}$	+0.920

$AgSCN + e^- \rightarrow Ag + SCN^-$	$+0.8951$
$HO_2^- + H_2O + 2e^- \rightarrow 3OH^-$	$+0.878$
$IrCl_6^{2-} + e^- \rightarrow IrCl_6^{3-}$	$+0.8665$
$Ag^+ + e^- \rightarrow Ag$	$+0.7996$
$Hg_2^{2+} + 2e^- \rightarrow 2Hg$	$+0.7973$
$AgF + e^- \rightarrow Ag + F^-$	$+0.779$
$Fe^{3+} + e^- \rightarrow Fe^{2+}$	$+0.771$
$IrCl_6^{3-} + 3e^- \rightarrow Ir + 6Cl^-$	$+0.77$
$ClO_2^- + 2H_2O + 4e^- \rightarrow Cl^- + 4OH^-$	$+0.76$
$2NO + H_2O + 2e^- \rightarrow N_2O + 2OH^-$	$+0.76$
$Po^{4+} + 4e^- \rightarrow Po$	$+0.76$
$Rh^{3+} + 3e^- \rightarrow Rh$	$+0.758$
$PtCl_4^{2-} + 2e^- \rightarrow Pt + 4Cl^-$	$+0.755$
$Tl^{3+} + 3e^- \rightarrow Tl$	$+0.741$
$Ag_2O_3 + H_2O + 2e^- \rightarrow 2AgO + 2OH^-$	$+0.739$
$O_2 + 2H^+ + 2e^- \rightarrow H_2O_2$	$+0.695$
$PtCl_6^{2-} + 2e^- \rightarrow PtCl_4^{2-} + 2Cl^-$	$+0.68$
$Ag_2SO_4 + 2e^- \rightarrow 2Ag + SO_4^{2-}$	$+0.654$
$ClO_3^- + 3H_2O + 6e^- \rightarrow Cl^- + 6OH^-$	$+0.62$
$Hg_2SO_4 + 2e^- \rightarrow 2Hg + SO_4^{2-}$	$+0.6125$
$2AgO + H_2O + 2e^- \rightarrow Ag_2O + 2OH^-$	$+0.607$
$Rh^{2+} + 4e^- \rightarrow Rh$	$+0.600$
$MnO_4^{2-} + 2H_2O + 2e^- \rightarrow MnO_2 + 4OH^-$	$+0.60$
$MnO_4^- + 2H_2O + 3e^- \rightarrow MnO_2 + 4OH^-$	$+0.595$
$PdCl_4^{2-} + 2e^- \rightarrow Pd + 4Cl^-$	$+0.591$
$Te^{4+} + 4e^- \rightarrow Te$	$+0.568$
$S_2O_6^{2-} + 4H^+ + 2e^- \rightarrow 2H_2SO_3$	$+0.564$
$H_3AsO_4 + 2H^+ + 2e^- \rightarrow HAsO_2 + 2H_2O$	$+0.560$
$Mn_2O_4^- + e^- \rightarrow MnO_2^-$	$+0.558$
$I_2 + 2e^- \rightarrow 2I^-$	$+0.5355$
$Cu^+ + e^- \rightarrow Cu$	$+0.521$
$Bi^+ + e^- \rightarrow Bi$	$+0.5$
$NiO_2 + 2H_2O + 2e^- \rightarrow Ni(OH)_2 + 2OH^-$	$+0.490$
$Ag_2CO_3 + 2e^- \rightarrow 2Ag + CO_3^{2-}$	$+0.47$
$Ag_2C_2O_4 + 2e^- \rightarrow 2Ag + C_2O_4^{2-}$	$+0.4647$
$Ru^{2+} + 2e^- \rightarrow Ru$	$+0.455$
$H_2SO_3 + 4H^+ + 4e^- \rightarrow S + 3H_2O$	$+0.449$
$Ag_2CrO_4 + 2e^- \rightarrow 2Ag + CrO_4^{2-}$	$+0.4470$
$O_2 + 2H_2O + 4e^- \rightarrow 4OH^-$	$+0.401$
$Tc^{2+} + 2e^- \rightarrow Tc$	$+0.400$
$ClO_4^- + H_2O + 2e^- \rightarrow ClO_3^- + 2OH^-$	$+0.36$
$Fe(CN)_6^{3-} + e^- \rightarrow Fe(CN)_6^{4-}$	$+0.358$
$AgIO + e^- \rightarrow Ag + IO_3^-$	$+0.354$
$Cu^{2+} + 2e^- \rightarrow Cu(Hg)$	$+0.345$
$Ag_2O + H_2O + 2e^- \rightarrow 2Ag + 2OH^-$	$+0.342$
$Cu^{2+} + 2e^- \rightarrow Cu$	$+0.3419$
$ClO_3^- + H_2O + 2e^- \rightarrow ClO_2^- + 2OH^-$	$+0.33$
$Bi^{3+} + 3e^- \rightarrow Bi$	$+0.308$
$Hg_2Cl_2 + 2e^- \rightarrow 2Hg + 2Cl^-$	$+0.26808$
$PbO_2 + H_2O + 2e^- \rightarrow PbO + 2OH^-$	$+0.247$
$Ge^{2+} + 2e^- \rightarrow Ge$	$+0.24$
$AgCl + e^- \rightarrow Ag + Cl^-$	$+0.22233$
$Bi^{3+} + 2e^- \rightarrow Bi^+$	$+0.2$
$SO_4^{2-} + 4H^+ + 2e^- \rightarrow H_2SO_3 + H_2O$	$+0.172$
$Fe_2O_3 + 4H^+ + 2e^- \rightarrow FeOH^+ + H_2O$	$+0.16$
$Cu^{2+} + e^- \rightarrow Cu^+$	$+0.153$
$Sn^{4+} + 2e^- \rightarrow Sn^{2+}$	$+0.151$
$Mn(OH)_3 + e^- \rightarrow Mn(OH)_2 + OH^-$	$+0.15$
$Ag_4Fe(CN)_6 + 4e^- \rightarrow 4Ag + Fe(CN)_6^{4-}$	$+0.1478$
$S + 2H^+ + 2e^- \rightarrow H_2S_{(aq)}$	$+0.142$
$Pt(OH)_2 + 2e^- \rightarrow Pt + 2OH^-$	$+0.14$
$Hg_2Br_2 + 2e^- \rightarrow 2Hg + 2Br^-$	$+0.13923$
$Ge^{4+} + 4e^- \rightarrow Ge$	$+0.124$
$Hg_2O + H_2O + 2e^- \rightarrow 2Hg + 2OH^-$	$+0.123$
$Co(NH_3)_6^{3+} + e^- \rightarrow Co(NH_3)_6^{2+}$	$+0.108$
$HgO + H_2O + 2e^- \rightarrow Hg + 2OH^-$	$+0.0977$
$N_2 + 2H_2O + 6H^+ \rightarrow 2NH_4OH$	$+0.092$
$S_4O_6^{2-} + 2e^- \rightarrow 2S_2O_3^{2-}$	$+0.08$
$AgBr + e^- \rightarrow Ag + Br^-$	$+0.07133$
$Pd(OH)_2 + 2e^- \rightarrow Pd + 2OH^-$	$+0.07$
$Tl_2O_3 + 3H_2O + 4e^- \rightarrow 2Tl^+ + 6OH^-$	$+0.02$
$NO_3^- + H_2O + 2e^- \rightarrow NO_2^- + 2OH^-$	$+0.01$
$CuI_2^- + e^- \rightarrow Cu + 2I^-$	$+0.00$
$\mathbf{2H^+ + 2e^- \rightarrow H_2}$	$\mathbf{0.00000}$
$2D^+ + 2e^- \rightarrow D_2$	-0.013
$AgCN + e^- \rightarrow Ag + CN^-$	-0.017
$Ag_2S + 2H^+ + 2e^- \rightarrow 2Ag + H_2S$	-0.0366
$Fe^{3+} + 3e^- \rightarrow Fe$	-0.037
$Hg_2I_2 + 2e^- \rightarrow 2Hg + 2I^-$	-0.0405
$Tl(OH)_3 + 2e^- \rightarrow TlOH + 2OH^-$	-0.05
$2H_2SO_3 + H^+ + 2e^- \rightarrow HS_2O_4^- + 2H_2O$	-0.056
$O_2 + H_2O + 2e^- \rightarrow HO_2^- + OH^-$	-0.076
$2Cu(OH)_2 + 2e^- \rightarrow Cu_2O + 2OH^- + H_2O$	-0.080
$SnO_2 + 4H^+ + 2e^- \rightarrow Sn^{2+} + 2H_2O$	-0.094

$SnO_2 + 4H^+ + 4e^- \rightarrow Sn + 2H_2O$	-0.117	$TlCl_2 + e^- \rightarrow Tl + Cl^-$	-0.5568
$Pb^{2+} + 2e^- \rightarrow Pb(Hg)$	-0.1205	$Fe(OH)_3 + e^- \rightarrow Fe(OH)_2 + OH^-$	-0.56
$Pb^{2+} + 2e^- \rightarrow Pb$	-0.1262	$2SO_3^{2-} + 3H_2O + 4e^- \rightarrow S_2O_3^{2-} + 6OH^-$	-0.571
$CrO_4^{2-} + 4H_2O + 3e^- \rightarrow Cr(OH)_3 + 5OH^-$	-0.13	$PbO + H_2O + 2e^- \rightarrow Pb + 2OH^-$	-0.580
$Sn^{2+} + 2e^- \rightarrow Sn$	-0.1375	$Ta^{3+} + 3e^- \rightarrow Ta$	-0.6
$O_2 + 2H_2O + 2e^- \rightarrow H_2O_2 + 2OH^-$	-0.146	$As + 3H^+ + 3e^- \rightarrow AsH_3$	-0.608
$AgI + e^- \rightarrow Ag + I^-$	-0.15224	$Cd(OH)_4^{2+} + 2e^- \rightarrow Cd + 4OH^-$	-0.658
$CO_2 + 2H^+ + 2e^- \rightarrow HCOOH$	-0.199	$AgS + 2e^- \rightarrow 2Ag + S^{2-}$	-0.691
$Mo^{3+} + 3e^- \rightarrow Mo$	-0.200	$Ni(OH)_2 + 2e^- \rightarrow Ni + 2OH^-$	$-+0.72$
$Ga^+ + e^- \rightarrow Ga$	-0.2	$Co(OH)_2 + 2e^- \rightarrow Co + 2OH^-$	-0.73
$2SO_2^{2-} + 4H^+ + 2e^- \rightarrow S_2O_6^{2-} + H_2O$	-0.22	$Zn^{2+} + 2e^- \rightarrow Zn$	-0.7618
$Cu(OH)_2 + 2e^- \rightarrow Cu + 2OH^-$	-0.222	$Zn^{2+} + 2e^- \rightarrow Zn(Hg)$	-0.7628
$CdSO_4 + 2e^- \rightarrow Cd + SO_4^{2-}$	-0.246	$CdO + H_2O + 2e^- \rightarrow Cd + 2OH^-$	-0.783
$Ni^{2+} + 2e^- \rightarrow Ni$	-0.257	$Cd(OH)_2 + 2e^- \rightarrow Cd(Hg) + 2OH^-$	-0.809
$PbCl_2 + 2e^- \rightarrow Pb + 2Cl^-$	-0.2675	$2H_2O + 2e^- \rightarrow H_2 + 2OH^-$	-0.8277
$H_3PO_4 + 2H^+ + 2e^- \rightarrow H_3PO_3 + H_2O$	-0.276	$Ti^{3+} + e^- \rightarrow Ti^{2+}$	-0.9
$Co^{2+} + 2e^- \rightarrow Co$	-0.28	$SnO_2 + 2H_2O + 4e^- \rightarrow Sn + 4OH^-$	-0.945
$PbBr_2 + 2e^- \rightarrow Pb + 2Br^-$	-0.284	$PO_4^{3-} + 2H_2O + 2e^- \rightarrow HPO_3^{2-} + 3OH^-$	-1.05
$Tl^+ + e^- \rightarrow Tl(Hg)$	-0.3338	$Nb^{3+} + 3e^- \rightarrow Nb$	-1.099
$Tl^+ + e^- \rightarrow Tl$	-0.336	$2SO_3^{2-} + 2H_2O + 2e^- \rightarrow S_2O_4^{2-} + 4OH^-$	-1.12
$In^{3+} + 3e^- \rightarrow In$	-0.3382	$V^{2+} + 2e^- \rightarrow V$	-1.175
$TlOH + e^- \rightarrow Tl + OH^-$	-0.34	$Mn^{2+} + 2e^- \rightarrow Mn$	-1.185
$PbF_2 + 2e^- \rightarrow Pb + 2F^-$	-0.3444	$Zn(OH)_4^{2-} + 2e^- \rightarrow Zn + 4OH^-$	-1.199
$PbSO_4 + 2e^- \rightarrow Pb(Hg) + SO_4^{2-}$	-0.3505	$CrO_2 + 2H_2O + 4e^- \rightarrow Cr + 4OH^-$	-1.2
$Cd^{2+} + 2e^- \rightarrow Cd(Hg)$	-0.3521	$No^{3+} + 3e^- \rightarrow No$	-1.20
$PbSO_4 + 2e^- \rightarrow Pb + SO_4^{2-}$	-0.3588	$ZnO_2^- + 2H_2O + 2e^- \rightarrow Zn + 4OH^-$	-1.215
$Cu_2O + H_2O + 2e^- \rightarrow 2Cu + 2OH^-$	-0.360	$Zn(OH)_2 + 2e^- \rightarrow Zn + 2OH^-$	-1.249
$PbI_2 + 2e^- \rightarrow Pb + 2I^-$	-0.365	$ZnO + H_2O + 2e^- \rightarrow Zn + 2OH^-$	-1.260
$In^{2+} + e^- \rightarrow In^+$	-0.40	$Pa^{3+} + 3e^- \rightarrow Pa$	-1.34
$Cd^{2+} + 2e^- \rightarrow Cd$	-0.403	$Ti^{3+} + 3e^- \rightarrow Ti$	-1.37
$In^{3+} + 2e^- \rightarrow In^+$	-0.443	$Zr^{4+} + 4e^- \rightarrow Zr$	-1.45
$Fe^{2+} + 2e^- \rightarrow Fe$	-0.447	$Cr(OH)_3 + 3e^- \rightarrow Cr + 3OH^-$	-1.48
$Bi_2O_3 + 3H_2O + 6e^- \rightarrow 2Bi + 6OH^-$	-0.46	$Pa^{4+} + 4e^- \rightarrow Pa$	-1.49
$S + 2e^- \rightarrow S^{2-}$	-0.47627	$Hf^{4+} + 4e^- \rightarrow Hf$	-1.55
$S + H_2O + 2e^- \rightarrow HS^- + OH^-$	-0.478	$ZrO_2 + 4H^+ + 4e^- \rightarrow Zr + 2H_2O$	-1.553
$ZnOH^+ + H^+ + 2e^- \rightarrow Zn + H_2O$	-0.497	$Mn(OH)_2 + 2e^- \rightarrow Mn + 2OH^-$	-1.56
$H_3PO_3 + 2H^+ + 2e^- \rightarrow H_3PO_2 + H_2O$	-0.499	$Ba^{2+} + 2e^- \rightarrow Ba(Hg)$	-1.570
$TiO_2 + 4H^+ + 2e^- \rightarrow Ti^{2+} + 2H_2O$	-0.502	$Ti^{2+} + 2e^- \rightarrow Ti$	-1.630
$H_3PO_2 + H^+ + e^- \rightarrow P + 2H_2O$	-0.508	$HPO_3^{2-} + 2H_2O + 2e^- \rightarrow H_2PO_2^- + 3OH^-$	-1.65
$Sb + 3H^+ + 3e^- \rightarrow SbH_3$	-0.510	$Al^{3+} + 3e^- \rightarrow Al$	-1.662
$HPbO_2 + H_2O + 2e^- \rightarrow Pb + 3OH^-$	-0.537	$Sr^{2+} + 2e^- \rightarrow Sr(Hg)$	-1.793
$Ga^{3+} + 3e^- \rightarrow Ga$	-0.549	$U^{3+} + 3e^- \rightarrow U$	-1.798

$Be^{2+} + 2e^- \rightarrow Be$	-1.847	$Ca^{2+} + 2e^- \rightarrow Ca$	-2.868
$Th^{4+} + 4e^- \rightarrow Th$	-1.899	$Sr(OH)_2 + 2e^- \rightarrow Sr + 2OH^-$	-2.88
$AlF_6^{3-} + 3e^- \rightarrow Al + 6F^-$	-2.069	$La(OH)_3 + 3e^- \rightarrow La + 3OH^-$	-2.90
$H_2 + 2e^- \rightarrow 2H^-$	-2.23	$Ba^{2+} + 2e^- \rightarrow Ba$	-2.912
$Al(OH)_3 + 3e^- \rightarrow Al + 3OH^-$	-2.31	$K^+ + e^- \rightarrow K$	-2.931
$H_2AlO_3^- + H_2O + 3e^- \rightarrow Al + 4OH^-$	-2.33	$Rb^+ + e^- \rightarrow Rb$	-2.98
$ZrO(OH)_2 + H_2O + 4e^- \rightarrow Zr + 4OH^-$	-2.36	$Ba(OH)_2 + 2e^- \rightarrow Ba + 2OH^-$	-2.99
$Mg^{2+} + 2e^- \rightarrow Mg$	-2.372	$Ca(OH)_2 + 2e^- \rightarrow Ca + 2OH^-$	-3.02
$La^{3+} + 3e^- \rightarrow La$	-2.379	$Cs^+ + e^- \rightarrow Cs$	-3.026
$Mg(OH)_2 + 2e^- \rightarrow Mg + 2OH^-$	-2.690	$Li^+ + e^- \rightarrow Li$	-3.0401
$Mg^+ + e^- \rightarrow Mg$	-2.70	$Ca^+ + e^- \rightarrow Ca$	-3.80
$Na^+ + e^- \rightarrow Na$	-2.71	$Sr^+ + e^- \rightarrow Sr$	-4.10

Source: *CRC Handbook of Chemistry and Physics, 84th Ed., D.R. Lide, Editor-in-Chief, CRC Press, Boca Raton, FL, U.S.A., 2003, 8-23~8-33.*

APPENDIX 10(B). STANDARD REDUCTION POTENTIALS (V *vs. SHE*) IN AQUEOUS SYSTEMS AT 25°C AND ONE ATM PRESSURE: <u>ALPHABETICAL</u>

$Ag^+ + e^- \rightarrow Ag$	+0.7996	$BrO_3^- + 6H^+ + 5e^- \rightarrow \frac{1}{2}Br_2 + 3H_2O$	+1.482
$Ag^{2+} + e^- \rightarrow Ag^+$	+1.980	$BrO_3^- + 6H^+ + 6e^- \rightarrow Br^- + 3H_2O$	+1.423
$AgBr + e^- \rightarrow Ag + Br^-$	+0.07133	$Ca^+ + e^- \rightarrow Ca$	−3.80
$AgCl + e^- \rightarrow Ag + Cl^-$	+0.22233	$Ca^{2+} + 2e^- \rightarrow Ca$	−2.868
$AgCN + e^- \rightarrow Ag + CN^-$	−0.017	$Ca(OH)_2 + 2e^- \rightarrow Ca + 2OH^-$	−3.02
$Ag_2CO_3 + 2e^- \rightarrow 2Ag + CO_3^{2-}$	+0.47	$Cd^{2+} + 2e^- \rightarrow Cd$	−0.403
$Ag_2C_2O_4 + 2e^- \rightarrow 2Ag + C_2O_4^{2-}$	+0.4647	$Cd^{2+} + 2e^- \rightarrow Cd(Hg)$	−0.3521
$Ag_2CrO_4 + 2e^- \rightarrow 2Ag + CrO_4^{2-}$	+0.4470	$CdO + H_2O + 2e^- \rightarrow Cd + 2OH^-$	−0.783
$AgF + e^- \rightarrow Ag + F^-$	+0.779	$Cd(OH)_2 + 2e^- \rightarrow Cd(Hg) + 2OH^-$	−0.809
$Ag_4Fe(CN)_6 + 4e^- \rightarrow 4Ag + Fe(CN)_6^{4-}$	+0.1478	$Cd(OH)_4^{2+} + 2e^- \rightarrow Cd + 4OH^-$	−0.658
$AgI + e^- \rightarrow Ag + I^-$	−0.15224	$CdSO_4 + 2e^- \rightarrow Cd + SO_4^{2-}$	−0.246
$AgIO + e^- \rightarrow Ag + IO_3^-$	+0.354	$Ce^{4+} + e^- \rightarrow Ce^{3+}$	+1.72
$2AgO + H_2O + 2e^- \rightarrow Ag_2O + 2OH^-$	+0.607	$Cl_{2(g)} + 2e^- \rightarrow 2Cl^-$	+1.35827
$Ag_2O + H_2O + 2e^- \rightarrow 2Ag + 2OH^-$	+0.342	$ClO_{2(aq)} + e^- \rightarrow ClO_2^-$	+0.954
$Ag_2O_3 + H_2O + 2e^- \rightarrow 2AgO + 2OH^-$	+0.739	$ClO_2 + H^+ + 2e^- \rightarrow HClO_2$	+1.277
$AgS + 2e^- \rightarrow 2Ag + S^{2-}$	−0.691	$ClO_2^- + 2H_2O + 4e^- \rightarrow Cl^- + 4OH^-$	+0.76
$Ag_2S + 2H^+ + 2e^- \rightarrow 2Ag + H_2S$	−0.0366	$ClO_3^- + 2H^+ + e^- \rightarrow ClO_2 + H_2O$	+1.152
$AgSCN + e^- \rightarrow Ag + SCN^-$	+0.8951	$ClO_3^- + 3H^+ + 2e^- \rightarrow HClO_2 + H_2O$	+1.214
$Ag_2SO_4 + 2e^- \rightarrow 2Ag + SO_4^{2-}$	+0.654	$ClO_3^- + 6H^+ + 5e^- \rightarrow \frac{1}{2}Cl_2 + 3H_2O$	+1.47
$Al^{3+} + 3e^- \rightarrow Al$	−1.662	$ClO_3^- + 3H_2O + 6e^- \rightarrow Cl^- + 6OH^-$	+0.62
$AlF_6^{3-} + 3e^- \rightarrow Al + 6F^-$	−2.069	$ClO_3^- + H_2O + 2e^- \rightarrow ClO_2^- + 2OH^-$	+0.33
$Al(OH)_3 + 3e^- \rightarrow Al + 3OH^-$	−2.31	$ClO_4^- + 2H^+ + 2e^- \rightarrow ClO_3^- + H_2O$	+1.189
$As + 3H^+ + 3e^- \rightarrow AsH_3$	−0.608	$ClO_4^- + 8H^+ + 7e^- \rightarrow \frac{1}{2}Cl_2 + 4H_2O$	+1.39
$Au^+ + e^- \rightarrow Au$	+1.692	$ClO_4^- + 8H^+ + 8e^- \rightarrow Cl^- + 4H_2O$	+1.389
$Au^{2+} + e^- \rightarrow Au^+$	+1.8	$ClO_4^- + H_2O + 2e^- \rightarrow ClO_3^- + 2OH^-$	+0.36
$Au^{3+} + 2e^- \rightarrow Au^+$	+1.401	$Co^{2+} + 2e^- \rightarrow Co$	−0.28
$Au^{3+} + 3e^- \rightarrow Au$	+1.498	$Co^{3+} + e^- \rightarrow Co^{2+}$	+1.92
$AuOH^{2+} + H^+ + 2e^- \rightarrow Au^+ + H_2O$	+1.32	$Co(NH_3)_6^{3+} + e^- \rightarrow Co(NH_3)_6^{2+}$	+0.108
$Au(OH)_3 + 3H^+ + 3e^- \rightarrow Au^- + 3H_2O$	+1.45	$Co(OH)_2 + 2e^- \rightarrow Co + 2OH^-$	−0.73
$Ba^{2+} + 2e^- \rightarrow Ba$	−2.912	$CO_2 + 2H^+ + 2e^- \rightarrow HCOOH$	−0.199
$Ba^{2+} + 2e^- \rightarrow Ba(Hg)$	−1.570	$Cu^+ + e^- \rightarrow Cu$	+0.521
$Ba(OH)_2 + 2e^- \rightarrow Ba + 2OH^-$	−2.99	$Cu^{2+} + e^- \rightarrow Cu^+$	+0.153
$Be^{2+} + 2e^- \rightarrow Be$	−1.847	$Cu^{2+} + 2e^- \rightarrow Cu$	+0.3419
$Bi^+ + e^- \rightarrow Bi$	+0.5	$Cu^{2+} + 2e^- \rightarrow Cu(Hg)$	+0.345
$Bi^{3+} + 2e^- \rightarrow Bi^+$	+0.2	$Cu^{3+} + e^- \rightarrow Cu^{2+}$	+2.4
$Bi^{3+} + 3e^- \rightarrow Bi$	+0.308	$Cu^{2+} + 2CN^- + e^- \rightarrow Cu(CN)_2^-$	+1.103
$Bi_2O_3 + 3H_2O + 6e^- \rightarrow 2Bi + 6OH^-$	−0.46	$CuI_2^- + e^- \rightarrow Cu + 2I^-$	+0.00
$Br_{2(aq)} + 2e^- \rightarrow 2Br^-$	+1.0873	$Cu_2O + H_2O + 2e^- \rightarrow 2Cu + 2OH^-$	−0.360

$Cu(OH)_2 + 2e^- \rightarrow Cu + 2OH^-$	-0.222		$HPO_3^{2-} + 2H_2O + 2e^- \rightarrow H_2PO_2^- + 3OH^-$	-1.65
$2Cu(OH)_2 + 2e^- \rightarrow Cu_2O + 2OH^- + H_2O$	-0.080		$H_3PO_2 + H^+ + e^- \rightarrow P + 2H_2O$	-0.508
$CrO_2 + 2H_2O + 4e^- \rightarrow Cr + 4OH^-$	-1.2		$H_3PO_3 + 2H^+ + 2e^- \rightarrow H_3PO_2 + H_2O$	-0.499
$CrO_4^{2-} + 4H_2O + 3e^- \rightarrow Cr(OH)_3 + 5OH^-$	-0.13		$H_3PO_4 + 2H^+ + 2e^- \rightarrow H_3PO_3 + H_2O$	-0.276
$Cr_2O_7^{2-} + 14H^+ + 6e^- \rightarrow 2Cr^{3+} + 7H_2O$	$+1.232$		$HPbO_2 + H_2O + 2e^- \rightarrow Pb + 3OH^-$	-0.537
$Cr(OH)_3 + 3e^- \rightarrow Cr + 3OH^-$	-1.48		$2H_2SO_3 + H^+ + 2e^- \rightarrow HS_2O_4^- + 2H_2O$	-0.056
$Cs^+ + e^- \rightarrow Cs$	-3.026		$H_2SO_3 + 4H^+ + 4e^- \rightarrow S + 3H_2O$	$+0.449$
$2D^+ + 2e^- \rightarrow D_2$	-0.013		$Hf^{4+} + 4e^- \rightarrow Hf$	-1.55
$F_2 + 2e^- \rightarrow 2F^-$	$+2.866$		$2Hg^+ + 2e^- \rightarrow Hg_2^{2+}$	$+0.920$
$F_2 + 2H^+ + 2e^- \rightarrow 2HF$	$+3.053$		$Hg_2^{2+} + 2e^- \rightarrow 2Hg$	$+0.7973$
$Fe^{2+} + 2e^- \rightarrow Fe$	-0.447		$Hg_2Br_2 + 2e^- \rightarrow 2Hg + 2Br^-$	$+0.13923$
$Fe^{3+} + e^- \rightarrow Fe^{2+}$	$+0.771$		$Hg_2Cl_2 + 2e^- \rightarrow 2Hg + 2Cl^-$	$+0.26808$
$Fe^{3+} + 3e^- \rightarrow Fe$	-0.037		$Hg_2I_2 + 2e^- \rightarrow 2Hg + 2I^-$	-0.0405
$Fe(CN)_6^{3-} + e^- \rightarrow Fe(CN)_6^{4-}$	$+0.358$		$HgO + H_2O + 2e^- \rightarrow Hg + 2OH^-$	$+0.0977$
$Fe_2O_3 + 4H^+ + 2e^- \rightarrow FeOH^+ + H_2O$	$+0.16$		$Hg_2O + H_2O + 2e^- \rightarrow 2Hg + 2OH^-$	$+0.123$
$Fe(OH)_3 + e^- \rightarrow Fe(OH)_2 + OH^-$	-0.56		$Hg(OH)_2 + 2H^+ + 2e^- \rightarrow Hg + 2H_2O$	$+1.034$
$Ga^+ + e^- \rightarrow Ga$	-0.2		$Hg_2SO_4 + 2e^- \rightarrow 2Hg + SO_4^{2-}$	$+0.6125$
$Ga^{3+} + 3e^- \rightarrow Ga$	-0.549		$I_2 + 2e^- \rightarrow 2I^-$	$+0.5355$
$Ge^{2+} + 2e^- \rightarrow Ge$	$+0.24$		$IO_3^- + 6H^+ + 6e^- \rightarrow I^- + 3H_2O$	$+1.085$
$Ge^{4+} + 4e^- \rightarrow Ge$	$+0.124$		$3IO_3^- + 6H^+ + 6e^- \rightarrow Cl^- + 3H_2O$	$+1.451$
$2H^+ + 2e^- \rightarrow H_2$	0.00000		$2IO_3^- + 12H^+ + 10e^- \rightarrow I_2 + 6H_2O$	$+1.195$
$H_2 + 2e^- \rightarrow 2H^-$	-2.23		$In^{2+} + e^- \rightarrow In^+$	-0.40
$H_2AlO_3^- + H_2O + 3e^- \rightarrow Al + 4OH^-$	-2.33		$In^{3+} + 2e^- \rightarrow In^+$	-0.443
$H_3AsO_4 + 2H^+ + 2e^- \rightarrow HAsO_2 + 2H_2O$	$+0.560$		$In^{3+} + 3e^- \rightarrow In$	-0.3382
$HClO + H^+ + e^- \rightarrow \frac{1}{2}Cl_2 + H_2O$	$+1.611$		$Ir^{3+} + 3e^- \rightarrow Ir$	$+1.156$
$HClO + H^+ + 2e^- \rightarrow Cl^- + H_2O$	$+1.482$		$IrCl_6^{2-} + e^- \rightarrow IrCl_6^{3-}$	$+0.8665$
$HClO_2 + 2H^+ + 2e^- \rightarrow HClO + H_2O$	$+1.645$		$IrCl_6^{3-} + 3e^- \rightarrow Ir + 6Cl^-$	$+0.77$
$HClO_2 + 3H^+ + 3e^- \rightarrow \frac{1}{2}Cl_2 + 2H_2O$	$+1.628$		$K^+ + e^- \rightarrow K$	-2.931
$HClO_2 + 3H^+ + 4e^- \rightarrow Cl^- + 2H_2O$	$+1.570$		$La^{3+} + 3e^- \rightarrow La$	-2.379
$HCrO_4^- + 7H^+ + 3e^- \rightarrow Cr^{3+} + 4H_2O$	$+1.350$		$La(OH)_3 + 3e^- \rightarrow La + 3OH^-$	-2.90
$HFeO_4^- + 4H^+ + 3e^- \rightarrow FeOOH + 2H_2O$	$+2.08$		$Li^+ + e^- \rightarrow Li$	-3.0401
$HFeO_4^- + 7H^+ + 3e^- \rightarrow Fe^{3+} + 4H_2O$	$+2.07$		$Mg^+ + e^- \rightarrow Mg$	-2.70
$HFeO_4^- + 8H^+ + 3e^- \rightarrow Fe^{3+} + 4H_2O$	$+2.20$		$Mg^{2+} + 2e^- \rightarrow Mg$	-2.372
$2HFeO_4^- + 8H^+ + 6e^- \rightarrow Fe_2O_3 + 5H_2O$	$+2.09$		$Mg(OH)_2 + 2e^- \rightarrow Mg + 2OH^-$	-2.690
$2HIO + 2H^+ + 2e^- \rightarrow I_2 + 2H_2O$	$+1.439$		$Mn^{2+} + 2e^- \rightarrow Mn$	1.185
$HNO_2 + H^+ + e^- \rightarrow NO + H_2O$	$+0.983$		$Mn^{3+} + e^- \rightarrow Mn^{2+}$	1.5415
$2HNO_2 + 4H^+ + 4e^- \rightarrow N_2O + 3H_2O$	$+1.297$		$MnO_2 + 4H^+ + 2e^- \rightarrow Mn^{2+} + 2H_2O$	$+1.224$
$HO_2 + H^+ + e^- \rightarrow H_2O_2$	$+1.495$		$MnO_4^- + 4H^+ + 3e^- \rightarrow MnO_2 + 2H_2O$	1.679
$HO_2^- + H_2O + 2e^- \rightarrow 3OH^-$	$+0.878$		$MnO_4^- + 8H^+ + 5e^- \rightarrow Mn^{2+} + 4H_2O$	1.507
$2H_2O + 2e^- \rightarrow H_2 + 2OH^-$	-0.8277		$MnO_4^{2-} + 2H_2O + 2e^- \rightarrow MnO_2 + 4OH^-$	$+0.60$
$H_2O_2 + 2H^+ + 2e^- \rightarrow 2H_2O$	$+1.776$		$MnO_4^- + 2H_2O + 3e^- \rightarrow MnO_2 + 4OH^-$	$+0.595$

Reaction	$E°$	Reaction	$E°$
$Mn_2O_3 + 6H^+ + e^- \rightarrow 2Mn^{2+} + 3H_2O$	+1.485	$PbSO_4 + 2e^- \rightarrow Pb(Hg) + SO_4^{2-}$	−0.3505
$Mn_2O_4^- + e^- \rightarrow MnO_2^-$	+0.558	$Pd^{2+} + 2e^- \rightarrow Pd$	+0.951
$Mn(OH)_3 + e^- \rightarrow Mn(OH)_2 + OH^-$	+0.15	$PdCl_4^{2-} + 2e^- \rightarrow Pd + 4Cl^-$	+0.591
$Mn(OH)_2 + 2e^- \rightarrow Mn + 2OH^-$	1.56	$PdCl_6^{2-} + 2e^- \rightarrow PdCl_4^{2-} + 2Cl^-$	+1.288
$Mo^{3+} + 3e^- \rightarrow Mo$	0.200	$Pd(OH)_2 + 2e^- \rightarrow Pd + 2OH^-$	+0.07
$N_2 + 2H_2O + 6H^+ \rightarrow 2NH_4OH$	+0.092	$Po^{4+} + 4e^- \rightarrow Po$	+0.76
$N_2O + 2H^+ + 2e^- \rightarrow N_2 + H_2O$	+1.766	$Pt^{2+} + 2e^- \rightarrow Pt$	+1.18
$2NO + 2H^+ + 2e^- \rightarrow N_2O + H_2O$	+1.591	$PtCl_4^{2-} + 2e^- \rightarrow Pt + 4Cl^-$	+0.755
$2NO + H_2O + 2e^- \rightarrow N_2O + 2OH^-$	+0.76	$PtCl_6^{2-} + 2e^- \rightarrow PtCl_4^{2-} + 2Cl^-$	+0.68
$NO_3^- + 3H^+ + 2e^- \rightarrow HNO_2 + H_2O$	+0.934	$PtO_2 + 2H^+ + 2e^- \rightarrow PtO + 2H_2O$	+1.01
$NO_3^- + 4H^+ + 3e^- \rightarrow NO + 2H_2O$	+0.957	$PtO_2 + 4H^+ + 4e^- \rightarrow Pt + 2H_2O$	+1.00
$NO_3^- + H_2O + 2e^- \rightarrow NO_2^- + 2OH^-$	+0.01	$PtO_3 + 4H^+ + 2e^- \rightarrow Pt(OH)_2^{2+} + H_2O$	+1.5
$Na^+ + e^- \rightarrow Na$	−2.71	$PtOH^+ + H^+ + 2e^- \rightarrow Pt + 2H_2O$	+1.2
$Nb^{3+} + 3e^- \rightarrow Nb$	−1.099	$Pt(OH)_2 + 2e^- \rightarrow Pt + 2OH^-$	+0.14
$Ni^{2+} + 2e^- \rightarrow Ni$	−0.257	$Pu^{5+} + e^- \rightarrow Pu^{4+}$	+1.099
$NiO_2 + 4H^+ + 2e^- \rightarrow Ni^{2+} + 2H_2O$	+1.678	$Pu^{4+} + e^- \rightarrow Pu^{3+}$	+1.006
$NiO_2 + 2H_2O + 2e^- \rightarrow Ni(OH)_2 + 2OH^-$	+0.490	$Rb^+ + e^- \rightarrow Rb$	−2.98
$No^{3+} + 3e^- \rightarrow No$	−1.20	$Rh^{2+} + 4e^- \rightarrow Rh$	+0.600
$O_{(g)} + 2H^+ + 2e^- \rightarrow H_2O$	+2.421	$Rh^{3+} + 3e^- \rightarrow Rh$	+0.758
$O_2 + 2H^+ + 2e^- \rightarrow H_2O_2$	+0.695	$Ru^{2+} + 2e^- \rightarrow Ru$	+0.455
$O_2 + 4H^+ + 4e^- \rightarrow 2H_2O$	+1.229	$S + 2e^- \rightarrow S^{2-}$	−0.47627
$O_2 + H_2O + 2e^- \rightarrow HO_2^- + OH^-$	−0.076	$S + 2H^+ + 2e^- \rightarrow H_2S_{(aq)}$	+0.142
$O_2 + 2H_2O + 2e^- \rightarrow H_2O_2 + 2OH^-$	−0.146	$S + H_2O + 2e^- \rightarrow HS^- + OH^-$	−0.478
$O_2 + 2H_2O + 4e^- \rightarrow 4OH^-$	+0.401	$2SO_2^- + 4H^+ + 2e^- \rightarrow S_2O_6^{2-} + H_2O$	−0.22
$O_3 + 2H^+ + 2e^- \rightarrow O_2 + H_2O$	+2.076	$2SO_3^{2-} + 2H_2O + 2e^- \rightarrow S_2O_4^{2-} + 4OH^-$	−1.12
$O_3 + H_2O + 2e^- \rightarrow O_2 + 2OH^-$	+1.24	$2SO_3^{2-} + 3H_2O + 4e^- \rightarrow S_2O_3^{2-} + 6OH^-$	−0.571
$OH + e^- \rightarrow OH^-$	+2.02	$SO_4^{2-} + 4H^+ + 2e^- \rightarrow H_2SO_3 + H_2O$	+0.172
$PO_4^{3-} + 2H_2O + 2e^- \rightarrow HPO_3^{2-} + 3OH^-$	−1.05	$S_2O_6^{2-} + 4H^+ + 2e^- \rightarrow 2H_2SO_3$	+0.564
$Pa^{3+} + 3e^- \rightarrow Pa$	−1.34	$S_2O_8^{2-} + 2e^- \rightarrow 2SO_4^{2-}$	+2.010
$Pa^{4+} + 4e^- \rightarrow Pa$	−1.49	$S_2O_8^{2-} + 2H^+ + 2e^- \rightarrow 2HSO_4^-$	+2.123
$Pb^{2+} + 2e^- \rightarrow Pb$	−0.1262	$S_4O_6^{2-} + 2e^- \rightarrow 2S_2O_3^{2-}$	+0.08
$Pb^{2+} + 2e^- \rightarrow Pb(Hg)$	−0.1205	$Sb + 3H^+ + 3e^- \rightarrow SbH_3$	−0.510
$PbBr_2 + 2e^- \rightarrow Pb + 2Br^-$	−0.284	$Sn^{2+} + 2e^- \rightarrow Sn$	−0.1375
$PbCl_2 + 2e^- \rightarrow Pb + 2Cl^-$	−0.2675	$Sn^{4+} + 2e^- \rightarrow Sn^{2+}$	+0.151
$PbF_2 + 2e^- \rightarrow Pb + 2F^-$	−0.3444	$SnO_2 + 4H^+ + 2e^- \rightarrow Sn^{2+} + 2H_2O$	−0.094
$PbI_2 + 2e^- \rightarrow Pb + 2I^-$	−00.365	$SnO_2 + 4H^+ + 4e^- \rightarrow Sn + 2H_2O$	−0.117
$PbO + H_2O + 2e^- \rightarrow Pb + 2OH^-$	−0.580	$SnO_2 + 2H_2O + 4e^- \rightarrow Sn + 4OH^-$	−0.945
$PbO_2 + 4H^+ + 2e^- \rightarrow Pb^{2+} + 2H_2O$	+1.455	$Sr^+ + e^- \rightarrow Sr$	−4.10
$PbO_2 + H_2O + 2e^- \rightarrow PbO + 2OH^-$	+0.247	$Sr^{2+} + 2e^- \rightarrow Sr(Hg)$	−1.793
$PbO_2 + SO_4^{2-} + 4H^+ + 2e^- \rightarrow PbSO_4 + 2H_2O$	1.6913	$Sr(OH)_2 + 2e^- \rightarrow Sr + 2OH^-$	−2.88
$PbSO_4 + 2e^- \rightarrow Pb + SO_4^{2-}$	−0.3588	$Ta^{3+} + 3e^- \rightarrow Ta$	−0.6

$Tb^{4+} + e^- \rightarrow Tb^{3+}$	$+3.1$	$Tl(OH)_3 + 2e^- \rightarrow TlOH + 2OH^-$	-0.05
$Tc^{2+} + 2e^- \rightarrow Tc$	$+0.400$	$U^{3+} + 3e^- \rightarrow U$	-1.798
$Te^{4+} + 4e^- \rightarrow Te$	$+0.568$	$V^{2+} + 2e^- \rightarrow V$	-1.175
$Th^{4+} + 4e^- \rightarrow Th$	-1.899	$XeF + e^- \rightarrow Xe + F^-$	$+3.4$
$Ti^{2+} + 2e^- \rightarrow Ti$	-1.630	$Zn^{2+} + 2e^- \rightarrow Zn$	-0.7618
$Ti^{3+} + e^- \rightarrow Ti^{2+}$	-0.9	$Zn^{2+} + 2e^- \rightarrow Zn(Hg)$	-0.7628
$Ti^{3+} + 3e^- \rightarrow Ti$	-1.37	$ZnO + H_2O + 2e^- \rightarrow Zn + 2OH^-$	-1.260
$TiO_2 + 4H^+ + 2e^- \rightarrow Ti^{2+} + 2H_2O$	-0.502	$ZnO_2^- + 2H_2O + 2e^- \rightarrow Zn + 4OH^-$	-1.215
$Tl^+ + e^- \rightarrow Tl$	-0.336	$ZnOH^+ + H^+ + 2e^- \rightarrow Zn + H_2O$	-0.497
$Tl^+ + e^- \rightarrow Tl(Hg)$	-0.3338	$Zn(OH)_2 + 2e^- \rightarrow Zn + 2OH^-$	-1.249
$Tl^{3+} + 2e^- \rightarrow Tl$	$+1.252$	$Zn(OH)_4^{2-} + 2e^- \rightarrow Zn + 4OH^-$	-1.199
$Tl^{3+} + 3e^- \rightarrow Tl$	$+0.741$	$Zr^{4+} + 4e^- \rightarrow Zr$	-1.45
$TlCl_2 + e^- \rightarrow Tl + Cl^-$	-0.5568	$ZrO_2 + 4H^+ + 4e^- \rightarrow Zr + 2H_2O$	-1.553
$Tl_2O_3 + 3H_2O + 4e^- \rightarrow 2Tl^+ + 6OH^-$	$+0.02$	$ZrO(OH)_2 + H_2O + 4e^- \rightarrow Zr + 4OH^-$	-2.36
$TlOH + e^- \rightarrow Tl + OH^-$	-0.34		

APPENDIX 11. ANSWERS TO EXERCISES

CHAPTER 1

[1-1] 576 m^3. **[1-2]** 74.73% Cl^{35}, 25.27% Cl^{37}.

CHAPTER 2

[2-1] 19.3 cm. **[2-2]** 11.0 atm. **[2-3]** 0.52 m. **[2-4]** 0.92. **[2-5]** 0.13 m^3. **[2-6]** 37.0°C, 310.15 K. **[2-7]** (a) –71.0°C (b) –56.7°C (c) +56.7°C.

CHAPTER 3

[3-1] 0.033 mol; 369 K. **[3-2]** 3.0. **[3-3]** ~30 m.

CHAPTER 4

[4-1] 957 m s^{-1}, 1058 m s^{-1}. **[4-2]** $121. **[4-3]** 6.75 bar. **[4-4]** (a) 1.77 km s^{-1} (b) 3.66 km s^{-1} (c) 629 m s^{-1}. **[4-5]** 3.6×10^{-9}%.

CHAPTER 5

[5-1] -1.48×10^{-4} m^3 mol^{-1}. **[5-2]** (a) 227 bar (b) 284 bar (c) 266 bar.

CHAPTER 6

[6-1] 5.0 GW. **[6-2]** 13 g. **[6-3]** Zero.

CHAPTER 7

[7-1] (a) 255 J (b) 0 (c) –235 J. **[7-2]** 19.6°C. **[7-3]** ~3 days.

CHAPTER 8

[8-1] (a) +0.71 J (b) +6.8 J.

CHAPTER 9

[9-1] +19.907 kJ mol^{-1}. **[9-2]** (a) +20.42 kJ mol^{-1} (b) endothermic. **[9-3]** –1234.78 kJ mol^{-1}. **[9-4]** +5.46 kJ mol^{-1}.

CHAPTER 10

[10-1] (a) 1.00 kJ (b) +1.00 kJ (c) +66.9 J K^{-1}. **[10-2]** +26.40 J K^{-1}. **[10-3]** +48.77 J K^{-1}.

CHAPTER 11

[11-1] 46.56 J K^{-1} mol^{-1}. **[11-2]** +47.7 J K^{-1} mol^{-1}. **[11-3]** –326.68 J K^{-1} mol^{-1}; yes.

CHAPTER 12

[12-1] 0. **[12-2]** –257.8 kJ mol^{-1}; –1094.3 kJ mol^{-1}. **[12-3]** (a) $17 520 (b) 2.68 tonne.

CHAPTER 13

[13-1] (a) Spontaneous; $\Delta G_R = -216.29$ kJ mol^{-1} (b) less spontaneous than in part (a); $\Delta G_R = -147.57$ kJ mol^{-1}.

CHAPTER 14

[14-1] 28.7%. **[14-2]** 0.17. **[14-3]** (a) not spontaneous at 25°C (b) ~152°C.

[14-4] (a) 7.58×10^{-21} (b) 2.48×10^{-5}%. **[14-5]** (a) 0.2537 (b) $P_{H_2} = 0.6074$ bar; $P_{HI} = 0.3926$ bar. **[14-6]** (a) 0.135 (b) 2.07.

CHAPTER 15

[15-1] 321 kg m^{-3}; 0.023 kg m^{-3}. **[15-2]** 337.2 K, 152.5 kPa.

CHAPTER 16

[16-1] 1.83×10^3 °C. **[16-2]** (a) -0.00741 K bar^{-1} (b) -0.084°C. **[16-3]** 78.0°C.

CHAPTER 17

[17-1] $x_B = 0.658$; $y_B = 0.829$. **[17-2]** $a_{H_2O} = 0.970$; $f_{H_2O} = 0.980$. **[17-3]** $P_A = 9.62, 84.55, 173.2$ Torr; $P_E = 420.9, 241.1, 28.4$ Torr. **[17-4]** 1.11×10^{-5} mol L^{-1}. **[17-5]** 0.201.

[17-6] *One phase:* 1, 2 = liquid; $x = 0$ = solid CaCl$_2$; $x = 0.5$ = solid KCaCl$_3$; $x = 1$ = solid KCl. *Two phases:* 3 = solid KCl + liquid 1; 4 = solid KCaCl$_3$ + liquid 1; 5 = solid KCaCl$_3$ + liquid 2; 6 = solid KCl + liquid 2; 7 = solid CaCl$_2$ + solid KCaCl$_3$; 8 = solid KCaCl$_3$ + solid KCl; *Three phase lines:* CaCl$_2$ + liquid 1 + KCaCl$_3$; KCaCl$_3$ + liquid 2 + KCl.

CHAPTER 18

[18-1] (a) -58.9 J mol^{-1} (b) -74 Pa. **[18-2]** (a) 100.095°C (b) ~85%. **[18-3]** 246 g mol^{-1}. **[18-4]** 33.5%. **[18-5]** 1.77×10^5 g mol^{-1}.

CHAPTER 19

[19-1] 1036.6 g L^{-1}. **[19-2]** 1.77×10^{-7}. **[19-3]** 0.50, 1.50, 7.50. **[19-4]** 3.31×10^{-8} mol kg^{-1}. **[19-5]** (a) 3.71 (b) 3.73.

CHAPTER 20

[20-1] 3.31. **[20-2]** 6.98. **[20-3]** [H$^+$] = 2.0496×10^{-6}; [OH$^-$] = 4.8790×10^{-9}

[HCO$_3^-$] = 2.0496×10^{-6}; [CO$_3^=$] = 4.688×10^{-11}; [H$_2$CO$_3$] = 9.4264×10^{-6}; $pH = 5.69$

CHAPTER 21

[21-1] $pH \leq 5.5$. **[21-2]** 4.97. **[21-3]** 6.59.

CHAPTER 22

[22-1] (a) 4.76 (b) 4.78 (c) 4.70; yes. **[22-2]** 9.32. **[22-3]** 9.42; decrease by 0.21 pH units.

CHAPTER 23

[23-1] (a) 1.9×10^{-3} (b) 1.2×10^{-4} (c) 2.3×10^{-5}. **[23-2]** 0.025 g. **[23-3]** (a) 0.027 M
(b) 0.094 M. **[23-4]** (a) 2.15×10^{-4} M (b) 1.00×10^{-5} M (c) 4.0×10^{-9} M. **[23-5]** (a) 4×10^{-4} M
(b) 0.00109 M. **[23-6]** (a) $AgIO_3$ (b) 0.00405 mol solid $AgIO_3$, 0.00297 mol solid Ag_2CrO_4,
$[Ag^+] = 1.056 \times 10^{-5}$ M, $[IO_3^-] = 9.47 \times 10^{-4}$, $[Cr_2O_4^{2-}] = 0.01703$ M.

CHAPTER 24

[24-1] 0, 0, +3, –2, +2, –2, +4, –2, +7, –2, +1, +5, –2, +1, +6, –2, +1, +6, +1, –4/7, –2, +1, –1, +1, –4, –2, +2, –2, +4.

[24-2]:

(a) $2\,CrO_4^{2-} + 3\,HSnO_2^- + 11H_2O \rightarrow 2\,Cr(OH)_3 + 3\,Sn(OH)_6^{2-} + OH^-$

(b) $2\,ClO_2 + 2\,OH^- \rightarrow ClO_3^- + ClO_2^- + H_2O$

(c) $5\,AsO_3^{3-} + 2\,MnO_4^- + 6H^+ \rightarrow 5\,AsO_4^{3-} + 2Mn^{2+} + 3H_2O$

(d) $5SO_2 + 2\,MnO_4^- + 2H_2O \rightarrow 5\,SO_4^{2-} + 2Mn^{2+} + 4H^+$

(e) $3Cl_2 + 6\,OH^- \rightarrow 5\,Cl^- + ClO_3^- + 3H_2O$

(f) $2\,CrO_4^{2-} + 3H_2S + 10H^+ \rightarrow 2Cr^{3+} + 3S + 8H_2O$

(g) $Pb + PbO_2 + 2\,SO_4^{2-} + 4H^+ \rightarrow 2PbSO_4 + 2H_2O$

(h) $2\,Cr(OH)_4^- + 3\,HO_2^- \rightarrow 2\,CrO_4^{2-} + 5H_2O + OH^-$

(i) $3Ag + NO_3^- + 4H^+ \rightarrow 3Ag^+ + NO + 2H_2O$

(j) $6\,I^- + 2\,MnO_4^- + 4H_2O \rightarrow 3I_2 + 2MnO_2 + 8OH^-$

(k) $4Zn + NO_3^- + 10H^+ \rightarrow 4Zn^{2+} + NH_4^+ + 3H_2O$

(l) $4Fe(OH)_2 + O_2 + 2H_2O \rightarrow 4Fe(OH)_3$

(m) $PbO_2 + 4HCl \rightarrow PbCl_2 + Cl_2 + 2H_2O$

(n) $6\,MnO_4^- + CH_3OH + 8\,OH^- \rightarrow 6\,MnO_4^{2-} + CO_3^{2-} + 6H_2O$

(o) $3H_2C_2O_4 + 2\,MoO_4^{2-} + 10H^+ \rightarrow 6CO_2 + 2Mo^{3+} + 8H_2O$

(p) $2\,MnO_4^- + 5H_2O_2 + 6H^+ \rightarrow 2Mn^{2+} + 5O_2 + 8H_2O$

(q) $Zn + 2\,OH^- + 2H_2O \rightarrow H_2 + Zn(OH)_4^{2-}$

(r) $3\,Cu(NH_3)_4^{2+} + S_2O_4^{2-} + 8\,OH^- \rightarrow 3Cu + 2\,SO_4^{2-} + 12NH_3 + 4H_2O$

CHAPTER 25

[25-1] –261.87 kJ mol^{-1}. **[25-2]** +0.1578 V. **[25-3]** 1.141 V. **[25-4]** 4.3×10^{21}.

[25-5] (a) +241.2 kJ mol^{-1} (b) 124.2 kJ mol^{-1} (c) –386 J K^{-1} mol^{-1}. **[25-6]** $K_{SP} = 2.9 \times 10^{-10}$.

CHAPTER 26

[26-1] (a) first (b) 3.72×10^{-3} min^{-1} (c) 373 min. **[26-2]** (a) 62.6% (b) 13.9%.

[26-3] $0.8284/\left(k\sqrt{c_o}\right)$. **[26-4]** 5.04×10^{-4} M. **[26-5]** (a) 8.25 s (b) 669 s. **[26-6]** Depends on the temperature: $\Delta G_a^*(25°C) = 51.2$ kJ mol^{-1}; $\Delta G_a^* (50°C) = 60.2$ kJ mol^{-1}. **[26-7]** 107 s. **[26-8]** 334 kJ mol^{-1}. **[26-9]** 3.77×10^{11} L mol^{-1} s^{-1}; 85.7 kJ mol^{-1}. **[26-10]** 54.3 ks.

APPENDIX 12.   ANSWERS TO PROBLEMS

CHAPTER 1

[1-1] 17.1 tonnes. **[1-2]** (a) 7.05 g (b) 30.7 mL. **[1-3]** 35.1 m^3. **[1-4]** $C_{10}H_{14}N_2$.
[1-5] (a) PCl_3 (b) 7.40 g; 9.88 g. **[1-6]** (a) $FeSO_4$ = 16.6 g L^{-1} (b) $Fe_2(SO_4)_3$ = 11.7 g L^{-1}.
[1-7] (a) BH_3 (b) B_2H_6 (c) 27.67 (d) 1.50 g. **[1-8]** 4.56 g. **[1-9]** 0.74. **[1-10]** 95.2%.
[1-11] 45.0%. **[1-12]** CH_2, C_3H_6. **[1-13]** (a) $Al + 3HCl \rightarrow AlCl_3 + \frac{3}{2}H_2$ (b) 0.830 g.
[1-14] (a) 28.81g mol^{-1} (b) 118.70 g mol^{-1}. **[1-15]** 48.25%. **[1-16]** 65.39 g mol^{-1}.
[1-17] C_2H_6O. **[1-18]** 163 kg. **[1-19]** 8.02 L, 5.05 kg. **[1-20]** 0.849. **[1-21]** 24 300 t/day,
850 t/day. **[1-22]** (a) 72.7% (b) 0.383 kg. **[1-23]** $C_5H_8O_2$. **[1-24]** 32.17% $NaCl$, 20.20%
Na_2SO_4, 47.65% $NaNO_3$.

CHAPTER 2

[2-1] 100.0419 g. **[2-2]** (a) 1.153 m (b) $12.86. **[2-3]** 80.2 L. **[2-4]** 38.9 kPa.
[2-5] 0.54 bar. **[2-6]** 41.9 kmol. **[2-7]** 8.17 N. **[2-8]** (a) 9.18 kPa (b) 17.0 kPa.
[2-9] 201 kN. **[2-10]** Level rises. **[2-11]** Level falls. **[2-12]** 882 kg. **[2-13]** 35.3 m.
[2-14] 798 kg m^{-3}. **[2-15]** 0.552 cm. **[2-16]** 750 kg m^{-3}.

CHAPTER 3

[3-1] (a) 0.479 atm (b) 1.57 atm **[3-2]** 39.9 mol % CH_4. **[3-3]** 108 kPa. **[3-4]** 0.448 g.
[3-5] (a) 2.00 (b) 16.89 kPa. **[3-6]** (a) 33.3% (b) 15.1 kg. **[3-7]** 43.0 mol % O_2.
[3-8] (a) 15.82 kPa (b) 0.13 kg (c) 0.952 kg m^{-3}. **[3-9]** 0.0448 atm. **[3-10]** (a) 96.22 m^3
(b) 22.7 kg. **[3-11]** 0.133 m^3. **[3-12]** (a) 24.79 L (b) 25.56 L (c) 0.2033 bar. **[3-13]** Yes.
[3-14] 1.17 g L^{-1}. **[3-15]** 0.400. **[3-16]** (a) 2.26 atm (b) 0.530 g. **[3-17]** 1.12 kg.
[3-18] 0.644 kg m^{-3}. **[3-19]** 53.0%.

CHAPTER 4

[4-1] 2.11×10^{-4} g/week. **[4-2]** 88.1 mol % He, 11.9 mol % Ar. **[4-3]** 412 m s^{-1}, 380 m s^{-1}.
[4-4] (a) 374 m s^{-1} (b) 505 m s^{-1}. **[4-5]** (a) 25.8% (b) 67.0% (c) 81.5%. **[4-6]** 8.2×10^{-8}%.
[4-10] 0.071%. **[4-12]** (a) 1.77% (b) 21.1%.

CHAPTER 5

[5-1] (a) 84.3 bar (b) 72.3 bar. **[5-2]** (a) 3.10 L mol^{-1} (b) 3.00 L mol^{-1}. **[5-3]** 95.5 kg.
[5-5] (a) 0.130 Pa m^6 mol^{-2} (b) 53.0 bar. **[5-6]** (a) –5.61 MPa (b) –18.2 Mpa. **[5-7]** 0.422 L
mol^{-1}. **[5-9]** (a) 111 MPa (b) 63.9 MPa. **[5-10]** (a) 2.479 MPa (b) 2.501 MPa (c) –1.16%
(ideal gas); –0.28% (van der Waals). **[5-11]** carbon monoxide. **[5-12]** (a) 1.38 mol (b) 1.404
mol. **[5-13]** (a) 99.8 kPa (b) 96.7 kPa.

CHAPTER 6

[6-1] (a) –0.895 kJ (b) –0.899 kJ. **[6-2]** (a) –88.0 J (b) –167 J. **[6-3]** (a) 118.71 g mol^{-1}
(b) –0.133 kJ. **[6-4]** (a) 24.75 $ d^{-1} (b) (i) Impossible; blunder; mistake; too low.
(ii) Impossible; blunder; mistake; too low. (iii) Good pumps; top efficiency; reasonable charge.
(iv) Too high; bad pumps; mistake. **[6-5]** 2579 Cal d^{-1}. **[6-6]** (a) 282 kJ. (b) 4.15 kJ/$. (c) 12
MJ/$. **[6-7]** 31.2 × 10^6 hp. **[6-8]** 0.1013 J. **[6-9]** Net work input of 7.55 kJ. **[6-10]** (a) 194
kPa (b) 0.459 m (c) 2.79 kJ. **[6-11]** 461 m. **[6-12]** 467 MJ. **[6-13]** –2.16 kJ. **[6-14]** 3052
kg. **[6-15]** 10.0 MJ. **[6-16]** (a) 1.20 kJ (b) 0.98 kJ. **[6-17]** (a) 99.77 bar (b) +14.97 kJ.
[6-18] (a) 49.5 bar (b) +7.43 kJ. **[6-19]** (a) 9.8 kJ (b) 39.2 kJ. **[6-20]** (a) +392 J (b) +273 J.
[6-21] (a) 99.0 MJ (b) 18.0 MJ (c) 8.65 MJ (d) 4.605 MJ. **[6-22]** (a) 4.01 kJ (b) 3.98 kJ.
[6-23] (a) +9.98 kJ (b) +9.82 kJ. **[6-24]** 1000 kg.

CHAPTER 7

[7-1] (a) +33.86 kJ (b) –0.324 J (c) +33.86 kJ. **[7-2]** (a) +8.42 kJ (b) –2.40 kJ (c) +6.02 kJ.
[7-3] (a) 40.1 L (b) 243 K (c) –1.50 kJ (d) –2.50 kJ. **[7-4]** (a) 2.8 ¢/MJ$_{elec}$ (b) 0.51 ¢/MJ$_{oil}$
(c) 9.7 ¢/MJ$_{oats}$ (d) 1.15 ¢/MJ$_{wood}$. **[7-5]** 17.8 MJ kg^{-1}. **[7-6]** (a) +3.14 kJ (b) +5.71 kJ.
[7-7] 5.07 × 10^{-8}%. **[7-8]** (a) 4.45 × 10^9 kg s^{-1} (b) 1.43 × 10^{11} y. **[7-9]** Mass decreases by
3.2 × 10^{-11}%. **[7-10]** (a) $4.80 /day (b) $5.40 /day. **[7-11]** (a) (i) +2.69 MJ (ii) 50.7°C
(b) 43.8°C. **[7-12]** (a) 24.0 $/day (b) 2.40 $/day (c) 24.0 $/day (d) 0 $/day. **[7-13]** *A* is correct.
[7-14] 1.84 kJ s^{-1}. **[7-15]** (a) 47.5 ¢/day (b) $3.17/day. **[7-16]** Decrease by 27.9 kWh d^{-1}.
[7-17] (a) 0 (b) –173 kJ (c) –247 kJ (d) –272 kJ. **[7-18]** (a) –270 kJ (b) evolved.
[7-19] (a) 0 (b) 350 K. **[7-20]** $75. **[7-21]** (a) Yes and no (b) ΔU = +144.0 kJ, Q_A = +144.0
kJ, W_A = 0 kJ (c) ΔU_B = +144.0 kJ, Q_B = +115.2 kJ, W_B = +28.8 kJ. **[7-22]** (a) 1.40 kJ (b) 1.78
kJ (c) –0.379 kJ. **[7-23]** (a) +0.748 kJ (b) +1.48 kJ (c) +1.62 kJ (d) 0 kJ. **[7-24]** (a) 1.03 kJ
(b) 0 (c) 3.46 kJ (d) 700 kJ. **[7-25]** (a) 0 J (b) 220 J (c) 659 J (d) 1.14 kJ (e) 5.17 kg (f) 1.32 kJ
(g) 1.57 kJ. **[7-26]** (a) 0 (b) +1.13 kJ. **[7-27]** (a) 1.13 kJ (b) 2.27 kJ (c) 1.32 kJ (d) 1.89 kJ
(e) –567 J. **[7-28]** (a) 9.173 kJ (b) 6.575 kJ (c) 21.04 J mol^{-1} K^{-1}. **[7-29]** 164 kJ.
[7-30] +14.77 kJ. **[31]** Not feasible.

CHAPTER 8

[8-1] (a) +2.205 kJ (b) +2.205 kJ (c) +1.581 kJ. **[8-2]** (a) +39.0 kJ (b) –3.12 kJ (c) +39.0 kJ
(d) +35.9 kJ. **[8-3]** (a) +466 kJ (b) +333 kJ. **[8-4]** (a) –0.218 kJ (b) +3.20 kJ (c) +3.20 kJ
(d) +2.98 kJ. **[8-5]** +69.90 kJ mol^{-1}. **[8-6]** 34.88 kJ mol^{-1}. **[8-7]** (a) 3.38 h (b) 3.78 h.
[8-8] –2.84 MJ mol^{-1}. **[8-9]** –1.366 MJ mol^{-1}. **[8-10]** (a) +0.165 J mol^{-1} (b) $\Delta H = \Delta U$ =
+6.01 kJ mol^{-1}. **[8-11]** –76 ¢/100 kg of fish. **[8-12]** (a) *Step 1*: (i) –9.36 MJ (ii) +9.36 MJ,
(iii) 0.0 (iv) 0.0. *Step 2*: (i) 0.0 (ii) –5.51 MJ (iii) –7.69 MJ (iv) –5.51 MJ. *Step 3*: (i) +0.728 MJ

(ii) –2.56 MJ (iii) –2.56 MJ (iv) –1.84 MJ. *Step 4*: (i) +5.90 MJ (ii) 0.0 (iii) +8.24 MJ (iv) +5.90 MJ. *Step 5*: (i) –0.571 MJ (ii) +2.01 MJ (iii) +2.01 MJ (iv) +1.44 MJ. (b) *Cycle*: (i) –3.30 MJ (ii) +3.30 MJ (iii) 0.0 (iv) 0.0. **[8-13]** $Q = +383.4$ kJ, $W = –58.2$ kJ, $\Delta U = +325.2$ kJ; $\Delta H = +383.4$ kJ. **[8-14]** $Q = +325.2$ kJ, $W = 0$, $\Delta U = +325.2$ kJ; $\Delta H = +383.4$ kJ. **[8-15]** In kJ mol^{-1}: (a) $Q = –12.2$, $W = +3.33$, $\Delta U = –8.86$, $\Delta H = –12.2$ (b) $Q = –8.86$, $W = 0$, $\Delta U = –8.86$, $\Delta H = –12.2$. **[8-16]** 29.48°C. **[8-17]** (a) +50 kJ (b) –10 kJ (c) +10 kJ (d) 0 (e) 0. **[8-19]** (a) +555 J (b) 492 J (c) 0 (d) –555 J (e) 0. **[8-21]** 1000 kg. **[8-22]** +14.77 kJ.

CHAPTER 9

[9-1] +25.27 kJ mol^{-1}. **[9-2]** –74.81 kJ mol^{-1}. **[9-3]** 1.10 MW cooling. **[9-4]** (a) –229.2 kJ (b) –231.3 kJ mol^{-1} (c) –2.09 kJ mol^{-1}. **[9-5]** –432 kJ. **[9-6]** +68.15 kJ mol^{-1}. **[9-7]** –74.81 kJ mol^{-1}. **[9-8]** +226.730 kJ mol^{-1}; endothermic. **[9-9]** +49.22 kJ mol^{-1}. **[9-10]** +105.597 kJ mol^{-1}. **[9-11]** +14.36 GJ tonne^{-1}. **[9-12]** 717 kJ mol^{-1}. **[9-13]** +90.124 kJ. mol^{-1}. **[9-14]** (a) –385.1 kJ mol^{-1} (b) –390.1 kJ mol^{-1}. **[9-15]** +40.9 kJ mol^{-1}. **[9-16]** (a) –41.17 kJ mol^{-1} (b) –38.59 kJ mol^{-1} (c) –18.68 kJ mol^{-1}. (d) –12.05 kJ mol^{-1}. **[9-17]** –5.47 kJ mol^{-1}. **[9-18]** +6.60 kJ mol^{-1}. **[9-19]** (a) 1.88 kJ mol^{-1} (b) 1.91 kJ mol^{-1} (c) 4.24 J g^{-1} K^{-1}. **[9-20]** 0. **[9-21]** –799.5 kJ mol^{-1}. **[9-22]** +2.32 kJ mol^{-1}. **[9-23]** (a) $\Delta H_T^o = –295\ 894.27 + 34.68T – 0.0031T^2$ (b) –284.033 kJ mol^{-1}. **[9-24]** 1.06×10^{-7} J g^{-1} K^{-4}. **[9-25]** 50.34 MJ. **[9-26]** +226.730 kJ mol^{-1}. **[9-27]** +86.23 kJ mol^{-1}. **[9-28]** –676.485 kJ mol^{-1}. **[9-29]** –74.81 kJ mol^{-1}. **[9-30]** 17.6%. **[9-31]** –1.45 MJ mol^{-1}. **[9-32]** +52.26 kJ mol^{-1}. **[9-33]** ~2890°C.

CHAPTER 10

[10-1] (a) +3.01 kJ (b) +8.57 J K^{-1}. **[10-2]** (a) 69.6°C (b) +20.3 J K^{-1} (c) +20.3 J K^{-1}. **[10-3]** (a) 377°C (b) +66.8 J K^{-1}. **[10-4]** (b) same (c) +191.0 J K^{-1} (d) +191.0 J K^{-1}. **[10-6]** +2.55 J K^{-1}. **[10-7]** +4.15 J K^{-1}. **[10-8]** +1.84 J K^{-1}. **[10-9]** +31.9 J K^{-1}. **[10-10]** (a) +4.16 J K^{-1} (b) +4.16 J K^{-1}. **[10-11]** (a) 0 (b) +64.8 kJ K^{-1}. **[10-12]** +7.46 J K^{-1}. **[10-13]** +45.4 J K^{-1}. **[10-14]** (a) 283 K (b) 22.1 J K^{-1}. **[10-15]** –3.51 J K^{-1} mol^{-1}. **[10-16]** (a) 2.32 kJ mol^{-1} (b) 3.21 J K^{-1} mol^{-1}. **[10-17]** –5.84 J K^{-1} mol^{-1}. **[10-18]** (a) 0 (b) +0.222 J K^{-1} mol^{-1} (c) spontaneous. **[10-19]** (a) +10.0 W K^{-1} (b) +1.77 W K^{-1} (c) +0.393 W K^{-1}. **[10-20]** +0.473 J K^{-1}. **[10-21]** +19.4 kJ K^{-1}. **[10-22]** (a) 0 (b) 0 (c) +25.0 kJ K^{-1} (d) +25.0 kJ K^{-1}. **[10-23]** (a) 0 (b) +7.06 J K^{-1}. **[10-24]** (a) 36.4°C (b) +252.2 J K^{-1}. **[10-25]** +0.524 J K^{-1}. **[10-26]** +2.57 J K^{-1}. **[10-27]** +226 J K^{-1}. **[10-28]** 0. **[10-29]** (In order, in J K^{-1} mol^{-1}): 86.8; 87.0; 85.4; 79.8; 93.6; 90.8; 88.9; 108.32. **[10-30]** +2.00 J K^{-1}. **[10-31]** (a) 18.0°C (b) 1.115 J K^{-1}. **[10-32]** (a) 40.25 kJ mol^{-1}, 107.89 J K^{-1} mol^{-1}, –1.084 kJ mol^{-1} (b) 41.05 kJ mol^{-1}, 110.02 J K^{-1} mol^{-1}, +1.095 kJ mol^{-1}. **[10-33]** +3.77 J K^{-1}, not spontaneous. **[10-34]** +524.0 J K^{-1} mol^{-1}. **[10-35]** +177 J K^{-1}.

CHAPTER 11

[11-1] (a) +5.40 J K^{-1} mol^{-1} (b) –4.43 J K^{-1} mol^{-1} (c) +0.97 J K^{-1} mol^{-1}. **[11-2]** No! ΔS_{univ} = –387.5 J K^{-1} mol^{-1}. **[11-3]** –242.877 J K^{-1} mol^{-1}; Yes. **[11-4]** No; ΔS_{univ} = –4.639 J K^{-1} mol^{-1} **[11-5]** No; ΔS_{univ} = –309.931 J K^{-1} mol^{-1}. **[11-6]** 32.52 J K^{-1} mol^{-1}. **[11-7]** 191.02 J K^{-1} mol^{-1}.

CHAPTER 12

[12-1] –112.09 J K^{-1} mol^{-1}. **[12-2]** 120 kJ mol^{-1}. **[12-3]** 122 kJ kg^{-1}. **[12-4]** (a) +91.99 J K^{-1} mol^{-1} (b) +3.65 kJ mol^{-1}. **[12-5]** +67.3 J mol^{-1}. **[12-6]** 12.6 MW. **[12-7]** (a) +1.67 J K^{-1} g^{-1} (b) +0.13 J K^{-1} g^{-1}, spontaneous (c) No. **[12-8]** +11.7 J g^{-1}, not spontanous. **[12-9]** 0.0; equilibrium, no process. **[12-10]** –141.05 kJ mol^{-1}. **[12-11]** –141.06 kJ mol^{-1}.
[12-12] (a) +1.88 kJ (b) –0.0304 J (c) +1.88 kJ (d)+5.580 J K^{-1} (e) +0.535 J K^{-1} (f) –1.98 kJ (g) spontaneous. **[12-13]** +8.50 kJ mol^{-1}. **[12-14]** +762 J, –762 J, 0, –2.56 J K^{-1}, +762 J.
[12-15] 0. **[12-16]** 2.20 kJ mol^{-1}. **[12-17]** 305 kJ. **[12-18]** 126 kJ. **[12-19]** +39.42 kJ.
[12-20] (a) 137.16 J K^{-1} mol^{-1} (b) –10.10 kJ mol^{-1}. **[12-21]** +22.9 J. **[12-22]** (a) Q = 0; W = +12.0 kJ; ΔU = +12.0 kJ; ΔH = +18.0 kJ (b) 7.48 kJ mol^{-1} (c) +60.00 J K^{-1}.
[12-23] (a) –46.3 kJ (b) Yes. **[12-24]** Only *Method 3* is possible.

CHAPTER 13

[13-1] –1.102 kJ mol^{-1}. **[13-2]** +2.302 kJ mol^{-1}. **[13-3]** +238.7 kJ. **[13-4]** (a) 597 J mol^{-1} (b) 1020 J mol^{-1}. **[13-5]** +10.0 kJ. **[13-6]** $\Delta G = \left(k' + c_P\right)(T_2 - T_1) - c_P \left(T_2 \ln T_2 - T_1 \ln T_1\right)$ $+ RT_2 \ln\left(P_2 / P\right)$, where $k' = R \ln P_1 - s^o_{298.15} + c_P \ln (298.15) - R \ln P°$ and P_1 is in bar.
[13-7] –9.51 MJ. **[13-8]** (a) 0.0352 (b) 0.188 (c) 5.33. **[13-9]** 6.165 kPa.

CHAPTER 14

[14-1] 8.90×10^{-4}. **[14-2]** (a) 3.906×10^{-3} (b) (i) NH_3 = 14.3%, N_2 = 35.7%, H_2 = 50.0% (c) 10.7 atm. **[14-3]** 1.0 mol. **[14-4]** 4.76×10^{-3}. **[14-5]** 266. **[14-6]** 0.0308 mol.
[14-7] (a) 76.8% (b) 11.5 kg/100 kg (c) 1.032 atm. **[14-8]** 2.05%. **[14-9]** 4.02×10^{-17}.
[14-10] 1.46×10^{-10} bar. **[14-11]** –2.78 kJ mol^{-1}. **[14-12]** +52.9 kJ mol^{-1}. **[14-13]** (a) 1.02 (b) 1.51. **[14-14]** (a) 1.02 (b) $\Delta H^o_T = -4000 + 10T + 0.01T^2$ (c) 1.26. **[14-15]** 5.7×10^{-21}.
[14-16] 1.85. **[14-17]** (a) 8.64×10^3 (b) –18.42 kJ mol^{-1} (c) –115.1 kJ mol^{-1} (d) –376 J K^{-1} mol^{-1}. **[14-18]** (a) 0.067 (b) 1.64. **[14-19]** 7.8×10^{-8}. **[14-20]** (a) 8.99×10^{-41} (b) 1.30×10^{-7} (c) 0.32%. **[14-21]** 5.22×10^7. **[14-22]** (a) 43.2 mbar (b) 1.56 mbar. **[14-23]** K = 4.00.

[14-24] (a) 0.0586 mol (b) 0.341 mol (c) 3.72 bar. **[14-25]** 43.5% dissociated; $K = 0.237$ (b) 2.52 kg m^{-3}. **[14-26]** No; $Q = 0.0210$. **[14-27]** (a) Yes; $Q = 3.84$ (b) $y_{CO} = 0.205$, $y_{CO_2} = 0.477$, $y_{N_2} = 0.318$. **[14-28]** (a) 25.6 bar (b) $y_{N_2} = 0.395$, $y_{O_2} = 0.593$, $y_{NO} = y_{NO_2} = 0.00593$. **[14-29]** +10.37 kJ mol^{-1}; not spontaneous. **[14-30]** (i) decrease (ii) large increase (iii) small decrease (iv) increase (v) negligible change (vi) decrease. **[14-31]** 1110 K. **[14-32]** 1119 K.

CHAPTER 15

[15-1] (b) +2676 J K^{-1} (c) +0.24 J K^{-1}. **[15-5]** (a) $P = 2$, $C = 1$, $F = 1$ (b) $P = 2$, $C = 2$, $F = 2$. **[15-6]** (a) $P = 3$, $C = 2$, $F = 1$ (b) $P = 3$, $C = 2$, $F = 1$ (c) $P = 3$, $C = 2$, $F = 1$.
[15-7] (a) $P = 2$, $C = 1$, $F = 1$ (b) $P = 2$, $C = 2$, $F = 2$. **[15-8]** (a) $P = 2$, $C = 2$, $F = 1$.
[15-9] (a) $F = 2$ (b) $F = 2$. **[15-10]** (a) 4 (b) 2 (c) H_2SO_4 concentration and *either* temperature *or* pressure.

CHAPTER 16

[16-1] 59.4 kJ mol^{-1}; 357°C. **[16-2]** 50.9 kJ mol^{-1}. **[16-3]** 5.80 bar; about –34°C.
[16-4] 141 kJ mol^{-1}. **[16-5]** (a) 1.34 K (b) +2.705 K kbar^{-1}. **[16-6]** +0.038 K.
[16-7] $A = (298.15\Delta C_P - \Delta H^o_{298.15})/R$; $B = \Delta C_P/R$. **[16-8]** 150°C. **[16-9]** +0.51°C.
[16-10] 1538°C. **[16-11]** 53.4 J g^{-1}. **[16-12]** 2883 atm. **[16-13]** (a) 195 K (b) 25.5 kJ mol^{-1}. **[16-14]** 168.6 J K^{-1} mol^{-1}. **[16-15]** 274°C. **[16-16]** 91.0°C. **[16-17]** (a) 29.95 kJ mol^{-1} (b) –79.0°C. **[16-18]** (a) 269°C (b) 344°C (c) 101 J mol^{-1} K^{-1} (d) 45.7 kJ mol^{-1}.
[16-19] 3.9 MPa. **[16-20]** (a) 44.1 kJ mol^{-1} (b) 42.3 kJ mol^{-1}. **[16-21]** 41.35 kJ mol^{-1}.
[16-22] 2.473 GJ. **[16-23]** –1.97 J g^{-1} K^{-1}. **[16-24]** 70.6 kPa. **[16-25]** (a) 3.281 kJ K^{-1} (b) 4.079 kJ K^{-1} (c) Clapeyron Eqn. **[16-26]** 7.02 km. **[16-27]** (a) 10.48 kJ mol^{-1} (b) 6.87°C, 5.26 kPa. **[16-28]** –20.5 J K^{-1} mol^{-1}. **[16-29]** (a) 38.0 kJ mol^{-1}, 112.9 J K^{-1} mol^{-1}, 4.34 kJ mol^{-1} (b) 36.5 kJ mol^{-1}, 108.3 J K^{-1} mol^{-1}, 0. **[16-30]** (a) 42.7 kJ mol^{-1} (b) –3.01 kJ mol^{-1} (c) 0.0726 or 7.26% (d) 18.6 bar. **[16-31]** (a) 196 K (b) 8.51 kJ mol^{-1}. **[16-32]** 1.45 MPa.

CHAPTER 17

[17-1] (a) 8.62 mol L^{-1}; 16.0 mol kg^{-1}; 0.224 (b) 1.61 h. **[17-2]** $x_B = 0.439$; $P = 65.8$ kPa.
[17-3] $y_B = 0.776$. **[17-4]** [N$_2$] = 0.83 mM; [O$_2$] = 0.46 mM. **[17-5]** 891 kg.
[17-6] 0.160. **[17-7]** 1.35 mol. **[17-8]** (a) $x_C = 0.749$ (b) $y_C = 0.850$ (c) $y_C = 0.656$.
[17-9] 0.422%. **[17-11]** (a) 18.62 cm^3 mol^{-1} (b) 18.06 cm^3 mol^{-1}. **[17-12]** (a) –4.16 cm^3 mol^{-1} (b) 18.04 cm^3 mol^{-1}. **[17-13]** (a) 189.1 cm^3 (b) 36.4 cm^3 mol^{-1} (c) 18.0 cm^3 mol^{-1}.
[17-14] (b) (i) 70.45 cm^3 mol^{-1} (ii) 18.01 cm^3 mol^{-1} (c) 20.63 L, 1.24 %. **[17-15]** 16.5%.
[17-16] (a) 5.136% (b) 0.009845 (c) 0.9845 (d) 0.5388 (e) 1.078 (f) 0.5519. **[17-17]** 54.5% H_2 and 45.5% O_2. **[17-18]** 0.678 mmol L^{-1}. **[17-19]** $P^{\bullet}_A = 450$ Torr, $P^{\bullet}_B = 150$ Torr.

[17-20] $x = 0.254$, $m = 18.93$, $M = 10.27$. **[17-21]** (a) 0.0612 (b) 7.74 mol m^{-3}.
[17-22] 1.52 g cm^{-3}. **[17-23]** (a) 63.3% (b) 64.3%. **[17-24]** 0.545 mol. **[17-25]** 814 kg.
[17-26] (a) $P_B = 231$ Torr, $P_T = 55.6$ Torr (b) 287 Torr (c) 0.194. **[17-27]** $x_B = 0.575$.
[17-28] (a) $x_E = 0.788$, $x_P = 0.212$ (b) 107.7 kPa. **[17-29]** (a) $x_B = 0.356$ (b) $y_x = 0.0784$.
[17-30] (a) $x_B = 0.411$ (b) $y_B = 0.632$ (c) 24.6 mol. **[17-31]** (a) 33.3 mol (b) 21.9 mol.
[17-32] (a) 55.9 kPa (b) 60.8 kPa (c) 69.55 kPa. **[17-33]** (a) No (b) $y_{CD} = 0.281$ (c) $K_H = 281$
kPa.

CHAPTER 18

[18-1] 255.6 g mol^{-1}. **[18-2]** ~86%. **[18-3]** 1.46 bar. **[18-4]** 5.39 bar. **[18-5]** 2.45 MPa.
[18-6] (a) 3161.8 Pa (b) 100.047°C (c) –0.170°C (d) 2.25 bar. **[18-7]** ~83%. **[18-8]** 20.52 K
(mol/kg)$^{-1}$. **[18-9]** (a) –0.055°C (b) 100.015°C (c) 71.60 kPa (d) 101.27 kPa (e) 2337.6 Pa.
[18-10] (a) 0.734 mol kg^{-1} (b) 100.38°C. **[18-11]** 178.235 g mol^{-1}. **[18-12]** 83.5 g mol^{-1}.
[18-13] 59.5 g mol^{-1}. **[18-14]** S$_8$. **[18-15]** (a) 687 g (b) 1331 g. **[18-16]** (a) 12.9%
(b) 1.9×10^{-3}. **[18-17]** 93.5%. **[18-18]** CaCl$_2$. **[18-19]** 34.4×10^4 g mol^{-1}. **[18-20]** 0.148
mol L^{-1}. **[18-21]** (a) 172 kPa (b) 314 g mol^{-1}. **[18-22]** –3.66 J mol^{-1}. **[18-23]** 251 m.
[18-24] –0.112°C, 157 g. **[18-25]** 94.18 g mol^{-1}.

CHAPTER 19

[19-1] (a) 0.790; 0.390 (b) 0.822; 0.457. **[19-2]** (a) 0.899 (b) 8.075×10^{-5} (c) 2.05.
[19-3] (a) 0.833 (b) 1.85×10^{-8}. **[19-4]** (a) 3.02 (b) 3.12. **[19-5]** 0.618.
[19-6] (a) 1.086×10^{-5} (b) 12.9% dissociated. **[19-7]** Using molarity, $\gamma_\pm = 0.9635$; using
molality, $\gamma_\pm = 0.9635$. **[19-8]** (a) (kg/mol)$^{1/2}$ (b) 0.4918, 0.5108, 0.5959 (c) 9.31×10^{-7},
9.28×10^{-7}, 9.17×10^{-7} (d) 3.016, 3.016, 3.019.

CHAPTER 20

[20-1] (a) 5.249×10^{-9} M (b) 2.105×10^{-6} (c) 5.267×10^{-9} M. **[20-2]** (a) $[H^+] = 2.504 \times 10^{-6}$
M; $pH = 5.601$ (b) 5.025×10^{-4} (c) 5.613. **[20-3]** (a) 5.15 (b) *Reaction [i]* (c) $[H^+] = 10^{-12}$ M;
$[OH^-] = 10^{-2}$ M; $[HS^-] = 0.05$ M; $[H_2S] = 5 \times 10^{-5}$ M; $[S^{2-}] = 5 \times 10^{-4}$ M (d) 0.0610 mol.
[20-4] 7.21. **[20-5]** 3.92. **[20-6]** (a) 1.752×10^{-5} (b) 1.00, 0.969, 0.942. **[20-7]** (a) 3.67%
(b) 3.67×10^{-4} (c) 0.977 (d) 1.34×10^{-5}]. **[20-8]** 6.23%. **[20-9]** 2.88. **[20-10]** 7.301,
1.5×10^{-7} mol L^{-1}. **[20-11]** 3.015×10^{-5} moles, 7.52. **[20-12]** (a) 2.88 (b) 0.979.
[20-13] 3.45.

CHAPTER 21

[21-1] (a) 2.88 (b) 1.54 (c) 8.88 (d) 11.12 (e) 5.12. **[21-2]** 6.98. **[21-3]** $[H^+] = 1.0117$;
0.1098; 0.01452; 0.001866; 1.984×10^{-4}; 1.998×10^{-5}. **[21-4]** (b) +54.7, +195, +59.5.

[21-5] (a) 1.37 (b) 5.27 (c) 8.85. **[21-6]** x must be within about $\pm 1.15\%$ of $[H^+]$.

CHAPTER 22

[22-1] 40.4 mL. **[22-2]** (a) 2.16 (b) 21.35 moles (c) 2.36. **[22-3]** 4.45. **[22-4]** (a) 9.42 (b) −0.26 pH units. **[22-5]** 70.3 mL. **[22-6]** 62.9 mL. **[22-7]** 27.7 mL. **[22-8]** (a) 3.75 (b) 4.27. **[22-9]** (a) 3.49 (b) 5.10. **[22-10]** $[HAc] = 0.079$ M, $[Ac^{2+}] = 0.221$ M. **[22-11]** (a) 1.77×10^{-3} M, 1.50×10^{-3} M, 0.50×10^{-3} M, 2.75 (b) 8.87×10^{-4} M, 1.20×10^{-3} M, 7.98×10^{-3} M, 3.05. **[22-12]** (a) $[CO_2] = 0.001209$ M, $[HCO_3^-] = 0.0240$ M (b) 7.38.

CHAPTER 23

[23-1] (a) 0.0429 M (b) 0.0284 M (c) 0.663 (d) 0.742 (e) 5.3×10^{-6} (f) 3.95×10^{-5} (g) −48%. **[23-2]** (a) 9.26 (b) 9.76 (c) 10.76. **[23-3]** 1.23×10^{-7} M, 7.00. **[23-4]** (a) 0.884 (b) 0.701. **[23-5]** 6.15 mg L^{-1}. **[23-6]** (a) 4.0325×10^{-5} (b) 0.0118 mol kg^{-1}. **[23-7]** (a) 1.45×10^{-18} (b) 6.92×10^{-16} mol kg^{-1}. **[23-8]** 1.26×10^{-5} M. **[23-9]** 8.3×10^{-3} mol L^{-1}. **[23-10]** 7.7×10^{-13} g. **[23-11]** (a) 8.218×10^{-9} mol L^{-1} (b) 6.988. **[23-12]** 9.48×10^{-4} mol L^{-1}.

CHAPTER 24

[24-1] $(CN)_2 + 2OH^- \rightarrow CN^- + OCN^- + H_2O$.
[24-2] Cathode: $2H_2O + 2e^- \rightarrow H_2 + 2OH^-$; Anode: $2OH^- \rightarrow H_2O + \frac{1}{2}O_2 + 2e^-$.
[24-3] Cathode: $2MnO_2 + H_2O + 2e^- \rightarrow Mn_2O_3 + 2OH^-$; Anode: $Zn + 2OH^- \rightarrow ZnO + H_2O + 2e^-$.

CHAPTER 25

[25-1] 1.67×10^{-8}. **[25-2]** 15.0. **[25-3]** +38.7 mV. **[25-4]** (a) −1.20 V (b) −1.286 V. **[25-5]** (a) −13.13 kJ (b) +60.2 J K^{-1} (c) +4.82 kJ. **[25-6]** (a) $2AgCl_{(s)} + H_{2(g)} \rightarrow 2Ag_{(s)} + 2H^+_{(aq)} + 2Cl^-_{(aq)}$ (b) 0.885 (c) 3.3×10^7. **[25-7]** (a) 97.0% (b) 111 tonne (c) $15 360/y (d) 13.8¢/kg. **[25-8]** (a) $2Ag_{(s)} + 2I^-_{(aq)} + Cl_{2(g)} \xrightarrow{2F} 2AgI_{(s)} + 2Cl^-_{(aq)}$ (b) −302.5 kJ mol^{-1} (c) −0.1522 V (d) −291.5 kJ mol^{-1} (e) 8.15×10^{-17} (f) −183.3 J K^{-1} mol^{-1} (g) −357.2 kJ mol^{-1}. **[25-9]** 2.00 V. **[25-10]** 1.008×10^{-14}. **[25-11]** (a) $Ag_{(pure\ Ag)} \rightarrow Ag_{(alloy)}$ (b) 0.1201 (c) 0.300. **[25-12]** 5.885. **[25-13]** 12.55 Pa. **[25-14]** (a) Cd (b) 7×10^{-15} mol kg^{-1}. **[25-15]** 0.9059 V. **[25-16]** 0.152 V. **[25-17]** (a) $Au_{(s)} + H^+_{(aq)} + I^-_{(aq)} \rightarrow AuI_{(s)} + \frac{1}{2}H_{2(g)}$ (b) 0.496 V (c) 1.78 (d) 6.10×10^{-21}. **[25-18]** 0.5513. **[25-19]** $E^o_{cell} = +0.1262$ V; $\Delta G^o = -24.35$ kJ mol^{-1}; $K = 1.85 \times 10^4$. **[25-20]** −1.2505 V. **[25-21]** (a) $4Al + 4OH^- + 3O_2 + 2H_2O \rightarrow 4H_2AlO_3^-$ (b) 2.731 V (c) 29.3 MJ kg^{-1} (d) 9.3×10^{553} (e) 0. **[25-22]** 1.74×10^{-10}. **[25-23]** 9.3×10^{-20} M.

CHAPTER 26

[26-1] (a) 5.5452×10^{-10} y^{-1} (b) 4.4 billion years. **[26-2]** (a) 1.25×10^9 y
(b) $t' = \left\{ 1/(k_1 + k_2) \right\} ln \left[\left\{ (k_1 + k_2)/k_1 \right\} \left\{ \left(N_{Ar}/N_K \right) + 1 \right\} \right]$ (c) 4.5 billion years. **[26-3]** (a) First order
(b) 4.19×10^{-5} s^{-1} (c) 4.60 h (d) 2.66% . **[26-4]** (a) Zeroth order (b) 1.34×10^{-6} mol L^{-1} s^{-1}
(c) 746 s (d) 6.70×10^{-7} mol L^{-1} s^{-1}. **[26-5]** (a) $Br^- =$ first order; $BrO_3^- =$ second order;
$H^+ =$ second order (b) different for Br^- and H^+ ; same for BrO_3^- (c) 1.37 (mol/L)$^{-3}$ s^{-1} (d) 8.21
(mol/L)$^{-3}$ s^{-1}. **[26-6]** (a) -6.79×10^{10} s^{-1} (b) 3.23 (c) Yes. **[26-7]** (a) $[O_3]_{10} = 3.34 \times 10^{-7}$ mol
L^{-1}; $[O_2]_{10} = 7.52 \times 10^{-7}$ mol L^{-1} (b) $d[O_3]/dt = -2.00 \times 10^{-8}$ (mol L^{-1}) s^{-1};
$d[O_2]/dt = +3.00 \times 10^{-8}$ (mol L^{-1}) s^{-1}]. **[26-8]** ~31.5 y. **[26-9]** (a) ~54.7% (b) ~7.5%.
[26-10] 0.44 min. **[26-11]** (a) 52.95 kJ mol^{-1}; 5.15×10^6 bar^{-1} s^{-1} (b) 1.89×10^{-3} bar^{-1} s^{-1}.
[26-12] Reaction [b] is 2.07 times faster than reaction [a]. **[26-13]** (a) 0.239 (mol/L)$^{-1}$ min^{-1}
(b) $z_A z_B = +2$. **[26-14]** (b) (i) 196×10^3 (ii) 42×10^6 (iii) 76×10^6 (iv) 97×10^6 (v) 135×10^6
(vi) 4.48×10^9 (vii) 9.92×10^9 (viii) 8.25×10^9 (ix) 9.92×10^9 (x) 33.9×10^9.

INDEX

That's all, foulkes!!